Recent Developments of Function Spaces and Their Applications I

Recent Developments of Function Spaces and Their Applications I

Editors

Dachun Yang
Wen Yuan

MDPI • Basel • Beijing • Wuhan • Barcelona • Belgrade • Manchester • Tokyo • Cluj • Tianjin

Editors
Dachun Yang
Beijing Normal University
China

Wen Yuan
Beijing Normal University
China

Editorial Office
MDPI
St. Alban-Anlage 66
4052 Basel, Switzerland

This is a reprint of articles from the Special Issue published online in the open access journal *Mathematics* (ISSN 2227-7390) (available at: https://www.mdpi.com/journal/mathematics/special_issues/Recent_Developments_of_Function_Spaces_and_Their_Applications_I).

For citation purposes, cite each article independently as indicated on the article page online and as indicated below:

LastName, A.A.; LastName, B.B.; LastName, C.C. Article Title. *Journal Name* **Year**, *Volume Number*, Page Range.

ISBN 978-3-0365-4017-7 (Hbk)
ISBN 978-3-0365-4018-4 (PDF)

Contents

Preface to "Recent Developments of Function Spaces and Their Applications I" . . . . **vii**

Jun Liu, Long Huang and Chenlong Yue
Molecular Characterizations of Anisotropic Mixed-Norm Hardy Spacesand Their Applications
Reprinted from: *Mathematics* **2021**, *9*, 2216, doi:10.3390/math9182216 . . . . **1**

Zhenzhen Lou, Qixiang Yang, Jianxun He and Kaili He
Wavelets and Real Interpolation of Besov Spaces
Reprinted from: *Mathematics* **2021**, *9*, 2235, doi:10.3390/math9182235 . . . . **25**

Jin Tao, Dachun Yang and Wen Yuan
A Survey on Function Spaces of John–Nirenberg Type
Reprinted from: *Mathematics* **2021**, *9*, 2264, doi:10.3390/math9182264 . . . . **37**

Daimei Chen, Yanping Chen and Teng Wang
Weighted Estimates for Iterated Commutators of Riesz Potential on Homogeneous Groups
Reprinted from: *Mathematics* **2021**, *9*, 2421, doi:10.3390/math9192421 . . . . **95**

Xing Fu
Boundedness of Some Paraproducts onSpaces of Homogeneous Type
Reprinted from: *Mathematics* **2021**, *9*, 2591, doi:10.3390/math9202591 . . . . **111**

Ziwei Li, Dachun Yang and Wen Yuan
Lebesgue Points of Besov and Triebel–Lizorkin Spaces with Generalized Smoothness
Reprinted from: *Mathematics* **2021**, *9*, 2724, doi:10.3390/math9212724 . . . . **137**

Eiichi Nakai and Yoshihiro Sawano
Spaces of Pointwise Multipliers on Morrey Spaces and Weak Morrey Spaces
Reprinted from: *Mathematics* **2021**, *9*, 2754, doi:10.3390/math9212754 . . . . **183**

Kwok-Pun Ho
Calderón Operator on Local Morrey Spaces with Variable Exponents
Reprinted from: *Mathematics* **2021**, *9*, 2977, doi:10.3390/math9222977 . . . . **201**

Aiting Wang, Wenhua Wang and Baode Li
Maximal Function Characterizations of Hardy Spaces on R^n with Pointwise Variable Anisotropy
Reprinted from: *Mathematics* **2021**, *9*, 3246, doi:10.3390/ math9243246 . . . . **213**

Shaoxiong Hou
An Optimal Estimate for the Anisotropic Logarithmic Potential
Reprinted from: *Mathematics* **2022**, *10*, 261, doi:10.3390/math10020261 . . . . **233**

Ali Behzadan and Michael Holst
Sobolev-Slobodeckij Spaces on Compact Manifolds, Revisited
Reprinted from: *Mathematics* **2022**, *10*, 522, doi:10.3390/math10030522 . . . . **247**

Jun Cao, Yongyang Jin, Yuanyuan Li and Qishun Zhang
Hardy Inequalities and Interrelations of FractionalTriebel–Lizorkin Spaces in a Bounded Uniform Domain
Reprinted from: *Mathematics* **2022**, *10*, 637, doi:10.3390/math10040637 . . . . **351**

Tianjun Shen and Bo Li
Schrödinger Harmonic Functions with Morrey Traces on Dirichlet Metric Measure Spaces
Reprinted from: *Mathematics* **2022**, *10*, 1112, doi:10.3390/math10071112 . . . . **377**

Preface to "Recent Developments of Function Spaces and Their Applications I"

As one of the central topics of modern harmonic analysis, the theory of function spaces has found wide applications in various branches of mathematics, such as harmonic analysis, partial differential equations, geometric analysis, and potential analysis, and has, for a long time, received a lot of attention. The development of various function spaces on different underlying spaces provides many new working spaces and research tools for the study of other related analysis fields.

This book contains 13 papers from the Special Issue "Recent Developments in Function Spaces and Their Applications I", including 12 research articles and 1 survey article. These papers concern some of the recent progress in the theory of various function spaces, such as Morrey and weak Morrey spaces, Hardy-type spaces, John–Nirenberg spaces, Sobolev spaces,and Besov and Triebel–Lizorkin spaces, as well as their applications in harmonic analysis, the boundedness of operators, potential analysis, and partial differential equations.

As the guest editors of the Special Issue, we hope that this book will be interesting and useful to researchers and graduate students in harmonic analysis, function spaces, and related areas.

Dachun Yang and Wen Yuan
Editors

mathematics

Article

Molecular Characterizations of Anisotropic Mixed-Norm Hardy Spaces and Their Applications

Jun Liu [1,*], Long Huang [2] and Chenlong Yue [1]

[1] School of Mathematics, China University of Mining and Technology, Xuzhou 221116, China; yclong@cumt.edu.cn
[2] School of Mathematics and Information Science, Key Laboratory of Mathematics and Interdisciplinary Sciences of the Guangdong Higher Education Institute, Guangzhou University, Guangzhou 510006, China; longhuang@gzhu.edu.cn
* Correspondence: junliu@cumt.edu.cn

Abstract: Let $\vec{p} \in (0,\infty)^n$ be an exponent vector and A be a general expansive matrix on $\mathbb{R}^n$. Let $H_A^{\vec{p}}(\mathbb{R}^n)$ be the anisotropic mixed-norm Hardy spaces associated with A defined via the non-tangential grand maximal function. In this article, using the known atomic characterization of $H_A^{\vec{p}}(\mathbb{R}^n)$, the authors characterize this Hardy space via molecules with the best possible known decay. As an application, the authors establish a criterion on the boundedness of linear operators from $H_A^{\vec{p}}(\mathbb{R}^n)$ to itself, which is used to explore the boundedness of anisotropic Calderón–Zygmund operators on $H_A^{\vec{p}}(\mathbb{R}^n)$. In addition, the boundedness of anisotropic Calderón–Zygmund operators from $H_A^{\vec{p}}(\mathbb{R}^n)$ to the mixed-norm Lebesgue space $L^{\vec{p}}(\mathbb{R}^n)$ is also presented. The obtained boundedness of these operators positively answers a question mentioned by Cleanthous et al. All of these results are new, even for isotropic mixed-norm Hardy spaces on $\mathbb{R}^n$.

Keywords: expansive matrix; (mixed-norm) Hardy space; molecule; Calderón–Zygmund operator

Citation: Liu, J.; Huang, L.; Yue, C. Molecular Characterizations of Anisotropic Mixed-Norm Hardy Spaces and Their Applications. *Mathematics* **2021**, *9*, 2216. https://doi.org/10.3390/math9182216

Academic Editor: Jay Jahangiri

Received: 19 August 2021
Accepted: 6 September 2021
Published: 9 September 2021

1. Introduction

This article is devoted to exploring the molecular characterization of the anisotropic mixed-norm Hardy space $H_A^{\vec{p}}(\mathbb{R}^n)$ from [1], where $\vec{p} \in (0,\infty)^n$ is an exponent vector and A is a general expansive matrix on $\mathbb{R}^n$ (see Definition 1 below). Recall that, as a generalization of the classical Lebesgue space $L^p(\mathbb{R}^n)$, the mixed-norm Lebesgue space $L^{\vec{p}}(\mathbb{R}^n)$, in which the constant exponent p is replaced by an exponent vector $\vec{p} \in [1,\infty]^n$, was studied by Benedek and Panzone [2] in 1961, which can be traced back to Hörmander [3]. Moreover, based on the mixed-norm Lebesgue space, the real-variable theory of various mixed-norm function spaces has rapidly developed over the last two decades; as can be seen, for instance, in ref. [4] on mixed-norm α-modulation spaces, in ref. [5] on mixed-norm Morrey spaces, in refs. [1,6–12] on mixed-norm Hardy spaces, as well as in [13–17] on mixed-norm Besov spaces and mixed-norm Triebel–Lizorkin spaces. For more details on the progress made with regard to the theory of mixed-norm function spaces, we refer the reader to [18–27] as well as to the survey article [28]. In particular, Cleanthous et al. [6] first introduced the anisotropic mixed-norm Hardy space $H_{\vec{a}}^{\vec{p}}(\mathbb{R}^n)$ associated with an anisotropic quasi-homogeneous norm $|\cdot|_{\vec{a}}$, where $\vec{a} \in [1,\infty)^n$ and $\vec{p} \in (0,\infty)^n$, via the non-tangential grand maximal function, and then established its various maximal function characterizations. Later on, Huang et al. [10,11] further completed the real-variable theory of $H_{\vec{a}}^{\vec{p}}(\mathbb{R}^n)$.

On the other hand, motivated by the important role of discrete groups of dilations in wavelet theory, Bownik [29] originally introduced the anisotropic Hardy space $H_A^p(\mathbb{R}^n)$, where $p \in (0,\infty)$. Nowadays, the anisotropic setting has proved useful not only in developing the function spaces arising in harmonic analysis, but also in some other areas such as the wavelet theory (see, for instance [29–32]) and partial differential equations (see, for instance [33,34]). Very recently, inspired by the previous works on both the Hardy spaces $H_{\vec{a}}^{\vec{p}}(\mathbb{R}^n)$ and $H_A^p(\mathbb{R}^n)$, Huang et al. [1] introduced the anisotropic mixed-norm Hardy space $H_A^{\vec{p}}(\mathbb{R}^n)$ associated with A, via the non-tangential grand maximal function, and established its various real-variable characterizations, respectively, by means of the radial or the non-tangential maximal functions, atoms, finite atoms, the Lusin area function, the Littlewood–Paley g-function or g_λ^*-function. The space $H_A^{\vec{p}}(\mathbb{R}^n)$ includes the aforementioned Hardy space $H_{\vec{a}}^{\vec{p}}(\mathbb{R}^n)$ as a special case; see Remark 1(i) below.

However, the molecular characterization of $H_A^{\vec{p}}(\mathbb{R}^n)$, which can be conveniently used to study the boundedness of many important operators (for instance, Calderón–Zygmund operators) on the space $H_A^{\vec{p}}(\mathbb{R}^n)$, is still missing. Thus, to further complete the real-variable theory of anisotropic mixed-norm Hardy spaces $H_A^{\vec{p}}(\mathbb{R}^n)$, in this article, we characterize the space $H_A^{\vec{p}}(\mathbb{R}^n)$ via molecules, in which the range of the decay index ε is in a sense the best possible known decay (see Remark 1(iv) below). As an application, we then obtain a criterion on the boundedness of linear operators on $H_A^{\vec{p}}(\mathbb{R}^n)$ (see Theorem 3 below), which is used to prove the boundedness of anisotropic Calderón–Zygmund operators on $H_A^{\vec{p}}(\mathbb{R}^n)$. In addition, the boundedness of anisotropic Calderón–Zygmund operators from $H_A^{\vec{p}}(\mathbb{R}^n)$ to the mixed-norm Lebesgue space $L^{\vec{p}}(\mathbb{R}^n)$ is also presented. When A is as in (6) below, the obtained boundedness of these Calderón–Zygmund operators positively answers a question mentioned by Cleanthous et al. in [6] (p. 2760); see [1,10] and Remark 2(iv) for more details. All these results are new, even for the isotropic mixed-norm Hardy spaces on $\mathbb{R}^n$. Here, we should point out that a molecular characterization of $H_A^{\vec{p}}(\mathbb{R}^n)$ has also been independently established in [35], in which the range of the decay index ε is just a proper subset of that from the present article. In this sense, the molecular characterization obtained in [35] is covered by the corresponding result of the present article.

The remainder of this article is organized as follows.

In Section 2, we present some notions on expansive matrices, homogeneous quasi-norms, the mixed-norm Lebesgue space $L^{\vec{p}}(\mathbb{R}^n)$ and the anisotropic mixed-norm Hardy space $H_A^{\vec{p}}(\mathbb{R}^n)$ (see Definitions 3 and 5 below).

Section 3 is devoted to characterizing the space $H_A^{\vec{p}}(\mathbb{R}^n)$ via molecules (see Theorem 1 below). To do this, we first give the notion of the anisotropic mixed-norm molecular Hardy space $H_A^{\vec{p},r,s,\varepsilon}(\mathbb{R}^n)$ (see Definition 7 below). Then, by the known atomic characterization of $H_A^{\vec{p}}(\mathbb{R}^n)$ from [1] (Theorem 4.7) (see also Lemma 2 below), we have $H_A^{\vec{p}}(\mathbb{R}^n) \subset H_A^{\vec{p},r,s,\varepsilon}(\mathbb{R}^n)$ with continuous inclusion. Therefore, to complete the proof of Theorem 1, we only need to show $H_A^{\vec{p},r,s,\varepsilon}(\mathbb{R}^n) \subset H_A^{\vec{p}}(\mathbb{R}^n)$ and the inclusion is continuous. Observe that, to obtain the inclusion of this type, the general method is to decompose a molecule into an infinite linear combination of the related atoms (see, for instance [36] (7.4) or [37] (3.23)), which does not work in the present article since the uniformly upper bound estimate of the dual-bases of the natural projection of each molecule on the infinite annuli of a dilated ball (see [36] (7.2) or [37] (3.18)) is still unclear due to its anisotropic structure. To overcome this difficulty, the main idea is to directly estimate the non-tangential maximal function of a molecule on the infinite annuli of a dilated ball (see (16) below), in which we need fully use the integral size condition of a molecule (see Definition 6(i) below). Then, we prove that $H_A^{\vec{p},r,s,\varepsilon}(\mathbb{R}^n)$ is continuously embedded into $H_A^{\vec{p}}(\mathbb{R}^n)$, which completes the proof of Theorem 1.

As applications, in Section 4, we present the boundedness of anisotropic Calderón–Zygmund operators from $H_A^{\vec{p}}(\mathbb{R}^n)$ to the mixed-norm Lebesgue space $L^{\vec{p}}(\mathbb{R}^n)$ (see Theorem 2 below) or to itself (see Theorem 3 below). For this purpose, by the known finite atomic characterization of $H_A^{\vec{p}}(\mathbb{R}^n)$, we first give the proof of Theorem 2. To prove Theorem 3, we then obtain a technical lemma, which shows that, if T is an anisotropic Calderón-Zygmund operator of order ℓ as in Definition 11, then, for any $(\vec{p}, r, \ell)$-atom $\widetilde{a}$, $T(\widetilde{a})$ is a harmless constant multiple of a $(\vec{p}, q, s_0, \varepsilon)$-molecule with s_0 and ε, respectively, as in Definition 11 and (24) below; see Lemma 8 below. In addition, the density of $H_{A,\text{fin}}^{\vec{p},\infty,s}(\mathbb{R}^n) \cap C(\mathbb{R}^n)$ in $H_A^{\vec{p}}(\mathbb{R}^n)$ is also presented in Lemma 9 below. Using this density and the molecular characterization of $H_A^{\vec{p}}(\mathbb{R}^n)$ from Section 3, we establish a useful criterion on the boundedness of linear operators on $H_A^{\vec{p}}(\mathbb{R}^n)$ (see Theorem 4 below), which shows that, if a linear operator T maps each atom to a related molecule, then T has a unique bounded linear extension on $H_A^{\vec{p}}(\mathbb{R}^n)$. Applying this criterion and Lemma 8, we then prove Theorem 3.

Finally, we make some conventions on notations. Let $\mathbf{0}$ be the origin of $\mathbb{R}^n$, $\mathbb{N} := \{1, 2, \ldots\}$ and $\mathbb{Z}_+ := \{0\} \cup \mathbb{N}$. We always use C to denote a positive constant which is independent of the main parameters, but may vary from line to line. The notation $f \lesssim g$ means $f \le Cg$ and if $f \lesssim g \lesssim f$, then we write $f \sim g$. We also use the following convention: if $f \le Cg$ and $g = h$ or $g \le h$, then we write $f \lesssim g \sim h$ or $f \lesssim g \lesssim h$, rather than $f \lesssim g = h$ or $f \lesssim g \le h$. For each multi-index $\beta := (\beta_1, \ldots, \beta_n) \in (\mathbb{Z}_+)^n =: \mathbb{Z}_+^n$, let $|\beta| := \beta_1 + \cdots + \beta_n$ and

$$\partial^\beta := \left(\frac{\partial}{\partial x_1}\right)^{\beta_1} \cdots \left(\frac{\partial}{\partial x_n}\right)^{\beta_n}.$$

For each $r \in [1, \infty]$, we denote by r' its conjugate index, namely $1/r + 1/r' = 1$. Moreover, if $\vec{r} := (r_1, \ldots, r_n) \in [1, \infty]^n$, we denote by $\vec{r}' := (r_1', \ldots, r_n')$ its conjugate index. In addition, for each set $\Omega \subset \mathbb{R}^n$, we denote by $\Omega^\complement$ the set $\mathbb{R}^n \setminus \Omega$, by $\mathbf{1}_\Omega$ its characteristic function, and by $|\Omega|$ its n-dimensional Lebesgue measure. For any $s \in \mathbb{R}$, we denote by $\lfloor s \rfloor$ the largest integer not greater than s. Throughout this article, the symbol $C^\infty(\mathbb{R}^n)$ denotes the set of all infinitely differentiable functions on $\mathbb{R}^n$.

2. Preliminaries

In this section, we present some notions on expansive matrices, mixed-norm Lebesgue spaces and anisotropic mixed-norm Hardy spaces (see, for instance [1,2,29]).

We begin with recalling the notion of expansive matrices from [29] (p. 5, Definition 2.1).

Definition 1. *An expansive matrix, i.e., a dilation, is a real $n \times n$ matrix A satisfying:*

$$\min_{\lambda \in \sigma(A)} |\lambda| > 1,$$

and here and thereafter, $\sigma(A)$ denotes the set of all eigenvalues of A.

Let $b := |\det A|$. Then, by [29] (p. 6, (2.7)), it is easy to see that $b \in (1, \infty)$. By [29] (p. 5, Lemma 2.2), we know that there exists an open ellipsoid Δ, with $|\Delta| = 1$, and $r \in (1, \infty)$ such that $\Delta \subset r\Delta \subset A\Delta$. This further implies that, for any $j \in \mathbb{Z}$, $B_j := A^j\Delta$ is open, $B_j \subset rB_j \subset B_{j+1}$ and $|B_j| = b^j$. For each $x \in \mathbb{R}^n$ and $j \in \mathbb{Z}$, an ellipsoid $x + B_j$ is called a dilated ball. Hereinafter, we always use $\mathfrak{B}$ to denote the collection of all such dilated balls, namely:

$$\mathfrak{B} := \{x + B_j : x \in \mathbb{R}^n \text{ and } j \in \mathbb{Z}\} \tag{1}$$

and:

$$\omega := \inf\left\{i \in \mathbb{Z} : r^i \ge 2\right\}. \tag{2}$$

The following notion of the homogeneous quasi-norm is just [29] (p. 6, Definition 2.3).

Definition 2. *For any given dilation A, a homogeneous quasi-norm, with respect to A, is a measurable mapping $\rho : \mathbb{R}^n \to [0, \infty)$ satisfying:*

(i) *If $x \ne \mathbf{0}$, then $\rho(x) \in (0, \infty)$;*
(ii) *For any $x \in \mathbb{R}^n$, $\rho(Ax) = b\rho(x)$;*
(iii) *There exists some $R \in [1, \infty)$ such that, for any $x, y \in \mathbb{R}^n$, $\rho(x + y) \le R[\rho(x) + \rho(y)]$.*

Throughout this article, for a fixed dilation A, by [29] (p. 6, Lemma 2.4), we can use the following step homogeneous quasi-norm ρ defined by setting for any $x \in \mathbb{R}^n$:

$$\rho(x) := \sum_{j\in\mathbb{Z}} b^j \mathbf{1}_{B_{j+1}\setminus B_j}(x) \quad \text{when } x \neq \mathbf{0}, \quad \text{or else} \quad \rho(\mathbf{0}) := 0 \tag{3}$$

for both simplicity and convenience.

For any $\vec{p} := (p_1, \ldots, p_n) \in (0,\infty)^n$, let:

$$p_- := \min\{p_1, \ldots, p_n\}, \quad p_+ := \max\{p_1, \ldots, p_n\} \quad \text{and} \quad \underline{p} \in (0, \min\{p_-, 1\}). \tag{4}$$

The following definition of mixed-norm Lebesgue spaces is from [2].

Definition 3. *Let $\vec{p} := (p_1, \ldots, p_n) \in (0,\infty]^n$. The mixed-norm Lebesgue space $L^{\vec{p}}(\mathbb{R}^n)$ is defined to be the set of all measurable functions f on $\mathbb{R}^n$ such that:*

$$\|f\|_{L^{\vec{p}}(\mathbb{R}^n)} := \left\{ \int_{\mathbb{R}} \cdots \left[\int_{\mathbb{R}} |f(x_1, \ldots, x_n)|^{p_1} \, dx_1 \right]^{\frac{p_2}{p_1}} \cdots dx_n \right\}^{\frac{1}{p_n}} < \infty$$

with the usual modifications made when $p_i = \infty$ for some $i \in \{1, \ldots, n\}$.

Obviously, when $\vec{p} := (\overbrace{p, \ldots, p}^{n \text{ times}})$ with some $p \in (0,\infty]$, the space $L^{\vec{p}}(\mathbb{R}^n)$ is just the classical Lebesgue space $L^p(\mathbb{R}^n)$.

Recall that a Schwartz function is a $C^\infty(\mathbb{R}^n)$ function φ satisfying that, for any $\nu \in \mathbb{Z}_+$ and multi-index $\gamma \in \mathbb{Z}_+^n$,

$$\|\varphi\|_{\gamma,\nu} := \sup_{x\in\mathbb{R}^n} [\rho(x)]^\nu |\partial^\gamma \varphi(x)| < \infty.$$

Denote by $\mathcal{S}(\mathbb{R}^n)$ the collection of all Schwartz functions as above, equipped with the topology determined by $\{\|\cdot\|_{\gamma,\nu}\}_{\gamma\in\mathbb{Z}_+^n, \nu\in\mathbb{Z}_+}$, and $\mathcal{S}'(\mathbb{R}^n)$ its dual space, equipped with the weak-$*$ topology. For any $N \in \mathbb{Z}_+$, denote by $\mathcal{S}_N(\mathbb{R}^n)$ the following set:

$$\left\{ \varphi \in \mathcal{S}(\mathbb{R}^n) : \|\varphi\|_{\mathcal{S}_N(\mathbb{R}^n)} := \sup_{\gamma\in\mathbb{Z}_+^n, |\gamma|\le N} \sup_{x\in\mathbb{R}^n} \left[|\partial^\gamma \varphi(x)| \max\left\{1, [\rho(x)]^N\right\} \right] \le 1 \right\}.$$

Hereinafter, for any $\varphi \in \mathcal{S}(\mathbb{R}^n)$ and $j \in \mathbb{Z}$, let: $\varphi_j(\cdot) := b^{-j}\varphi(A^{-j}\cdot)$.

Let $\lambda_-, \lambda_+ \in (1,\infty)$ be two numbers such that:

$$\lambda_- \le \min\{|\lambda| : \lambda \in \sigma(A)\} \le \max\{|\lambda| : \lambda \in \sigma(A)\} \le \lambda_+.$$

We should point out that if A is diagonalizable over $\mathbb{C}$, then we can let:

$$\lambda_- := \min\{|\lambda| : \lambda \in \sigma(A)\} \quad \text{and} \quad \lambda_+ := \max\{|\lambda| : \lambda \in \sigma(A)\}.$$

Otherwise, we may choose them sufficiently close to these equalities in accordance with what we need in our arguments.

Definition 4. *For any fixed $N \in \mathbb{N}$, the non-tangential grand maximal function $M_N(f)$ of $f \in \mathcal{S}'(\mathbb{R}^n)$ is defined by setting, for any $x \in \mathbb{R}^n$:*

$$M_N(f)(x) := \sup_{\varphi\in\mathcal{S}_N(\mathbb{R}^n)} \sup_{y\in x+B_j, j\in\mathbb{Z}} |f * \varphi_j(y)|.$$

We now recall the notion of anisotropic mixed-norm Hardy spaces as follows; see [1] (Definition 2.5).

Definition 5. *Let $\vec{p} \in (0,\infty)^n$ and $N \in \mathbb{N} \cap [\lfloor(\frac{1}{\min\{1,p_-\}} - 1)\frac{\ln b}{\ln \lambda_-}\rfloor + 2, \infty)$ with p_- as in (4). The anisotropic mixed-norm Hardy space $H_A^{\vec{p}}(\mathbb{R}^n)$ is defined as the set of all $f \in \mathcal{S}'(\mathbb{R}^n)$ such that $M_N(f) \in L^{\vec{p}}(\mathbb{R}^n)$. Moreover, for any $f \in H_A^{\vec{p}}(\mathbb{R}^n)$, let:*

$$\|f\|_{H_A^{\vec{p}}(\mathbb{R}^n)} := \|M_N(f)\|_{L^{\vec{p}}(\mathbb{R}^n)}.$$

Observe that, by [1] (Theorem 4.7), we know that the Hardy space $H_A^{\vec{p}}(\mathbb{R}^n)$ is independent of the choice of N as in Definition 5.

3. Molecular Characterization of $H_A^{\vec{p}}(\mathbb{R}^n)$

In this section, we characterize $H_A^{\vec{p}}(\mathbb{R}^n)$ via molecules. Recall that, for any $r \in (0,\infty]$ and measurable set $\Omega \subset \mathbb{R}^n$, the Lebesgue space $L^r(E)$ is defined as the set of all measurable functions g on Ω such that, when $r \in (0,\infty)$,

$$\|g\|_{L^r(\Omega)} := \left[\int_\Omega |g(x)|^r\,dx\right]^{1/r} < \infty$$

and

$$\|g\|_{L^\infty(\Omega)} := \operatorname*{ess\,sup}_{x\in\Omega} |g(x)| < \infty.$$

We now introduce the notion of anisotropic mixed-norm $(\vec{p}, r, s, \varepsilon)$-molecules as follows.

Definition 6. *Let $\vec{p} \in (0,\infty)^n$, $r \in (1,\infty]$:*

$$s \in \left[\left\lfloor\left(\frac{1}{p_-} - 1\right)\frac{\ln b}{\ln \lambda_-}\right\rfloor, \infty\right) \cap \mathbb{Z}_+ \tag{5}$$

and $\varepsilon \in (0,\infty)$, where p_- is as in (4). An anisotropic mixed-norm $(\vec{p}, r, s, \varepsilon)$-molecule, associated with some dilated ball $B := x_0 + B_{k_0} \in \mathfrak{B}$ with $x_0 \in \mathbb{R}^n$, $k_0 \in \mathbb{Z}$ and $\mathfrak{B}$ as in (1), is a measurable function m satisfying the following two conditions:

(i) *For any $k \in \mathbb{Z}_+$, $\|m\|_{L^r(U_k(B))} \le b^{-k\varepsilon}|B|^{1/r}\|\mathbf{1}_B\|^{-1}_{L^{\vec{p}}(\mathbb{R}^n)}$, where $U_0(B) := B$ and, for any $k \in \mathbb{N}$,*

$$U_k(B) = U_k(x_0 + B_{k_0}) := x_0 + (A^k B_{k_0}) \setminus (A^{k-1} B_{k_0});$$

(ii) *For any multi-index $\gamma \in \mathbb{Z}_+^n$ with $|\gamma| \le s$, $\int_{\mathbb{R}^n} m(x)x^\gamma\,dx = 0$.*

Henceforth, we call an anisotropic mixed-norm $(\vec{p}, r, s, \varepsilon)$-molecule simply by a $(\vec{p}, r, s, \varepsilon)$-molecule. Via $(\vec{p}, r, s, \varepsilon)$-molecules, we give the following notion of anisotropic mixed-norm molecular Hardy spaces $H_A^{\vec{p},r,s,\varepsilon}(\mathbb{R}^n)$.

Definition 7. *Let $\vec{p} \in (0,\infty)^n$, $r \in (1,\infty]$, s be as in (5) and $\varepsilon \in (0,\infty)$. The anisotropic mixed-norm molecular Hardy space $H_A^{\vec{p},r,s,\varepsilon}(\mathbb{R}^n)$ is defined to be the set of all $f \in \mathcal{S}'(\mathbb{R}^n)$ satisfying that there exists a sequence $\{\lambda_k\}_{k\in\mathbb{N}} \subset \mathbb{C}$ and a sequence of $(\vec{p}, r, s, \varepsilon)$-molecules, $\{m_k\}_{k\in\mathbb{N}}$, associated, respectively, with $\{B^{(k)}\}_{k\in\mathbb{N}} \subset \mathfrak{B}$ such that:*

$$f = \sum_{k\in\mathbb{N}} \lambda_k m_k \quad \text{in} \quad \mathcal{S}'(\mathbb{R}^n).$$

Moreover, for any $f \in H_A^{\vec{p},r,s,\varepsilon}(\mathbb{R}^n)$, *let:*

$$\|f\|_{H_A^{\vec{p},r,s,\varepsilon}(\mathbb{R}^n)} := \inf\left\|\left\{\sum_{k\in\mathbb{N}}\left[\frac{|\lambda_k|\mathbf{1}_{B^{(k)}}}{\|\mathbf{1}_{B^{(k)}}\|_{L^{\vec{p}}(\mathbb{R}^n)}}\right]^{\underline{p}}\right\}^{1/\underline{p}}\right\|_{L^{\vec{p}}(\mathbb{R}^n)},$$

where the infimum is taken over all decompositions of f as above and $\underline{p}$ *as in* (4).

The main result of this section is the subsequent Theorem 1.

Theorem 1. *Let* $\vec{p} \in (0,\infty)^n$, $r \in (\max\{p_+,1\},\infty]$ *with* p_+ *as in* (4), *s be as in* (5):

$$N \in \mathbb{N} \cap \left[\left\lfloor\left(\frac{1}{\min\{1,p_-\}}-1\right)\frac{\ln b}{\ln \lambda_-}\right\rfloor + 2, \infty\right) \text{ with } p_- \text{ as in (4)},$$

and $\varepsilon \in ((s+1)\log_b(\lambda_+/\lambda_-),\infty)$. *Then,* $H_A^{\vec{p}}(\mathbb{R}^n) = H_A^{\vec{p},r,s,\varepsilon}(\mathbb{R}^n)$ *with equivalent quasi-norms.*

Remark 1. *(i)* *When:*

$$A := \begin{pmatrix} 2^{a_1} & 0 & \cdots & 0 \\ 0 & 2^{a_2} & \cdots & 0 \\ \vdots & \vdots & & \vdots \\ 0 & 0 & \cdots & 2^{a_n} \end{pmatrix} \tag{6}$$

with $\vec{a} := (a_1,\dots,a_n) \in [1,\infty)^n$, *the Hardy space* $H_A^{\vec{p}}(\mathbb{R}^n)$ *and the anisotropic mixed-norm Hardy space* $H_{\vec{a}}^{\vec{p}}(\mathbb{R}^n)$ *from [6] coincide with equivalent quasi-norms; see [1] (Remark 2(iv)). In this case, Theorem 1 is new. Moreover, if* $A := d\,\mathrm{I}_{n\times n}$ *for some* $d \in \mathbb{R}$ *with* $|d| \in (1,\infty)$, *here and thereafter,* $\mathrm{I}_{n\times n}$ *denotes the* $n \times n$ *unit matrix, then* $H_A^{\vec{p}}(\mathbb{R}^n)$ *becomes the classical isotropic mixed-norm Hardy space from [7] which is just a special case of* $H_{\vec{a}}^{\vec{p}}(\mathbb{R}^n)$ *from [6]; see [10] Remark 4.4(i) for more details. Even in this case, Theorem 1 is still new;*

(ii) *Let* $\varphi : \mathbb{R}^n \times [0,\infty) \to [0,\infty)$ *be an anisotropic growth function (see, for instance, ref. [38] (Definition 2.5)). Recall that, in [38] (Theorem 3.12), the authors established a molecular characterization of the anisotropic Musielak–Orlicz Hardy space* $H_A^{\varphi}(\mathbb{R}^n)$; *see also [37,39] for the special cases. It follows from [40] (Remark 2.5(iii)), that the anisotropic Musielak–Orlicz Hardy space* $H_A^{\varphi}(\mathbb{R}^n)$ *and anisotropic mixed-norm Hardy space* $H_A^{\vec{p}}(\mathbb{R}^n)$ *in this article cannot cover each other, and hence neither do [38] (Theorem 3.12) and Theorem 1;*

(iii) *Let* $p(\cdot) : \mathbb{R}^n \to (0,\infty]$ *be a variable exponent function satisfying the so-called globally log-Hölder continuous condition (see [40] (2.5) and (2.6))). Very recently, the molecular characterization of the variable anisotropic Hardy space* $H_A^{p(\cdot)}(\mathbb{R}^n)$ *was established by Liu [41] (Theorem 3.1) and, independently, by Wang et al. [42] (Theorem 2.9) with some stronger assumptions on the decay of molecules. As pointed out in [1] (Introduction), the variable anisotropic Hardy space* $H_A^{p(\cdot)}(\mathbb{R}^n)$ *in [41] or [42] and the anisotropic mixed-norm Hardy space* $H_A^{\vec{p}}(\mathbb{R}^n)$ *in this article cannot cover each other. Thus, Theorem 1 cannot be covered by [41] (Theorem 3.1) or [42] (Theorem 2.9);*

(iv) *When* $A := d\,\mathrm{I}_{n\times n}$ *for some* $d \in \mathbb{R}$ *with* $|d| \in (1,\infty)$ *and* $\vec{p} := (\overbrace{p,\dots,p}^{n \text{ times}})$ *with some* $p \in (0,\infty)$, *the space* $H_A^{\vec{p}}(\mathbb{R}^n)$ *becomes the classical isotropic Hardy space* $H^p(\mathbb{R}^n)$ *and* $\log_b(\lambda_+/\lambda_-) = 0$. *In this case, Theorem 1 gives a molecular characterization of* $H^p(\mathbb{R}^n)$ *with the best possible known decay of molecules, namely,* $\varepsilon \in (0,\infty)$.

To show Theorem 1, we need several technical lemmas. First, Lemma 1 is just [1] (Lemma 4.5).

Lemma 1. *Let $\vec{p} \in (0,\infty)^n$, $i \in \mathbb{Z}$ and $r \in [1,\infty] \cap (p_+,\infty]$ with p_+ as in (4). Assume that $\{t_k\}_{k\in\mathbb{N}} \subset \mathbb{C}$, $\{B^{(k)}\}_{k\in\mathbb{N}} := \{x_k + B_{\ell_k}\}_{k\in\mathbb{N}} \subset \mathfrak{B}$ and $\{a_k\}_{k\in\mathbb{N}} \subset L^r(\mathbb{R}^n)$ satisfy that, for any $k \in \mathbb{N}$, $\operatorname{supp} a_k \subset x_k + A^i B_{\ell_k}$:*

$$\|a_k\|_{L^r(\mathbb{R}^n)} \le \frac{|B^{(k)}|^{1/r}}{\|\mathbf{1}_{B^{(k)}}\|_{L^{\vec{p}}(\mathbb{R}^n)}}$$

and:

$$\left\| \left\{ \sum_{k\in\mathbb{N}} \left[\frac{|t_k|\mathbf{1}_{B^{(k)}}}{\|\mathbf{1}_{B^{(k)}}\|_{L^{\vec{p}}(\mathbb{R}^n)}} \right]^{\underline{p}} \right\}^{1/\underline{p}} \right\|_{L^{\vec{p}}(\mathbb{R}^n)} < \infty,$$

where $\underline{p}$ is as in (4). Then:

$$\left\| \left[\sum_{k\in\mathbb{N}} |t_k a_k|^{\underline{p}} \right]^{1/\underline{p}} \right\|_{L^{\vec{p}}(\mathbb{R}^n)} \le C \left\| \left\{ \sum_{k\in\mathbb{N}} \left[\frac{|t_k|\mathbf{1}_{B^{(k)}}}{\|\mathbf{1}_{B^{(k)}}\|_{L^{\vec{p}}(\mathbb{R}^n)}} \right]^{\underline{p}} \right\}^{1/\underline{p}} \right\|_{L^{\vec{p}}(\mathbb{R}^n)},$$

where C is a positive constant independent of $\{t_k\}_{k\in\mathbb{N}}$, $\{B^{(k)}\}_{k\in\mathbb{N}}$ and $\{a_k\}_{k\in\mathbb{N}}$.

The following notions of anisotropic mixed-norm $(\vec{p}, r, s)$-atoms and anisotropic mixed-norm atomic Hardy spaces $H_A^{\vec{p},r,s}(\mathbb{R}^n)$ are from [1].

Definition 8. *Let $\vec{p} \in (0,\infty)^n$, $r \in (1,\infty]$ and s be as in (5).*

(i) *A measurable function a on $\mathbb{R}^n$ is called an anisotropic mixed-norm $(\vec{p}, r, s)$-atom if:*
 - (i)$_1$ *$\operatorname{supp} a \subset B$ with some $B \in \mathfrak{B}$, where $\mathfrak{B}$ is as in (1);*
 - (i)$_2$ *$\|a\|_{L^r(\mathbb{R}^n)} \le \frac{|B|^{1/r}}{\|\mathbf{1}_B\|_{L^{\vec{p}}(\mathbb{R}^n)}}$;*
 - (i)$_3$ *For any $\alpha \in \mathbb{Z}_+^n$ with $|\alpha| \le s$, $\int_{\mathbb{R}^n} a(x) x^\alpha \, dx = 0$.*

(ii) *The anisotropic mixed-norm atomic Hardy space $H_A^{\vec{p},r,s}(\mathbb{R}^n)$ is defined to be the set of all $f \in \mathcal{S}'(\mathbb{R}^n)$ satisfying that there exists a sequence $\{\lambda_k\}_{k\in\mathbb{N}} \subset \mathbb{C}$ and a sequence of $(\vec{p}, r, s)$-atoms, $\{a_k\}_{k\in\mathbb{N}}$, supported, respectively, in $\{B^{(k)}\}_{k\in\mathbb{N}} \subset \mathfrak{B}$ such that:*

$$f = \sum_{k\in\mathbb{N}} \lambda_k a_k \quad \text{in} \quad \mathcal{S}'(\mathbb{R}^n).$$

Furthermore, for any $f \in H_A^{\vec{p},r,s}(\mathbb{R}^n)$, let:

$$\|f\|_{H_A^{\vec{p},r,s}(\mathbb{R}^n)} := \inf \left\| \left\{ \sum_{k\in\mathbb{N}} \left[\frac{|\lambda_k|\mathbf{1}_{B^{(k)}}}{\|\mathbf{1}_{B^{(k)}}\|_{L^{\vec{p}}(\mathbb{R}^n)}} \right]^{\underline{p}} \right\}^{1/\underline{p}} \right\|_{L^{\vec{p}}(\mathbb{R}^n)},$$

where the infimum is taken over all decompositions of f as above.

We also need the atomic characterization of $H_A^{\vec{p}}(\mathbb{R}^n)$ obtained in [1] (Theorem 4.7).

Lemma 2. *Let $\vec{p}$, r, s and N be as in Theorem 1. Then:*

$$H_A^{\vec{p}}(\mathbb{R}^n) = H_A^{\vec{p},r,s}(\mathbb{R}^n)$$

with equivalent quasi-norms.

In addition, by [29] (p. 8, (2.11), p. 5, (2.1) and (2.2) and p. 17, Proposition 3.10), we have the following conclusions.

Lemma 3. *Let A be some fixed dilation. Then:*

(i) *For any $i \in \mathbb{Z}$:*

$$B_i + B_i \subset B_{i+\omega} \quad \text{and} \quad B_i + (B_{i+\omega})^{\complement} \subset (B_i)^{\complement},$$

where ω is as in (2);

(ii) *There exists a positive constant C such that, for any $x \in \mathbb{R}^n$, when $k \in \mathbb{Z}_+$:*

$$\frac{1}{C}(\lambda_-)^k|x| \le |A^k x| \le C(\lambda_+)^k|x|$$

and, when $k \in \mathbb{Z} \setminus \mathbb{Z}_+$:

$$\frac{1}{C}(\lambda_+)^k|x| \le |A^k x| \le C(\lambda_-)^k|x|;$$

(iii) *For any given $N \in \mathbb{N}$, there exists a constant $C_{(N)} \in (0,\infty)$, depending on N, such that, for any $f \in \mathcal{S}'(\mathbb{R}^n)$ and $x \in \mathbb{R}^n$,*

$$M_N^0(f)(x) \le M_N(f)(x) \le C_{(N)} M_N^0(f)(x),$$

where $M_N^0(f)$ denotes the radial grand maximal function of $f \in \mathcal{S}'(\mathbb{R}^n)$ defined by setting, for any $x \in \mathbb{R}^n$,

$$M_N^0(f)(x) := \sup_{\varphi \in \mathcal{S}_N(\mathbb{R}^n)} \sup_{k \in \mathbb{Z}} |f * \varphi_k(x)|.$$

Denote by $L^1_{\text{loc}}(\mathbb{R}^n)$ the set of all locally integrable functions on $\mathbb{R}^n$. Recall that the anisotropic Hardy–Littlewood maximal function $M_{\text{HL}}(f)$ of $f \in L^1_{\text{loc}}(\mathbb{R}^n)$ is defined by setting, for any $x \in \mathbb{R}^n$:

$$M_{\text{HL}}(f)(x) := \sup_{k \in \mathbb{Z}} \sup_{y \in x + B_k} \frac{1}{|B_k|} \int_{y+B_k} |f(z)|\,dz = \sup_{x \in B \in \mathfrak{B}} \frac{1}{|B|} \int_B |f(z)|\,dz, \tag{7}$$

where $\mathfrak{B}$ is as in (1).

The two following lemmas are, respectively, from [1] (Lemma 4.4) and [16] (p. 188).

Lemma 4. *Let $\vec{p} \in (1,\infty)^n$ and $u \in (1,\infty]$. Then, there exists a positive constant C such that, for any sequence $\{f_k\}_{k\in\mathbb{N}}$ of measurable functions:*

$$\left\| \left\{ \sum_{k\in\mathbb{N}} [M_{\text{HL}}(f_k)]^u \right\}^{1/u} \right\|_{L^{\vec{p}}(\mathbb{R}^n)} \le C \left\| \left(\sum_{k\in\mathbb{N}} |f_k|^u \right)^{1/u} \right\|_{L^{\vec{p}}(\mathbb{R}^n)}$$

with the usual modification made when $u = \infty$, where M_{HL} denotes the Hardy–Littlewood maximal operator as in (7).

Lemma 5. *Let $\vec{p} \in (0,\infty]^n$. Then, for any $t \in (0,\infty)$ and $f \in L^{\vec{p}}(\mathbb{R}^n)$:*

$$\| |f|^t \|_{L^{\vec{p}}(\mathbb{R}^n)} = \|f\|^t_{L^{t\vec{p}}(\mathbb{R}^n)}.$$

In addition, for any $\mu \in \mathbb{C}$, $t \in [0, \min\{p_-, 1\}]$ and f, $g \in L^{\vec{p}}(\mathbb{R}^n)$, $\|\mu f\|_{L^{\vec{p}}(\mathbb{R}^n)} = |\mu| \|f\|_{L^{\vec{p}}(\mathbb{R}^n)}$ and:

$$\|f+g\|^t_{L^{\vec{p}}(\mathbb{R}^n)} \le \|f\|^t_{L^{\vec{p}}(\mathbb{R}^n)} + \|g\|^t_{L^{\vec{p}}(\mathbb{R}^n)}.$$

We now prove Theorem 1.

Proof of Theorem 1. Let $\vec{p} \in (0,\infty)^n$, $r \in (\max\{p_+,1\},\infty]$ with p_+ as in (4) and s be as in (5). Then, by the fact that a $(\vec{p},r,s)$-atom is a $(\vec{p},r,s,\varepsilon)$-molecule for any $\varepsilon \in (0,\infty)$, as well as the notions of both $H_A^{\vec{p},r,s}(\mathbb{R}^n)$ and $H_A^{\vec{p},r,s,\varepsilon}(\mathbb{R}^n)$, it is easy to see that $H_A^{\vec{p},r,s}(\mathbb{R}^n) \subset H_A^{\vec{p},r,s,\varepsilon}(\mathbb{R}^n)$ with continuous inclusion. In addition, by Lemma 2, we have $H_A^{\vec{p}}(\mathbb{R}^n) = H_A^{\vec{p},r,s}(\mathbb{R}^n)$ with equivalent quasi-norms. Therefore, $H_A^{\vec{p}}(\mathbb{R}^n) \subset H_A^{\vec{p},r,s,\varepsilon}(\mathbb{R}^n)$ and this inclusion is continuous.

Thus, to complete the proof of Theorem 1, it suffices to prove that:

$$H_A^{\vec{p},r,s,\varepsilon}(\mathbb{R}^n) \subset H_A^{\vec{p}}(\mathbb{R}^n) \tag{8}$$

holds true with continuous inclusion. For this purpose, without loss of generality, for any $f \in H_A^{\vec{p},r,s,\varepsilon}(\mathbb{R}^n)$, we may assume that f is not the zero element of $H_A^{\vec{p},r,s,\varepsilon}(\mathbb{R}^n)$. Then, by Definition 7, we find that there exists a sequence $\{\lambda_k\}_{k\in\mathbb{N}} \subset \mathbb{C}$ and a sequence of $(\vec{p},r,s,\varepsilon)$-molecules $\{m_k\}_{k\in\mathbb{N}}$, associated, respectively, to $\{B^{(k)}\}_{k\in\mathbb{N}} \subset \mathfrak{B}$ such that:

$$f = \sum_{k\in\mathbb{N}} \lambda_k m_k \quad \text{in} \quad \mathcal{S}'(\mathbb{R}^n), \tag{9}$$

and:

$$\|f\|_{H_A^{\vec{p},r,s,\varepsilon}(\mathbb{R}^n)} \sim \left\| \left\{ \sum_{k\in\mathbb{N}} \left[\frac{|\lambda_k| \mathbf{1}_{B^{(k)}}}{\|\mathbf{1}_{B^{(k)}}\|_{L^{\vec{p}}(\mathbb{R}^n)}} \right]^{\underline{p}} \right\}^{1/\underline{p}} \right\|_{L^{\vec{p}}(\mathbb{R}^n)} \tag{10}$$

with $\underline{p}$ as in (4). Take two sequences $\{x_k\}_{k\in\mathbb{N}} \subset \mathbb{R}^n$ and $\{i_k\}_{k\in\mathbb{N}} \subset \mathbb{Z}$ such that, for any $k \in \mathbb{N}$, $x_k + B_{i_k} = B^{(k)}$. From (9), we deduce that, for any $N \in \mathbb{N} \cap [\lfloor (\frac{1}{\underline{p}} - 1)\frac{\ln b}{\ln \lambda_-} \rfloor + 2, \infty)$ and $x \in \mathbb{R}^n$:

$$\begin{aligned} M_N(f)(x) &\leq \sum_{k\in\mathbb{N}} |\lambda_k| M_N(m_k)(x) \mathbf{1}_{x_k + A^\omega B_{i_k}}(x) + \sum_{k\in\mathbb{N}} |\lambda_k| M_N(m_k)(x) \mathbf{1}_{(x_k + A^\omega B_{i_k})^\complement}(x) \\ &=: \mathrm{J}_1 + \mathrm{J}_2, \end{aligned} \tag{11}$$

where ω is an integer as in (2).

For the term J_1, by the boundedness of M_N on $L^q(\mathbb{R}^n)$ with $q \in (1,\infty]$ (see [43] (Remark 2.10)) and the definition of $(\vec{p},r,s,\varepsilon)$-molecules, we conclude that, for any $\varepsilon \in ((s+1)\log_b(\lambda_+/\lambda_-),\infty)$ and $k \in \mathbb{N}$:

$$\begin{aligned} \|M_N(m_k)\|_{L^r(\mathbb{R}^n)} &\lesssim \|m_k\|_{L^r(\mathbb{R}^n)} \lesssim \sum_{\ell\in\mathbb{Z}_+} \|m_k\|_{L^r(U_\ell(B^{(k)}))} \\ &\lesssim \sum_{\ell\in\mathbb{Z}_+} b^{-\ell\varepsilon} \frac{|B^{(k)}|^{1/r}}{\|\mathbf{1}_{B^{(k)}}\|_{L^{\vec{p}}(\mathbb{R}^n)}} \sim \frac{|B^{(k)}|^{1/r}}{\|\mathbf{1}_{B^{(k)}}\|_{L^{\vec{p}}(\mathbb{R}^n)}}, \end{aligned}$$

where $U_0(B^{(k)}) := B^{(k)}$ and, for each $\ell \in \mathbb{N}$:

$$U_\ell(B^{(k)}) = U_\ell(x_k + B_{i_k}) := x_k + (A^\ell B_{i_k}) \setminus (A^{\ell-1} B_{i_k}).$$

This, together with the well-known inequality that, for any $\{\alpha_k\}_{k\in\mathbb{N}} \subset \mathbb{C}$ and $t \in (0,1]$:

$$\left[\sum_{k\in\mathbb{N}} |\alpha_k|\right]^t \leq \sum_{k\in\mathbb{N}} |\alpha_k|^t$$

as well as Lemma 1 and (10), implies thatL

$$\begin{aligned}\|\mathrm{J}_1\|_{L^{\vec{p}}(\mathbb{R}^n)} &\lesssim \left\| \left\{ \sum_{k\in\mathbb{N}} \left[|\lambda_k| M_N(m_k) \mathbf{1}_{x_k+A^\omega B_{i_k}} \right]^{\underline{p}} \right\}^{1/\underline{p}} \right\|_{L^{\vec{p}}(\mathbb{R}^n)} \\ &\lesssim \left\| \left\{ \sum_{k\in\mathbb{N}} \left[\frac{|\lambda_k| \mathbf{1}_{B^{(k)}}}{\|\mathbf{1}_{B^{(k)}}\|_{L^{\vec{p}}(\mathbb{R}^n)}} \right]^{\underline{p}} \right\}^{1/\underline{p}} \right\|_{L^{\vec{p}}(\mathbb{R}^n)} \\ &\sim \|f\|_{H_A^{\vec{p},r,s,\varepsilon}(\mathbb{R}^n)}. \end{aligned} \tag{12}$$

Then, we deal with J_2. To this end, we assume that Q is a polynomial with a degree not greater than s. Then, from Definition 6 and the Hölder inequality, it follows that, for any $N \in \mathbb{N}$, $\varphi \in \mathcal{S}_N(\mathbb{R}^n)$, $\nu \in \mathbb{Z}$ and $x \in (x_k + B_{i_k+\omega})^{\complement}$ with $k \in \mathbb{N}$:

$$\begin{aligned} &|(m_k * \varphi_\nu)(x)| \\ &= b^{-\nu} \left| \int_{\mathbb{R}^n} m_k(z) \varphi(A^{-\nu}(x-z))\, dz \right| \\ &\le b^{-\nu} \sum_{\ell\in\mathbb{Z}_+} \left| \int_{U_\ell(x_k+B_{i_k})} m_k(z) \left[\varphi(A^{-\nu}(x-z)) - Q(A^{-\nu}(x-z)) \right] dz \right| \\ &\le b^{-\nu} \sum_{\ell\in\mathbb{Z}_+} \sup_{z\in A^{-\nu}(x-x_k)+A^\ell B_{i_k-\nu}} |\varphi(z) - Q(z)| \int_{U_\ell(x_k+B_{i_k})} |m_k(z)|\, dz \\ &\lesssim b^{i_k/r'-\nu} \sum_{\ell\in\mathbb{Z}_+} b^{\ell/r'} \sup_{z\in A^{-\nu}(x-x_k)+A^\ell B_{i_k-\nu}} |\varphi(z) - Q(z)| \|m_k\|_{L^r(U_\ell(x_k+B_{i_k}))} \\ &\lesssim b^{i_k-\nu} \left\| \mathbf{1}_{x_k+B_{i_k}} \right\|_{L^{\vec{p}}(\mathbb{R}^n)}^{-1} \sum_{\ell\in\mathbb{Z}_+} b^{(1/r'-\varepsilon)\ell} \sup_{z\in A^{-\nu}(x-x_k)+A^\ell B_{i_k-\nu}} |\varphi(z) - Q(z)|. \end{aligned} \tag{13}$$

For any $k \in \mathbb{N}$ and $x \in (x_k + B_{i_k+\omega})^{\complement}$, it is easy to see that there exists some $j \in \mathbb{Z}_+$ such that $x \in [x_k + (B_{i_k+\omega+j+1} \setminus B_{i_k+\omega+j})]$. Then, for any $\nu \in \mathbb{Z}$ and $\ell \in \mathbb{Z}_+$, by Lemma 3(i), we have:

$$\begin{aligned} A^{-\nu}(x-x_k) + A^\ell B_{i_k-\nu} &\subset A^{-\nu+\ell}(B_{i_k+\omega+j+1} \setminus B_{i_k+\omega+j}) + A^\ell B_{i_k-\nu} \\ &= A^{i_k-\nu+\ell}([B_{\omega+j+1} \setminus B_{\omega+j}] + B_0) \subset A^{i_k-\nu+\ell}(B_j)^{\complement}. \end{aligned} \tag{14}$$

When $i_k \ge \nu$, we pick $Q \equiv 0$. Then, by (14), the fact that $\varphi \in \mathcal{S}_N(\mathbb{R}^n)$ and (3), we find that, for any $N \in \mathbb{N}$ and $\ell \in \mathbb{Z}_+$:

$$\begin{aligned} \sup_{z\in A^{-\nu}(x-x_k)+A^\ell B_{i_k-\nu}} |\varphi(z) - Q(z)| &\le \sup_{z\in A^{i_k-\nu+\ell}(B_j)^{\complement}} \min\left\{1, \rho(z)^{-N}\right\} \\ &\le b^{-N(i_k-\nu+\ell+j)}. \end{aligned} \tag{15}$$

When $i_k < \nu$, we let Q be the Taylor expansion of φ at the point $A^{-\nu}(x-x_k)$ with order s. Then, from the Taylor remainder theorem, Lemma 3(ii) and (14), we deduce that, for any $N \in \mathbb{N} \cap [s+1, \infty)$ and $\ell \in \mathbb{Z}_+$:

$$
\begin{aligned}
&\sup_{z\in A^{-\nu}(x-x_k)+A^{\ell}B_{i_k-\nu}} |\varphi(z)-Q(z)| \\
&\quad \lesssim (\lambda_+)^{\ell(s+1)}(\lambda_-)^{(s+1)(i_k-\nu)} \sup_{z\in A^{-\nu}(x-x_k)+A^{\ell}B_{i_k-\nu}} \min\left\{1,\, \rho(z)^{-N}\right\} \\
&\quad \lesssim b^{\ell(s+1)\log_b(\lambda_+)}(\lambda_-)^{(s+1)(i_k-\nu)} \sup_{z\in A^{i_k-\nu+\ell}(B_j)^{\complement}} \min\left\{1,\, \rho(z)^{-N}\right\} \\
&\quad \lesssim b^{\ell(s+1)\log_b(\lambda_+)}(\lambda_-)^{(s+1)(i_k-\nu)} \min\left\{1,\, b^{-N(i_k-\nu+\ell+j)}\right\}.
\end{aligned}
$$

This, combined with Lemma 3(iii), (13) and (15), further implies that, for any $k\in\mathbb{N}$, $N\in\mathbb{N}\cap[s+1,\infty)$ and $x\in[x_k+(B_{i_k+\omega+j+1}\setminus B_{i_k+\omega+j})]$ with some $j\in\mathbb{Z}_+$:

$$
\begin{aligned}
&M_N(m_k)(x) \\
&\quad \sim \sup_{\varphi\in\mathcal{S}_N(\mathbb{R}^n)} \sup_{\nu\in\mathbb{Z}} |(m_k*\varphi_\nu)(x)| \\
&\quad \lesssim \left\|\mathbf{1}_{x_k+B_{i_k}}\right\|_{L^{\vec{p}}(\mathbb{R}^n)}^{-1} \sum_{\ell\in\mathbb{Z}_+} b^{(1/r'-\varepsilon)\ell} \max\left\{ \sup_{\nu\in\mathbb{Z},\,\nu\le i_k} b^{i_k-\nu} b^{-N(i_k-\nu+\ell+j)}, \right. \\
&\qquad \left. \sup_{\nu\in\mathbb{Z},\,\nu>i_k} b^{i_k-\nu} b^{\ell(s+1)\log_b(\lambda_+)}(\lambda_-)^{(s+1)(i_k-\nu)} \min\left\{1,\, b^{-N(i_k-\nu+\ell+j)}\right\}\right\}.
\end{aligned}
$$

Notice that the supremum over $\nu\le i_k$ has the largest value when $\nu=i_k$. Without loss of generality, we can take $s=\lfloor(1/\min\{1,p_-\}-1)\ln b/\ln\lambda_-\rfloor$ and $N=s+2$, which implies that $b\lambda_-^{s+1}\le b^N$ and the above supremum over $\nu>i_k$ is attained when $i_k-\nu+\ell+j=0$. By this and the fact that $\varepsilon\in((s+1)\log_b(\lambda_+/\lambda_-),\infty)$, we conclude that:

$$
\begin{aligned}
M_N(m_k)(x) &\lesssim \left\|\mathbf{1}_{x_k+B_{i_k}}\right\|_{L^{\vec{p}}(\mathbb{R}^n)}^{-1} \\
&\quad \times \sum_{\ell\in\mathbb{Z}_+}\left\{b^{-\ell\varepsilon}+b^{-\ell[\varepsilon-(s+1)\log_b(\lambda_+/\lambda_-)]}\right\}\max\left\{b^{-Nj},\,\left[b(\lambda_-)^{s+1}\right]^{-j}\right\} \\
&\lesssim \left\|\mathbf{1}_{x_k+B_{i_k}}\right\|_{L^{\vec{p}}(\mathbb{R}^n)}^{-1}\left[b(\lambda_-)^{s+1}\right]^{-j} \\
&\sim \left\|\mathbf{1}_{x_k+B_{i_k}}\right\|_{L^{\vec{p}}(\mathbb{R}^n)}^{-1} b^{-j} b^{-(s+1)j\frac{\ln\lambda_-}{\ln b}} \\
&\lesssim \left\|\mathbf{1}_{x_k+B_{i_k}}\right\|_{L^{\vec{p}}(\mathbb{R}^n)}^{-1} b^{i_k[(s+1)\frac{\ln\lambda_-}{\ln b}+1]} b^{-(i_k+\omega+j)[(s+1)\frac{\ln\lambda_-}{\ln b}+1]} \\
&\lesssim \left\|\mathbf{1}_{x_k+B_{i_k}}\right\|_{L^{\vec{p}}(\mathbb{R}^n)}^{-1} \frac{|x_k+B_{i_k}|^{\sigma}}{[\rho(x-x_k)]^{\sigma}} \\
&\lesssim \left\|\mathbf{1}_{x_k+B_{i_k}}\right\|_{L^{\vec{p}}(\mathbb{R}^n)}^{-1}\left[M_{\mathrm{HL}}\left(\mathbf{1}_{x_k+B_{i_k}}\right)(x)\right]^{\sigma} \\
&\sim \left\|\mathbf{1}_{B^{(k)}}\right\|_{L^{\vec{p}}(\mathbb{R}^n)}^{-1}\left[M_{\mathrm{HL}}\left(\mathbf{1}_{B^{(k)}}\right)(x)\right]^{\sigma},
\end{aligned} \tag{16}
$$

where:

$$
\sigma := \left(\frac{\ln b}{\ln\lambda_-}+s+1\right)\frac{\ln\lambda_-}{\ln b} > \frac{1}{\min\{1,p_-\}}.
$$

By this and Lemmas 4 and 5, we obtain:

$$\begin{aligned}
\|J_2\|_{L^{\vec{p}}(\mathbb{R}^n)} &\lesssim \left\| \sum_{k\in\mathbb{N}} \frac{|\lambda_k|}{\|\mathbf{1}_{B^{(k)}}\|_{L^{\vec{p}}(\mathbb{R}^n)}} \left[M_{\mathrm{HL}}\left(\mathbf{1}_{B^{(k)}}\right)\right]^{\sigma} \right\|_{L^{\vec{p}}(\mathbb{R}^n)} \\
&\sim \left\| \left\{ \sum_{k\in\mathbb{N}} \frac{|\lambda_k|}{\|\mathbf{1}_{B^{(k)}}\|_{L^{\vec{p}}(\mathbb{R}^n)}} \left[M_{\mathrm{HL}}(\mathbf{1}_{B^{(k)}})\right]^{\sigma} \right\}^{1/\sigma} \right\|_{L^{\sigma\vec{p}}(\mathbb{R}^n)}^{\sigma} \\
&\lesssim \left\| \sum_{k\in\mathbb{N}} \frac{|\lambda_k|\mathbf{1}_{B^{(k)}}}{\|\mathbf{1}_{B^{(k)}}\|_{L^{\vec{p}}(\mathbb{R}^n)}} \right\|_{L^{\vec{p}}(\mathbb{R}^n)} \\
&\lesssim \left\| \left\{ \sum_{k\in\mathbb{N}} \left[\frac{|\lambda_k|\mathbf{1}_{B^{(k)}}}{\|\mathbf{1}_{B^{(k)}}\|_{L^{\vec{p}}(\mathbb{R}^n)}} \right]^{\underline{p}} \right\}^{1/\underline{p}} \right\|_{L^{\vec{p}}(\mathbb{R}^n)} \\
&\sim \|f\|_{H_A^{\vec{p},r,s,\varepsilon}(\mathbb{R}^n)}.
\end{aligned}$$

This, together with (11), (12) and Lemma 5 again, implies that:

$$\|f\|_{H_A^{\vec{p}}(\mathbb{R}^n)} = \|M_N(f)\|_{L^{\vec{p}}(\mathbb{R}^n)} \lesssim \|f\|_{H_A^{\vec{p},r,s,\varepsilon}(\mathbb{R}^n)},$$

which completes the proof of (8) and hence of Theorem 1. □

4. Some Applications

In this section, as applications, we establish a criterion on the boundedness of linear operators on $H_A^{\vec{p}}(\mathbb{R}^n)$, which further implies the boundedness of anisotropic Calderón–Zygmund operators on $H_A^{\vec{p}}(\mathbb{R}^n)$. Moreover, the boundedness of these operators from $H_A^{\vec{p}}(\mathbb{R}^n)$ to the mixed-norm Lebesgue space $L^{\vec{p}}(\mathbb{R}^n)$ is also obtained.

We begin with the definition the notion of anisotropic Calderón–Zygmund operators from [29] (p. 60, Definition 9.1).

Definition 9. *An anisotropic Calderón–Zygmund standard kernel is a locally integrable function $\mathcal{K}$ on $E := \{(x,y) \in \mathbb{R}^n \times \mathbb{R}^n : x \neq y\}$ satisfying that there exist two positive constants C and τ such that, for any $(x_1, y_1), (x_1, y_2), (x_2, y_1) \in E$:*

$$|\mathcal{K}(x_1, y_1)| \le \frac{C}{\rho(x_1 - y_1)},$$

$$|\mathcal{K}(x_1,y_1) - \mathcal{K}(x_1,y_2)| \le C\frac{[\rho(y_1-y_2)]^{\tau}}{[\rho(x_1-y_1)]^{1+\tau}} \quad \text{when} \quad \rho(x_1-y_1) \ge b^{2\omega}\rho(y_1-y_2),$$

and:

$$|\mathcal{K}(x_1,y_1) - \mathcal{K}(x_2,y_1)| \le C\frac{[\rho(x_1-x_2)]^{\tau}}{[\rho(x_1-y_1)]^{1+\tau}} \quad \text{when} \quad \rho(x_1-y_1) \ge b^{2\omega}\rho(x_1-x_2),$$

with ω as in (2). Moreover, an anisotropic Calderón–Zygmund operator is a linear operator T satisfying that it is bounded on $L^2(\mathbb{R}^n)$ and there exists an anisotropic Calderón–Zygmund standard kernel $\mathcal{K}$ such that, for any $f \in L^2(\mathbb{R}^n)$ with compact support and $x \notin \operatorname{supp} f$,

$$T(f)(x) = \int_{\operatorname{supp} f} \mathcal{K}(x,z) f(z)\, dz.$$

Hereinafter, for each $\ell \in \mathbb{N}$, let $C^{\ell}(\mathbb{R}^n)$ be the collection of all functions on $\mathbb{R}^n$ whose derivatives with order not greater than ℓ are continuous. The following notion of anisotropic Calderón–Zygmund operator of order ℓ originates from [29] (p. 61, Definition 9.2).

Definition 10. *Let $\ell \in \mathbb{N}$. An anisotropic Calderón–Zygmund operator of order ℓ is an anisotropic Calderón–Zygmund operator T whose kernel $\mathcal{K}$ is a $C^{\ell}(\mathbb{R}^n)$ function with respect to the second variable y and satisfying that there exists a positive constant C such that, for any $\gamma \in \mathbb{Z}_+^n$ with $1 \le |\gamma| \le \ell$, $t \in \mathbb{Z}$ and $(x, y) \in E$ with $\rho(x-y) \sim b^t$:*

$$\left|\partial_y^{\gamma} \widetilde{\mathcal{K}}(x, A^{-t}y)\right| \le C[\rho(x-y)]^{-1} \sim Cb^{-t}, \tag{17}$$

where the implicit equivalent positive constants are independent of x, y, t and, for any $x, y \in \mathbb{R}^n$ with $x \neq A^t y$, $\widetilde{\mathcal{K}}(x,y) := \mathcal{K}(x, A^t y)$.

Then, we first have the boundedness of anisotropic Calderón–Zygmund operators of order ℓ from $H_A^{\vec{p}}(\mathbb{R}^n)$ to $L^{\vec{p}}(\mathbb{R}^n)$.

Theorem 2. *Let $\vec{p} \in (0,\infty)^n$ and T be an anisotropic Calderón–Zygmund operator of order ℓ with $\ell \in [s_0+1,\infty)$, where $s_0 := \lfloor (1/p_- - 1)\ln b/\ln \lambda_- \rfloor$ and p_- is as in (4). Then, there exists a positive constant C such that, for any $f \in H_A^{\vec{p}}(\mathbb{R}^n)$:*

$$\|T(f)\|_{L^{\vec{p}}(\mathbb{R}^n)} \le C\|f\|_{H_A^{\vec{p}}(\mathbb{R}^n)}. \tag{18}$$

To prove this theorem, we need the finite atomic characterization of anisotropic mixed-norm Hardy spaces $H_{A,\mathrm{fin}}^{\vec{p},r,s}(\mathbb{R}^n)$; see [1] (Theorem 5.3). Denote by $C(\mathbb{R}^n)$ the set of all continuous functions on $\mathbb{R}^n$.

Lemma 6. *Let $\vec{p} \in (0,\infty)^n$ and s be as in (5):*

(i) *If $r \in (\max\{p_+,1\},\infty)$ with p_+ as in (4), then $\|\cdot\|_{H_{A,\mathrm{fin}}^{\vec{p},r,s}(\mathbb{R}^n)}$ and $\|\cdot\|_{H_A^{\vec{p}}(\mathbb{R}^n)}$ are two equivalent quasi-norms on $H_{A,\mathrm{fin}}^{\vec{p},r,s}(\mathbb{R}^n)$;*

(ii) *$\|\cdot\|_{H_{A,\mathrm{fin}}^{\vec{p},\infty,s}(\mathbb{R}^n)}$ and $\|\cdot\|_{H_A^{\vec{p}}(\mathbb{R}^n)}$ are two equivalent quasi-norms on $H_{A,\mathrm{fin}}^{\vec{p},\infty,s}(\mathbb{R}^n) \cap C(\mathbb{R}^n)$.*

Here and thereafter, $H_{A,\mathrm{fin}}^{\vec{p},r,s}(\mathbb{R}^n)$ denotes the anisotropic mixed-norm finite atomic Hardy space, namely the set of all $f \in \mathcal{S}'(\mathbb{R}^n)$ satisfying that there exists $K \in \mathbb{N}$, $\{\lambda_k\}_{k\in[1,K]\cap\mathbb{N}} \subset \mathbb{C}$ and a finite sequence of $(\vec{p},r,s)$-atoms, $\{a_k\}_{k\in[1,K]\cap\mathbb{N}}$, supported, respectively, in $\{B^{(k)}\}_{k\in[1,K]\cap\mathbb{N}} \subset \mathfrak{B}$ such that:

$$f = \sum_{k=1}^{K} \lambda_k a_k \quad \text{in} \quad \mathcal{S}'(\mathbb{R}^n).$$

Moreover, for any $f \in H_{A,\mathrm{fin}}^{\vec{p},r,s}(\mathbb{R}^n)$, let:

$$\|f\|_{H_{A,\mathrm{fin}}^{\vec{p},r,s}(\mathbb{R}^n)} := \inf\left\|\left\{\sum_{k=1}^{K}\left[\frac{|\lambda_k|\mathbf{1}_{B^{(k)}}}{\|\mathbf{1}_{B^{(k)}}\|_{L^{\vec{p}}(\mathbb{R}^n)}}\right]^{\underline{p}}\right\}^{1/\underline{p}}\right\|_{L^{\vec{p}}(\mathbb{R}^n)},$$

where $\underline{p}$ is as in (4) and the infimum is taken over all decompositions of f as above.

In addition, let $\vec{p} \in (1,\infty)^n$ and $i \in \mathbb{Z}_+$. Then, by Lemma 4 and the fact that, for any dilated ball $B \in \mathfrak{B}$ and $\epsilon \in (0, \underline{p})$, $\mathbf{1}_{A^i B} \le b^{\frac{i}{\epsilon}} [M_{\mathrm{HL}}(\mathbf{1}_B)]^{\frac{1}{\epsilon}}$, we know that there exists a positive constant C such that, for any sequence $\{B^{(k)}\}_{k\in\mathbb{N}} \subset \mathfrak{B}$:

$$\left\| \sum_{k\in\mathbb{N}} \mathbf{1}_{A^i B^{(k)}} \right\|_{L^{\vec{p}}(\mathbb{R}^n)} \le C b^{\frac{i}{\epsilon}} \left\| \sum_{k\in\mathbb{N}} \mathbf{1}_{B^{(k)}} \right\|_{L^{\vec{p}}(\mathbb{R}^n)}. \tag{19}$$

Now, we show Theorem 2.

Proof of Theorem 2. Let $\vec{p}$, r and s be as in Lemma 6(i). We next prove this theorem in two steps.

Step (1). In this step, we prove that (18) holds true for any $f \in H^{\vec{p},r,s}_{A,\mathrm{fin}}(\mathbb{R}^n)$. For this purpose, for any $f \in H^{\vec{p},r,s}_{A,\mathrm{fin}}(\mathbb{R}^n)$, by Lemma 6, we can find some $K \in \mathbb{N}$, three finite sequences $\{\lambda_k\}_{k\in[1,K]\cap\mathbb{N}} \subset \mathbb{C}$, $\{x_k\}_{k\in[1,K]\cap\mathbb{N}} \subset \mathbb{R}^n$ and $\{i_k\}_{k\in[1,K]\cap\mathbb{N}} \subset \mathbb{Z}$, and a finite sequence of $(\vec{p},r,s)$-atoms, $\{a_k\}_{k\in[1,K]\cap\mathbb{N}}$, supported, respectively, in $\{x_k + B_{i_k}\}_{k\in[1,K]\cap\mathbb{N}} \subset \mathfrak{B}$ such that $f = \sum_{k=1}^K \lambda_k a_k$ in $\mathcal{S}'(\mathbb{R}^n)$ and:

$$\|f\|_{H^{\vec{p},r,s}_{A,\mathrm{fin}}(\mathbb{R}^n)} \sim \left\| \left\{ \sum_{k=1}^K \left[\frac{|\lambda_k| \mathbf{1}_{x_k+B_{i_k}}}{\|\mathbf{1}_{x_k+B_{i_k}}\|_{L^{\vec{p}}(\mathbb{R}^n)}} \right]^{\underline{p}} \right\}^{1/\underline{p}} \right\|_{L^{\vec{p}}(\mathbb{R}^n)}. \tag{20}$$

From the linearity of T and Lemma 5, we obtain:

$$\begin{aligned} \|T(f)\|_{L^{\vec{p}}(\mathbb{R}^n)} &\lesssim \left\| \sum_{k=1}^K |\lambda_k| T(a_k) \mathbf{1}_{x_k+B_{i_k+\omega}} \right\|_{L^{\vec{p}}(\mathbb{R}^n)} + \left\| \sum_{k=1}^K |\lambda_k| T(a_k) \mathbf{1}_{(x_k+B_{i_k+\omega})^\complement} \right\|_{L^{\vec{p}}(\mathbb{R}^n)} \\ &=: \mathrm{J}_1 + \mathrm{J}_2. \end{aligned} \tag{21}$$

We first deal with J_2. To do this, by a similar argument to that used in the proof of [44] (4.13), we conclude that, for each $k \in [1,K]\cap\mathbb{N}$ and $x \in (x_k + B_{i_k+\omega})^\complement$:

$$T(a_k)(x) \lesssim \left\| \mathbf{1}_{x_k+B_{i_k}} \right\|^{-1}_{L^{\vec{p}}(\mathbb{R}^n)} \left[M_{\mathrm{HL}}\left(\mathbf{1}_{x_k+B_{i_k}}\right)(x) \right]^u,$$

where:

$$u := \left(\frac{\ln b}{\ln \lambda_-} + s_0 + 1 \right) \frac{\ln \lambda_-}{\ln b} > \frac{1}{\underline{p}}.$$

This, together with Lemmas 5 and 4, and (20), implies that:

$$\begin{aligned} \|\mathrm{J}_2\|_{L^{\vec{p}}(\mathbb{R}^n)} &\lesssim \left\| \sum_{k=1}^K \frac{|\lambda_k|}{\|\mathbf{1}_{x_k+B_{i_k}}\|_{L^{\vec{p}}(\mathbb{R}^n)}} \left[M_{\mathrm{HL}}\left(\mathbf{1}_{x_k+B_{i_k}}\right) \right]^u \right\|_{L^{\vec{p}}(\mathbb{R}^n)} \\ &\sim \left\| \left\{ \sum_{k=1}^K \frac{|\lambda_k|}{\|\mathbf{1}_{x_k+B_{i_k}}\|_{L^{\vec{p}}(\mathbb{R}^n)}} \left[M_{\mathrm{HL}}(\mathbf{1}_{x_k+B_{i_k}}) \right]^u \right\}^{1/u} \right\|^u_{L^{u\vec{p}}} \\ &\lesssim \left\| \sum_{k=1}^K \frac{|\lambda_k| \mathbf{1}_{x_k+B_{i_k}}}{\|\mathbf{1}_{x_k+B_{i_k}}\|_{L^{\vec{p}}(\mathbb{R}^n)}} \right\|_{L^{\vec{p}}(\mathbb{R}^n)} \\ &\lesssim \left\| \left\{ \sum_{k=1}^K \left[\frac{|\lambda_k| \mathbf{1}_{x_k+B_{i_k}}}{\|\mathbf{1}_{x_k+B_{i_k}}\|_{L^{\vec{p}}(\mathbb{R}^n)}} \right]^{\underline{p}} \right\}^{1/\underline{p}} \right\|_{L^{\vec{p}}(\mathbb{R}^n)} \\ &\sim \|f\|_{H^{\vec{p},r,s}_{A,\mathrm{fin}}(\mathbb{R}^n)}. \end{aligned} \tag{22}$$

For J_1, take $g \in L^{(\vec{p}/\underline{p})'}(\mathbb{R}^n)$ such that $\|g\|_{L^{(\vec{p}/\underline{p})'}(\mathbb{R}^n)} \le 1$ and:

$$\left\| \sum_{k=1}^{K} |\lambda_k|^{\underline{p}} [T(a_k)]^{\underline{p}} \mathbf{1}_{x_k+B_{i_k+\omega}} \right\|_{L^{\vec{p}/\underline{p}}(\mathbb{R}^n)} = \int_{\mathbb{R}^n} \sum_{k=1}^{K} |\lambda_k|^{\underline{p}} [T(a_k)(x)]^{\underline{p}} \mathbf{1}_{x_k+B_{i_k+\omega}}(x) g(x)\, dx.$$

From this, Lemma 5 and the Hölder inequality, it follows that, for any $q \in (1,\infty)$ satisfying $p_+ < q\underline{p} < r$:

$$\begin{aligned}
(\mathrm{J}_1)^{\underline{p}} &\lesssim \left\| \sum_{k=1}^{K} |\lambda_k|^{\underline{p}} [T(a_k)]^{\underline{p}} \mathbf{1}_{x_k+B_{i_k+\omega}} \right\|_{L^{\vec{p}/\underline{p}}(\mathbb{R}^n)} \\
&\sim \int_{\mathbb{R}^n} \sum_{k=1}^{K} |\lambda_k|^{\underline{p}} [T(a_k)(x)]^{\underline{p}} \mathbf{1}_{x_k+B_{i_k+\omega}}(x) g(x)\, dx. \\
&\lesssim \sum_{k=1}^{K} |\lambda_k|^{\underline{p}} \left\| [T(a_k)]^{\underline{p}} \mathbf{1}_{x_k+B_{i_k+\omega}} \right\|_{L^q(\mathbb{R}^n)} \left\| \mathbf{1}_{x_k+B_{i_k+\omega}} g \right\|_{L^{q'}(\mathbb{R}^n)} \\
&\lesssim \sum_{k=1}^{K} |\lambda_k|^{\underline{p}} \|T(a_k)\|_{L^r(\mathbb{R}^n)}^{\underline{p}} \left\| \mathbf{1}_{x_k+B_{i_k+\omega}} \right\|_{L^{r/(r-q\underline{p})}(\mathbb{R}^n)}^{1/q} \left\| \mathbf{1}_{x_k+B_{i_k+\omega}} g \right\|_{L^{q'}(\mathbb{R}^n)}.
\end{aligned}$$

This, combined with the boundedness of T on $L^t(\mathbb{R}^n)$ for any $t \in (1,\infty)$ (see [29] (p. 60)), Definition 8(i) and the Hölder inequality again, further implies that:

$$\begin{aligned}
(\mathrm{J}_1)^{\underline{p}} &\lesssim \sum_{k=1}^{K} |\lambda_k|^{\underline{p}} \left\| \mathbf{1}_{x_k+B_{i_k}} \right\|_{L^{\vec{p}}(\mathbb{R}^n)}^{-\underline{p}} |B_{i_k}|^{\underline{p}/r} |B_{i_k+\omega}|^{(r-q\underline{p})/rq} \left\| \mathbf{1}_{x_k+B_{i_k+\omega}} g \right\|_{L^{q'}(\mathbb{R}^n)} \\
&\sim \sum_{k=1}^{K} |\lambda_k|^{\underline{p}} \left\| \mathbf{1}_{x_k+B_{i_k}} \right\|_{L^{\vec{p}}(\mathbb{R}^n)}^{-\underline{p}} |B_{i_k+\omega}| \left[\frac{1}{|B_{i_k+\omega}|} \int_{x_k+B_{i_k+\omega}} [g(x)]^{q'}\, dx \right]^{1/q'} \\
&\lesssim \sum_{k=1}^{K} |\lambda_k|^{\underline{p}} \left\| \mathbf{1}_{x_k+B_{i_k}} \right\|_{L^{\vec{p}}(\mathbb{R}^n)}^{-\underline{p}} \int_{\mathbb{R}^n} \mathbf{1}_{x_k+B_{i_k+\omega}}(x) \left[M_{\mathrm{HL}}\left(g^{q'}\right)(x) \right]^{1/q'} dx \\
&\lesssim \left\| \sum_{k=1}^{K} |\lambda_k|^{\underline{p}} \left\| \mathbf{1}_{x_k+B_{i_k}} \right\|_{L^{\vec{p}}(\mathbb{R}^n)}^{-\underline{p}} \mathbf{1}_{x_k+B_{i_k+\omega}} \right\|_{L^{\vec{p}/\underline{p}}(\mathbb{R}^n)} \left\| \left[M_{\mathrm{HL}}\left(g^{q'}\right) \right]^{1/q'} \right\|_{L^{(\vec{p}/\underline{p})'}(\mathbb{R}^n)}.
\end{aligned}$$

Note that $p_+/\underline{p} \in (0,q)$, we know that $(\vec{p}/\underline{p})' \in (q',\infty]$. By this, (19), the boundedness of M_{HL} on $L^{\vec{v}}(\mathbb{R}^n)$ with $\vec{v} \in (1,\infty]^n$ (see [10] (Lemma 3.5)), Lemma 5, the fact that $\|g\|_{L^{(\vec{p}/\underline{p})'}(\mathbb{R}^n)} \le 1$ and (20), we conclude that:

$$\begin{aligned}
\mathrm{J}_1 &\lesssim \left\| \sum_{k=1}^{K} |\lambda_k|^{\underline{p}} \left\| \mathbf{1}_{x_k+B_{i_k}} \right\|_{L^{\vec{p}}(\mathbb{R}^n)}^{-\underline{p}} \mathbf{1}_{x_k+B_{i_k}} \right\|_{L^{\vec{p}/\underline{p}}(\mathbb{R}^n)}^{1/\underline{p}} \|g\|_{L^{(\vec{p}/\underline{p})'}(\mathbb{R}^n)}^{1/\underline{p}} \\
&\lesssim \left\| \left\{ \sum_{k=1}^{K} \left[\frac{|\lambda_k| \mathbf{1}_{x_k+B_{i_k}}}{\left\| \mathbf{1}_{x_k+B_{i_k}} \right\|_{L^{\vec{p}}(\mathbb{R}^n)}} \right]^{\underline{p}} \right\}^{1/\underline{p}} \right\|_{L^{\vec{p}}(\mathbb{R}^n)} \\
&\sim \|f\|_{H_{A,\mathrm{fin}}^{\vec{p},r,s}(\mathbb{R}^n)}.
\end{aligned}$$

From this, (22), (21) and Lemma 6(i), we deduce that (18) holds true for any $f \in H_{A,\mathrm{fin}}^{\vec{p},r,s}(\mathbb{R}^n)$, which completes the proof of Step (1).

Step (2). This step aims to show that (18) holds true for any $f \in H_A^{\vec{p}}(\mathbb{R}^n)$. To this end, for any $f \in H_A^{\vec{p}}(\mathbb{R}^n)$, by the obvious density of $H_{A,\mathrm{fin}}^{\vec{p},r,s}(\mathbb{R}^n)$ in $H_A^{\vec{p}}(\mathbb{R}^n)$, with respect to the quasi-norm $\|\cdot\|_{H_A^{\vec{p}}(\mathbb{R}^n)}$, we find that there exists a Cauchy sequence $\{f_i\}_{i\in\mathbb{N}} \subset H_{A,\mathrm{fin}}^{\vec{p},r,s}(\mathbb{R}^n)$ such that

$$\lim_{i\to\infty}\|f_i - f\|_{H_A^{\vec{p}}(\mathbb{R}^n)} = 0.$$

By this and the linearity of T, it is easy to see that, as $i, \iota \to \infty$:

$$\begin{aligned}\|T(f_i) - T(f_\iota)\|_{H_A^{\vec{p}}(\mathbb{R}^n)} &= \|T(f_i - f_\iota)\|_{H_A^{\vec{p}}(\mathbb{R}^n)}\\ &\lesssim \|f_i - f_\iota\|_{H_A^{\vec{p}}(\mathbb{R}^n)} \to 0.\end{aligned}$$

Therefore, $\{T(f_i)\}_{i\in\mathbb{N}}$ is also a Cauchy sequence in $H_A^{\vec{p}}(\mathbb{R}^n)$. By this and the completeness of $H_A^{\vec{p}}(\mathbb{R}^n)$, we know that there exists some $h \in H_A^{\vec{p}}(\mathbb{R}^n)$ such that $h = \lim_{i\to\infty} T(f_i)$ in $H_A^{\vec{p}}(\mathbb{R}^n)$. Let $T(f) := h$. Then, for any $f \in H_A^{\vec{p}}(\mathbb{R}^n)$:

$$\begin{aligned}\|T(f)\|_{H_A^{\vec{p}}(\mathbb{R}^n)} &\lesssim \limsup_{i\to\infty}\left[\|T(f) - T(f_i)\|_{H_A^{\vec{p}}(\mathbb{R}^n)} + \|T(f_i)\|_{H_A^{\vec{p}}(\mathbb{R}^n)}\right]\\ &\sim \limsup_{i\to\infty}\|T(f_i)\|_{H_A^{\vec{p}}(\mathbb{R}^n)} \lesssim \lim_{i\to\infty}\|f_i\|_{H_A^{\vec{p}}(\mathbb{R}^n)} \sim \|f\|_{H_A^{\vec{p}}(\mathbb{R}^n)}.\end{aligned} \tag{23}$$

This finishes the proof of Step (2) and hence of Theorem 2. □

Motivated by [29] (p. 64, Definition 9.4), we introduce the vanishing moment condition as follows.

Definition 11. *Let $\vec{p} \in (0,\infty)^n$, $\ell \in \mathbb{N}$ satisfy:*

$$\frac{1}{p_-} - 1 < \frac{(\ln\lambda_-)^2}{\ln b \ln\lambda_+}\ell$$

and $s_0 := \lfloor (1/p_- - 1)\ln b/\ln\lambda_- \rfloor$, where p_- is as in (4). An anisotropic Calderón–Zygmund operator T of order ℓ is said to satisfy $T^*(x^\gamma) = 0$ for any $\gamma \in \mathbb{Z}_+^n$ with $|\gamma| \le s_0$ if, for any $g \in L^2(\mathbb{R}^n)$ with compact support and satisfying that, for each $\beta \in \mathbb{Z}_+^n$ with $|\beta| \le \ell$, $\int_{\mathbb{R}^n} g(x)x^\beta\,dx = 0$, the equality $\int_{\mathbb{R}^n} T(g)(x)x^\gamma\,dx = 0$ holds true for each $\gamma \in \mathbb{Z}_+^n$ satisfying $|\gamma| \le s_0$.

We have the following boundedness of anisotropic Calderón–Zygmund operators on $H_A^{\vec{p}}(\mathbb{R}^n)$.

Theorem 3. *Let $\vec{p}$, ℓ, s_0 be as in Definition 11. Assume that T is an anisotropic Calderón–Zygmund operator of order ℓ and satisfies $T^*(x^\gamma) = 0$ for any $\gamma \in \mathbb{Z}_+^n$ with $|\gamma| \le s_0$. Then, there exists a positive constant C such that, for any $f \in H_A^{\vec{p}}(\mathbb{R}^n)$,*

$$\|T(f)\|_{H_A^{\vec{p}}(\mathbb{R}^n)} \le C\|f\|_{H_A^{\vec{p}}(\mathbb{R}^n)}.$$

By [1] (Lemma 6.8) and [45] (Lemma 2.3), we easily obtain the succeeding Lemma 7; the details are omitted.

Lemma 7. *Assume that $E \subset \mathbb{R}^n$, $F \in \mathfrak{B}$ with $\mathfrak{B}$ as in (1), $E \subset F$ and there exists a constant $c_0 \in (0,1]$ such that $|E| \ge c_0|F|$. Then, for any $\vec{p} \in (0,\infty)^n$, there exists a positive constant C, independent of E and F, such that:*

$$\frac{\|\mathbf{1}_F\|_{L^{\vec{p}}(\mathbb{R}^n)}}{\|\mathbf{1}_E\|_{L^{\vec{p}}(\mathbb{R}^n)}} \le C.$$

To prove Theorem 3, we need the following technical lemma, which is motivated by [44] (Lemma 4.10) and [39] (Lemma 4.13).

Lemma 8. *Let $\vec{p}$, ℓ, s_0 be as in Definition 11. Assume that $r \in (1, \infty]$ and T is an anisotropic Calderón–Zygmund operator of order ℓ satisfying $T^*(x^\gamma) = 0$ for any $\gamma \in \mathbb{Z}_+^n$ with $|\gamma| \le s_0$. Then, there exists a positive constant C such that, for any $(\vec{p}, r, \ell)$-atom $\widetilde{a}$ supported in some dilated ball $x_0 + B_{i_0} \in \mathfrak{B}$ with $x_0 \in \mathbb{R}^n$, $i_0 \in \mathbb{Z}$ and $\mathfrak{B}$ as in (1), $\frac{1}{C}T(\widetilde{a})$ is a $(\vec{p}, r, s_0, \varepsilon)$-molecule associated with $x_0 + B_{i_0+\omega}$, where:*

$$\varepsilon := \ell \log_b(\lambda_-) + 1/r' \tag{24}$$

and ω is as in (2).

Proof. Let T be an anisotropic Calderón–Zygmund operator of order ℓ satisfying:

$$T^*(x^\gamma) = 0 \text{ for any } \gamma \in \mathbb{Z}_+^n \text{ with } |\gamma| \le s_0.$$

For any $(\vec{p}, r, \ell)$-atom $\widetilde{a}$ supported in some dilated ball $x_0 + B_{i_0} \in \mathfrak{B}$, without losing generality, we may assume that $x_0 = \mathbf{0}$. Then, by the vanishing moments of $\widetilde{a}$ and Definition 11, we find that $T(\widetilde{a})$ has vanishing moments up to an order of s_0.

Let $U_0(B_{i_0}) := B_{i_0+\omega}$ and, for any $k \in \mathbb{N}$:

$$U_k(B_{i_0}) := (A^k B_{i_0+\omega}) \setminus (A^{k-1} B_{i_0+\omega}).$$

To show that $T(\widetilde{a})$ is a harmless constant multiple of a $(\vec{p}, r, s_0, \varepsilon)$-molecule associated with $B_{i_0+\omega}$, it suffices to prove that, for any $k \in \mathbb{Z}_+$:

$$\|T(\widetilde{a})\|_{L^r(U_k(B_{i_0}))} \lesssim \frac{b^{-k\varepsilon}|B_{i_0+\omega}|^{1/r}}{\|\mathbf{1}_{B_{i_0+\omega}}\|_{L^{\vec{p}}(\mathbb{R}^n)}}, \tag{25}$$

where ε is as in (24).

Indeed, from the boundedness of T on L^r, the fact that supp $\widetilde{a} \subset B_{i_0}$, the size condition of $\widetilde{a}$ and Lemma 7, it follows that:

$$\|T(\widetilde{a})\|_{L^r(U_k(B_{i_0}))} \lesssim \|\widetilde{a}\|_{L^r(B_{i_0})} \lesssim \frac{|B_{i_0}|^{1/r}}{\|\mathbf{1}_{B_{i_0}}\|_{L^{\vec{p}}(\mathbb{R}^n)}} \lesssim \frac{|B_{i_0+\omega}|^{1/r}}{\|\mathbf{1}_{B_{i_0+\omega}}\|_{L^{\vec{p}}(\mathbb{R}^n)}}$$

and hence (25) holds true for $k = 0$.

On another hand, for any $(\vec{p}, r, \ell)$-atom $\widetilde{a}$, $k \in \mathbb{N}$, $x \in U_k(B_{i_0})$ and $y \in B_{i_0}$, by Lemma 3(i), we know that $x - y \in B_{i_0+k+2\omega} \setminus B_{i_0+k-1}$, which implies that $\rho(x-y) \sim b^{i_0+k}$. From this and (17), we deduce that, for any $\gamma \in \mathbb{Z}_+^n$ with $1 \le |\gamma| \le \ell$:

$$\left|\partial_y^\gamma \left[\mathcal{K}(\cdot, A^{i_0+k}\cdot)\right]\left(x, A^{-i_0-k}y\right)\right| \lesssim [\rho(x-y)]^{-1} \lesssim b^{-i_0-k}. \tag{26}$$

Note that supp $\widetilde{a} \subset B_{i_0}$. Then, we have:

$$T(\widetilde{a})(x) = \int_{B_{i_0}} \mathcal{K}(x,y)\widetilde{a}(y)\,dy = \int_{B_{i_0}} \widetilde{\mathcal{K}}(x, A^{-i_0-k}y)\widetilde{a}(y)\,dy, \tag{27}$$

where $\widetilde{\mathcal{K}}(x,y) := \mathcal{K}(x, A^{i_0+k}y)$ for any x, $y \in \mathbb{R}^n$ with $x \ne A^{i_0+k}y$. Moreover, by Taylor expansion theorem for the variable y at the point $(x, \mathbf{0})$, we easily obtain:

$$\widetilde{\mathcal{K}}(x, \widetilde{y}) = \sum_{\gamma \in \mathbb{Z}_+^n, |\gamma| \le \ell-1} \frac{\partial_y^\gamma \widetilde{\mathcal{K}}(x, \mathbf{0})}{\gamma!}(\widetilde{y})^\gamma + R_\ell(\widetilde{y}), \tag{28}$$

where $\widetilde{y} := A^{-i_0-k}y$ for any $y \in B_{i_0}$. This, combined with (26), further implies that:

$$|R_\ell(\widetilde{y})| \lesssim \sup_{t\in B_{-k}} \sup_{\gamma\in\mathbb{Z}_+^n, |\gamma|=\ell} \left|\partial_y^\gamma \widetilde{\mathcal{K}}(x,t)\right| |\widetilde{y}|^\ell \lesssim b^{-i_0-k} \sup_{t\in B_{-k}} |t|^\ell.$$

By the fact that, for any $t \in B_{-k}$, $\rho(t) < b^{-k} < 1$ and [29] (p. 11, Lemma 3.2), we conclude that, for any $\ell \in \mathbb{N}$ as in Definition 11,

$$\sup_{t\in B_{-k}} |t|^\ell \lesssim \sup_{t\in B_{-k}} [\rho(t)]^{\ell \frac{\ln \lambda_-}{\ln b}} \lesssim b^{-k\ell \log_b(\lambda_-)}.$$

Thus, we have:

$$|R_\ell(\widetilde{y})| \lesssim b^{-i_0-k} b^{-k\ell \log_b(\lambda_-)}.$$

From this, (27), (28), the vanishing moments of atoms and the Hölder inequality, it follows that, for any $(\vec{p}, r, \ell)$-atom $\widetilde{a}$, $k \in \mathbb{N}$ and $x \in U_k(B_{i_0})$:

$$\begin{aligned}
|T(\widetilde{a})(x)| &\le \int_{B_{i_0}} \left|R_\ell(A^{-i_0-k}y)\widetilde{a}(y)\right| dy \\
&\lesssim b^{-i_0-k} b^{-k\ell\log_b(\lambda_-)} \int_{B_{i_0}} |\widetilde{a}(y)|\, dy \\
&\lesssim b^{-i_0-k} b^{-k\ell\log_b(\lambda_-)} |B_{i_0}|^{1/r'} \|\widetilde{a}\|_{L^r(B_{i_0})} \\
&\sim b^{-k[1+\ell\log_b(\lambda_-)]} b^{-i_0/r} \|\widetilde{a}\|_{L^r(B_{i_0})}.
\end{aligned}$$

This, together with the size condition of $\widetilde{a}$, (24) and Lemma 7, imply that, for any $k \in \mathbb{N}$:

$$\begin{aligned}
\|T(\widetilde{a})\|_{L^r(U_k(B_{i_0}))} &\lesssim b^{-k[1+\ell\log_b(\lambda_-)]} b^{-i_0/r} \|\widetilde{a}\|_{L^r(B_{i_0})} |B_{i_0+k+\omega}|^{1/r} \\
&\lesssim b^{-k[1+\ell\log_b(\lambda_-)]} b^{k/r} \frac{|B_{i_0}|^{1/r}}{\|\mathbf{1}_{B_{i_0}}\|_{L^{\vec{p}}(\mathbb{R}^n)}} \\
&\lesssim \frac{b^{-k\varepsilon}|B_{i_0+\omega}|^{1/r}}{\|\mathbf{1}_{B_{i_0+\omega}}\|_{L^{\vec{p}}(\mathbb{R}^n)}},
\end{aligned}$$

which completes the proof of (25) for $k \in \mathbb{N}$ and hence of Lemma 8. □

In addition, we also need the subsequent density of $H_A^{\vec{p}}(\mathbb{R}^n)$.

Lemma 9. *Let $\vec{p} \in (0,\infty)^n$. Then:*

(i) *$H_A^{\vec{p}}(\mathbb{R}^n) \cap C_c^\infty(\mathbb{R}^n)$ is dense in $H_A^{\vec{p}}(\mathbb{R}^n)$; here and thereafter, $C_c^\infty(\mathbb{R}^n)$ denotes the set of all infinitely differentiable functions with compact support on $\mathbb{R}^n$;*

(ii) *For any s as in (5), $H_{A,\mathrm{fin}}^{\vec{p},\infty,s}(\mathbb{R}^n) \cap C(\mathbb{R}^n)$ is dense in $H_A^{\vec{p}}(\mathbb{R}^n)$.*

Proof. To prove (i), we first show that, for any $\varphi \in \mathcal{S}(\mathbb{R}^n)$ with $\int_{\mathbb{R}^n} \varphi(x)\,dx \neq 0$ and $f \in H_A^{\vec{p}}(\mathbb{R}^n)$, as $k \to -\infty$,

$$f * \varphi_k \to f \quad \text{in} \quad H_A^{\vec{p}}(\mathbb{R}^n). \tag{29}$$

For this purpose, we first assume that $f \in H_A^{\vec{p}}(\mathbb{R}^n) \cap L^2(\mathbb{R}^n)$. In this case, to prove (29), we only need to show that, for almost every $x \in \mathbb{R}^n$, as $k \to -\infty$:

$$M_N(f * \varphi_k - f)(x) \to 0 \quad \text{for almost every } x \in \mathbb{R}^n \text{ as } k \to -\infty \tag{30}$$

where $N := N_{\vec{p}} + 2$ with $N_{\vec{p}} := \lfloor (\frac{1}{\min\{1,p_-\}} - 1)\frac{\ln b}{\ln \lambda_-} \rfloor + 2$. Indeed, note that, for any $k \in \mathbb{Z}$, $f * \varphi_k - f \in L^2(\mathbb{R}^n)$. Then, by [29] (p. 13, Theorem 3.6), we know that, for any $k \in \mathbb{Z}$, $M_N(f * \varphi_k - f) \in L^2(\mathbb{R}^n)$. From this, ref. [29] (p. 39, Lemma 6.6), (30) and the Lebesgue-dominated convergence theorem, it follows that, (29) holds true for any $f \in H_A^{\vec{p}}(\mathbb{R}^n) \cap L^2(\mathbb{R}^n)$.

Subsequently, we prove (30). To this end, let g be a continuous function with compact support. Then, g is uniformly continuous on $\mathbb{R}^n$. Thus, for any $\delta \in (0, \infty)$, there exists some $\eta \in (0, \infty)$ such that, for any $y \in \mathbb{R}^n$ satisfying $\rho(y) < \eta$ and $x \in \mathbb{R}^n$,

$$|g(x-y) - g(x)| < \frac{\delta}{2\|\varphi\|_{L^1(\mathbb{R}^n)}}.$$

Without loss of generality, we can assume that $\int_{\mathbb{R}^n} \varphi(x)\,dx = 1$. Then, for any $k \in \mathbb{Z}$ and $x \in \mathbb{R}^n$, we have:

$$\begin{aligned}|g * \varphi_k(x) - g(x)| &\le \int_{\rho(y)<\eta} |g(x-y) - g(x)||\varphi_k(y)|\,dy + \int_{\rho(y)\ge\eta} \cdots \\ &< \frac{\delta}{2} + 2\|g\|_{L^\infty(\mathbb{R}^n)} \int_{\rho(y)\ge b^{-k}\eta} |\varphi(y)|\,dy. \end{aligned} \tag{31}$$

By the integrability of φ, we can find a $K \in \mathbb{Z}$ such that, for any $k \in (-\infty, K] \cap \mathbb{Z}$:

$$2\|g\|_{L^\infty(\mathbb{R}^n)} \int_{\rho(y)\ge b^{-k}\eta} |\varphi(y)|\,dy < \frac{\delta}{2}.$$

From this and (31), we deduce that, for any $x \in \mathbb{R}^n$:

$$\lim_{k\to-\infty} |g * \varphi_k(x) - g(x)| = 0 \quad \text{holds true uniformly.}$$

Therefore, $\|g * \varphi_k - g\|_{L^\infty(\mathbb{R}^n)} \to 0$ as $k \to -\infty$. This, together with [29] (p. 13, Theorem 3.6), again implies that:

$$\|M_N(g * \varphi_k - g)\|_{L^\infty(\mathbb{R}^n)} \lesssim \|g * \varphi_k - g\|_{L^\infty(\mathbb{R}^n)} \to 0 \quad \text{as } k \to -\infty. \tag{32}$$

For any given $\epsilon \in (0, \infty)$, there exists a continuous function g with compact support such that:

$$\|f - g\|^2_{L^2(\mathbb{R}^n)} < \epsilon.$$

By (32) and [29] (p. 39, Lemma 6.6), we again know that there exists a positive constant κ such that, for any $x \in \mathbb{R}^n$:

$$\begin{aligned}&\limsup_{k\to-\infty} M_N(f * \varphi_k - f)(x) \\ &\quad \le \sup_{k\in\mathbb{Z}} M_N((f-g) * \varphi_k)(x) + \limsup_{k\to-\infty} M_N(g * \varphi_k - g)(x) + M_N(g-f)(x) \\ &\quad \le \kappa M_{N_{\vec{p}}}(g-f)(x).\end{aligned}$$

Thus, for any $\lambda \in (0, \infty)$, we have:

$$\begin{aligned}&\left|\left\{x \in \mathbb{R}^n : \limsup_{k\to-\infty} M_N(f * \varphi_k - f)(x) > \lambda\right\}\right| \\ &\quad \le \left|\left\{x \in \mathbb{R}^n : M_{N_{\vec{p}}}(g-f)(x) > \frac{\lambda}{\kappa}\right\}\right| \lesssim \frac{\|f-g\|^2_{L^2(\mathbb{R}^n)}}{\lambda^2} \lesssim \frac{\epsilon}{\lambda^2}.\end{aligned}$$

This implies that, for any $f \in H_A^{\vec{p}}(\mathbb{R}^n) \cap L^2(\mathbb{R}^n)$, (30) holds true.

When $f \in H_A^{\vec{p}}(\mathbb{R}^n)$, by an argument similar to that used in [43] (p. 1700), it is easy to see that (29) also holds true.

Moreover, if $f \in H_{A,\,\mathrm{fin}}^{\vec{p},r,s}(\mathbb{R}^n)$ and $\varphi \in C_c^\infty(\mathbb{R}^n)$ with $\int_{\mathbb{R}^n} \varphi(x)\,dx \neq 0$, then, for any $k \in \mathbb{Z}$,

$$f * \varphi_k \in C_c^\infty(\mathbb{R}^n) \cap H_A^{\vec{p}}(\mathbb{R}^n)$$

and, by (29),

$$f * \varphi_k \to f \quad \text{in} \quad H_A^{\vec{p}}(\mathbb{R}^n) \quad \text{as} \quad k \to -\infty.$$

This, combined with the density of the set $H_{A,\,\mathrm{fin}}^{\vec{p},r,s}(\mathbb{R}^n)$ in $H_A^{\vec{p}}(\mathbb{R}^n)$, further implies that $C_c^\infty(\mathbb{R}^n) \cap H_A^{\vec{p}}(\mathbb{R}^n)$ is dense in $H_A^{\vec{p}}(\mathbb{R}^n)$, which completes the proof of (i).

We now prove (ii). By (i) and the proof of [43] (Theorem 6.13 (ii)) with some slight modifications, we conclude that $H_{A,\,\mathrm{fin}}^{\vec{p},\infty,s}(\mathbb{R}^n) \cap C(\mathbb{R}^n)$ is dense in $H_A^{\vec{p}}(\mathbb{R}^n)$. This finishes the proof of (ii) and hence of Lemma 9. □

Applying Lemmas 6, 7 and 9 as well as Theorem 1, we obtain a criterion on the boundedness of linear operators on $H_A^{\vec{p}}(\mathbb{R}^n)$ as follows, which plays a key role in the proof of Theorem 3.

Theorem 4. *Let T be a linear operator defined on the set of all measurable functions. Assume that $\vec{p} \in (0,\infty)^n$, $r \in (\max\{p_+,1\},\infty]$ with p_+ as in (4) and $\widetilde{s}$ is as in (5) with s replaced by $\widetilde{s}$. If there exists some $i_0 \in \mathbb{Z}$ and a positive constant C such that, for any $(p(\cdot),r,\widetilde{s})$-atom $\widetilde{a}$ supported in some dilated ball $x_0 + B_{k_0} \in \mathfrak{B}$ with $x_0 \in \mathbb{R}^n$, $k_0 \in \mathbb{Z}$ and $\mathfrak{B}$ as in (1), $\frac{1}{C}T(\widetilde{a})$ is a $(p(\cdot),r,s,\varepsilon)$-molecule associated with $x_0 + B_{k_0+i_0}$, where s and ε are as in Theorem 1, then T has a unique bounded linear extension on $H_A^{\vec{p}}(\mathbb{R}^n)$.*

Proof. Let $\vec{p} \in (0,\infty)^n$, $r \in (\max\{p_+,1\},\infty]$ and

$$\widetilde{s} \in \left[\left\lfloor \left(\frac{1}{p_-} - 1\right)\frac{\ln b}{\ln \lambda_-}\right\rfloor, \infty\right) \cap \mathbb{Z}_+$$

with p_- as in (4). We next show Theorem 4 by considering two cases.

Case (1). $r \in (\max\{p_+,1\},\infty)$. For this case, let $f \in H_{A,\,\mathrm{fin}}^{\vec{p},r,\widetilde{s}}(\mathbb{R}^n)$. Then, by the notion of $H_{A,\,\mathrm{fin}}^{\vec{p},r,\widetilde{s}}(\mathbb{R}^n)$ in Lemma 6, we find that there exists some $K \in \mathbb{N}$, three finite sequences $\{\lambda_k\}_{k\in[1,K]\cap\mathbb{N}} \subset \mathbb{C}$, $\{x_k\}_{k\in[1,K]\cap\mathbb{N}} \subset \mathbb{R}^n$ and $\{i_k\}_{k\in[1,K]\cap\mathbb{N}} \subset \mathbb{Z}$, and a finite sequence of $(\vec{p},r,s)$-atoms, $\{a_k\}_{k\in[1,K]\cap\mathbb{N}}$, supported, respectively, in $\{x_k + B_{i_k}\}_{k\in[1,K]\cap\mathbb{N}} \subset \mathfrak{B}$ such that:

$$f = \sum_{k=1}^{K} \lambda_k a_k \quad \text{in} \quad \mathcal{S}'(\mathbb{R}^n) \tag{33}$$

and:

$$\|f\|_{H_{A,\,\mathrm{fin}}^{\vec{p},r,s}(\mathbb{R}^n)} \sim \left\| \left\{ \sum_{k=1}^{K} \left[\frac{|\lambda_k| \mathbf{1}_{x_k+B_{i_k}}}{\|\mathbf{1}_{x_k+B_{i_k}}\|_{L^{\vec{p}}(\mathbb{R}^n)}} \right]^{\underline{p}} \right\}^{1/\underline{p}} \right\|_{L^{\vec{p}}(\mathbb{R}^n)}. \tag{34}$$

This, together with (33) and the linearity of T, implies that $T(f) = \sum_{k=1}^{K} \lambda_k T(a_k)$ in $\mathcal{S}'(\mathbb{R}^n)$, where, for any $k \in [1,K] \cap \mathbb{N}$, $\frac{1}{C}T(a_k)$ with C being a positive constant independent of k is a $(\vec{p},r,s,\varepsilon)$-molecule associated with $x_k + B_{i_k+i_0}$ with s, ε and i_0 as in Theorem 4.

From this, Theorem 1, Definition 7, as well as Lemmas 7, 4 and 5, (34) and Lemma 6, we further deduce that, for any $f \in H_{A,\mathrm{fin}}^{\vec{p},r,\vec{s}}(\mathbb{R}^n)$:

$$\begin{aligned}\|T(f)\|_{H_A^{\vec{p}}(\mathbb{R}^n)} &\sim \|T(f)\|_{H_A^{\vec{p},r,s,\varepsilon}(\mathbb{R}^n)} \\ &\lesssim \left\| \left\{ \sum_{k=1}^{K} \left[\frac{|\lambda_k| \mathbf{1}_{x_k+B_{i_k+i_0}}}{\|\mathbf{1}_{x_k+B_{i_k+i_0}}\|_{L^{\vec{p}}(\mathbb{R}^n)}} \right]^{\underline{p}} \right\}^{1/\underline{p}} \right\|_{L^{\vec{p}}(\mathbb{R}^n)} \\ &\lesssim b^{i_0/\tau} \left\| \left[\sum_{k=1}^{K} \left\{ \frac{|\lambda_k| [M_{\mathrm{HL}}(\mathbf{1}_{x_k+B_{i_k}})]^{1/\tau}}{\|\mathbf{1}_{x_k+B_{i_k}}\|_{L^{\vec{p}}(\mathbb{R}^n)}} \right\}^{\underline{p}} \right]^{1/\underline{p}} \right\|_{L^{\vec{p}}(\mathbb{R}^n)} \\ &\sim b^{i_0/\tau} \left\| \left\{ \sum_{k=1}^{K} \left[\frac{|\lambda_k|^\tau M_{\mathrm{HL}}(\mathbf{1}_{x_k+B_{i_k}})}{\|\mathbf{1}_{x_k+B_{i_k}}\|_{L^{\vec{p}}(\mathbb{R}^n)}^\tau} \right]^{\underline{p}/\tau} \right\}^{\tau/\underline{p}} \right\|_{L^{\vec{p}/\tau}(\mathbb{R}^n)}^{1/\tau} \\ &\lesssim \left\| \left\{ \sum_{k=1}^{K} \left[\frac{|\lambda_k| \mathbf{1}_{x_k+B_{i_k}}}{\|\mathbf{1}_{x_k+B_{i_k}}\|_{L^{\vec{p}}(\mathbb{R}^n)}} \right]^{\underline{p}} \right\}^{1/\underline{p}} \right\|_{L^{\vec{p}}(\mathbb{R}^n)} \\ &\sim \|f\|_{H_{A,\mathrm{fin}}^{\vec{p},r,\vec{s}}(\mathbb{R}^n)} \sim \|f\|_{H_A^{\vec{p}}(\mathbb{R}^n)},\end{aligned} \tag{35}$$

where $\tau \in (0, \underline{p})$ is a constant.

Moreover, by the obvious density of $H_{A,\mathrm{fin}}^{p(\cdot),r,\vec{s}}(\mathbb{R}^n)$ in $H_A^{\vec{p}}(\mathbb{R}^n)$ with respect to the quasi-norm $\|\cdot\|_{H_A^{\vec{p}}(\mathbb{R}^n)}$ and a proof similar to the estimation of (23), we conclude that, for any $f \in H_A^{\vec{p}}(\mathbb{R}^n)$, (35) also holds true. This finishes the proof of Theorem 4 in Case (1).

Case (2). $r = \infty$. In this case, by Lemma 9(ii), we know that $H_{A,\mathrm{fin}}^{\vec{p},\infty,\vec{s}}(\mathbb{R}^n) \cap C(\mathbb{R}^n)$ is dense in $H_A^{\vec{p}}(\mathbb{R}^n)$. From this, repeating the proof of Case (1) with some slight modifications, it follows that Theorem 4 also holds true when $r = \infty$, which completes the proof of Theorem 4. □

We now prove Theorem 3.

Proof of Theorem 3. Indeed, Theorem 3 is an immediate corollary of Theorem 4 and Lemma 8. This finishes the proof of Theorem 3. □

Remark 2. (i) *Assume that $\ell \in \mathbb{N}$, $p \in (0,1]$ and:*

$$\frac{1}{p} - 1 \le \frac{(\ln \lambda_-)^2}{\ln b \ln \lambda_+} \ell. \tag{36}$$

When $\vec{p} := (\overbrace{p,\ldots,p}^{n \text{ times}})$ with some $p \in (0,\infty)$, the spaces $H_A^{\vec{p}}(\mathbb{R}^n)$ and $L^{\vec{p}}(\mathbb{R}^n)$ are just, respectively, the anisotropic Hardy space $H_A^p(\mathbb{R}^n)$ of Bownik [29] and the Lebesgue space $L^p(\mathbb{R}^n)$. In this case, Theorems 2 and 3 implies that, for any $\ell \in \mathbb{N}$ and $p \in (0,1]$ as in (36), the anisotropic Calderón–Zygmund operator of order ℓ (see Definition 10) is bounded from $H_A^p(\mathbb{R}^n)$ to $L^p(\mathbb{R}^n)$ (or to itself), which are just, respectively, ref. [29] (p. 69, Theorem 9.9 and p. 68, Theorem 9.8). Moreover, let $A := d\,\mathrm{I}_{n\times n}$ for some $d \in \mathbb{R}$ with $|d| \in (1,\infty)$, $\ell = 1$. Then, $\frac{(\ln \lambda_-)^2}{\ln b \ln \lambda_+}\ell = \frac{1}{n}$ and $H_A^{\vec{p}}(\mathbb{R}^n)$ becomes the classical isotropic Hardy space $H^p(\mathbb{R}^n)$. In this case, by Theorems 2 and 3 and [37] ((i) and (ii) of Remark 4.4), we further know that, for any $p \in (\frac{n}{n+1}, 1]$, the classical Calderón–Zygmund operator is bounded from $H^p(\mathbb{R}^n)$ to $L^p(\mathbb{R}^n)$ (or to itself), which is a well-known result (see, for instance [46]).

(ii) *When $A := d\,\mathrm{I}_{n\times n}$ for some $d \in \mathbb{R}$ with $|d| \in (1,\infty)$, the space $H_A^{\vec{p}}(\mathbb{R}^n)$ becomes the mixed-norm Hardy space $H^{\vec{p}}(\mathbb{R}^n)$ (see [7]). In this case, Theorems 2 and 3 are new.*

(iii) *Very recently, Bownik et al. [47] introduced a kind of more general anisotropic Calderón–Zygmund operators (see [47] (Definition 5.4)) and established the boundedness of these operators from the anisotropic Hardy space $H^p(\Theta)$ to the Lebesgue space $L^p(\mathbb{R}^n)$ or to itself (see, respectively, ref. [47] (Theorems 5.12 and 5.11)), where Θ is a continuous multi-level ellipsoid cover of $\mathbb{R}^n$ (see [47] (Definition 2.1)). Here, we should point out that the space $H_A^{\vec{p}}(\mathbb{R}^n)$, in this article, is not covered by the space $H^p(\Theta)$, since the exponent p in $H^p(\Theta)$ is only a constant. Thus, Theorems 2 and 3 are covered by neither [47] (Theorems 5.12 or 5.11).*

(iv) *Recall that Huang et al. also introduced another sort of anisotropic non-convolutional β-order Calderón–Zygmund operators (see [1] (Definition 8.3)) and obtained the boundedness of these Calderón–Zygmund operators from $H_A^{\vec{p}}(\mathbb{R}^n)$ to the mixed-norm Lebesgue space $L^{\vec{p}}(\mathbb{R}^n)$ (or to itself), where $\beta \in (0,\infty)$ and $\vec{p} \in (0,2)^n$ with:*

$$p_- \in \left(\frac{\ln b}{\ln b + \beta \ln \lambda_-}, \frac{\ln b}{\ln b + (\lceil \beta \rceil - 1)\ln \lambda_-}\right],$$

where the symbol $\lceil \beta \rceil$ denotes the least integer not less than β; see [1] (Theorem 8.5). Observe that the Calderón–Zygmund operator in [1] (Definition 8.3) is different from the one used in the present article (see Definition 10) and ref. [1] (Theorem 8.5) requires the integrable exponent $\vec{p}$ which belongs to $(0,2)^n$; however, this restriction is removed in Theorems 2 and 3. Thus, Theorems 2 and 3 cannot be covered by [1] (Theorem 8.5).

5. Conclusions

In this article, we characterize the anisotropic mixed-norm Hardy space $H_A^{\vec{p}}(\mathbb{R}^n)$ via molecules, in which the range of the decay index ε is the known best possible in some sense. As an application, we then obtain a criterion on the boundedness of linear operators on $H_A^{\vec{p}}(\mathbb{R}^n)$, which is used to prove the boundedness of the anisotropic Calderón–Zygmund operators on $H_A^{\vec{p}}(\mathbb{R}^n)$. In addition, the boundedness of anisotropic Calderón–Zygmund operators from $H_A^{\vec{p}}(\mathbb{R}^n)$ to the mixed-norm Lebesgue space $L^{\vec{p}}(\mathbb{R}^n)$ is also presented. When A is as in (6), the obtained boundedness of these Calderón–Zygmund operators positively answers a question formulated by Cleanthous et al. in [6] (p. 2760). All these results are new, even for the isotropic mixed-norm Hardy spaces on $\mathbb{R}^n$.

Author Contributions: Conceptualization, J.L. and L.H.; methodology, J.L.; writing—original draft preparation, L.H. and C.Y.; writing—review and editing, J.L. All authors have read and agreed to the published version of the manuscript.

Funding: This research was supported by the Fundamental Research Funds for the Central Universities (Grant No. 2020QN21), the Natural Science Foundation of Jiangsu Province (Grant No. BK20200647), the National Natural Science Foundation of China (Grant No. 12001527) and the Project Funded by China Postdoctoral Science Foundation (Grant No. 2021M693422).

Institutional Review Board Statement: Not applicable.

Informed Consent Statement: Not applicable.

Data Availability Statement: Not applicable.

Acknowledgments: The authors would like to thank the referees for their careful reading and helpful comments which indeed improved the presentation of this article.

Conflicts of Interest: The authors declare no conflict of interest.

References

1. Huang, L.; Liu, J.; Yang, D.; Yuan, W. Real-variable characterizations of new anisotropic mixed-norm Hardy spaces. *Commun. Pure Appl. Anal.* **2020**, *19*, 3033–3082. [CrossRef]
2. Benedek, A.; Panzone, R. The space L^p, with mixed norm. *Duke Math. J.* **1961**, *28*, 301–324. [CrossRef]
3. Hörmander, L. Estimates for translation invariant operators in L^p spaces. *Acta Math.* **1960**, *104*, 93–140. [CrossRef]
4. Cleanthous, G.; Georgiadis, A.G. Mixed-norm α-modulation spaces. *Trans. Am. Math. Soc.* **2020**, *373*, 3323–3356. [CrossRef]

5. Nogayama, T. Mixed Morrey spaces. *Positivity* **2019**, *23*, 961–1000. [CrossRef]
6. Cleanthous, G.; Georgiadis, A.G.; Nielsen, M. Anisotropic mixed-norm Hardy spaces. *J. Geom. Anal.* **2017**, *27*, 2758–2787. [CrossRef]
7. Hart, J.; Torres, R.H.; Wu, X. Smoothing properties of bilinear operators and Leibniz-type rules in Lebesgue and mixed Lebesgue spaces. *Trans. Am. Math. Soc.* **2018**, *370*, 8581–8612. [CrossRef]
8. Ho, K.-P. Sublinear operators on mixed-norm Hardy spaces with variable exponents. *Atti Accad. Naz. Lincei Rend. Lincei Mat. Appl.* **2020**, *31*, 481–502. [CrossRef]
9. Huang, L.; Chang, D.-C.; Yang, D. Fourier transform of anisotropic mixed-norm Hardy spaces. *Front. Math. China* **2021**, *16*, 119–139. [CrossRef]
10. Huang, L.; Liu, J.; Yang, D.; Yuan, W. Atomic and Littlewood–Paley characterizations of anisotropic mixed-norm Hardy spaces and their applications. *J. Geom. Anal.* **2019**, *29*, 1991–2067. [CrossRef]
11. Huang, L.; Liu, J.; Yang, D.; Yuan, W. Dual spaces of anisotropic mixed-norm Hardy spaces. *Proc. Am. Math. Soc.* **2019**, *147*, 1201–1215. [CrossRef]
12. Huang, L.; Liu, J.; Yang, D.; Yuan, W. Identification of anisotropic mixed-norm Hardy spaces and certain homogeneous Triebel–Lizorkin spaces. *J. Approx. Theory* **2020**, *258*, 105459. [CrossRef]
13. Georgiadis, A.G.; Nielsen, M. Pseudodifferential operators on mixed-norm Besov and Triebel–Lizorkin spaces. *Math. Nachr.* **2016**, *289*, 2019–2036. [CrossRef]
14. Cleanthous, G.; Georgiadis, A.G.; Nielsen, M. Discrete decomposition of homogeneous mixed-norm Besov spaces. In *Functional Analysis, Harmonic Analysis, and Image Processing: A Collection of Papers in Honor of Björn Jawerth*; Contemporary Mathematics, 693; American Mathematical Society: Providence, RI, USA, 2017; pp. 167–184.
15. Johnsen, J.; Munch Hansen, S.; Sickel, W. Characterisation by local means of anisotropic Lizorkin–Triebel spaces with mixed norms. *Z. Anal. Anwend.* **2013**, *32*, 257–277. [CrossRef]
16. Johnsen, J.; Sickel, W. A direct proof of Sobolev embeddings for quasi-homogeneous Lizorkin–Triebel spaces with mixed norms. *J. Funct. Spaces Appl.* **2007**, *5*, 183–198. [CrossRef]
17. Johnsen, J.; Munch Hansen, S.; Sickel, W. Anisotropic Lizorkin–Triebel spaces with mixed norms—Traces on smooth boundaries. *Math. Nachr.* **2015**, *288*, 1327–1359. [CrossRef]
18. Ho, K.-P. Strong maximal operator on mixed-norm spaces. *Ann. Univ. Ferrara Sez. VII Sci. Mat.* **2016**, *62*, 275–291. [CrossRef]
19. Ho, K.-P. Mixed norm Lebesgue spaces with variable exponents and applications. *Riv. Math. Univ. Parma (N.S.)* **2018**, *9*, 21–44.
20. Chen, T.; Sun, W. Hardy–Littlewood–Sobolev inequality on mixed-norm Lebesgue spaces. *arXiv* **2019**, arXiv:1912.03712.
21. Chen, T.; Sun, W. Iterated and mixed weak norms with applications to geometric inequalities. *J. Geom. Anal.* **2020**, *304*, 268–4323.
22. Chen, T.; Sun, W. Extension of multilinear fractional integral operators to linear operators on Lebesgue spaces with mixed norms. *Math. Ann.* **2021**, *379*, 1089–1172. [CrossRef]
23. Cleanthous, G.; Georgiadis, A.G.; Nielsen, M. Molecular decomposition of anisotropic homogeneous mixed-norm spaces with applications to the boundedness of operators. *Appl. Comput. Harmon. Anal.* **2019**, *47*, 447–480. [CrossRef]
24. Huang, L.; Weisz, F.; Yang, D.; Yuan, W. Summability of Fourier transforms on mixed-norm Lebesgue spaces via associated Herz spaces. *Anal. Appl. (Singap.)* **2021**. [CrossRef]
25. Cleanthous, G.; Georgiadis, A.G.; Nielsen, M. Fourier multipliers on anisotropic mixed-norm spaces of distributions. *Math. Scand.* **2019**, *124*, 289–304. [CrossRef]
26. Georgiadis, A.G.; Johnsen, J.; Nielsen, M. Wavelet transforms for homogeneous mixed-norm Triebel–Lizorkin spaces. *Monatsh. Math.* **2017**, *183*, 587–624. [CrossRef]
27. Zhang, J.; Xue, Q. Multilinear strong maximal operators on weighted mixed norm spaces. *Publ. Math. Debr.* **2020**, *96*, 347–361. [CrossRef]
28. Huang, L.; Yang, D. On function spaces with mixed norms—A survey. *J. Math. Study* **2021**, *54*, 262–336.
29. Bownik, M. Anisotropic Hardy Spaces and Wavelets. *Mem. Am. Math. Soc.* **2003**, *164*, 122. [CrossRef]
30. Barrios, B.; Betancor, J.J. Anisotropic weak Hardy spaces and wavelets. *J. Funct. Spaces Appl.* **2012**, *2012*, 809121.
31. Bownik, M.; Rzeszotnik, Z. Construction and reconstruction of tight framelets and wavelets via matrix mask functions. *J. Funct. Anal.* **2009**, *256*, 1065–1105. [CrossRef]
32. Dekel, S.; Han, Y.; Petrushev, P. Anisotropic meshless frames on $\mathbb{R}^n$. *J. Fourier Anal. Appl.* **2009**, *15*, 634–662. [CrossRef]
33. Jakab, T.; Mitrea, M. Parabolic initial boundary value problems in nonsmooth cylinders with data in anisotropic Besov spaces. *Math. Res. Lett.* **2006**, *13*, 825–831. [CrossRef]
34. Bownik, M.; Wang, L.-A.D. A PDE characterization of anisotropic Hardy spaces. *arXiv* **2020**, arXiv: 2011.10651.
35. Wang, A.; Wang, W. The atomic and molecular decompositions of anisotropic mixed-norm Hardy spaces and their application. *J. Korean Math. Soc.* under review.
36. Lu, S.-Z. *Four Lectures on Real H^p Spaces*; World Scientific Publishing Co., Inc.: River Edge, NJ, USA, 1995.
37. Liao, M.; Li, J.; Li, B.; Li, B. A new molecular characterization of diagonal anisotropic Musielak–Orlicz Hardy spaces. *Bull. Sci. Math.* **2021**, *166*, 102939. [CrossRef]
38. Liu, J.; Haroske, D.D.; Yang, D. New molecular characterizations of anisotropic Musielak–Orlicz Hardy spaces and their applications. *J. Math. Anal. Appl.* **2019**, *475*, 1341–1366. [CrossRef]

39. Li, B.; Fan, X.; Fu, Z.; Yang, D. Molecular characterization of anisotropic Musielak–Orlicz Hardy spaces and their applications. *Acta Math. Sin. (Engl. Ser.)* **2016**, *32*, 1391–1414. [CrossRef]
40. Liu, J.; Weisz, F.; Yang, D; Yuan, W. Variable anisotropic Hardy spaces and their applications. *Taiwan J. Math.* **2018**, *22*, 1173–1216. [CrossRef]
41. Liu, J. Molecular characterizations of variable anisotropic Hardy spaces with applications to boundedness of Calderón–Zygmund operators. *Banach J. Math. Anal.* **2021**, *15*, 1–24. [CrossRef]
42. Wang, W.; Liu, X.; Wang, A.; Li, B. Molecular decomposition of anisotropic Hardy spaces with variable exponents. *Indian J. Pure Appl. Math.* **2020**, *51*, 1471–1495. [CrossRef]
43. Liu, J.; Yang, D; Yuan, W. Anisotropic Hardy–Lorentz spaces and their applications. *Sci. China Math.* **2016**, *59*, 1669–1720. [CrossRef]
44. Liu, X.; Qiu, X.; Li, B. Molecular characterization of anisotropic variable Hardy–Lorentz spaces. *Tohoku Math. J.* **2020**, *72*, 211–233. [CrossRef]
45. Bownik, M.; Li, B.; Yang, D.; Zhou, Y. Weighted anisotropic product Hardy spaces and boundedness of sublinear operators. *Math. Nachr.* **2010**, *283*, 392–442. [CrossRef]
46. Stein, E.M. *Harmonic Analysis: Real-Variable Methods, Orthogonality, and Oscillatory Integrals;* Princeton Mathematical Series 43, Monographs in Harmonic Analysis III; Princeton University Press: Princeton, NJ, USA, 1993.
47. Bownik, M.; Li, B.; Li, J. Variable anisotropic singular integral operators. *arXiv* **2020**, arXiv:2004.09707.

mathematics

Article

Wavelets and Real Interpolation of Besov Spaces

Zhenzhen Lou [1,2], Qixiang Yang [3], Jianxun He [1,*] and Kaili He [1]

1 School of Mathematics and Information Sciences, Guangzhou University, Guangzhou 510006, China; zhenzhenlou@e.gzhu.edu.cn (Z.L.); kailihe@e.gzhu.edu.cn (K.H.)
2 School of Mathematics and Statistics, Qujing Normal University, Qujing 655011, China
3 School of Mathematics and Statistics, Wuhan University, Wuhan 430072, China; qxyang@whu.edu.cn
* Correspondence: hejianxun@gzhu.edu.cn

Abstract: In view of the importance of Besov space in harmonic analysis, differential equations, and other fields, Jaak Peetre proposed to find a precise description of $(B^{s_0,q_0}_{p_0}, B^{s_1,q_1}_{p_1})_{\theta,r}$. In this paper, we come to consider this problem by wavelets. We apply Meyer wavelets to characterize the real interpolation of homogeneous Besov spaces for the crucial index p and obtain a precise description of $(\dot{B}^{s,q}_{p_0}, \dot{B}^{s,q}_{p_1})_{\theta,r}$.

Keywords: real interpolation; besov space; meyer wavelet

Citation: Lou, Z.; Yang, Q.; He, J.; He, K. Wavelets and Real Interpolation of Besov Spaces. *Mathematics* **2021**, *9*, 2235. https://doi.org/10.3390/math9182235

Academic Editor: Palle E.T. Jorgensen

Received: 24 July 2021
Accepted: 10 September 2021
Published: 12 September 2021

1. Introduction

Since the middle of 20th century, the study of interpolation space has greatly promoted the development of function space, operator theory, and developed a set of perfect mathematical theories. It greatly enriches the theory of harmonic analysis, see [1–4]. However, for a long time, only the real interpolation spaces of Lebesgue spaces have been studied thoroughly, their forms are known as Lorentz spaces, and there are a lot of literature about Lorentz spaces, see [2,5–9].

For the real interpolation of Besov spaces, we can refer to [9–16]. When the index p is fixed, it has been shown that $(B^{s_0,q_0}_{p}, B^{s_1,q_1}_{p})_{\theta,r}$ are still Besov spaces, see [4,9,16]. The interpolation for the index p is very different to which for the indices s and q. If $p_0 \neq p_1$, then $(B^{s,q}_{p_0}, B^{s,q}_{p_1})_{\theta,r}$ will fall outside of the scale of Besov spaces. J. Peetre proposed to consider the real interpolation of Besov spaces in [4]. For more than forty years, due to some inherent difficulties, little progress has been made in this regard.

In this paper, we consider the interpolation problem introduced in [4] for the crucial index p. Wavelets have localization of both frequency and spatial position, which provides a powerful tool for the study of the interpolation of Besov spaces. In this paper, we obtain a precise description of $(\dot{B}^{s,q}_{p_0}, \dot{B}^{s,q}_{p_1})_{\theta,r}$ by Meyer wavelets. Further, as $q = r$, we prove that $(\dot{B}^{s,q}_{p_0}, \dot{B}^{s,q}_{p_1})_{\theta,q}$ can fall into the Besov–Lorentz spaces in [17].

For Besov and Triebel–Lizorkin spaces, we use the characterization based on the Littlewood–Paley decomposition, see [9,18,19]. Given a function φ, such that its Fourier transform $\widehat{\varphi}(\xi) \in C_0^\infty(\mathbb{R}^n)$ and satisfies

$$\text{supp } \widehat{\varphi} \subset \{\xi \in \mathbb{R}^n : |\xi| \leq 2\} \text{ and } \hat{\varphi}(\xi) = 1, \text{ if } |\xi| \leq \frac{1}{2}.$$

For $u \in \mathbb{Z}$, we define φ_u by

$$\varphi_u(x) = 2^{n(u+1)}\varphi(2^{u+1}x) - 2^{nu}\varphi(2^u x).$$

These functions $\{\varphi_u(x)\}_{u\in\mathbb{Z}}$ satisfy

$$\begin{cases} \text{supp } \hat{\varphi}_u \subset \{\xi \in \mathbb{R}^n, \frac{1}{2} \leq 2^{-u}|\xi| \leq 2\}; \\ |\hat{\varphi}_u(\xi)| \geq C > 0, \text{ if } \frac{1}{2} < C_1 \leq 2^{-u}|\xi| \leq C_2 < 2; \\ |\partial^k \hat{\varphi}_u(\xi)| \leq C_k 2^{-u|k|}, \text{ for all } k \in \mathbb{N}^n; \\ \sum_{u=-\infty}^{+\infty} \hat{\varphi}_u(\xi) = 1, \text{ for any } \xi \in \mathbb{R}^n. \end{cases}$$

Denote the space of all Schwartz functions on $\mathbb{R}^n$ by $\mathcal{S}(\mathbb{R}^n)$. The dual space of $\mathcal{S}(\mathbb{R}^n)$, namely, the space of all tempered distributions on $\mathbb{R}^n$, equipped with the weak-$*$ topology, is denoted by $\mathcal{S}'(\mathbb{R}^n)$. Denote the space of all polynomials on $\mathbb{R}^n$ by $P(\mathbb{R}^n)$. Let $f \in \mathcal{S}'(\mathbb{R}^n)\backslash P(\mathbb{R}^n)$. Define $f_u = \varphi_u * f$, the f_u is called the u-th dyadic block of the Littlewood–Paley decomposition of f. We recall the definition of $\dot{B}_p^{s,q}$ and $\dot{F}_p^{s,q}$.

Definition 1. *Given $s \in \mathbb{R}$, $0 < q \leq \infty$ and $u \in \mathbb{Z}$. For $f \in S'(\mathbb{R}^n)\backslash P(\mathbb{R}^n)$, we define*

(i) For $0 < p \leq \infty$, $f \in \dot{B}_p^{s,q}$, if $\left(\sum_u 2^{usq}\|f_u(x)\|_{L^p}^q\right)^{\frac{1}{q}} < \infty$.

(ii) For $0 < p < \infty$, $f \in \dot{F}_p^{s,q}$, if $\left\|\left(\sum_u 2^{usq}|f_u(x)|^q\right)^{\frac{1}{q}}\right\|_{L^p} < \infty$.

As $q = \infty$, it should be replaced by the supremum norm.

The definition of the above two spaces are independent of the selection of the functions φ, see [9].

Then, we recall some notations of Meyer wavelets. Let Ψ^0 be an even function in $C_0^\infty([-\frac{4\pi}{3}, \frac{4\pi}{3}])$ satisfying

$$\begin{cases} 0 \leq \Psi^0(\xi) \leq 1; \\ \Psi^0(\xi) = 1, \text{ for } |\xi| \leq \frac{2\pi}{3}. \end{cases}$$

Let

$$\Omega(\xi) = \sqrt{(\Psi^0(\frac{\xi}{2}))^2 - (\Psi^0(\xi))^2}.$$

Then, $\Omega(\xi)$ is an even function in $C_0^\infty([-\frac{8\pi}{3}, \frac{8\pi}{3}])$. It is easy to get

$$\begin{cases} \Omega(\xi) = 0, \text{ for } |\xi| \leq \frac{2\pi}{3}; \\ \Omega^2(\xi) + \Omega^2(2\xi) = 1 = \Omega^2(\xi) + \Omega^2(2\pi - \xi), \text{ for } \frac{2\pi}{3} \leq \xi \leq \frac{4\pi}{3}. \end{cases}$$

Denote $\Psi^1(\xi) := \Omega(\xi)e^{-\frac{i\xi}{2}}$. For all $\epsilon = (\epsilon_1, \cdots, \epsilon_n) \in \{0,1\}^n$, define

$$\hat{\Phi}^\epsilon(\xi) := \prod_{i=1}^n \Psi^{\epsilon_i}(\xi_i).$$

Furthermore, $\Gamma := \{(\epsilon, k), \epsilon \in \{0,1\}^n \setminus \{(0, ..., 0)\}, k \in \mathbb{Z}^n\}$ and

$$\Lambda := \{(\epsilon, j, k) : \epsilon \in \{0,1\}^n \setminus \{(0, ..., 0)\}, j \in \mathbb{Z}, k \in \mathbb{Z}^n\}.$$

For $(\epsilon, j, k) \in \Lambda$, denote

$$\Phi_{j,k}^\epsilon(x) := 2^{\frac{jn}{2}} \Phi^\epsilon(2^j x - k).$$

For $f \in \mathcal{S}'$, let $a^{\epsilon}_{j,k} = \langle f, \Phi^{\epsilon}_{j,k}\rangle$. The following results are well-known, see [17,18,20].

Lemma 1. *The Meyer wavelets $\{\Phi^{\epsilon}_{j,k}\}_{(\epsilon,j,k)\in\Lambda}$ form an orthogonal basis in $L^2(\mathbb{R}^n)$, hence, for all $f \in L^2(\mathbb{R}^n)$, the following wavelet decomposition holds in L^2 sense,*

$$f = \sum_{(\epsilon,j,k)\in\Lambda} a^{\epsilon}_{j,k}\Phi^{\epsilon}_{j,k}.$$

In this paper, we first give some precise descriptions of $(\dot{B}^{s,q}_{p_0}, \dot{B}^{s,q}_{p_1})_{\theta,r}$ with wavelets. Let $\chi(x)$ be the characteristic function on the unit cube $[0,1)^n$. For Borel set F in $\mathbb{R}^n$, denote $|F|$ the Lebesgue measure of F. Suppose that $j, u \in \mathbb{Z}$, $1 < p_0 < p_1 < \infty$ and $\frac{1}{\alpha} = \frac{1}{p_0} - \frac{1}{p_1}$, denote

$$c_{j,n}(\tau) := \inf\left\{\lambda : \left|\left\{x \in \mathbb{R}^n : \sum_{(\epsilon,k)\in\Gamma} |a^{\epsilon}_{j,k}|\chi(2^j x - k) > 2^{-\frac{nj}{2}}\lambda\right\}\right| \leq \tau\right\},$$

$$b^{p_0,p_1}_{j,n,u} := \left(\int_0^{2^{u\alpha}} (c_{j,n}(\tau))^{p_0} d\tau\right)^{\frac{1}{p_0}} + 2^u\left(\int_{2^{u\alpha}}^{\infty} (c_{j,n}(\tau))^{p_1} d\tau\right)^{\frac{1}{p_1}}.$$

Theorem 1. *Given $\theta \in (0,1)$, $s \in \mathbb{R}$, $1 < p_0 < p_1 < \infty$, $0 < q, r \leq \infty$ and $\frac{1}{p} = \frac{1-\theta}{p_0} + \frac{\theta}{p_1}$. For $f = \sum_{(\epsilon,j,k)\in\Lambda} a^{\epsilon}_{j,k}\Phi_{j,k}$, we have*

(i) $f \in (\dot{B}^{s,q}_{p_0}, \dot{B}^{s,q}_{\infty})_{\theta,r}$ *if, and only if,*

$$\sum_u 2^{-ur\theta}\left\{\sum_j 2^{jsq}\left[\int_0^{2^{up_0}} (c_{j,n}(\tau))^{p_0} d\tau\right]^{\frac{q}{p_0}}\right\}^{\frac{r}{q}} < \infty;$$

(ii) $f \in (\dot{B}^{s,q}_{p_0}, \dot{B}^{s,q}_{p_1})_{\theta,r}$ *if and only if*

$$\sum_u 2^{-ur\theta}\left\{\sum_j 2^{jsq}\left[b^{p_0,p_1}_{j,n,u}\right]^q\right\}^{\frac{r}{q}} < \infty.$$

The above wavelet characterization is slightly complicated. Yang-Cheng-Peng [17] introduced Besov–Lorentz spaces. Further, when $q = r$, we can prove that $(\dot{B}^{s,q}_{p_0}, \dot{B}^{s,q}_{p_1})_{\theta,q}$ are just the Besov–Lorentz spaces defined in [17]. We have

Theorem 2. *Let $\theta \in (0,1)$, $s \in \mathbb{R}$, $0 < q \leq \infty$, $1 < p_0 < p_1 < \infty$, $\frac{1}{p} = \frac{1-\theta}{p_0} + \frac{\theta}{p_1}$, $u \in \mathbb{Z}$ and $f = \sum_{(\epsilon,j,k)\in\Lambda} a^{\epsilon}_{j,k}\Phi^{\epsilon}_{j,k}$. Then the following conditions are equivalent.*

(i) $f \in (\dot{B}^{s,q}_{p_0}, \dot{B}^{s,q}_{p_1})_{\theta,q}$ *if, and only if,*

$$\sum_j 2^{jsq}\left\{\sum_u 2^{uq}\left|\left\{x \in \mathbb{R}^n : \left|\sum_{(\epsilon,k)\in\Gamma} a^{\epsilon}_{j,k}\Phi^{\epsilon}_{j,k}(x)\right| > 2^u\right\}\right|^{\frac{q}{p}}\right\} < \infty.$$

(ii) $f \in (\dot{B}^{s,q}_{p_0}, \dot{B}^{s,q}_{p_1})_{\theta,q}$ *if, and only if,*

$$\sum_j 2^{jsq}\left(\sum_u 2^{uq}\left|\left\{x \in \mathbb{R}^n : 2^{\frac{nj}{2}} \sum_{(\epsilon,k)\in\Gamma} |a^{\epsilon}_{j,k}|\chi(2^j x - k) > 2^u\right\}\right|^{\frac{q}{p}}\right) < \infty.$$

Although the above main results still not solve the problem proposed by J. Peetre [4] thoroughly, we obtain a precise description of $(\dot{B}^{s,q}_{p_0}, \dot{B}^{s,q}_{p_1})_{\theta,r}$ by Meyer wavelets. The wavelet characterization of real interpolation spaces of Besov spaces provides people with an effective means to study the continuity of linear operators and bilinear operators on such spaces. We are using this point to study the well-posedness of non-linear fluid equations.

The plan of this paper is the following. In Section 2, we recall the general background of the real interpolation method and Lorentz spaces. Then we review wavelet characterization of $\dot{B}^{s,q}_p$ and $\dot{F}^{s,q}_p$. In Section 3, we give the proof of Theorem 1. Finally, in Section 4 we prove Theorem 2.

In this paper, $A \lesssim B$ means the estimation of the form $A \leq CB$ with some constant C independent of the main parameters, C may vary from line to line. $A \sim B$ means $A \lesssim B$ and $B \lesssim A$.

2. Preliminaries on Real Interpolation and Wavelets

In this section, we present some preliminaries on real interpolation and wavelets.

2.1. K-Functional and Real Interpolation

The K-functional was introduced by J. Peetre in the process of dealing with real interpolation spaces, see [1,4]. If (A_0, A_1) is a pair of quasi-normed spaces which are continuously embedded in a Hausdorff space X, then the K-functional

$$K(t, f, A_0, A_1) := \inf_{f=f_0+f_1} \{\|f_0\|_{A_0} + t\|f\|_{A_1}\}$$

is defined for all $f = f_0 + f_1$, where $f_0 \in A_0$, $f_1 \in A_1$.

Definition 2. *Let $0 < \theta < 1$ and $0 < q < \infty$. We define*

$$(A_0, A_1)_{\theta,q,K} =: \left\{ f : f \in A_0 + A_1, \|f\|_{(A_0,A_1)_{\theta,q,K}} = \left\{ \int_0^\infty [t^{-\theta} K(t, f, A_0, A_1)]^q \frac{dt}{t} \right\}^{\frac{1}{q}} < \infty \right\}. \quad (1)$$

Further, we define

$$(A_0, A_1)_{\theta,\infty,K} =: \left\{ f : f \in A_0 + A_1, \|f\|_{(A_0,A_1)_{\theta,\infty,K}} = \sup_t t^{-\theta} K(t, f, A_0, A_1) < \infty \right\}. \quad (2)$$

Bergh-Löfström [1] has shown that the norms of the spaces $(A_0, A_1)_{\theta,q,K}$ in (1) and (2) have the following discrete representation.

Lemma 2. *Let $0 < \theta < 1$. Then,*

$$\|f\|_{(A_0,A_1)_{\theta,q,K}} \sim \begin{cases} \left[\sum_{j\in\mathbb{Z}} 2^{-jq\theta} K(2^j, f, A_0, A_1)^q \right]^{\frac{1}{q}}, & 0 < q < \infty; \\ \sup_{j\in\mathbb{Z}} 2^{-j\theta} K(2^j, f, A_0, A_1), & q = \infty. \end{cases} \quad (3)$$

In the following part, we always use this form. For $x \in \mathbb{R}^n$ and function $f(x)$, the distribution function $\sigma_f(\lambda)$ and rearrangement function $f^*(\tau)$ are defined in the following way

$$\sigma_f(\lambda) = |\{x : |f(x)| > \lambda\}| \text{ and } f^*(\tau) = \inf\{\lambda : \sigma_f(\lambda) \leq \tau\}.$$

We review some results about K-functional, see [3].

Lemma 3. *Suppose that* $0 < p < \infty$ *and* $f \in L^p + L^\infty$. *Then*

$$K(t, f, L^p, L^\infty) \sim \left[\int_0^{t^p} (f^*(\tau))^p d\tau\right]^{\frac{1}{p}}.$$

Lemma 4. *If* $0 < p_0 < p_1 < \infty$ *and* $\frac{1}{\alpha} = \frac{1}{p_0} - \frac{1}{p_1}$, *then*

$$K(t, f, L^{p_0}, L^{p_1}) \sim \left[\int_0^{t^\alpha} (f^*(\tau))^{p_0} d\tau\right]^{\frac{1}{p_0}} + t\left[\int_{t^\alpha}^{\infty} (f^*(\tau))^{p_1} d\tau\right]^{\frac{1}{p_1}}.$$

For $0 < p < \infty$, K-functional can be replaced to K_p functional, see [21]. Define K_p functional by

$$K_p := K_p(t, f, A_0, A_1) = \inf_{f=f_0+f_1} (\|f_0\|_{A_0}^p + t^p \|f_1\|_{A_1}^p)^{\frac{1}{p}},$$

and

$$\|f\|_{(A_0,A_1)_{\theta,q,K_p}} = \left[\int_0^\infty [t^{-\theta} K_p(t, f, A_0, A_1)]^q \frac{dt}{t}\right]^{\frac{1}{q}}.$$

We recall an important lemma about $K_p(t, f, A_0, A_1)$, see [21].

Lemma 5. *Let* (A_0, A_1) *be a couple of quasi-normed spaces. For any* $0 < p < \infty$, *we have*

$$\|f\|_{(A_0,A_1)_{\theta,q,K}} \sim \|f\|_{(A_0,A_1)_{\theta,q,K_p}}.$$

2.2. Lorentz Spaces and Lebesgue Spaces

In this subsection, we present first the definition of Lorentz spaces which are the generalization of Lebesgue spaces and then some relative lemmas.

Definition 3. *For* $1 \le p < \infty$ *and* $0 < r < \infty$, *the Lorentz spaces* $L^{p,r}$ *are defined as follows*

$$L^{p,r} = \left\{ f : \|f\|_{p,r} = \left[\frac{r}{p}\int_0^\infty \left(\tau^{\frac{1}{p}} f^*(\tau)\right)^r \frac{d\tau}{\tau}\right]^{\frac{1}{r}} < \infty \right\}.$$

For $r = \infty$,

$$L^{p,\infty} = \left\{ f : \|f\|_{p,\infty} = \sup_\tau \tau^{\frac{1}{p}} f^*(\tau) < \infty \right\}.$$

It is easy to see that $L^{p,p} = L^p$. Further, $L^{p,\infty}$ corresponds to the weak L^p spaces. The above definition depends on the rearrangement function $f^*(\tau)$. These spaces can be characterized by distribution function $\sigma_f(\lambda)$ also, see [2].

Lemma 6. *Let* $1 \le p < \infty$ *and* $0 < r \le \infty$. *Then, for any* $f \in L^{p,r}$, *one has*

$$\|f\|_{p,r} \sim \left[r\int_0^\infty \left(\lambda \sigma_f^{\frac{1}{p}}(\lambda)\right)^r \frac{d\lambda}{\lambda}\right]^{\frac{1}{r}} \text{ and } \|f\|_{p,\infty} \sim \sup_\lambda \lambda \sigma_f^{\frac{1}{p}}(\lambda).$$

The above continuous integral can be written as the following discrete form, see [17].

Lemma 7. *Suppose that* $1 \le p < \infty$ *and* $0 < r < \infty$. *Then* $f \in L^{p,r}$, *if*

$$\left(\sum_u 2^{ru} |\{x \in \mathbb{R}^n : |f(x)| > 2^u\}|^{\frac{r}{p}}\right)^{\frac{1}{r}} < \infty,$$

as $r = \infty$, the L^r-norm should be replaced by the L^∞-norm.

The above Lorentz spaces are in fact real interpolation of Lebesgue spaces L^p, see [1].

Lemma 8. *Assume that $0 < p_0 < p_1 \leq \infty$, $0 < r \leq \infty$, $0 < \theta < 1$ and $\frac{1}{p} = \frac{1-\theta}{p_0} + \frac{\theta}{p_1}$. Then*

$$(L^{p_0}, L^{p_1})_{\theta,r} = L^{p,r}, \text{ with } \frac{1}{p} = \frac{1-\theta}{p_0} + \frac{\theta}{p_1}.$$

By Lemma 8, we get another characterization of $L^{p,r}$ as below.

Corollary 1. *Let all parameters be as defined in Lemma 8. Then,*

$$\|f\|_{p,r} \sim \left[\int_0^\infty \left(t^{-\theta} K(t, f, L^{p_0}, L^{p_1})\right)^r \frac{dt}{t}\right]^{\frac{1}{r}}.$$

2.3. Wavelet Characterization of $\dot{B}_p^{s,q}$ and $\dot{F}_p^{s,q}$

For any function $f(x)$ in $\dot{B}_p^{s,q}$ or $\dot{F}_p^{s,q}$ in Definition 1, the following wavelet decomposition holds in the sense of distribution,

$$f = \sum_{(\epsilon,j,k)\in\Lambda} a_{j,k}^\epsilon \Phi_{j,k}^\epsilon.$$

We recall the wavelet characterization of $\dot{B}_p^{s,q}$ and $\dot{F}_p^{s,q}$ in this subsection, see [16–18,20]. For any $s \in \mathbb{R}$ and $0 < q \leq \infty$, denote

$$S_{s,q}f(x) := \left(\sum_{(\epsilon,j,k)\in\Lambda} 2^{qj(s+\frac{n}{2})} |a_{j,k}^\epsilon|^q \chi(2^j x - k)\right)^{\frac{1}{q}}.$$

When $s = 0$ *and* $q = 2$, we denote $Sf := S_{s,q}f$.

Lemma 9. *Let $s \in \mathbb{R}$ and $0 < q \leq \infty$.*

(i) For $0 < p < \infty$, $f \in \dot{F}_p^{s,q}(\mathbb{R}^n)$ if, and only if,

$$\|S_{s,q}f\|_{L^p} < +\infty.$$

(ii) For $0 < p \leq \infty$, $f \in \dot{B}_p^{s,q}(\mathbb{R}^n)$ if, and only if,

$$\left[\sum_{j\in\mathbb{Z}} 2^{qj(s-\frac{n}{p}+\frac{n}{2})} \left(\sum_{(\epsilon,k)\in\Gamma} |a_{j,k}^\epsilon|^p\right)^{\frac{q}{p}}\right]^{\frac{1}{q}} < \infty.$$

It is easy to see that $\dot{F}_p^{0,2} = L^p$. In [17], Yang-Cheng-Peng proved the wavelet characterization of Lorentz spaces $L^{p,r}$.

Lemma 10. *Suppose that $1 \leq p < \infty$, $0 < r < \infty$ and $u \in \mathbb{Z}$. Then $f \in L^{p,r}$, if*

$$\left(\sum_u 2^{ru} |\{x \in \mathbb{R}^n : |Sf(x)| > 2^u\}|^{\frac{r}{p}}\right)^{\frac{1}{r}} < \infty,$$

as $r = \infty$, the L^r-norm should be replaced by the L^∞-norm.

Remark 1. *f and Sf can control each other by using good$-\lambda$ inequality. When the Fourier transform of f is supported on a ring, f and Sf can control each other. The distribution function $\sigma_f(\lambda)$ and rearrangement function $f^*(\tau)$ can be replaced by $\sigma_{Sf}(\lambda)$ and $(Sf)^*(\tau)$, see [17]. Without affecting the proof, these notations are not strictly distinguished in this paper.*

3. Proof of Theorem 1

In this section, we characterize $(\dot{B}^{s,q}_{p_0}, \dot{B}^{s,q}_{\infty})_{\theta,r}$, and $(\dot{B}^{s,q}_{p_0}, \dot{B}^{s,q}_{p_1})_{\theta,r}$ with wavelets. Now we come to prove Theorem 1.

Proof. Denote

$$\|f\|_p := \|f\|_{L^p}, \|f\|_{(A_0,A_1)_{\theta,q}} := \|f\|_{(A_0,A_1)_{\theta,q,K}}.$$

For any function f in $\dot{B}^{s,q}_p$, the following wavelet decomposition holds in the sense of distribution,

$$f = \sum_{(\epsilon,j,k)\in\Lambda} a^{\epsilon}_{j,k}\Phi^{\epsilon}_{j,k}.$$

From Lemma 9, it follows that

$$K_q(t,f) := K_q(t,f,\dot{B}^{s,q}_{p_0},\dot{B}^{s,q}_{p_1}) = \inf\left[\sum_j 2^{jq(s-\frac{n}{p_0}+\frac{n}{2})}\left(\sum_{(\epsilon,k)\in\Gamma}|x^{\epsilon}_{j,k}|^{p_0}\right)^{\frac{q}{p_0}}\right.$$
$$\left. + t^q\sum_j 2^{jq(s-\frac{n}{p_1}+\frac{n}{2})}\left(\sum_{(\epsilon,k)\in\Gamma}|a^{\epsilon}_{j,k}-x^{\epsilon}_{j,k}|^{p_1}\right)^{\frac{q}{p_1}}\right]^{\frac{1}{q}}.$$

Denote

$$x_j = \sum_{(\epsilon,k)\in\Gamma} x^{\epsilon}_{j,k}\Phi^{\epsilon}_{j,k}(x)\ ,\ a_j = \sum_{(\epsilon,k)\in\Gamma} a^{\epsilon}_{j,k}\Phi^{\epsilon}_{j,k}(x).$$

By Lemma 9, we deduce that

$$\begin{aligned}
\|x_j\|_{p_0} &= 2^{j(-\frac{n}{p_0}+\frac{n}{2})}\left\{\sum_k\left(\sum_{\epsilon}|x^{\epsilon}_{j,k}|^2\right)^{\frac{p_0}{2}}\right\}^{\frac{1}{p_0}} \\
&\sim 2^{j(-\frac{n}{p_0}+\frac{n}{2})}\left\{\sum_{(\epsilon,k)\in\Gamma}|x^{\epsilon}_{j,k}|^{p_0}\right\}^{\frac{1}{p_0}}, \\
\|a_j - x_j\|_{p_1} &= 2^{j(-\frac{n}{p_1}+\frac{n}{2})}\left\{\sum_k\left(\sum_{\epsilon}|a^{\epsilon}_{j,k}-x^{\epsilon}_{j,k}|^2\right)^{\frac{p_1}{2}}\right\}^{\frac{1}{p_1}} \\
&\sim 2^{j(-\frac{n}{p_1}+\frac{n}{2})}\left\{\sum_{(\epsilon,k)\in\Gamma}|a^{\epsilon}_{j,k}-x^{\epsilon}_{j,k}|^{p_1}\right\}^{\frac{1}{p_1}}.
\end{aligned}$$

Hence,

$$K_q(t,a_j) \sim \left[\sum_j 2^{jsq} \inf\Big(\|x_j\|_{p_0}^q + t^q\|a_j - x_j\|_{p_1}^q\Big)\right]^{\frac{1}{q}}$$

$$\sim \left\{\sum_j 2^{jsq}\left[\inf(\|x_j\|_{p_0} + t\|a_j - x_j\|_{p_1})\right]^q\right\}^{\frac{1}{q}}$$

$$= \left\{\sum_j 2^{jsq}\left[K(t,a_j,L^{p_0},L^{p_1})\right]^q\right\}^{\frac{1}{q}}.$$

Consequently,

$$\|f\|_{(\dot{B}^{s,q}_{p_0},\dot{B}^{s,q}_{p_1})_{\theta,r}} \sim \left\{\int_0^\infty [t^{-\theta}K_q(t,a)]^r \frac{dt}{t}\right\}^{\frac{1}{r}}$$

$$\sim \left\{\int_0^\infty \left[t^{-\theta}\left\{\sum_j 2^{jsq}\left[K(t,a_j,L^{p_0},L^{p_1})\right]^q\right\}^{\frac{1}{q}}\right]^r \frac{dt}{t}\right\}^{\frac{1}{r}}. \tag{4}$$

If $1 < p_0 < p_1 < \infty$, then $L^{p_0} = \dot{F}^{0,2}_{p_0}$ and $L^{p_1} = \dot{F}^{0,2}_{p_1}$. Applying Remark 1, we have

$$(S^*f)(\tau) := (S^*_{0,2}f)(\tau) = \inf\{\lambda : |\{x \in \mathbb{R}^n : S_{0,2}f(x) > \lambda\}| \le \tau\}.$$

For $a_j = \sum_{(\epsilon,k)\in\Gamma} a^\epsilon_{j,k}\Phi^\epsilon_{j,k}(x)$, we have

$$Sa_j(x) := S_{0,2}a_j(x) = \left(\sum_{(\epsilon,k)\in\Gamma} 2^{2j(0+\frac{n}{2})}|a^\epsilon_{j,k}|^2\chi(2^jx-k)\right)^{\frac{1}{2}}$$

$$= \left(\sum_{(\epsilon,k)\in\Gamma} 2^{jn}|a^\epsilon_{j,k}|^2\chi(2^jx-k)\right)^{\frac{1}{2}}$$

$$= 2^{\frac{jn}{2}}\sum_k\left(\sum_\epsilon |a^\epsilon_{j,k}|^2\right)^{\frac{1}{2}}\chi(2^jx-k)$$

$$\sim 2^{\frac{jn}{2}}\sum_{(\epsilon,k)\in\Gamma}|a^\epsilon_{j,k}|\chi(2^jx-k).$$

Thus,

$$Sa_j(x) = 2^{\frac{jn}{2}}\sum_{(\epsilon,k)\in\Gamma}|a^\epsilon_{j,k}|\chi(2^jx-k). \tag{5}$$

By (5), we deduce that

$$(S^*a_j)(\tau) = \inf\{\lambda : |\{x \in \mathbb{R}^n : Sa_j(x) > \lambda\}| \le \tau\}$$

$$= \inf\left\{\lambda : \left|\left\{x \in \mathbb{R}^n : 2^{\frac{jn}{2}}\sum_{(\epsilon,k)\in\Gamma}|a^\epsilon_{j,k}|\chi(2^jx-k) > \lambda\right\}\right| \le \tau\right\}$$

$$= \inf\left\{\lambda : \left|\left\{x \in \mathbb{R}^n : \sum_{(\epsilon,k)\in\Gamma}|a^\epsilon_{j,k}|\chi(2^jx-k) > 2^{-\frac{jn}{2}}\lambda\right\}\right| \le \tau\right\}.$$

Denote

$$c_{j,n}(\tau) := \inf\left\{\lambda : \left|\left\{x \in \mathbb{R}^n : \sum_{(\epsilon,k)\in\Gamma} |a^{\epsilon}_{j,k}|\chi(2^j x - k) > 2^{-\frac{jn}{2}}\lambda\right\}\right| \leq \tau\right\}. \tag{6}$$

Hence,

$$(S^* a_j)(\tau) = c_{j,n}(\tau). \tag{7}$$

Let us prove the theorem in two cases.

(i) For $p_1 = \infty$, by Remark 1 and Lemma 3, we have

$$K(t, a_j, L^{p_0}, L^{\infty}) \sim \left[\int_0^{t^{p_0}} (a_j^*(\tau))^{p_0} d\tau\right]^{\frac{1}{p_0}} \sim \left[\int_0^{t^{p_0}} [(S^* a_j)(\tau)]^{p_0} d\tau\right]^{\frac{1}{p_0}}.$$

By (6) and (7), we get

$$K(t, a_j, L^{p_0}, L^{\infty}) \sim \left[\int_0^{t^{p_0}} (c_{j,n}(\tau))^{p_0} d\tau\right]^{\frac{1}{p_0}}. \tag{8}$$

Applying (4), (8) and the discrete representation of the spaces $(A_0, A_1)_{\theta,q,K}$ which is described in Remark 3, we obtain

$$\|f\|^r_{(\dot{B}^{s,q}_{p_0}, \dot{B}^{s,q}_{\infty})_{\theta,r}} \sim \sum_u 2^{-ur\theta} \left\{\sum_j 2^{jsq} \left[\int_0^{2^{up_0}} (c_{j,n}(\tau))^{p_0} d\tau\right]^{\frac{q}{p_0}}\right\}^{\frac{r}{q}}.$$

(ii) For $1 < p_0 < p_1 < \infty$, by Lemma 4, similar as we did in (i), we have

$$K(t, a_j, L^{p_0}, L^{p_1}) \sim \left[\int_0^{t^{\alpha}} (c_{j,n}(\tau))^{p_0} d\tau\right]^{\frac{1}{p_0}} + t\left[\int_{t^{\alpha}}^{\infty} (c_{j,n}(\tau))^{p_1} d\tau\right]^{\frac{1}{p_1}}, \tag{9}$$

where $\frac{1}{\alpha} = \frac{1}{p_0} - \frac{1}{p_1}$. Denote

$$b^{p_0,p_1}_{j,n,u,} := \left(\int_0^{2^{u\alpha}} (c_{j,\lambda}(\tau))^{p_0} d\tau\right)^{\frac{1}{p_0}} + 2^u \left(\int_{2^{u\alpha}}^{\infty} (c_{j,\lambda}(\tau))^{p_1} d\tau\right)^{\frac{1}{p_1}}.$$

Combining (4) with (9) and using the discrete representation of the spaces $(A_0, A_1)_{\theta,q,K}$ which is described in Remark 3, we know that

$$\|f\|^r_{(\dot{B}^{s,q}_{p_0}, \dot{B}^{s,q}_{p_1})_{\theta,r}} \sim \sum_u 2^{-ur\theta} \left\{\sum_j 2^{jsq} \left[b^{p_0,p_1}_{j,n,u}\right]^q\right\}^{\frac{r}{q}}.$$

The proof of Theorem 1 is complete. □

4. Proof of Theorem 2

Now we come to prove Theorem 2.

Proof. Applying Lemma 5, the same as we did in the proof of Theorem 1, we can also get

$$\|f\|_{(\dot{B}^{s,q}_{p_0}, \dot{B}^{s,q}_{p_1})_{\theta,r}} \sim \left\{\int_0^{\infty} \left[t^{-\theta}\left\{\sum_j 2^{jsq} [K(t, a_j, L^{p_0}, L^{p_1})]^q\right\}^{\frac{1}{q}}\right]^r \frac{dt}{t}\right\}^{\frac{1}{r}},$$

where $f = \sum_{(\epsilon,j,k)\in\Lambda} a^{\epsilon}_{j,k}\Phi^{\epsilon}_{j,k}$, $a_j = \sum_{(\epsilon,k)\in\Gamma} a^{\epsilon}_{j,k}\Phi^{\epsilon}_{j,k}$. As $r = q$, we can write

$$\begin{aligned}
\|f\|_{(\dot{B}^{s,q}_{p_0},\dot{B}^{s,q}_{p_1})_{\theta,q}} &\sim \left\{\int_0^{\infty}\left[t^{-\theta q}\sum_j 2^{jsq}[K(t,a_j,L^{p_0},L^{p_1})]^q\right]\frac{dt}{t}\right\}^{\frac{1}{q}} \\
&\sim \left\{\sum_j 2^{jsq}\left[\int_0^{\infty} t^{-\theta q}[K(t,a_j,L^{p_0},L^{p_1})]^q\frac{dt}{t}\right]\right\}^{\frac{1}{q}} \\
&= \left\{\sum_j 2^{jsq}\left(\left[\int_0^{\infty} t^{-\theta q}[K(t,a_j,L^{p_0},L^{p_1})]^q\frac{dt}{t}\right]^{\frac{1}{q}}\right)^q\right\}^{\frac{1}{q}} \\
&\sim \left\{\sum_j 2^{jsq}\|a_j\|^q_{(L^{p_0},L^{p_1})_{\theta,q}}\right\}^{\frac{1}{q}}.
\end{aligned}$$

Thus,

$$\|f\|_{(\dot{B}^{s,q}_{p_0},\dot{B}^{s,q}_{p_1})_{\theta,q}} \sim \left\{\sum_j 2^{jsq}\|a_j\|^q_{(L^{p_0},L^{p_1})_{\theta,q}}\right\}^{\frac{1}{q}}. \tag{10}$$

We will prove the theorem in two cases.

(i) For $a_j = \sum_{(\epsilon,k)\in\Gamma} a^{\epsilon}_{j,k}\Phi^{\epsilon}_{j,k}(x)$ and $\frac{1}{p} = \frac{1-\theta}{p_0} + \frac{\theta}{p_1}$, using Lemma 7, we have

$$\|a_j\|_{(L^{p_0},L^{p_1})_{\theta,q}} = \|a_j\|_{p,q} = \left\{\sum_u 2^{uq}\left|\left\{x\in\mathbb{R}^n : \left|\sum_{(\epsilon,k)\in\Gamma} a^{\epsilon}_{j,k}\Phi^{\epsilon}_{j,k}(x)\right| > 2^u\right\}\right|^{\frac{q}{p}}\right\}^{\frac{1}{q}}. \tag{11}$$

From (10) and (11), it follows that

$$\|f\|^q_{(\dot{B}^{s,q}_{p_0},\dot{B}^{s,q}_{p_1})_{\theta,q}} \sim \sum_j 2^{jsq}\left\{\sum_u 2^{uq}\left|\left\{x\in\mathbb{R}^n : \left|\sum_{(\epsilon,k)\in\Gamma} a^{\epsilon}_{j,k}\Phi^{\epsilon}_{j,k}(x)\right| > 2^u\right\}\right|^{\frac{q}{p}}\right\}.$$

(ii) Applying Lemma 10, we obtain another equivalent form of $\|a_j\|_{p,q}$,

$$\begin{aligned}
\|a_j\|_{(L^{p_0},L^{p_1})_{\theta,q}} = \|a_j\|_{p,q} &= \left(\sum_u 2^{ru}|\{x\in\mathbb{R}^n : |Sa_j(x)| > 2^u\}|^{\frac{q}{p}}\right)^{\frac{1}{q}} \\
&= \left\{\sum_u 2^{uq}\left|\left\{x\in\mathbb{R}^n : 2^{\frac{nj}{2}}\sum_{(\epsilon,k)\in\Gamma}|a^{\epsilon}_{j,k}|\chi(2^jx-k) > 2^u\right\}\right|^{\frac{q}{p}}\right\}^{\frac{1}{q}}.
\end{aligned} \tag{12}$$

Applying (10) and (12), we obtain that

$$\|f\|^q_{(\dot{B}^{s,q}_{p_0},\dot{B}^{s,q}_{p_1})_{\theta,q}} \sim \sum_j 2^{jsq}\left\{\sum_u 2^{uq}\left|\left\{x\in\mathbb{R}^n : 2^{\frac{nj}{2}}\sum_{(\epsilon,k)\in\Gamma}|a^{\epsilon}_{j,k}|\chi(2^jx-k) > 2^u\right\}\right|^{\frac{q}{p}}\right\}.$$

We finish the proof of Theorem 2. □

Author Contributions: Conceptualization, Q.Y., J.H., Z.L. and K.H.; methodology, Q.Y. and Z.L.; formal analysis, Z.L. and J.H.; investigation, Z.L., J.H. and K.H.; supervision, J.H. All authors have read and agreed to the published version of the manuscript.

Funding: This work was partially supported by the National Natural Science Foundation of China under Grant Nos. 12071229, 11571261, and the Project of Guangzhou Scientific and Technological Bureau Grant No. 202102010402.

Institutional Review Board Statement: Not applicable.

Informed Consent Statement: Not applicable.

Data Availability Statement: Not applicable.

Conflicts of Interest: The authors declare no conflicts of interest.

References

1. Bergh, J.; Löfström, J. *Interpolation Spaces, An Introduction;* Springer: New York, NY, USA, 1976.
2. Cheng, M.; Deng, D.; Long, R. *Real Analysis*, 2rd ed.; High Eduacatin Press: Beijing, China, 2008.
3. Holmstedt, T. Interpolation of quasi-normed spaces. *Math. Scand.* **1970**, *10*, 177–199. [CrossRef]
4. Peetre, J. *New Thoughts on Besov Space;* Math Series; Duke University: Durham, NC, USA, 1976.
5. Blozinski, A.P. On a convolution theorem for $L(p,q)$ spaces. *Trans. Am. Math. Soc.* **1972**, *164*, 255–265.
6. Hunt, R.A. On $L(p,q)$ spaces. *Enseign. Math.* **1966**, *12*, 249–276.
7. Lorentz, G.G. Some new functional spaces. *Ann. Math.* **1950**, *51*, 37–55. [CrossRef]
8. O'Neil, R. Convolution operators and $L(p,q)$ spaces. *Duke Math. J.* **1963**, *30*, 129–142. [CrossRef]
9. Triebel, H. *Theory of Function Spaces;* Birkhauser Verlag: Boston, MA, USA, 1983.
10. Asekritova, I.; Kruglyak, N. Interpolation of Besov spaces in the nondiagonal case. *St. Petersb. Math. J.* **2007**, *18*, 511–516. [CrossRef]
11. Besoy, B.F.; Cobos, F.; Triebel, H. On Function Spaces of Lorentz-Sobolev type. *Math. Ann.* **2021**, 1–33. [CrossRef]
12. Devore, R.A.; Popov, V.A. Interpolation of Besov Spaces. *Trans. Am. Math. Soc.* **1988**, *305*, 397–414. [CrossRef]
13. Cobos, F.; Fernandez, D. *Hardy-Sobolev Spaces and Besov Spaces with a Function Parameter;* Springer: Berlin, Germany, 1988.
14. Devore, R.A.; Yu, X. K-functionals for Besov spaces. *J. Approx. Theory* **1991**, *67*, 38–50. [CrossRef]
15. Triebel, H. Spaces of distributions of Besov type on Euclidean n-space. Duality, interpolation. *Ark. Mat.* **1973**, *11*, 13–64. [CrossRef]
16. Yang, D. Real interpolations for Besov and Triebel-Lizorkin spaces on spaces of homogeneous type. *Math. Nachr.* **2004**, *273*, 96–113. [CrossRef]
17. Yang, Q.; Cheng, Z.; Peng, L. Uniform characterization of function spaces by wavelets. *Acta Math. Sci. Ser. A Chin. Ed.* **2005**, *25*, 130–144.
18. Yang, Q. *Wavelet and Distribution;* Beijing Science and Technology Press: Beijing, China, 2002.
19. Yang, Q. *Introduction to Harmonic Analysis and Wavelets;* Wuhan University Press: Wuhan, China, 2012.
20. Meyer, Y. *Ondelettes et Opérateur, I et II;* Hermann: Paris, France, 1990.
21. Holmstedt, T.; Peetre, J. On certain functionals arising in the theory of interpolation spaces. *J. Funct. Anal.* **1969**, *4*, 88–94. [CrossRef]

mathematics MDPI

Systematic Review
A Survey on Function Spaces of John–Nirenberg Type

Jin Tao, Dachun Yang * and Wen Yuan

Laboratory of Mathematics and Complex Systems (Ministry of Education of China), School of Mathematical Sciences, Beijing Normal University, Beijing 100875, China; jintao@mail.bnu.edu.cn (J.T.); wenyuan@bnu.edu.cn (W.Y.)
* Correspondence: dcyang@bnu.edu.cn

Abstract: In this systematic review, the authors give a survey on the recent developments of both the John–Nirenberg space JN_p and the space BMO as well as their vanishing subspaces such as VMO, XMO, CMO, VJN_p, and CJN_p on $\mathbb{R}^n$ or a given cube $Q_0 \subset \mathbb{R}^n$ with finite side length. In addition, some related open questions are also presented.

Keywords: Euclidean space; cube; congruent cube; BMO; JN_p; (localized) John–Nirenberg–Campanato space; Riesz–Morrey space; vanishing John–Nirenberg space; duality; commutator

Citation: Tao, J.; Yang, D.; Yuan, W. A Survey on Function Spaces of John–Nirenberg Type. *Mathematics* **2021**, *9*, 2264. https://doi.org/10.3390/math9182264

Academic Editor: Juan Benigno Seoane-Sepúlveda

Received: 25 July 2021
Accepted: 9 September 2021
Published: 15 September 2021

1. Introduction

In this article, a *cube* Q means that it has finite side length and all its sides parallel to the coordinate axes, but Q is not necessarily open or closed. Moreover, we always let $\mathcal{X}$ be $\mathbb{R}^n$ or a given cube of $\mathbb{R}^n$. Recall that the *Lebesgue space* $L^q(\mathcal{X})$ with $q \in [1, \infty]$ is defined to be the set of all measurable functions f on $\mathcal{X}$ such that

$$\|f\|_{L^q(\mathcal{X})} := \begin{cases} \left[\int_{\mathcal{X}} |f(x)|^q \, dx\right]^{\frac{1}{q}} & \text{when} \quad q \in [1, \infty), \\ \operatorname*{ess\,sup}_{x \in \mathcal{X}} |f(x)| & \text{when} \quad q = \infty \end{cases}$$

is finite. In what follows, we use $\mathbf{1}_E$ to denote the *characteristic function* of a set $E \subset \mathbb{R}^n$, and for any given $q \in [1, \infty)$, $L^q_{\rm loc}(\mathcal{X})$ to denote the set of all measurable functions f on $\mathcal{X}$ such that $f\mathbf{1}_E \in L^q(\mathcal{X})$ for any bounded measurable set $E \subset \mathcal{X}$.

It is well known that $L^p(\mathcal{X})$ with $p \in [1, \infty]$ plays a leading role in the modern analysis of mathematics. In particular, when $p \in (1, \infty)$, the space $L^p(\mathcal{X})$ enjoys some elegant properties, such as the reflexivity and the separability, which no longer hold true in $L^\infty(\mathcal{X})$. Thus, many studies related to $L^p(\mathcal{X})$ need some modifications when $p = \infty$: for instance, the boundedness of Calderón–Zygmund operators. Recall that the Calderón–Zygmund operator T is bounded on $L^p(\mathbb{R}^n)$ for any given $p \in (1, \infty)$, but not bounded on $L^\infty(\mathbb{R}^n)$. Indeed, T maps $L^\infty(\mathbb{R}^n)$ into the space $\mathrm{BMO}(\mathbb{R}^n)$ which was introduced by John and Nirenberg [1] in 1961 to study the functions of *bounded mean oscillation*; here and thereafter,

$$\mathrm{BMO}(\mathcal{X}) := \left\{ f \in L^1_{\rm loc}(\mathcal{X}) : \|f\|_{\mathrm{BMO}(\mathcal{X})} := \sup_{\text{cube } Q \subset \mathcal{X}} \fint_Q \left|f(x) - f_Q\right| dx < \infty \right\}$$

with

$$f_Q := \fint_Q f(y)\, dy := \frac{1}{|Q|} \int_Q f(y)\, dy$$

and the supremum taken over all cubes Q of $\mathcal{X}$. This implies that $\mathrm{BMO}(\mathcal{X})$ is a fine substitute of $L^\infty(\mathcal{X})$. Furthermore, it should be mentioned that, in the sense modulo constants, $\mathrm{BMO}(\mathcal{X})$ is a Banach space, but, for simplicity, we regard $f \in \mathrm{BMO}(\mathcal{X})$ as a function rather than an equivalent class $f + \mathbb{C} := \{f + c : c \in \mathbb{C}\}$ if there exists no confusion. Moreover, the space BMO $(\mathcal{X})$ and its numerous variants as well as their vanishing subspaces have attracted a lot of attention since 1961. For instance, Fefferman and Stein [2] proved that the dual space of the Hardy space $H^1(\mathbb{R}^n)$ is BMO $(\mathbb{R}^n)$; Coifman et al. [3] showed an equivalent characterization of the boundedness of Calderón–Zygmund commutators via BMO $(\mathbb{R}^n)$; Coifman and Weiss [4,5] introduced the space of homogeneous type and studied the Hardy space and the BMO space in this context; Sarason [6] obtained the equivalent characterization of VMO $(\mathbb{R}^n)$, the closure in BMO $(\mathbb{R}^n)$ of uniformly continuous functions, and used it to study stationary stochastic processes satisfying the strong mixing condition and the algebra $H^\infty + C$; Uchiyama [7] established an equivalent characterization of the compactness of Calderón–Zygmund commutators via CMO $(\mathbb{R}^n)$ which is defined to be the closure in BMO $(\mathbb{R}^n)$ of infinitely differentiable functions on $\mathbb{R}^n$ with compact support; Nakai and Yabuta [8] studied pointwise multipliers for functions on $\mathbb{R}^n$ of bounded mean oscillation; and Iwaniec [9] used the compactness theorem in Uchiyama [7] to study linear complex Beltrami equations and the $L^p(\mathbb{C})$ theory of quasiregular mappings. All these classical results have wide generalizations as well as applications and have inspired a myriad of further studies in recent years: see, for instance, the References [10–13] for their applications in singular integral operators as well as their commutators, the References [14–19] for their applications in pointwise multipliers, the References [20–22] for their applications in partial differential equations, and the References [23–28] for more variants and properties of BMO $(\mathbb{R}^n)$. In particular, we refer the reader to Chang and Sadosky [29] for an instructive survey on functions of bounded mean oscillation and also Chang et al. [25] for BMO spaces on the Lipschitz domain of $\mathbb{R}^n$.

Naturally, BMO $(\mathcal{X})$ extends $L^\infty(\mathcal{X})$, in the sense that $L^\infty(\mathcal{X}) \subsetneqq$ BMO $(\mathcal{X})$ and, moreover, $\|\cdot\|_{\mathrm{BMO}\,(\mathcal{X})} \le 2\|\cdot\|_{L^\infty(\mathcal{X})}$. Similarly, such extension exists for any $L^p(\mathcal{X})$ with $p \in (1,\infty)$. Indeed, John and Nirenberg [1] also introduced a generalized version of the BMO condition which was subsequently used to define the so-called John–Nirenberg space $JN_p(Q_0)$ with exponent $p \in (1,\infty)$ and Q_0 being any given cube of $\mathbb{R}^n$. Recall that for any given $p \in (1,\infty)$ and any given cube Q_0 of $\mathbb{R}^n$, the *John–Nirenberg space* $JN_p(Q_0)$ is defined to be the set of all $f \in L^1(Q_0)$ such that

$$\|f\|_{JN_p(Q_0)} := \sup\left[\sum_i |Q_i|\left\{\fint_{Q_i} |f(x) - f_{Q_i}|\,dx\right\}^p\right]^{\frac{1}{p}} < \infty, \tag{1}$$

where the supremum is taken over all collections of *interior pairwise disjoint* cubes $\{Q_i\}_i$ of Q_0. It is easy to see that the limit of $JN_p(Q_0)$ when $p \to \infty$ is just BMO (Q_0) (see also Corollary 2 below). Moreover, the John–Nirenberg space is closely related to the Lebesgue space $L^p(Q_0)$ and the weak Lebesgue space $L^{p,\infty}(Q_0)$ which is defined in Definition 1 below. Precisely, let $p \in (1,\infty)$. On the one hand, the inequality obtained in ([1], Lemma 3) (see also Theorem 2 below) implies that $JN_p(Q_0) \subset L^{p,\infty}(Q_0)$; additionally, by ([30], Example 3.5), we further know that $JN_p(Q_0) \subsetneqq L^{p,\infty}(Q_0)$. On the other hand, it is obvious that $L^p(Q_0) \subset JN_p(Q_0)$ with $\|\cdot\|_{JN_p(Q_0)} \le 2\|\cdot\|_{L^p(Q_0)}$, but the striking nontriviality was shown very recently by Dafni et al. ([31], Proposition 3.2 and Corollary 4.2), who say that $L^p(Q_0) \subsetneqq JN_p(Q_0)$. Combining these facts, we conclude that

$$L^p(Q_0) \subsetneqq JN_p(Q_0) \subsetneqq L^{p,\infty}(Q_0). \tag{2}$$

Therefore, John–Nirenberg spaces are new spaces between Lebesgue spaces and weak Lebesgue spaces, which motivates us to study the properties of JN_p. Furthermore, various John–Nirenberg-type spaces have also attracted a lot of attention in recent years (see, for instance, [31–37] for the Euclidean space case and [30,38–40] for the metric measure space case).

It should be mentioned that the mean oscillation truly makes a difference in both BMO and JN_p; for instance,

(i) Via the characterization of distribution functions, we know that BMO is closely related to the space $L_{\exp}$ whose definition (see (6) below) is similar to an equivalent expression of BMO but with $f - f_Q$ replaced by f (see Proposition 3 below);

(ii) There exists an interesting observation presented by Riesz [41], which says that in (1), if we replace $f - f_{Q_i}$ by f, then $JN_p(Q_0)$ turns to be $L^p(Q_0)$. Moreover, this conclusion also holds true when Q_0 is replaced by $\mathbb{R}^n$ (see Proposition 28 below).

The main purpose of this article is to give a survey on some recent developments of both the John–Nirenberg space JN_p and the space BMO, including their several generalized (or related) spaces and some vanishing subspaces. We begin in Section 2 by recalling some definitions and basic properties of BMO and JN_p. Section 3 summarizes some recent developments of the John–Nirenberg–Campanato space, the localized John–Nirenberg–Campanato space, and the special John–Nirenberg–Campanato space via congruent cubes. Section 4 focuses on the Riesz-type space, which differs from the John–Nirenberg space in subtracting integral means, and its congruent counterpart. In Section 5, we pay attention to some vanishing subspaces of the aforementioned John–Nirenberg-type spaces, such as VMO, XMO, CMO, VJN_p, and CJN_p on $\mathbb{R}^n$ or any given cube Q_0 of $\mathbb{R}^n$. In addition, several related open questions are also summarized in this survey.

More precisely, the remainder of this survey is organized as follows.

Section 2 is split into two subsections. In Section 2.1, via recalling the definitions of distribution functions and some related function spaces (including the weak Lebesgue space, the Morrey space, and the space $L_{\exp}$), we present the relation

$$L^\infty(Q_0) \subsetneqq \mathrm{BMO}(Q_0) \subsetneqq L_{\exp}(Q_0)$$

in Proposition 2 below, which is a counterpart of (2) above, and also show two equivalent Orlicz-type norms on $\mathrm{BMO}(\mathbb{R}^n)$ in Proposition 3 below; moreover, the corresponding results for the localized BMO space are also obtained in Corollary 1 below. Section 2.2 is devoted to some significant results of JN_p, including the famous John–Nirenberg inequality (see Theorem 2 below), and the accurate relations of JN_p and L^p as well as $L^{p,\infty}$ (see Remark 2 below). Furthermore, some recent progress of JN_p is also briefly listed at the end of this subsection.

Section 3 is split into three subsections. In Section 3.1, we first recall the notions of the John–Nirenberg–Campanato space (for short, JNC space), the corresponding Hardy-type space, and their basic properties, which include the limit results and the relations with other classical spaces. Then we review the dual theorem between these two spaces and the independence over the second sub-index of JNC spaces and Hardy-type spaces. Section 3.2 is devoted to the localized counterpart of Section 3.1. The aim of Section 3.3 is the summary of the special JNC space defined via congruent cubes (for short, congruent JNC space), including their basic properties corresponding to those in Section 3.1. Furthermore, some applications about the boundedness of operators on congruent spaces are mentioned as well.

In Section 4, via subtracting integral means in the JNC space, we first give the definition of the Riesz-type space appearing in [37] and then present some basic facts about this space in Section 4.1. Moreover, the predual space (namely, the block-type space) and the corresponding dual theorem of the Riesz-type space are also displayed in this subsection. Section 4.2 is devoted to the congruent counterpart of the Riesz-type space and the boundedness of some important operators.

Section 5 is split into three subsections. Section 5.1 is devoted to several vanishing subspaces of $\mathrm{BMO}(\mathbb{R}^n)$, including $\mathrm{VMO}(\mathbb{R}^n)$, $\mathrm{CMO}(\mathbb{R}^n)$, $\mathrm{MMO}(\mathbb{R}^n)$, $\mathrm{XMO}(\mathbb{R}^n)$, and $\mathrm{X_1MO}(\mathbb{R}^n)$. We first recall their definitions and then review their (except $\mathrm{MMO}(\mathbb{R}^n)$) mean oscillation characterizations, respectively, in Theorems 11–13 below. Meanwhile, an open question on the corresponding equivalent characterization of $\mathrm{MMO}(\mathbb{R}^n)$ is also listed in Question 11 below. Then, we further review the compactness theorems of the Calderón–Zygmund commutators $[b, T]$, where b belongs to the vanishing subspaces $\mathrm{CMO}(\mathbb{R}^n)$ as well as $\mathrm{XMO}(\mathbb{R}^n)$, and propose an open question on $[b, T]$ with

$b \in \mathrm{XMO}(\mathbb{R}^n)$. Moreover, the characterizations via Riesz transforms of $\mathrm{BMO}(\mathbb{R}^n)$, $\mathrm{VMO}(\mathbb{R}^n)$, and $\mathrm{CMO}(\mathbb{R}^n)$, as well as the localized results of these vanishing subspaces, are presented. Furthermore, some open questions are listed in this subsection. Section 5.2 devotes to the vanishing subspaces of JNC spaces. We first recall the definition of the vanishing JNC space on cubes in Definition 17 and then review its equivalent characterization as well as its dual result, respectively, in Theorems 19 and 20. Moreover, for the case of $\mathbb{R}^n$, we review the corresponding results for $VJN_p(\mathbb{R}^n)$ and $CJN_p(\mathbb{R}^n)$, which are, respectively, counterparts of $\mathrm{VMO}(\mathbb{R}^n)$ and $\mathrm{CMO}(\mathbb{R}^n)$ (see Theorems 21 and 22 below). As before, some open questions are also listed at the end of this subsection. Section 5.3 is devoted to the congruent counterpart of Section 5.2, and some similar conclusions are listed in this subsection; meanwhile, some open questions on the JNC space have affirmative answers in the congruent setting (see Proposition 32 below).

Finally, we make some conventions on notation. Let $\mathbb{N} := \{1,2,\ldots\}$, $\mathbb{Z}_+ := \mathbb{N} \cup \{0\}$, and $\mathbb{Z}_+^n := (\mathbb{Z}_+)^n$. We always denote by C and $\widetilde{C}$ positive constants which are independent of the main parameters, but they may vary from line to line. Moreover, we use $C_{(\gamma,\,\beta,\,\ldots)}$ to denote a positive constant depending on the indicated parameters $\gamma,\ \beta,\ldots$ Constants with subscripts, such as C_0 and A_1, do not change in different occurrences. Moreover, the symbol $f \lesssim g$ represents that $f \le Cg$ for some positive constant C. If $f \lesssim g$ and $g \lesssim f$, we then write $f \sim g$. If $f \le Cg$ and $g = h$ or $g \le h$, we then write $f \lesssim g \sim h$ or $f \lesssim g \lesssim h$, *rather than* $f \lesssim g = h$ or $f \lesssim g \le h$. For any $p \in [1,\infty]$, let p' be its *conjugate index*, that is, p' satisfies $1/p + 1/p' = 1$. We use $\mathbf{1}_E$ to denote the *characteristic function* of a set $E \subset \mathbb{R}^n$, $|E|$ to denote the *Lebesgue measure* when $E \subset \mathbb{R}^n$ is measurable, and $\mathbf{0}$ to denote the *origin* of $\mathbb{R}^n$. For any function f on $\mathbb{R}^n$, let $\mathrm{supp}\,(f) := \{x \in \mathbb{R}^n :\ f(x) \neq 0\}$. Let $\mathbb{X}$ be a normed linear space. We use $(\mathbb{X})^*$ to denote its dual space.

2. BMO and JN_p

It is well known that the space BMO has played an important role in harmonic analysis, partial differential equations, and other mathematical fields since it was introduced by John and Nirenberg in their celebrated article [1]. However, in the same article [1], another mysterious space appeared as well, which is now called the John–Nirenberg space JN_p. Indeed, BMO can be viewed as the limit space of JN_p as $p \to \infty$ (see Proposition 6 and Corollary 2 below with $\alpha := 0$). To establish the relations of BMO and JN_p, and also to summarize some recent works of John–Nirenberg-type spaces, we first recall some basic properties of BMO and JN_p in this section.

This section is devoted to some well-known results of $\mathrm{BMO}(\mathcal{X})$ and $JN_p(\mathcal{X})$, respectively, in Sections 2.1 and 2.2. In addition, it is trivial to find that all the results in Section 2.1 also hold true with the cube Q_0 replaced by the ball B_0 of $\mathbb{R}^n$.

2.1. (Localized) BMO and $L_{\exp}$

This subsection is devoted to several equivalent norms of the spaces BMO and localized BMO. To this end, we begin with the *distribution function*

$$\mathfrak{D}(f;\mathcal{X})(t) := |\{x \in \mathcal{X} :\ |f(x)| > t\}|, \tag{3}$$

where $f \in L^1_{\mathrm{loc}}(\mathcal{X})$ and $t \in (0,\infty)$. Recall that the distribution function is closely related to the following weak Lebesgue space.

Definition 1. *Let $p \in (0,\infty)$. The* weak Lebesgue space $L^{p,\infty}(\mathcal{X})$ *is defined by setting*

$$L^{p,\infty}(\mathcal{X}) := \left\{f \text{ is measurable on } \mathcal{X} :\ \|f\|_{L^{p,\infty}(\mathcal{X})} < \infty\right\},$$

where, for any measurable function f on $\mathcal{X}$,

$$\|f\|_{L^{p,\infty}(\mathcal{X})} := \sup_{t\in(0,\infty)} \left[t|\{x \in \mathcal{X} :\ |f(x)| > t\}|^{\frac{1}{p}}\right].$$

Moreover, the distribution function also features BMO $(\mathcal{X})$, which is exactly the famous result obtained by John and Nirenberg ([1], Lemma 1'): there exist positive constants C_1 and C_2, depending only on the dimension n, such that, for any given $f \in \mathrm{BMO}(\mathcal{X})$, any given cube $Q \subset \mathcal{X}$, and any $t \in (0, \infty)$,

$$\left|\left\{x \in Q : |f(x) - f_Q| > t\right\}\right| \leq C_1 e^{-\frac{C_2}{\|f\|_{\mathrm{BMO}(\mathcal{X})}} t} |Q|. \tag{4}$$

The main tool used in the proof of (4) is the following well-known *Calderón–Zygmund decomposition* (see, for instance, [42], p. 34, Theorem 2.11, and also [43], p. 150, Lemma 1).

Theorem 1. *For a given function f which is integrable and non-negative on $\mathcal{X}$, and a given positive number λ, there exists a sequence $\{Q_j\}_j$ of disjoint dyadic cubes of $\mathcal{X}$ such that*

(i) $f(x) \leq \lambda$ *for almost every* $x \in \mathcal{X} \setminus \bigcup_j Q_j$;
(ii) $|\bigcup_j Q_j| \leq \frac{1}{\lambda}\|f\|_{L^1(\mathcal{X})}$;
(iii) $\lambda < \fint_{Q_j} f(x)\, dx \leq 2^n \lambda$.

As an application of (4), we find that for any given $q \in (1, \infty)$, $f \in \mathrm{BMO}(\mathbb{R}^n)$ if and only if $f \in L^1_{\mathrm{loc}}(\mathbb{R}^n)$ and

$$\|f\|_{\mathrm{BMO}_q(\mathbb{R}^n)} := \sup_{\text{cube } Q \subset \mathbb{R}^n} \left[\fint_Q |f(x) - f_Q|^q \, dx\right]^{\frac{1}{q}} < \infty.$$

Meanwhile, $\|\cdot\|_{\mathrm{BMO}(\mathbb{R}^n)} \sim \|\cdot\|_{\mathrm{BMO}_q(\mathbb{R}^n)}$ (see, for instance, [42], p. 125, Corollary 6.12).

Recently, Bényi et al. [44] gave a comprehensive approach for the boundedness of weighted commutators via a new equivalent Orlicz-type norm

$$\|f\|_{\mathcal{BMO}(\mathcal{X})} := \sup_{\text{cube } Q \subset \mathcal{X}} \|f - f_Q\|_{L_{\exp}(Q)}. \tag{5}$$

This equivalence is proved in Proposition 3 below. Here and thereafter, for any given cube Q of $\mathbb{R}^n$ and any measurable function g, the *locally normalized Orlicz norm* $\|g\|_{L_{\exp}(Q)}$ is defined by setting

$$\|g\|_{L_{\exp}(Q)} := \inf\left\{\lambda \in (0, \infty) : \fint_Q \left[e^{\frac{|g(x)|}{\lambda}} - 1\right] dx \leq 1\right\}. \tag{6}$$

Moreover, for any given cube Q of $\mathbb{R}^n$, the *space* $L_{\exp}(Q)$ is defined by setting

$$L_{\exp}(Q) := \left\{f \text{ is measurable on } Q : \ \exists\, \lambda \in (0, \infty) \text{ such that } \fint_Q e^{\frac{|f(x)|}{\lambda}} \, dx < \infty\right\}.$$

The space $L_{\exp}(Q)$ was studied in the interpolation of operators (see, for instance, [45], p. 243), and it is closely related to the space BMO (Q) (see Proposition 3 below).

On the Orlicz function in (6), we have the following properties.

Lemma 1. *For any $t \in [0, \infty)$, let $\Phi(t) := e^t - 1$. Then,*

(i) *Φ is of* lower type 1, *namely for any $s \in (0, 1)$ and $t \in (0, \infty)$,*

$$\Phi(st) \leq s\Phi(t);$$

(ii) *Φ is of* critical lower type 1, *namely there exists no $p \in (1, \infty)$, such that for any $s \in (0, 1)$ and $t \in (0, \infty)$,*

$$\Phi(st) \leq Cs^p \Phi(t)$$

holds true for some constant $C \in [1, \infty)$ independent of s and t.

Proof. We first show (i). For any $s \in (0,1)$ and $t \in (0,\infty)$, let

$$h(s,t) := \Phi(st) - s\Phi(t) = e^{st} - 1 - s(e^t - 1).$$

Then,

$$\frac{\partial}{\partial t} h(s,t) = se^{st} - se^t = s(e^{st} - e^t).$$

From this and $s \in (0,1)$, we deduce that for any $t \in (0,\infty)$, $\frac{\partial}{\partial t} h(s,t) < 0$, and hence $h(s,t) \le h(s,0) = 0$, which shows that Φ is of lower type 1 and hence completes the proof of (i).

Next, we show that Φ is of critical lower type 1. Suppose that there exist a $p \in (1,\infty)$ and a constant $C \in [1,\infty)$, such that for any $s \in (0,1)$ and $t \in (0,\infty)$, $\Phi(st) \le Cs^p\Phi(t)$, namely

$$e^{st} - 1 \le Cs^p(e^t - 1). \tag{7}$$

From $p \in (1,\infty)$ and the L'Hospital rule, we deduce that

$$\lim_{s \to 0^+} \frac{\Phi(st)}{s^p\Phi(t)} = \lim_{s \to 0^+} \frac{e^{st} - 1}{s^p(e^t - 1)} = \lim_{s \to 0^+} \frac{te^{st}}{ps^{p-1}(e^t - 1)} = \infty,$$

which contradicts (7), and hence Φ is of critical lower type 1. Here and thereafter, $s \to 0^+$ means $s \in (0,1)$ and $s \to 0$. This finishes the proof of (ii) and hence of Lemma 1. □

Before showing the equivalent Orlicz-type norms of $\mathrm{BMO}\,(\mathcal{X})$, we first prove the following equivalent characterizations of $\mathrm{BMO}\,(\mathcal{X})$. These characterizations might be well known. However, to the best of our knowledge, we did not find a complete proof. For the convenience of the reader, we present the details here.

Proposition 1. *The following three statements are mutually equivalent:*

(i) $f \in \mathrm{BMO}\,(\mathcal{X})$;

(ii) $f \in L^1_{\mathrm{loc}}(\mathcal{X})$ *and there exist positive constants* C_3 *and* C_4*, such that for any cube* $Q \subset \mathcal{X}$ *and any* $t \in (0,\infty)$,

$$\left|\left\{x \in Q : |f(x) - f_Q| > t\right\}\right| \le C_3 e^{-C_4 t}|Q|;$$

(iii) $f \in L^1_{\mathrm{loc}}(\mathcal{X})$ *and there exists a* $\lambda \in (0,\infty)$*, such that*

$$\sup_{\text{cube } Q \subset \mathcal{X}} \fint_Q e^{\frac{|f(x) - f_Q|}{\lambda}}\,dx < \infty.$$

Proof. We prove this proposition via showing (i) $\Longrightarrow$ (ii) $\Longrightarrow$ (iii) $\Longrightarrow$ (i).

First, the implication (i) $\Longrightarrow$ (ii) was proved by John and Nirenberg in [1], Lemma 1′ (see (4) above).

Next, we show the implication (ii) $\Longrightarrow$ (iii). Suppose that f satisfies (ii). Then, there exist positive constants C_3 and C_4, such that for any cube $Q \subset \mathcal{X}$ and any $t \in (0,\infty)$,

$$\left|\left\{x \in Q : |f(x) - f_Q| > t\right\}\right| \le C_3 e^{-C_4 t}|Q|$$

and hence

$$\begin{aligned} &\fint_Q e^{\frac{C_4}{2}|f(x) - f_Q|}\,dx \\ &= \frac{1}{|Q|}\int_0^\infty \left|\left\{x \in Q : e^{\frac{C_4}{2}|f(x) - f_Q|} > t\right\}\right| dt \end{aligned}$$

$$
\begin{aligned}
&= \frac{1}{|Q|}\left(\int_0^1 + \int_1^\infty\right)\left|\left\{x \in Q : e^{\frac{C_4}{2}|f(x)-f_Q|} > t\right\}\right| dt \\
&\le 1 + \frac{1}{|Q|}\int_1^\infty \left|\left\{x \in Q : |f(x) - f_Q| > 2C_4^{-1}\log t\right\}\right| dt \\
&\le 1 + \frac{1}{|Q|}\int_1^\infty C_3 e^{-C_4 2C_4^{-1}\log t}|Q|\, dt \\
&= 1 + C_3\int_1^\infty t^{-2}\, dt = 1 + C_3,
\end{aligned} \tag{8}
$$

which implies that f satisfies (iii). This shows the implication (ii) $\Longrightarrow$ (iii).

Finally, we show the implication (iii) $\Longrightarrow$ (i). Suppose that f satisfies (iii). Then, there exists a $\lambda \in (0, \infty)$, such that

$$
\sup_{Q \subset \mathcal{X}} \fint_Q e^{\frac{|f(x)-f_Q|}{\lambda}}\, dx < \infty.
$$

From this and the basic inequality $x \le e^x - 1$ for any $x \in \mathbb{R}$, we deduce that

$$
\sup_{\text{cube } Q \subset \mathcal{X}} \fint_Q \left|f(x) - f_Q\right| dx \le \lambda \sup_{\text{cube } Q \subset \mathcal{X}} \fint_Q \left[e^{\frac{|f(x)-f_Q|}{\lambda}} - 1\right] dx < \infty,
$$

which implies that f satisfies (i), and hence the implication (iii) $\Longrightarrow$ (i) holds true. This finishes the proof of Proposition 1. □

In what follows, for any normed space $\mathbb{Y}(\mathcal{X})$, equipped with the norm $\|\cdot\|_{\mathbb{Y}(\mathcal{X})}$, whose elements are measurable functions on $\mathcal{X}$, let

$$
\mathbb{Y}(\mathcal{X})/\mathbb{C} := \left\{f \text{ is measurable on } \mathcal{X} : \|f\|_{\mathbb{Y}(\mathcal{X})/\mathbb{C}} := \inf_{c \in \mathbb{C}} \|f + c\|_{\mathbb{Y}(\mathcal{X})} < \infty\right\}.
$$

Proposition 2. *Let Q_0 be a given cube of $\mathbb{R}^n$. Then,*

$$
[L^\infty(Q_0)/\mathbb{C}] \subsetneqq \mathrm{BMO}\,(Q_0) \subsetneqq \left[L_{\exp}(Q_0)/\mathbb{C}\right].
$$

Proof. Indeed, on the one hand, from

$$
\fint_Q \left|f(x) - f_Q\right| dx \le 2\fint_Q \left|f(x) + c\right| dx \le 2\|f + c\|_{L^\infty(Q_0)}
$$

for any $c \in \mathbb{C}$, we deduce that $[L^\infty(Q_0)/\mathbb{C}] \subset \mathrm{BMO}\,(Q_0)$. Moreover, let $g(\cdot) := \log|\cdot - c_0|$, where c_0 is the center of Q_0. Then, $g \in \mathrm{BMO}\,(Q_0) \setminus [L^\infty(Q_0)/\mathbb{C}]$ (see [46], Example 3.1.3, for this fact).

On the other hand, by Proposition 1(iii), we easily find that $\mathrm{BMO}\,(Q_0) \subset [L_{\exp}(Q_0)/\mathbb{C}]$. Moreover, without loss of generality, we may assume that $Q_0 := (-1, 1)$ and let

$$
g(x) := \begin{cases} -\log(-x), & x \in (-1, 0), \\ 0, & x = 0, \\ \log(x), & x \in (0, 1). \end{cases}
$$

We claim that $g \in [L_{\exp}(Q_0)/\mathbb{C}] \setminus \mathrm{BMO}\,(Q_0)$. Indeed, for any $\epsilon \in (0, 1)$, let $I_\epsilon := (-\epsilon, \epsilon)$. Then,

$$
\fint_{I_\epsilon} \left|g(x) - g_{I_\epsilon}\right| dx = \frac{1}{2\epsilon}\int_{-\epsilon}^{\epsilon} \left|\log|x|\right| dx = -\frac{1}{\epsilon}\int_0^\epsilon \log(x)\, dx = 1 - \log(\epsilon) \to \infty
$$

as $\epsilon \to 0^+$, which implies that $g \notin \mathrm{BMO}(Q_0)$. However,

$$\int_{Q_0} e^{\frac{1}{2}|g(x)|}\,dx = 2\int_0^1 e^{-\frac{1}{2}\log(x)}\,dx = 2\int_0^1 x^{-\frac{1}{2}}\,dx = 4 < \infty,$$

which implies that $g \in L_{\exp}(Q_0)$. Therefore, $\mathrm{BMO}(Q_0) \subsetneqq [L_{\exp}(Q_0)/\mathbb{C}]$, which completes the proof of Proposition 2. □

Now, we show that the two Orlicz-type norms, (5) and

$$\|f\|_{\widetilde{L_{\exp}}(\mathcal{X})} := \inf\left\{\lambda \in (0,\infty) : \sup_{\text{cube } Q\subset\mathcal{X}} \fint_Q \left[e^{\frac{|f(x)-f_Q|}{\lambda}} - 1\right] dx \le 1\right\}$$

for any $f \in L^1_{\mathrm{loc}}(\mathcal{X})$, are equivalent norms of $\mathrm{BMO}(\mathcal{X})$.

Proposition 3. *The following three statements are mutually equivalent:*

(i) $f \in \mathrm{BMO}(\mathcal{X})$;
(ii) $f \in L^1_{\mathrm{loc}}(\mathcal{X})$ *and* $\|f\|_{\mathcal{BMO}(\mathcal{X})} < \infty$;
(iii) $f \in L^1_{\mathrm{loc}}(\mathcal{X})$ *and* $\|f\|_{\widetilde{L_{\exp}}(\mathcal{X})} < \infty$.

Moreover, $\|\cdot\|_{\mathrm{BMO}(\mathcal{X})} \sim \|\cdot\|_{\mathcal{BMO}(\mathcal{X})} \sim \|\cdot\|_{\widetilde{L_{\exp}}(\mathcal{X})}$.

Proof. To prove this proposition, we only need to prove that for any $f \in L^1_{\mathrm{loc}}(\mathcal{X})$,

$$\|f\|_{\mathrm{BMO}(\mathcal{X})} \sim \|f\|_{\mathcal{BMO}(\mathcal{X})} \sim \|f\|_{\widetilde{L_{\exp}}(\mathcal{X})}.$$

We first show that for any $f \in L^1_{\mathrm{loc}}(\mathcal{X})$, $\|f\|_{\mathrm{BMO}(\mathcal{X})} \le \|f\|_{\mathcal{BMO}(\mathcal{X})}$ and $\|f\|_{\mathrm{BMO}(\mathcal{X})} \le \|f\|_{\widetilde{L_{\exp}}(\mathcal{X})}$. To this end, let $f \in L^1_{\mathrm{loc}}(\mathcal{X})$. For any cube $Q \subset \mathcal{X}$ and any $\lambda \in (0,\infty)$, by $t \le e^t - 1$ for any $t \in (0,\infty)$, we have

$$\fint_Q |f(x) - f_Q|\,dx \le \lambda \fint_Q \left[e^{\frac{|f(x)-f_Q|}{\lambda}} - 1\right] dx \le \lambda,$$

which implies that

$$\fint_Q |f(x) - f_Q|\,dx \le \|f - f_Q\|_{L_{\exp}(Q)}$$

and hence

$$\|f\|_{\mathrm{BMO}(\mathcal{X})} \le \|f\|_{\mathcal{BMO}(\mathcal{X})}.$$

Moreover, to show $\|f\|_{\mathrm{BMO}(\mathcal{X})} \le \|f\|_{\widetilde{L_{\exp}}(\mathcal{X})}$, it suffices to assume that $f \in \widetilde{L_{\exp}}(\mathcal{X})$; otherwise, $\|f\|_{\widetilde{L_{\exp}}(\mathcal{X})} = \infty$, and hence the desired inequality automatically holds true. Then, by $t \le e^t - 1$ for any $t \in (0,\infty)$, we conclude that for any $n \in \mathbb{N}$ and any cube $Q \subset \mathcal{X}$,

$$\fint_Q \frac{|f(x) - f_Q|}{\|f\|_{\widetilde{L_{\exp}}(\mathcal{X})} + \frac{1}{n}}\,dx \le \fint_Q \left[e^{\frac{|f(x)-f_Q|}{\|f\|_{\widetilde{L_{\exp}}(\mathcal{X})} + \frac{1}{n}}} - 1\right] dx. \tag{9}$$

From the definition of $\|\cdot\|_{\widetilde{L_{\exp}}(\mathcal{X})}$, we deduce that for any $n \in \mathbb{N}$, there exists a

$$\lambda_n \in \left(\|f\|_{\widetilde{L_{\exp}}(\mathcal{X})}, \|f\|_{\widetilde{L_{\exp}}(\mathcal{X})} + \frac{1}{n}\right)$$

such that

$$\sup_{\text{cube } Q\subset\mathcal{X}} \fint_Q \left[e^{\frac{|f(x)-f_Q|}{\lambda_n}} - 1\right] dx \le 1.$$

By this, (9), and the monotonicity of $e^{(\cdot)} - 1$, we conclude that, for any $n \in \mathbb{N}$ and any cube $Q \subset \mathcal{X}$,

$$\fint_Q \frac{|f(x) - f_Q|}{\|f\|_{\widetilde{L_{\exp}}(\mathcal{X})} + \frac{1}{n}} dx \le 1$$

and hence

$$\fint_Q |f(x) - f_Q| \, dx \le \|f\|_{\widetilde{L_{\exp}}(\mathcal{X})} + \frac{1}{n}.$$

Letting $n \to \infty$, we then obtain

$$\|f\|_{\mathrm{BMO}(\mathcal{X})} = \sup_{\text{cube } Q\subset\mathcal{X}} \fint_Q |f(x) - f_Q| \, dx \le \|f\|_{\widetilde{L_{\exp}}(\mathcal{X})}.$$

To summarize, we have, for any $f \in L^1_{\mathrm{loc}}(\mathcal{X})$,

$$\|f\|_{\mathrm{BMO}(\mathcal{X})} \le \|f\|_{\mathcal{BMO}(\mathcal{X})} \text{ and } \|f\|_{\mathrm{BMO}(\mathcal{X})} \le \|f\|_{\widetilde{L_{\exp}}(\mathcal{X})}. \tag{10}$$

Next, we show that the reverse inequalities hold true for any $f \in L^1_{\mathrm{loc}}(\mathcal{X})$, respectively. In fact, we may assume that $f \in \mathrm{BMO}(\mathcal{X})$ because, otherwise, the desired inequalities automatically hold true. Now, let $f \in \mathrm{BMO}(\mathcal{X})$. Then, for any cube $Q \subset \mathcal{X}$ and any $\lambda \in (C_2^{-1}\|f\|_{\mathrm{BMO}(\mathcal{X})}, \infty)$, by (4) and the calculation of (8), we obtain

$$\begin{aligned}
&\fint_Q e^{\frac{|f(x)-f_Q|}{\lambda}} dx \\
&\le 1 + \frac{1}{|Q|}\int_1^\infty \left|\left\{x \in Q : |f(x) - f_Q| > \lambda \log t\right\}\right| dt \\
&\le 1 + \frac{1}{|Q|}\int_1^\infty C_1 e^{-\frac{C_2}{\|f\|_{\mathrm{BMO}(\mathcal{X})}}\lambda \log t} |Q| \, dt \\
&= 1 + C_1 \int_1^\infty t^{-\frac{C_2\lambda}{\|f\|_{\mathrm{BMO}(\mathcal{X})}}} dt = 1 + C_1
\end{aligned}$$

and hence

$$\fint_Q \left[e^{\frac{|f(x)-f_Q|}{\lambda}} - 1\right] dx \le C_1,$$

where $C_1 \in (1, \infty)$ is as in (4). From this and Lemma 1(i) with s replaced by $1/C_1$, we deduce that

$$\fint_Q \left[e^{\frac{|f(x)-f_Q|}{\lambda C_1}} - 1\right] dx \le \frac{1}{C_1} \fint_Q \left[e^{\frac{|f(x)-f_Q|}{\lambda}} - 1\right] dx \le 1. \tag{11}$$

On the one hand, by (11) and

$$\frac{C_1}{C_2}\|f\|_{\mathrm{BMO}(\mathcal{X})} < \lambda C_1 < \infty,$$

we conclude that

$$\|f - f_Q\|_{L_{\exp}(Q)} = \inf\left\{\widetilde{\lambda} > 0 : \fint_Q \left[e^{\frac{|f(x)-f_Q|}{\widetilde{\lambda}}} - 1\right] dx \le 1\right\}$$

$$\leq \frac{C_1}{C_2} \|f\|_{\mathrm{BMO}\,(\mathcal{X})}$$

and hence

$$\|f\|_{\mathcal{BMO}(\mathcal{X})} = \sup_{\text{cube } Q \subset \mathcal{X}} \|f - f_Q\|_{L_{\exp}(Q)} \leq \frac{C_1}{C_2} \|f\|_{\mathrm{BMO}\,(\mathcal{X})}. \tag{12}$$

On the other hand, by (11), we conclude that

$$\sup_{\text{cube } Q \subset \mathcal{X}} \fint_Q \left[e^{\frac{|f(x) - f_Q|}{\lambda C_1}} - 1 \right] dx \leq 1.$$

From this and

$$\frac{C_1}{C_2} \|f\|_{\mathrm{BMO}\,(\mathcal{X})} < \lambda C_1 < \infty,$$

we deduce that

$$\|f\|_{\widetilde{L_{\exp}}(\mathcal{X})} = \inf\left\{ \lambda \in (0, \infty) : \sup_{\text{cube } Q \subset \mathcal{X}} \fint_Q \left[e^{\frac{|f(x) - f_Q|}{\lambda}} - 1 \right] dx \leq 1 \right\}$$
$$\leq \frac{C_1}{C_2} \|f\|_{\mathrm{BMO}\,(\mathcal{X})}.$$

Combining this with (12), we have, for any $f \in \mathrm{BMO}\,(\mathcal{X})$,

$$\|f\|_{\mathcal{BMO}(\mathcal{X})} \leq \frac{C_1}{C_2} \|f\|_{\mathrm{BMO}\,(\mathcal{X})} \text{ and } \|f\|_{\widetilde{L_{\exp}}(\mathcal{X})} \leq \frac{C_1}{C_2} \|f\|_{\mathrm{BMO}\,(\mathcal{X})}.$$

This, together with (10), then finishes the proof of Proposition 3. □

Remark 1. *There exists another norm on $L_{\exp}(Q_0)$, defined by the distribution functions as follows. Let f be a measurable function on Q_0. The* decreasing rearrangement *f^* of f is defined by setting, for any $u \in [0, \infty)$,*

$$f^*(u) := \inf\{t \in (0, \infty) : |\{x \in Q_0 : |f(x)| > t\}| \leq u\}.$$

Moreover, for any $v \in (0, \infty)$, let

$$f^{**}(v) := \frac{1}{v} \int_0^v f^*(u)\, du.$$

Then, $f \in L_{\exp}(Q_0)$ if and only if f is measurable on Q_0 and

$$\|f\|_{L^*_{\exp}(Q_0)} := \sup_{v \in (0, |Q_0|]} \frac{f^{**}(v)}{1 + \log(\frac{|Q_0|}{v})} < \infty.$$

Meanwhile, $\|\cdot\|_{L^*_{\exp}(Q_0)}$ is a norm of $L_{\exp}(Q_0)$ (see [45], p. 246, Theorem 6.4, for more details). Furthermore, from [45] (p. 7, Corollary 1.9), we deduce that $\|\cdot\|_{L^*_{\exp}(Q_0)}$ and $\|\cdot\|_{L_{\exp}(Q_0)}$ are equivalent. Notice that f^* and f^{**} are fundamental tools in the theory of Lorentz spaces *(see [47], p. 48, for more details).*

Recently, Izuki et al. [48] obtained both the John–Nirenberg inequality and the equivalent characterization of $\mathrm{BMO}\,(\mathbb{R}^n)$ on the ball Banach function space which contains Morrey spaces, (weighted, mixed-norm, variable) Lebesgue spaces, and Orlicz-slice spaces as special cases (see [48], Definition 2.8, and also [49], for the related definitions). Precisely, let X be a ball Banach function space satisfying the additional assumption that the Hardy–

Littlewood maximal operator M is bounded on X' (the associated space of X; see [48], Definition 2.9, for its definition), and for any $b \in L^1_{\mathrm{loc}}(\mathbb{R}^n)$,

$$\|b\|_{\mathrm{BMO}_X} := \sup_B \frac{1}{\|\mathbf{1}_B\|_X} \left\| |b - b_B| \mathbf{1}_B \right\|_X,$$

where the supremum is taken over all balls B of $\mathbb{R}^n$. It is obvious that $\|\cdot\|_{\mathrm{BMO}_{L^1(\mathbb{R}^n)}} = \|\cdot\|_{\mathrm{BMO}(\mathbb{R}^n)}$. Moreover, in [48] (Theorem 1.2), Izuki et al. showed that under the above assumption of X, $b \in \mathrm{BMO}(\mathbb{R}^n)$ if and only if $b \in L^1_{\mathrm{loc}}(\mathbb{R}^n)$ and $\|b\|_{\mathrm{BMO}_X} < \infty$; meanwhile,

$$\|\cdot\|_{\mathrm{BMO}_X} \sim \|\cdot\|_{\mathrm{BMO}(\mathbb{R}^n)}.$$

Furthermore, the John–Nirenberg inequality on X was also obtained in [48] (Theorem 3.1), which shows that there exists some positive constant $\widetilde{C}$, such that for any ball $B \subset \mathbb{R}^n$ and any $\tau \in [0, \infty)$,

$$\left\| \mathbf{1}_{\{x \in B:\ |b(x) - b_B| > \tau 2^{n+2} \|b\|_{\mathrm{BMO}(\mathbb{R}^n)}\}} \right\|_X \le \widetilde{C} 2^{-\frac{\tau}{1 + 2^{n+4} \|M\|_{X' \to X'}}} \|\mathbf{1}_B\|_X,$$

where $\|M\|_{X' \to X'}$ denotes the *operator norm* of M on X'. Later, these results were applied in [49] to establish the compactness characterization of commutators on ball Banach function spaces.

Now, we come to the localized counterpart. The local space $\mathrm{BMO}(\mathbb{R}^n)$, denoted by $\mathrm{bmo}(\mathbb{R}^n)$, was originally introduced by Goldberg [50]. In the same article, Goldberg also introduced the localized Campanato space $\Lambda_\alpha(\mathbb{R}^n)$ with $\alpha \in (0, \infty)$, which proves the dual space of the localized Hardy space. Later, Jonsson et al. [51] constructed the localized Hardy space and the localized Campanato space on the subset of $\mathbb{R}^n$; Chang [52] studied the localized Campanato space on bounded Lipschitz domains; Chang et al. [20] studied the localized Hardy space and its dual space on smooth domains as well as their applications to boundary value problems; and Dafni and Liflyand [53] characterized the localized Hardy space in the sense of Goldberg, respectively, by means of the localized Hilbert transform and localized molecules. In what follows, for any cube Q of $\mathbb{R}^n$, we use $\ell(Q)$ to denote its side length, and let $\ell(\mathbb{R}^n) := \infty$. Recall that

$$\mathrm{bmo}(\mathcal{X}) := \left\{ f \in L^1_{\mathrm{loc}}(\mathcal{X}) :\ \|f\|_{\mathrm{bmo}(\mathcal{X})} < \infty \right\},$$

where

$$\|f\|_{\mathrm{bmo}(\mathcal{X})} := \sup_Q \fint_Q \left| f(x) - f_{Q,c_0} \right| dx$$

with

$$f_{Q,c_0} := \begin{cases} f_Q & \text{if } \ell(Q) \in (0, c_0), \\ 0 & \text{if } \ell(Q) \in [c_0, \ell(\mathcal{X})) \end{cases} \tag{13}$$

for some given $c_0 \in (0, \ell(\mathcal{X}))$, and the supremum taken over all cubes Q of $\mathcal{X}$. Furthermore, a well-known fact is that $\mathrm{bmo}(\mathcal{X})$ is independent of the choice of c_0 (see, for instance, [54], Lemma 6.1).

Proposition 4. *Let $\mathcal{X}$ be $\mathbb{R}^n$ or a cube Q_0 of $\mathbb{R}^n$. Then,*

$$[L^\infty(\mathcal{X})/\mathbb{C}] \subset [\mathrm{bmo}(\mathcal{X})/\mathbb{C}] \subset \mathrm{BMO}(\mathcal{X}) \tag{14}$$

and

$$\|\cdot\|_{\mathrm{BMO}(\mathcal{X})} \le 2 \inf_{c \in \mathbb{C}} \|\cdot + c\|_{\mathrm{bmo}(\mathcal{X})} \le 4 \inf_{c \in \mathbb{C}} \|\cdot + c\|_{L^\infty(\mathcal{X})}. \tag{15}$$

Moreover,

$$[L^\infty(\mathbb{R}^n)/\mathbb{C}] \subsetneqq [\mathrm{bmo}\,(\mathbb{R}^n)/\mathbb{C}] \subsetneqq \mathrm{BMO}\,(\mathbb{R}^n) \tag{16}$$

and, for any cube Q_0 *of* $\mathbb{R}^n$*,*

$$[L^\infty(Q_0)/\mathbb{C}] \subsetneqq [\mathrm{bmo}\,(Q_0)/\mathbb{C}] = \mathrm{BMO}\,(Q_0) \subsetneqq \left[L_{\exp}(Q_0)/\mathbb{C}\right] \tag{17}$$

with

$$\|\cdot\|_{\mathrm{BMO}\,(Q_0)} \le 2 \inf_{c\in\mathbb{C}} \|\cdot+c\|_{\mathrm{bmo}\,(Q_0)} \le 4\|\cdot\|_{\mathrm{BMO}\,(Q_0)}.$$

Proof. First, we prove (15). To this end, let $f \in L^1_{\mathrm{loc}}(\mathcal{X})$. Then, for any $c \in \mathbb{C}$ and any cube Q of $\mathcal{X}$,

$$\begin{aligned} \fint_Q \left|f(x) - f_Q\right| dx &= \fint_Q \left|[f(x)+c] - (f+c)_Q\right| dx \\ &\le 2\fint_Q \left|f(x)+c\right| dx \le 2\|f+c\|_{L^\infty(Q)}. \end{aligned}$$

From this and the definitions of $\|\cdot\|_{\mathrm{BMO}\,(\mathcal{X})}$ and $\|\cdot\|_{\mathrm{bmo}\,(\mathcal{X})}$, it follows that (15) holds true, which further implies (14).

We now show (16). Indeed, let

$$g_1(x) := \begin{cases} \log(|x|) & \text{if } x \in \mathbb{R}^n \setminus \{\mathbf{0}\}, \\ 0 & \text{if } x = \mathbf{0}. \end{cases}$$

From [46] (Example 3.1.3), we deduce that $g_1 \in \mathrm{BMO}\,(\mathbb{R}^n)$. However, $g_1 \notin \mathrm{bmo}\,(\mathbb{R}^n)$ because, for any $M > \max\{c_0, 1\}$, by the sphere coordinate changing method, we have

$$\fint_{B(\mathbf{0},M)} \left|\log(|x|)\right| dx \sim \log(M),$$

which tends to infinity as $M \to \infty$. Thus, $g_1 \in \mathrm{BMO}\,(\mathbb{R}^n) \setminus [\mathrm{bmo}\,(\mathbb{R}^n)/\mathbb{C}]$, and hence we have $[\mathrm{bmo}\,(\mathbb{R}^n)/\mathbb{C}] \subsetneqq \mathrm{BMO}\,(\mathbb{R}^n)$. Moreover, define

$$g_2(x) := \begin{cases} \log(|x|) & \text{if } |x| \in (0,1), \\ 0 & \text{if } |x| \in \{0\} \bigcup [1,\infty). \end{cases}$$

Notice that $g_2 \notin L^\infty(\mathbb{R}^n)$ and $g_2 = \max\{g_1, 0\} \in \mathrm{BMO}\,(\mathbb{R}^n)$. Then, for any cube $Q \subset \mathbb{R}^n$, if $\ell(Q) \in (0, c_0)$, then

$$\fint_Q \left|g_2(x) - (g_2)_Q\right| dx \le \|g_2\|_{\mathrm{BMO}\,(\mathbb{R}^n)};$$

if $\ell(Q) \in [c_0, \infty)$, then

$$\fint_Q \left|g_2(x)\right| dx \le \fint_{B(\mathbf{0},1)} \log(|x|)\, dx \sim \|g_2\|_{L^1(\mathbb{R}^n)} \sim 1.$$

To summarize, $\|g_2\|_{\mathrm{bmo}\,(\mathbb{R}^n)} \lesssim 1 + \|g_2\|_{\mathrm{BMO}\,(\mathbb{R}^n)}$, which implies that $g_2 \in \mathrm{bmo}\,(\mathbb{R}^n)$ and hence $L^\infty(\mathbb{R}^n) \subsetneqq \mathrm{bmo}\,(\mathbb{R}^n)$. This shows (16).

We next prove (17). By the above example g_2, we conclude that $L^\infty(Q_0) \subsetneqq \mathrm{bmo}\,(Q_0)$. Meanwhile, $\mathrm{BMO}\,(Q_0) \subsetneqq [L_{\exp}(Q_0)/\mathbb{C}]$ was obtained in Proposition 2. Moreover, for any given $f \in \mathrm{BMO}\,(Q_0)$, we have $f \in L^1(Q_0)$ and hence

$$\inf_{c\in\mathbb{C}} \|f-c\|_{\mathrm{bmo}\,(Q_0)}$$

$$= \begin{cases} \fint_Q \left|f(x) - f_Q\right| dx \le \|f\|_{\mathrm{BMO}(Q_0)} & \text{if } \ell(Q) \in (0, c_0), \\ \inf\limits_{c\in\mathbb{C}} \fint_Q |f(x) - c|\, dx \le 2\|f\|_{\mathrm{BMO}(Q_0)} & \text{if } \ell(Q) \in [c_0, \ell(Q_0)), \end{cases}$$
$$\le 2\|f\|_{\mathrm{BMO}(Q_0)}.$$

Combining this with the observations that $[\mathrm{bmo}(Q_0)/\mathbb{C}] \subset \mathrm{BMO}(Q_0)$ and that, for any $c \in \mathbb{C}$,

$$\|f\|_{\mathrm{BMO}(Q_0)} = \|f + c\|_{\mathrm{BMO}(Q_0)} \le 2\|f + c\|_{\mathrm{bmo}(Q_0)},$$

we find that $[\mathrm{bmo}(Q_0)/\mathbb{C}] = \mathrm{BMO}(Q_0)$ and

$$\|f\|_{\mathrm{BMO}(Q_0)} \le 2 \inf_{c\in\mathbb{C}} \|f + c\|_{\mathrm{bmo}(Q_0)} \le 4\|f\|_{\mathrm{BMO}(Q_0)}.$$

To summarize, we obtain (17). This finishes the proof of Proposition 4. □

Let $f \in L^1_{\mathrm{loc}}(\mathcal{X})$. Similar to Proposition 3, let

$$\|f\|_{\mathrm{bmo}_1(\mathcal{X})} := \sup_{\text{cube } Q\subset\mathcal{X}} \left\|f - f_{Q,c_0}\right\|_{L_{\exp}(Q)} \tag{18}$$

and

$$\|f\|_{\mathrm{bmo}_2(\mathcal{X})} := \inf\left\{\lambda \in (0,\infty) : \sup_{\text{cube } Q\subset\mathcal{X}} \fint_Q \left[e^{\frac{|f(x)-f_{Q,c_0}|}{\lambda}} - 1\right] dx \le 1\right\}, \tag{19}$$

where $c_0 \in (0, \ell(\mathcal{X}))$, and f_{Q,c_0} is as in (13). To show that they are equivalent norms of $\mathrm{bmo}(\mathcal{X})$, we first establish the following John–Nirenberg inequality for $\mathrm{bmo}(\mathcal{X})$, namely Proposition 5 below. In what follows, for any given cube Q of $\mathbb{R}^n$, $(a_1, \ldots, a_n)$ denotes the *left and lower vertex* of Q, which means that for any $(x_1, \ldots, x_n) \in Q$, $x_i \ge a_i$ for any $i \in \{1, \ldots, n\}$. Recall that for any given cube Q of $\mathbb{R}^n$, the *dyadic system* $\mathscr{D}_Q$ of Q is defined by setting

$$\mathscr{D}_Q := \bigcup_{j=0}^{\infty} \mathscr{D}_Q^{(j)}, \tag{20}$$

where, for any $j \in \{0, 1, \ldots\}$, $\mathscr{D}_Q^{(j)}$ denotes the set of all $(x_1, \ldots, x_n) \in Q$, such that for any $i \in \{1, \ldots, n\}$, either

$$x_i \in \left[a_i + k_i 2^{-j}\ell(Q), a_i + (k_i + 1)2^{-j}\ell(Q)\right)$$

for some $k_i \in \{0, 1, \ldots, 2^j - 2\}$ or

$$x_i \in \left[a_i + (1 - 2^{-j})\ell(Q), a_i + \ell(Q)\right].$$

Proposition 5. *Let $f \in \mathrm{bmo}(\mathcal{X})$ and $c_0 \in (0, \ell(\mathcal{X}))$. Then, there exist positive constants C_5 and C_6, such that for any given cube $Q \subset \mathcal{X}$ and any $t \in (0, \infty)$,*

$$\left|\left\{x \in Q : |f(x) - f_{Q,c_0}| > t\right\}\right| \le C_5 e^{-\frac{C_6}{\|f\|_{\mathrm{bmo}(\mathcal{X})}} t} |Q|. \tag{21}$$

Proof. Indeed, this proof is a slight modification of the proof of [1] (Lemma 1) or [42] (Theorem 6.11). We give some details here, again for the sake of completeness.

Let $f \in \mathrm{bmo}\,(\mathcal{X})$. Then, from Proposition 4, we deduce that $f \in \mathrm{BMO}\,(\mathcal{X})$ with $\|f\|_{\mathrm{BMO}\,(\mathcal{X})} \le 2\|f\|_{\mathrm{bmo}\,(\mathcal{X})}$, which further implies that for any cube $Q \subset \mathcal{X}$ with $\ell(Q) < c_0$ and any $t \in (0, \infty)$,

$$\begin{aligned}\mathfrak{D}\big(f - f_{Q,c_0}; Q\big)(t) = \mathfrak{D}\big(f - f_Q; Q\big)(t) &\le C_1 e^{-\frac{C_2}{\|f\|_{\mathrm{BMO}\,(\mathcal{X})}} t}|Q| \\ &\le C_1 e^{-\frac{C_2}{2\|f\|_{\mathrm{bmo}\,(\mathcal{X})}} t}|Q|,\end{aligned}$$

where C_1 and C_2 are as in (4), and the distribution function $\mathfrak{D}$ is defined as in (3). Therefore, to show (21), it remains to prove that for any given cube Q with $\ell(Q) \ge c_0$, and any $t \in (0, \infty)$,

$$\Big|\{x \in Q : |f(x)| > t\}\Big| \le C_5 e^{-\frac{C_6}{\|f\|_{\mathrm{bmo}\,(\mathcal{X})}} t}|Q|.$$

Notice that, in this case, there exists a unique $m_0 \in \mathbb{Z}_+$ such that $2^{-(m_0+1)}\ell(Q) < c_0 \le 2^{-m_0}\ell(Q)$. Moreover, since inequality (21) is not altered when we multiply both f and t by the same constant, without loss of generality, we may assume that $\|f\|_{\mathrm{bmo}\,(\mathcal{X})} = 1$. Let Q_0 be any given dyadic subcube of Q with level m_0, namely $Q_0 \in \mathscr{D}_Q^{(m_0)}$. Then, by $c_0 \le 2^{-m_0}\ell(Q) = \ell(Q_0)$ and the definition of $\|f\|_{\mathrm{bmo}\,(\mathcal{X})}$, we have

$$\fint_{Q_0} |f(x)|\,dx \le \|f\|_{\mathrm{bmo}\,(\mathcal{X})} = 1. \tag{22}$$

From the Calderón–Zygmund decomposition (namely Theorem 1) of f with height $\lambda := 2$, we deduce that there exists a family $\{Q_{1,j}\}_j \subset \mathscr{D}_{Q_0}^{(1)}$, such that for any j,

$$2 < \fint_{Q_{1,j}} |f(x)|\,dx \le 2^{n+1}$$

and $|f(x)| \le 2$ when $x \in Q \setminus \bigcup_j Q_{1,j}$. By this and (22), we conclude that

$$\sum_j |Q_{1,j}| \le \frac{1}{2}\sum_j \int_{Q_{1,j}} |f(x)|\,dx \le \frac{1}{2}\int_{Q_0} |f(x)|\,dx \le \frac{1}{2}|Q_0|$$

and, for any j,

$$\left|f_{Q_{1,j}}\right| \le \left|\fint_{Q_{1,j}} f(x)\,dx\right| \le 2^{n+1}.$$

Moreover, for any j, from the Calderón–Zygmund decomposition of $f - f_{Q_{1,j}}$ with height 2, we deduce that there exists a family $\{Q_{1,j,k}\}_k \subset \mathscr{D}_{Q_{1,j}}^{(1)}$, such that for any k,

$$2 < \fint_{Q_{1,j,k}} |f(x) - f_{Q_{1,j}}|\,dx \le 2^{n+1}$$

and $|f(x) - f_{Q_{1,j}}| \le 2$ when $x \in Q \setminus \bigcup_k Q_{1,j,k}$. Meanwhile, by the construction of $\{Q_{1,j}\}_j$, we know that $\ell(Q_{1,j}) = \frac{1}{2}\ell(Q_0) = 2^{-(m_0+1)}\ell(Q)$, which, combined with the facts $\|f\|_{\mathrm{bmo}\,(\mathcal{X})} = 1$ and $2^{-(m_0+1)}\ell(Q) < c_0$, further implies that

$$\fint_{Q_{1,j}} \left|f(x) - f_{Q_{1,j}}\right| dx \le \|f\|_{\mathrm{bmo}\,(\mathcal{X})} = 1.$$

Thus, we obtain, for any j,

$$\sum_k |Q_{1,j,k}| \le \frac{1}{2} \sum_j \int_{Q_{1,j,k}} |f(x) - f_{Q_{1,j}}|\, dx$$
$$\le \frac{1}{2} \int_{Q_{1,j}} |f(x) - f_{Q_{1,j}}|\, dx \le \frac{1}{2} |Q_{1,j}|$$

and, for any k,

$$\left| f_{Q_{1,j,k}} - f_{Q_{1,j}} \right| \le \fint_{Q_{1,j,k}} |f(x) - f_{Q_{1,j}}|\, dx \le 2^{n+1}.$$

Rewrite $\bigcup_{j,k}\{Q_{1,j,k}\} =: \bigcup_j \{Q_{2,j}\}$. Then, we have

$$\sum_j |Q_{2,j}| \le \frac{1}{2} \sum_j |Q_{1,j}| \le \frac{1}{4} |Q_0|$$

and, for any $x \in Q \setminus \bigcup_j Q_{2,j}$,

$$|f(x)| \le \left| f(x) - f_{Q_{1,j}} \right| + \left| f_{Q_{1,j}} \right| \le 2 + 2^{n+1} \le 2 \cdot 2^{n+1}.$$

Repeating this process, then, for any $T \in \mathbb{N}$, we obtain a family $\{Q_{T,j}\}_j \subset \mathscr{D}_{Q_0}$ of disjoint dyadic cubes, such that

$$\sum_j |Q_{T,j}| \le 2^{-T} |Q_0|$$

and, for any $x \in Q_0 \setminus \bigcup_j Q_{T,j}$,

$$|f(x)| \le T 2^{n+1}.$$

Notice that, for any $t \in [2^{n+1}, \infty)$, there exists a unique $T \in \mathbb{N}$, such that $T2^{n+1} \le t < (T+1)2^{n+1} \le T2^{n+2}$. Therefore, we obtain

$$|\{x \in Q_0 : |f(x)| > t\}| \le \sum_j |Q_{T,j}| \le 2^{-T} |Q_0|$$
$$= e^{-T \log 2} |Q_0| \le e^{-C_6 t} |Q_0|, \tag{23}$$

where $C_6 := 2^{-(n+2)} \log 2$. Furthermore, observe that if $t \in (0, 2^{n+1})$, then $C_6 t < 2^{-1} \log 2$ and hence

$$|\{x \in Q_0 : |f(x)| > t\}| \le |Q_0| \le e^{2^{-1} \log 2 - C_6 t} |Q_0| = C_5 e^{-C_6 t} |Q_0|,$$

where $C_5 := \sqrt{2}$. By this, (23), and the arbitrariness of $Q_0 \in \mathscr{D}_Q^{(m_0)}$, we conclude that for any $t \in (0, \infty)$,

$$|\{x \in Q : |f(x)| > t\}| = \sum_{Q_0 \in \mathscr{D}_Q^{(m_0)}} |\{x \in Q_0 : |f(x)| > t\}|$$
$$\le C_5 e^{-C_6 t} \sum_{Q_0 \in \mathscr{D}_Q^{(m_0)}} |Q_0| = C_5 e^{-C_6 t} |Q|$$

and hence (21) holds true. This finishes the proof of Proposition 5. □

As a corollary of Proposition 5, we have the following result: namely, $\|\cdot\|_{\mathrm{bmo}_1(\mathcal{X})}$ in (18) and $\|\cdot\|_{\mathrm{bmo}_2(\mathcal{X})}$ in (19) are equivalent norms of bmo $(\mathcal{X})$. The proof of Corollary 1 is just a repetition of the proof of Proposition 3 with (4) replaced by (21); we omit the details here.

Corollary 1. *The following three statements are mutually equivalent:*

(i) $f \in \mathrm{bmo}\,(\mathcal{X})$;

(ii) $f \in L^1_{\mathrm{loc}}(\mathcal{X})$ *and* $\|f\|_{\mathrm{bmo}_1(\mathcal{X})} < \infty$;

(iii) $f \in L^1_{\mathrm{loc}}(\mathcal{X})$ *and* $\|f\|_{\mathrm{bmo}_2(\mathcal{X})} < \infty$.

Moreover, $\|\cdot\|_{\mathrm{bmo}\,(\mathcal{X})} \sim \|\cdot\|_{\mathrm{bmo}_1(\mathcal{X})} \sim \|\cdot\|_{\mathrm{bmo}_2(\mathcal{X})}$.

2.2. John–Nirenberg Space JN_p

Although there exist many fruitful studies of the space BMO in recent years, as was mentioned before, the structure of JN_p is largely a mystery, and there still exist many unsolved problems on JN_p. The first well-known property of JN_p is the following *John–Nirenberg inequality* obtained in [1] (Lemma 3), which says that $JN_p(Q_0)$ is embedded into the weak Lebesgue space $L^{p,\infty}(Q_0)$ (see Definition 1).

Theorem 2 (John–Nirenberg). *Let $p \in (1,\infty)$ and Q_0 be a given cube of $\mathbb{R}^n$. If $f \in JN_p(Q_0)$, then $f - f_{Q_0} \in L^{p,\infty}(Q_0)$, and there exists a positive constant $C_{(n,p)}$, depending only on n and p, but independent of f, such that*

$$\left\|f - f_{Q_0}\right\|_{L^{p,\infty}(Q_0)} \le C_{(n,p)}\|f\|_{JN_p(Q_0)}.$$

It should be mentioned that the proof of Theorem 2 relies on the Calderón–Zygmund decomposition (namely Theorem 1) as well. Moreover, as an application of Theorem 2, Dafni et al. recently showed in [31] (Proposition 5.1) that for any given $p \in (1,\infty)$ and $q \in [1,p)$, $f \in JN_p(Q_0)$ if and only if $f \in L^1(Q_0)$ and

$$\|f\|_{JN_{p,q}(Q_0)} := \sup\left[\sum_i |Q_i|\left(\fint_{Q_i} |f(x) - f_{Q_i}|^q\,dx\right)^{\frac{p}{q}}\right]^{\frac{1}{p}} < \infty,$$

where the supremum is taken in the same way as in (1); meanwhile, $\|\cdot\|_{JN_p(Q_0)} \sim \|\cdot\|_{JN_{p,q}(Q_0)}$. Furthermore, in [31] (Proposition 5.1), Dafni et al. also showed that for any given $p \in (1,\infty)$ and $q \in [p,\infty)$, the spaces $JN_{p,q}(Q_0)$ and $L^q(Q_0)$ coincide as sets.

Remark 2.

(i) *As a counterpart of Proposition 2, for any given $p \in (1,\infty)$ and any given cube Q_0 of $\mathbb{R}^n$, we have*

$$L^p(Q_0) \subsetneqq JN_p(Q_0) \subsetneqq L^{p,\infty}(Q_0).$$

Indeed, $L^p(Q_0) \subset JN_p(Q_0)$ is obvious from their definitions; $JN_p(Q_0) \subset L^{p,\infty}(Q_0)$ is just Theorem 2; $JN_p(Q_0) \subsetneqq L^{p,\infty}(Q_0)$ was shown in [30] (Example 3.5); and the desired function is just $x^{-1/p}$ on $[0,2]$. However, the fact $L^p(Q_0) \subsetneqq JN_p(Q_0)$ is extremely non-trivial and was obtained in [31] (Proposition 3.2 and Corollary 4.2) via constructing a nice fractal function based on skillful dyadic techniques. Moreover, in [31] (Theorem 1.1 and Remark 2.4), Dafni et al. showed that for any given $p \in (1,\infty)$ and any given interval $I_0 \subset \mathbb{R}$, no matter whether bounded or not, monotone functions are in $JN_p(I_0)$ if and only if they are also in $L^p(I_0)$. Thus, $JN_p(\mathcal{X})$ may be very "close" to $L^p(\mathcal{X})$ for any given $p \in (1,\infty)$.

(ii) *$JN_1(Q_0)$ coincides with $L^1(Q_0)$. To be precise, let Q_0 be any given cube of $\mathbb{R}^n$, and*

$$JN_1(Q_0) := \left\{f \in L^1(Q_0) : \|f\|_{JN_1(Q_0)} < \infty\right\},$$

where $\|f\|_{JN_1(Q_0)}$ is defined as in (1) with p replaced by 1. Then, we claim that $JN_1(Q_0) = [L^1(Q_0)/\mathbb{C}]$ with equivalent norms. Indeed, for any $f \in JN_1(Q_0)$, by the definition of $\|f\|_{JN_1(Q_0)}$, we have

$$\|f\|_{JN_1(Q_0)} \ge \left\|f - f_{Q_0}\right\|_{L^1(Q_0)} \ge \inf_{c\in\mathbb{C}} \|f + c\|_{L^1(Q_0)} =: \|f\|_{L^1(Q_0)/\mathbb{C}}.$$

Conversely, for any given $f \in L^1(Q_0)$ and any $c \in \mathbb{C}$, we have

$$\begin{aligned}\|f\|_{JN_1(Q_0)} &= \sup \sum_i \int_{Q_i} \left|f(x) - f_{Q_i}\right| dx \\ &\le 2 \sup \sum_i \int_{Q_i} \left|f(x) + c\right| dx \\ &\le 2\|f + c\|_{L^1(Q_0)},\end{aligned}$$

which implies that $\|f\|_{JN_1(Q_0)} \lesssim \|f\|_{L^1(Q_0)/\mathbb{C}}$ and hence the above claim holds true. Moreover, the relation between $JN_1(\mathbb{R})$ and $L^1(\mathbb{R})$ was studied in [33] (Proposition 2).

(iii) *Garsia and Rodemich in [55] (Theorem 7.4) showed that for any given $p \in (1, \infty)$, $f \in L^{p,\infty}(Q_0)$ if and only if $f \in L^1(Q_0)$ and*

$$\|f\|_{\mathrm{GaRo}_p(Q_0)} := \sup \frac{1}{(\sum_i |Q_i|)^{1/p'}} \sum_i \frac{1}{|Q_i|} \int_{Q_i} \int_{Q_i} \left|f(x) - f(y)\right| dx\, dy < \infty,$$

where the supremum is taken in the same way as in (1); meanwhile,

$$\|\cdot\|_{L^{p,\infty}(Q_0)} \sim \|\cdot\|_{\mathrm{GaRo}_p(Q_0)};$$

(see also [35], Theorem 5(ii), for this equivalence). Moreover, in [35] (Theorem 5(i)), Milman showed that $\|\cdot\|_{\mathrm{GaRo}_p(Q_0)} \le 2\|\cdot\|_{JN_p(Q_0)}$.

Recall that the predual space of $\mathrm{BMO}(\mathcal{X})$ is the Hardy space $H^1(\mathcal{X})$ (see, for instance, [5], Theorem B). Similar to this duality, Dafni et al. [31] also obtained the predual space of $JN_p(Q_0)$ for any given $p \in (1, \infty)$, which is denoted by the Hardy kind space $HK_{p'}(Q_0)$, here and thereafter $1/p + 1/p' = 1$. Later, these properties, including equivalent norms and duality, were further studied on several John–Nirenberg-type spaces, such as John–Nirenberg–Campanato spaces, localized John–Nirenberg–Campanato spaces, congruent John–Nirenberg–Campanato spaces (see Section 3 for more details), and Riesz-type spaces (see Section 4 for more details).

Finally, let us briefly recall some other related studies concerning the John–Nirenberg space JN_p, which will not be stated in detail in this survey, although all of them are quite instructive:

- Stampacchia [56] introduced the space $N^{(p,\lambda)}$, which coincides with $JN_{(p,1,0)_\alpha}(Q_0)$ in Definitions 3 if we write $\lambda = p\alpha$ with $p \in (1, \infty)$ and $\alpha \in (-\infty, \infty)$, and applied them to the context of interpolation of operators.
- Campanato [57] also used the John–Nirenberg spaces to study the interpolation of operators.
- In the context of doubling metric spaces, JN_p and median-type JN_p were studied, respectively, by Aalto et al. in [30] and Myyryläinen in [58].
- Hurri-Syrjänen et al. [34] established a local-to-global result for the space $JN_p(\Omega)$ on an open subset Ω of $\mathbb{R}^n$. More precisely, it was proved that the norm $\|\cdot\|_{JN_p(\Omega)}$ is dominated by its local version $\|\cdot\|_{JN_{p,\tau}(\Omega)}$ modulus constants; here, $\tau \in [1, \infty)$; for any open subset Ω of $\mathbb{R}^n$, the related "norm" $\|\cdot\|_{JN_p(\Omega)}$ is defined in the same way as $\|\cdot\|_{JN_p(Q_0)}$ in (1) with Q_0 replaced by Ω; and $\|\cdot\|_{JN_{p,\tau}(\Omega)}$ is defined in the same way as $\|\cdot\|_{JN_p(\Omega)}$ with an additional requirement $\tau Q \subset \Omega$ for all chosen cubes Q in the definition of $\|\cdot\|_{JN_p(\Omega)}$.
- Marola and Saari [40] studied the corresponding results of Hurri-Syrjänen et al. [34] on metric measure spaces and obtained the equivalence between the local and the global JN_p norms. Moreover, in both articles [34,40], a global John–Nirenberg inequality for $JN_p(\Omega)$ was established.

- Berkovits et al. [32] applied the dyadic variant of $JN_p(Q_0)$ in the study of self-improving properties of some Poincaré-type inequalities. Later, the dyadic $JN_p(Q_0)$ was further studied by Kinnunen and Myyryläinen in [59].
- A. Brudnyi and Y. Brudnyi [60] introduced a class of function spaces $V_\kappa([0,1]^n)$ which coincides with $JN_{(p,q,s)_\alpha}([0,1]^n)$, defined below for suitable range of indices (see [61], Proposition 2.9, for more details). Very recently, Domínguez and Milman [62] further introduced and studied sparse Brudnyi and John–Nirenberg spaces.
- Blasco and Espinoza-Villalva [33] computed the concrete value of $\|\mathbf{1}_A\|_{JN_p(\mathbb{R})}$ for any given $p \in [1,\infty]$ and any measurable set $A \subset \mathbb{R}$ of positive and finite Lebesgue measure, where $JN_\infty(\mathbb{R}) := \mathrm{BMO}\,(\mathbb{R})$.
- The $JN_p(Q_0)$-type norm $\|\cdot\|_{\mathrm{GaRo}_p(Q_0)}$ in Remark 2(iii) was further generalized and studied in Astashkin and Milman [63] via the Strömberg–Jawerth–Torchinsky local maximal operator.

3. John–Nirenberg–Campanato Space

The main target of this section is to summarize the main results of John–Nirenberg–Campanato spaces, localized John–Nirenberg–Campanato spaces, and congruent John–Nirenberg–Campanato spaces obtained, respectively, in [36,61,64]. Moreover, at the end of each part, we list some open questions which are still unsolved so far. Now, we first recall some definitions of some basic function spaces.

- For any $s \in \mathbb{Z}_+$ (the set of all non-negative integers), let $\mathcal{P}_s(Q)$ denote the set of all polynomials of degree not greater than s on the cube Q, and $P_Q^{(s)}(f)$ denote the unique polynomial of degree not greater than s, such that

$$\int_Q \left[f(x) - P_Q^{(s)}(f)(x)\right] x^\gamma \, dx = 0, \quad \forall\, |\gamma| \le s, \tag{24}$$

  where $\gamma := (\gamma_1, \ldots, \gamma_n) \in \mathbb{Z}_+^n := (\mathbb{Z}_+)^n$, $|\gamma| := \gamma_1 + \cdots + \gamma_n$, and $x^\gamma := x_1^{\gamma_1} \cdots x_n^{\gamma_n}$ for any $x := (x_1, \ldots, x_n) \in \mathbb{R}^n$.
- Let $q \in [1,\infty]$ and Q_0 be a given cube of $\mathbb{R}^n$. For any measurable function f, let

$$\|f\|_{L^q(Q_0,|Q_0|^{-1}dx)} := \left[\fint_{Q_0} |f(x)|^q \, dx\right]^{\frac{1}{q}}.$$

- Let $q \in (1,\infty)$, $s \in \mathbb{Z}_+$, and Q_0 be a given cube of $\mathbb{R}^n$. The *space* $L^q(Q_0,|Q_0|^{-1}dx)/\mathcal{P}_s(Q_0)$ is defined by setting

$$L^q(Q_0,|Q_0|^{-1}dx)/\mathcal{P}_s(Q_0) := \left\{f \in L^q(Q_0) : \|f\|_{L^q(Q_0,|Q_0|^{-1}dx)/\mathcal{P}_s(Q_0)} < \infty\right\},$$

  where

$$\|f\|_{L^q(Q_0,|Q_0|^{-1}dx)/\mathcal{P}_s(Q_0)} := \inf_{m \in \mathcal{P}_s(Q_0)} \|f + m\|_{L^q(Q_0,|Q_0|^{-1}dx)}.$$

- For any given $v \in [1,\infty]$ and $s \in \mathbb{Z}_+$, and any measurable subset $E \subset \mathbb{R}^n$, let

$$L_s^v(E) := \left\{f \in L^v(E) : \int_E f(x) x^\gamma \, dx = 0,\ \forall\, \gamma \in \mathbb{Z}_+^n,\ |\gamma| \le s\right\}.$$

Let Q be any given cube of $\mathbb{R}^n$. It is well known that $P_Q^{(0)}(f) = f_Q$, and for any $s \in \mathbb{Z}_+$, there exists a constant $C_{(s)} \in [1,\infty)$, independent of f and Q, such that

$$\left|P_Q^{(s)}(f)(x)\right| \le C_{(s)} \fint_Q |f(x)| \, dx, \quad \forall\, x \in Q. \tag{25}$$

Indeed, let $\{\varphi_Q^{(\gamma)} : \gamma \in \mathbb{Z}_+^n, |\gamma| \le s\}$ denote the Gram–Schmidt orthonormalization of $\{x^\gamma : \gamma \in \mathbb{Z}_+^n, |\gamma| \le s\}$ on the cube Q with respect to the weight $1/|Q|$, namely for any γ, ν, $\mu \in \mathbb{Z}_+^n$ with $|\gamma| \le s$, $|\nu| \le s$, and $|\mu| \le s$, $\varphi_Q^{(\gamma)} \in \mathcal{P}_s(Q)$ and

$$\langle \varphi_Q^{(\nu)}, \varphi_Q^{(\mu)} \rangle := \frac{1}{|Q|} \int_Q \varphi_Q^{(\nu)}(x) \varphi_Q^{(\mu)}(x)\, dx = \begin{cases} 1, & \nu = \mu, \\ 0, & \nu \ne \mu. \end{cases}$$

Then,

$$P_Q^{(s)}(f)(x) := \sum_{\{\gamma \in \mathbb{Z}_+^n :\, |\gamma| \le s\}} \langle \varphi_Q^{(\gamma)}, f \rangle \varphi_Q^{(\gamma)}(x), \quad \forall\, x \in Q,$$

and we can choose $C_{(s)} := \sum_{\{\gamma \in \mathbb{Z}_+^n :\, |\gamma| \le s\}} \|\varphi_Q^{(\gamma)}\|_{L^\infty(Q)}^2$ satisfying (25) (see [65], p. 83, and [66], p. 54, Lemma 4.1, for more details).

3.1. *John–Nirenberg–Campanato Spaces*

In this subsection, we first recall the definitions of Campanato spaces, John–Nirenberg–Campanato spaces (for short, JNC spaces), and Hardy-type spaces, respectively, in Definitions 2, 3, and 6 below. Moreover, we review some properties of JNC spaces and Hardy-type spaces, including their limit spaces (Proposition 6 and Corollary 2 below), relations with the Lebesgue space (Propositions 7 and 8 below), the dual result (Theorem 3 below), the monotonicity over the first sub-index (Proposition 9 below), the John–Nirenberg-type inequality (Theorem 4 below), and the equivalence over the second sub-index (Propositions 10 and 11 below).

A general dual result for Hardy spaces was given by Coifman and Weiss [5] who proved that for any given $p \in (0, 1]$ and $q \in [1, \infty]$, and s being a non-negative integer not smaller than $n(\frac{1}{p} - 1)$, the dual space of the Hardy space $H^p(\mathbb{R}^n)$ is the Campanato space $C_{\frac{1}{p}-1, q, s}(\mathbb{R}^n)$, which was introduced by Campanato [67] and coincides with $\mathrm{BMO}(\mathbb{R}^n)$ when $p = 1$.

Definition 2. *Let $\alpha \in [0, \infty)$, $q \in [1, \infty)$, and $s \in \mathbb{Z}_+$.*

(i) *The* Campanato space $C_{\alpha,q,s}(\mathcal{X})$ *is defined by setting*

$$C_{\alpha,q,s}(\mathcal{X}) := \left\{ f \in L^q_{\mathrm{loc}}(\mathcal{X}) : \|f\|_{C_{\alpha,q,s}(\mathcal{X})} < \infty \right\},$$

where

$$\|f\|_{C_{\alpha,q,s}(\mathcal{X})} := \sup |Q|^{-\alpha} \left[\fint_Q \left| f - P_Q^{(s)}(f) \right|^q \right]^{\frac{1}{q}}$$

and the supremum is taken over all cubes Q of $\mathcal{X}$. In addition, the "norm" $\|\cdot\|_{C_{\alpha,q,s}(\mathcal{X})}$ of polynomials is zero, and for simplicity, the space $C_{\alpha,q,s}(\mathcal{X})$ is regarded as the quotient space $C_{\alpha,q,s}(\mathcal{X})/\mathcal{P}_s(\mathcal{X})$.

(ii) *The* dual space $(C_{\alpha,q,s}(\mathcal{X}))^*$ *of $C_{\alpha,q,s}(\mathcal{X})$ is defined to be the set of all continuous linear functionals on $C_{\alpha,q,s}(\mathcal{X})$ equipped with the weak-$*$ topology.*

In what follows, for any $\ell \in (0, \infty)$, $Q(\mathbf{0}, \ell)$ denotes the cube centered at the origin $\mathbf{0}$ with side length ℓ.

Remark 3. *Let $0 < q \le p \le \infty$. The* Morrey space $M_q^p(\mathbb{R}^n)$, *introduced by Morrey in [68], is defined by setting*

$$M_q^p(\mathbb{R}^n) := \left\{ f \in L^q_{\mathrm{loc}}(\mathbb{R}^n) : \|f\|_{M_q^p(\mathbb{R}^n)} < \infty \right\},$$

where, for any $f \in L^q_{\rm loc}(\mathbb{R}^n)$,

$$\|f\|_{M^p_q(\mathbb{R}^n)} := \sup_{\text{cube } Q\subset\mathbb{R}^n} |Q|^{\frac{1}{p}}\left[\fint_Q |f(y)|^q\,dy\right]^{\frac{1}{q}}.$$

From Campanato ([67], Theorem 6.II), it follows that for any given $q \in [1,\infty)$ *and* $\alpha \in [-\frac{1}{q}, 0)$, *and any* $f \in C_{q,\alpha,0}(\mathcal{X})$,

$$\|f\|_{C_{q,\alpha,0}(\mathcal{X})} \sim \left\|f - \sigma(f)\right\|_{M_q^{-1/\alpha}(\mathcal{X})}, \tag{26}$$

where the positive equivalence constants are independent of f, *and*

$$\sigma(f) := \begin{cases} \lim\limits_{\ell\to\infty} \dfrac{1}{|Q(\mathbf{0},\ell)|}\displaystyle\int_{Q(\mathbf{0},\ell)} f(x)\,dx & \text{if } \mathcal{X} = \mathbb{R}^n,\\ \dfrac{1}{|Q_0|}\displaystyle\int_{Q_0} f(x)\,dx & \text{if } \mathcal{X} = Q_0;\end{cases}$$

see also Nakai [16], Theorem 2.1 and Corollary 2.3, for this conclusion on spaces of homogeneous type. In addition, a surprising result says that in the definition of supremum $\|\cdot\|_{M^p_q(\mathbb{R}^n)}$, *if "cubes" were changed into "measurable sets", then the Morrey norm* $\|\cdot\|_{M^p_q(\mathbb{R}^n)}$ *becomes an equivalent norm of the weak Lebesgue space (see Definition 1). To be precise, for any given* $0 < q < p < \infty$, $f \in L^{p,\infty}(\mathbb{R}^n)$ *if and only if* $f \in L^q_{\rm loc}(\mathbb{R}^n)$ *and*

$$\|f\|_{\widetilde{M^p_q}(\mathbb{R}^n)} := \sup_{A\subset\mathbb{R}^n,\,|A|\in(0,\infty)} |A|^{\frac{1}{p}}\left[\fint_A |f(y)|^q\,dy\right]^{\frac{1}{q}} < \infty;$$

moreover,

$$\|\cdot\|_{L^{p,\infty}(\mathbb{R}^n)} \le \|\cdot\|_{\widetilde{M^p_q}(\mathbb{R}^n)} \le \left(\frac{p}{p-q}\right)^{\frac{1}{q}}\|\cdot\|_{L^{p,\infty}(\mathbb{R}^n)};$$

see, for instance, [69], p. 485, Lemma 2.8. Another interesting JN_p*-type equivalent norm of the weak Lebesgue space was presented in Remark 2(iii).*

Inspired by the relation between BMO and the Campanato space, as well as the relation between BMO and JN_p, Tao et al. [61] introduced a Campanato-type space $JN_{(p,q,s)_\alpha}(\mathcal{X})$ in the spirit of the John–Nirenberg space $JN_p(Q_0)$, which contains $JN_p(Q_0)$ as a special case. This John–Nirenberg–Campanato space is defined not only on any cube Q_0 but also on the whole space $\mathbb{R}^n$.

Definition 3. *Let* $p, q \in [1,\infty)$, $s \in \mathbb{Z}_+$, *and* $\alpha \in \mathbb{R}$.

(i) *The* John–Nirenberg–Campanato space *(for short,* JNC space*)* $JN_{(p,q,s)_\alpha}(\mathcal{X})$ *is defined by setting*

$$JN_{(p,q,s)_\alpha}(\mathcal{X}) := \left\{f \in L^q_{\rm loc}(\mathcal{X}) : \|f\|_{JN_{(p,q,s)_\alpha}(\mathcal{X})} < \infty\right\},$$

where

$$\|f\|_{JN_{(p,q,s)_\alpha}(\mathcal{X})} := \sup\left\{\sum_i |Q_i|\left[|Q_i|^{-\alpha}\left\{\fint_{Q_i}\left|f(x) - P^{(s)}_{Q_i}(f)(x)\right|^q dx\right\}^{\frac{1}{q}}\right]^p\right\}^{\frac{1}{p}},$$

$P^{(s)}_{Q_i}(f)$ *for any* i *is as in (24) with* Q *replaced by* Q_i, *and the supremum is taken over all collections of interior pairwise disjoint cubes* $\{Q_i\}_i$ *of* $\mathcal{X}$. *Furthermore, the "norm"* $\|\cdot\|_{JN_{(p,q,s)_\alpha}(\mathcal{X})}$

of polynomials is zero, and for simplicity, the space $JN_{(p,q,s)_\alpha}(\mathcal{X})$ *is regarded as the quotient space* $JN_{(p,q,s)_\alpha}(\mathcal{X})/\mathcal{P}_s(\mathcal{X})$.

(ii) *The* dual space $(JN_{(p,q,s)_\alpha}(\mathcal{X}))^*$ *of* $JN_{(p,q,s)_\alpha}(\mathcal{X})$ *is defined to be the set of all continuous linear functionals on* $JN_{(p,q,s)_\alpha}(\mathcal{X})$ *equipped with the weak-∗ topology.*

Remark 4. *In [61], the JNC space was introduced only for any given* $\alpha \in [0,\infty)$ *to study its relation with the Campanato space in Definition 2, and for any given* $p \in (1,\infty)$ *due to Remark 2(ii). However, many results in [61] also hold true when* $\alpha \in \mathbb{R}$ *and* $p = 1$*, just with some slight modifications of their proofs. Thus, in this survey, we introduce the JNC space for any given* $\alpha \in \mathbb{R}$ *and* $p \in [1,\infty)$ *and naturally extend some related results with some identical proofs omitted.*

The following proposition, which is just [61] (Proposition 2.6), means that the classical Campanato space serves as a limit space of $JN_{(p,q,s)_\alpha}(\mathcal{X})$, similar to the Lebesgue spaces $L^\infty(\mathcal{X})$ and $L^p(\mathcal{X})$ when $p \to \infty$.

Proposition 6. *Let* $\alpha \in [0,\infty)$, $q \in [1,\infty)$, *and* $s \in \mathbb{Z}_+$. *Then,*

$$\lim_{p\to\infty} JN_{(p,q,s)_\alpha}(\mathcal{X}) = C_{\alpha,q,s}(\mathcal{X})$$

in the following sense: for any $f \in \bigcup_{r\in[1,\infty)} \bigcap_{p\in[r,\infty)} JN_{(p,q,s)_\alpha}(\mathcal{X})$,

$$\lim_{p\to\infty} \|f\|_{JN_{(p,q,s)_\alpha}(\mathcal{X})} = \|f\|_{C_{\alpha,q,s}(\mathcal{X})}.$$

In Proposition 6, if we take $\mathcal{X} = Q_0$, we then have the following corollary, which is just [61] (Corollary 2.8).

Corollary 2. *Let* $q \in [1,\infty)$, $\alpha \in [0,\infty)$, $s \in \mathbb{Z}_+$, *and* Q_0 *be a given cube of* $\mathbb{R}^n$. *Then,*

$$C_{\alpha,q,s}(Q_0) = \left\{ f \in \bigcap_{p\in[1,\infty)} JN_{(p,q,s)_\alpha}(Q_0) : \lim_{p\to\infty} \|f\|_{JN_{(p,q,s)_\alpha}(Q_0)} < \infty \right\}$$

and for any $f \in C_{\alpha,q,s}(Q_0)$,

$$\|f\|_{C_{\alpha,q,s}(Q_0)} = \lim_{p\to\infty} \|f\|_{JN_{(p,q,s)_\alpha}(Q_0)}.$$

Remark 5.

(i) *Let* $p \in (1,\infty)$ *and* Q_0 *be a given cube of* $\mathbb{R}^n$. *It is easy to show that*

$$\mathrm{BMO}\,(Q_0) \subset JN_p(Q_0).$$

However, we claim that

$$\mathrm{BMO}\,(\mathbb{R}^n) \not\subset JN_p(\mathbb{R}^n).$$

Indeed, for the simplicity of the presentation, without loss of generality, we may show this claim only in $\mathbb{R}$. *Let* $g(x) := \log(|x|)$ *for any* $x \in \mathbb{R}\setminus\{0\}$, *and* $g(0) := 0$. *Then,* $g \in \mathrm{BMO}\,(\mathbb{R})$ *due to [46] (Example 3.1.3), and hence it suffices to prove that* $g \notin JN_p(\mathbb{R})$ *for any given* $p \in (1,\infty)$. *To do this, let* $I_t := (0,t)$ *for any* $t \in (0,\infty)$. *Then, by some simple calculations, we obtain*

$$g_{I_t} = \fint_{I_t} g(x)\,dx = \frac{1}{t}\int_0^t \log(x)\,dx = \log(t) - 1$$

and hence

$$\left|\left\{x \in I_t : \left|g(x) - g_{I_t}\right| > \frac{1}{2}\right\}\right|$$

$$= \left|\left\{x \in (0,t) : \left|\log(x) - [\log(t) - 1]\right| > \frac{1}{2}\right\}\right|$$
$$\geq t - te^{-\frac{1}{2}} = t\left(1 - e^{-\frac{1}{2}}\right) \to \infty$$

as $t \to \infty$. However, the John–Nirenberg inequality of $JN_p(I_t)$ in Theorem 2 implies that for any $t \in (0,\infty)$,

$$\left|\left\{x \in I_t : \left|g(x) - g_{I_t}\right| > \frac{1}{2}\right\}\right| \lesssim \left[\frac{\|g\|_{JN_p(I_t)}}{\frac{1}{2}}\right]^p \lesssim \|g\|^p_{JN_p(\mathbb{R})}$$

with the implicit positive constants depending only on p. Thus, $g \notin JN_p(\mathbb{R})$, and hence the above claim holds true.

(ii) *The predual counterpart of Corollary 2 is still unclear so far (see Question 2 below for more details).*

Obviously, $JN_{(p,q,0)_0}(Q_0)$ is just $JN_{p,q}(Q_0)$. From this and [31] (Proposition 5.1), we deduce that when $p \in (1,\infty)$ and $q \in [1,p)$, $JN_{(p,q,0)_0}(Q_0)$ coincides with $JN_p(Q_0)$ in the sense of equivalent norms, and when $p \in (1,\infty)$ and $q \in [p,\infty)$, $JN_{(p,q,0)_0}(Q_0)$ and $L^q(Q_0)$ coincide as sets. Moreover, by adding a particular weight of $|Q_0|$, the authors of this article showed that the aforementioned coincidence (as sets) can be modified into equivalent norms (see Proposition 7 below, which is just [61], Proposition 2.5). In what follows, for any given positive constant A and any given function space $(\mathbb{X}, \|\cdot\|_{\mathbb{X}})$, we write $A\mathbb{X} := \{Af : f \in \mathbb{X}\}$ with its norm defined by setting, for any $Af \in A\mathbb{X}$, $\|Af\|_{A\mathbb{X}} := A\|f\|_{\mathbb{X}}$.

Proposition 7. *Let $p \in [1,\infty)$, $q \in [p,\infty)$, $s \in \mathbb{Z}_+$, $\alpha = 0$, and Q_0 be a given cube of $\mathbb{R}^n$. Then,*

$$\left[|Q_0|^{-\frac{1}{p}} JN_{(p,q,s)_\alpha}(Q_0)\right] = \left[L^q(Q_0, |Q_0|^{-1}dx)/\mathcal{P}_s(Q_0)\right]$$

with equivalent norms, namely

$$\|f\|_{L^q(Q_0,|Q_0|^{-1}dx)/\mathcal{P}_s(Q_0)} \leq |Q_0|^{-\frac{1}{p}} \|f\|_{JN_{(p,q,s)_0}(Q_0)}$$
$$\leq 2^{p-\frac{p}{q}}\left[1 + C_{(s)}\right]^{\frac{p}{q}} \|f\|_{L^q(Q_0,|Q_0|^{-1}dx)/\mathcal{P}_s(Q_0)},$$

where $C_{(s)}$ is as in (25).

It is a very interesting open question to find a counterpart of Proposition 7 when $\alpha \in \mathbb{R} \setminus \{0\}$ (see Question 1 below for more details).

Now, we review the predual of the John–Nirenberg–Campanato space via introducing atoms, polymers, and Hardy-type spaces in order, which coincide with the same notation as in [31] when $u \in (1,\infty)$, $v \in (u,\infty]$, and $\alpha = 0 = s$ (see [61], Remarks 3.4 and 3.8, for more details). In particular, when $\alpha = 0$, the $(u,v,s)_0$-atom below is just the classic atom of the Hardy space (see [61], Remark 3.2).

Definition 4. *Let $u, v \in [1,\infty]$, $s \in \mathbb{Z}_+$, and $\alpha \in \mathbb{R}$. A function a is called a $(u,v,s)_\alpha$-*atom *on a cube Q if*

(i) $\mathrm{supp}\,(a) := \{x \in \mathbb{R}^n : a(x) \neq 0\} \subset Q$;

(ii) $\|a\|_{L^v(Q)} \leq |Q|^{\frac{1}{v}-\frac{1}{u}-\alpha}$;

(iii) $\int_Q a(x)x^\gamma\,dx = 0$ *for any $\gamma \in \mathbb{Z}^n_+$ with $|\gamma| \leq s$.*

In what follows, for any $u \in [1, \infty]$, let u' denote its *conjugate index*, namely $1/u + 1/u' = 1$, and for any $\{\lambda_j\}_j \subset \mathbb{C}$, let

$$\left\|\{\lambda_j\}_j\right\|_{\ell^u} := \begin{cases} \left(\sum_j |\lambda_j|^u\right)^{\frac{1}{u}} & \text{when } u \in [1, \infty), \\ \sup_j |\lambda_j| & \text{when } u = \infty. \end{cases} \tag{27}$$

Definition 5. *Let $u, v \in [1, \infty]$, $s \in \mathbb{Z}_+$, and $\alpha \in \mathbb{R}$. The* space of $(u, v, s)_\alpha$-polymers, *denoted by $\widetilde{HK}_{(u,v,s)_\alpha}(\mathcal{X})$, is defined to be the set of all $g \in (JN_{(u',v',s)_\alpha}(\mathcal{X}))^*$ satisfying that there exist $(u, v, s)_\alpha$-atoms $\{a_j\}_j$ supported, respectively, in interior pairwise disjoint cubes $\{Q_j\}_j$ of $\mathcal{X}$, and $\{\lambda_j\}_j \subset \mathbb{C}$ with $|\lambda_j|^u < \infty$, such that*

$$g = \sum_j \lambda_j a_j$$

*in $(JN_{(u',v',s)_\alpha}(\mathcal{X}))^*$. Moreover, any $g \in \widetilde{HK}_{(u,v,s)_\alpha}(\mathcal{X})$ is called a $(u, v, s)_\alpha$-*polymer *with its norm $\|g\|_{\widetilde{HK}_{(u,v,s)_\alpha}(\mathcal{X})}$ defined by setting*

$$\|g\|_{\widetilde{HK}_{(u,v,s)_\alpha}(\mathcal{X})} := \inf\left\|\{\lambda_j\}_j\right\|_{\ell^u},$$

where the infimum is taken over all decompositions of g as above.

Definition 6. *Let $u, v \in [1, \infty]$, $s \in \mathbb{Z}_+$, and $\alpha \in \mathbb{R}$. The* Hardy-type space *$HK_{(u,v,s)_\alpha}(\mathcal{X})$ is defined by setting*

$$HK_{(u,v,s)_\alpha}(\mathcal{X}) := \left\{ g \in (JN_{(u',v',s)_\alpha}(\mathcal{X}))^* : \ g = \sum_i g_i \text{ in } (JN_{(u',v',s)_\alpha}(\mathcal{X}))^*, \right.$$
$$\left. \{g_i\}_i \subset \widetilde{HK}_{(u,v,s)_\alpha}(\mathcal{X}), \text{ and } \sum_i \|g_i\|_{\widetilde{HK}_{(u,v,s)_\alpha}(\mathcal{X})} < \infty \right\}$$

and for any $g \in HK_{(u,v,s)_\alpha}(\mathcal{X})$, let

$$\|g\|_{HK_{(u,v,s)_\alpha}(\mathcal{X})} := \inf \sum_i \|g_i\|_{\widetilde{HK}_{(u,v,s)_\alpha}(\mathcal{X})},$$

where the infimum is taken over all decompositions of g as above. Moreover, the finite atomic Hardy-type space *$HK^{\text{fin}}_{(u,v,s)_\alpha}(\mathcal{X})$ is defined to be the set of all finite summations $\sum_{m=1}^M \lambda_m a_m$, where $M \in \mathbb{N}$, $\{\lambda_m\}_{m=1}^M \subset \mathbb{C}$, and $\{a_m\}_{m=1}^M$ are $(u, v, s)_\alpha$-atoms.*

The significant dual relation between $JN_{(p,q,s)_\alpha}(\mathcal{X})$ and $HK_{(p',q',s)_\alpha}(\mathcal{X})$ reads as follows, which is just [61] (Theorem 3.9) with $\alpha \in [0, \infty)$ replaced by $\alpha \in \mathbb{R}$ (this makes sense because the crucial lemma ([61], Lemma 3.12) still holds true with the corresponding replacement).

Theorem 3. *Let $p, q \in (1, \infty)$, $1/p = 1/p' = 1 = 1/q + 1/q'$, $s \in \mathbb{Z}_+$, and $\alpha \in \mathbb{R}$. Then, $(HK_{(p',q',s)_\alpha}(\mathcal{X}))^* = JN_{(p,q,s)_\alpha}(\mathcal{X})$ in the following sense:*

(i) *If $f \in JN_{(p,q,s)_\alpha}(\mathcal{X})$, then f induces a linear functional $\mathcal{L}_f$ on $HK_{(p',q',s)_\alpha}(\mathcal{X})$ and*

$$\|\mathcal{L}_f\|_{(HK_{(p',q',s)_\alpha}(\mathcal{X}))^*} \le C\|f\|_{JN_{(p,q,s)_\alpha}(\mathcal{X})},$$

where C is a positive constant independent of f.

(ii) *If* $\mathcal{L} \in (HK_{(p',q',s)_\alpha}(\mathcal{X}))^*$, *then there exists an* $f \in JN_{(p,q,s)_\alpha}(\mathcal{X})$, *such that for any* $g \in HK^{\text{fin}}_{(p',q',s)_\alpha}(\mathcal{X})$,

$$\mathcal{L}(g) = \int_{\mathcal{X}} f(x)g(x)\,dx,$$

and

$$\|\mathcal{L}\|_{(HK_{(p',q',s)_\alpha}(\mathcal{X}))^*} \sim \|f\|_{JN_{(p,q,s)_\alpha}(\mathcal{X})}$$

with the positive equivalence constants independent of f.

When $\mathcal{X} := Q_0$, $\alpha = 0 = s$, and $q \in [1, p)$, by [61] (Remark 3.10 and Proposition 10), we know that Theorem 3 in this case coincides with [31] (Theorem 6.6). As an application of Theorem 3, the authors obtained the following atomic characterization of $L^{q'}_s(Q_0)$ for any given $q' \in (1, \infty)$ and $s \in \mathbb{Z}_+$, which is just [61] (Corollary 3.13).

Proposition 8. *Let* $p \in (1, \infty)$, $q \in [p, \infty)$, $1/p = 1/p' = 1 = 1/q + 1/q'$, $s \in \mathbb{Z}_+$, *and* Q_0 *be a given cube of* $\mathbb{R}^n$. *Then,*

$$L^{q'}_s(Q_0, |Q_0|^{q'-1}dx) = |Q_0|^{\frac{1}{p}} HK_{(p',q',s)_0}(Q_0)$$

with equivalent norms.

From Theorem 2 and [47] (p. 14, Exercise 1.1.11), we deduce that for any $1 < p_1 < p_2 < \infty$,

$$JN_{p_2}(Q_0) \subset L^{p_2,\infty}(Q_0) \subset L^{p_1}(Q_0) \subset JN_{p_1}(Q_0).$$

Moreover, it is easy to show the following monotonicity over the first sub-index of both $JN_{(p,q,s)_\alpha}(Q_0)$ and $HK_{(u,v,s)_\alpha}(Q_0)$.

Proposition 9. *Let* $s \in \mathbb{Z}_+$ *and* Q_0 *be a given cube of* $\mathbb{R}^n$.

(i) *Let* $1 < u_1 < u_2 < \infty$. *If* $v \in (1, \infty)$ *and* $\alpha \in \mathbb{R}$, *or* $v = \infty$ *and* $\alpha \in [0, \infty)$, *then*

$$HK_{(u_2,v,s)_\alpha}(Q_0) \subset HK_{(u_1,v,s)_\alpha}(Q_0)$$

and

$$\|\cdot\|_{HK_{(u_1,v,s)_\alpha}(Q_0)} \le |Q_0|^{\frac{1}{u_1} - \frac{1}{u_2}} \|\cdot\|_{HK_{(u_2,v,s)_\alpha}(Q_0)}.$$

(ii) *Let* $1 < p_1 < p_2 < \infty$. *If* $q \in (1, \infty)$ *and* $\alpha \in \mathbb{R}$, *or* $q = 1$ *and* $\alpha \in [0, \infty)$, *then*

$$JN_{(p_2,q,s)_\alpha}(Q_0) \subset JN_{(p_1,q,s)_\alpha}(Q_0)$$

and there exists some positive constant C, such that

$$\|\cdot\|_{JN_{(p_1,q,s)_\alpha}(Q_0)} \le C|Q_0|^{\frac{1}{p_1} - \frac{1}{p_2}} \|\cdot\|_{JN_{(p_2,q,s)_\alpha}(Q_0)}.$$

Proof. (i) is a direct corollary of the fact that for any $(u_2, v, s)_\alpha$-atom a on the cube Q,

$$|Q|^{\frac{1}{v_2} - \frac{1}{v_1}} a$$

is a $(u_1, v, s)_\alpha$-atom (see [36], Remark 5.5, for more details).

(ii) is a direct consequence of the Jensen inequality (see, for instance, [61], Remark 4.2(ii)). This finishes the proof of Proposition 9. □

Now, we consider the independence over the second sub-index, which strongly relies on the John–Nirenberg inequality as in the BMO case. The following John–Nirenberg-type inequality is just [61] (Theorem 4.3), which coincides with Theorem 2 when $\alpha = 0 = s$.

Theorem 4. *Let $p \in (1,\infty)$, $s \in \mathbb{Z}_+$, $\alpha \in [0,\infty)$, and Q_0 be a given cube of $\mathbb{R}^n$. If $f \in JN_{(p,1,s)_\alpha}(Q_0)$, then $f - P^{(s)}_{Q_0}(f) \in L^{p,\infty}(Q_0)$, and there exists a positive constant $C_{(n,p,s)}$, depending only on n, p, and s, but independent of f, such that*

$$\left\|f - P^{(s)}_{Q_0}(f)\right\|_{L^{p,\infty}(Q_0)} \le C_{(n,p,s)}|Q_0|^{\alpha}\|f\|_{JN_{(p,1,s)_\alpha}(Q_0)}.$$

It should be mentioned that the main tool used in the proof of Theorem 4 is the following *good-λ inequality* (namely, Lemma 2 below), which is just [61] (Lemma 4.6) (see also [30], Lemma 4.5, when $s = 0$). Recall that for any given cube Q_0 of $\mathbb{R}^n$, the *dyadic maximal operator* $\mathcal{M}^{(\mathrm{d})}_{Q_0}$ is defined by setting, for any given $g \in L^1(Q_0)$ and any $x \in Q_0$,

$$\mathcal{M}^{(\mathrm{d})}_{Q_0}(g)(x) := \sup_{Q\in\mathcal{D}_{Q_0},\, Q\ni x} \frac{1}{|Q|}\int_Q |g(x)|\,dx,$$

where $\mathcal{D}_{Q_0}$ is as in (20) with Q replaced by Q_0, and the supremum is taken over all dyadic cubes $Q \in \mathcal{D}_{Q_0}$ and $Q \ni x$.

Lemma 2. *Let $p \in (1,\infty)$, $s \in \mathbb{Z}_+$, $C_{(s)} \in [1,\infty)$ be as in (25), $\theta \in (0, 2^{-n}C^{-1}_{(s)})$, Q_0 be a given cube of $\mathbb{R}^n$, and $f \in JN_{(p,1,s)_0}(Q_0)$. Then, for any real number $\lambda > \frac{1}{\theta}\fint_{Q_0}|f - P^{(s)}_{Q_0}(f)|$,*

$$\begin{aligned}&\left|\left\{x \in Q_0 :\ \mathcal{M}^{(\mathrm{d})}_{Q_0}\left(f - P^{(s)}_{Q_0}(f)\right)(x) > \lambda\right\}\right|\\ &\quad\le \frac{\|f\|_{JN_{(p,1,s)_0}(Q_0)}}{[1-2^n\theta C_{(s)}]\lambda}\left|\left\{x \in Q_0 :\ \mathcal{M}^{(\mathrm{d})}_{Q_0}\left(f - P^{(s)}_{Q_0}(f)\right)(x) > \theta\lambda\right\}\right|^{\frac{1}{p'}}.\end{aligned}$$

Moreover, based on Theorem 4 in [61] (Proposition 4.1), Tao et al. further obtained the following independence over the second sub-index of $JN_{(p,q,s)_\alpha}(\mathcal{X})$.

Proposition 10. *Let $1 \le q < p < \infty$, $s \in \mathbb{Z}_+$, and $\alpha \in [0,\infty)$. Then,*

$$JN_{(p,q,s)_\alpha}(\mathcal{X}) = JN_{(p,1,s)_\alpha}(\mathcal{X})$$

with equivalent norms.

Furthermore, the following independence over the second sub-index of $HK_{(u,v,s)_\alpha}(\mathcal{X})$ is just [61] (Proposition 4.7), whose proof is based on Theorem 3 and Proposition 10.

Proposition 11. *Let $1 < u < v \le \infty$, $s \in \mathbb{Z}_+$, and $\alpha \in [0,\infty)$. Then,*

$$HK_{(u,v,s)_\alpha}(\mathcal{X}) = HK_{(u,\infty,s)_\alpha}(\mathcal{X})$$

with equivalent norms.

In particular, when $\alpha = 0 = s$, Propositions 10 and 11 were obtained, respectively, in [31] (Propositions 5.1 and 6.4).

Combining Theorem 3 and Propositions 10 and 11, we immediately have the following corollary; we omit the details here.

Corollary 3. *Let $p \in (1,\infty)$, $s \in \mathbb{Z}_+$, and $\alpha \in [0,\infty)$. Then, $(HK_{(p',\infty,s)_\alpha}(\mathcal{X}))^* = JN_{(p,1,s)_\alpha}(\mathcal{X})$.*

Finally, we list some open questions.

Question 1. *For any given cube Q_0 of $\mathbb{R}^n$, by [61] (Remark 4.2(ii)) with slight modifications, we know that*

(i) *for any given $p \in [1,\infty)$ and $s \in \mathbb{Z}_+$,*

$$JN_{(p,q,s)_0}(Q_0) = \begin{cases} JN_{(p,1,s)_0}(Q_0), & q \in [1,p), \\ JN_{(q,q,s)_0}(Q_0), & q \in [p,\infty); \end{cases}$$

(ii) *for any given $p \in [1,\infty)$, $q \in [p,\infty)$, $s \in \mathbb{Z}_+$, and $\alpha \in \mathbb{R}$,*

$$JN_{(q,q,s)_\alpha}(Q_0) \subset JN_{(p,q,s)_\alpha}(Q_0)$$

and

$$\left[|Q_0|^{-\frac{1}{p}}\|f\|_{JN_{(p,q,s)_\alpha}(Q_0)}\right] \le \left[|Q_0|^{-\frac{1}{q}}\|f\|_{JN_{(q,q,s)_\alpha}(Q_0)}\right];$$

(iii) *for any given $p \in [1,\infty)$, $q \in [p,\infty)$, $s \in \mathbb{Z}_+$, and $\alpha \in (\frac{s+1}{n},\infty)$,*

$$JN_{(q,q,s)_\alpha}(Q_0) = \mathcal{P}_s(Q_0) = JN_{(p,q,s)_\alpha}(Q_0).$$

However, letting $RM_{p,q,\alpha}(\mathcal{X})$ denote the Riesz–Morrey space in Definition 14, it is still unknown whether or not

(i) *for any given $p \in [1,\infty)$, $q \in [p,\infty)$, $s \in \mathbb{Z}_+$, and $\alpha \in (-\infty,\frac{s+1}{n}] \setminus \{0\}$,*

$$JN_{(p,q,s)_\alpha}(Q_0) = JN_{(q,q,s)_\alpha}(Q_0) \text{ or } JN_{(p,q,s)_\alpha}(Q_0) = \left[RM_{p,q,\alpha}(Q_0)/\mathcal{P}_s(Q_0)\right]$$

holds true;

(ii) *for any given $p \in [1,\infty)$, $q \in [p,\infty)$, $s \in \mathbb{Z}_+$, and $\alpha \in \mathbb{R}$,*

$$JN_{(p,q,s)_\alpha}(\mathbb{R}^n) = JN_{(q,q,s)_\alpha}(\mathbb{R}^n) \text{ or } JN_{(p,q,s)_\alpha}(\mathbb{R}^n) = \left[RM_{p,q,\alpha}(\mathbb{R}^n)/\mathcal{P}_s(\mathbb{R}^n)\right]$$

holds true, where $\mathcal{P}_s(\mathbb{R}^n)$ denotes the set of all polynomials of degree not greater than s on $\mathbb{R}^n$.

Question 2. *Let $1 < u_1 < u_2 < \infty$, $v \in (1,\infty]$, $s \in \mathbb{Z}_+$, and Q_0 be a given cube of $\mathbb{R}^n$. From Proposition 9(i), we deduce that*

$$HK_{(u_2,v,s)_0}(Q_0) \subset HK_{(u_1,v,s)_0}(Q_0)$$

and

$$\|\cdot\|_{HK_{(u_1,v,s)_0}(Q_0)} \le \left[|Q_0|^{\frac{1}{u_1}-\frac{1}{u_2}}\|\cdot\|_{HK_{(u_2,v,s)_0}(Q_0)}\right].$$

Moreover, by [61] (Remark 4.2(iii)) and [36] (Proposition 5.7), we find that for any $u \in [1,\infty)$,

$$HK_{(u,v,s)_0}(Q_0) \subset H^{1,v,s}_{\mathrm{at}}(Q_0)$$

and for any $g \in \bigcup_{u\in[1,\infty)} HK_{(u,v,s)_0}(Q_0)$,

$$\|g\|_{H^{1,v,s}_{\mathrm{at}}(Q_0)} \le \liminf_{u\to 1^+} \|g\|_{HK_{(u,v,s)_0}(Q_0)},$$

where $H^{1,v,s}_{\mathrm{at}}(\mathcal{X})$ denotes the atomic Hardy space *(see Coifman and Weiss [5], and also [61], Remark 3.2(ii), for its definition). Here and thereafter, $u \to 1^+$ means $u \in (1,\infty)$ and $u \to 1$. However, for any given $v \in (1,\infty]$, $s \in \mathbb{Z}_+$, $\alpha \in [0,\infty)$, and any given cube Q_0 of $\mathbb{R}^n$,*

(i) *it is still unknown whether or not for any $g \in \bigcup_{u\in[1,\infty)} HK_{(u,v,s)_\alpha}(Q_0)$,*

$$\|g\|_{H^{\frac{1}{\alpha+1},v,s}_{\mathrm{at}}(Q_0)} = \lim_{u\to 1^+} \|g\|_{HK_{(u,v,s)_\alpha}(Q_0)}$$

holds true;

(ii) *it is interesting to clarify the relation between* $\bigcup_{u\in[1,\infty)} HK_{(u,v,s)_\alpha}(Q_0)$ *and* $H_{\rm at}^{\frac{1}{\alpha+1},v,s}(Q_0)$.

The last question in this subsection is on an interpolation result in [56]. We first recall some notation in [56]. Let $p \in (1,\infty)$, $\lambda \in \mathbb{R}$, and Q_0 be a given cube of $\mathbb{R}^n$. The *space* $N^{(p,\lambda)}(Q_0)$ is defined by setting

$$N^{(p,\lambda)}(Q_0) := \left\{u \in L^1(Q_0) : \ [u]_{N^{(p,\lambda)}(Q_0)} < \infty\right\},$$

where

$$[u]_{N^{(p,\lambda)}}(Q_0) := \sup\left\{\sum_i \left|\int_{Q_i} |u(x) - u_{Q_i}|\,dx\right|^p |Q_i|^{1-p-\lambda}\right\}^{1/p}$$

and the supremum is taken over all collections of interior pairwise disjoint cubes $\{Q_i\}_i$ of Q_0, and u_{Q_i} is the mean of u over Q_i for any i. Let $\mathcal{F}(Q_0)$ denote the set of all simple functions on Q_0.

Definition 7 ([56], Definition 3.1). *A linear operator T defined on $\mathcal{F}(Q_0)$ is said to be of* strong type $N[p,(q,\mu)]$ *if there exists a positive constant K, such that for any $u \in \mathcal{F}(Q_0)$,*

$$[Tu]_{N^{(q,\mu)}(Q_0)} \le K\|u\|_{L^p(Q_0)};$$

the smallest of the constant K for which the above inequality holds true is called the strong $N[p,(q,\mu)]$-norm.

Theorem 5 ([56], Theorem 3.1). *Let $[p_i, q_i, \mu_i]$ be real numbers, such that p_i, $q_i \in [1,\infty)$ for any $i \in \{1,2\}$. If T is a linear operator which is simultaneously of strong type $N[p_i,(q_i,\mu_i)]$ with respective norms K_i $(i \in \{1,2\})$, then T is of strong type $N[p_t,(q_t,\mu)]$, where*

$$\begin{cases} \dfrac{1}{p_t} := \dfrac{1-t}{p_1} + \dfrac{t}{p_2}, \quad \dfrac{1}{q_t} := \dfrac{1-t}{q_1} + \dfrac{t}{q_2} \\ \dfrac{\mu}{q} = (1-t)\dfrac{\mu_1}{q_1} \quad for \quad t \in [0,1]. \end{cases}$$

Moreover, for any $t \in [0,1]$,

$$[Tu]_{N[p_t,(q_t,\mu)]} \le K_1^{1-t}K_2^t\|u\|_{L^p(Q_0)}.$$

The theorem also holds true in the limit case $p_1 = \infty$ and $\frac{1}{q_1} = \mu_1 = 0$.

Question 3. *In the proof of Theorem 5, lines 1–3 of [56] (p. 454), the author applied [56] (Lemma 2.3) with*

$$F[u,v,S] := \sum_i \int_{Q_i} \left[u(y) - u_{Q_i}\right] v\,dy |Q_i|^{-\lambda/p_t}$$

replaced by

$$\Phi(S,t) := \sum_i \int_{Q_i} \left[T(\widetilde{u}(y,t)) - (T\widetilde{u})_{Q_i}\right]\widetilde{v}(y,t)\,dy|Q_i|^{-\mu(t)\beta(t)}.$$

Therefore, by the proof of [56] (Lemma 2.3), we need to choose a function $\widetilde{v}$ satisfying that for any i, there exists some constant c_i, such that

$$\widetilde{v}(y,t) = c_i \overline{\left\{\mathrm{sign}\left[T(\widetilde{u}(y,t)) - (T\widetilde{u})_{Q_i}\right]\right\}} \tag{28}$$

in Q_i. Meanwhile, from the definition of $\widetilde{v}$ (see line 3 of [56], p. 452), it follows that

$$\widetilde{v}(y,t) = |v(y)|^{[1-\beta(t)]q_t'} e^{i \arg v(y)} \tag{29}$$

for some simple function $v \in \mathcal{F}(Q_0)$, where $1/q_t + 1/q_t' = 1$. To summarize, we need to find a simple function v, such that both (28) and (29) hold true, which seems unreasonable because $T\widetilde{u}$ may behave so badly even though both u and $\widetilde{u}$ are simple functions. Thus, the proof of Theorem 5 in [56] seems problematic. It is interesting to check whether or not Theorem 5 is really true.

3.2. Localized John–Nirenberg–Campanato Spaces

As a combination of the JNC space and the localized BMO space in Section 2.1, Sun et al. [36] studied the localized John–Nirenberg–Campanato space, which is new even in a special case: localized John–Nirenberg spaces. Now, we recall the definition of the localized Campanato space, which was first introduced by Goldberg in [50] (Theorem 5). In what follows, for any $s \in \mathbb{Z}_+$ and $c_0 \in (0, \ell(\mathcal{X}))$, let

$$P_{Q,c_0}^{(s)}(f) := \begin{cases} P_Q^{(s)}(f), & \ell(Q) < c_0, \\ 0, & \ell(Q) \geq c_0, \end{cases}$$

where $P_Q^{(s)}(f)$ is as in (24).

Definition 8. *Let $q \in [1, \infty)$, $s \in \mathbb{Z}_+$, and $\alpha \in [0, \infty)$. Fix $c_0 \in (0, \ell(\mathcal{X}))$. The* local Campanato space *$\Lambda_{(\alpha,q,s)}(\mathcal{X})$ is defined to be the set of all functions $f \in L^q_{\mathrm{loc}}(\mathcal{X})$, such that*

$$\|f\|_{\Lambda_{(\alpha,q,s)}(\mathcal{X})} := \sup |Q|^{-\alpha} \left[\fint_Q \left| f(x) - P_{Q,c_0}^{(s)}(f)(x) \right|^q dx \right]^{\frac{1}{q}} < \infty,$$

where the supremum is taken over all cubes Q of $\mathcal{X}$.

Fix the constant $c_0 \in (0, \ell(\mathcal{X}))$. In Definition 3, if $P_{Q_j}^{(s)}(f)$ were replaced by $P_{Q_j,c_0}^{(s)}(f)$, then we obtain the following localized John–Nirenberg–Campanato space. As was mentioned in Remark 4, we naturally extend the ranges of α and p, similar to Section 3.1; we omit some identical proofs.

Definition 9. *Let p, $q \in [1, \infty)$, $s \in \mathbb{Z}_+$, and $\alpha \in \mathbb{R}$. Fix the constant $c_0 \in (0, \ell(\mathcal{X}))$. The* local John–Nirenberg–Campanato space *$jn_{(p,q,s)_{\alpha,c_0}}(\mathcal{X})$ is defined to be the set of all functions $f \in L^q_{\mathrm{loc}}(\mathcal{X})$, such that*

$$\|f\|_{jn_{(p,q,s)_{\alpha,c_0}}(\mathcal{X})} := \sup \left[\sum_{j \in \mathbb{N}} |Q_j| \left\{ |Q_j|^{-\alpha} \left[\fint_{Q_j} \left| f(x) - P_{Q_j,c_0}^{(s)}(f)(x) \right|^q dx \right]^{\frac{1}{q}} \right\}^p \right]^{\frac{1}{p}}$$

is finite, where the supremum is taken over all collections of interior pairwise disjoint cubes $\{Q_j\}_{j \in \mathbb{N}}$ of $\mathcal{X}$. Moreover, the dual space *$(jn_{(p,q,s)_{\alpha,c_0}}(\mathcal{X}))^*$ of $jn_{(p,q,s)_{\alpha,c_0}}(\mathcal{X})$ is defined to be the set of all continuous linear functionals on $jn_{(p,q,s)_{\alpha,c_0}}(\mathcal{X})$ equipped with the weak-$*$ topology.*

Remark 6. *Notice that the Campanato space and the John–Nirenberg–Campanato space are quotient spaces, while their localized versions are not.*

Furthermore, in [36] (Proposition 2.5), Sun et al. showed that $jn_{(p,q,s)_{\alpha,c_0}}(\mathcal{X})$ in Definition 9 is independent of the choice of the positive constant c_0. Therefore, in what follows, we write

$$jn_{(p,q,s)_\alpha}(\mathcal{X}) := jn_{(p,q,s)_{\alpha,c_0}}(\mathcal{X}).$$

In particular, if $q = 1$ and $s = 0 = \alpha$, then $jn_{(p,q,s)_\alpha}(\mathcal{X})$ becomes the *local John–Nirenberg space*

$$jn_p(\mathcal{X}) := jn_{(p,1,0)_0}(\mathcal{X}).$$

The following Banach structure of $jn_{(p,q,s)_\alpha}(\mathcal{X})$ is just [36] (Proposition 2.7).

Proposition 12. *Let $p, q \in [1, \infty)$, $s \in \mathbb{Z}_+$, and $\alpha \in \mathbb{R}$. Then, $jn_{(p,q,s)_\alpha}(\mathcal{X})$ is a Banach space.*

In what follows, the *space* $jn_{(p,q,s)_\alpha}(Q_0)/\mathcal{P}_s(Q_0)$ is defined by setting

$$jn_{(p,q,s)_\alpha}(Q_0)/\mathcal{P}_s(Q_0) := \left\{f \in jn_{(p,q,s)_\alpha}(Q_0) : \|f\|_{jn_{(p,q,s)_\alpha}(Q_0)/\mathcal{P}_s(Q_0)} < \infty\right\},$$

where

$$\|f\|_{jn_{(p,q,s)_\alpha}(Q_0)/\mathcal{P}_s(Q_0)} := \inf_{a \in \mathcal{P}_s(Q_0)} \|f + a\|_{jn_{(p,q,s)_\alpha}(Q_0)};$$

the *space* $JN_{(p,q,s)_\alpha}(\mathcal{X}) \cap L^p(\mathcal{X})$ is defined by setting

$$JN_{(p,q,s)_\alpha}(\mathcal{X}) \cap L^p(\mathcal{X}) := \left\{f \in L^1_{\text{loc}}(\mathcal{X}) : \|f\|_{JN_{(p,q,s)_\alpha}(\mathcal{X}) \cap L^p(\mathcal{X})} < \infty\right\},$$

where

$$\|f\|_{JN_{(p,q,s)_\alpha}(\mathcal{X}) \cap L^p(\mathcal{X})} := \max\left\{\|f\|_{JN_{(p,q,s)_\alpha}(\mathcal{X})}, \|f\|_{L^p(\mathcal{X})}\right\}.$$

Moreover, the relations between $jn_{(p,q,s)_\alpha}(\mathcal{X})$ and $JN_{(p,q,s)_\alpha}(\mathcal{X})$, namely the following Propositions 13 and 14, are just [36] (Propositions 2.9 and 2.10), respectively.

Proposition 13. *Let $p, q \in [1, \infty)$, $s \in \mathbb{Z}_+$, and $\alpha \in \mathbb{R}$. Then,*

(i) $jn_{(p,q,s)_\alpha}(\mathcal{X}) \subset JN_{(p,q,s)_\alpha}(\mathcal{X})$;

(ii) *if Q_0 is a given cube of $\mathbb{R}^n$, then $JN_{(p,q,s)_\alpha}(Q_0) = jn_{(p,q,s)_\alpha}(Q_0)/\mathcal{P}_s(Q_0)$ with equivalent norms;*

(iii) $L^p(\mathbb{R}) \subsetneqq jn_p(\mathbb{R}) \subsetneqq JN_p(\mathbb{R})$ *if* $p \in (1, \infty)$.

Proposition 14. *Let $p \in [1, \infty)$, $q \in [1, p]$, $s \in \mathbb{Z}_+$, and $\alpha \in (0, \infty)$. Then,*

$$jn_{(p,q,s)_\alpha}(\mathcal{X}) = \left[JN_{(p,q,s)_\alpha}(\mathcal{X}) \cap L^p(\mathcal{X})\right] \tag{30}$$

with equivalent norms.

Furthermore, observe that Proposition 14 is the counterpart of [51] (Theorem 4.1), which says that for any $\alpha \in (0, \infty)$, $q \in [1, \infty)$, and $s \in \mathbb{Z}_+$,

$$\Lambda_{(\alpha,q,s)}(\mathcal{X}) = \left[C_{(\alpha,q,s)}(\mathcal{X}) \cap L^\infty(\mathcal{X})\right].$$

However, the case $q \in [p, \infty)$ in Proposition 14 is unclear so far (see Question 5 below). As an application of Propositions 13(ii) and 14, we have the following result.

Proposition 15. *Let $p \in [1, \infty)$, $q \in [1, p]$, $s \in \mathbb{Z}_+$, $\alpha \in (0, \infty)$, and Q_0 be a given cube of $\mathbb{R}^n$. Then,*

$$JN_{(p,q,s)_\alpha}(Q_0) \subset [L^p(Q_0)/\mathcal{P}_s(Q_0)].$$

Proof. Let p, q, s, α, and Q_0 be as in this proposition. Then, by Propositions 13(ii) and 14, we obtain

$$\begin{aligned} JN_{(p,q,s)_\alpha}(Q_0) &= \left[jn_{(p,q,s)_\alpha}(Q_0)/\mathcal{P}_s(Q_0)\right] \\ &= \left\{JN_{(p,q,s)_\alpha}(Q_0) \cap [L^p(Q_0)/\mathcal{P}_s(Q_0)]\right\} \end{aligned}$$

and

$$\|\cdot\|_{JN_{(p,q,s)_\alpha}(Q_0)} \sim \inf_{a\in\mathcal{P}_s(Q_0)} \|\cdot + a\|_{jn_{(p,q,s)_\alpha}(Q_0)}$$
$$\sim \max\left\{\|\cdot\|_{JN_{(p,q,s)_\alpha}(Q_0)}, \inf_{a\in\mathcal{P}_s(Q_0)} \|\cdot + a\|_{L^p(Q_0)}\right\}.$$

This implies that $JN_{(p,q,s)_\alpha}(Q_0) \subset [L^p(Q_0)/\mathcal{P}_s(Q_0)]$ with

$$\inf_{a\in\mathcal{P}_s(Q_0)} \|\cdot + a\|_{L^p(Q_0)} \lesssim \|\cdot\|_{JN_{(p,q,s)_\alpha}(Q_0)},$$

which completes the proof of Proposition 15. □

Propositions 16 and 17 below are just, respectively, [36] (Propositions 2.12 and 2.13), which show that the localized Campanato space is the limit of the localized John–Nirenberg–Campanato space.

Proposition 16. *Let $q \in [1,\infty)$, $s \in \mathbb{Z}_+$, $\alpha \in [0,\infty)$, and Q_0 be a given cube of $\mathbb{R}^n$. Then, for any $f \in L^1(Q_0)$,*

$$\|f\|_{\Lambda_{(\alpha,q,s)}(Q_0)} = \lim_{p\to\infty} \|f\|_{jn_{(p,q,s)_\alpha}(Q_0)}.$$

Moreover,

$$\Lambda_{(\alpha,q,s)}(Q_0) = \left\{f \in \bigcap_{p\in[1,\infty)} jn_{(p,q,s)_\alpha}(Q_0) : \lim_{p\to\infty} \|f\|_{jn_{(p,q,s)_\alpha}(Q_0)} < \infty\right\}.$$

Proposition 17. *Let $q \in [1,\infty)$, $s \in \mathbb{Z}_+$, and $\alpha \in [0,\infty)$. Then,*

$$\lim_{p\to\infty} jn_{(p,q,s)_\alpha}(\mathbb{R}^n) = \Lambda_{(\alpha,q,s)}(\mathbb{R}^n)$$

in the following sense: if $f \in jn_{(p,q,s)_\alpha}(\mathbb{R}^n) \cap \Lambda_{(\alpha,q,s)}(\mathbb{R}^n)$, then

$$f \in \bigcap_{r\in[p,\infty)} jn_{(r,q,s)_\alpha}(\mathbb{R}^n)$$

and

$$\|f\|_{\Lambda_{(\alpha,q,s)}(\mathbb{R}^n)} = \lim_{r\to\infty} \|f\|_{jn_{(r,q,s)_\alpha}(\mathbb{R}^n)}.$$

As in Proposition 10, the following invariance of $jn_{(p,q,s)_\alpha}(\mathcal{X})$ on its indices in the appropriate range is just [36] (Proposition 3.1).

Proposition 18. *Let $p \in (1,\infty)$, $q \in [1,p)$, $s \in \mathbb{Z}_+$, and $\alpha \in [0,\infty)$. Then,*

$$jn_{(p,q,s)_\alpha}(\mathcal{X}) = jn_{(p,1,s)_\alpha}(\mathcal{X})$$

with equivalent norms.

In other ranges of indices, namely $q \geq p$, the following relation between $jn_{(p,q,s)_\alpha}(\mathcal{X})$ and the Lebesgue space is just [36] (Proposition 3.4).

Proposition 19. *Let $s \in \mathbb{Z}_+$ and Q_0 be a given cube of $\mathbb{R}^n$.*

(i) *If $1 \leq p \leq q < \infty$, then $[|Q_0|^{\frac{1}{q}-\frac{1}{p}} jn_{(p,q,s)_0}(Q_0)] = L^q(Q_0)$ with equivalent norms.*
(ii) *If $p \in [1,\infty)$, then $jn_{(p,p,s)_0}(\mathbb{R}^n) = L^p(\mathbb{R}^n)$ with equivalent norms.*
(iii) *If $p, q \in [1,\infty)$, $\alpha \in (-\infty, \frac{1}{p} - \frac{1}{q})$, and $f \in jn_{(p,q,s)_\alpha}(\mathbb{R}^n)$, then $f = 0$ almost everywhere.*

Using the localized atom, Sun et al. [36] introduced the localized Hardy-type space and showed that this space is the predual of the localized John–Nirenberg–Campanato space. First, recall the definitions of localized atoms, localized polymers, and localized Hardy-type spaces in order as follows.

Definition 10. *Let v, $w \in [1,\infty]$, $s \in \mathbb{Z}_+$, and $\alpha \in \mathbb{R}$. Fix $c_0 \in (0, \ell(\mathcal{X}))$, and let Q denote a cube of $\mathbb{R}^n$. Then, a function a on $\mathbb{R}^n$ is called a* local $(v,w,s)_{\alpha,c_0}$-atom *supported in Q if*

(i) $\operatorname{supp}(a) := \{x \in \mathbb{R}^n : a(x) \neq 0\} \subset Q$;

(ii) $\|a\|_{L^w(Q)} \leq |Q|^{\frac{1}{w}-\frac{1}{v}-\alpha}$;

(iii) *when $\ell(Q) < c_0$, $\int_Q a(x)x^\beta dx = 0$ for any $\beta \in \mathbb{Z}_+^n$ and $|\beta| \leq s$.*

Definition 11. *Let v, $w \in [1,\infty]$, $s \in \mathbb{Z}_+$, $\alpha \in \mathbb{R}$, and $c_0 \in (0, \ell(\mathcal{X}))$. The* space $\widetilde{hk}_{(v,w,s)_{\alpha,c_0}}(\mathcal{X})$ *is defined to be the set of all $g \in (jn_{(v',w',s)_{\alpha,c_0}}(\mathcal{X}))^*$, such that*

$$g = \sum_{j\in\mathbb{N}} \lambda_j a_j$$

in $(jn_{(v',w',s)_{\alpha,c_0}}(\mathcal{X}))^*$, where $1/v + 1/v' = 1 = 1/w + 1/w'$, $\{a_j\}_{j\in\mathbb{N}}$ are local $(v,w,s)_{\alpha,c_0}$-atoms supported, respectively, in interior pairwise disjoint subcubes $\{Q_j\}_{j\in\mathbb{N}}$ of $\mathcal{X}$, and $\{\lambda_j\}_{j\in\mathbb{N}} \subset \mathbb{C}$ with $\|\{\lambda_j\}_{j\in\mathbb{N}}\|_{\ell^v} < \infty$ (see (27) for the definition of $\|\cdot\|_{\ell^v}$). Any $g \in \widetilde{hk}_{(v,w,s)_{\alpha,c_0}}(\mathcal{X})$ is called a local $(v,w,s)_{\alpha,c_0}$-polymer *on $\mathcal{X}$, and let*

$$\|g\|_{\widetilde{hk}_{(v,w,s)_{\alpha,c_0}}(\mathcal{X})} := \inf\left\|\{\lambda_j\}_{j\in\mathbb{N}}\right\|_{\ell^v},$$

where the infimum is taken over all decompositions of g as above.

Definition 12. *Let v, $w \in [1,\infty]$, $s \in \mathbb{Z}_+$, $\alpha \in \mathbb{R}$, and $c_0 \in (0, \ell(\mathcal{X}))$. The* local Hardy-type space $hk_{(v,w,s)_{\alpha,c_0}}(\mathcal{X})$ *is defined to be the set of all $g \in (jn_{(v',w',s)_{\alpha,c_0}}(\mathcal{X}))^*$, such that there exists a sequence $\{g_i\}_{i\in\mathbb{N}} \subset \widetilde{hk}_{(v,w,s)_{\alpha,c_0}}(\mathcal{X})$ satisfying that $\sum_{i\in\mathbb{N}} \|g_i\|_{\widetilde{hk}_{(v,w,s)_{\alpha,c_0}}(\mathcal{X})} < \infty$ and*

$$g = \sum_{i\in\mathbb{N}} g_i \tag{31}$$

in $(jn_{(v',w',s)_{\alpha,c_0}}(\mathcal{X}))^*$. For any $g \in hk_{(v,w,s)_{\alpha,c_0}}(\mathcal{X})$, let

$$\|g\|_{hk_{(v,w,s)_{\alpha,c_0}}(\mathcal{X})} := \inf \sum_{i\in\mathbb{N}} \|g_i\|_{\widetilde{hk}_{(v,w,s)_{\alpha,c_0}}(\mathcal{X})},$$

where the infimum is taken over all decompositions of g as in (31).

Correspondingly, $hk_{(v,w,s)_{\alpha,c_0}}(\mathcal{X})$ is independent of the choice of the positive constant c_0 as well, which is just [36] (Proposition 4.7).

Proposition 20. *Let $v \in (1,\infty)$, $w \in (1,\infty]$, $s \in \mathbb{Z}_+$, $\alpha \in \mathbb{R}$, and $0 < c_1 < c_2 < \ell(\mathcal{X})$. Then, $hk_{(v,w,s)_{\alpha,c_1}}(\mathcal{X}) = hk_{(v,w,s)_{\alpha,c_2}}(\mathcal{X})$ with equivalent norms.*

Henceforth, we simply write

$$\text{local } (v,w,s)_{\alpha,c_0}\text{–atoms, } \widetilde{hk}_{(v,w,s)_{\alpha,c_0}}(\mathcal{X}), \text{ and } hk_{(v,w,s)_{\alpha,c_0}}(\mathcal{X}),$$

respectively, as

$$\text{local } (v,w,s)_{\alpha}\text{–atoms, } \widetilde{hk}_{(v,w,s)_{\alpha}}(\mathcal{X}), \text{ and } hk_{(v,w,s)_{\alpha}}(\mathcal{X}).$$

The corresponding dual theorem (namely Theorem 6 below) is just [36] (Theorem 4.11). In what follows, the *space* $hk^{\rm fin}_{(v,w,s)_\alpha}(\mathcal{X})$ is defined to be the set of all finite linear combinations of local $(v,w,s)_\alpha$-atoms supported, respectively, in cubes of $\mathcal{X}$.

Theorem 6. *Let* $v, w \in (1,\infty)$, $1/v + 1/v' = 1 = 1/w + 1/w' = 1$, $s \in \mathbb{Z}_+$, *and* $\alpha \in \mathbb{R}$. *Then,* $jn_{(v',w',s)_\alpha}(\mathcal{X}) = (hk_{(v,w,s)_\alpha}(\mathcal{X}))^*$ *in the following sense:*

(i) *For any given* $f \in jn_{(v',w',s)_\alpha}(\mathcal{X})$, *the linear functional*

$$\mathcal{L}_f : g \longmapsto \langle \mathcal{L}_f, g \rangle := \int_{\mathcal{X}} f(x)g(x)\,dx, \qquad \forall\, g \in hk^{\rm fin}_{(v,w,s)_\alpha}(\mathcal{X})$$

can be extended to a bounded linear functional on $hk_{(v,w,s)_\alpha}(\mathcal{X})$. *Moreover, it holds true that* $\|\mathcal{L}_f\|_{(hk_{(v,w,s)_\alpha}(\mathcal{X}))^*} \le \|f\|_{jn_{(v',w',s)_\alpha}(\mathcal{X})}$.

(ii) *Any bounded linear functional* $\mathcal{L}$ *on* $hk_{(v,w,s)_\alpha}(\mathcal{X})$ *can be represented by a function* $f \in jn_{(v',w',s)_\alpha}(\mathcal{X})$ *in the following sense:*

$$\langle \mathcal{L}, g \rangle = \int_{\mathcal{X}} f(x)g(x)\,dx, \quad \forall\, g \in hk^{\rm fin}_{(v,w,s)_\alpha}(\mathcal{X}).$$

Moreover, there exists a positive constant C, depending only on s, such that $\|f\|_{jn_{(v',w',s)_\alpha}(\mathcal{X})} \le C\|\mathcal{L}\|_{(hk_{(v,w,s)_\alpha}(\mathcal{X}))^*}$.

As a corollary of Theorem 6, as well as a counterpart of Proposition 18, for any admissible (v,s,α), Proposition 21, which is just [36] (Proposition 5.1), shows that $hk_{(v,w,s)_\alpha}(\mathcal{X})$ is invariant on $w \in (v,\infty]$.

Proposition 21. *Let* $v \in (1,\infty)$, $w \in (v,\infty]$, $s \in \mathbb{Z}_+$, *and* $\alpha \in [0,\infty)$. *Then,*

$$hk_{(v,w,s)_\alpha}(\mathcal{X}) = hk_{(v,\infty,s)_\alpha}(\mathcal{X})$$

with equivalent norms.

The following proposition, which is just [36] (Proposition 5.6), might be viewed as a counterpart of Proposition 19.

Proposition 22. *Let* $v \in (1,\infty)$ *and* $s \in \mathbb{Z}_+$.

(i) *If* $w \in (1,v]$, *and* Q_0 *is a given cube of* $\mathbb{R}^n$, *then* $hk_{(v,w,s)_0}(Q_0) = |Q_0|^{\frac{1}{v}-\frac{1}{w}}L^w(Q_0)$ *with equivalent norms.*

(ii) $L^v(\mathbb{R}^n) = hk_{(v,v,s)_0}(\mathbb{R}^n)$ *with equivalent norms.*

Finally, the following relation between $hk_{(v,w,s)_\alpha}(\mathcal{X})$ and the atomic localized Hardy space is just [36] (Proposition 5.7).

Proposition 23. *Let* $w \in (1,\infty]$ *and* Q_0 *be a given cube of* $\mathbb{R}^n$. *Then,*

$$\bigcup_{v\in[1,\infty)} hk_{(v,w,0)_0}(Q_0) \subset h^{1,w}_{at}(Q_0).$$

Moreover, if $g \in \bigcup_{v\in[1,\infty)} hk_{(v,w,0)_0}(Q_0)$, *then*

$$\|g\|_{h^{1,w}_{at}(Q_0)} \le \liminf_{v\to 1^+} \|g\|_{hk_{(v,w,0)_0}(Q_0)},$$

where $v \to 1^+$ *means that* $v \in (1,\infty)$ *and* $v \to 1$.

We also list some open questions at the end of this subsection.

Question 4. *There still exists something* unclear *in Proposition 13(iii). Precisely, let* $p \in (1, \infty)$,

$$jn_p(\mathbb{R})/\mathbb{C} := \left\{ f \in L^1_{\mathrm{loc}}(\mathbb{R}) : \|f\|_{jn_p(\mathbb{R})/\mathbb{C}} := \inf_{c \in \mathbb{C}} \|f + c\|_{jn_p(\mathbb{R})} < \infty \right\}$$

and

$$L^p(\mathbb{R})/\mathbb{C} := \left\{ f \in L^1_{\mathrm{loc}}(\mathbb{R}) : \|f\|_{L^p(\mathbb{R})/\mathbb{C}} := \inf_{c \in \mathbb{C}} \|f + c\|_{L^p(\mathbb{R})} < \infty \right\}.$$

Then, it is still unknown whether or not

$$\left[jn_p(\mathbb{R})/\mathbb{C} \right] \subsetneqq JN_p(\mathbb{R})$$

holds true; namely, it is still unknown whether or not there exists some non-constant *function* h*, such that* $h \in JN_p(\mathbb{R})$ *but* $h \notin jn_p(\mathbb{R})$. *Moreover, it is still unknown whether or not*

$$[L^p(\mathbb{R}^n)/\mathbb{C}] \subsetneqq \left[jn_p(\mathbb{R}^n)/\mathbb{C} \right] \subsetneqq JN_p(\mathbb{R}^n)$$

holds true.

The following question is on the case $q > p$ corresponding to Proposition 14.

Question 5. *Let* $p \in [1, \infty)$, $q \in (p, \infty)$, $s \in \mathbb{Z}_+$, *and* $\alpha \in (0, \infty)$. *Then, it is still unknown whether or not*

$$jn_{(p,q,s)_\alpha}(\mathcal{X}) = \left[JN_{(p,q,s)_\alpha}(\mathcal{X}) \cap L^p(\mathcal{X}) \right]$$

still holds true.

Furthermore, the corresponding localized cases of Questions 1 and 2 are listed as follows. The following Question 6 is a modification of [36] (Remark 3.5), and Question 7 is just [36] (Remark 5.8).

Question 6. *Let* $p \in [1, \infty)$, $q \in [1, \infty)$, $s \in \mathbb{Z}_+$, *and* $\alpha \in [\frac{1}{p} - \frac{1}{q}, \infty)$. *Then, the relation between* $jn_{(p,q,s)_\alpha}(\mathbb{R}^n)$ *and the Riesz–Morrey space* $RM_{p,q,\alpha}(\mathbb{R}^n)$ *(see Section 4.1 for its definition) is still* unclear, *except the identity*

$$jn_{(p,p,s)_0}(\mathbb{R}^n) = L^p(\mathbb{R}^n) = RM_{p,p,0}(\mathbb{R}^n)$$

due to Proposition 19(ii) and Theorem 8(ii), and the inclusion

$$jn_{(p,q,s)_\alpha}(\mathbb{R}^n) \supset RM_{p,q,\alpha}(\mathbb{R}^n) \quad \text{with} \quad \|\cdot\|_{jn_{(p,q,s)_\alpha}(\mathbb{R}^n)} \lesssim \|\cdot\|_{RM_{p,q,\alpha}(\mathbb{R}^n)}$$

due to (25) and their definitions, where the implicit positive constant is independent of the functions under consideration.

Question 7. *Let* $v \in (1, \infty)$, $w \in (1, \infty]$, *and* Q_0 *be a given cube of* $\mathbb{R}^n$.

(i) *It is interesting to clarify the relation between* $\bigcup_{v \in (1,\infty)} hk_{(v,w,0)_0}(Q_0)$ *and* $h^{1,w}_{at}(Q_0)$, *and to find the condition on* g, *such that* $\|g\|_{h^{1,w}_{at}(Q_0)} = \lim_{v \to 1^+} \|g\|_{hk_{(v,w,0)_0}(Q_0)}$.

(ii) *Let* $\alpha \in (0, \infty)$ *and* $s \in \mathbb{Z}_+$. *As* $v \to 1^+$, *the relation between the localized atomic Hardy space (see [50] for the definition) and* $hk_{(v,w,s)_\alpha}(Q_0)$ *is still unknown.*

3.3. *Congruent John–Nirenberg–Campanato Spaces*

Inspired by the JNC space (see Section 3.1) and the space $\mathcal{B}$ (introduced and studied by Bourgain et al. [70]), Jia et al. [64] introduced the special John–Nirenberg–Campanato spaces via congruent cubes, which are of some amalgam features. This subsection is devoted to the main properties and some applications of congruent JNC spaces.

In what follows, for any $m \in \mathbb{Z}$, $\mathcal{D}_m(\mathbb{R}^n)$ denotes the set of all subcubes of $\mathbb{R}^n$ with side length 2^{-m}, $\mathcal{D}_m(Q_0)$ the set of all subcubes of Q_0 with side length $2^{-m}\ell(Q_0)$ for any given $m \in \mathbb{Z}_+$, and $\mathcal{D}_m(Q_0) := \emptyset$ for any given $m \in \mathbb{Z} \setminus \mathbb{Z}_+$; here and thereafter, $\ell(Q_0)$ denotes the side length of Q_0.

Definition 13. *Let $p, q \in [1,\infty)$, $s \in \mathbb{Z}_+$, and $\alpha \in \mathbb{R}$. The* special John–Nirenberg–Campanato space via congruent cubes *(for short,* congruent JNC space*) $JN^{\rm con}_{(p,q,s)_\alpha}(\mathcal{X})$ is defined to be the set of all $f \in L^1_{\rm loc}(\mathcal{X})$, such that*

$$\|f\|_{JN^{\rm con}_{(p,q,s)_\alpha}(\mathcal{X})} := \sup_{m\in\mathbb{Z}}\left\{[f]^{(m)}_{(p,q,s)_\alpha,\mathcal{X}}\right\} < \infty,$$

where, for any $m \in \mathbb{Z}$, $[f]^{(m)}_{(p,q,s)_\alpha,\mathcal{X}}$ is defined to be

$$\sup_{\{Q_j\}_j\subset\mathcal{D}_m(\mathcal{X})}\left[\sum_j |Q_j|\left\{|Q_j|^{-\alpha}\left[\fint_{Q_j}\left|f(x)-P^{(s)}_{Q_j}(f)(x)\right|^q dx\right]^{\frac{1}{q}}\right\}^p\right]^{\frac{1}{p}}$$

with $P^{(s)}_{Q_j}(f)$ for any j as in (24) via Q replaced by Q_j and the supremum taken over all collections of interior pairwise disjoint cubes $\{Q_j\}_j \subset \mathcal{D}_m(\mathcal{X})$. In particular, let

$$JN^{\rm con}_{p,q}(\mathcal{X}) := JN^{\rm con}_{(p,q,0)_0}(\mathcal{X}).$$

Remark 7. *Let $p, q \in [1,\infty)$, $s \in \mathbb{Z}_+$, and $\alpha \in \mathbb{R}$. There exist some useful equivalent norms on $JN^{\rm con}_{(p,q,s)_\alpha}(\mathcal{X})$ as follows.*

(i) *(non-dyadic side length) $f \in JN^{\rm con}_{(p,q,s)_\alpha}(\mathcal{X})$ if and only if $f \in L^1_{\rm loc}(\mathcal{X})$ and*

$$\|f\|_{\widetilde{JN}^{\rm con}_{(p,q,s)_\alpha}(\mathcal{X})} := \sup\left[\sum_j |Q_j|\left\{|Q_j|^{-\alpha}\left[\fint_{Q_j}\left|f(x)-P^{(s)}_{Q_j}(f)(x)\right|^q dx\right]^{\frac{1}{q}}\right\}^p\right]^{\frac{1}{p}} < \infty$$

if and only if $f \in L^1_{\rm loc}(\mathcal{X})$ and

$$\|f\|_{\widehat{JN}^{\rm con}_{(p,q,s)_\alpha}(\mathcal{X})} := \sup\left[\sum_j |Q_j|\left\{|Q_j|^{-\alpha}\inf_{P\in\mathcal{P}_s(Q_j)}\left[\fint_{Q_j}|f(x)-P(x)|^q dx\right]^{\frac{1}{q}}\right\}^p\right]^{\frac{1}{p}} < \infty, \quad (32)$$

where the suprema are taken over all collections of interior pairwise disjoint cubes $\{Q_j\}_j$ of $\mathcal{X}$ with the same side length; moreover, $\|\cdot\|_{JN^{\rm con}_{(p,q,s)_\alpha}(\mathcal{X})} \sim \|\cdot\|_{\widetilde{JN}^{\rm con}_{(p,q,s)_\alpha}(\mathcal{X})} \sim \|\cdot\|_{\widehat{JN}^{\rm con}_{(p,q,s)_\alpha}(\mathcal{X})}$; see [64] (Remark 1.6(ii) and Propositions 2.6 and 2.7).

(ii) *(integral representation) In what follows, for any $y \in \mathbb{R}^n$ and $r \in (0,\infty)$, let*

$$B(y,r) := \{x \in \mathbb{R}^n : |x-y| < r\}.$$

Then $f \in JN^{\rm con}_{(p,q,s)_\alpha}(\mathbb{R}^n)$ if and only if $f \in L^1_{\rm loc}(\mathbb{R}^n)$ and

$$\|f\|_* := \sup_{r\in(0,\infty)}\left[\int_{\mathbb{R}^n}\left\{|B(y,r)|^{-\alpha}\left[\fint_{B(y,r)}\left|f(x)-P^{(s)}_{B(y,r)}(f)(x)\right|^q dx\right]^{\frac{1}{q}}\right\}^p dy\right]^{\frac{1}{p}} < \infty;$$

moreover, $\|\cdot\|_{JN^{\rm con}_{(p,q,s)_\alpha}(\mathbb{R}^n)} \sim \|\cdot\|_*$; *see [64] (Proposition 2.2) for this equivalence, which plays an essential role when establishing the boundedness of operators on congruent JNC spaces (see [71–73] for more details).*

The following proposition is just [64] (Proposition 2.10).

Proposition 24. *Let* $s \in \mathbb{Z}_+$, $\alpha \in \mathbb{R}$, *and* Q_0 *be a given cube of* $\mathbb{R}^n$.

(i) *For any given* $p \in [1,\infty)$ *and* $q \in [1,\infty)$,

$$JN^{\rm con}_{(p,q,s)_\alpha}(Q_0) \subset \left[|Q_0|^{\frac{1}{p}-\frac{1}{q}-\alpha} L^q(Q_0)/\mathcal{P}_s(Q_0)\right].$$

Moreover, for any $f \in JN^{\rm con}_{(p,q,s)_\alpha}(Q_0)$,

$$\|f\|_{|Q_0|^{\frac{1}{p}-\frac{1}{q}-\alpha} L^q(Q_0)/\mathcal{P}_s(Q_0)} \le \|f\|_{JN^{\rm con}_{(p,q,s)_\alpha}(Q_0)}.$$

(ii) *If* $\alpha \in (-\infty, 0]$, *then, for any given* $p \in [1,\infty)$ *and* $q \in [p,\infty)$,

$$JN^{\rm con}_{(p,q,s)_\alpha}(Q_0) = \left[|Q_0|^{\frac{1}{p}-\frac{1}{q}-\alpha} L^q(Q_0)/\mathcal{P}_s(Q_0)\right]$$

with equivalent norms.

(iii) *If* $q \in [1,\infty)$ *and* $1 \le p_1 \le p_2 < \infty$, *then* $JN^{\rm con}_{(p_2,q,s)_\alpha}(Q_0) \subset JN^{\rm con}_{(p_1,q,s)_\alpha}(Q_0)$. *Moreover, for any* $f \in JN^{\rm con}_{(p_2,q,s)_\alpha}(Q_0)$,

$$|Q_0|^{-\frac{1}{p_1}} \|f\|_{JN^{\rm con}_{(p_1,q,s)_\alpha}(Q_0)} \le |Q_0|^{-\frac{1}{p_2}} \|f\|_{JN^{\rm con}_{(p_2,q,s)_\alpha}(Q_0)}.$$

(iv) *If* $p \in [1,\infty)$ *and* $1 \le q_1 \le q_2 < \infty$, *then* $JN^{\rm con}_{(p,q_2,s)_\alpha}(\mathcal{X}) \subset JN^{\rm con}_{(p,q_1,s)_\alpha}(\mathcal{X})$. *Moreover, for any* $f \in JN^{\rm con}_{(p,q_2,s)_\alpha}(\mathcal{X})$,

$$\|f\|_{JN^{\rm con}_{(p,q_1,s)_\alpha}(\mathcal{X})} \le \|f\|_{JN^{\rm con}_{(p,q_2,s)_\alpha}(\mathcal{X})}.$$

The relation of congruent JNC spaces and Campanato spaces is similar to Proposition 6 and Corollary 2, and hence we omit the statement here; see [64] (Proposition 2.11) for details. The relation of congruent JNC spaces and the space $\mathcal{B}$ was discussed in [64] (Proposition 2.20 and Remark 2.21). Recall that the *local Sobolev space* $W^{1,p}_{\rm loc}(\mathbb{R}^n)$ is defined by setting

$$W^{1,p}_{\rm loc}(\mathbb{R}^n) := \left\{f \in L^p_{\rm loc}(\mathbb{R}^n) : |\nabla f| \in L^p_{\rm loc}(\mathbb{R}^n)\right\},$$

here and thereafter, $\nabla f := (\partial_1 f, \ldots, \partial_n f)$, where for any $i \in \{1,\ldots,n\}$, $\partial_i f$ denotes the *weak derivative* of f, namely a locally integrable function on $\mathbb{R}^n$, such that for any $\varphi \in C^\infty_{\rm c}(\mathbb{R}^n)$ (the set of all infinitely differentiable functions on $\mathbb{R}^n$ with compact support),

$$\int_{\mathbb{R}^n} f(x)\partial_i\varphi(x)\,dx = -\int_{\mathbb{R}^n} \varphi(x)\partial_i f(x)\,dx.$$

The following proposition is just [64] (Proposition 2.13).

Proposition 25. *Let* $p \in (1,\infty)$ *and* $f \in L^p_{\rm loc}(\mathbb{R}^n)$. *Then,* $|\nabla f| \in L^p(\mathbb{R}^n)$ *if and only if*

$$\liminf_{m\to\infty} [f]^{(m)}_{(p,p,0)_{1/n},\mathbb{R}^n} < \infty,$$

where $[f]^{(m)}_{(p,p,0)_{1/n},\mathbb{R}^n}$ *is as in Definition 13. Moreover, for any given* $p \in [1,\infty)$*, there exists a constant* $C_{(n,p)} \in [1,\infty)$*, such that for any* $f \in W^{1,p}_{\mathrm{loc}}(\mathbb{R}^n)$*,*

$$\frac{1}{C_{(n,p)}}\left[\int_{\mathbb{R}^n} |\nabla f(x)|^p \, dx\right]^{\frac{1}{p}} \le \liminf_{m\to\infty} [f]^{(m)}_{(p,p,0)_{1/n},\mathbb{R}^n} \le C_{(n,p)}\left[\int_{\mathbb{R}^n} |\nabla f(x)|^p \, dx\right]^{\frac{1}{p}}.$$

Remark 8. *Fusco et al. studied BMO-type seminorms and Sobolev functions in [74]. Indeed, in [74] (Theorem 2.2), Fusco et al. showed that Proposition 25 still holds true with cubes* $\{Q_j\}_j$*, in the supremum of* $[f]^{(m)}_{(p,p,0)_{1/n},\mathbb{R}^n}$*, having the same side length but an* arbitrary orientation. *Later, the main results of [74] were further extended by Di Fratta and Fiorenza in [75], via replacing a family of open cubes by a broader class of* tessellations *(from pentagonal and hexagonal tilings to space-filling polyhedrons and creative tessellations).*

The following nontriviality is just [64] (Propositions 2.16 and 2.19).

Proposition 26. *Let* $p \in (1,\infty)$ *and* $q \in [1,p)$*.*

(i) *Let* I_0 *be a given bounded interval of* $\mathbb{R}$*. Then,*

$$JN_{p,q}(I_0) \subsetneqq JN^{\mathrm{con}}_{p,q}(I_0) \quad \text{and} \quad JN_{p,q}(\mathbb{R}) \subsetneqq JN^{\mathrm{con}}_{p,q}(\mathbb{R}).$$

(ii) *Let* Q_0 *be a given cube of* $\mathbb{R}^n$*. Then,*

$$JN_{p,q}(Q_0) \subsetneqq JN^{\mathrm{con}}_{p,q}(Q_0).$$

Similar to Theorem 3, the following dual result is just [64] (Theorem 4.10). Recall that the congruent Hardy-type space $HK^{\mathrm{con}}_{(u,v,s)_\alpha}(\mathcal{X})$ is defined as in Definition 6 with the additional condition that all cubes of the polymer have the same side length (see [64], Definition 4.7, for more details).

Theorem 7. *Let* p*,* $q \in (1,\infty)$*,* $1/p = 1/p' = 1 = 1/q + 1/q'$*,* $s \in \mathbb{Z}_+$*, and* $\alpha \in \mathbb{R}$*. If* $JN^{\mathrm{con}}_{(p,q,s)_\alpha}(\mathcal{X})$ *is equipped with the norm* $\|\cdot\|_{\widetilde{JN}^{\mathrm{con}}_{(p,q,s)_\alpha}(\mathcal{X})}$ *in (32), then*

$$\left(HK^{\mathrm{con}}_{(p',q',s)_\alpha}(\mathcal{X})\right)^* = JN^{\mathrm{con}}_{(p,q,s)_\alpha}(\mathcal{X})$$

with equivalent norms in the following sense:

(i) *Any* $f \in JN^{\mathrm{con}}_{(p,q,s)_\alpha}(\mathcal{X})$ *induces a linear functional* $\mathcal{L}_f$ *which is given by setting, for any* $g \in HK^{\mathrm{con}}_{(p',q',s)_\alpha}(\mathcal{X})$ *and* $\{g_i\}_i \subset \widetilde{HK}^{\mathrm{con}}_{(p',q',s)_\alpha}(\mathcal{X})$ *with* $g = \sum_i g_i$ *in* $(JN^{\mathrm{con}}_{(p,q,s)_\alpha}(\mathcal{X}))^*$*,*

$$\mathcal{L}_f(g) := \langle g, f\rangle = \sum_i \langle g_i, f\rangle.$$

Moreover, for any $g \in HK^{\mathrm{con-fin}}_{(p',q',s)_\alpha}(\mathcal{X})$*,*

$$\mathcal{L}(g) = \int_{\mathcal{X}} f(x)g(x)\, dx \quad \text{and} \quad \|\mathcal{L}_f\|_{(HK^{\mathrm{con}}_{(p',q',s)_\alpha}(\mathcal{X}))^*} \le \|f\|_{\widetilde{JN}^{\mathrm{con}}_{(p,q,s)_\alpha}(\mathcal{X})}.$$

(ii) *Conversely, for any continuous linear functional* $\mathcal{L}$ *on* $HK^{\mathrm{con}}_{(p',q',s)_\alpha}(\mathcal{X})$*, there exists a unique* $f \in JN^{\mathrm{con}}_{(p,q,s)_\alpha}(\mathcal{X})$*, such that for any* $g \in HK^{\mathrm{con-fin}}_{(p',q',s)_\alpha}(\mathcal{X})$*,*

$$\mathcal{L}(g) = \int_{\mathcal{X}} f(x)g(x)\, dx \quad \text{and} \quad \|f\|_{\widetilde{JN}^{\mathrm{con}}_{(p,q,s)_\alpha}(\mathcal{X})} \le \|\mathcal{L}\|_{(HK^{\mathrm{con}}_{(p',q',s)_\alpha}(\mathcal{X}))^*}.$$

Moreover, when $\mathcal{X} = Q_0$, we further have the VMO-H^1-type duality for the congruent Hardy-type space (see Theorem 25 below).

Recall that Essén et al. [76] introduced and studied the Q space on $\mathbb{R}^n$, which generalizes the space BMO $(\mathbb{R}^n)$. Later, the Q space proved very useful in harmonic analysis, potential analysis, partial differential equations, and closely related fields (see, for instance, [77–79]). Thus, it is natural to consider some "new Q space" corresponding to the John–Nirenberg space JN_p. Based on Remark 7(ii), Tao et al. [80] introduced and studied the *John–Nirenberg-Q space* on $\mathbb{R}^n$ via congruent cubes, which contains the congruent John–Nirenberg space on $\mathbb{R}^n$ as special cases and also sheds some light on the mysterious John–Nirenberg space.

4. Riesz-Type Space

Observe that if we partially subtract integral means (or polynomials for high order cases) in $\|f\|_{JN_{(p,q,s)_\alpha}(\mathcal{X})}$, namely dropping $P^{(s)}_{Q_i}(f)$ in

$$\left\{\sum_i |Q_i|\left[|Q_i|^{-\alpha}\left\{\fint_{Q_i}\left|f(x) - P^{(s)}_{Q_i}(f)(x)\right|^q dx\right\}^{\frac{1}{q}}\right]^p\right\}^{\frac{1}{p}}$$

for any i satisfying $\ell(Q_i) \geq c_0$, then we obtain the localized JNC space as in Definition 9. Thus, a natural question arises: what if we thoroughly drop all $\{P^{(s)}_{Q_i}(f)\}_i$ in $\|f\|_{JN_{(p,q,s)_\alpha}(\mathcal{X})}$? In this section, we study the space with such a norm (subtracting all $\{P^{(s)}_{Q_i}(f)\}_i$ in the norm of the JNC space). As a bridge connecting Lebesgue and Morrey spaces via Riesz norms, it is called the "Riesz–Morrey space". For more studies on the well-known Morrey space, we refer the reader to, for instance, [81–84] and, in particular, the recent monographs by Sawano et al. [85,86].

4.1. Riesz–Morrey Spaces

As a suitable substitute of $L^\infty(\mathcal{X})$, the space BMO $(\mathcal{X})$ proves very useful in harmonic analysis and partial differential equations. Recall that

$$\|f\|_{\mathrm{BMO}(\mathcal{X})} := \sup_{\text{cube } Q\subset\mathcal{X}} \fint_Q \left|f(x) - f_Q\right| dx.$$

Indeed, the only difference between them exists in subtracting integral means, which is just the following proposition. In what follows, for any $q \in (0, \infty)$ and any measurable function f, let

$$\|f\|_{L^q_*(\mathcal{X})} := \sup_{\text{cube } Q\subset\mathcal{X}} \left[\fint_Q |f(x)|^q dx\right]^{\frac{1}{q}}.$$

Proposition 27. *Let $q \in (0, \infty)$. Then, $f \in L^\infty(\mathcal{X})$ if and only if $f \in L^q_{\mathrm{loc}}(\mathcal{X})$ and $\|f\|_{L^q_*(\mathcal{X})} < \infty$. Moreover,*

$$\|\cdot\|_{L^\infty(\mathcal{X})} = \|\cdot\|_{L^q_*(\mathcal{X})}.$$

Proof. For the simplicity of the presentation, we only consider the case $q = 1$. On the one hand, for any $f \in L^\infty(\mathcal{X})$, it is easy to see that $f \in L^1_{\mathrm{loc}}(\mathcal{X})$ and

$$\|f\|_{L^1_*(\mathcal{X})} = \sup_{Q \subset \mathcal{X}} \fint_Q |f(x)|\,dx \le \sup_{Q \subset \mathcal{X}} \|f\|_{L^\infty(\mathcal{X})} = \|f\|_{L^\infty(\mathcal{X})}.$$

On the other hand, for any $f \in L^1_{\mathrm{loc}}(\mathcal{X})$ and $\|f\|_{L^1_*(\mathcal{X})} < \infty$, let x be any Lebesgue point of f. Then, from the Lebesgue differentiation theorem, we deduce that

$$|f(x)| = \lim_{|Q| \to 0^+,\, Q \ni x} \fint_Q |f(y)|\,dy \le \sup_{Q \subset \mathcal{X}} \fint_Q |f(y)|\,dy = \|f\|_{L^1_*(\mathcal{X})},$$

which, together with the Lebesgue differentiation theorem again, further implies that

$$\|f\|_{L^\infty(\mathcal{X})} \le \|f\|_{L^1_*(\mathcal{X})}$$

and hence $f \in L^\infty(\mathcal{X})$. Moreover, we have $\|\cdot\|_{L^\infty(\mathcal{X})} = \|\cdot\|_{L^1_*(\mathcal{X})}$. This finishes the proof of Proposition 27. □

Furthermore, if we remove integral means in the $JN_p(Q_0)$-norm

$$\|f\|_{JN_p(Q_0)} = \sup\left[\sum_i |Q_i| \left(\fint_{Q_i} \left|f(x) - f_{Q_i}\right| dx\right)^p\right]^{\frac{1}{p}},$$

where the supremum is taken over all collections of cubes $\{Q_i\}_i$ of Q_0 with pairwise disjoint interiors, then we obtain

$$\sup\left[\sum_i |Q_i| \left(\fint_{Q_i} |f(x)|\,dx\right)^p\right]^{\frac{1}{p}} =: \|f\|_{R_p(Q_0)}$$

which coincides with $\|f\|_{L^p(Q_0)}$ due to Riesz [41]. Corresponding to the JNC space, the following triple index Riesz-type space $R_{p,q,\alpha}(\mathcal{X})$, called the Riesz–Morrey space, was introduced and studied in [37] and, independently, by Fofana et al. [87] when $\mathcal{X} = \mathbb{R}^n$.

Definition 14. *Let $p \in [1,\infty]$, $q \in [1,\infty]$, and $\alpha \in \mathbb{R}$. The* Riesz–Morrey space *$RM_{p,q,\alpha}(\mathcal{X})$ is defined by setting*

$$RM_{p,q,\alpha}(\mathcal{X}) := \left\{f \in L^q_{\mathrm{loc}}(\mathcal{X}) : \|f\|_{RM_{p,q,\alpha}(\mathcal{X})} < \infty\right\},$$

where

$$\|f\|_{RM_{p,q,\alpha}(\mathcal{X})} := \begin{cases} \sup\left[\sum_i |Q_i|^{1-p\alpha-\frac{p}{q}} \|f\|^p_{L^q(Q_i)}\right]^{\frac{1}{p}} & \text{if } p \in [1,\infty),\ q \in [1,\infty], \\ \sup\sup_i |Q_i|^{-\alpha-\frac{1}{q}} \|f\|_{L^q(Q_i)} & \text{if } p = \infty,\ q \in [1,\infty] \end{cases}$$

and the suprema are taken over all collections of subcubes $\{Q_i\}_i$ of $\mathcal{X}$ with pairwise disjoint interiors. In addition, $R_{p,q,0}(\mathcal{X}) =: R_{p,q}(\mathcal{X})$.

Observe that the Riesz–Morrey norm $\|\cdot\|_{RM_{p,q,\alpha}(\mathcal{X})}$ is different from the JNC norm $\|\cdot\|_{JN_{(p,q,s)_\alpha}(\mathcal{X})}$ with $s = 0$, only in subtracting mean oscillations (see [37], Remark 2, for more details). It is easy to see that $\|\cdot\|_{R_{p,1,0}(Q_0)} = \|\cdot\|_{R_p(Q_0)}$, and, as a generalization of the above equivalence in Riesz [41], the following proposition is just [37] (Proposition 1).

Proposition 28. *Let $p \in [1,\infty]$ and $q \in [1,p]$. Then, $f \in L^p(\mathcal{X})$ if and only if $f \in R_{p,q}(\mathcal{X})$. Moreover, $L^p(\mathcal{X}) = R_{p,q}(\mathcal{X})$ with equivalent norms, namely, for any $f \in L^q_{\rm loc}(\mathcal{X})$, $\|f\|_{L^p(\mathcal{X})} = \|f\|_{R_{p,q}(\mathcal{X})}$.*

As for the case $1 \le p < q \le \infty$, by [37] (Remark 2.3), we know that

$$R_{p,q}(\mathbb{R}^n) = \{0\} \ne L^q(\mathbb{R}^n) = R_{q,q}(\mathbb{R}^n),$$

and

$$\left[|Q_0|^{-\frac{1}{p}} R_{p,q}(Q_0)\right] = \left[|Q_0|^{-\frac{1}{q}} L^q(Q_0)\right] = \left[|Q_0|^{-\frac{1}{q}} R_{q,q}(Q_0)\right]$$

with equivalent norms.

Moreover, it is shown in [37] (Theorem 1 and Corollary 1) that the endpoint spaces of Riesz–Morrey spaces are Lebesgue spaces or Morrey spaces. In this sense, we regard the Riesz–Morrey space as a bridge connecting the Lebesgue space and the Morrey space. Thus, a natural question arises: whether or not Riesz–Morrey spaces are truly new spaces different from Lebesgue spaces or Morrey spaces. Very recently, Zeng et al. [88] gave an *affirmative* answer to this question via constructing two nontrivial functions over $\mathbb{R}^n$ and any given cube Q of $\mathbb{R}^n$. It should be pointed out that the nontrivial function on the cube Q is geometrically similar to the striking function constructed by Dafni et al. in the proof of [31] (Proposition 3.2). Furthermore, we have the following classifications of Riesz–Morrey spaces, which are just [88] (Corollary 3.7).

Theorem 8.

(i) *Let $p \in (1,\infty]$ and $q \in [1,p)$. Then,*

$$RM_{p,q,\alpha}(\mathbb{R}^n)\begin{cases} = L^q(\mathbb{R}^n) & \text{if } \alpha = \frac{1}{p} - \frac{1}{q}, \\ \supsetneqq L^{\frac{p}{1-p\alpha}}(\mathbb{R}^n) & \text{if } \alpha \in \left(\frac{1}{p} - \frac{1}{q}, 0\right), \\ = L^p(\mathbb{R}^n) & \text{if } \alpha = 0, \\ = \{0\} & \text{if } \alpha \in \left(-\infty, \frac{1}{p} - \frac{1}{q}\right) \cup (0,\infty). \end{cases}$$

In particular, if $\alpha \in (-\frac{1}{q}, 0)$, then $RM_{\infty,q,\alpha}(\mathbb{R}^n) = M_q^{-1/\alpha}(\mathbb{R}^n)$, which is just the Morrey space defined in Remark 3.

(ii) *Let $p \in [1,\infty]$ and $q \in [p,\infty]$. Then,*

$$RM_{p,q,\alpha}(\mathbb{R}^n)\begin{cases} = L^q(\mathbb{R}^n) & \text{if } \alpha = \frac{1}{p} - \frac{1}{q} = 0, \\ = \{0\} & \text{if } \alpha = \frac{1}{p} - \frac{1}{q} \ne 0, \\ = \{0\} & \text{if } \alpha \in \mathbb{R} \setminus \left\{\frac{1}{p} - \frac{1}{q}\right\}. \end{cases}$$

(iii) *Let $p \in (1,\infty]$, $q \in [1,p)$, and Q_0 be a given cube of $\mathbb{R}^n$. Then,*

$$RM_{p,q,\alpha}(Q_0)\begin{cases} = L^q(Q_0) & \text{if } \alpha = \left(-\infty, \frac{1}{p} - \frac{1}{q}\right], \\ \supsetneqq L^{\frac{p}{1-p\alpha}}(Q_0) & \text{if } \alpha \in \left(\frac{1}{p} - \frac{1}{q}, 0\right), \\ = L^p(Q_0) & \text{if } \alpha = 0, \\ = \{0\} & \text{if } \alpha \in (0,\infty). \end{cases}$$

In particular, $RM_{\infty,q,\alpha}(Q_0) = M_q^{-1/\alpha}(Q_0)$ if $\alpha \in (-\frac{1}{q}, 0)$.

(iv) *Let $p \in [1,\infty]$, $q \in [p,\infty]$, and Q_0 be a given cube of $\mathbb{R}^n$. Then,*

$$RM_{p,q,\alpha}(Q_0)\begin{cases} = L^q(Q_0) & \text{if } \alpha \in (-\infty, 0], \\ = \{0\} & \text{if } \alpha \in (0,\infty). \end{cases}$$

Recall that by [89] (Theorem 1), the predual space of the Morrey space is the so-called block space. Combining this with the duality of John–Nirenberg–Campanato spaces in [61] (Theorem 3.9), the authors in [37] introduced the block-type space which proves the predual of the Riesz–Morrey space. Observe that every (∞, v, α)-block in Definition 15(i) is exactly a $(v, \frac{\alpha}{n})$-*block* introduced in [89].

Definition 15. *Let* $u,\ v \in [1,\infty]$, $\frac{1}{u} + \frac{1}{u'} = 1 = \frac{1}{v} + \frac{1}{v'}$, *and* $\alpha \in \mathbb{R}$. *Let* $(RM_{u',v',\alpha}(\mathcal{X}))^*$ *be the* dual space *of* $RM_{u',v',\alpha}(\mathcal{X})$ *equipped with the weak-∗ topology.*

(i) *A function b is called a* (u, v, α)-block *if*

$$\operatorname{supp}(b) := \{x \in \mathcal{X} : b(x) \neq 0\} \subset Q \quad and \quad \|b\|_{L^v(Q)} \leq |Q|^{\frac{1}{v}-\frac{1}{u}-\alpha}.$$

(ii) *The* space of (u, v, α)-chains, $\widetilde{B}_{u,v,\alpha}(\mathcal{X})$, *is defined by setting*

$$\widetilde{B}_{u,v,\alpha}(\mathcal{X}) := \left\{ h \in (RM_{u',v',\alpha}(\mathcal{X}))^* : \ h = \sum_j \lambda_j b_j \ and \ \left\|\{\lambda_j\}_j\right\|_{\ell^u} < \infty \right\},$$

where $\{b_j\}_j$ *are* (u, v, α)*-blocks supported, respectively, in subcubes* $\{Q_j\}$ *of* $\mathcal{X}$ *with pairwise disjoint interiors, and* $\{\lambda_j\}_j \subset \mathbb{C}$ *with* $\|\{\lambda_j\}_j\|_{\ell^u} < \infty$ *(see (27) for the definition of* $\|\cdot\|_{\ell^u}$*). Moreover, any* $h \in \widetilde{B}_{u,v,\alpha}(\mathcal{X})$ *is called a* (u, v, α)-chain, *and its norm is defined by setting*

$$\|h\|_{\widetilde{B}_{u,v,\alpha}(\mathcal{X})} := \inf \left\|\{\lambda_j\}_j\right\|_{\ell^u},$$

where the infimum is taken over all decompositions of h as above.

(iii) *The* block-type space $B_{u,v,\alpha}(\mathcal{X})$ *is defined by setting*

$$B_{u,v,\alpha}(\mathcal{X}) := \left\{ g \in (RM_{u',v',\alpha}(\mathcal{X}))^* : \ g = \sum_i h_i \ and \ \sum_i \|h_j\|_{\widetilde{B}_{u,v,\alpha}(\mathcal{X})} < \infty \right\},$$

where $\{h_i\}_i$ *are* (u, v, α)*-chains. Moreover, for any* $g \in B_{u,v,\alpha}(\mathcal{X})$,

$$\|g\|_{B_{u,v,\alpha}(\mathcal{X})} := \inf \sum_i \|h_j\|_{\widetilde{B}_{u,v,\alpha}(\mathcal{X})},$$

where the infimum is taken over all decompositions of g as above.

(iv) *The* finite block-type space $B^{\mathrm{fin}}_{u,v,\alpha}(\mathcal{X})$ *is defined to be the set of all finite summations*

$$\sum_{m=1}^{M} \lambda_m b_m,$$

where $M \in \mathbb{N}$, $\{\lambda_m\}_{m=1}^M \subset \mathbb{C}$, *and* $\{b_m\}_{m=1}^M$ *are* (u, v, α)*-blocks.*

The following dual theorem is just [37] (Theorem 2).

Theorem 9. *Let* $p,\ q \in (1,\infty)$, $1/p + 1/p' = 1 = 1/q + 1/q'$, *and* $\alpha \in \mathbb{R}$. *Then,* $(B_{p',q',\alpha}(\mathcal{X}))^* = RM_{p,q,\alpha}(\mathcal{X})$ *in the following sense:*

(i) *If* $f \in RM_{p,q,\alpha}(\mathcal{X})$, *then f induces a linear functional* $\mathcal{L}_f$ *on* $B_{p',q',\alpha}(\mathcal{X})$ *with*

$$\|\mathcal{L}_f\|_{(B_{p',q',\alpha}(\mathcal{X}))^*} \leq C\|f\|_{RM_{p,q,\alpha}(\mathcal{X})},$$

where C is a positive constant independent of f.

(ii) *If* $\mathcal{L} \in (B_{p',q',\alpha}(\mathcal{X}))^*$, *then there exists some* $f \in RM_{p,q,\alpha}(\mathcal{X})$, *such that for any* $g \in B^{\mathrm{fin}}_{p',q',\alpha}(\mathcal{X})$,

$$\mathcal{L}(g) = \int_{\mathcal{X}} f(x)g(x)\,dx,$$

and

$$\|\mathcal{L}\|_{(B_{p',q',\alpha}(\mathcal{X}))^*} \sim \|f\|_{RM_{p,q,\alpha}(\mathcal{X})}$$

with the positive equivalence constants independent of f.

Furthermore, for the Riesz–Morrey space, there exist three open questions unsolved so far. The first question is on the relation between the Riesz–Morrey space and the weak Lebesgue space.

Question 8. *Let* $p \in (1,\infty)$, $q \in [1,p)$, *and* $\alpha \in (\frac{1}{p} - \frac{1}{q}, 0)$. *Then, Zeng et al. ([88], Remark 3.4) showed that*

$$RM_{p,q,\alpha}(\mathbb{R}^n) \not\subseteq L^{\frac{p}{1-p\alpha},\infty}(\mathbb{R}^n) \not\subseteq RM_{p,q,\alpha}(\mathbb{R}^n),$$

which implies that on $\mathbb{R}^n$, *the Riesz–Morrey space and the weak Lebesgue space do not cover each other. Furthermore, for a given cube* Q_0 *of* $\mathbb{R}^n$, *Zeng et al. ([88], Remark 3.6) showed that*

$$L^{\frac{p}{1-p\alpha},\infty}(Q_0) \not\subseteq RM_{p,q,\alpha}(Q_0).$$

However, it is still unknown whether or not

$$RM_{p,q,\alpha}(Q_0) \not\subseteq L^{\frac{p}{1-p\alpha},\infty}(Q_0)$$

still holds true. This question was posed in [88] (Remark 3.6), and is still unclear.

The following Questions 9 and 10 are just [37] (Remarks 4 and 5), respectively.

Question 9. *As a counterpart of* (26), *for any given* $p \in [1,\infty)$, $q \in [1,p)$, $s \in \mathbb{Z}_+$, *and* $\alpha \in [\frac{1}{p} - \frac{1}{q}, 0)$, *it is interesting to ask whether or not*

$$JN_{(p,q,s)_\alpha}(\mathcal{X}) = \left[RM_{p,q,\alpha}(\mathcal{X})/\mathcal{P}_s(\mathcal{X})\right]$$

and, for any $f \in JN_{(p,q,s)_\alpha}(\mathcal{X})$,

$$\|f\|_{JN_{(p,q,s)_\alpha}(\mathcal{X})} \sim \left\|f - \sigma(f)\right\|_{RM_{p,q,\alpha}(\mathcal{X})},$$

with the positive equivalence constants independent of f, *still hold true. This is still* unclear.

Question 10. *Recall that for any given* $f \in L^1_{\mathrm{loc}}(\mathcal{X})$ *and any* $x \in \mathcal{X}$, *the* Hardy–Littlewood maximal function $\mathcal{M}(f)(x)$ *is defined by setting*

$$\mathcal{M}(f)(x) := \sup_{Q \ni x} \fint_Q |f(y)|\,dy, \tag{33}$$

where the supremum is taken over all cubes Q *containing* x. *Meanwhile,* $\mathcal{M}$ *is called the* Hardy–Littlewood maximal operator. *It is well known that* $\mathcal{M}$ *is bounded on* $L^q(\mathcal{X})$ *for any given* $q \in (1,\infty]$ *(see, for instance, [42], p. 31, Theorem 2.5). Moreover,* $\mathcal{M}$ *is also bounded on* $M_q^{-1/\alpha}(\mathcal{X})$ *for any given* $q \in (1,\infty]$ *and* $\alpha \in [-\frac{1}{q}, 0]$ *(see, for instance, [90], Theorem 1). To summarize, the boundedness of* $\mathcal{M}$ *on endpoint spaces of Riesz–Morrey spaces (Lebesgue spaces and Morrey spaces) has already been obtained. Therefore, it is very interesting to ask whether or not* $\mathcal{M}$ *is* *bounded on the Riesz–Morrey space $RM_{p,q,\alpha}(\mathcal{X})$ with $p \in (1,\infty]$, $q \in [1,p)$, and $\alpha \in (\frac{1}{p}-\frac{1}{q},0)$. This is a challenging and important problem which is still open.*

4.2. Congruent Riesz–Morrey Spaces

To obtain the boundedness of several important operators, we next consider a special Riesz–Morrey space via congruent cubes, denoted by $RM_{p,q,\alpha}(\mathbb{R}^n)$, as in Section 3.3. In this subsection, we first recall the definition of $RM^{\rm con}_{p,q,\alpha}(\mathbb{R}^n)$, and then review the boundedness of the Hardy–Littlewood maximal operator on this space.

Definition 16. *Let p, $q \in [1,\infty]$, and $\alpha \in \mathbb{R}$. The* special Riesz–Morrey space via congruent cubes *(for short,* congruent Riesz–Morrey space*) $RM^{\rm con}_{p,q,\alpha}(\mathbb{R}^n)$ is defined to be the set of all locally integrable functions f on $\mathbb{R}^n$, such that*

$$\|f\|_{RM^{\rm con}_{p,q,\alpha}(\mathbb{R}^n)} := \begin{cases} \sup\left[\sum_j |Q_j|^{1-p\alpha-\frac{p}{q}}\|f\|^p_{L^q(Q_j)}\right]^{\frac{1}{p}}, & p\in[1,\infty), \\ \sup\limits_{\text{cube } Q\subset\mathbb{R}^n} |Q|^{-\alpha-\frac{1}{q}}\|f\|_{L^q(Q)}, & p=\infty \end{cases}$$

is finite, where the first supremum is taken over all collections of interior pairwise disjoint cubes $\{Q_j\}_j$ of $\mathbb{R}^n$ with the same side length.

Remark 9.

(i) *If we do not require that $\{Q_j\}_j$ has the same size in the definition of congruent Riesz–Morrey spaces, then it is just the Riesz–Morrey space $RM_{p,q,\alpha}(\mathbb{R}^n)$ in Section 4.1.*

(ii) *If $p=\infty$, $q\in(0,\infty)$, and $\alpha\in[-\frac{1}{q},0)$, then $RM^{\rm con}_{p,q,\alpha}(\mathbb{R}^n)$ in Definition 16 coincides with the Morrey space $M^{-1/\alpha}_q(\mathbb{R}^n)$ in Remark 3.*

(iii) *Similar to Remark 7, for any given p, $q\in[1,\infty)$, and $\alpha\in\mathbb{R}$, $f\in RM^{\rm con}_{p,q,\alpha}(\mathbb{R}^n)$ if and only if $f\in L^1_{\rm loc}(\mathbb{R}^n)$ and*

$$\|f\|_{\widetilde{RM}^{\rm con}_{p,q,\alpha}(\mathbb{R}^n)} := \sup_{r\in(0,\infty)}\left[\int_{\mathbb{R}^n}\left\{|B(y,r)|^{-\alpha}\left[\fint_{B(y,r)}|f(x)|^q dx\right]^{\frac{1}{q}}\right\}^p dy\right]^{\frac{1}{p}}$$

is finite; moreover,

$$\|\cdot\|_{RM^{\rm con}_{p,q,\alpha}(\mathbb{R}^n)} \sim \|\cdot\|_{\widetilde{RM}^{\rm con}_{p,q,\alpha}(\mathbb{R}^n)};$$

see [71] for more details. Recall that for any $y\in\mathbb{R}^n$ and $r\in(0,\infty)$,

$$B(y,r) := \{x\in\mathbb{R}^n : |x-y|<r\}.$$

(iv) *If $1\le q<\alpha<p\le\infty$, then the space $RM^{\rm con}_{p,q,\alpha}(\mathbb{R}^n)$ coincides with the amalgam space $(L^q,\ell^p)^{\frac{p}{1-p\alpha}}(\mathbb{R}^n)$, which was introduced by Fofana [91]. (See [87,92–96] for more studies on the amalgam space.)*

The following boundedness of the Hardy–Littlewood maximal operator on congruent Riesz–Morrey spaces was obtained in [71].

Theorem 10. *Let p, $q\in(1,\infty)$, $\alpha\in\mathbb{R}$, and $\mathcal{M}$ be the Hardy–Littlewood maximal operator as in (33). Then $\mathcal{M}$ is bounded on $RM^{\rm con}_{p,q,\alpha}(\mathbb{R}^n)$.*

Moreover, via Theorem 10, Jia et al. [71] also established the boundedness of Calderón–Zygmund operators on congruent Riesz–Morrey spaces.

Finally, since a congruent Riesz–Morrey space is a *ball Banach function space*, we refer the reader to [49] for the equivalent characterizations of the boundedness and the compactness of Calderón–Zygmund commutators on ball Banach function spaces. It should be mentioned that a crucial assumption in [49] is the boundedness of $\mathcal{M}$, and hence Theorem 10 provides an essential tool when studying the boundedness of operators on congruent Riesz–Morrey spaces.

5. Vanishing Subspace

In this section, we focus on several vanishing subspaces of aforementioned John–Nirenberg-type spaces. In what follows, $C^\infty(\mathbb{R}^n)$ denotes the set of all infinitely differentiable functions on $\mathbb{R}^n$; $\mathbf{0}$ denotes the origin of $\mathbb{R}^n$; for any $\alpha := (\alpha_1, \dots, \alpha_n) \in \mathbb{Z}_+^n := (\mathbb{Z}_+)^n$, let $\partial^\alpha := (\frac{\partial}{\partial x_1})^{\alpha_1} \cdots (\frac{\partial}{\partial x_n})^{\alpha_n}$; for any given normed linear space $\mathcal{Y}$ and any given its subset $\mathcal{X}$, $\overline{\mathcal{X}}^{\mathcal{Y}}$ denotes the *closure* of the set $\mathcal{X}$ in $\mathcal{Y}$ in terms of the topology of $\mathcal{Y}$; and if $\mathcal{Y} = \mathbb{R}^n$, we then denote $\overline{\mathcal{X}}^{\mathcal{Y}}$ simply by $\overline{\mathcal{X}}$.

5.1. Vanishing BMO Spaces

We now recall several vanishing subspaces of the space $\mathrm{BMO}\,(\mathbb{R}^n)$.

- $\mathrm{VMO}\,(\mathbb{R}^n)$, introduced by Sarason [6], is defined by setting

$$\mathrm{VMO}\,(\mathbb{R}^n) := \overline{C_{\mathrm{u}}(\mathbb{R}^n) \cap \mathrm{BMO}\,(\mathbb{R}^n)}^{\mathrm{BMO}\,(\mathbb{R}^n)},$$

where $C_{\mathrm{u}}(\mathbb{R}^n)$ denotes the set of all uniformly continuous functions on $\mathbb{R}^n$.

- $\mathrm{CMO}\,(\mathbb{R}^n)$, announced in Neri [97], is defined by setting

$$\mathrm{CMO}\,(\mathbb{R}^n) := \overline{C_{\mathrm{c}}^\infty(\mathbb{R}^n)}^{\mathrm{BMO}\,(\mathbb{R}^n)},$$

where $C_{\mathrm{c}}^\infty(\mathbb{R}^n)$ denotes the set of all infinitely differentiable functions on $\mathbb{R}^n$ with compact support. In addition, by approximations of the identity, it is easy to find that

$$\mathrm{CMO}\,(\mathbb{R}^n) = \overline{C_{\mathrm{c}}(\mathbb{R}^n)}^{\mathrm{BMO}\,(\mathbb{R}^n)} = \overline{C_0(\mathbb{R}^n)}^{\mathrm{BMO}\,(\mathbb{R}^n)}, \tag{34}$$

where $C_{\mathrm{c}}(\mathbb{R}^n)$ denotes the set of all functions on $\mathbb{R}^n$ with compact support, and $C_0(\mathbb{R}^n)$ denotes the set of all continuous functions on $\mathbb{R}^n$ which vanish at the infinity.

- $\mathrm{MMO}\,(\mathbb{R}^n)$, introduced by Torres and Xue [98], is defined by setting

$$\mathrm{MMO}\,(\mathbb{R}^n) := \overline{A_\infty(\mathbb{R}^n)}^{\mathrm{BMO}\,(\mathbb{R}^n)},$$

where

$$A_\infty(\mathbb{R}^n) := \left\{b \in C^\infty(\mathbb{R}^n) \cap L^\infty(\mathbb{R}^n) :\ \forall\, \alpha \in \mathbb{Z}_+^n \setminus \{\mathbf{0}\},\ \lim_{|x|\to\infty} \partial^\alpha b(x) = 0\right\}.$$

- $\mathrm{XMO}\,(\mathbb{R}^n)$, introduced by Torres and Xue [98], is defined by setting

$$\mathrm{XMO}\,(\mathbb{R}^n) := \overline{B_\infty(\mathbb{R}^n)}^{\mathrm{BMO}\,(\mathbb{R}^n)},$$

where

$$B_\infty(\mathbb{R}^n) := \left\{b \in C^\infty(\mathbb{R}^n) \cap \mathrm{BMO}\,(\mathbb{R}^n) :\ \forall\, \alpha \in \mathbb{Z}_+^n \setminus \{\mathbf{0}\},\ \lim_{|x|\to\infty} \partial^\alpha b(x) = 0\right\}.$$

- $X_1MO(\mathbb{R}^n)$, introduced by Tao et al. [99], is defined by setting

$$X_1MO(\mathbb{R}^n) := \overline{B_1(\mathbb{R}^n)}^{BMO(\mathbb{R}^n)},$$

where

$$B_1(\mathbb{R}^n) := \left\{ b \in C^1(\mathbb{R}^n) \cap BMO(\mathbb{R}^n) : \lim_{|x|\to\infty} |\nabla b(x)| = 0 \right\}$$

with $C^1(\mathbb{R}^n)$ being the set of all functions f on $\mathbb{R}^n$ whose gradients $\nabla f := (\frac{\partial f}{\partial x_1}, \dots, \frac{\partial f}{\partial x_n})$ are continuous.

The relation of these vanishing subspaces reads as follows.

Proposition 29. $CMO(\mathbb{R}^n) \subsetneqq MMO(\mathbb{R}^n) \subsetneqq XMO(\mathbb{R}^n) = X_1MO(\mathbb{R}^n) \subsetneqq VMO(\mathbb{R}^n)$.

Indeed,

$$CMO(\mathbb{R}^n) \subsetneqq MMO(\mathbb{R}^n) \subsetneqq XMO(\mathbb{R}^n)$$

was obtained in [98] (p. 5). Moreover,

$$XMO(\mathbb{R}^n) = X_1MO(\mathbb{R}^n) \subsetneqq VMO(\mathbb{R}^n)$$

was obtained in [99] (Corollary 1.3), which completely answered the open question proposed in [98] (p. 6).

Next, we investigate the mean oscillation characterizations of these vanishing subspaces. Recall that, for any cube Q of $\mathbb{R}^n$, and any $f \in L^1_{loc}(\mathbb{R}^n)$, the *mean oscillation* $O(f;Q)$ is defined by setting

$$O(f;Q) := \fint_Q |f(x) - f_Q|\, dx = \frac{1}{|Q|}\int_Q \left| f(x) - \frac{1}{|Q|}\int_Q f(y)\, dy \right| dx.$$

The earliest results of $VMO(\mathbb{R}^n)$ were obtained by Sarason in [6], and Theorem 11 below is a part of [6] (Theorem 1). In what follows, $a \to 0^+$ means $a \in (0,\infty)$ and $a \to 0$.

Theorem 11. $f \in VMO(\mathbb{R}^n)$ *if and only if* $f \in BMO(\mathbb{R}^n)$ *and*

$$\lim_{a\to 0^+} \sup_{|Q|=a} O(f;Q) = 0.$$

The following equivalent characterization of $CMO(\mathbb{R}^n)$ is just Uchiyama ([7], p. 166).

Theorem 12. $f \in CMO(\mathbb{R}^n)$ *if and only if* $f \in BMO(\mathbb{R}^n)$ *and satisfies the following three conditions:*

(i) $\lim_{a\to 0^+} \sup_{|Q|=a} O(f;Q) = 0$;

(ii) *for any cube* Q *of* $\mathbb{R}^n$, $\lim_{|x|\to\infty} O(f;Q+x) = 0$;

(iii) $\lim_{a\to \infty} \sup_{|Q|=a} O(f;Q) = 0$.

Very recently, Tao et al. obtained the following equivalent characterization of $XMO(\mathbb{R}^n)$ and $X_1MO(\mathbb{R}^n)$, which is just [99] (Theorem 1.2).

Theorem 13. *The following statements are mutually equivalent:*

(i) $f \in X_1MO(\mathbb{R}^n)$;

(ii) $f \in BMO(\mathbb{R}^n)$ *and enjoys the properties that*

  a) $\lim_{a\to 0^+} \sup_{|Q|=a} O(f;Q) = 0$;

b) *for any cube Q of* $\mathbb{R}^n$, $\lim_{|x|\to\infty} \mathcal{O}(f; Q + x) = 0$.

(iii) $f \in \mathrm{XMO}(\mathbb{R}^n)$.

Remark 10. *Proposition 12*(ii) *can be replaced by*

(ii′) $\lim_{M\to\infty} \sup_{Q\cap Q(\mathbf{0},M)=\emptyset} \mathcal{O}(f; Q) = 0,$

where $Q(\mathbf{0}, M)$ *denotes the cube centered at* $\mathbf{0}$ *with the side length M. However,* $(\mathrm{ii})_2$ *of Theorem 13*(ii) *can not be replaced by* (ii′) *(see [99], Proposition 2.5, for more details).*

However, the equivalent characterization of $\mathrm{MMO}(\mathbb{R}^n)$ is still unknown (see [99], Proposition 2.5 and Remark 2.6, for more details on the following open question.)

Question 11. *It is interesting to find the equivalent characterization of* $\mathrm{MMO}(\mathbb{R}^n)$, *as well as its localized counterpart (see Question 14), via the mean oscillations.*

As for the applications of these vanishing subspaces, we know that the commutator $[b, T]$, generated by $b \in \mathrm{BMO}(\mathbb{R}^n)$ and the Calderón–Zygmund operator T, plays an important role in harmonic analysis, complex analysis, partial differential equations, and other fields in mathematics. Here, we only list several typical *bilinear* results; other *linear* and *multi-linear* results can be found, for instance, in [22,100,101] and their references.

In what follows, let $\mathbb{Z}_+^{3n} := (\mathbb{Z}_+)^{3n}$ and $L_c^\infty(\mathbb{R}^n)$ denote the set of all functions $f \in L^\infty(\mathbb{R}^n)$ with compact support. We now consider the following particular type of bilinear Calderón–Zygmund operator T, whose kernel K satisfies

(i) The standard *size* and *regularity* conditions: for any multi-index $\alpha := (\alpha_1, \dots, \alpha_{3n}) \in \mathbb{Z}_+^{3n}$ with $|\alpha| := \alpha_1 + \cdots + \alpha_{3n} \le 1$, there exists a positive constant $C_{(\alpha)}$, depending on α, such that for any $x,\ y,\ z \in \mathbb{R}^n$ with $x \neq y$ or $x \neq z$,

$$|\partial^\alpha K(x, y, z)| \le C_{(\alpha)}(|x - y| + |x - z|)^{-2n-|\alpha|}. \tag{35}$$

Here and thereafter, $\partial^\alpha := (\frac{\partial}{\partial x_1})^{\alpha_1} \cdots (\frac{\partial}{\partial x_{3n}})^{\alpha_{3n}}$.

(ii) The additional decay condition: there exist positive constants C and δ, such that for any $x,\ y,\ z \in \mathbb{R}^n$ with $|x - y| + |x - z| > 1$,

$$|K(x, y, z)| \le C(|x - y| + |x - z|)^{-2n-2-\delta}, \tag{36}$$

and for any $f,\ g \in L_c^\infty(\mathbb{R}^n)$ and $x \notin \mathrm{supp}(f) \cap \mathrm{supp}(g)$, T is supposed to have the following usual representation:

$$T(f, g)(x) = \int_{\mathbb{R}^{2n}} K(x, y, z) f(y) g(z)\, dy\, dz,$$

here and thereafter, $\mathrm{supp}(f) := \{x \in \mathbb{R}^n : f(x) \neq 0\}$. Notice that the (inhomogeneous) Coifman–Meyer bilinear Fourier multipliers and the bilinear pseudodifferential operators with certain symbols satisfy the above two conditions (see, for instance, [98] and references therein).

Recall that, usually, a non-negative measurable function w on $\mathbb{R}^n$ is called a *weight* on $\mathbb{R}^n$. For any given $\mathbf{p} := (p_1, p_2) \in (1, \infty) \times (1, \infty)$, let p satisfy $\frac{1}{p} = \frac{1}{p_1} + \frac{1}{p_2}$. Following [10], we call $\mathbf{w} := (w_1, w_2)$ a *vector* $\mathbf{A_p}(\mathbb{R}^n)$ *weight*, denoted by $\mathbf{w} := (w_1, w_2) \in \mathbf{A_p}(\mathbb{R}^n)$, if

$$[\mathbf{w}]_{\mathbf{A_p}(\mathbb{R}^n)} := \sup_Q \left[\frac{1}{|Q|}\int_Q w(x)\, dx\right] \left\{\frac{1}{|Q|}\int_Q [w_1(x)]^{1-p_1'}\, dx\right\}^{\frac{p}{p_1'}} \times \left\{\frac{1}{|Q|}\int_Q [w_2(x)]^{1-p_2'}\, dx\right\}^{\frac{p}{p_2'}} < \infty,$$

where $w := w_1^{p/p_1} w_2^{p/p_2}$, $1/p_1 + 1/p_1' = 1 = 1/p_2 + 1/p_2'$, and the supremum is taken over all cubes Q of $\mathbb{R}^n$. In what follows, for any given weight w on $\mathbb{R}^n$ and any measurable subset $E \subsetneqq \mathbb{R}^n$, the *symbol* $L_w^p(E)$, with $p \in (0,\infty)$, denotes the set of all measurable functions f on E, such that

$$\|f\|_{L_w^p(E)} := \left[\int_E |f(x)|^p w(x)\,dx\right]^{\frac{1}{p}} < \infty,$$

and, when $w \equiv 1$, we write $L_w^p(E) =: L^p(E)$. Furthermore, $\|\cdot\|_{L^\infty(E)}$ represents the essential supremum on E.

In addition, recall that the *bilinear commutators* $[b,T]_1$ and $[b,T]_2$ are defined, respectively, by setting, for any $f,\ g \in L_c^\infty(\mathbb{R}^n)$ and $x \notin \operatorname{supp}(f) \cap \operatorname{supp}(g)$,

$$\begin{aligned} [b,T]_1(f,g)(x) &:= (bT(f,g) - T(bf,g))(x) \\ &= \int_{\mathbb{R}^{2n}} [b(x)-b(y)]K(x,y,z)f(y)g(z)\,dy\,dz \end{aligned} \tag{37}$$

and

$$\begin{aligned} [b,T]_2(f,g)(x) &:= (bT(f,g) - T(f,bg))(x) \\ &= \int_{\mathbb{R}^{2n}} [b(x)-b(z)]K(x,y,z)f(y)g(z)\,dy\,dz. \end{aligned} \tag{38}$$

The following theorem, obtained in [11] (Theorem 1) for any given $p \in (1,\infty)$ and in [102] (Theorem 1) for any given $p \in (\frac{1}{2}, 1]$, showed that the bilinear commutators $\{[b,T]_i\}_{i=1,2}$ are compact for $b \in \mathrm{CMO}(\mathbb{R}^n)$.

Theorem 14. *Let $(p_1,p_2) \in (1,\infty)\times(1,\infty)$, $p \in (\frac{1}{2},\infty)$ with $\frac{1}{p} = \frac{1}{p_1} + \frac{1}{p_2}$, $b \in \mathrm{CMO}(\mathbb{R}^n)$, and T be a bilinear Calderón–Zygmund operator whose kernel satisfies (35). Then, for any $i \in \{1,2\}$, the bilinear commutator $[b,T]_i$ as in (37) or (38) is compact from $L^{p_1}(\mathbb{R}^n)\times L^{p_2}(\mathbb{R}^n)$ to $L^p(\mathbb{R}^n)$.*

If we require an extra additional decay (36) for the Calderón–Zygmund kernel in Theorem 14, we can then replace $\mathrm{CMO}(\mathbb{R}^n)$ by $\mathrm{XMO}(\mathbb{R}^n)$, that is, delete condition (iii) in Theorem 12 of $\mathrm{CMO}(\mathbb{R}^n)$. This new compactness result was first obtained in [98] (Theorem 1.1) and then generalized into the weighted case, namely the following Theorem 15, which is just [99] (Theorem 1.4).

Theorem 15. *Let $\mathbf{p} := (p_1,p_2) \in (1,\infty)\times(1,\infty)$, $p \in (\frac{1}{2},\infty)$ with $\frac{1}{p} = \frac{1}{p_1} + \frac{1}{p_2}$, $\mathbf{w} := (w_1,w_2) \in \mathbf{A_p}(\mathbb{R}^n)$, $w := w_1^{p/p_1} w_2^{p/p_2}$, $b \in \mathrm{XMO}(\mathbb{R}^n)$, and T be a bilinear Calderón–Zygmund operator whose kernel satisfies (35) and (36). Then, for any $i \in \{1,2\}$, the bilinear commutator $[b,T]_i$ as in (37) or (38) is compact from $L_{w_1}^{p_1}(\mathbb{R}^n)\times L_{w_2}^{p_2}(\mathbb{R}^n)$ to $L_w^p(\mathbb{R}^n)$.*

On the other hand, if the kernel behaves "good", such as the *Riesz transforms* $\{\mathcal{R}_j\}_{j=1}^n$:

$$\mathcal{R}_j(f)(x) := \text{p.v.}\,\pi^{-\frac{n+1}{2}}\Gamma\!\left(\frac{n+1}{2}\right)\int_{\mathbb{R}^n} \frac{y_j}{|y|^{n+1}} f(x-y)\,dy,$$

then the reverse of Theorem 14 holds true as well (see, for instance, the following Theorem 16, which is just [103], Theorem 3.1). Moreover, it should be mentioned that the linear case of Theorem 16 was obtained by Uchiyama ([7], Theorem 2).

Theorem 16. *Let $(p_1,p_2) \in (1,\infty)\times(1,\infty)$ and $p \in (\frac{1}{2},\infty)$ with $\frac{1}{p} = \frac{1}{p_1} + \frac{1}{p_2}$. Then, for any $i \in \{1,2\}$ and $j \in \{1,\dots,n\}$, the bilinear commutator $[b,\mathcal{R}_j]_i$ is compact from $L^{p_1}(\mathbb{R}^n)\times L^{p_2}(\mathbb{R}^n)$ to $L^p(\mathbb{R}^n)$ if and only if $b \in \mathrm{CMO}(\mathbb{R}^n)$.*

However, the corresponding equivalent characterization of XMO $(\mathbb{R}^n)$ is still unknown. For simplicity, we state this question in the unweighted case.

Question 12. *Let* $(p_1, p_2) \in (1, \infty) \times (1, \infty)$, *and* $p \in (\frac{1}{2}, \infty)$ *be such that* $\frac{1}{p} = \frac{1}{p_1} + \frac{1}{p_2}$. *Then, it is interesting to find some bilinear Calderón–Zygmund operator* T, *such that for any* $i \in \{1, 2\}$, *the bilinear commutator* $[b, T]_i$ *is compact from* $L^{p_1}(\mathbb{R}^n) \times L^{p_2}(\mathbb{R}^n)$ *to* $L^p(\mathbb{R}^n)$ if and only if $b \in$ XMO $(\mathbb{R}^n)$.

Next, recall the Riesz transform characterizations of BMO $(\mathbb{R}^n)$ and its vanishing subspaces.

Theorem 17. *Let* $f \in L^1_{\text{loc}}(\mathbb{R}^n)$. *Then,*

(i) *([2], Theorem 3)* $f \in$ BMO $(\mathbb{R}^n)$ *if and only if there exist functions* $\{f_j\}_{j=0}^n \subset L^\infty(\mathbb{R}^n)$, *such that*

$$f = f_0 + \sum_{j=1}^{n} \mathcal{R}_j(f_j)$$

and

$$C^{-1}\|f\|_{\mathrm{BMO}\,(\mathbb{R}^n)} \le \sum_{j=0}^{n} \|f_j\|_{L^\infty(\mathbb{R}^n)} \le C\|f\|_{\mathrm{BMO}\,(\mathbb{R}^n)} \tag{39}$$

for some positive constant C *independent of* f *and* $\{f_j\}_{j=0}^n$.

(ii) *([6], Theorem 1)* $f \in$ VMO $(\mathbb{R}^n)$ *if and only if there exist functions* $\{f_j\}_{j=0}^n \subset [C_u(\mathbb{R}^n) \cap L^\infty(\mathbb{R}^n)]$, *such that*

$$f = f_0 + \sum_{j=1}^{n} \mathcal{R}_j(f_j)$$

and (39) holds true in this case.

(iii) *([97], p. 185)* $f \in$ CMO $(\mathbb{R}^n)$ *if and only if there exist functions* $\{f_j\}_{j=0}^n \subset C_0(\mathbb{R}^n)$, *such that*

$$f = f_0 + \sum_{j=1}^{n} \mathcal{R}_j(f_j)$$

and (39) holds true in this case.

Question 13. *Since the Riesz transform is well defined on* $L^\infty(\mathbb{R}^n)$, *it is interesting to find the counterpart of Theorem 17 when* $f \in$ MMO $(\mathbb{R}^n)$. *Moreover, since the Riesz transform characterization is useful when proving the duality of the CMO-*H^1 *type, it is also interesting to find the dual spaces of* MMO $(\mathbb{R}^n)$ *and* XMO $(\mathbb{R}^n)$.

When $\mathbb{R}^n$ is replaced by some cube Q_0 with finite side length, we then have VMO (Q_0) = CMO (Q_0) (see [104] for more details). Moreover, the vanishing subspace on the *spaces of homogeneous type*, denoted by $\mathfrak{X}$, was studied in Coifman et al. [5], and they proved $(\mathcal{VMO}(\mathfrak{X}))^* = H^1(\mathfrak{X})$, where $\mathcal{VMO}(\mathfrak{X})$ denotes the closure in BMO $(\mathfrak{X})$ of continuous functions on $\mathfrak{X}$ with compact support. Notice that when $\mathfrak{X} = \mathbb{R}^n$, by (34), we have $\mathcal{VMO}(\mathfrak{X}) = \mathcal{VMO}(\mathbb{R}^n) =$ CMO $(\mathbb{R}^n)$.

Finally, we consider the localized version of these vanishing subspaces. The following characterization of local VMO $(\mathbb{R}^n)$ is a part of [105] (Theorem 1).

Proposition 30. *Let* $\mathrm{vmo}\,(\mathbb{R}^n)$ *be the closure of* $C_u(\mathbb{R}^n) \cap \mathrm{bmo}\,(\mathbb{R}^n)$ *in* $\mathrm{bmo}\,(\mathbb{R}^n)$. *Then,* $f \in \mathrm{vmo}\,(\mathbb{R}^n)$ *if and only if* $f \in \mathrm{bmo}\,(\mathbb{R}^n)$ *and*

$$\lim_{a\to 0^+} \sup_{|Q|=a} O(f;Q) = 0.$$

Moreover, the following localized result of $\mathrm{CMO}\,(\mathbb{R}^n)$ is just Dafni ([104], Theorem 6) (see also [105], Theorem 3).

Theorem 18. *Let* $\mathrm{cmo}\,(\mathbb{R}^n)$ *be the closure of* $C_0(\mathbb{R}^n)$ *in* $\mathrm{bmo}\,(\mathbb{R}^n)$. *Then,* $f \in \mathrm{cmo}\,(\mathbb{R}^n)$ *if and only if* $f \in \mathrm{bmo}\,(\mathbb{R}^n)$ *and*

$$\lim_{a\to 0^+} \sup_{|Q|=a} O(f;Q) = 0 = \lim_{M\to\infty} \sup_{|Q|>1,\, Q\cap Q(\mathbf{0},M)=\emptyset} \fint_Q |f|.$$

In addition, the localized version of Theorem 17 can be found in [50] (Corollary 1) for $\mathrm{bmo}\,(\mathbb{R}^n)$, and in [105] (Theorems 1 and 3) for $\mathrm{vmo}\,(\mathbb{R}^n)$ and $\mathrm{cmo}\,(\mathbb{R}^n)$, respectively.

Question 14. *Let* $\mathrm{mmo}\,(\mathbb{R}^n)$, $\mathrm{xmo}\,(\mathbb{R}^n)$, *and* $\mathrm{x_1mo}\,(\mathbb{R}^n)$ *be, respectively, the closure in* $\mathrm{bmo}\,(\mathbb{R}^n)$ *of* $A_\infty(\mathbb{R}^n)$, $B_\infty(\mathbb{R}^n)$, *and* $B_1(\mathbb{R}^n)$. *It is interesting to find the counterparts of*

(i) *Theorem 18 with* $\mathrm{cmo}\,(\mathbb{R}^n)$ *replaced by* $\mathrm{xmo}\,(\mathbb{R}^n)$;
(ii) *Theorem 13 with* $\mathrm{XMO}\,(\mathbb{R}^n)$ *and* $\mathrm{X_1MO}\,(\mathbb{R}^n)$ *replaced, respectively, by* $\mathrm{xmo}\,(\mathbb{R}^n)$ *and* $\mathrm{x_1mo}\,(\mathbb{R}^n)$;
(iii) *Question 13 with* $\mathrm{MMO}\,(\mathbb{R}^n)$ *replaced by* $\mathrm{mmo}\,(\mathbb{R}^n)$;
(iv) *The dual result* $(\mathrm{cmo}\,(\mathbb{R}^n))^* = h^1(\mathbb{R}^n)$, *in ([104], Theorem 9), with* $\mathrm{cmo}\,(\mathbb{R}^n)$ *replaced by* $\mathrm{mmo}\,(\mathbb{R}^n)$ *or* $\mathrm{xmo}\,(\mathbb{R}^n)$, *where* $h^1(\mathbb{R}^n)$ *is the* localized Hardy space;
(v) *The equivalent characterizations for* $\mathrm{mmo}\,(\mathbb{R}^n)$ *and* $\mathrm{xmo}\,(\mathbb{R}^n)$ *via* localized Riesz transforms.

Remark 11. *For the studies of vanishing Morrey spaces, we refer the reader to [106–109].*

5.2. Vanishing John–Nirenberg–Campanato Spaces

Very recently, the vanishing subspaces of John–Nirenberg spaces were also studied in [60,110]. Indeed, as a counterpart of Section 5.1, the vanishing subspaces of JNC spaces enjoy similar characterizations, which are summarized in this subsection.

Definition 17. *Let* $p \in (1,\infty)$, $q \in [1,\infty)$, $s \in \mathbb{Z}_+$, *and* $\alpha \in \mathbb{R}$. *The* vanishing subspace $VJN_{(p,q,s)_\alpha}(\mathcal{X})$ *is defined by setting*

$$VJN_{(p,q,s)_\alpha}(\mathcal{X}) := \left\{ f \in JN_{(p,q,s)_\alpha}(\mathcal{X}) : \limsup_{a\to 0^+} \sup_{\mathrm{size}\le a} \widetilde{O}_{(p,q,s)_\alpha}(f;\{Q_i\}_i) = 0 \right\},$$

where

$$\widetilde{O}_{(p,q,s)_\alpha}(f;\{Q_i\}_i) := \left\{ \sum_i |Q_i| \left[|Q_i|^{-\alpha} \left\{ \fint_{Q_i} \left| f(x) - P^{(s)}_{Q_i}(f)(x) \right|^q dx \right\}^{\frac{1}{q}} \right]^p \right\}^{\frac{1}{p}}$$

and the supremum is taken over all collections of interior pairwise disjoint cubes $\{Q_i\}_i$ *of* $\mathcal{X}$ *with side lengths no more than a. To simplify the notation, write* $VJN_{p,q}(\mathcal{X}) := VJN_{(p,q,0)_0}(\mathcal{X})$ *and* $VJN_p(\mathcal{X}) := VJN_{p,1}(\mathcal{X})$.

On the unit cube $[0,1]^n$, the space $VJN_{(p,q,s)_\alpha}([0,1]^n)$ was studied by A. Brudnyi and Y. Brudnyi in [60] with different symbols. The following characterization (Theorem 19) and duality (Theorem 20) are just, respectively, [60] (Theorem 3.14 and 3.7). Notice that when $\alpha \geq \frac{s+1}{n}$, from [60] (Lemma 4.1), we deduce that $JN_{(p,q,s)_\alpha}([0,1]^n) = \mathcal{P}_s([0,1]^n)$ is trivial.

Theorem 19. *Let $p, q \in [1,\infty)$, $s \in \mathbb{Z}_+$, and $\alpha \in (-\infty, \frac{s+1}{n})$. Then,*

$$VJN_{(p,q,s)_\alpha}([0,1]^n) = \overline{C^\infty([0,1]^n) \cap JN_{(p,q,s)_\alpha}([0,1]^n)}^{JN_{(p,q,s)_\alpha}([0,1]^n)},$$

where $C^\infty([0,1]^n) := C^\infty(\mathbb{R}^n)|_{[0,1]^n}$ denotes the restriction of infinitely differentiable functions from $\mathbb{R}^n$ to $[0,1]^n$.

Theorem 20. *Let $p, q \in (1,\infty)$, $s \in \mathbb{Z}_+$, and $\alpha \in (-\infty, \frac{s+1}{n})$. Then,*

$$\left(VJN_{(p,q,s)_\alpha}([0,1]^n)\right)^* = HK_{(p',q',s)_\alpha}([0,1]^n),$$

where $\frac{1}{p} + \frac{1}{p'} = 1 = \frac{1}{q} + \frac{1}{q'}$.

It is obvious that Theorems 19 and 20 hold true with $[0,1]^n$ replaced by a given cube Q_0 of $\mathbb{R}^n$. As an application of the duality, Tao et al. ([110], Proposition 5.7) showed that for any $p \in (1,\infty)$ and any given cube Q_0 of $\mathbb{R}^n$,

$$[L^p(Q_0)/\mathbb{C}] \subsetneqq VJN_p(Q_0)$$

which proves the *nontriviality* of $VJN_p(Q_0)$, here and thereafter,

$$L^p(\mathcal{X})/\mathbb{C} := \left\{f \in L^1_{\mathrm{loc}}(\mathcal{X}) : \|f\|_{L^p(\mathcal{X})/\mathbb{C}} < \infty\right\}$$

with

$$\|f\|_{L^p(\mathcal{X})/\mathbb{C}} := \inf_{c\in\mathbb{C}} \|f + c\|_{L^p(\mathcal{X})}.$$

Remark 12. *There exists a gap in the proof of [110] (Proposition 5.7): we cannot deduce*

$$\left(VJN_p(Q_0)\right)^{**} = JN_p(Q_0), \tag{40}$$

namely [110] (5.2), directly from Theorems 20 and 3 because, in the statements of these dual theorems, q cannot equal 1. Indeed, (40) still holds true due to the equivalence of $JN_{p,q}(Q_0)$ with $q \in [1,p)$. Precisely, let $p \in (1,\infty)$ and $q \in (1,p)$. By Theorems 20 and 3, we obtain

$$\left(VJN_{p,q}(Q_0)\right)^{**} = JN_{p,q}(Q_0),$$

which, together with Theorems 10 and 21 below, further implies that

$$\left(VJN_p(Q_0)\right)^{**} = \left(VJN_{p,q}(Q_0)\right)^{**} = JN_{p,q}(Q_0) = JN_p(Q_0),$$

and hence (40) holds true. This fixes the gap in the proof of [110] (5.2).

Next, we consider the case $\mathcal{X} = \mathbb{R}^n$. The following proposition indicates that the convolution is a suitable tool when approximating functions in $JN_p(\mathbb{R}^n)$, which is a counterpart of [6] (Lemma 1). Indeed, the approximate functions in the proofs of both Theorems 21 and 22 are constructed via the convolution (see [110] for more details).

Proposition 31. *Let $p \in (1,\infty)$ and $\varphi \in L^1(\mathbb{R}^n)$ with compact support. If $f \in JN_p(\mathbb{R}^n)$, then $f * \varphi \in JN_p(\mathbb{R}^n)$ and*

$$\|f * \varphi\|_{JN_p(\mathbb{R}^n)} \leq 2\|\varphi\|_{L^1(\mathbb{R}^n)}\|f\|_{JN_p(\mathbb{R}^n)}.$$

Proof. Let p, φ, and f be as in this lemma. Then, for any cube Q of $\mathbb{R}^n$, by the Fubini theorem, we have

$$
\begin{aligned}
O(f*\varphi;Q) &= \fint_Q \left|f*\varphi(x)-(f*\varphi)_Q\right|dx \\
&= \fint_Q \left|\fint_Q \int_{\mathbb{R}^n} \varphi(z)[f(x-z)-f(y-z)]\,dz\,dy\right|dx \\
&\le \int_{\mathbb{R}^n} \fint_Q \fint_Q |\varphi(z)||f(x-z)-f(y-z)|\,dy\,dx\,dz \\
&= \int_{\mathbb{R}^n} |\varphi(z)| \fint_{Q-z} \fint_{Q-z} |f(x)-f(y)|\,dy\,dx\,dz \\
&\le 2\int_{\mathbb{R}^n} |\varphi(z)| O(f;Q-z)\,dz, \qquad (41)
\end{aligned}
$$

where $Q-z := \{w-z:\ w\in Q\}$. Therefore, for any interior pairwise disjoint subcubes $\{Q_i\}_i$ of $\mathbb{R}^n$, by (41) and the generalized Minkowski integral inequality, we conclude that

$$
\begin{aligned}
&\left\{\sum_i |Q_i|[O(f*\varphi;Q_i)]^p\right\}^{\frac{1}{p}} \\
&\quad\le 2\left\{\sum_i |Q_i|\left[\int_{\mathbb{R}^n} |\varphi(z)|O(f;Q-z)\,dz\right]^p\right\}^{\frac{1}{p}} \\
&\quad= 2\left\{\sum_i \left[\int_{\mathbb{R}^n} |Q_i|^{\frac{1}{p}} |\varphi(z)|O(f;Q_i-z)\,dz\right]^p\right\}^{\frac{1}{p}} \\
&\quad\le 2\int_{\mathbb{R}^n} \left\{\sum_i \left[|Q_i|^{\frac{1}{p}} |\varphi(z)|O(f;Q_i-z)\right]^p\right\}^{\frac{1}{p}} dz \\
&\quad= 2\int_{\mathbb{R}^n} |\varphi(z)| \left\{\sum_i |Q_i-z|[O(f;Q_i-z)]^p\right\}^{\frac{1}{p}} dz \\
&\quad\le 2\|\varphi\|_{L^1(\mathbb{R}^n)}\|f\|_{JN_p(\mathbb{R}^n)},
\end{aligned}
$$

where $Q_i-z := \{w-z:\ w\in Q_i\}$ for any i. This further implies that

$$
\|f*\varphi\|_{JN_p(\mathbb{R}^n)} \le 2\|\varphi\|_{L^1(\mathbb{R}^n)}\|f\|_{JN_p(\mathbb{R}^n)}
$$

and hence finishes the proof of Proposition 31. □

The following equivalent characterization is just [110] (Theorem 3.2).

Theorem 21. *Let $p\in(1,\infty)$. Then, the following three statements are mutually equivalent:*

(i) $f\in \overline{D_p(\mathbb{R}^n)\cap JN_p(\mathbb{R}^n)}^{JN_p(\mathbb{R}^n)} =: VJN_p(\mathbb{R}^n)$, *where*

$$
D_p(\mathbb{R}^n) := \{f\in C^\infty(\mathbb{R}^n):\ |\nabla f|\in L^p(\mathbb{R}^n)\}
$$

and ∇f denotes the gradient of f;

(ii) $f\in JN_p(\mathbb{R}^n)$ *and, for any given $q\in[1,p)$,*

$$
\lim_{a\to 0^+} \sup_{\{\{Q_i\}_i:\ \ell(Q_i)\le a,\ \forall i\}} \left\{\sum_i |Q_i|\left[\fint_{Q_i} |f(x)-f_{Q_i}|^q\,dx\right]^{\frac{p}{q}}\right\}^{\frac{1}{p}} = 0,
$$

where the supremum is taken over all collections $\{Q_i\}_i$ *of interior pairwise disjoint subcubes of* $\mathbb{R}^n$ *with side lengths no more than a;*

(iii) $f \in JN_p(\mathbb{R}^n)$ *and*

$$\lim_{a\to 0^+} \sup_{\{\{Q_i\}_i:\ \ell(Q_i)\le a,\ \forall i\}} \left\{\sum_i |Q_i| \left[\fint_{Q_i} \left| f(x) - f_{Q_i} \right| dx \right]^p \right\}^{\frac{1}{p}} = 0,$$

where the supremum is taken over all collections $\{Q_i\}_i$ *of interior pairwise disjoint subcubes of* $\mathbb{R}^n$ *with side lengths no more than a.*

Now, we recall another vanishing subspace of $JN_p(\mathbb{R}^n)$ introduced in [110], which is of the CMO type.

Definition 18. *Let* $p \in (1,\infty)$*. The* vanishing subspace $CJN_p(\mathbb{R}^n)$ *of* $JN_p(\mathbb{R}^n)$ *is defined by setting*

$$CJN_p(\mathbb{R}^n) := \overline{C_c^\infty(\mathbb{R}^n)}^{JN_p(\mathbb{R}^n)},$$

where $C_c^\infty(\mathbb{R}^n)$ *denotes the set of all infinitely differentiable functions on* $\mathbb{R}^n$ *with compact support.*

The following theorem is just [110] (Theorem 4.3).

Theorem 22. *Let* $p \in (1,\infty)$*. Then,* $f \in CJN_p(\mathbb{R}^n)$ *if and only if* $f \in JN_p(\mathbb{R}^n)$*, and* f *satisfies the following two conditions:*

(i)

$$\lim_{a\to 0^+} \sup_{\{\{Q_i\}_i:\ \ell(Q_i)\le a,\ \forall i\}} \left\{\sum_i |Q_i| \left[\fint_{Q_i} \left| f(x) - f_{Q_i} \right| dx \right]^p \right\}^{\frac{1}{p}} = 0,$$

where the supremum is taken over all collections $\{Q_i\}_i$ *of interior pairwise disjoint subcubes of* $\mathbb{R}^n$ *with side lengths* $\{\ell(Q_i)\}_i$ *no more than a;*

(ii)

$$\lim_{a\to \infty} \sup_{\{Q\subset\mathbb{R}^n:\ \ell(Q)\ge a\}} |Q|^{1/p} \fint_Q \left| f(x) - f_Q \right| dx = 0,$$

where the supremum is taken over all cubes Q of $\mathbb{R}^n$ *with side lengths* $\ell(Q)$ *no less than a.*

Moreover, Tao et al. ([110], Theorem 4.4) showed that Theorem 22(ii) can be replaced by the following statement:

$$\lim_{a\to \infty} \sup_{\{\{Q_i\}_i:\ \ell(Q_i)\ge a,\ \forall i\}} \left\{\sum_i |Q_i| \left[\fint_{Q_i} \left| f(x) - f_{Q_i} \right| dx \right]^p \right\}^{\frac{1}{p}} = 0,$$

where the supremum is taken over all collections $\{Q_i\}_i$ of interior pairwise disjoint subcubes of $\mathbb{R}^n$ with side lengths $\{\ell(Q_i)\}_i$ greater than a.

Furthermore, Tao et al. ([110], Corollary 4.5) showed that Theorem 22 holds true with

$$\fint_Q \left| f(x) - f_Q \right| dx \quad \text{and} \quad \fint_{Q_i} \left| f(x) - f_{Q_i} \right| dx$$

in (i) and (ii) replaced, respectively, by

$$\left[\fint_Q \left| f(x) - f_Q \right|^q dx \right]^{\frac{1}{q}} \quad \text{and} \quad \left[\fint_{Q_i} \left| f(x) - f_{Q_i} \right|^q dx \right]^{\frac{1}{q}}$$

for any $q \in [1,p)$.

However, there still exist some unsolved questions on the vanishing John–Nirenberg space. The first question is on the case $p = 1$.

Question 15. *The proof of [110] (Theorem 3.2) indicates that (i) and (iii) of Theorem 21 are equivalent when $p = 1$. However, the corresponding equivalent characterization of $CJN_1(\mathbb{R}^n)$ is still* unclear.

The following question is just [110] (Question 5.5).

Question 16.

(i) *It is still unknown whether or not Theorems 21 and 22 hold true with $JN_p(\mathbb{R}^n)$ replaced by $JN_{(p,q,s)_\alpha}(\mathbb{R}^n)$ when $p, q \in [1, \infty)$, $s \in \mathbb{Z}_+$, and $\alpha \in \mathbb{R} \setminus \{0\}$.*

(ii) *It is interesting to ask whether or not for any given $p \in (1, \infty)$, $q \in [1, \infty)$, $s \in \mathbb{Z}_+$, and $\alpha \in \mathbb{R}$,*

$$\left(CJN_{(p,q,s)_\alpha}(\mathbb{R}^n)\right)^* = HK_{(p',q',s)_\alpha}(\mathbb{R}^n) \quad \text{or} \quad \left(CJN_{(p,q,s)_\alpha}(\mathbb{R}^n)\right)^{**} = JN_{(p,q,s)_\alpha}(\mathbb{R}^n)$$

still holds true, where $1/p + 1/p' = 1 = 1/q + 1/q'$, $CJN_{(p,q,s)_\alpha}(\mathbb{R}^n)$ denotes the closure of $C_c^\infty(\mathbb{R}^n)$ in $JN_{(p,q,s)_\alpha}(\mathbb{R}^n)$, and $HK_{(p',q',s)_\alpha}(\mathbb{R}^n)$ the Hardy-type space introduced in [61] (Definition 3.6).

Obviously, $[L^p(\mathbb{R}^n)/\mathbb{C}] \subset CJN_p(\mathbb{R}^n) \subset VJN_p(\mathbb{R}^n) \subset JN_p(\mathbb{R}^n)$. Then, the last question naturally arises, which is just [110] (Questions 5.6 and 5.8).

Question 17. *Let $p \in (1, \infty)$. It is interesting to ask whether or not*

$$[L^p(\mathbb{R}^n)/\mathbb{C}] \subsetneqq CJN_p(\mathbb{R}^n) \subsetneqq VJN_p(\mathbb{R}^n) \subsetneqq JN_p(\mathbb{R}^n)$$

holds true. This is still unclear.

5.3. Vanishing Congruent John–Nirenberg–Campanato Spaces

As a counterpart of Section 5.2, the vanishing subspace of congruent John–Nirenberg–Campanato spaces $VJN^{\rm con}_{(p,q,s)_\alpha}(\mathcal{X})$ was studied in [64].

Definition 19. *Let $p, q \in [1, \infty)$, $s \in \mathbb{Z}_+$, and $\alpha \in \mathbb{R}$. The* space *$VJN^{\rm con}_{(p,q,s)_\alpha}(\mathcal{X})$ is defined by setting*

$$VJN^{\rm con}_{(p,q,s)_\alpha}(\mathcal{X}) := \overline{D_p(\mathcal{X}) \cap JN^{\rm con}_{(p,q,s)_\alpha}(\mathcal{X})}^{JN^{\rm con}_{(p,q,s)_\alpha}(\mathcal{X})},$$

where

$$D_p(\mathcal{X}) := \{f \in C^\infty(\mathcal{X}) : \ |\nabla f| \in L^p(\mathcal{X})\}.$$

Furthermore, simply write $VJN^{\rm con}_{p,q}(\mathcal{X}) := VJN^{\rm con}_{(p,q,0)_0}(\mathcal{X})$ and $VJN^{\rm con}_p(\mathcal{X}) := VJN^{\rm con}_{p,1}(\mathcal{X})$.

Remark 13. *Let $p, q \in [1, \infty)$, $s \in \mathbb{Z}_+$, $\alpha \in \mathbb{R}$, and Q_0 be a given cube of $\mathbb{R}^n$. Then, the observation $D_p(Q_0) = C^\infty(Q_0)$ implies that*

$$VJN^{\rm con}_{(p,q,s)_\alpha}(Q_0) = \overline{C^\infty(Q_0) \cap JN^{\rm con}_{(p,q,s)_\alpha}(Q_0)}^{JN^{\rm con}_{(p,q,s)_\alpha}(Q_0)}.$$

Recall that $\mathcal{D}_m(\mathcal{X})$ with $m \in \mathbb{Z}$ is defined in the beginning of Section 3.3. The following characterizations, namely Theorems 23 and 24, are just [64] (Theorems 3.5 and 3.9, respectively).

Theorem 23. *Let $p, q \in [1, \infty)$, $s \in \mathbb{Z}_+$, $\alpha \in (-\infty, \frac{s+1}{n})$, and Q_0 be a given cube of $\mathbb{R}^n$. Then, $f \in VJN^{\rm con}_{(p,q,s)_\alpha}(Q_0)$ if and only if $f \in L^q(Q_0)$ and*

$$\limsup_{m\to\infty} \sup_{\{Q_j\}_j \subset \mathcal{D}_m(Q_0)} \left[\sum_j |Q_j| \left\{ |Q_j|^{-\alpha} \left[\fint_{Q_j} \left| f - P^{(s)}_{Q_j}(f) \right|^q \right]^{\frac{1}{q}} \right\}^p \right]^{\frac{1}{p}} = 0, \tag{42}$$

where the second supremum is taken over all collections of interior pairwise disjoint cubes $\{Q_j\}_j \subset \mathcal{D}_m(Q_0)$ for any $m \in \mathbb{Z}$.

Corollary 4. *Let $p = 1$, $q \in [1, \infty)$, $s \in \mathbb{Z}_+$, $\alpha = 0$, and Q_0 be a given cube of $\mathbb{R}^n$. Then, (42) holds true for any $f \in L^q(Q_0)$.*

Proof. By Proposition 24(ii) and the definition of $VJN^{\rm con}_{(p,q,s)_\alpha}(Q_0)$, we have

$$[L^q(Q_0)/\mathcal{P}_s(Q_0)] = VJN^{\rm con}_{(p,q,s)_\alpha}(Q_0) = JN^{\rm con}_{(p,q,s)_\alpha}(Q_0),$$

which, combined with Theorem 23, then completes the proof of Corollary 4. □

Theorem 24. *Let $p \in [1, \infty)$ and $q \in [1, p]$. Then, $f \in VJN^{\rm con}_{p,q}(\mathbb{R}^n)$ if and only if $f \in JN^{\rm con}_{p,q}(\mathbb{R}^n)$ and*

$$\limsup_{m\to\infty} \sup_{\{Q_j\}_j \subset \mathcal{D}_m(\mathbb{R}^n)} \left[\sum_j |Q_j| \left(\fint_{Q_j} \left| f - f_{Q_j} \right|^q \right)^{\frac{p}{q}} \right]^{\frac{1}{p}} = 0,$$

where the second supremum is taken over all collections of interior pairwise disjoint cubes $\{Q_j\}_j \subset \mathcal{D}_m(\mathbb{R}^n)$ for any $m \in \mathbb{Z}$.

We can partially answer Question 17 in the congruent JNC space as follows.

Proposition 32. *Let I_0 be a given bounded interval of $\mathbb{R}$, and Q_0 a given cube of $\mathbb{R}^n$.*

(i) *([64], Proposition 3.11) If $p \in (1, \infty)$ and $q \in [1, p)$, then $[L^p(\mathbb{R})/\mathbb{C}] \subsetneqq VJN^{\rm con}_{p,q}(\mathbb{R})$.*
(ii) *([64], Proposition 3.12) If $p \in (1, \infty)$ and $q \in [1, p)$, then $VJN^{\rm con}_{p,q}(\mathbb{R}) \subsetneqq JN^{\rm con}_{p,q}(\mathbb{R})$ and $VJN^{\rm con}_{p,q}(I_0) \subsetneqq JN^{\rm con}_{p,q}(I_0)$.*
(iii) *([64], Proposition 4.40) If $p \in (1, \infty)$ and $q \in (1, p)$, then $[L^p(Q_0)/\mathbb{C}] \subsetneqq VJN^{\rm con}_{p,q}(Q_0)$.*

Furthermore, it is easy to show that $[L^1(Q_0)/\mathbb{C}] = VJN^{\rm con}_1(Q_0) = JN^{\rm con}_1(Q_0)$ (see Remark 2(ii)).

The following VMO-H^1-type duality is just [64] (Theorem 4.39).

Theorem 25. *Let $p, q \in (1, \infty)$, $s \in \mathbb{Z}_+$, $\frac{1}{p} + \frac{1}{p'} = 1 = \frac{1}{q} + \frac{1}{q'}$, $\alpha \in (-\infty, \frac{s+1}{n})$, and Q_0 be a given cube of $\mathbb{R}^n$. Then,*

$$\left(VJN^{\rm con}_{(p,q,s)_\alpha}(Q_0)\right)^* = HK^{\rm con}_{(p',q',s)_\alpha}(Q_0)$$

in the following sense: there exists an isometric isomorphism

$$K:\ HK^{\rm con}_{(p',q',s)_\alpha}(Q_0) \longrightarrow \left(VJN^{\rm con}_{(p,q,s)_\alpha}(Q_0)\right)^*$$

such that for any $g \in HK^{\rm con}_{(p',q',s)_\alpha}(Q_0)$ and $f \in VJN^{\rm con}_{(p,q,s)_\alpha}(Q_0)$,

$$\langle Kg, f\rangle = \langle g, f\rangle.$$

Similar to Question 16(ii), the following question, posed in [64] (Remark 4.41), is still unsolved.

Question 18. *For any given $p, q \in (1,\infty)$, $s \in \mathbb{Z}_+$, and $\alpha \in (-\infty, \frac{s+1}{n})$, it is interesting to ask whether or not*

$$\left(CJN^{\mathrm{con}}_{(p,q,s)_\alpha}(\mathbb{R}^n)\right)^* = HK^{\mathrm{con}}_{(p',q',s)_\alpha}(\mathbb{R}^n) \quad \text{and} \quad \left(CJN^{\mathrm{con}}_{(p,q,s)_\alpha}(\mathbb{R}^n)\right)^{**} = JN^{\mathrm{con}}_{(p,q,s)_\alpha}(\mathbb{R}^n)$$

hold true, where $CJN^{\mathrm{con}}_{(p,q,s)_\alpha}(\mathbb{R}^n)$ denotes the closure of $C_c^\infty(\mathbb{R}^n)$ in $JN^{\mathrm{con}}_{(p,q,s)_\alpha}(\mathbb{R}^n)$ and $\frac{1}{p}+\frac{1}{p'} = 1 = \frac{1}{q}+\frac{1}{q'}$. This is still unclear.

Author Contributions: Conceptualization, J.T., D.Y. and W.Y.; methodology, J.T., D.Y. and W.Y.; software, J.T., D.Y. and W.Y.; validation, J.T., D.Y. and W.Y.; formal analysis, J.T., D.Y. and W.Y.; investigation, J.T., D.Y. and W.Y.; resources, J.T., D.Y. and W.Y.; data curation, J.T., D.Y. and W.Y.; writing—original draft preparation, J.T., D.Y. and W.Y.; writing—review and editing, J.T., D.Y. and W.Y.; visualization, J.T., D.Y. and W.Y.; supervision, J.T., D.Y. and W.Y.; project administration, J.T., D.Y. and W.Y.; funding acquisition, J.T., D.Y. and W.Y. All authors have read and agreed to the published version of the manuscript.

Funding: This research was funded by the National Natural Science Foundation of China (Grant Nos. 11971058, 12071197, 12122102, and 11871100) and the National Key Research and Development Program of China (Grant No. 2020YFA0712900).

Institutional Review Board Statement: Not applicable.

Informed Consent Statement: Not applicable.

Data Availability Statement: Not applicable.

Acknowledgments: Jin Tao would like to thank Hongchao Jia and Jingsong Sun for some useful discussions on this survey. The authors would also like to thank the referees for their carefully reading and valuable remarks, which improved the presentation of this article.

Conflicts of Interest: The authors declare no conflict of interest.

References

1. John, F.; Nirenberg, L. On functions of bounded mean oscillation. *Commun. Pure Appl. Math.* **1961**, *14*, 415–426. [CrossRef]
2. Fefferman, C.; Stein, E.M. H^p spaces of several variables. *Acta Math.* **1972**, *129*, 137–193. [CrossRef]
3. Coifman, R.R.; Rochberg, R.; Weiss, G. Factorization theorems for Hardy spaces in several variables. *Ann. Math.* **1976**, *103*, 611–635. [CrossRef]
4. Coifman, R.R.; Weiss, G. *Analyse Harmonique Non-Commutative sur Certains Espaces Homogènes. (French) Étude de Certaines Intégrales Singulières*; Lecture Notes in Mathematics; Springer: Berlin, Germany; New York, NY, USA, 1971; Volume 242.
5. Coifman, R.R.; Weiss, G. Extensions of Hardy spaces and their use in analysis. *Bull. Am. Math. Soc.* **1977**, *83*, 569–645. [CrossRef]
6. Sarason, D. Functions of vanishing mean oscillation. *Trans. Am. Math. Soc.* **1975**, *207*, 391–405. [CrossRef]
7. Uchiyama, A. On the compactness of operators of Hankel type. *Tôhoku Math. J.* **1978**, *30*, 163–171. [CrossRef]
8. Nakai, E.; Yabuta, K. Pointwise multipliers for functions of bounded mean oscillation. *J. Math. Soc. Jpn.* **1995**, *37*, 207–218.
9. Iwaniec, T. L^p-theory of quasiregular mappings. In *Quasiconformal Space Mappings*; Lecture Notes in Mathematics; Springer: Berlin/Heidelberg, Germany, 1992; Volume 1508, pp. 39–64.
10. Bényi, A.; Damián, W.; Moen, K.; Torres, R.H. Compact bilinear commutators: The weighted case. *Mich. Math. J.* **2015**, *64*, 39–51. [CrossRef]
11. Bényi, A.; Torres, R.H. Compact bilinear operators and commutators. *Proc. Am. Math. Soc.* **2013**, *141*, 3609–3621. [CrossRef]
12. Chang, D.-C.; Li, S.-Y. On the boundedness of multipliers, commutators and the second derivatives of Green's operators on H^1 and BMO. *Ann. Scuola Norm. Sup. Pisa Cl. Sci.* **1999**, *28*, 341–356.
13. Janson, S. Mean oscillation and commutators of singular integral operators. *Ark. Mat.* **1978**, *16*, 263–270. [CrossRef]
14. Li, W.; Nakai, E.; Yang, D. Pointwise multipliers on BMO spaces with non-doubling measures. *Taiwan. J. Math.* **2018**, *22*, 183–203. [CrossRef]
15. Nakai, E. Pointwise multipliers on several functions spaces—A survey. *Linear Nonlinear Anal.* **2017**, *3*, 27–59.
16. Nakai, E. The Campanato, Morrey and Hölder spaces on spaces of homogeneous type. *Stud. Math.* **2006**, *176*, 1–19. [CrossRef]
17. Nakai, E. Pointwise multipliers on weighted BMO spaces. *Stud. Math.* **1997**, *125*, 35–56. [CrossRef]
18. Nakai, E. Pointwise multipliers for functions of weighted bounded mean oscillation. *Stud. Math.* **1993**, *105*, 105–119. [CrossRef]
19. Nakai, E.; Yabuta, K. Pointwise multipliers for functions of weighted bounded mean oscillation on spaces of homogeneous type. *Math. Japon.* **1997**, *46*, 15–28.

20. Chang, D.-C.; Dafni, G.; Stein, E.M. Hardy spaces, BMO, and boundary value problems for the Laplacian on a smooth domain in $\mathbb{R}^n$. *Trans. Am. Math. Soc.* **1999**, *351*, 1605–1661. [CrossRef]
21. Nakai, E.; Yoneda, T. Bilinear estimates in dyadic BMO and the Navier–Stokes equations. *J. Math. Soc. Jpn.* **2012**, *64*, 399–422. [CrossRef]
22. Tao, J.; Yang, D.; Yang, D. Beurling–Ahlfors commutators on weighted Morrey spaces and applications to Beltrami equations. *Potential Anal.* **2020**, *53*, 1467–1491. [CrossRef]
23. Ambrosio, L.; Bourgain, J.; Brezis, H.; Figalli, A. BMO-type norms related to the perimeter of sets. *Commun. Pure Appl. Math.* **2016**, *69*, 1062–1086. [CrossRef]
24. Butaev, A.; Dafni, G. Approximation and extension of functions of vanishing mean oscillation. *J. Geom. Anal.* **2021**, *31*, 6892–6921. [CrossRef]
25. Chang, D.-C.; Dafni, G.; Sadosky, C. A div-curl lemma in BMO on a domain. In *Harmonic Analysis, Signal Processing, and Complexity*; Progress in Mathematics; Birkhäuser Boston: Boston, MA, USA, 2005; Volume 238.
26. Dafni, G.; Gibara, R. BMO on shapes and sharp constants. In *Advances in Harmonic Analysis and Partial Differential Equations*; Contemporary Mathematics; American Mathematical Society: Providence, RI, USA, 2020; Volume 748, pp. 1–33.
27. Dafni, G.; Gibara, R.; Lavigne, A. BMO and the John–Nirenberg inequality on measure spaces. *Anal. Geom. Metr. Spaces* **2020**, *8*, 335–362. [CrossRef]
28. Dafni, G.; Gibara, R.; Yue, H. Geometric maximal operators and BMO on product bases. *J. Geom. Anal.* **2021**, *31*, 5740–5765. [CrossRef]
29. Chang, D.-C.; Sadosky, C. Functions of bounded mean oscillation. *Taiwan. J. Math.* **2006**, *10*, 573–601. [CrossRef]
30. Aalto, D.; Berkovits, L.; Kansanen, O.E.; Yue, H. John–Nirenberg lemmas for a doubling measure. *Stud. Math.* **2011**, *204*, 21–37. [CrossRef]
31. Dafni, G.; Hytönen, T.; Korte, R.; Yue, H. The space JN_p: Nontriviality and duality. *J. Funct. Anal.* **2018**, *275*, 577–603. [CrossRef]
32. Berkovits, L.; Kinnunen, J.; Martell, J.M. Oscillation estimates, self-improving results and good-λ inequalities. *J. Funct. Anal.* **2016**, *270*, 3559–3590. [CrossRef]
33. Blasco, O.; Espinoza-Villalva, C. The norm of the characteristic function of a set in the John–Nirenberg space of exponent p. *Math. Methods Appl. Sci.* **2020**, *43*, 9327–9336. [CrossRef]
34. Hurri-Syrjänen, R.; Marola, N.; Vähäkangas, A.V. Aspects of local-to-global results. *Bull. Lond. Math. Soc.* **2014**, *46*, 1032–1042. [CrossRef]
35. Milman, M. Marcinkiewicz spaces, Garsia–Rodemich spaces and the scale of John–Nirenberg self improving inequalities. *Ann. Acad. Sci. Fenn. Math.* **2016**, *41*, 491–501. [CrossRef]
36. Sun, J.; Xie, G.; Yang, D. Localized John–Nirenberg–Campanato spaces. *Anal. Math. Phys.* **2021**, *11*, 29. [CrossRef]
37. Tao, J.; Yang, D.; Yuan, W. A bridge connecting Lebesgue and Morrey spaces via Riesz norms. *Banach J. Math. Anal.* **2021**, *15*, 20. [CrossRef]
38. Franchi, B.; Pérez, C.; Wheeden, R.L. Self-improving properties of John–Nirenberg and Poincaré inequalities on spaces of homogeneous type. *J. Funct. Anal.* **1998**, *153*, 108–146. [CrossRef]
39. MacManus, P.; Pérez, C. Generalized Poincaré inequalities: Sharp self-improving properties. *Internat. Math. Res. Notices* **1998**, *1998*, 101–116. [CrossRef]
40. Marola, N.; Saari, O. Local to global results for spaces of BMO type. *Math. Z.* **2016**, *282*, 473–484. [CrossRef]
41. Riesz, F. Untersuchungen über systeme integrierbarer funktionen. *Math. Ann.* **1910**, *69*, 449–497. (In German) [CrossRef]
42. Duoandikoetxea, J. *Fourier Analysis*; Graduate Studies in Mathematics; American Mathematical Society: Providence, RI, USA, 2001; Volume 29.
43. Stein, E.M. *Harmonic Analysis: Real-Variable Methods, Orthogonality, and Oscillatory Integrals*; Princeton Mathematical Series, Monographs in Harmonic Analysis III; Princeton University Press: Princeton, NJ, USA, 1993; Volume 43.
44. Bényi, A.; Martell, J.M.; Moen, K.; Stachura, E.; Torres, R.H. Boundedness results for commutators with BMO functions via weighted estimates: A comprehensive approach. *Math. Ann.* **2020**, *376*, 61–102. [CrossRef]
45. Bennett, C.; Sharpley, R. *Interpolation of Operators*; Pure and Applied Mathematics; Academic Press Inc.: Boston, MA, USA, 1988; Volume 129.
46. Grafakos, L. *Morden Fourier Analysis*, 3rd ed.; Graduate Texts in Mathematics; Springer: New York, NY, USA, 2014; Volume 250.
47. Grafakos, L. *Classical Fourier Analysis*, 3rd ed.; Graduate Texts in Mathematics; Springer: New York, NY, USA, 2014; Volume 249.
48. Izuki, M.; Noi, T.; Sawano, Y. The John–Nirenberg inequality in ball Banach function spaces and application to characterization of BMO. *J. Inequal. Appl.* **2019**, *2019*, 268. [CrossRef]
49. Tao, J.; Yang, D.; Yuan, W.; Zhang, Y. Compactness characterizations of commutators on ball Banach function spaces. *Potential Anal.* **2021**, in press.
50. Goldberg, D. A local version of real Hardy spaces. *Duke Math. J.* **1979**, *46*, 27–42. [CrossRef]
51. Jonsson, A.; Sjögren, P.; Wallin, H. Hardy and Lipschitz spaces on subsets of $\mathbb{R}^n$. *Stud. Math.* **1984**, *80*, 141–166. [CrossRef]
52. Chang, D.-C. The dual of Hardy spaces on a bounded domain in $\mathbb{R}^n$. *Forum Math.* **1994**, *6*, 65–81. [CrossRef]
53. Dafni, G.; Liflyand, E. A local Hilbert transform, Hardy's inequality and molecular characterization of Goldberg's local Hardy space. *Complex Anal. Synerg.* **2019**, *5*, 10. [CrossRef]

54. Dafni, G.; Yue, H. Some characterizations of local bmo and h^1 on metric measure spaces. *Anal. Math. Phys.* **2012**, *2*, 285–318. [CrossRef]
55. Garsia, A.M.; Rodemich, E. Monotonicity of certain functionals under rearrangement. *Ann. Inst. Fourier (Grenoble)* **1974**, *24*, 67–116. [CrossRef]
56. Stampacchia, G. The spaces $L^{(p,\lambda)}$, $N^{(p,\lambda)}$ and interpolation. *Ann. Scuola Norm. Sup. Pisa Cl. Sci.* **1965**, *19*, 443–462.
57. Campanato, S. Su un teorema di interpolazione di G. Stampacchia. *Ann. Scuola Norm. Sup. Pisa Cl. Sci.* **1966**, *20*, 649–652.
58. Myyryläinen, K. Median-type John–Nirenberg space in metric measure spaces. *arXiv* **2021**, arXiv:2104.05380.
59. Kinnunen, J.; Myyryläinen, K. Dyadic John–Nirenberg space. *arXiv* **2021**, arXiv:2107.00492.
60. Brudnyi, A.; Brudnyi, Y. On the Banach structure of multivariate BV spaces. *Diss. Math.* **2020**, *548*, 1–52. [CrossRef]
61. Tao, J.; Yang, D.; Yuan, W. John–Nirenberg–Campanato spaces. *Nonlinear Anal.* **2019**, *189*, 111584. [CrossRef]
62. Domínguez, Ó.; Milman, M. Sparse Brudnyi and John–Nirenberg spaces. *arXiv* **2021**, arXiv:2107.05117.
63. Astashkin, S.; Milman, M. Garsia–Rodemich spaces: Local maximal functions and interpolation. *Stud. Math.* **2020**, *255*, 1–26. [CrossRef]
64. Jia, H.; Tao, J.; Yuan, W.; Yang, D.; Zhang, Y. Special John–Nirenberg–Campanato spaces via congruent cubes. *Sci. China Math.* **2021**. [CrossRef]
65. Taibleson, M.H.; Weiss, G. The molecular characterization of certain Hardy spaces. In *Representation Theorems for Hardy Spaces, Astérisque 77*; Société Mathématique de France: France, Paris, 1980; pp. 67–149.
66. Lu, S. *Four Lectures on Real H^p Spaces*; World Scientific Publishing Co., Inc.: River Edge, NJ, USA, 1995.
67. Campanato, S. Proprietà di una famiglia di spazi funzionali. *Ann. Scuola Norm. Sup. Pisa Cl. Sci.* **1964**, *18*, 137–160.
68. Morrey, C.B. On the solutions of quasi-linear elliptic partial differential equations. *Trans. Am. Math. Soc.* **1938**, *43*, 126–166. [CrossRef]
69. García-Cuerva, J.; Rubio de Francia, J. *Weighted Norm Inequalities and Related Topics*; North-Holland Mathematics Studies, Notas de Matemática [Mathematical Notes]; North-Holland Publishing Co.: Amsterdam, The Netherlands, 1985; Volume 116.
70. Bourgain, J.; Brezis, H.; Mironescu, P. A new function space and applications. *J. Eur. Math. Soc. (JEMS)* **2015**, *17*, 2083–2101. [CrossRef]
71. Jia, H.; Tao, J.; Yuan, W.; Yang, D.; Zhang, Y. Boundedness of Calderón–Zygmund operators on special John–Nirenberg–Campanato and Hardy-type spaces via congruent cubes. *Anal. Math. Phys.* **2021**, under review.
72. Jia, H.; Tao, J.; Yuan, W.; Yang, D.; Zhang, Y. Boundedness of fractional integrals on special John–Nirenberg–Campanato and Hardy-type spaces via congruent cubes. *Front. Math. China* **2021**, under review.
73. Jia, H.; Yuan, W.; Yang, D.; Zhang, Y. Estimates for Littlewood–Paley operators on special John–Nirenberg–Campanato spaces via congruent cubes. *arXiv* **2021**, arXiv:2108.01559.
74. Fusco, N.; Moscariello, G.; Sbordone, C. BMO-type seminorms and Sobolev functions. *ESAIM Control Optim. Calc. Var.* **2018**, *24*, 835–847. [CrossRef]
75. Di Fratta, G.; Fiorenza, A. BMO-type seminorms from Escher-type tessellations. *J. Funct. Anal.* **2020**, *279*, 108556. [CrossRef]
76. Essén, M.; Janson, S.; Peng, L.; Xiao, J. Q spaces of several real variables. *Indiana Univ. Math. J.* **2000**, *49*, 575–615. [CrossRef]
77. Koskela, P.; Xiao, J.; Zhang, Y.; Zhou, Y. A quasiconformal composition problem for the Q-spaces. *J. Eur. Math. Soc. (JEMS)* **2017**, *19*, 1159–1187. [CrossRef]
78. Xiao, J. *Q_α Analysis on Euclidean Spaces*; Advances in Analysis and Geometry 1; De Gruyter: Berlin, Germany, 2019.
79. Zhuo, C.; Sickel, W.; Yang, D.; Yuan, W. Characterizations of Besov-type and Triebel–Lizorkin-type spaces via averages on balls. *Canad. Math. Bull.* **2017**, *60*, 655–672. [CrossRef]
80. Tao, J.; Yang, Z.; Yuan, W. John–Nirenberg-Q spaces via congruent cubes. 2021, submitted.
81. Hakim, D.I.; Nakamura, S.; Sawano, Y. Complex interpolation of smoothness Morrey subspaces. *Constr. Approx.* **2017**, *46*, 489–563. [CrossRef]
82. Hakim, D.I.; Sawano, Y. Complex interpolation of various subspaces of Morrey spaces. *Sci. China Math.* **2020**, *63*, 937–964. [CrossRef]
83. Mastyło, M.; Sawano, Y. Complex interpolation and Calderón–Mityagin couples of Morrey spaces. *Anal. PDE* **2019**, *12*, 1711–1740. [CrossRef]
84. Mastyło, M.; Sawano, Y.; Tanaka, H. Morrey-type space and its Köthe dual space. *Bull. Malays. Math. Sci. Soc.* **2018**, *41*, 1181–1198. [CrossRef]
85. Sawano, Y.; Di Fazio, G.; Hakim, D. *Morrey Spaces: Introduction and Applications to Integral Operators and PDE's*; Chapman and Hall/CRC: New York, NY, USA, 2020; Volume I. Available online: https://www.taylorfrancis.com/books/mono/10.1201/9780429085925/morrey-spaces-yoshihiro-sawano-giuseppe-di-fazio-denny-ivanal-hakim (accessed on 8 September 2021).
86. Sawano, Y.; Di Fazio, G.; Hakim, D. *Morrey Spaces: Introduction and Applications to Integral Operators and PDE's*; Chapman and Hall/CRC: New York, NY, USA, 2020; Volume II. Available online: https://www.taylorfrancis.com/books/mono/10.1201/9781003029076/morrey-spaces-yoshihiro-sawano-giuseppe-di-fazio-denny-ivanal-hakim (accessed on 8 September 2021).
87. Fofana, I.; Faléa, F.R.; Kpata, B.A. A class of subspaces of Morrey spaces and norm inequalities on Riesz potential operators. *Afr. Mat.* **2015**, *26*, 717–739. [CrossRef]
88. Zeng, Z.; Chang, D.-C.; Tao, J.; Yang, D. Nontriviality of Riesz–Morrey spaces. *Appl. Anal.* **2021**. 021.1932836. [CrossRef]

89. Blasco, O.; Ruiz, A.; Vega, L. Non-interpolation in Morrey–Campanato and block spaces. *Ann. Scuola Norm. Sup. Pisa Cl. Sci.* **1999**, *28*, 31–40.
90. Chiarenza, F.; Frasca, M. Morrey spaces and Hardy–Littlewood maximal function. *Rend. Mat. Appl.* **1987**, *7*, 273–279.
91. Fofana, I. Étude d'une classe d'espace de fonctions contenant les espaces de Lorentz. *Afr. Mat.* **1988**, *1*, 29–50.
92. Adama, A.; Feuto, J.; Fofana, I. A weighted inequality for potential type operators. *Adv. Pure Appl. Math.* **2019**, *10*, 413–426. [CrossRef]
93. Coulibaly, S.; Fofana, I. On Lebesgue integrability of Fourier transforms in amalgam spaces. *J. Fourier Anal. Appl.* **2019**, *25*, 184–209. [CrossRef]
94. Diarra, N.; Fofana, I. On preduals and Köthe duals of some subspaces of Morrey spaces. *J. Math. Anal. Appl.* **2021**, *496*, 124842. [CrossRef]
95. Dosso, M.; Fofana, I.; Sanogo, M. On some subspaces of Morrey–Sobolev spaces and boundedness of Riesz integrals. *Ann. Polon. Math.* **2013**, *108*, 133–153. [CrossRef]
96. Feuto, J.; Fofana, I.; Koua, K. Integrable fractional mean functions on spaces of homogeneous type. *Afr. Diaspora J. Math. (N.S.)* **2010**, *9*, 8–30.
97. Neri, U. Fractional integration on the space H^1 and its dual. *Stud. Math.* **1975**, *53*, 175–189. [CrossRef]
98. Torres, R.H.; Xue, Q. On compactness of commutators of multiplication and bilinear pesudodifferential operators and a new subspace of BMO. *Rev. Mat. Iberoam.* **2020**, *36*, 939–956. [CrossRef]
99. Tao, J.; Xue, Q.; Yang, D.; Yuan, W. XMO and weighted compact bilinear commutators. *J. Fourier Anal. Appl.* **2021**, *27*, 60. [CrossRef]
100. Li, K. Multilinear commutators in the two-weight setting. *arXiv* **2020**, arXiv:2006.09071.
101. Tao, J.; Yang, D.; Yang, D. Boundedness and compactness characterizations of Cauchy integral commutators on Morrey spaces. *Math. Methods Appl. Sci.* **2019**, *42*, 1631–1651. [CrossRef]
102. Torres, R.H.; Xue, Q.; Yan, J. Compact bilinear commutators: The quasi-Banach space case. *J. Anal.* **2018**, *26*, 227–234. [CrossRef]
103. Chaffee, L.; Chen, P.; Han, Y.; Torres, R.H.; Ward, L.A. Characterization of compactness of commutators of bilinear singular integral operators. *Proc. Am. Math. Soc.* **2018**, *146*, 3943–3953. [CrossRef]
104. Dafni, G. Local VMO and weak convergence in h^1. *Canad. Math. Bull.* **2002**, *45*, 46–59. [CrossRef]
105. Bourdaud, G. Remarques sur certains sous-espaces de $\mathrm{BMO}(\mathbb{R}^n)$ et de $\mathrm{bmo}(\mathbb{R}^n)$. (French. English, French summary) [Remarks on some subspaces of $\mathrm{BMO}(\mathbb{R}^n)$ and $\mathrm{bmo}(\mathbb{R}^n)$]. *Ann. Inst. Fourier (Grenoble)* **2002**, *52*, 1187–1218. [CrossRef]
106. Alabalik, A.; Almeida, A.; Samko, S. On the invariance of certain vanishing subspaces of Morrey spaces with respect to some classical operators. *Banach J. Math. Anal.* **2020**, *14*, 987–1000. [CrossRef]
107. Alabalik, A.; Almeida, A.; Samko, S. Preservation of certain vanishing properties of generalized Morrey spaces by some classical operators. *Math. Methods Appl. Sci.* **2020**, *43*, 9375–9386. [CrossRef]
108. Almeida, A. Maximal commutators and commutators of potential operators in new vanishing Morrey spaces. *Nonlinear Anal.* **2020**, *192*, 111684. [CrossRef]
109. Almeida, A.; Samko, S. Approximation in Morrey spaces. *J. Funct. Anal.* **2017**, *272*, 2392–2411. [CrossRef]
110. Tao, J.; Yang, D.; Yuan, W. Vanishing John–Nirenberg spaces. *Adv. Calc. Var.* **2021**. [CrossRef]

mathematics

MDPI

Article

Weighted Estimates for Iterated Commutators of Riesz Potential on Homogeneous Groups

Daimei Chen , Yanping Chen * and Teng Wang

Department of Applied Mathematics, School of Mathematics and Physics, University of Science and Technology Beijing, Beijing 100083, China; daimeich@126.com (D.C.); tengwustb@126.com (T.W.)
* Correspondence: yanpingch@ustb.edu.cn

Abstract: In this paper, we study the two weight commutators theorem of Riesz potential on an arbitrary homogeneous group $\mathbb{H}$ of dimension N. Moreover, in accordance with the results in the Euclidean space, we acquire the quantitative weighted bound on homogeneous group.

Keywords: commutators; Riesz potential; homogeneous group

Citation: Chen, D.; Chen, Y.; Wang, T. Weighted Estimates for Iterated Commutators of Riesz Potential on Homogeneous Groups. *Mathematics* **2021**, 9, 2421. https://doi.org/10.3390/math9192421

Academic Editor: Christopher Goodrich

Received: 24 August 2021
Accepted: 23 September 2021
Published: 29 September 2021

1. Introduction and Main Results

Suppose $\mathbb{H}$ is a nilpotent Lie group, which has the multiplication, inverse, expansion and norm configurations $(x,y) \mapsto xy, x \mapsto x^{-1}, (t,x) \mapsto t \circ x, x \mapsto \rho(x)$ for $x, y \in \mathbb{H}, t > 0$, respectively, then we call $\mathbb{H}$ being a homogeneous group (see [1] or [2]). The multiplication and inverse operations are polynomials and t-action is an automorphism of the group structure, where t is of the form

$$t \circ (x_1, \ldots, x_n) = (t^{\beta_1} x_1, \ldots, t^{\beta_n} x_n)$$

for some constants $0 < \beta_1 \le \beta_2 \le \ldots \le \beta_n$. Besides, $\rho(x) := \max_{1 \le j \le n} \{|x_j|^{1/\beta_j}\}$ is a norm linked to the expansion configuration. We call the value $N = \sum_{j=1}^{n} \beta_j$ the dimensionality of $\mathbb{H}$. In addition to the Euclidean structure, $\mathbb{H}$ is equipped with a homogeneous nilpotent Lie group structure, where Lebesgue measure is a bi-invariant Haar measure, the identity is the origin $0, x^{-1} = -x$ and multiplication $xy, x, y \in \mathbb{H}$, satisfies

(1) $(ax)(bx) = ax + bx, x \in \mathbb{H}, a, b \in \mathbb{R}$;

(2) $t \circ (xy) = (t \circ x)(t \circ y), x, y \in \mathbb{H}, t > 0$;

(3) if $z = xy$, then $z_k = P_k(x,y)$, where $P_1(x,y) = x_1 + y_1$ and $P_k(x,y) = x_k + y_k + P_k(x,y)$ for $k \ge 2$ with a polynomial $P_k(x,y)$ depending only on $x_1, \cdots, x_{k-1}, y_1, \cdots, y_{k-1}$.

Finally, the Heisenberg group on $\mathbb{R}^3$ is an example of a homogeneous group. If we define the multiplication

$$(x,y,u)(x',y',u') = (x + x', y + y', u + u' + (xy' - yx')/2),$$

$(x,y,u)(x',y',u') \in \mathbb{R}^3$, the $\mathbb{R}^3$ with this group law is the Heisenberg group $\mathbb{H}_1$; a dilation is defined by $t \circ (x,y,u) = (tx, ty, t^2 u)$, that is the parameters $\beta_1 = 1, \beta_2 = 1, \beta_3 = 2$.

Definition 1. *Let $w(x)$ is a function on $\mathbb{H}$, which is non-negative locally integrable. For $1 < p < \infty$, we call that w is an A_p weight, denoted by $w \in A_p$, if*

$$[w]_{A_p} := \sup_B \left(\frac{1}{|B|}\int_B w(x)dx\right)\left(\frac{1}{|B|}\int_B \left(\frac{1}{w}\right)^{\frac{1}{p-1}} dx\right)^{p-1} < \infty,$$

The supremum here is taken over of all balls $B \subset \mathbb{H}$. We call that the quantity $[w]_{A_p}$ is the A_p constant of w. For $p = 1$, if $M(w)(x) \le cw(x)$ for a.e.$x \in \mathbb{H}$, then we say that w is an A_1 weight, *denoted by $w \in A_1$, where M represents the Hardy-Littlewood maximal function. In addition, let $A_\infty := \cup_{1 \le p \le \infty} A_p$, then we have*

$$[w]_{A_\infty} := \sup_B \left(\frac{1}{|B|} \int_B w dx \right) \exp\left(\frac{1}{|B|} \int_B log(\frac{1}{w}) dx \right) < \infty.$$

Definition 2. *Let $x \in \mathbb{H}$, and $w(x)$ be a non-negative locally integrable function. For $1 < p < q < \infty$, $w \in A_{p,q}$ if*

$$[w]_{A_{p,q}} := \sup_B \left(\frac{1}{|B|} \int_B w^q \right) \left(\frac{1}{|B|} \int_B w^{-p'} \right)^{\frac{q}{p'}} < \infty,$$

where p' is the conjugate exponent of p, that is $\frac{1}{p} + \frac{1}{p'} = 1$.

Definition 3. *Suppose $w \in A_\infty$. Let $b \in L^1_{loc}(\mathbb{H})$, then $b(x) \in BMO_w(\mathbb{H})$ if*

$$\|b\|_{BMO_w(\mathbb{H})} := \sup_B \frac{1}{w(B)} \int_B |b(x) - b_B| dx < \infty,$$

where $b_B := \frac{1}{|B|} \int_B b(x) dx$ and the supremum is taken over of all balls $B \subset \mathbb{H}$.

We now review the definition of Riesz potential on homogeneous group. For $0 < \alpha < N$,

$$I_\alpha f(x) := \int_{\mathbb{H}} \frac{f(y)}{\rho(xy^{-1})^{N-\alpha}} dy,$$

and the corresponding associated maximal function M_α by

$$M_\alpha f(x) = \sup_{x \in B} \frac{1}{|B|^{1-\frac{\alpha}{N}}} \int_B |f(y)| dy.$$

The reason why we study the weighted estimates for these operators is because they have a wide range of applications in partial differential equations, Sobolev embeddings or quantum mechanics (see [3] or [4]).

Muckenhoupt and Wheeden [5] are the first scholars to study the Riesz potential. When $\mathbb{H}$ is an isotropic Euclidean space, Muckenhoupt and Wheeden [5] show that I_α is bounded from $L^p(w^p)$ to $L^q(w^q)$ for $1 < p < \frac{n}{\alpha}, \frac{1}{q} = \frac{1}{p} - \frac{\alpha}{n}, w \in A_{p,q}$. Moreover, the sharp constant in this inequality was given in [6]:

$$\|I_\alpha\|_{L^p(w^p) \to L^q(w^q)} \le C[w]_{A_{p,q}}^{(1-\frac{\alpha}{n}) \max(1, \frac{p'}{q})}.$$

Definition 4. *Suppose $b \in L^1_{loc}(\mathbb{H}), f \in L^p(\mathbb{H})$. Let $[b, I_\alpha]$ be the commutator defined by*

$$[b, I_\alpha] f(x) := b(x) I_\alpha(f)(x) - I_\alpha(bf)(x).$$

The iterative commutators $(I_\alpha)_b^m, m \in \mathbb{N}$, are defined naturally by

$$(I_\alpha)_b^m f(x) := [b, (I_\alpha)_b^{m-1}] f(x), (I_\alpha)_b^1 f(x) := [b, I_\alpha] f(x).$$

In 2016, Holmes, Rahm and Spencer [7] prove that

$$[b, I_\alpha] : L^p_{w^p}(\mathbb{R}^n) \to L^q_{\lambda^q}(\mathbb{R}^n) \Leftrightarrow b \in BMO_\mu(\mathbb{R}^n),$$

where $1 < p < \frac{n}{\alpha}, \frac{1}{q} = \frac{1}{p} - \frac{\alpha}{n}, w, \lambda \in A_{p,q}, \mu = \frac{w}{\lambda}$. Later, the quantitative estimates for iterated commutators of fractional integrals was obtained by N. Accomazzo, J. C. Martínez-Perales and I. P. Rivera-Ríos [8].

In 2013, Sato [9] gave the estimates for singular integrals on homogeneous groups. In [10], X. T. Duong, H. Q. Li and J. Li established the Bloom-type two weight estimates for the commutator of Riesz transform on stratified Lie groups. Moreover, Z. Fan and J. Li [11] obtained the quantitative weighted estimates for rough singular integrals on homogeneous groups.

Motivated by the above estimates, we investigate the quantitative weighted estimation for the higher order commutators of fractional integral operators on homogeneous groups.

In this paper, our main result is the follow theorem.

Theorem 1. *Let $0 < \alpha < N$ and $1 < p < \frac{N}{\alpha}$, q defined by $\frac{1}{q} + \frac{\alpha}{N} = \frac{1}{p}$, and m is a positive integer. Assume that $\mu, \lambda \in A_{p,q}$ and that $\nu = \frac{\mu}{\lambda}$.*

1. *If $b \in BMO_{\nu^{1/m}}(\mathbb{H})$, then*

$$\|(I_\alpha)_b^m f\|_{L^q_{\lambda^q}(\mathbb{H})} \le C_{m,N,\alpha,p} \|b\|^m_{BMO_{\nu^{1/m}}(\mathbb{H})} \kappa_m \|f\|_{L^p_{\mu^p}(\mathbb{H})}, \tag{1}$$

where

$$\kappa_m = \sum_{k=0}^{m} \binom{m}{k} \left([\lambda]^{\frac{k}{m}}_{A_{p,q}} [\mu]^{\frac{m-k}{m}}_{A_{p,q}} \right)^{(1-\frac{\alpha}{N})\max\{1,\frac{p'}{q}\}} A(m,k)B(m,k)$$

and

$$A(m,k) \le \left([\lambda^q]^{\frac{m+k+1}{2}}_{A_q} [\mu^q]^{\frac{m-k-1}{2}}_{A_q} \right)^{\frac{m-k}{m}\max\{1,\frac{1}{q-1}\}},$$

$$B(m,k) \le \left([\lambda^p]^{\frac{k-1}{2}}_{A_p} [\mu^p]^{m-\frac{k-1}{2}}_{A_p} \right)^{\frac{k}{m}\max\{1,\frac{1}{p-1}\}}.$$

2. *For every $b \in L^1_{loc}(\mathbb{H})$, if $(I_\alpha)_b^m$ is bounded from $L^p_{\mu^p}(\mathbb{H})$ to $L^q_{\lambda^q}(\mathbb{H})$, then $b \in BMO_{\nu^{1/m}}(\mathbb{H})$ with*

$$\|b\|^m_{BMO_{\nu^{1/m}}(\mathbb{H})} \lesssim \|(I_\alpha)_b^m\|_{L^p_{\mu^p}(\mathbb{H}) \to L^q_{\lambda^q}(\mathbb{H})}.$$

2. Domination of the Iterated Commutators by Sparse Operators

2.1. A System of Dyadic Cubes

We define a left-unchanged analogous-distance d on $\mathbb{H}$ by $d(x,y) = \rho(x^{-1}y)$, which signifies that there has a constant $A_0 \ge 1$ such that for any $x, y, z \in \mathbb{H}$,

$$d(x,y) \le A_0[d(x,z) + d(z,y)].$$

Next, let $B(x,r) := \{y \in \mathbb{H} : d(x,y) < r\}$ be the open ball which is centered on $x \in \mathbb{H}$ and $r > 0$ is the radius.

Let $\mathscr{A}_k$ be k-th denumerable index set. A denumerable class $\mathcal{D} := \cup_{k\in\mathbb{Z}}\mathcal{D}_k, \mathcal{D}_k := \{Q^k_\beta : \beta \in \mathscr{A}_k\}$, of Borel sets $Q^k_\beta \subseteq \mathbb{H}$ is known as a set of dyadic cubes with arguments $\delta \in (0,1)$ and $0 < a_1 \le A_1 < \infty$ if it has the characteristics below:

(1) $\mathbb{H} = \cup_{\beta\in\mathscr{A}_k} Q^k_\beta$ (disjoint union) for all $k \in \mathbb{Z}$;

(2) If $\ell \ge k$, then either $Q^\ell_\gamma \subseteq Q^k_\beta$ or $Q^k_\beta \cap Q^\ell_\gamma = \varnothing$;

(3) For arbitrary (k,β) and for any $\ell \le k$, there is a exclusive γ such that $Q^k_\beta \subseteq Q^\ell_\gamma$;

(4) For arbitrary (k,β) there exists no more that M (a settled geometric constant) γ such that $Q_\gamma^{k+1} \subseteq Q_\beta^k$, and $Q_\beta^k = \cup_{Q\in\mathcal{D}_{k+1}, Q\subseteq Q_\beta^k} Q$;

(5) $B(x_\beta^k, a_1\delta^k) \subseteq Q_\beta^k \subseteq B(x_\beta^k, A_1\delta^k) =: B(Q_\beta^k)$;

(6) If $\ell \geq k$ and $Q_\gamma^\ell \subseteq Q_\beta^k$, then $B(Q_\gamma^\ell) \subseteq B(Q_\beta^k)$. The set Q_β^k is called a dyadic cube of generation k with centre $x_\beta^k \in Q_\beta^k$ and side length $\ell(Q_\beta^k) = \delta^k$.

From the natures of the dyadic system above, for any Q_β^k, Q_γ^{k+1} and $Q_\gamma^{k+1} \subset Q_\beta^k$, we get that there is a constant $\widetilde{A}_0 > 0$ such that:

$$|Q_\gamma^{k+1}| \leq |Q_\beta^k| \leq \widetilde{A}_0 |Q_\gamma^{k+1}|.$$

2.2. Adjacent Systems of Dyadic Cubes

Let $\{\mathcal{D}^t : t = 1,2,\ldots,\mathcal{T}\}$ be a limited set of the dyadic families, then we call that it is a collection of neighbor systems of dyadic cubes with arguments $\delta \in (0,1)$, $0 < a_1 \leq A_1 < \infty$ and $1 \leq C_{adj} < \infty$ if it has the following two characteristics:

(1) For any $t \in \{1,2,\ldots,\mathcal{T}\}$, $\mathcal{D}^t$ is a system of dyadic cubes with arguments $\delta \in (0,1)$ and $0 < a_1 \leq A_1 < \infty$;

(2) For any ball $B(x,r) \subseteq \mathbb{H}$ with $\delta^{k+3} < r \leq \delta^{k+2}, k \in \mathbb{Z}$, there have $t \in \{1,2,\ldots,\mathcal{T}\}$ and $Q \in \mathcal{D}^t$ of generation k which is centered on ${}^t x_\beta^k$ such that $d(x, {}^t x_\beta^k) < 2A_0\delta^k$ and

$$B(x,r) \subseteq Q \subseteq B(x, C_{adj} r). \tag{2}$$

2.3. Sparse Operators

We review the concept of sparse family given in [12] on ordinary spaces of homogeneous description in the sense of Coifman and Weiss [13], which is also suitable in the case of homogeneous groups.

Definition 5. *Let $0 < \eta < 1$, for every $Q \in \mathcal{S}$, we call that the collection $\mathcal{S} \subset \mathcal{D}$ of dyadic cubes be a η-sparse, if there exists a measurable subset $E_Q \subset Q$ such that $|E_Q| \geq \eta|Q|$ and the sets $\{E_Q\}_{Q\in\mathcal{S}}$ have only limited overlap.*

Definition 6. *Given a sparse family, the sparse operator $\mathcal{A}_\mathcal{S}$ is defined by*

$$\mathcal{A}_\mathcal{S}(f)(x) = \sum_{Q\in\mathcal{S}} \langle f\rangle_Q \chi_Q(x),$$

where $\langle f\rangle_Q = \frac{1}{|Q|}\int_Q f(x)dx$.

In this subfraction, the primary target is to reveal the following quantitative edition of Lacey's pointwise domination inequality.

Proposition 1. *Let $0 < \alpha < N$. Let m be a nonnegative integer. For every $f \in C_c^\infty(\mathbb{H})$ and $b \in L^m_{loc}(\mathbb{H})$, there exits $\mathcal{T}$ dyadic systems $\mathcal{D}^t, t = 1,2,\ldots,\mathcal{T}$ and η-sparse families $\mathcal{S}_t \subset \mathcal{D}^t$ such that for a.e.$x \in \mathbb{H}$,*

$$|(I_\alpha)_b^m f| \leq C_{N,m,\alpha} \sum_{t=1}^{\mathcal{T}} \sum_{k=0}^{m} \binom{m}{k} \mathcal{A}^{m,k}_{\alpha,\mathcal{S}_t}(b,f)(x), \quad a.e. x \in \mathbb{H}, \tag{3}$$

where for a sparse family $\mathcal{S}$, $\mathcal{A}^{m,k}_{\alpha,\mathcal{S}}(b,\cdot)$ is the sparse operator given by

$$\mathcal{A}^{m,k}_{\alpha,\mathcal{S}}(b,f)(x) = \sum_{Q\in\mathcal{S}} |b(x) - b_Q|^{m-k} |Q|^{\frac{\alpha}{N}} \langle f(b-b_Q)^k\rangle_Q \chi_Q(x).$$

To show the Proposition 1, we need some auxiliary maximal operators. To begin with, let $\widetilde{j_0}$ be the smallest integer such that

$$2^{\widetilde{j_0}} > \max\{3A_0, 2A_0C_{adj}\} \tag{4}$$

and let $C_{\widetilde{j_0}} := 2^{\widetilde{j_0}+2}A_0$.

Next we define the grand maximal truncated operator $\mathcal{M}_{I_\alpha}$ as follows:

$$\mathcal{M}_{I_\alpha}f(x) = \sup_{x\in B} \operatorname*{ess\,sup}_{\xi\in B} |I_\alpha(f\chi_{\mathbb{H}\setminus C_{\widetilde{j_0}}B})(\xi)|,$$

where the first supremum is taken over of all balls $B \subset \mathbb{H}$ satisfying $x \in B$. We can know that this operator is of vital importance in the following proof, Given a ball $B_0 \subset \mathbb{H}$, for $x \in B_0$ we also define a local edition of $\mathcal{M}_{I_\alpha}$ by

$$\mathcal{M}_{I_\alpha,B_0}f(x) = \sup_{x\in B\subset B_0} \operatorname*{ess\,sup}_{\xi\in B} |I_\alpha(f\chi_{C_{\widetilde{j_0}}B_0\setminus C_{\widetilde{j_0}}B})(\xi)|.$$

Now, we claim that the following lemma is true.

Lemma 1. *Let $0 < \alpha < N$. The following pointwise estimates holds:*

1. *For a.e.$x \in B_0$,*

$$|I_\alpha(f\chi_{C_{\widetilde{j_0}}B_0})(x)| \leq \mathcal{M}_{I_\alpha,B_0}f(x).$$

2. *There exists a constant $C_{N,\alpha} > 0$ such that for a.e.$x \in \mathbb{H}$,*

$$\mathcal{M}_{I_\alpha}f(x) \leq C_{N,\alpha}\Big(M_\alpha f(x) + I_\alpha|f|(x)\Big).$$

Using the results of Lemma 1, we then prove the Proposition 1.

Proof of Proposition 1. In order to proof the Proposition 1, we refer to the thinking in [8] for this domination, which is adapted to our situation of homogeneous groups.

Firstly, we suppose that f is supported in a ball $B_0 := B(x_0, r) \subset \mathbb{H}$, next we disintegrate $\mathbb{H}$ which respect to this ball B_0. We can do it as follows. We start define the annuli $U_j := 2^{j+1}B_0 \setminus 2^jB_0, j \geq 0$ and select the minimum integer j_0 such that

$$j_0 > \widetilde{j_0} \quad \text{and} \quad 2^{j_0} > 4A_0 \tag{5}$$

Next, for any U_j, we select the balls

$$\{\widetilde{B}_{j,\ell}\}_{\ell-1}^{L_j}, \tag{6}$$

centred in U_j and with radius $2^{j-\widetilde{j_0}}r$ to cover U_j. From the doubling property [13], we obtain

$$\sup_j L_j \leq C_{A_0,\widetilde{j_0}}, \tag{7}$$

where $C_{A_0,\widetilde{j_0}}$ is an positive constant that only relates on A_0 and $\widetilde{j_0}$.

We now go over the characters of these $\widetilde{B}_{j,\ell}$. Denote $\widetilde{B}_{j,\ell} := B(x_{j,\ell}, 2^{j-\widetilde{j_0}}r)$, where $\widetilde{j_0}$ is defines as in (4). Then we have $C_{adj}\widetilde{B}_{j,\ell} := B(x_{j,\ell}, C_{adj}2^{j-\widetilde{j_0}}r)$, which was shown in the proof of Theorem 3.7 in [12] that

$$C_{adj}\widetilde{B}_{j,\ell} \cap U_{j+j_0} = \emptyset, \quad \forall j \geq 0 \quad \text{and} \quad \forall \ell = 1, 2, \ldots, L_j; \tag{8}$$

and

$$C_{adj}\widetilde{B}_{j,\ell} \cap U_{j-j_0} = \emptyset, \quad \forall j \geq j_0 \quad \text{and} \quad \forall \ell = 1,2,\ldots,L_j. \tag{9}$$

Now, because of the Equation (8) and (9), we see that each $C_{adj}\widetilde{B}_{j,\ell}$, at most overlap with $2j_0+1$ annuli U_j's. Moreover, for every j and ℓ, $C_{\tilde{j_0}}\widetilde{B}_{j,\ell}$ covers B_0.

Next by observing the (2), there is an integer $t_0 \in \{1,2,\ldots,\mathcal{T}\}$ and $Q_0 \in \mathcal{D}^{t_0}$ such that $B_0 \subseteq Q_0 \subseteq C_{adj}B_0$. Additionally, for this Q_0, as in Section 2.1 the ball that includes Q_0 and has comparable measure to Q_0 is represented by $B(Q_0)$. Consequently, B_0 is overwritten by $B(Q_0)$ and $|B(Q_0)| \lesssim |B_0|$, where the implicit constant relates only to C_{adj} and A_1.

Now we claim that there exists a $\frac{1}{2}$-sparse family $\mathcal{F}^{t_0} \subset \mathcal{D}^{t_0}(Q_0)$, the set of all dyadic cubes in t_0-th dyadic system that are contained in Q_0, such that for *a.e.* $x \in B_0$,

$$|(I_\alpha)_b^m(f\chi_{C_{\tilde{j_0}}B(Q_0)})(x)| \leq C_{N,m,\alpha}\sum_{k=0}^{m}\binom{m}{k}\mathcal{B}^{m,k}_{\alpha,\mathcal{F}^{t_0}}(b,f)(x), \tag{10}$$

where

$$\mathcal{B}^{m,k}_{\alpha,\mathcal{F}^{t_0}}(b,f)(x) = \sum_{Q\in\mathcal{F}^{t_0}}|b(x)-b_{R_Q}|^{m-k}|C_{\tilde{j_0}}B(Q)|^{\frac{\alpha}{N}}\langle f(b-b_{R_Q})^k\rangle_{C_{\tilde{j_0}}B(Q)}\chi_Q(x).$$

Here, R_Q is the dyadic cube in $\mathscr{D}^t$ for some $t \in \{1,2,\ldots,\mathcal{T}\}$ such that $C_{\tilde{j_0}}B(Q) \subset R_Q \subset C_{adj}\cdot C_{\tilde{j_0}}B(Q)$, where $B(Q)$ is defined as in Section 2.1, j_0 defined as in (5) and $\tilde{j_0}$ defined as in (4).

Assume that we have already proven the assertion (10). Let us take a partition of $\mathbb{H}$ as follows:

$$\mathbb{H} = \bigcup_{j=0}^{\infty} 2^j B_0.$$

We next consider the annuli $U_j := 2^{j+1}B_0 \setminus 2^jB_0$ for $j \geq 0$ and the covering $\{\widetilde{B}_{j,\ell}\}_{\ell=1}^{L_j}$ of U_j as in (6). We note that for each $\widetilde{B}_{j,\ell}$, there exist $t_{j,\ell} \in \{1,2,\ldots,\mathcal{T}\}$ and $\widetilde{Q}_{j,\ell} \in \mathcal{D}^{t_{j,\ell}}$ such that $\widetilde{B}_{j,\ell} \subseteq \widetilde{Q}_{j,\ell} \subseteq C_{adj}\widetilde{B}_{j,\ell}$. Therefore, we acquire that for each such $\widetilde{B}_{j,\ell}$, the enlargement $C_{\tilde{j_0}}B(\widetilde{Q}_{j,l})$ covers B_0 since $C_{\tilde{j_0}}\widetilde{B}_{j,\ell}$ covers B_0.

Next, we utilize (10) to each $\widetilde{B}_{j,\ell}$, then we acquire a $\frac{1}{2}$-sparse family $\widetilde{\mathcal{F}}_{j,\ell} \subset \mathcal{D}^{t_{j,\ell}}(\widetilde{Q}_{j,\ell})$ such that (10) can be established for *a.e.* $x \in \widetilde{B}_{j,\ell}$.

Now, set $\mathcal{F} := \cup_{j,\ell}\widetilde{\mathcal{F}}_{j,\ell}$. Then we observe that the balls $C_{adj}\widetilde{B}_{j,\ell}$ are overlapping not more than $C_{A_0,\tilde{j_0}}(2j_0+1)$ times, where $C_{A_0,\tilde{j_0}}$ is the constant in (7). Then, we can obtain that $\mathcal{F}$ is a $\frac{1}{2C_{A_0,\tilde{j_0}}(2j_0+1)}$-sparse family and for *a.e.* $c \in \mathbb{H}$,

$$\begin{aligned}&|(I_\alpha)_b^m(f)(x)| \\ &\leq C_{N,m,\alpha}\sum_{k=0}^{m}\binom{m}{k}\sum_{Q\in\mathcal{F}}\left(|b(x)-b_{R_Q}|^{m-k}|C_{\tilde{j_0}}B(Q)|^{\frac{\alpha}{N}}\langle f(b-b_{R_Q})^k\rangle_{C_{\tilde{j_0}}B(Q)}\right)\chi_Q(x).\end{aligned}$$

Since $C_{\tilde{j_0}}B(Q) \subset R_Q$, and it is clear that $|R_Q| \leq \overline{C}|C_{\tilde{j_0}}B(Q)|$ ($\overline{C}$ depends only on C_{adj}), we obtain that $\langle f\rangle_{C_{\tilde{j_0}}B(Q)} \leq \overline{C}\langle f\rangle_{R_Q}$. Now, we set $\mathcal{S}_t := \{R_Q \in \mathcal{D}^t : Q \in \mathcal{F}\}$, $t \in \{1,2,\ldots,\mathcal{T}\}$, then since the fact that $\mathcal{F}$ is $\frac{1}{2C_{A_0,\tilde{j_0}}(2j_0+1)}$-sparse, we can acquire that each family $\mathcal{S}_t$ is $\frac{1}{2C_{A_0,\tilde{j_0}}(2j_0+1)\overline{c}}$-sparse.

Now, we let

$$\eta := \frac{1}{2C_{A_0,\tilde{j}_0}(2j_0+1)\bar{c}},$$

where $\bar{c}$ is a constant relating only on $\overline{C}$, $C_{\tilde{j}_0}$. Then it follows that (3) holds, which finishes the proof. □

Proof of the Assertion (10). To demonstrate the assertion it suffice to attest the following recursive computation: there exist the cubes $P_j \in \mathcal{D}^{t_0}(Q_0)$ that does not intersect each other such that $\sum_j |P_j| \le \frac{1}{2}|Q_0|$ and for *a.e.* $x \in B_0$,

$$\begin{aligned}
&|(I_\alpha)_b^m(f\chi_{C_{\tilde{j}_0}B(Q_0)})(x)|\chi_{Q_0}(x)\\
&\le C_{N,m,\alpha}\sum_{k=0}^{m}\binom{m}{k}|b(x)-b_{R_{Q_0}}|^{m-k}|C_{\tilde{j}_0}B(Q_0)|^{\frac{\alpha}{N}}\langle f(b-b_{R_{Q_0}})^k\rangle_{C_{\tilde{j}_0}B(Q_0)}\chi_{Q_0}(x)\\
&+\sum_j|(I_\alpha)_b^m(f\chi_{C_{\tilde{j}_0}B(P_j)})(x)|\chi_{P_j}(x).
\end{aligned}$$

Iterating this estimate, we acquire (10) with $\mathcal{F}^{t_0}$ being the union of all the families $\{P_j^k\}$, where $\{P_j^0\}=\{Q_0\}$, $\{P_j^1\}=\{P_j\}$ as mentioned above, and $\{P_j^k\}$ are the cubes acquired at the k-th stage of the iterative approach. Clearly $\mathcal{F}^{t_0}$ is a $\frac{1}{2}$-sparse family, since let

$$E_{P_j^k} = P_j^k \setminus \cup_j P_j^{k+1}.$$

Now we prove the recursive estimate. For any countable family $\{P_j\}_j$ of disjoint cubes $P_j \subset \mathcal{D}^{t_0}(Q_0)$, we have that

$$\begin{aligned}
&|(I_\alpha)_b^m(f\chi_{C_{\tilde{j}_0}B(Q_0)})(x)\chi_{Q_0}(x)\\
&\le |(I_\alpha)_b^m(f\chi_{C_{\tilde{j}_0}B(Q_0)})(x)\chi_{Q_0\setminus\cup_j P_j}(x)+\sum_j|(I_\alpha)_b^m(f\chi_{C_{\tilde{j}_0}B(Q_0)})(x)\chi_{P_j}(x)\\
&\le |(I_\alpha)_b^m(f\chi_{C_{\tilde{j}_0}B(Q_0)})(x)\chi_{Q_0\setminus\cup_j P_j}(x)+\sum_j|(I_\alpha)_b^m(f\chi_{C_{\tilde{j}_0}B(Q_0)\setminus C_{\tilde{j}_0}B(P_j)})(x)\chi_{P_j}(x)\\
&+\sum_j|(I_\alpha)_b^m(f\chi_{C_{\tilde{j}_0}B(P_j)})(x)\chi_{P_j}(x)
\end{aligned}$$

So we just have to reveal that we can opt for a family of pairwise disjoint cubes $\{P_j\} \subset \mathcal{D}^{t_0}(Q_0)$ such that $\sum_j |P_j| \le \frac{1}{2}|Q_0|$ and that for *a.e.* $x \in B_0$,

$$\begin{aligned}
&|(I_\alpha)_b^m(f\chi_{C_{\tilde{j}_0}B(Q_0)})(x)|\chi_{Q_0\setminus\cup_j P_j}(x)+\sum_j|(I_\alpha)_b^m(f\chi_{C_{\tilde{j}_0}B(Q_0)\setminus C_{\tilde{j}_0}B(P_j)})(x)|\chi_{P_j}(x)\\
&\le C_{N,m,\alpha}\sum_{k=0}^{m}\binom{m}{k}|b(x)-b_{R_{Q_0}}|^{m-k}|C_{\tilde{j}_0}B(Q_0)|^{\frac{\alpha}{N}}\langle f(b-b_{R_{Q_0}})^k\rangle_{C_{\tilde{j}_0}B(Q_0)}\chi_{Q_0}(x).
\end{aligned}$$

Using that $(I_\alpha)_b^m f = (I_\alpha)_{b-c}^m f$ for any $c \in \mathbb{R}$, and also that

$$(I_\alpha)_{b-c}^m f = \sum_{k=0}^{m}(-1)^k\binom{m}{k}I_\alpha((b-c)^k f)(b-c)^{m-k},$$

it follows that

$$
\begin{aligned}
&|(I_\alpha)_b^m(f\chi_{C_{\tilde{j_0}}B(Q_0)})(x)|\chi_{Q_0\setminus\cup_jP_j(x)}+\sum_j|(I_\alpha)_b^m(f\chi_{C_{\tilde{j_0}}B(Q_0)\setminus C_{\tilde{j_0}}B(P_j)})(x)|\chi_{P_j}(x)\\
&\leq\sum_{k=0}^m\binom{m}{k}|b(x)-b_{R_{Q_0}}|^{m-k}|I_\alpha((b-b_{R_{Q_0}})^kf\chi_{C_{\tilde{j_0}}B(Q_0)})(x)|\chi_{Q_0\setminus\cup_jP_j}(x)\\
&+\sum_{k=0}^m\binom{m}{k}|b(x)-b_{R_{Q_0}}|^{m-k}\sum_j|I_\alpha((b-b_{R_{Q_0}})^kf\chi_{C_{\tilde{j_0}}B(Q_0)\setminus C_{\tilde{j_0}}B(P_j)})(x)|\chi_{P_j}(x)\\
&=:W_1+W_2.
\end{aligned}
$$

Now we define the set $E=\cup_{k=0}^mE_k$, where

$$E_k=\{x\in B_0:\mathcal{M}_{I_\alpha,B_0}((b-b_{R_{Q_0}})^kf)(x)>C_{N,m,\alpha}|C_{\tilde{j_0}}B(Q_0)|^{\frac{\alpha}{N}}\langle(b-b_{R_{Q_0}})^kf\rangle_{C_{\tilde{j_0}}B(Q_0)}\},$$

with $C_{N,m,\alpha}$ being a positive number to be chosen.

From [8], we can choose $C_{N,m,\alpha}$ big enough (depending on $C_{\tilde{j_0}}$, C_{adj}, and A_1) such that

$$|E|\leq\frac{1}{4\widetilde{A_0}}|B_0|,$$

where $\widetilde{A_0}$ is defined in Section 2.1. We now utilize the Calderón-Zygmund decomposition to the function χ_E on B_0 at the height $\lambda:=\frac{1}{2\widetilde{A_0}}$, to acquire pairwise disjoint cubes $\{P_j\}\subset\mathcal{D}^{t_0}(Q_0)$ such that

$$\frac{1}{2\widetilde{A_0}}|P_j|\leq|P_j\cap E|\leq\frac{1}{2}|P_j|$$

and $|E\setminus\cup_jP_j|=0$. This implies that

$$\sum_j|P_j|\leq\frac{1}{2}|B_0|\quad\text{and}\quad P_j\cap E^c\neq\emptyset.$$

Fix some j. Since we have $P_j\cap E^c\neq\emptyset$, we observe that

$$\mathcal{M}_{I_\alpha,B_0}((b-b_{R_{Q_0}})^kf)(x)\leq C_{N,m,\alpha}|C_{\tilde{j_0}}B(Q_0)|^{\frac{\alpha}{N}}\langle(b-b_{R_{Q_0}})^kf\rangle_{C_{\tilde{j_0}}B(Q_0)},$$

which allows us to control the summation in W_2 by considering the cube P_j.

Now by (i) in Lemma 1, we know that

$$|I_\alpha((b-b_{R_{Q_0}})^kf\chi_{C_{\tilde{j_0}}B(Q_0)})(x)|\leq\mathcal{M}_{I_\alpha,B_0}((b-b_{R_{Q_0}})^kf)(x),\quad\text{for a.e. }x\in B_0.$$

Since $|E\setminus\cup_jP_j|=0$, we have that

$$
\begin{aligned}
&\mathcal{M}_{I_\alpha,B_0}((b-b_{R_{Q_0}})^kf)(x)\\
&\leq C_{N,m,\alpha}|C_{\tilde{j_0}}B(Q_0)|^{\frac{\alpha}{N}}\langle(b-b_{R_{Q_0}})^kf\rangle_{C_{\tilde{j_0}}B(Q_0)},\quad\text{for a.e. }x\in B_0\setminus\cup_jP_j.
\end{aligned}
$$

Consequently,

$$
\begin{aligned}
&|I_\alpha((b-b_{R_{Q_0}})^kf\chi_{C_{\tilde{j_0}}B(Q_0)})(x)|\\
&\leq C_{N,m,\alpha}|C_{\tilde{j_0}}B(Q_0)|^{\frac{\alpha}{N}}\langle(b-b_{R_{Q_0}})^kf\rangle_{C_{\tilde{j_0}}B(Q_0)},\quad\text{for a.e. }x\in B_0\setminus\cup_jP_j.
\end{aligned}
$$

These estimates allow us to control the remaining terms in W_1, so we are done. □

Proof of Lemma 1. Now we give the proof process of Lemma 1.

The result in the Euclidean space case can be referred to as [8]. Now, we can adapt the proof in [8] to our setting of homogeneous groups.

(i) Let r is close enough to 0 such that $B(x,r) \subset B_0$. Then,

$$\begin{aligned} |I_\alpha(f\chi_{C_{\tilde{j}0}B_0})(x)| &\le |I_\alpha(f\chi_{C_{\tilde{j}0}B(x,r)})(x)| + |I_\alpha(f\chi_{C_{\tilde{j}0}B_0\setminus C_{\tilde{j}0}B(x,r)})(x)| \\ &\le |I_\alpha(f\chi_{C_{\tilde{j}0}B(x,r)})(x)| + \mathcal{M}_{I_\alpha,B_0}f(x), \end{aligned}$$

the estimate for the first term follows by standard computations involving a dyadic annuli-type decomposition of the $B(x,r)$.

$$\begin{aligned} |I_\alpha(f\chi_{C_{\tilde{j}0}B(x,r)})(x)| &= \left|\int_{\mathbb{H}} \frac{f(y)\chi_{C_{\tilde{j}0}B(x,r)}}{d(x,y)^{N-\alpha}}dy\right| \\ &\le \int_{B(x,C_{\tilde{j}0}r)} \frac{|f(y)|}{d(x,y)^{N-\alpha}}dy \\ &= \sum_{i=-\infty}^{1} \int_{B(x,C_{\tilde{j}0}{}^{i}r)\setminus B(x,C_{\tilde{j}0}{}^{i-1}r)} \frac{|f(y)|}{d(x,y)^{N-\alpha}}dy \\ &\le \sum_{i=-\infty}^{1} (C_{\tilde{j}0}{}^{i-1}r)^{\alpha-N} \int_{B(x,C_{\tilde{j}0}{}^{i}r)} |f(y)|dy \\ &= \sum_{i=-\infty}^{1} \left(\frac{1}{C_{\tilde{j}0}}\right)^{\alpha-N} (C_{\tilde{j}0}{}^{i}r)^{\alpha} \frac{1}{(C_{\tilde{j}0}{}^{i}r)^{N}} \int_{B(x,C_{\tilde{j}0}{}^{i}r)} |f(y)|dy \\ &\le C_{N,\alpha,C_{\tilde{j}0}} r^\alpha Mf(x). \end{aligned}$$

Then,

$$|I_\alpha(f\chi_{C_{\tilde{j}0}B_0})(x)| \le C_{N,\alpha,C_{\tilde{j}0}} r^\alpha Mf(x) + \mathcal{M}_{I_\alpha,B_0}f(x), \tag{11}$$

the estimate in (i) is settled letting $r \to 0$ in (11).

(ii) Let $x, \xi \in B := B(x_0, r)$. Let B_x be the closed ball with radius $4(A_0 + C_{\tilde{j}0})r$, which centered at x. Then $C_{\tilde{j}0}B \subset B_x$, and we acquire

$$\begin{aligned} |I_\alpha(f\chi_{\mathbb{H}\setminus C_{\tilde{j}0}B})(\xi)| &= |I_\alpha(f\chi_{\mathbb{H}\setminus B_x})(\xi) + I_\alpha(f\chi_{B_x\setminus C_{\tilde{j}0}B})(\xi)| \\ &\le |I_\alpha(f\chi_{\mathbb{H}\setminus B_x})(\xi) - I_\alpha(f\chi_{\mathbb{H}\setminus B_x})(x)| \\ &\quad + |I_\alpha(f\chi_{B_x\setminus C_{\tilde{j}0}B})(\xi)| + |I_\alpha(f\chi_{\mathbb{H}\setminus B_x})(x)| \end{aligned}$$

For the first term, since ρ is homogeneous of degree $\alpha - N$, and by using the Proposition 1.7 in [1], we get

$$
\begin{aligned}
&|I_\alpha(f\chi_{\mathbb{H}\setminus B_x})(\xi) - I_\alpha(f\chi_{\mathbb{H}\setminus B_x})(x)| \\
&\leq \int_{\mathbb{H}\setminus B_x} |f(y)| \left| \frac{1}{d(y,\xi)^{N-\alpha}} - \frac{1}{d(x,y)^{N-\alpha}} \right| dy \\
&\leq C_{N,\alpha} \int_{\mathbb{H}\setminus B_x} \frac{2r}{d(x,y)^{N-\alpha+1}} |f(y)| dy \\
&= C_{N,\alpha} \sum_{i=1}^{\infty} \int_{2^i B_x \setminus 2^{i-1} B_x} \frac{2r}{d(x,y)^{N-\alpha+1}} |f(y)| dy \\
&\leq C_{N,\alpha} \sum_{i=1}^{\infty} \frac{2r}{\left(2^{i-1}|B_x|^{\frac{1}{N}}\right)^{N-\alpha+1}} \int_{2^i B_x} |f(y)| dy \\
&= C_{N,\alpha} \sum_{i=1}^{\infty} \frac{2r}{\left(2^{i-1}2^2 r(A_0 + C_{\tilde{j}_0})\right)^{N-\alpha+1}} \int_{2^i B_x} |f(y)| dy \\
&= C_{N,\alpha} \sum_{i=1}^{\infty} \frac{2r}{2^{i+1} r(A_0 + C_{\tilde{j}_0})} \cdot \frac{1}{\left(2^{i+1} r(A_0 + C_{\tilde{j}_0})\right)^{N-\alpha}} \int_{2^i B_x} |f(y)| dy \\
&\leq C_{N,\alpha} M_\alpha f(x).
\end{aligned}
$$

Next, for $\xi \in B, y \in B_x \setminus C_{\tilde{j}_0} B$, we have $d(y,\xi) > 2^{\tilde{j}_0} r$. Then we have

$$
\begin{aligned}
|I_\alpha(f\chi_{B_x \setminus C_{\tilde{j}_0} B})(\xi)| &\leq \int_{B_x \setminus C_{\tilde{j}_0} B} \frac{1}{d(y,\xi)^{N-\alpha}} |f(y)| dy \\
&\leq \frac{1}{|2^{\tilde{j}_0} r|^{N-\alpha}} \int_{B_x} |f(y)| dy \\
&= C_{N,\alpha} \frac{1}{|4(A_0 + C_{\tilde{j}_0}) r|^{N-\alpha}} \int_{B_x} |f(y)| dy \\
&\leq C_{N,\alpha} M_\alpha f(x).
\end{aligned}
$$

Finally, we observe that

$$
\begin{aligned}
|I_\alpha(f\chi_{\mathbb{H}\setminus B_x})(x)| &= |\int_{\mathbb{H}\setminus B_x} \frac{f(y)}{d(x,y)^{N-\alpha}} dx| \\
&\leq \int_{\mathbb{H}} \frac{|f(y)|}{d(x,y)^{N-\alpha}} dx \\
&= I_\alpha |f|(x),
\end{aligned}
$$

which finishes the proof of (ii). □

Next, we review that the dyadic weighted BMO space associated with the system $\mathcal{D}^t$ is defined as

$$
BMO_{\eta,\mathcal{D}^t}(\mathbb{H}) := \{b \in L^1_{loc}(\mathbb{H}) : \|b\|_{BMO_{\eta,\mathcal{D}^t}} < \infty\},
$$

where $\|b\|_{BMO_{\eta,\mathcal{D}^t}} = \sup_{Q \in \mathcal{D}^t} \frac{1}{\eta(Q)} \int_Q |b(x) - b_Q| dx$. Then according to the dyadic structure theorem studies in [14], one has

$$
BMO_\eta(\mathbb{H}) = \bigcap_{t=1}^{\mathcal{T}} BMO_{\eta,\mathcal{D}^t}(\mathbb{H}).
$$

Now, to verify a function b is in $BMO_\eta(\mathbb{H})$, it suffices to verify it belongs to each weighted dyadic BMO space $BMO_{\eta,\mathcal{D}^t}(\mathbb{H})$. Given a dyadic cube $Q \in \mathcal{D}^t$ with $t = 1,2,\ldots,\mathcal{T}$, and a measurable function f on $\mathbb{H}$, we define the local mean oscillation of f on Q by

$$\omega_\lambda(f;Q) = \inf_{c\in\mathbb{R}} \left((f-c)\chi_Q\right)^*(\lambda|Q|), \quad 0<\lambda<1,$$

where

$$\left((f-c)\chi_Q\right)^*(\lambda|Q|) = \sup_{E\subset Q, |E|=\lambda|Q|} \inf_{x\in E} |(f-c)(x)|.$$

With these notation and dyadic structure theorem above, following the same proof in [10], we also acquire that for any weight $\eta \in A_2$, we have

$$\|b\|_{BMO_\eta(\mathbb{H})} \le C\sum_{t=1}^{\mathcal{T}} \sup_{Q\in\mathcal{D}^t} \omega_\lambda(b;Q)\frac{|Q|}{\eta(Q)}, \quad 0<\lambda\le 2^{N+1}, \tag{12}$$

where C depends on η.

Proposition 2. *Suppose that $\mathbb{H}$ is a homogeneous group with dimension $N, b \in L^1_{loc}(\mathbb{H})$. Then for any cube $Q \subset \mathbb{H}$, there exist measurable set $F_i \subset Q$ with $i = 1,2$, such that*

$$\omega_{\frac{1}{2^{N+2}}}(b;Q) \le b(x) - b(y), \ \forall (x,y) \in F_1 \times F_2.$$

Proof. We take ideas from N. Accomazzo, J. C. Martínez-Perales and I. P. Rivera-Ríos [8]. In [8], for any cube $Q \in \mathcal{D}^t$ with $t = 1,2,\ldots,\mathcal{T}$, there exists a subset $E \subset Q$ with $|E| = \frac{1}{2^{N+2}}|Q|$ such that for every $x \in E$,

$$\omega_{\frac{1}{2^{N+2}}}(b;Q) \le |b(x) - m_b(Q)|,$$

where $m_b(Q)$ is a not necessarily unique number that satisfies

$$\max\left\{|\{x\in Q: b(x) > m_b(Q)\}|, |\{x \in Q : b(x) < m_b(Q)\}|\right\} \le \frac{|Q|}{2}.$$

Let $E_1 \subset Q$ with $|E| = \frac{1}{2}|Q|$ and such that $b(x) \ge m_b(Q)$ for every $x \in E_1$. Further let $E_2 = Q \setminus E_1$, then $|E_2| = \frac{1}{2}|Q|$ and for every $x \in E_2, b(x) \le m_b(Q)$.

We obtain that at least half of the set E is contained either in E_1 or in E_2 since Q is the disjoint union of E_1 and E_2. Without loss of generality, we assume that half of E is in E_1, then we let $F_1 = E \cap E_1, F_2 = E_2 \cap (E \cap E_1)^c$, we have

$$|F_1| = |E| - |E \cap (E\cap E_1)^C| \ge |E| - \frac{|E|}{2} = \frac{|Q|}{2^{N+3}},$$

and

$$|F_2| = |E_2| - |E_2 \cap (E\cap E_1)| \ge \frac{1}{2}|Q| - \frac{1}{2^{N+3}}|Q| = (\frac{1}{2} - \frac{1}{2^{N+3}})|Q|.$$

Then if $x \in F_1$ and $y \in F_2$, we have that

$$\omega_{\frac{1}{2^{N+2}}}(b;Q) \le b(x) - m_b(Q) \le b(x) - b(y),$$

which shows that Proposition 2 holds. □

Given a dyadic grid $\mathcal{D}$, define the dyadic Riesz potential operator

$$I_\alpha^{\mathcal{D}} f(x) = \sum_{Q \in \mathcal{D}} \frac{1}{|Q|^{1-\frac{\alpha}{N}}} \int_Q |f(y)| dy \chi_Q(x).$$

Proposition 3. *Given $0 < \alpha < N$, then for any dyadic grid $\mathcal{D}$,*

$$I_\alpha^{\mathcal{D}} f(x) \lesssim I_\alpha f(x). \tag{13}$$

Proof. The result in the Euclidean setting is from the Proposition 2.1 in [15]. Here, we can adapt the proof in [15] to our setting of spaces of homogeneous type. □

3. Proof of Theorem 1

To proof (i), we are following the ideas in [16] or [8].

Let $\mathcal{D}$ be a dyadic system in $\mathbb{H}$ and let $\mathcal{S}$ be a sparse family from $\mathcal{D}$. We know

$$\mathcal{A}_{\alpha,\mathcal{S}}^{m,k}(b,f)(x) = \sum_{Q \in \mathcal{S}} |b(x) - b_Q|^{m-k} |Q|^{\frac{\alpha}{N}} \langle (b - b_Q)^k f \rangle_Q \chi_Q(x),$$

by duality, we have that

$$\|\mathcal{A}_{\alpha,\mathcal{S}}^{m,k}(b,f)(x)\|_{L^q_{\lambda^q}(\mathbb{H})} \leq \sup_{g:\|g\|_{L^{q'}_{\lambda^{q'}}(\mathbb{H})}=1} \sum_{Q \in \mathcal{S}} \left(\int_Q |g(x)\lambda^q| |b(x) - b_Q|^{m-k} dx \right) |Q|^{\frac{\alpha}{N}}$$
$$\times \left(\frac{1}{|Q|} \int_Q |b(x) - b_Q|^k |f(x)| dx \right).$$

By Lemma 3.5 in [12], there exists a sparse family $\widetilde{\mathcal{S}} \subset \mathcal{D}$ such that $\mathcal{S} \subset \widetilde{\mathcal{S}}$ and for every cube $Q \in \widetilde{\mathcal{S}}$, for *a.e.* $x \in Q$,

$$|b(x) - b_Q| \leq C_N \sum_{P \in \widetilde{\mathcal{S}}, P \subset Q} \Omega(b,P) \chi_P(x),$$

where $\Omega(b,P) = \frac{1}{|P|} \int_P |b(x) - b_P| dx$

Assume that $b \in BMO_\eta(\mathbb{H})$ with η to be chosen, then we have for *a.e.* $x \in Q$,

$$|b(x) - b_Q| \leq C_N \sum_{P \in \widetilde{\mathcal{S}}, P \subset Q} \frac{1}{\eta(P)} \int_P |b(x) - b_P| dx \cdot \frac{\eta(P)}{|P|} \chi_P(x)$$
$$\leq C_N \|b\|_{BMO_\eta(\mathbb{H})} \sum_{P \in \widetilde{\mathcal{S}}, P \subset Q} \frac{\eta(P)}{|P|} \chi_P(x).$$

Then, we further have

$$\|\mathcal{A}_{\alpha,\mathcal{S}}^{m,k}(b,f)(x)\|_{L^q_{\lambda^q}(\mathbb{H})}$$
$$\leq C_N \|b\|^m_{BMO_\eta(\mathbb{H})} \sup_{g:\|g\|_{L^{q'}_{\lambda^{q'}}(\mathbb{H})}=1} \sum_{Q \in \mathcal{S}} \left(\frac{1}{|Q|} \int_Q |g(x)\lambda^q| \Big(\sum_{P \in \widetilde{\mathcal{S}}, P \subset Q} \frac{\eta(P)}{|P|} \chi_P(x) \Big)^{m-k} dx \right)$$
$$\times \left(\frac{1}{|Q|} \int_Q \Big(\sum_{P \in \widetilde{\mathcal{S}}, P \subset Q} \frac{\eta(P)}{|P|} \chi_P(x) \Big)^k |f(x)| dx \right) \cdot |Q| \cdot |Q|^{\frac{\alpha}{N}}.$$

Next, note that for each $\ell \in \mathbb{N}$, from [12], for an arbitrary function h, we have

$$\int_Q |h(x)| \Big(\sum_{Q\in\tilde{\mathcal{S}}, P\subset Q} \frac{\eta(P)}{|P|} \chi_P(x) \Big)^{\ell} dx$$
$$\lesssim \int_Q \mathcal{A}^{\ell}_{\tilde{\mathcal{S}},\eta}(|h|)(x)dx,$$

where $\mathcal{A}_{\tilde{\mathcal{S}},\eta}(|h|)(x) := \mathcal{A}_{\tilde{\mathcal{S}}}(|h|)\eta$, $\mathcal{A}_{\tilde{\mathcal{S}}}(h) := \sum_{Q\in\tilde{\mathcal{S}}} h_Q\chi_Q$ and $\mathcal{A}^{\ell}_{\tilde{\mathcal{S}},\eta}f$ stands for the ℓ-th iteration of $\mathcal{A}_{\tilde{\mathcal{S}},\eta}$.

Then we have

$$\begin{aligned}
&\|\mathcal{A}^{m,k}_{\alpha,\mathcal{S}}(b,f)(x)\|_{L^q_{\lambda^q}(\mathbb{H})} \\
&\le C_N \|b\|^m_{BMO_\eta(\mathbb{H})} \sup_{g:\|g\|_{L^{q'}_{\lambda^{q'}}(\mathbb{H})}=1} \sum_{Q\in\mathcal{S}} \Big(\int_Q \mathcal{A}^{m-k}_{\tilde{\mathcal{S}},\eta}(|g|\lambda^q) \Big) \cdot \frac{1}{|Q|^{1-\frac{\alpha}{N}}} \Big(\int_Q \mathcal{A}^{k}_{\tilde{\mathcal{S}},\eta}(|f|) \Big) \\
&\le C_N \|b\|^m_{BMO_\eta(\mathbb{H})} \sup_{g:\|g\|_{L^{q'}_{\lambda^{q'}}(\mathbb{H})}=1} \int_{\mathbb{H}} \sum_{Q\in\mathcal{S}} \frac{1}{|Q|^{1-\frac{\alpha}{N}}} \Big(\int_Q \mathcal{A}^{k}_{\tilde{\mathcal{S}},\eta}(|f|)\chi_Q(x) \Big) \cdot \mathcal{A}^{m-k}_{\tilde{\mathcal{S}},\eta}(|g|\lambda^q) \\
&= C_N \|b\|^m_{BMO_\eta(\mathbb{H})} \sup_{g:\|g\|_{L^{q'}_{\lambda^{q'}}(\mathbb{H})}=1} \int_{\mathbb{H}} I^{\alpha}_{\mathcal{S}} \Big(\mathcal{A}^{k}_{\tilde{\mathcal{S}},\eta}(|f|) \Big)(x) \Big(\mathcal{A}^{m-k}_{\tilde{\mathcal{S}},\eta}(|g|\lambda^q) \Big)(x)dx,
\end{aligned}$$

where $I^{\alpha}_{\mathcal{S},\eta}f := I^{\alpha}_{\mathcal{S}}(f)\eta$ and $I^{\alpha}_{\mathcal{S}}f(x) = \sum_{Q\in\mathcal{S}} \frac{1}{|Q|^{1-\frac{\alpha}{N}}} \int_Q |f|\chi_Q(x)$.

From (13) and the boundedness of $I_\alpha f$, if p, q, α are as in the hypothesis of Theorem 1.1 and $w \in A_{p,q}$, $\mathcal{S} \subset \mathcal{D}$, then

$$\|I^{\alpha}_{\mathcal{S}}\|_{L^p_{w^p}(\mathbb{H})\to L^q_{w^q}(\mathbb{H})} \le C_{N,p,q,\alpha}[w]^{(1-\frac{\alpha}{N})\max\{1,\frac{p'}{q}\}}_{A_{p,q}}. \tag{14}$$

Observe that $\mathcal{A}_{\tilde{\mathcal{S}}}$ is self-adjoint, then

$$\int_{\mathbb{H}} I^{\alpha}_{\mathcal{S}} \Big(\mathcal{A}^{k}_{\tilde{\mathcal{S}},\eta}(|f|) \Big) \Big(\mathcal{A}^{m-k}_{\tilde{\mathcal{S}},\eta}(|g|\lambda^q) \Big) = \int_{\mathbb{H}} \mathcal{A}_{\tilde{\mathcal{S}}} \mathcal{A}^{m-k-1}_{\tilde{\mathcal{S}},\eta} \Big[I^{\alpha}_{\mathcal{S},\eta} \Big(\mathcal{A}^{K}_{\tilde{\mathcal{S}},\eta}(|f|) \Big) \Big] |g|\lambda^q.$$

By Hölder inequality, we have that

$$\|\mathcal{A}^{m,k}_{\alpha,\mathcal{S}}(b,f)(x)\|_{L^q_{\lambda^q}(\mathbb{H})} \le C_N \|b\|^m_{BMO_\eta(\mathbb{H})} \|\mathcal{A}_{\tilde{\mathcal{S}}} \mathcal{A}^{m-k-1}_{\tilde{\mathcal{S}},\eta} I^{\alpha}_{\mathcal{S},\eta} \mathcal{A}^{k}_{\tilde{\mathcal{S}},\eta}(|f|)\|_{L^q_{\lambda^q}(\mathbb{H})}.$$

Applying that $\|\mathcal{A}_{\tilde{\mathcal{S}}}\|_{L^p_w(\mathbb{H})} \le C_{N,p}[w]^{\max\{1,\frac{1}{p-1}\}}_{A_p}$ (see, e.g., [17]),

$$\begin{aligned}
&\|\mathcal{A}_{\tilde{\mathcal{S}}} \mathcal{A}^{m-k-1}_{\tilde{\mathcal{S}},\eta} I^{\alpha}_{\mathcal{S},\eta} \mathcal{A}^{k}_{\tilde{\mathcal{S}},\eta}(|f|)\|_{L^q_{\lambda^q}(\mathbb{H})} \\
&\le C_{N,p}[\lambda^q]^{\max\{1,\frac{1}{q-1}\}}_{A_q} \|\mathcal{A}_{\tilde{\mathcal{S}}} \mathcal{A}^{m-k-2}_{\tilde{\mathcal{S}},\eta} I^{\alpha}_{\mathcal{S},\eta} \mathcal{A}^{k}_{\tilde{\mathcal{S}},\eta}(|f|)\|_{L^q_{\lambda^q\eta^q}(\mathbb{H})} \\
&\le C_{N,p}[\lambda^q]^{\max\{1,\frac{1}{q-1}\}}_{A_q} [\lambda^q\eta^q]^{\max\{1,\frac{1}{q-1}\}}_{A_q} \|\mathcal{A}^{m-k-2}_{\tilde{\mathcal{S}},\eta} I^{\alpha}_{\mathcal{S},\eta} \mathcal{A}^{k}_{\tilde{\mathcal{S}},\eta}(|f|)\|_{L^q_{\lambda^q\eta^q}(\mathbb{H})} \\
&\le C_{N,p} \Big(\prod_{i=0}^{m-k-1} [\lambda^q\eta^{iq}]_{A_q} \Big)^{\max\{1,\frac{1}{q-1}\}} \|I^{\alpha}_{\mathcal{S},\eta} \mathcal{A}^{k}_{\tilde{\mathcal{S}},\eta}(|f|)\|_{L^q_{\lambda^q\eta^{(m-k-1)q}}(\mathbb{H})}.
\end{aligned}$$

Using (14), we have that

$$\|I^{\alpha}_{\mathcal{S},\eta}\mathcal{A}^{k}_{\tilde{\mathcal{S}},\eta}(|f|)\|_{L^q_{\lambda^q\eta^{(m-k-1)q}}(\mathbb{H})} = \|I^{\alpha}_{\mathcal{S}}\mathcal{A}^{k}_{\tilde{\mathcal{S}},\eta}(|f|)\|_{L^q_{\lambda^q\eta^{(m-k)q}}(\mathbb{H})}$$
$$\leq C_{N,p,\alpha}[\lambda\eta^{m-k}]_{A_{p,q}}^{(1-\frac{\alpha}{N})\max\{1,\frac{p'}{q}\}}\|\mathcal{A}^{k}_{\tilde{\mathcal{S}},\eta}(|f|)\|_{L^p_{\lambda^p\eta^{(m-k)p}}(\mathbb{H})},$$

and applying again $\|\mathcal{A}_{\tilde{\mathcal{S}}}\|_{L^p_w(\mathbb{H})} \leq C_{N,p}[w]_{A_p}^{\max\{1,\frac{1}{p-1}\}}$,

$$\|\mathcal{A}^{k}_{\tilde{\mathcal{S}},\eta}(|f|)\|_{L^p_{\lambda^p\eta^{(m-k)p}}(\mathbb{H})} \leq C_{N,p}\Big(\prod_{i=m-k+1}^{m}[\lambda^p\eta^{ip}]_{A_p}\Big)^{\max\{1,\frac{1}{p-1}\}}\|f\|_{L^p_{\lambda^p\eta^{mp}}(\mathbb{H})},$$

which, along with the previous estimate, yields

$$\|\mathcal{A}^{m,k}_{\alpha,\mathcal{S}}(b,f)(x)\|_{L^q_{\lambda^q}(\mathbb{H})}$$
$$\leq C_{N,p,\alpha}\|b\|^m_{BMO_\eta(\mathbb{H})}A(m,k)B(m,k)[\lambda\eta^{m-k}]^{(1-\frac{\alpha}{N})\max\{1,\frac{p'}{q}\}}\|f\|_{L^p_{\lambda^p\eta^{mp}}(\mathbb{H})},$$

where

$$A(m,k) = \Big(\prod_{i=0}^{m-k-1}[\lambda^q\eta^{iq}]_{A_q}\Big)^{\max\{1,\frac{1}{q-1}\}},$$

and

$$B(m,k) = \Big(\prod_{i=m-k+1}^{\infty}[\lambda^p\eta^{ip}]_{A_p}\Big)^{\max\{1,\frac{1}{p-1}\}}.$$

Hence, setting $\eta = \nu^{1/m}$, where $\nu = (\frac{\mu}{\lambda})^{1/p}$, it reading follows from Hölder's inequality

$$[\lambda^s\nu^{s\frac{i}{m}}]_{A_s} \leq [\lambda^s]_{A_s}^{\frac{m-i}{m}}[\mu^s]_{A_s}^{\frac{i}{m}}, \quad s = p,q.$$

Thus, we acquire that

$$A(m,k) \leq \Big(\prod_{i=0}^{m-k-1}[\lambda^q]_{A_q}^{\frac{m-i}{m}}[\mu^q]_{A_q}^{\frac{i}{m}}\Big)^{\max\{1,\frac{1}{q-1}\}} \leq \Big([\lambda^q]_{A_q}^{\frac{m+k+1}{2}}[\mu^q]_{A_q}^{\frac{m-k-1}{2}}\Big)^{\frac{m-k}{m}\max\{1,\frac{1}{q-1}\}},$$

and

$$B(m,k) \leq \Big(\prod_{i=m-k+1}^{m}[\lambda^p]_{A_p}^{\frac{m-i}{m}}[\mu^p]_{A_p}^{\frac{i}{m}}\Big)^{\max\{1,\frac{1}{p-1}\}} \leq \Big([\lambda^p]_{A_p}^{\frac{k-1}{2}}[\mu^p]_{A_p}^{m-\frac{k-1}{2}}\Big)^{\frac{k}{m}\max\{1,\frac{1}{p-1}\}}.$$

Combining all the preceding estimates obtains (i).

To proof (ii), we are going to follow ideas in [10]. Based on (12), it suffices to show that there exists a positive constant C such that for all dyadic cubes $Q \in \mathcal{D}^t$,

$$\omega_{\frac{1}{2^{N+2}}}(b;Q)^m \leq C\Big(\frac{\nu^{1/m}(Q)}{|Q|}\Big)^m\|(I_\alpha)^m_b\|_{L^p_{\mu^p}(\mathbb{H})\to L^q_{\lambda^q}(\mathbb{H})} \quad (15)$$

Using Proposition 2 and Hölder inequality implies that

$$\begin{aligned}
\omega_{\frac{1}{2^{N+2}}}(b;Q)^m|F_1||F_2| &\le \int_{F_1}\int_{F_2}\Big(b(x)-b(y)\Big)^m dxdy \\
&\le dima(Q)^{N-\alpha}\int_{F_1}\int_{F_2}\frac{\Big(b(x)-b(y)\Big)^m}{d(x,y)^{N-\alpha}}dxdy \\
&= dima(Q)^{N-\alpha}\int_{F_1}(I_\alpha)_b^m(\chi_{F_2})(x)dx \\
&\le C|Q|^{1-\frac{\alpha}{N}}\Big(\int_Q \lambda^{-q'}\Big)^{\frac{1}{q'}}\cdot\Big(\int_{\mathbb{H}}[(I_\alpha)_b^m(\chi_{F_2})]^q\lambda^q dx\Big)^{\frac{1}{q}} \\
&\le C|Q|^{1-\frac{\alpha}{N}}\Big(\int_Q \lambda^{-q'}\Big)^{\frac{1}{q'}}\cdot\Big(\int_Q \mu^p\Big)^{\frac{1}{p}}\|(I_\alpha)_b^m\|_{L^p_{\mu^p}(\mathbb{H})\to L^q_{\lambda^q}(\mathbb{H})} \\
&= C|Q|^2\Big(\frac{1}{|Q|}\int_Q \lambda^{-q'}\Big)^{\frac{1}{q'}}\cdot\Big(\frac{1}{|Q|}\int_Q \mu^p\Big)^{\frac{1}{p}}\|(I_\alpha)_b^m\|_{L^p_{\mu^p}(\mathbb{H})\to L^q_{\lambda^q}(\mathbb{H})},
\end{aligned}$$

where we used that $\frac{1}{q}+\frac{\alpha}{N}=\frac{1}{p}$.

Further, this yields

$$\omega_{\frac{1}{2^{N+2}}}(b;Q)^m \le C\Big(\frac{1}{|Q|}\int_Q \lambda^{-q'}\Big)^{\frac{1}{q'}}\cdot\Big(\frac{1}{|Q|}\int_Q \mu^p\Big)^{\frac{1}{p}}\|(I_\alpha)_b^m\|_{L^p_{\mu^p}(\mathbb{H})\to L^q_{\lambda^q}(\mathbb{H})}.$$

Then from [8], we have

$$\Big(\frac{1}{|Q|}\int_Q \mu^p\Big)^{\frac{1}{p}} \le C\Big(\frac{1}{|Q|}\int_Q \nu^{1/m}\Big)^m\Big(\frac{1}{|Q|}\int_Q \lambda^p\Big)^{\frac{1}{p}},$$

so the

$$\begin{aligned}
&\omega_{\frac{1}{2^{N+2}}}(b;Q)^m \\
&\le C\Big(\frac{1}{|Q|}\int_Q \nu^{1/m}\Big)^m\Big(\frac{1}{|Q|}\int_Q \lambda^{-q'}\Big)^{\frac{1}{q'}}\Big(\frac{1}{|Q|}\int_Q \lambda^p\Big)^{\frac{1}{p}}\|(I_\alpha)_b^m\|_{L^p_{\mu^p}(\mathbb{H})\to L^q_{\lambda^q}(\mathbb{H})}.
\end{aligned}$$

Now we observe that since $q>p$ then by Hölder inequality,

$$\Big(\frac{1}{|Q|}\int_Q \lambda^p\Big)^{\frac{1}{p}} \le \Big(\frac{1}{|Q|}\int_Q \lambda^q\Big)^{\frac{1}{q}} \quad\text{and}\quad \Big(\frac{1}{|Q|}\int_Q \lambda^{-q'}\Big)^{\frac{1}{q'}} \le \Big(\frac{1}{|Q|}\int_Q \lambda^{-p'}\Big)^{\frac{1}{p'}},$$

then

$$\Big(\frac{1}{|Q|}\int_Q \lambda^{-q'}\Big)^{\frac{1}{q'}}\Big(\frac{1}{|Q|}\int_Q \lambda^p\Big)^{\frac{1}{p}} \le \Big[\Big(\frac{1}{|Q|}\int_Q \lambda^q\Big)\Big(\frac{1}{|Q|}\int_Q \lambda^{-p'}\Big)^{\frac{q}{p'}}\Big]^{\frac{1}{q}}.$$

Consequently, since $\lambda\in A_{p,q}$, we finally get

$$\omega_{\frac{1}{2^{N+2}}}(b;Q)^m \le C\Big(\frac{1}{|Q|}\int_Q \nu^{1/m}\Big)^m\|(I_\alpha)_b^m\|_{L^p_{\mu^p}(\mathbb{H})\to L^q_{\lambda^q}(\mathbb{H})}.$$

Thus, (15) holds and hence, the proof of (ii) is complete.

Therefore, we complete the proof of Theorem 1.

Author Contributions: Writing—original draft preparation, D.C.; methodology, Y.C.; check, T.W. All authors have read and agreed to the published version of the manuscript.

Funding: The project was in part supported by: Yanping Chen's National Natural Sciience Foundation of China (# 11871096, # 11471033).

Institutional Review Board Statement: Not applicable.

Informed Consent Statement: Not applicable.

Data Availability Statement: Data sharing not applicable to this article as no datasets were generated or analysed during the current study.

Conflicts of Interest: The authors declare no conflict of interest.

Abbreviation

BMO Bounded Mean Oscillation

References

1. Folland, G.B.; Stein, E.M. *Hardy Spaces on Homogeneous Groups*; Princetion University Press: Princeton, NJ, USA; Univerity of Tokyo: Tokyo, Japan, 1982.
2. Stein, E.M. *Harmonic Analysis: Real-Variable Methods, Orthogonality, and Oscillatory Integrals*; Princeton University Press: Priceton, NJ, USA, 1993.
3. Frazier, M.; Jawerth, B.; Weiss, G. *Littlewood-Paley Theory and the Study of Function Spaces*; American Mathematical Society: Providence, RI, USA, 1991; Volume 79. [CrossRef]
4. Sawyer, E.; Wheeden, R.L. Weighted Inequalities for Fractional Integrals on Euclidean and Homogeneous Spaces. *Am. J. Math.* **1992**, *114*, 813–874. [CrossRef]
5. Muckenhoupt, B.; Wheeden, R. Weighted Norm Inequalities for Fractional Integrals. *Trans. Am. Math. Soc.* **1974**, *192*, 261–274. [CrossRef]
6. Lacey, M.T.; Moen, K.; Pérez, C.; Torres, R.H. Sharp Weighted Bounds for Fractional Integral Operators. *J. Funct. Anal.* **2010**, *259*, 1073–1097. [CrossRef]
7. Holmes, I.; Rahm, R.; Spencer, S. Commutators with Fractional Integral Operators. *Stud. Math.* **2016**, *233*, 279–291. [CrossRef]
8. Accomazzo, N.; Martínez-Perales, J.C.; Rivera-Ríos, I.P. On Bloom-Type Estimates for Iterated Commutators of Fractional Integrals. *Indiana Univ. Math. J.* **2020**, *69*, 1207–1230. [CrossRef]
9. Sato, S. Estimate for Singular Integrals on Homegeneous Groups. *J. Math. Anal. Appl.* **2013**, *400*, 311–330. [CrossRef]
10. Duong, X.T.; Li, H.Q.; Li, J.; Wick, B.D.; Wu, Q.Y. Lower Bound of Riesz Transform Kernels Revisited and Commutators on Stratified Lie Groups. *arXiv* **2018**, arXiv:1803.01301.
11. Fan, Z.; Li, J. Quantitative Weighted Estimates for Rough Singular Integrals on Homogeneous Groups. *arXiv* **2020**, arXiv:2009.02433.
12. Duong, X.T.; Gong, R.; Kuffner, M.S.; Li, J.; Wick, B.D.; Yang, D. Two Weight Commutators on Spaces of Homogeneous Type and Applications. *J. Geom. Anal.* **2021**, *31*, 980–1038. [CrossRef]
13. Coifman, R.R.; Weiss, G. Analyse Harmonique Non-Commutative Sur Certains Espaces Homogènes: Étude De Certaines Intégrales Singulières. In *Lecture Notes in Mathematics*; Springer: Heidelberg/Berlin, Germany, 1971; Volume 242.
14. Kairema, A.; Li, J.; Pereyra, M.C.; Ward, L.A. Haar Bases on Quasi-Metric Measure Spaces, and Dyadic Structure Theorems for Function Spaces on Product Spaces of Homogeneous Type. *J. Funct. Anal.* **2016**, *271*, 1793–1843. [CrossRef]
15. Cruz-Uribe, D.; Moen, K. One and two weighted norm inequalities for Riesz potentials. *Ill. J. Math.* **2013**, *57*, 295–323.
16. Lerner, A.K.; Ombrosi, S.; Rivera-Ríos, I.P. Commutators of Singular Integrals Revisited. *Bull. Lond. Math. Soc.* **2019**, *51*, 107–119. [CrossRef]
17. Cruz-Uribe, D.; Martell, J.M.; Pérez, C. Sharp Weighted Estimates for Classical Operators. *Adv. Math.* **2012**, *229*, 408–441. [CrossRef]

mathematics MDPI

Article

Boundedness of Some Paraproducts on Spaces of Homogeneous Type

Xing Fu

Hubei Key Laboratory of Applied Mathematics, Faculty of Mathematics and Statistics, Hubei University, Wuhan 430062, China; xingfu@hubu.edu.cn

Abstract: Let $(\mathcal{X}, d, \mu)$ be a space of homogeneous type in the sense of Coifman and Weiss. In this article, the author develops a partial theory of paraproducts $\{\Pi_j\}_{j=1}^3$ defined via approximations of the identity with exponential decay (and integration 1), which are extensions of paraproducts defined via regular wavelets. Precisely, the author first obtains the boundedness of Π_3 on Hardy spaces and then, via the methods of interpolation and the well-known $T(1)$ theorem, establishes the endpoint estimates for $\{\Pi_j\}_{j=1}^3$. The main novelty of this paper is the application of the Abel summation formula to the establishment of some relations among the boundedness of $\{\Pi_j\}_{j=1}^3$, which has independent interests. It is also remarked that, throughout this article, μ is not assumed to satisfy the reverse doubling condition.

Keywords: space of homogeneous type; paraproduct; $T(1)$ theorem; hardy space; bilinear estimate

Citation: Fu, X. Boundedness of Some Paraproducts on Spaces of Homogeneous Type. *Mathematics* **2021**, *9*, 2591. https://doi.org/10.3390/math9202591

Academic Editor: Jay Jahangiri

Received: 1 August 2021
Accepted: 11 October 2021
Published: 15 October 2021

1. Introduction

Classical paraproducts defined via convolutions are kinds of non-commutative bilinear operators, which are useful tools in the decompositions of products of functions. The prototypes of paraproducts can be found, for examples, in the work of Fujita and Kato [1] and Kato [2] on the study of mild solutions of Navier–Stokes equations and in the investigation of pseudo-differential operators and para-differential operators by Meyer and Coifman [3–5]. The formal notion of paraproducts has been introduced in 1981 by Bony for the study of the nonlinear hyperbolic partial differential equations in [6]. Since then the theory of papraproducts has been developed rapidly, which plays an essential role in both harmonic analysis and partial differential equations. For applications of paraproducts in harmonic analysis, we refer the reader to [7–16]. See also [17,18] for more applications of paraproducts in mathematical physics. The paraproducts defined via wavelets was first investigated by Grafakos and Torres [19] and then studied by Bonami et al. [20], which play crucial roles in both the bilinear decompositions of products of functions in [20,21], the (sub-)bilinear decompositions of commutators and the endpoint estimates of commutators in [22,23]. See the survey [24] and the monographs [25,26] for more information.

In 1970s, Coifman and Weiss [27,28] introduced the notion of the space of homogeneous type which has been proven to be a natural background for extensions of many classical results on Euclidean spaces. Recall that a *quasi-metric space* $(\mathcal{X}, d)$ is a non-empty set $\mathcal{X}$ equipped with a *quasi-metric* d such that, for any x, y, $z \in \mathcal{X}$,

(i) $d(x,y) = 0$ if and only if $x = y$;
(ii) $d(x,y) = d(y,x)$;
(iii) the *quasi-triangle inequality* $d(x,y) \le A_0[d(x,z) + d(z,y)]$ holds true, where $A_0 \in [1,\infty)$ is called the *quasi-triangle constant* which is independent of x, y and z.

The triple $(\mathcal{X}, d, \mu)$ is called a *space of homogeneous type* if μ is a non-negative measure satisfying the following *doubling condition*: there exists a positive constant $C_{(\mathcal{X})} \in [1,\infty)$, depending on $\mathcal{X}$, such that, for any $r \in (0,\infty)$ and $x \in \mathcal{X}$,

$$\mu(B(x,2r)) \le C_{(\mathcal{X})}\, \mu(B(x,r))$$

or, equivalently, there exists a positive constant C such that, for any $\lambda \in [1, \infty)$, $r \in (0, \infty)$ and $x \in \mathcal{X}$,

$$\mu(B(x, \lambda r)) \leq C\lambda^n \mu(B(x, r)), \tag{1}$$

where $B(x, r) := \{y \in \mathcal{X} : d(y, x) < r\}$ and $n := \log_2 C_{(\mathcal{X})}$ represents the "upper dimension" of $\mathcal{X}$.

As in ([29], Section 1) (see also ([30], Section 1)), throughout the whole article, we *always* assume that $(\mathcal{X}, d, \mu)$ is a space of homogeneous type satisfying the following additional assumptions:

(i) Suppose that, for any given $x \in \mathcal{X}$, the sequence of balls, $\{B(x, r)\}_{r \in (0,\infty)}$, in $\mathcal{X}$ is a basis of open neighborhoods around x.
(ii) Assume that μ satisfies that all the open sets are measurable and, for any measurable set $A \subset \mathcal{X}$, there exists a Borel set $E \supset A$ such that $\mu(A) = \mu(E)$, which is called *Borel regular*.
(iii) Suppose that, for any $x \in \mathcal{X}$ and $r \in (0, \infty)$, $\mu(B(x, r)) \in (0, \infty)$.
(iv) For the sake of the presentation simplicity, without loss of generality, we always assume that $\operatorname{diam}(\mathcal{X}) := \sup\{d(x, y) : x, y \in \mathcal{X}\} = \infty$ and $(\mathcal{X}, d, \mu)$ is *non-atomic*, that is, for any $x \in \mathcal{X}$, $\mu(\{x\}) = 0$.

It was shown in ([31], Lemma 5.1) or ([32], Lemma 8.1) (see also ([30], Section 1)) that, under the above assumptions, $\operatorname{diam}(\mathcal{X}) = \infty$ if and only if $\mu(\mathcal{X}) = \infty$.

A space of homogeneous type, $(\mathcal{X}, d, \mu)$, is called an RD-*space* introduced by Han et al. [33] (see also [34]) if μ further satisfies the following *reverse doubling condition* (or, for brevity, RD-*condition*): there exist positive constants a_0, $\widetilde{C}_{(\mathcal{X})} \in (1, \infty)$, depending on $\mathcal{X}$, such that, for any $x \in \mathcal{X}$ and $r \in (0, \operatorname{diam}(\mathcal{X})/a_0)$,

$$\mu(B(x, a_0 r)) \geq \widetilde{C}_{(\mathcal{X})} \mu(B(x, r)). \tag{2}$$

Notice that the harmonic analysis on spaces of homogeneous type has a long history; see, for example, [27,28,35,36]. We refer the reader to [33,34,37–46] for the real-variable theory of some function spaces and Calderón–Zygmund operators on RD-spaces. Furthermore, for some recent developments on the real-variable theory of function spaces and its applications on spaces of homogeneous type, please see [29,47–61].

Some progress is also made on the boundedness of paraproducts on metric measure spaces. Let $(\mathcal{X}, d, \mu)$ be an RD-space. Han et al. ([33], Theorem 5.56) extended the celebrated $T(1)$-theorem of David and Journé [11] to the RD-space via paraproducts. Later, Grafakos et al. [43] introduced a kind of paraproducts on $\mathcal{X}$, which extends the corresponding notion of paraproducts in ([33], Theorem 5.56), and investigated their boundedness from $H^p(\mathcal{X}) \times H^q(\mathcal{X})$ into $H^r(\mathcal{X})$ by (in)homogeneous Calderón reproducing formulae, which also generalizes a classical result on Euclidean spaces obtained by Grafakos and Kalton [14]. Grafakos et al. [43] also studied the endpoint estimates of paraproducts on $\mathcal{X}$ via the theory of Calderón–Zygmund operators. Moreover, via the off-diagonal estimates of integral kernels, Grafakos et al. [42] showed that a kind of bilinear discrete paraproducts on $\mathcal{X}$ via the theory of multilinear Calderón–Zygmund operators established in [42], are bounded on weighted Lebesgue spaces, Triebel–Lizorkin spaces and Besov spaces. Recently, Chang et al. [30,62] showed that the aforementioned boundedness of paraproducts on RD-spaces remains true on spaces of homogeneous type, namely, without having recourse to the RD-condition (2).

A space of homogeneous type, $(\mathcal{X}, d, \mu)$, is called a *metric measure space of homogeneous type* if the quasi-triangle constant $A_0 = 1$. In this setting, Fu et al. [48] proved that $f \times g$ of $f \in H^1_{\mathrm{at}}(\mathcal{X})$ and $g \in \mathrm{BMO}(\mathcal{X})$ can be written into a sum of three bilinear operators $\{\Pi_j\}_{j=1}^3$, which are also called paraproducts. These paraproducts play important roles in the study on the endpoint boundedness of the (sub-)linear commutator $[b, T]$ of a (sub-)linear operator T and $b \in \mathrm{BMO}(\mathcal{X})$ on (local) Hardy spaces in [29,57,58]; see also the survey [63] for more details. A natural question is whether there exists a relatively complete boundedness theory for paraproducts $\{\Pi_j\}_{j=1}^3$ in [48] which enjoy the same boundedness as the paraproducts in [30,62].

In this article, we give a partial affirmative answer to this question with the paraproducts $\{\Pi_j\}_{j=1}^3$ in [48] replaced by more general forms via the exp-ATIs and 1-exp-ATIs from [53]. We obtained the boundedness of Π_3 on Hardy spaces and its endpoint estimates, and the endpoint estimates for Π_1 and Π_2. The boundedness of Π_1 and Π_2 on Hardy spaces may need different approaches and was left as an unsolved question.

In what follows, we always assume that $(\mathcal{X}, d, \mu)$ is a space of homogeneous type. The remainder of this article is organized as follows.

Section 2 is devoted to some preliminary notions and results which are needed to the proof of the main results Theorems 2–4 below. In particular, we recall the $T(1)$ theorem from ([32], Section 12) (see Lemma 3 below), and use the Abel summation formula to build some relations among the boundedness of $\{\Pi_j\}_{j=1}^3$ (see Theorem 1 below).

In Section 3, we prove Theorems 2–4 below. In precise, Theorem 2 is an easy consequence of the Hölder inequality and the definition of $H^p(\mathcal{X})$. To show (i)–(iv) of Theorem 3, we first fix an $f \in \mathrm{BMO}(\mathcal{X})$ and express the paraproduct Π_3 by an integral operator $K_f^{(3)}$. Then, via the methods of interpolation and the crucial estimates (11) and (12), we show that $K_f^{(3)}$ has the weak boundedness property $\mathrm{WBP}(\eta)$ with η as in Lemma 2 below. Next we prove that the kernel of $K_f^{(3)}$ is an η-Calderón–Zygmund kernel, which also relies on estimates (11) and (12). Moreover, we point out that $K_f^{(3)}(1)$, $(K_f^{(3)})^*(1) \in \mathrm{BMO}(\mathcal{X})$, which, together with the $T(1)$ theorem from ([32], Theorem 12.2) and the boundedness of Calderón–Zygmund operators, we finally finish the proof of (i)-(iv) of Theorem 3. In order to prove (v) and (vi) of Theorem 3, we first fix $g \in L^\infty(\mathcal{X})$ and write Π_3 as an integral operator $K_g^{(3)}$. By the fact that $L^\infty(\mathcal{X}) \subset \mathrm{BMO}(\mathcal{X})$ and some arguments used in the proof of (i)–(iv) of Theorem 3, we obtain the desired results and finish the proof of Theorem 3. The proof of (i)–(iv) of Theorem 4 is a consequence of the arguments and ideas from the proof of (i)–(iv) of Theorem 3. The main novelty of this paper lies in the proof of (v)–(vi) of Theorem 4, where we use the Abel summation formula to build some relations among the boundedness of $\{\Pi_j\}_{j=1}^3$ and then transform the same boundedness of Π_1 from $L^2(\mathcal{X}) \times L^\infty(\mathcal{X})$ into $L^2(\mathcal{X})$ into the same boundedness of Π_2 and Π_3. We also remark that, throughout this article, μ is not assumed to satisfy the reverse doubling condition (2).

Finally, we list some notation used throughout this article. Let $\mathbb{N} := \{1, 2, \ldots\}$ and $\mathbb{Z}_+ := \{0\} \cup \mathbb{N}$. We use C or c to denote a *positive constant* which may be different from line to line, but is independent of main parameters. In addition, we also use $C_{(\rho,\alpha,\ldots)}$ or $c_{(\rho,\alpha,\ldots)}$ to denote a positive constant depending on the indicated parameters ρ, α, ... For any two real functions f and g, we write $f \lesssim g$ when $f \le Cg$ and $f \sim g$ when $f \lesssim g \lesssim f$. For any subset E of $\mathcal{X}$, denote by $\mathbf{1}_E$ its *characteristic function*. For any x, $y \in \mathcal{X}$, r, $\rho \in (0, \infty)$ and ball $B := B(x,r) := \{y \in \mathcal{X} : d(y,x) < r\}$, define $\rho B := B(x, \rho r)$, $V(x,r) := \mu(B(x,r)) =: V_r(x)$, and $V(x,y) := \mu(B(x, d(x,y)))$. For any $p \in [1, \infty]$, let p' denote its *conjugate index*, namely, $1/p + 1/p' = 1$. For any a, $b \in \mathbb{R}$, let $a \wedge b := \min\{a, b\}$ and $a \vee b := \max\{a, b\}$. Finally, for any linear integral operator T, we keep the notation T for its integral kernel.

2. Preliminary Notions and Results

In this section, we mainly state some preliminary notions and results which are needed to the proof of the main results Theorems 2–4 below. In particular, we investigate some relations among the boundedness of $\{\Pi_j\}_{j=1}^3$.

We first recall the notions of some function spaces. Let $q \in (0, \infty]$. The *Lebesgue space* $L^q(\mathcal{X})$ is defined to be the set of all μ-measurable functions f on $\mathcal{X}$ such that, if $q \in (0, \infty)$,

$$\|f\|_{L^q(\mathcal{X})} := \left[\int_{\mathcal{X}} |f(x)|^q \, d\mu(x)\right]^{1/q} < \infty;$$

if $q = \infty$, $\|f\|_{L^\infty(\mathcal{X})} := \operatorname*{ess\,sup}_{x\in\mathcal{X}} |f(x)| < \infty$, where $\operatorname*{ess\,sup}_{x\in\mathcal{X}} |f(x)|$ denotes the *essential supremum* of $|f|$ on $\mathcal{X}$. Denote by $L^1_{\mathrm{loc}}(\mathcal{X})$ the *space of all locally integrable functions*.

Let $s \in (0,1]$ and denote by $C(\mathcal{X})$ the *space of all continuous functions on* $\mathcal{X}$. Then the *homogeneous* and *inhomogeneous spaces* $C^s(\mathcal{X})$ and $\dot{C}^s(\mathcal{X})$ *of s-Hölder continuous functions on* $\mathcal{X}$ are, respectively, defined by setting

$$C^s(\mathcal{X}) := \left\{f \in C(\mathcal{X}) : \|f\|_{C^s(\mathcal{X})} < \infty\right\} \quad\text{and}\quad \dot{C}^s(\mathcal{X}) := \left\{f \in C(\mathcal{X}) : \|f\|_{\dot{C}^s(\mathcal{X})} < \infty\right\}$$

with

$$\|f\|_{C^s(\mathcal{X})} := \|f\|_{L^\infty(\mathcal{X})} + \|f\|_{\dot{C}^s(\mathcal{X})} \quad\text{and}\quad \|f\|_{\dot{C}^s(\mathcal{X})} := \sup_{\{(x,y)\in\mathcal{X}\times\mathcal{X}:\, x\neq y\}} \frac{|f(x)-f(y)|}{[d(x,y)]^s}.$$

Moreover, the *space* $C^s_b(\mathcal{X})$ *of all s-Hölder continuous functions with bounded support on* $\mathcal{X}$ is defined by setting

$$C^s_b(\mathcal{X}) := \{f \in C^s(\mathcal{X}) : f \text{ has bounded support}\},$$

where we equip $C^s_b(\mathcal{X})$ with the usual strict inductive limit topology (see, for instance, ([36], p. 273) and ([33], p. 23)). A useful subspace $\mathring{C}^s_b(\mathcal{X})$ of $C^s_b(\mathcal{X})$ is defined by setting $\mathring{C}^s_b(\mathcal{X}) := \{f \in C^s_b(\mathcal{X}) : \int_{\mathcal{X}} f(x)\,d\mu(x) = 0\}$. Moreover, the *dual space* $(C^s_b(\mathcal{X}))'$ [resp., $(\mathring{C}^s_b(\mathcal{X}))'$] of $C^s_b(\mathcal{X})$ [resp., $\mathring{C}^s_b(\mathcal{X})$] is defined to be the set of all linear functionals on $C^s_b(\mathcal{X})$ [resp., on $\mathring{C}^s_b(\mathcal{X})$] equipped with the weak-$*$ topology.

Definition 1 ([27,32,35])**.** *Let $s \in (0,1]$. A function $K : (\mathcal{X}\times\mathcal{X}) \setminus \{(x,x) : x \in \mathcal{X}\} \to \mathbb{C}$ is called an* s-Calderón–Zygmund kernel *if there exists a positive constant $C_{(K)}$, depending on K, such that*

(i) *for any x, $y \in \mathcal{X}$ with $x \neq y$,*

$$|K(x,y)| \le C_{(K)} \frac{1}{V(x,y)}; \tag{3}$$

(ii) *for any x, $\widetilde{x}$, $y \in \mathcal{X}$ satisfying $d(x,\widetilde{x}) \le (2A_0)^{-1} d(x,y)$ with $x \neq y$,*

$$|K(x,y) - K(\widetilde{x},y)| \le C_{(K)} \left[\frac{d(x,\widetilde{x})}{d(x,y)}\right]^s \frac{1}{V(x,y)} \tag{4}$$

and

$$|K(y,x) - K(y,\widetilde{x})| \le C_{(K)} \left[\frac{d(x,\widetilde{x})}{d(x,y)}\right]^s \frac{1}{V(x,y)}. \tag{5}$$

A linear operator $T : C^s_b(\mathcal{X}) \to (C^s_b(\mathcal{X}))'$ is called an s-Calderón–Zygmund operator *if T can be extended to a bounded linear operator on $L^2(\mathcal{X})$ and if there exists an s-Calderón–Zygmund kernel K such that, for any $f \in C^s_b(\mathcal{X})$ and $x \notin \operatorname{supp} f$, $Tf(x) := \int_{\mathcal{X}} K(x,y) f(y)\,d\mu(y)$.*

Definition 2 ([28])**.** *Let $p \in (0,1]$ and $q \in [1,\infty] \cap (p,\infty]$. A function a on $\mathcal{X}$ is called a* (p,q)-atom supported on a ball B *if (i) $\operatorname{supp} a \subset B$; (ii) $\|a\|_{L^q(\mathcal{X})} \le [\mu(B)]^{1/q-1/p}$; (iii) $\int_{\mathcal{X}} a(x)\,d\mu(x) = 0$, here and thereafter, for any measurable function f, $\operatorname{supp} f := \{x \in \mathcal{X} : f(x) \neq 0\}$.*

A function $f \in (\mathrm{Lip}_{1/p-1}(\mathcal{X}))'$ when $p \in (0,1)$, or $f \in L^1(\mathcal{X})$ when $p = 1$, is said to belong to the atomic Hardy space $H^{p,q}_{\mathrm{at}}(\mathcal{X})$ if there exist (p,q)-atoms $\{a_j\}_{j=1}^\infty$ and numbers

$\{\lambda_j\}_{j=1}^{\infty} \subset \mathbb{C}$ such that $\sum_{j=1}^{\infty} |\lambda_j|^p < \infty$ and $f = \sum_{j=1}^{\infty} \lambda_j a_j$ in $(\mathrm{Lip}_{1/p-1}(\mathcal{X}))'$ when $p \in (0,1)$, or in $L^1(\mathcal{X})$ when $p = 1$. Moreover, the quasi-norm *of f in $H_{\mathrm{at}}^{p,q}(\mathcal{X})$ is defined by setting*

$$\|f\|_{H_{\mathrm{at}}^{p,q}(\mathcal{X})} := \inf\left\{\left[\sum_{j=1}^{\infty} |\lambda_j|^p\right]^{1/p}\right\},$$

where the infimum is taken over all decompositions of f as above.

Let $p \in (0,1]$. It was shown in ([28], Theorem A) that $H_{\mathrm{at}}^{p,q}(\mathcal{X})$ is independent of the choice of $q \in [1,\infty] \cap (p,\infty]$ and hence simply denoted by $H_{\mathrm{at}}^{p}(\mathcal{X})$.

Definition 3 (([34], Definition 2.2) and ([33], Definition 2.8))**.** *Let $x_1 \in \mathcal{X}$ be fixed, $r, \vartheta \in (0,\infty)$ and $\kappa \in (0,1]$. The space $\mathcal{G}(x_1, r, \kappa, \vartheta)$ of test functions is defined to be the set of all measurable functions f on $\mathcal{X}$ such that there exists a positive constant C such that*

(T1) for any $x \in \mathcal{X}$,

$$|f(x)| \le C \frac{1}{\mu(B(x_1, r)) + V(x_1, x)} \left[\frac{r}{r + d(x_1, x)}\right]^{\vartheta};$$

(T2) for any $x, y \in \mathcal{X}$ with $d(x,y) \le [r + d(x_1, x)]/(2A_0)$,

$$|f(x) - f(y)| \le C \left[\frac{d(x,y)}{r + d(x_1, x)}\right]^{\kappa} \frac{1}{\mu(B(x_1, r)) + V(x_1, x)} \left[\frac{r}{r + d(x_1, x)}\right]^{\vartheta}.$$

Moreover, the norm *of f in $\mathcal{G}(x_1, r, \kappa, \vartheta)$ is defined by setting*

$$\|f\|_{\mathcal{G}(x_1, r, \kappa, \vartheta)} := \inf\{C : C \text{ satisfies (T1) and (T2)}\}.$$

It was shown in ([33], pp. 18–20) that, for any $x \in \mathcal{X}$ and $r \in (0,\infty)$,

$$\mathcal{G}(x, r, \kappa, \vartheta) = \mathcal{G}(x_1, 1, \kappa, \vartheta)$$

with equivalent norms, but the positive equivalence constants may depend on x and r and that $\mathcal{G}(x_1, 1, \kappa, \vartheta)$ is a Banach space. In what follows, for short, we write $\mathcal{G}(\kappa, \vartheta) := \mathcal{G}(x_1, 1, \kappa, \vartheta)$ and let $\mathring{\mathcal{G}}(\kappa, \vartheta) := \{f \in \mathcal{G}(\kappa, \vartheta) : \int_{\mathcal{X}} f(x)\, d\mu(x) = 0\}$.

Let $\varepsilon \in (0,1]$, $\kappa, \vartheta \in (0, \varepsilon]$ and $\mathcal{G}_0^{\varepsilon}(\kappa, \vartheta)$ [resp., $\mathring{\mathcal{G}}_0^{\varepsilon}(\kappa, \vartheta)$] be the completion of the space $\mathcal{G}(\varepsilon, \varepsilon)$ [resp., $\mathring{\mathcal{G}}(\varepsilon, \varepsilon)$] in the $\mathcal{G}(\kappa, \vartheta)$ norm. Moreover, if $f \in \mathcal{G}_0^{\varepsilon}(\kappa, \vartheta)$, we then let

$$\|f\|_{\mathcal{G}_0^{\varepsilon}(\kappa, \vartheta)} := \|f\|_{\mathcal{G}(\kappa, \vartheta)}.$$

The dual space *$(\mathcal{G}_0^{\varepsilon}(\kappa, \vartheta))'$ [resp., $(\mathring{\mathcal{G}}_0^{\varepsilon}(\kappa, \vartheta))'$] is defined to be the set of all continuous linear functionals on $\mathcal{G}_0^{\varepsilon}(\kappa, \vartheta)$ [resp., $\mathring{\mathcal{G}}_0^{\varepsilon}(\kappa, \vartheta)$] and equipped with the weak-$*$ topology.*

We then recall the following system of dyadic cubes given in ([64], Theorem 2.2), which was formulated in ([53], Lemma 2.3).

Lemma 1. *Fix constants c_0, C_0 and δ such that $0 < c_0 \le C_0 < \infty$, $\delta \in (0,1)$, and $12A_0^3 C_0 \delta \le c_0$. Assume that a set of points, $\{z_\alpha^k : k \in \mathbb{Z}, \alpha \in \mathscr{A}_k\} \subset \mathcal{X}$ with*

$$\mathscr{A}_k \text{ being a countable set of indices for any } k \in \mathbb{Z}, \tag{6}$$

satisfies the following properties: for any $k \in \mathbb{Z}$, (i) *$d(z_\alpha^k, z_\beta^k) \ge c_0 \delta^k$ when $\alpha \ne \beta$;* (ii) *for any $x \in \mathcal{X}$, $\min_{\alpha \in \mathscr{A}_k} d(x, z_\alpha^k) \le C_0 \delta^k$.*

Then there exists a family of sets, $\{Q_\alpha^k : k \in \mathbb{Z}, \alpha \in \mathscr{A}_k\}$*, which is called the* system of half-open dyadic cubes*, satisfying*

(iii) $\mathcal{X} = \bigcup_{\alpha\in\mathscr{A}_k} Q_\alpha^k$ *with* $\{Q_\alpha^k : \alpha \in \mathscr{A}_k\}$ *mutually disjoint;*

(iv) if $\ell \geq k$, $\alpha \in \mathscr{A}_k$ *and* $\beta \in \mathscr{A}_\ell$*, then either* $Q_\beta^\ell \subset Q_\alpha^k$ *or* $Q_\alpha^k \cap Q_\beta^\ell = \emptyset$ *holds true;*

(v) for any $\alpha \in \mathscr{A}_k$, $B(z_\alpha^k, c_\natural \delta^k) \subset Q_\alpha^k \subset B(z_\alpha^k, C^\natural \delta^k)$ *with* $c_\natural := (3A_0^2)^{-1}c_0$, $C^\natural := 2A_0C_0$ *and* z_α^k *being called the* "center" *of* Q_α^k.

In what follows, for any $k \in \mathbb{Z}$, let

$$\mathcal{X}^k := \{z_\alpha^k\}_{\alpha\in\mathscr{A}_k}, \quad \mathscr{G}_k := \mathscr{A}_{k+1} \setminus \mathscr{A}_k, \qquad \mathcal{Y}^k := \mathcal{X}^{k+1} \setminus \mathcal{X}^k =: \{y_\beta^k\}_{\beta\in\mathscr{G}_k}, \tag{7}$$

and, for any $y \in \mathcal{X}$, let $d(y, \mathcal{Y}^k) := \inf_{z\in\mathcal{Y}^k} d(y,z)$.

Based on the set $\{z_\alpha^k\}_{k\in\mathbb{Z},\alpha\in\mathscr{A}_k}$ [with $\mathscr{A}_k$ as in (6)] of points as in Lemma 1 and its related dyadic cubes, Auscher and Hytönen ([32], Theorem 7.1) constructed the following notable system $\{\psi_\beta^k\}_{k\in\mathbb{Z},\beta\in\mathscr{G}_k}$ of regular wavelets on $\mathcal{X}$, which is an orthonormal basis of $L^2(\mathcal{X})$.

Lemma 2. *There exist constants* $C, \nu \in (0,\infty)$, $a \in (0,1]$, $\eta \in (0,1)$*, and regular wavelets* $\{\psi_\beta^k\}_{k\in\mathbb{Z},\beta\in\mathscr{G}_k}$*, with* $\mathscr{G}_k$ *as in* (7)*, satisfying*

(i) for any $k \in \mathbb{Z}$, $\beta \in \mathscr{G}_k$ *and* $x \in \mathcal{X}$,

$$\left|\psi_\beta^k(x)\right| \leq C\frac{1}{\sqrt{V_{\delta^k}(y_\beta^k)}}\exp\left\{-\nu\left[\frac{d(y_\beta^k,x)}{\delta^k}\right]^a\right\};$$

(ii) for any $k \in \mathbb{Z}$, $\beta \in \mathscr{G}_k$ *and* $x, y \in \mathcal{X}$ *with* $d(x,y) \leq \delta^k$,

$$\left|\psi_\beta^k(x) - \psi_\beta^k(y)\right| \leq C\left[\frac{d(x,y)}{\delta^k}\right]^\eta \frac{1}{\sqrt{V_{\delta^k}(y_\beta^k)}}\exp\left\{-\nu\left[\frac{d(y_\beta^k,x)}{\delta^k}\right]^a\right\};$$

(iii) for any $k \in \mathbb{Z}$ *and* $\beta \in \mathscr{G}_k$, $\int_{\mathcal{X}} \psi_\beta^k(x)\,d\mu(x) = 0$ *with* $\{y_\beta^k\}_{k\in\mathbb{Z},\beta\in\mathscr{G}_k}$ *as in* (7).

Moreover, the system of regular wavelets $\{\psi_\beta^k\}_{k\in\mathbb{Z},\beta\in\mathscr{G}_k}$ *is both an orthonormal basis of* $L^2(\mathcal{X})$ *and an unconditional basis of* $L^p(\mathcal{X})$ *for any given* $p \in (1,\infty)$.

Definition 4 (([54], Definition 2.7), ([53], Definition 2.4) and ([30], Definition 2.3)). *A sequence* $\{Q_k\}_{k\in\mathbb{Z}}$ *of bounded linear integral operators on* $L^2(\mathcal{X})$ *is called an* approximation of the identity with exponential decay *(for short,* exp-ATI*) if there exist constants* $C, \nu \in (0,\infty)$, $a \in (0,1]$ *and* $\eta \in (0,1)$ *such that, for any* $k \in \mathbb{Z}$*, the kernel of the operator* Q_k*, which is still denoted by* Q_k*, satisfies*

*(i) (*the identity condition*)* $\sum_{k=-\infty}^{\infty} Q_k = I$ *in* $L^2(\mathcal{X})$*, where I denotes the* identity operator *on* $L^2(\mathcal{X})$*;*

*(ii) (*the size condition*) for any* $x, y \in \mathcal{X}$,

$$|Q_k(x,y)| \leq CR_k(x,y)$$

with

$$R_k(x,y) := \frac{1}{\sqrt{V_{\delta^k}(x)V_{\delta^k}(y)}}\exp\left\{-\nu\left[\frac{d(x,y)}{\delta^k}\right]^a\right\} \times \exp\left\{-\nu\left[\frac{\max\{d(x,\mathcal{Y}^k),d(y,\mathcal{Y}^k)\}}{\delta^k}\right]^a\right\};$$

(iii) *(the* regularity condition*) for any x, $\widetilde{x}$, $y \in \mathcal{X}$ with $d(x, \widetilde{x}) \le \delta^k$,*

$$|Q_k(x,y) - Q_k(\widetilde{x},y)| + |Q_k(y,x) - Q_k(y,\widetilde{x})| \le C\left[\frac{d(x,\widetilde{x})}{\delta^k}\right]^\eta R_k(x,y);$$

(iv) *(the* second difference regularity condition*) for any x, $\widetilde{x}$, y, $\widetilde{y} \in \mathcal{X}$ with $d(x, \widetilde{x}) \le \delta^k$ and $d(y, \widetilde{y}) \le \delta^k$,*

$$|[Q_k(x,y) - Q_k(\widetilde{x},y)] - [Q_k(x,\widetilde{y}) - Q_k(\widetilde{x},\widetilde{y})]| \le C\left[\frac{d(x,\widetilde{x})}{\delta^k}\right]^\eta \left[\frac{d(y,\widetilde{y})}{\delta^k}\right]^\eta R_k(x,y);$$

(v) *(the* cancellation condition*) for any x, $y \in \mathcal{X}$, $\int_{\mathcal{X}} Q_k(x,y)\,d\mu(y) = 0 = \int_{\mathcal{X}} Q_k(x,y)\,d\mu(x)$.*

Remark 1. *Let $\{\psi^k_\beta\}_{k\in\mathbb{Z},\,\beta\in\mathscr{G}_k}$ be as in Lemma 2. For any $k \in \mathbb{Z}$ and x, $y \in \mathcal{X}$, let*

$$D_k(x,y) := \sum_{\beta\in\mathscr{G}_k} \psi^k_\beta(x)\psi^k_\beta(y).$$

It was shown in ([54], p. 291) that the sequence $\{D_k\}_{k\in\mathbb{Z}}$ of linear integral operators associated with kernels $\{D_k(\cdot,\cdot)\}_{k\in\mathbb{Z}}$ satisfies all conditions (i)–(v) of Definition 4.

Definition 5 ([53], Definition 2.8)**.** *A sequence $\{P_k\}_{k\in\mathbb{Z}}$ of bounded linear integral operators on $L^2(\mathcal{X})$ is called an* approximation of the identity with exponential decay and integration 1 *(for short,* 1-exp-ATI*) if $\{P_k\}_{k\in\mathbb{Z}}$ has the following properties:*

(i) *for any $k \in \mathbb{Z}$, P_k satisfies (ii), (iii), and (iv) of Definition 4, but without the exponential decay factor*

$$\exp\left\{-\nu\left[\frac{\max\{d(x,\mathcal{Y}^k), d(y,\mathcal{Y}^k)\}}{\delta^k}\right]^a\right\}$$

with $\mathcal{Y}^k$ as in (7);

(ii) *$\int_{\mathcal{X}} P_k(x,y)\,d\mu(y) = 1 = \int_{\mathcal{X}} P_k(y,x)\,d\mu(y)$ for any $k \in \mathbb{Z}$ and $x \in \mathcal{X}$;*

(iii) *Let $Q_k := P_k - P_{k-1}$ for any $k \in \mathbb{Z}$. Then $\{Q_k\}_{k\in\mathbb{Z}}$ is an* exp-ATI.

Remark 2.

(i) *The existence of the* 1-exp-ATI *is ensured by ([32], Lemma 10.1) (see also ([53], Remark 2.9)).*

(ii) *For any given $p \in [1,\infty]$, P_k and hence Q_k are bounded on $L^p(\mathcal{X})$ uniformly in $k \in \mathbb{Z}$; see, for instance, ([54], Proposition 2.2(iii)).*

(iii) *It was shown that $\lim_{k\to\infty} P_k = I$ on $L^2(\mathcal{X})$; see, for example, ([53], Remark 2.9).*

Definition 6 (([53], Section 3 and Theorem 5.10) and ([62], Definition 1.1))**.** *Let κ, $\vartheta \in (0,\eta)$ with η as in Lemma 2, $\{P_k\}_{k\in\mathbb{Z}}$ be a* 1-exp-ATI *and $Q_k := P_k - P_{k-1}$ for any $k \in \mathbb{Z}$. Then, for any $f \in (\mathcal{G}^\eta_0(\kappa,\vartheta))'$, the* non-tangential maximal function $\mathcal{M}_\rho(f)$ of f, *with* aperture $\rho \in (0,\infty)$, *is defined by setting, for any $x \in \mathcal{X}$,*

$$\mathcal{M}_\rho(f)(x) := \sup_{k\in\mathbb{Z}} \sup_{y\in B(x,\rho\delta^k)} |P_k f(y)|.$$

Moreover, for any $f \in (\mathring{\mathcal{G}}^\eta_0(\kappa,\vartheta))'$, the Littlewood–Paley g-function $g(f)$ *of f is defined by setting, for any $x \in \mathcal{X}$,*

$$g(f)(x) := \left[\sum_{k\in\mathbb{Z}} |Q_k f(x)|^2\right]^{1/2}.$$

Let $p \in (0,\infty]$ and $\rho \in (0,\infty)$. Then the Hardy spaces $H^p_\rho(\mathcal{X})$ *and* $H^p(\mathcal{X})$ *are defined, respectively, by setting*

$$H^p_\rho(\mathcal{X}) := \left\{ f \in \left(\mathcal{G}^\eta_0(\kappa,\vartheta)\right)' : \|f\|_{H^p_\rho(\mathcal{X})} := \|\mathcal{M}_\rho(f)\|_{L^p(\mathcal{X})} < \infty \right\}$$

and

$$H^p(\mathcal{X}) := \left\{ f \in \left(\mathring{\mathcal{G}}^\eta_0(\kappa,\vartheta)\right)' : \|f\|_{H^p(\mathcal{X})} := \|g(f)\|_{L^p(\mathcal{X})} < \infty \right\}.$$

Remark 3. *Let $\rho \in (0,\infty)$, $p \in (n/(n+\eta),1]$ and κ, $\vartheta \in (n(1/p-1),\eta)$. It was shown in ([62], Remark 1.2) and ([60], Theorem 6.1) that*

(i) *$H^p_\rho(\mathcal{X})$ and $H^p_{\rm at}(\mathcal{X})$ coincide with equivalent quasi-norms;*
(ii) *$H^p(\mathcal{X}) = H^p_{\rm at}(\mathcal{X})$ with equivalent quasi-norms as subspaces of $(\mathring{\mathcal{G}}^\eta_0(\kappa,\vartheta))'$;*
(ii) *for any given $p \in (1,\infty)$, $H^p_\rho(\mathcal{X}) = L^p(\mathcal{X}) = H^p(\mathcal{X})$ with equivalent norms.*

We now introduce the following notion of paraproducts on $\mathcal{X}$ adapted from ([48], (3.2)).

Definition 7. *Let κ, $\vartheta \in (0,\eta)$ with η as in Lemma 2. Let $\{P_j\}_{j\in\mathbb{Z}}$ be a 1-exp-ATI and $Q_j := P_j - P_{j-1}$ for any $j \in \mathbb{Z}$. Then the* paraproduct Π_3 *is formally defined by setting, for any $f \in \left(\mathring{\mathcal{G}}^\eta_0(\kappa,\vartheta)\right)'$, $g \in \left(\mathring{\mathcal{G}}^\eta_0(\kappa,\vartheta)\right)'$ and $x \in \mathcal{X}$,*

$$\Pi_3(f,g)(x) := \sum_{j\in\mathbb{Z}} Q_j(f)(x) Q_j(g)(x),$$

where the series converges in $\left(\mathcal{G}^\eta_0(\kappa,\vartheta)\right)'$.

Remark 4. *In Theorems 2 and 3 below, we prove that $\Pi_3(f,g)$ in Definition 7 is well defined for any $(f,g) \in H^p(\mathcal{X}) \times H^q(\mathcal{X})$ with p, $q \in (0,\infty)$ and any $(f,g) \in$ BMO$(\mathcal{X}) \times C^\eta_b(\mathcal{X})$.*

Definition 8. *Let κ, $\vartheta \in (0,\eta)$ with η be as in Lemma 2. Let $\{P_j\}_{j\in\mathbb{Z}}$ be a 1-exp-ATI and $Q_j := P_j - P_{j-1}$ for any $j \in \mathbb{Z}$. Then the* paraproducts Π_1 *and* Π_2 *are formally defined, respectively, by setting*

(i) *for any $f \in (\mathring{\mathcal{G}}^\eta_0(\kappa,\vartheta))'$, $g \in (\mathcal{G}^\eta_0(\kappa,\vartheta))'$ and $x \in \mathcal{X}$,*

$$\Pi_1(f,g)(x) := \sum_{j\in\mathbb{Z}} Q_j(f)(x) P_j(g)(x);$$

(ii) *for any $f \in (\mathcal{G}^\eta_0(\kappa,\vartheta))'$, $g \in (\mathring{\mathcal{G}}^\eta_0(\kappa,\vartheta))'$ and $x \in \mathcal{X}$,*

$$\Pi_2(f,g)(x) := \sum_{j\in\mathbb{Z}} P_j(f)(x) Q_j(g)(x),$$

where the above two series converge in $\left(\mathcal{G}^\eta_0(\kappa,\vartheta)\right)'$.

Remark 5.

(i) *In Theorem 4 below, we show that $\Pi_1(f,g)$ in Definition 8 is well defined for any $(f,g) \in L^\infty(\mathcal{X}) \times C^\eta_b(\mathcal{X})$.*
(ii) *Due to the fact that $\Pi_2(f,g) = \Pi_1(g,f)$ for any proper functions f and g, we conclude that Π_2 shares corresponding boundedness to Π_1 as in Theorem 4 below.*

To prove Theorem 3 below, we need to recall the $T(1)$ theorem from ([32], Section 12). Let $\sigma \in (0,1)$ and $s \in (0,\sigma]$. A linear continuous operator $T : C^s_b(\mathcal{X}) \to (C^s_b(\mathcal{X}))'$ is said to have *weak boundedness property* WBP(σ) if there exists a positive constant C_1such that, for any $f, g \in C_b^\sigma(\mathcal{X})$ normalized by $\|f\|_{L^\infty(\mathcal{X})} + r^\sigma\|f\|_{\dot{C}^\sigma(\mathcal{X})} \le 1$ and $\|g\|_{L^\infty(\mathcal{X})} + r^\sigma\|g\|_{\dot{C}^\sigma(\mathcal{X})} \le 1$, with support in some ball $B(x,r)$ ($x \in \mathcal{X}$ and $r \in (0,\infty)$),

$$|(Tf,g)| \le C_1 V(x,r).$$

As for $T(1)$ with T associated with the s-Calderón–Zygmund kernel, it is defined as a continuous linear functional on $\mathring{C}_b^s(\mathcal{X})$ by setting

$$\langle T(1), f\rangle := \langle T(g), f\rangle + \int_{\mathcal{X}} (1-g(x))T^*(f)(x)\,d\mu(x), \tag{8}$$

where $g : \mathcal{X} \to \mathbb{R}$ satisfies that there exists a ball $B(x_0,r) \supset \operatorname{supp} f$ such that, for any $x \in \mathcal{X}$, $\mathbf{1}_{B(x_0,r)}(x) \le g(x) \le \mathbf{1}_{B(x_0,2A_0r)}(x)$. It is not difficult to show that both of the two terms in the right hand side of (8) are well defined.

Lemma 3. *Let $\sigma \in (0,1)$, $s \in (0,\sigma]$, $(\mathcal{X},d,\mu)$ be any space of homogeneous type and T be associated to an s-Calderón–Zygmund kernel. Then T can be extended to a bounded operator on $L^2(\mathcal{X})$ if and only if T has* WBP(s) *and $T(1)$, $T^*(1) \in$ BMO $(\mathcal{X})$.*

At the end of this section, we use the Abel summation formula to make some links among the boundedness of Π_1, Π_2 and Π_3 in some sense, which plays an important role in the proof of Theorem 4 below. In what follows, for any $N \in \mathbb{Z}$ and suitable functions f and g,

$$\Pi_1^{(N)}(f,g) := \sum_{j=-N}^{N} P_j(f)Q_j(g), \quad \Pi_2^{(N)}(f,g) := \sum_{j=-N}^{N} Q_j(f)P_j(g)$$

and $\Pi_3^{(N)}(f,g) := \sum_{j=-N}^{N} Q_j(f)Q_j(g)$.

Theorem 1. *Assume that there exists a positive constant C such that, for any $N \in \mathbb{N}$, $f \in L^2(\mathcal{X})$ and $g \in L^\infty(\mathcal{X})$,*

$$\left\|\Pi_2^{(N)}(f,g)\right\|_{L^2(\mathcal{X})} + \left\|\Pi_3^{(N)}(f,g)\right\|_{L^2(\mathcal{X})} \le C\|f\|_{L^2(\mathcal{X})}\|g\|_{L^\infty(\mathcal{X})}. \tag{9}$$

Then Π_1 defined as in Definition 8 is bounded from $L^2(\mathcal{X}) \times L^\infty(\mathcal{X})$ into $L^2(\mathcal{X})$.

Proof. Let $f \in L^2(\mathcal{X})$ and $g \in L^\infty(\mathcal{X})$. For any $N \in \mathbb{N}$, by the Abel summation formula, we know that

$$\begin{aligned}
\Pi_1^{(N)}(f,g) &= \sum_{j=-N}^{N} P_j(f)Q_j(g) = \sum_{j=-N}^{N} P_j(f)\left[P_j(g) - P_{j-1}(g)\right] \\
&= \sum_{j=-N}^{N} P_j(f)P_j(g) - \sum_{j=-N}^{N} P_j(f)P_{j-1}(g) \\
&= \sum_{j=-N}^{N} P_j(f)P_j(g) - \sum_{j=-N-1}^{N-1} P_{j+1}(f)P_j(g) \\
&= P_{N+1}(f)P_N(g) - P_{-N}(f)P_{-N-1}(g) \\
&\quad + \sum_{j=-N}^{N} \left[P_j(f) - P_{j+1}(f)\right]P_j(g) \\
&= P_{N+1}(f)P_N(g) - P_{-N}(f)P_{-N-1}(g)
\end{aligned}$$

$$
\begin{aligned}
&- \sum_{j=-N}^{N} Q_{j+1}(f)P_j(g) \\
&= P_{N+1}(f)P_N(g) - P_{-N}(f)P_{-N-1}(g) \\
&\quad - \sum_{j=-N}^{N} Q_{j+1}(f)P_{j+1}(g) + \sum_{j=-N}^{N} Q_{j+1}(f)\left[P_{j+1}(g) - P_j(g)\right] \\
&= P_{N+1}(f)P_N(g) - P_{-N}(f)P_{-N-1}(g) \\
&\quad - \sum_{j=-N}^{N} Q_{j+1}(f)P_{j+1}(g) + \sum_{j=-N}^{N} Q_{j+1}(f)Q_{j+1}(g) \\
&= P_{N+1}(f)P_N(g) - P_{-N}(f)P_{-N-1}(g) \\
&\quad - \sum_{j=-N+1}^{N+1} Q_j(f)P_j(g) + \sum_{j=-N+1}^{N+1} Q_j(f)Q_j(g) \\
&= P_{N+1}(f)P_N(g) - P_{-N}(f)P_{-N-1}(g) + Q_{-N}(f)P_{-N}(g) \\
&\quad - Q_{N+1}(f)P_{N+1}(g) + Q_{N+1}(f)Q_{N+1}(g) - Q_{-N}(f)Q_{-N}(g) \\
&\quad - \Pi_2^{(N)}(f,g) + \Pi_3^{(N)}(f,g).
\end{aligned}
$$

From this, (9) and Remark 2(ii), we deduce that

$$
\begin{aligned}
\left\|\Pi_1^{(N)}(f,g)\right\|_{L^2(\mathcal{X})} &\le \|P_{N+1}(f)P_N(g)\|_{L^2(\mathcal{X})} + \|P_{-N}(f)P_{-N-1}(g)\|_{L^2(\mathcal{X})} \\
&\quad + \|Q_{-N}(f)P_{-N}(g)\|_{L^2(\mathcal{X})} + \|Q_{N+1}(f)P_{N+1}(g)\|_{L^2(\mathcal{X})} \\
&\quad + \|Q_{N+1}(f)Q_{N+1}(g)\|_{L^2(\mathcal{X})} + \|Q_{-N}(f)Q_{-N}(g)\|_{L^2(\mathcal{X})} \\
&\quad + \left\|\Pi_2^{(N)}(f,g)\right\|_{L^2(\mathcal{X})} + \left\|\Pi_3^{(N)}(f,g)\right\|_{L^2(\mathcal{X})} \\
&\lesssim \|f\|_{L^2(\mathcal{X})}\|g\|_{L^\infty(\mathcal{X})},
\end{aligned}
$$

which, combined with the Fatou lemma, implies that

$$
\|\Pi_1(f,g)\|_{L^2(\mathcal{X})} \le \limsup_{N\to\infty} \left\|\Pi_1^{(N)}(f,g)\right\|_{L^2(\mathcal{X})} \lesssim \|f\|_{L^2(\mathcal{X})}\|g\|_{L^\infty(\mathcal{X})}.
$$

This completes the proof of Theorem 1. □

3. Boundedness of Paraproducts $\{\Pi_j\}_{j=1}^3$

This section is devoted to the proofs of the main results of this article on the boundedness of paraproducts $\{\Pi_j\}_{j=1}^3$.

We now state the first main result of this article as follows.

Theorem 2. *Let η be as in Lemma 2, p, q, $r \in (n/(n+\eta),\infty)$ with $1/r = 1/p + 1/q$, and κ, $\vartheta \in (\max\{0, n(1/r-1)\},\eta)$. Then the paraproduct Π_3 as in Definition 7 is a bounded bilinear operator from $H^p(\mathcal{X}) \times H^q(\mathcal{X})$ into $L^r(\mathcal{X})$.*

Remark 6.

(i) *Theorem 2 is an extension of ([48], Lemma 3.3).*

(ii) *It is still unclear whether Π_1 and Π_2 can be extended to bounded operators from $H^p(\mathcal{X}) \times H^q(\mathcal{X})$ into $H^r(\mathcal{X})$ or not.*

The following result is an easy consequence of Theorem 2, we omit the details here.

Corollary 1. *Let $q \in (1,\infty)$ and $q' := q/(q-1)$. Then the paraproduct Π_3 as in Definition 7 is a bounded bilinear operator from $L^q(\mathcal{X}) \times L^{q'}(\mathcal{X})$ into $L^1(\mathcal{X})$.*

Now, we are ready to prove Theorem 2.

Proof of Theorem 2. Let p, q, r, η, κ, ϑ, and Π_3 be as in Theorem 2. For any $(f,g) \in H^p(\mathcal{X}) \times H^q(\mathcal{X})$, we know that $f,g \in \left(\mathring{\mathcal{G}}_0^\eta(\kappa,\vartheta)\right)'$. By the Hölder inequality, we immediately have

$$\begin{aligned}
\|\Pi_3(f,g)\|_{L^r(\mathcal{X})} &\le \left\| \sum_{j\in\mathbb{Z}} |Q_j(f)Q_j(g)| \right\|_{L^r(\mathcal{X})} \\
&\le \left\| \left[\sum_{j\in\mathbb{Z}} |Q_j(f)|^2\right]^{1/2} \left[\sum_{j\in\mathbb{Z}} |Q_j(g)|^2\right]^{1/2} \right\|_{L^r(\mathcal{X})} \\
&\le \left\| \left[\sum_{j\in\mathbb{Z}} |Q_j(f)|^2\right]^{1/2} \right\|_{L^p(\mathcal{X})} \left\| \left[\sum_{j\in\mathbb{Z}} |Q_j(g)|^2\right]^{1/2} \right\|_{L^q(\mathcal{X})} \\
&= \|f\|_{H^p(\mathcal{X})} \|g\|_{H^q(\mathcal{X})},
\end{aligned}$$

which completes the proof of Theorem 2. □

Then we state other two main results of this article, which give various endpoint estimates of Π_3 and Π_1. In what follows, the *weak Lebesgue space* $L^{1,\infty}(\mathcal{X})$ is defined to be the set of all μ-measurable functions f on $\mathcal{X}$ such that

$$\|f\|_{L^{1,\infty}(\mathcal{X})} := \sup_{\lambda\in(0,\infty)} [\lambda\mu(\{x \in \mathcal{X} : |f(x)| > \lambda\})] < \infty,$$

and the *space* BMO $(\mathcal{X})$ the set of all locally integrable functions f on $\mathcal{X}$ such that

$$\|f\|_{\mathrm{BMO}(\mathcal{X})} := \sup_{B\subset\mathcal{X}} \frac{1}{\mu(B)} \int_B |f(x) - m_B(f)|\, d\mu(x) < \infty,$$

where the supremum is taken over all balls of $\mathcal{X}$ and, here and thereafter, for any locally integrable function f and a ball $B \subset \mathcal{X}$, $m_B(f) := \frac{1}{\mu(B)} \int_B f(y)\, d\mu(y)$.

Theorem 3. *Let η be as in Lemma 2, $q \in (1,\infty)$, κ, $\vartheta \in (\max\{0, n(1/q-1)\}, \eta)$, and Π_3 be as in Definition 7. Assume that the* exp-ATI, *$\{Q_j\}_{j\in\mathbb{Z}}$, further satisfies*

(a) $Q_j^* = Q_j$ *and* $Q_j^2 = Q_j$ *on* $L^2(\mathcal{X})$ *for any* $j \in \mathbb{Z}$*, namely,* $\{Q_j\}_{j\in\mathbb{Z}}$ *are* projection operators on $L^2(\mathcal{X})$.

(b) $\sum_{j\in\mathbb{Z}} Q_j = I$ *in* $H^1_{\mathrm{at}}(\mathcal{X})$.

Then Π_3 *can be extended to a bounded bilinear operator*

(i) *from* BMO $(\mathcal{X}) \times L^q(\mathcal{X})$ *into* $L^q(\mathcal{X})$*;*

(ii) *from* BMO $(\mathcal{X}) \times H^1_{\mathrm{at}}(\mathcal{X})$ *into* $L^1(\mathcal{X})$*;*

(iii) *from* BMO $(\mathcal{X}) \times L^\infty(\mathcal{X})$ *into* BMO $(\mathcal{X})$*;*

(iv) *from* BMO $(\mathcal{X}) \times L^1(\mathcal{X})$ *into* $L^{1,\infty}(\mathcal{X})$*;*

(v) *from* $L^q(\mathcal{X}) \times L^\infty(\mathcal{X})$ *into* $L^q(\mathcal{X})$*;*

(vi) *from* $L^1(\mathcal{X}) \times L^\infty(\mathcal{X})$ *into* $L^{1,\infty}(\mathcal{X})$*.*

Remark 7. *From ([32], Section 10) and ([65], Theorem 3.10), it follows that the sequence* $\{D_k\}_{k\in\mathbb{Z}}$ *in Remark 1 still satisfies all the assumptions in Theorem 3. Thus, Theorem 3(ii) is an extension of ([48], Theorem 4.9).*

The following result is a variant of ([62], Theorem 7).

Theorem 4. *Let η be as in Lemma 2, $q \in (1,\infty)$, $p \in (\frac{n}{n+\eta},\infty)$, κ, $\vartheta \in (\max\{0, n(1/p - 1)\}, \eta)$, and Π_1 be as in Definition 8. Assume that the* exp-ATI, *$\{Q_j\}_{j\in\mathbb{Z}}$, further satisfies, for any $f \in L^\infty(\mathcal{X})$ and $(1,2)$-atom h supported on some ball B_0,*

$$\int_{\mathcal{X}} \Pi_1(f,h)(x)\,d\mu(x) = 0. \tag{10}$$

Then Π_1 can be extended to a bounded bilinear operator

(i) from $L^\infty(\mathcal{X}) \times H^p(\mathcal{X})$ into $H^p(\mathcal{X})$;
(ii) from $L^\infty(\mathcal{X}) \times H^1_{\mathrm{at}}(\mathcal{X})$ into $L^1(\mathcal{X})$;
(iii) from $L^\infty(\mathcal{X}) \times L^\infty(\mathcal{X})$ into $\mathrm{BMO}\,(\mathcal{X})$;
(iv) from $L^\infty(\mathcal{X}) \times L^1(\mathcal{X})$ into $L^{1,\infty}(\mathcal{X})$.
(v) from $L^q(\mathcal{X}) \times L^\infty(\mathcal{X})$ into $L^q(\mathcal{X})$;
(vi) from $L^1(\mathcal{X}) \times L^\infty(\mathcal{X})$ into $L^{1,\infty}(\mathcal{X})$.

Remark 8.

(i) Let $f \in L^\infty(\mathcal{X}) \subset \mathrm{BMO}\,(\mathcal{X})$, h be a $(1,2)$-atom and $Q_k := D_k$ ($k \in \mathbb{Z}$) be as in Remark 1. By ([48], p. 985, lines 1–3 from the bottom), we have $\Pi_1(f,h) = H + hm_{B_0}(f)$ with $H \in H^1_{\mathrm{at}}(\mathcal{X})$, which, together with the fact that, for any $G \in H^1_{\mathrm{at}}(\mathcal{X})$, $\int_{\mathcal{X}} G(x)\,d\mu(x) = 0$, further implies that $\int_{\mathcal{X}} \Pi_1(f,h)(x)\,d\mu(x) = 0$. Therefore, the sequence $\{D_k\}_{k\in\mathbb{Z}}$ in Remark 1 still satisfies all the assumptions in Theorem 4.
(ii) It is still unknown what happens if we replace $f \in L^\infty(\mathcal{X})$ (resp., $g \in L^\infty(\mathcal{X})$) by $f \in \mathrm{BMO}\,(\mathcal{X})$ (resp., $g \in \mathrm{BMO}\,(\mathcal{X})$) in Theorem 4.

As in ([30], Remark 3.3) or ([62], Remark 1.8), the following estimates are important to escape the dependence on the RD-condition (2). For any given a, $c \in (0,\infty)$, and, for any $r \in (0,\infty)$ and $x \in \mathcal{X}$,

$$\sum_{\{k\in\mathbb{Z}:\ \delta^k \geq r\}} \frac{1}{V_{\delta^k}(x)} \exp\left\{-c\left[\frac{d(x,\mathcal{Y}^k)}{\delta^k}\right]^a\right\} \lesssim \frac{1}{V_r(x)} \tag{11}$$

(see ([32], Lemma 8.3)) and, for any x, $y \in \mathcal{X}$ with $x \neq y$,

$$\sum_{k\in\mathbb{Z}} \frac{1}{V_{\delta^k}(x)} \exp\left\{-c\left[\frac{d(x,y)}{\delta^k}\right]^a\right\} \exp\left\{-c\left[\frac{d(x,\mathcal{Y}^k)}{\delta^k}\right]^a\right\} \lesssim \frac{1}{V(x,y)}, \tag{12}$$

where the implicit positive constant is independent of x and y (see ([54], Lemma 4.9)), which essentially connect the geometrical properties of $\mathcal{X}$ expressed via its equipped quasi-metric d, dyadic reference points and dyadic cubes.

Now, we are ready to prove Theorem 3.

Proof of Theorem 3. Without loss of generality, we may assume that the sum $\sum_{j\in\mathbb{Z}}$ in $\Pi_3(f,g)$ is a finite sum $\sum_{j=-N}^{N}$ for any fixed $N \in \mathbb{N}$, see ([66], pp. 302–305) for some details.

We first prove (i)–(iv) of this theorem. To this end, we temporarily fix an $f \in \mathrm{BMO}\,(\mathcal{X})$. For any $x \in \mathcal{X}$, we write

$$\begin{aligned}\Pi_3(f,g)(x) &= \sum_{j\in\mathbb{Z}} Q_j(f)(x)Q_j(g)(x) = \int_{\mathcal{X}}\left[\sum_{j\in\mathbb{Z}} Q_j(x,y)Q_j(f)(x)\right]g(y)\,d\mu(y)\\ &=: \int_{\mathcal{X}} K_f^{(3)}(x,y)g(y)\,d\mu(y) =: K_f^{(3)}(g)(x),\end{aligned}$$

where $K_f^{(3)}$ is an integral operator associated with the kernel defined by setting, for any $x, y \in \mathcal{X}$,

$$K_f^{(3)}(x,y) := \sum_{j\in\mathbb{Z}} Q_j(x,y)Q_j(f)(x).$$

To prove (i)–(iv) of this theorem, the key point is the proof of the boundedness of $K_f^{(3)}$ on $L^2(\mathcal{X})$, where we need some ideas from ([67], Remark 4.4.5).

We first claim that $K_f^{(3)}$ has WBP(η) and hence maps from $C_b^\eta(\mathcal{X})$ into $(C_b^\eta(\mathcal{X}))'$. Indeed, let g, $h \in C_b^\eta(\mathcal{X})$, supported on some ball $B(x_0, r_0)$ with $x_0 \in \mathcal{X}$ and $r_0 \in (0,\infty)$, be normalized by

$$\|g\|_{L^\infty(\mathcal{X})} + r_0^\eta \|g\|_{\dot{C}_b^\eta(\mathcal{X})} \le 1 \quad \text{and} \quad \|h\|_{L^\infty(\mathcal{X})} + r_0^\eta \|h\|_{\dot{C}_b^\eta(\mathcal{X})} \le 1.$$

Then, by the fact from ([62], (2.3)) that

$$\sup_{j\in\mathbb{Z}} \|Q_j(f)\|_{L^\infty(\mathcal{X})} \lesssim \|f\|_{\mathrm{BMO}(\mathcal{X})} \tag{13}$$

and the Hölder inequality, we conclude that

$$\begin{aligned}
\left|\left\langle K_f^{(3)}(g), h\right\rangle\right| &= |\langle \Pi_3(f,g), h\rangle| = \left|\int_{\mathcal{X}} \Pi_3(f,g)(x)h(x)\,d\mu(x)\right| \\
&\le \sum_{j\in\mathbb{Z}} \int_{\mathcal{X}} |Q_j(f)(x)||Q_j(g)(x)||h(x)|\,d\mu(x) \\
&\lesssim \|f\|_{\mathrm{BMO}(\mathcal{X})} \sum_{j\in\mathbb{Z}} \int_{\mathcal{X}} |Q_j(g)(x)||h(x)|\,d\mu(x) \\
&\lesssim \|f\|_{\mathrm{BMO}(\mathcal{X})} \sum_{j\in\mathbb{Z}} \|Q_j(g)\|_{L^2(\mathcal{X})} \|h\|_{L^2(\mathcal{X})} \\
&\lesssim \|f\|_{\mathrm{BMO}(\mathcal{X})} [V(x_0,r_0)]^{1/2} \sum_{j\in\mathbb{Z}} \|Q_j(g)\|_{L^2(\mathcal{X})}.
\end{aligned}$$

Thus, to prove the above claim, it suffices to show that

$$\sum_{j\in\mathbb{Z}} \|Q_j(g)\|_{L^2(\mathcal{X})} \lesssim [V(x_0,r_0)]^{1/2}. \tag{14}$$

We further consider the following two cases.

(Case 1) $\delta^j \ge r_0$. Choose a fixed $x_1 \in B(x_0, 2r_0) \setminus B(x_0, r_0)$. Then, by (v) and (ii) of Definition 4 and (1), we have

$$\begin{aligned}
&|Q_j(g)(x)| \\
&\quad = \left|\int_{B(x_0,r_0)} Q_j(x,y)[g(y) - g(x_1)]\,d\mu(y)\right| \\
&\quad \le \int_{B(x_0,r_0)} |Q_j(x,y)||g(y) - g(x_1)|\,d\mu(y) \\
&\quad \lesssim \int_{B(x_0,r_0)} \frac{1}{V_{\delta^j}(x)} \exp\left\{-\frac{\nu}{2}\left[\frac{d(x,y)}{\delta^j}\right]^a\right\} \exp\left\{-\nu\left[\frac{d(y,\mathcal{Y}^j)}{\delta^j}\right]^a\right\} [d(y,x_1)]^\eta \|g\|_{\dot{C}_b^\eta(\mathcal{X})}\,d\mu(y) \\
&\quad \lesssim r_0^\eta \|g\|_{\dot{C}_b^\eta(\mathcal{X})} \int_{B(x_0,r_0)} \frac{1}{V_{\delta^j}(x)} \exp\left\{-\frac{\nu}{2}\left[\frac{d(x,y)}{\delta^j}\right]^a\right\} \exp\left\{-\frac{\nu}{2}\left[\frac{d(y,x_0)}{\delta^j}\right]^a\right\} \\
&\qquad \times \exp\left\{-\nu\left[\frac{d(y,\mathcal{Y}^j)}{\delta^j}\right]^a\right\} \exp\left\{-\nu\left[\frac{d(y,x_0)}{\delta^j}\right]^a\right\} d\mu(y)
\end{aligned} \tag{15}$$

$$\lesssim \int_{B(x_0,r_0)} \frac{1}{V_{\delta^j}(x)} \exp\left\{-\frac{\nu}{2}\left[\frac{d(x,x_0)}{A_0\delta^j}\right]^a\right\} \exp\left\{-\nu\left[\frac{d(x_0,\mathcal{Y}^j)}{A_0\delta^j}\right]^a\right\} d\mu(y)$$
$$\lesssim \frac{V(x_0,r_0)}{V_{\delta^j}(x_0)} \exp\left\{-\frac{\nu}{4}\left[\frac{d(x,x_0)}{A_0\delta^j}\right]^a\right\} \exp\left\{-\nu\left[\frac{d(x_0,\mathcal{Y}^j)}{A_0\delta^j}\right]^a\right\},$$

which implies that

$$\|Q_j(g)\|_{L^\infty(\mathcal{X})} \lesssim \frac{V(x_0,r_0)}{V_{\delta^j}(x_0)} \exp\left\{-\nu\left[\frac{d(x_0,\mathcal{Y}^j)}{A_0\delta^j}\right]^a\right\}.$$

On the other hand, from (15), we deduce that

$$\|Q_j(g)\|_{L^1(\mathcal{X})} \lesssim V(x_0,r_0) \int_{\mathcal{X}} \frac{1}{V_{\delta^j}(x_0)} \exp\left\{-\frac{\nu}{4}\left[\frac{d(x,x_0)}{A_0\delta^j}\right]^a\right\} d\mu(x) \exp\left\{-\nu\left[\frac{d(x_0,\mathcal{Y}^j)}{A_0\delta^j}\right]^a\right\}$$
$$\lesssim V(x_0,r_0) \exp\left\{-\nu\left[\frac{d(x_0,\mathcal{Y}^j)}{A_0\delta^j}\right]^a\right\}.$$

Thus,

$$\|Q_j(g)\|_{L^2(\mathcal{X})} \le \|Q_j(g)\|_{L^\infty(\mathcal{X})}^{1/2} \|Q_j(g)\|_{L^1(\mathcal{X})}^{1/2} \lesssim V(x_0,r_0) \frac{1}{\sqrt{V_{\delta^j}(x_0)}} \exp\left\{-\nu\left[\frac{d(x_0,\mathcal{Y}^j)}{A_0\delta^j}\right]^a\right\},$$

which, combined with (11), further implies that

$$\sum_{\{j\in\mathbb{Z}:\delta^j\ge r_0\}} \|Q_j(g)\|_{L^2(\mathcal{X})} \lesssim V(x_0,r_0) \sum_{\{j\in\mathbb{Z}:\delta^j\ge r_0\}} \frac{1}{\sqrt{V_{\delta^j}(x_0)}} \exp\left\{-\nu\left[\frac{d(x_0,\mathcal{Y}^j)}{A_0\delta^j}\right]^a\right\} \tag{16}$$
$$\lesssim \sqrt{V(x_0,r_0)}.$$

(Case 2) $\delta^j < r_0$. In this case, for a fixed $x \in \mathcal{X}$, by Definition 4(v), we first write

$$|Q_j(g)(x)| \le \int_{\mathcal{X}} |Q_j(x,y)||g(y)-g(x)|\, d\mu(y)$$
$$= \int_{B(x,\delta^j)} |Q_j(x,y)||g(y)-g(x)|\, d\mu(y) + \int_{\mathcal{X}\setminus B(x,\delta^j)} \cdots := \mathrm{I}_1 + \mathrm{I}_2.$$

Indeed, by Definition 4(ii) and (1), we have

$$\mathrm{I}_1 \lesssim \int_{B(x,\delta^j)} \frac{1}{V_{\delta^j}(x)} \exp\left\{-\frac{\nu}{2}\left[\frac{d(x,y)}{\delta^j}\right]^a\right\} [d(y,x)]^\eta \|g\|_{\mathring{C}_b^\eta(\mathcal{X})}\, d\mu(y)$$
$$\lesssim \left(\frac{\delta^j}{r_0}\right)^\eta \int_{B(x,\delta^j)} \frac{1}{V_{\delta^j}(x)} \exp\left\{-\frac{\nu}{2}\left[\frac{d(x,y)}{\delta^j}\right]^a\right\} d\mu(y) \lesssim \left(\frac{\delta^j}{r_0}\right)^\eta.$$

and

$$\mathrm{I}_2 \lesssim \sum_{k=1}^\infty \int_{B(x,2^k\delta^j)\setminus B(x,2^{k-1}\delta^j)} \frac{1}{V_{\delta^j}(x)} \exp\left\{-\frac{\nu}{2}\left[\frac{d(x,y)}{\delta^j}\right]^a\right\} [d(y,x)]^\eta \|g\|_{\mathring{C}_b^\eta(\mathcal{X})}\, d\mu(y)$$
$$\lesssim \sum_{k=1}^\infty 2^{k\eta}\left(\frac{\delta^j}{r_0}\right)^\eta \exp\left\{-\frac{\nu}{4}2^{(k-1)a}\right\} \int_{B(x,2^k\delta^j)} \frac{1}{V_{\delta^j}(x)} \exp\left\{-\frac{\nu}{4}\left[\frac{d(x,y)}{\delta^j}\right]^a\right\} d\mu(y)$$
$$\lesssim \sum_{k=1}^\infty 2^{k\eta} \exp\left\{-\frac{\nu}{4}2^{(k-1)a}\right\}\left(\frac{\delta^j}{r_0}\right)^\eta \lesssim \left(\frac{\delta^j}{r_0}\right)^\eta.$$

Combining the estimates of I_1 and I_2, we obtain

$$\|Q_j(g)\|_{L^\infty(\mathcal{X})} \lesssim \left(\frac{\delta^j}{r_0}\right)^\eta. \tag{17}$$

Now we estimate $\|Q_j(g)\|_{L^1(\mathcal{X})}$. Indeed, we know that

$$\begin{aligned}|Q_j(g)(x)| &\le \int_{B(x_0,r_0)} |Q_j(x,y)||g(y)|\,d\mu(y) \le \|g\|_{L^\infty(\mathcal{X})} \int_{B(x_0,r_0)} |Q_j(x,y)|\,d\mu(y)\\ &\le \int_{B(x_0,r_0)} |Q_j(x,y)|\,d\mu(y).\end{aligned}$$

For any fixed $x \in \mathcal{X}$, we further consider the following two cases.
Case 1 $d(x, x_0) < 2A_0 r_0$. Observe that, by Definition 4(ii) and (1),

$$|Q_j(g)(x)| \lesssim 1.$$

Case 2 $d(x, x_0) \ge 2A_0 r_0$. In this case, we observe that, for any $y \in B(x_0, r_0)$,

$$d(x,y) \ge \frac{d(x,x_0)}{A_0} - d(y,x_0) \ge \frac{d(x,x_0)}{2A_0}$$

and hence, by Definition 4(ii) and (1),

$$\begin{aligned}|Q_j(g)(x)| &\lesssim \int_{B(x_0,r_0)} \frac{1}{V_{\delta^j}(x)} \exp\left\{-\frac{\nu}{2}\left[\frac{d(x,y)}{\delta^j}\right]^a\right\} d\mu(y)\\ &\lesssim \int_{B(x_0,r_0)} \frac{1}{V_{\delta^j}(x)} \exp\left\{-\frac{\nu}{2}\left[\frac{d(x,x_0))}{2A_0\delta^j}\right]^a\right\} d\mu(y)\\ &\lesssim \frac{V(x_0,r_0)}{V_{\delta^j}(x_0)} \exp\left\{-\frac{\nu}{4}\left[\frac{d(x,x_0))}{2A_0\delta^j}\right]^a\right\}.\end{aligned}$$

Combining Cases 1 and 2, we find that, for any $x \in \mathcal{X}$,

$$|Q_j(g)(x)| \lesssim \mathbf{1}_{B(x_0,2A_0r_0)}(x) + \mathbf{1}_{\mathcal{X}\setminus B(x_0,2A_0r_0)}(x) \frac{V(x_0,r_0)}{V_{\delta^j}(x_0)} \exp\left\{-\frac{\nu}{4}\left[\frac{d(x,x_0))}{2A_0\delta^j}\right]^a\right\},$$

which implies that

$$\|Q_j(g)\|_{L^1(\mathcal{X})} \lesssim V(x_0, r_0).$$

From this and and (17), it follows that

$$\|Q_j(g)\|_{L^2(\mathcal{X})} \le \|Q_j(g)\|^{1/2}_{L^\infty(\mathcal{X})} \|Q_j(g)\|^{1/2}_{L^1(\mathcal{X})} \lesssim \left(\frac{\delta^j}{r_0}\right)^{\eta/2} \sqrt{V(x_0,r_0)},$$

which further implies that

$$\sum_{\{j\in\mathbb{Z}:\ \delta^j<r_0\}} \|Q_j(g)\|_{L^2(\mathcal{X})} \lesssim r_0^{-\eta/2} \sum_{\{j\in\mathbb{Z}:\ \delta^j<r_0\}} \delta^{j\eta/2}\sqrt{V(x_0,r_0)} \lesssim \sqrt{V(x_0,r_0)}.$$

By this and (16), we conclude that

$$\sum_{j\in\mathbb{Z}} \|Q_j(g)\|_{L^2(\mathcal{X})} \lesssim \sqrt{V(x_0,r_0)},$$

which further completes the proof of the above claim.

Now we begin to show that $K_f^{(3)}(\cdot,\cdot)$ satisfies (3) through (5). To achieve this, by (13), Definition 4(ii), (1) and (12), we find that, for any $x,\ y \in \mathcal{X}$ with $x \neq y$,

$$\left|K_f^{(3)}(x,y)\right| \leq \sum_{j\in\mathbb{Z}} |Q_j(x,y)||Q_j(f)(x)| \lesssim \|f\|_{\mathrm{BMO}(\mathcal{X})} \sum_{j\in\mathbb{Z}} |Q_j(x,y)| \tag{18}$$
$$\lesssim \|f\|_{\mathrm{BMO}(\mathcal{X})} \sum_{j\in\mathbb{Z}} \frac{1}{V_{\delta^j}(x)} \exp\left\{-\frac{\nu}{2}\left[\frac{d(x,y)}{\delta^j}\right]^a\right\} \exp\left\{-\frac{\nu}{2}\left[\frac{d(x,\mathcal{Y}^j)}{\delta^j}\right]^a\right\}$$
$$\lesssim \|f\|_{\mathrm{BMO}(\mathcal{X})} \frac{1}{V(x,y)}.$$

This shows that $K_f^{(3)}(\cdot,\cdot)$ satisfies (3).

Then we prove that $K_f^{(3)}(\cdot,\cdot)$ satisfies (4). Indeed, let $x,\ \widetilde{x},\ y \in \mathcal{X}$ with $d(x,\widetilde{x}) \leq \frac{1}{2A_0}d(x,y)$ and $x \neq y$. We observe that

$$d(\widetilde{x},y) \geq \frac{d(x,y)}{A_0} - d(x,\widetilde{x}) \geq \frac{d(x,y)}{2A_0}. \tag{19}$$

From (13), it follows that

$$\left|K_f^{(3)}(x,y) - K_f^{(3)}(\widetilde{x},y)\right| \leq \sum_{j\in\mathbb{Z}} |Q_j(f)(y)||Q_j(x,y) - Q_j(\widetilde{x},y)|$$
$$\lesssim \|f\|_{\mathrm{BMO}(\mathcal{X})} \sum_{j\in\mathbb{Z}} |Q_j(x,y) - Q_j(\widetilde{x},y)|.$$

We further consider the following two cases.

Case (1) $d(x,\widetilde{x}) \leq \delta^j$. In this case, by Definition 4(iii) and (1), we have

$$|Q_j(x,y) - Q_j(\widetilde{x},y)| \lesssim \left[\frac{d(x,\widetilde{x})}{\delta^j}\right]^\eta \frac{1}{V_{\delta^j}(x)} \exp\left\{-\frac{\nu}{2}\left[\frac{d(x,y)}{\delta^j}\right]^a\right\} \exp\left\{-\nu\left[\frac{d(x,\mathcal{Y}^j)}{\delta^j}\right]^a\right\}$$
$$\lesssim \left[\frac{d(x,\widetilde{x})}{d(x,y)}\right]^\eta \frac{1}{V_{\delta^j}(x)} \exp\left\{-\frac{\nu}{4}\left[\frac{d(x,y)}{\delta^j}\right]^a\right\} \exp\left\{-\frac{\nu}{4}\left[\frac{d(x,\mathcal{Y}^j)}{\delta^j}\right]^a\right\},$$

which, together with (12), implies that

$$\sum_{\{j\in\mathbb{Z}:\ \delta^j \geq d(x,\widetilde{x})\}} |Q_j(x,y) - Q_j(\widetilde{x},y)| \lesssim \left[\frac{d(x,\widetilde{x})}{d(x,y)}\right]^\eta \sum_{\{j\in\mathbb{Z}:\ \delta^j \geq d(x,\widetilde{x})\}} \frac{1}{V_{\delta^j}(x)} \exp\left\{-\frac{\nu}{4}\left[\frac{d(x,y)}{\delta^j}\right]^a\right\}$$
$$\times \exp\left\{-\frac{\nu}{4}\left[\frac{d(x,\mathcal{Y}^j)}{\delta^j}\right]^a\right\}$$
$$\lesssim \left[\frac{d(x,\widetilde{x})}{d(x,y)}\right]^\eta \frac{1}{V(x,y)}.$$

Case (2) $d(x,\widetilde{x}) > \delta^j$. In this case, from Definition 4(ii), (1) and (19), we deduce that

$$|Q_j(x,y) - Q_j(\widetilde{x},y)| \leq |Q_j(x,y)| + |Q_j(\widetilde{x},y)|$$
$$\lesssim \frac{1}{V_{\delta^j}(y)} \exp\left\{-\frac{\nu}{2}\left[\frac{d(x,y)}{\delta^j}\right]^a\right\} \exp\left\{-\nu\left[\frac{d(y,\mathcal{Y}^j)}{\delta^j}\right]^a\right\}$$
$$+ \frac{1}{V_{\delta^j}(y)} \exp\left\{-\frac{\nu}{2}\left[\frac{d(\widetilde{x},y)}{\delta^j}\right]^a\right\} \exp\left\{-\nu\left[\frac{d(y,\mathcal{Y}^j)}{\delta^j}\right]^a\right\}$$

$$\lesssim \frac{1}{V_{\delta^j}(y)} \exp\left\{-\frac{\nu}{2}\left[\frac{d(x,y)}{2A_0\delta^j}\right]^a\right\} \exp\left\{-\nu\left[\frac{d(y,\mathcal{Y}^j)}{\delta^j}\right]^a\right\}$$
$$\lesssim \left[\frac{d(x,\widetilde{x})}{\delta^j}\right]^\eta \frac{1}{V_{\delta^j}(y)} \exp\left\{-\frac{\nu}{2}\left[\frac{d(x,y)}{2A_0\delta^j}\right]^a\right\} \exp\left\{-\nu\left[\frac{d(y,\mathcal{Y}^j)}{\delta^j}\right]^a\right\}$$
$$\lesssim \left[\frac{d(x,\widetilde{x})}{d(x,y)}\right]^\eta \frac{1}{V_{\delta^j}(y)} \exp\left\{-\frac{\nu}{4}\left[\frac{d(x,y)}{2A_0\delta^j}\right]^a\right\} \exp\left\{-\frac{\nu}{4}\left[\frac{d(y,\mathcal{Y}^j)}{2A_0\delta^j}\right]^a\right\}.$$

Thus, by (12), we obtain

$$\sum_{\{j\in\mathbb{Z}:\ \delta^j<d(x,\widetilde{x})\}} |Q_j(x,y)-Q_j(\widetilde{x},y)|$$
$$\lesssim \left[\frac{d(x,\widetilde{x})}{d(x,y)}\right]^a \sum_{\{j\in\mathbb{Z}:\ \delta^j<d(x,\widetilde{x})\}} \frac{1}{V_{\delta^j}(y)} \exp\left\{-\frac{\nu}{4}\left[\frac{d(x,y)}{2A_0\delta^j}\right]^a\right\} \exp\left\{-\frac{\nu}{4}\left[\frac{d(y,\mathcal{Y}^j)}{2A_0\delta^j}\right]^a\right\}$$
$$\lesssim \left[\frac{d(x,\widetilde{x})}{d(x,y)}\right]^\eta \frac{1}{V(x,y)}.$$

Combining the Cases (1) and (2), we have

$$\sum_{j\in\mathbb{Z}} |Q_j(x,y)-Q_j(\widetilde{x},y)| \lesssim \left[\frac{d(x,\widetilde{x})}{d(x,y)}\right]^\eta \frac{1}{V(x,y)}, \tag{20}$$

which further proves that $K_f^{(3)}(\cdot,\cdot)$ satisfies (4). By the arguments similar to those used in the proof of (20), we conclude that

$$\sum_{\{j\in\mathbb{Z}\}} |Q_j(y,x)-Q_j(y,\widetilde{x})| \lesssim \left[\frac{d(x,\widetilde{x})}{d(x,y)}\right]^\eta \frac{1}{V(x,y)}. \tag{21}$$

We further show that $K_f^{(3)}(\cdot,\cdot)$ satisfies (5). Indeed, let x, $\widetilde{x}, y \in \mathcal{X}$ with $d(x,\widetilde{x}) \le \frac{1}{2A_0}d(x,y)$ and $x \neq y$. From (13), Definition 4(v), (1) and (21), we deduce that

$$\left|K_f^{(3)}(y,x)-K_f^{(3)}(y,\widetilde{x})\right|$$
$$\le \sum_{j\in\mathbb{Z}} |Q_j(y,x)Q_j(f)(x)-Q_j(y,\widetilde{x})Q_j(f)(\widetilde{x})|$$
$$\le \sum_{j\in\mathbb{Z}} |Q_j(y,x)-Q_j(y,\widetilde{x})||Q_j(f)(x)| + \sum_{j\in\mathbb{Z}} |Q_j(y,\widetilde{x})||Q_j(f)(x)-Q_j(f)(\widetilde{x})|$$
$$\lesssim \|f\|_{\mathrm{BMO}(\mathcal{X})} \sum_{j\in\mathbb{Z}} |Q_j(y,x)-Q_j(y,\widetilde{x})|$$
$$+ \sum_{j\in\mathbb{Z}} |Q_j(y,\widetilde{x})| \int_{\mathcal{X}} |Q_j(x,z)-Q_j(\widetilde{x},z)| \left|f(z)-m_{B(x,\delta^j)}(f)\right| d\mu(z)$$
$$\lesssim \|f\|_{\mathrm{BMO}(\mathcal{X})} \left[\frac{d(x,\widetilde{x})}{d(y,x)}\right]^\eta \frac{1}{V(x,y)} + \mathrm{A},$$

where

$$\mathrm{A} := \sum_{j\in\mathbb{Z}} |Q_j(y,\widetilde{x})| \int_{\mathcal{X}} |Q_j(x,z)-Q_j(\widetilde{x},z)| \left|f(z)-m_{B(x,\delta^j)}(f)\right| d\mu(z).$$

To estimate A, we deal with the following two cases.

Case (i) $d(x,\widetilde{x}) \le \delta^j$. By (ii) and (iii) of Definition 4, (1), (19), some arguments similar to those used in the proof of ([62], (2.3)) and (12), we conclude that

$$\begin{aligned}
\mathrm{A} &\lesssim \sum_{j\in\mathbb{Z}} \frac{1}{V_{\delta^j}(y)} \exp\left\{-\frac{\nu}{2}\left[\frac{d(\widetilde{x},y)}{\delta^j}\right]^a\right\} \exp\left\{-\nu\left[\frac{d(y,\mathcal{Y}^j)}{\delta^j}\right]^a\right\} \\
&\quad\times \int_{\mathcal{X}} \left[\frac{d(x,\widetilde{x})}{\delta^j}\right]^\eta \frac{1}{\sqrt{V_{\delta^j}(x)V_{\delta^j}(\widetilde{x})}} \exp\left\{-\frac{\nu}{4}\left[\frac{d(x,z)}{\delta^j}\right]^a\right\} \exp\left\{-\frac{\nu}{4}\left[\frac{d(\widetilde{x},z)}{\delta^j}\right]^a\right\} \\
&\quad\times \left|f(z) - m_{B(x,\delta^j)}(f)\right| d\mu(z) \\
&\lesssim \sum_{j\in\mathbb{Z}} \frac{1}{V_{\delta^j}(y)} \exp\left\{-\frac{\nu}{2}\left[\frac{d(x,y)}{2A_0\delta^j}\right]^a\right\} \exp\left\{-\nu\left[\frac{d(y,\mathcal{Y}^j)}{\delta^j}\right]^a\right\} \\
&\quad\times \int_{\mathcal{X}} \left[\frac{d(x,\widetilde{x})}{\delta^j}\right]^\eta \frac{1}{\sqrt{V_{\delta^j}(x)V_{\delta^j}(\widetilde{x})}} \exp\left\{-\frac{\nu}{8}\left[\frac{d(x,z)}{\delta^j}\right]^a\right\} \exp\left\{-\frac{\nu}{8}\left[\frac{d(x,\widetilde{x})}{A_0\delta^j}\right]^a\right\} \\
&\quad\times \left|f(z) - m_{B(x,\delta^j)}(f)\right| d\mu(z) \\
&\lesssim \sum_{j\in\mathbb{Z}} \frac{1}{V_{\delta^j}(y)} \exp\left\{-\frac{\nu}{4}\left[\frac{d(x,y)}{2A_0\delta^j}\right]^a\right\} \exp\left\{-\nu\left[\frac{d(y,\mathcal{Y}^j)}{\delta^j}\right]^a\right\} \\
&\quad\times \int_{\mathcal{X}} \left[\frac{d(x,\widetilde{x})}{d(x,y)}\right]^\eta \frac{1}{V_{\delta^j}(x)} \exp\left\{-\frac{\nu}{8}\left[\frac{d(x,z)}{\delta^j}\right]^a\right\} \left|f(z) - m_{B(x,\delta^j)}(f)\right| d\mu(z) \\
&\lesssim \|f\|_{\mathrm{BMO}(\mathcal{X})} \left[\frac{d(x,\widetilde{x})}{d(x,y)}\right]^\eta \sum_{j\in\mathbb{Z}} \frac{1}{V_{\delta^j}(y)} \exp\left\{-\frac{\nu}{4}\left[\frac{d(x,y)}{2A_0\delta^j}\right]^a\right\} \exp\left\{-\frac{\nu}{4}\left[\frac{d(y,\mathcal{Y}^j)}{2A_0\delta^j}\right]^a\right\} \\
&\lesssim \|f\|_{\mathrm{BMO}(\mathcal{X})} \left[\frac{d(x,\widetilde{x})}{d(y,x)}\right]^\eta \frac{1}{V(x,y)}.
\end{aligned}$$

Case (ii) $d(x,\widetilde{x}) > \delta^j$. From Definition 4(ii), (1), (19), some arguments similar to those used in the proof of ([62], (2.3)) and (12), we deduce that

$$\begin{aligned}
\mathrm{A} &\lesssim \sum_{j\in\mathbb{Z}} \frac{1}{V_{\delta^j}(y)} \exp\left\{-\frac{\nu}{2}\left[\frac{d(\widetilde{x},y)}{\delta^j}\right]^a\right\} \exp\left\{-\nu\left[\frac{d(y,\mathcal{Y}^j)}{\delta^j}\right]^a\right\} \qquad (22) \\
&\quad\times \int_{\mathcal{X}} \left[\frac{d(x,\widetilde{x})}{\delta^j}\right]^\eta \left[\frac{1}{V_{\delta^j}(x)} \exp\left\{-\frac{\nu}{2}\left[\frac{d(x,z)}{\delta^j}\right]^a\right\} + \frac{1}{V_{\delta^j}(z)} \exp\left\{-\frac{\nu}{2}\left[\frac{d(\widetilde{x},z)}{\delta^j}\right]^a\right\}\right] \\
&\quad\times \left|f(z) - m_{B(x,\delta^j)}(f)\right| d\mu(z) \\
&\lesssim \sum_{j\in\mathbb{Z}} \frac{1}{V_{\delta^j}(y)} \exp\left\{-\frac{\nu}{4}\left[\frac{d(\widetilde{x},y)}{\delta^j}\right]^a\right\} \exp\left\{-\frac{\nu}{4}\left[\frac{d(x,y)}{2A_0\delta^j}\right]^a\right\} \exp\left\{-\nu\left[\frac{d(y,\mathcal{Y}^j)}{\delta^j}\right]^a\right\} \\
&\quad\times \int_{\mathcal{X}} \left[\frac{d(x,\widetilde{x})}{\delta^j}\right]^\eta \left[\frac{1}{V_{\delta^j}(x)} \exp\left\{-\frac{\nu}{2}\left[\frac{d(x,z)}{\delta^j}\right]^a\right\} + \frac{1}{V_{\delta^j}(z)} \exp\left\{-\frac{\nu}{2}\left[\frac{d(\widetilde{x},z)}{\delta^j}\right]^a\right\}\right] \\
&\quad\times \left|f(z) - m_{B(x,\delta^j)}(f)\right| d\mu(z) \\
&\lesssim \left[\frac{d(x,\widetilde{x})}{d(y,x)}\right]^\eta \sum_{j\in\mathbb{Z}} \frac{1}{V_{\delta^j}(y)} \exp\left\{-\frac{\nu}{8}\left[\frac{d(x,y)}{2A_0\delta^j}\right]^a\right\} \exp\left\{-\nu\left[\frac{d(y,\mathcal{Y}^j)}{\delta^j}\right]^a\right\} \\
&\quad\times \int_{\mathcal{X}} \frac{1}{V_{\delta^j}(x)} \exp\left\{-\frac{\nu}{2}\left[\frac{d(x,z)}{\delta^j}\right]^a\right\} \left|f(z) - m_{B(x,\delta^j)}(f)\right| d\mu(z) \\
&\quad+ \left[\frac{d(x,\widetilde{x})}{d(y,x)}\right]^\eta \sum_{j\in\mathbb{Z}} \frac{1}{V_{\delta^j}(y)} \exp\left\{-\frac{\nu}{16}\left[\frac{d(x,y)}{2A_0\delta^j}\right]^a\right\} \exp\left\{-\nu\left[\frac{d(y,\mathcal{Y}^j)}{\delta^j}\right]^a\right\}
\end{aligned}$$

$$\times \int_{\mathcal{X}} \frac{1}{V_{\delta^j}(z)} \exp\left\{-\frac{\nu}{16}\left[\frac{d(x,z)}{2A_0^2\delta^j}\right]^a\right\} \left|f(z) - m_{B(x,\delta^j)}(f)\right| d\mu(z)$$
$$\lesssim \|f\|_{\mathrm{BMO}\,(\mathcal{X})} \left[\frac{d(x,\widetilde{x})}{d(x,y)}\right]^\eta \sum_{j\in\mathbb{Z}} \frac{1}{V_{\delta^j}(y)} \exp\left\{-\frac{\nu}{8}\left[\frac{d(x,y)}{2A_0\delta^j}\right]^a\right\} \exp\left\{-\frac{\nu}{8}\left[\frac{d(y,\mathcal{Y}^j)}{2A_0\delta^j}\right]^a\right\}$$
$$\lesssim \|f\|_{\mathrm{BMO}\,(\mathcal{X})} \left[\frac{d(x,\widetilde{x})}{d(y,x)}\right]^\eta \frac{1}{V(x,y)}.$$

Combining Cases (i) and (ii), we know that $K_f^{(3)}(\cdot,\cdot)$ satisfies (5). These complete the proof of (3) through (5) for $K_f^{(3)}(\cdot,\cdot)$.

Next we show that $K_f^{(3)}(1)$, $(K_f^{(3)})^*(1) \in \mathrm{BMO}\,(\mathcal{X})$. Obviously, $K_f^{(3)}(1) = 0 \in \mathrm{BMO}\,(\mathcal{X})$. Now we prove that $(K_f^{(3)})^*(1) \in \mathrm{BMO}\,(\mathcal{X})$. It is easy to see that

$$\left(K_f^{(3)}\right)^*(x,y) = K_f^{(3)}(y,x) = \sum_{j\in\mathbb{Z}} Q_j(y,x)Q_j(f)(y).$$

For any $h \in \mathring{C}_b^\eta(\mathcal{X})$ with $\operatorname{supp} h \subset B(x_0,r_0)$ and any $N \in \mathbb{N}$, choose $\eta_N \in C_b^\eta(\mathcal{X})$ with $\eta_N \equiv 1$ on $B(x_0,2A_0Nr_0)$, $\operatorname{supp}(\eta_N) \subset B(x_0,4A_0Nr_0)$, and $0 \le \eta_N \le 1$. We write

$$\left\langle \left(K_f^{(3)}\right)^*(1),h\right\rangle = \left\langle \left(K_f^{(3)}\right)^*(\eta_N),h\right\rangle + \int_{\mathcal{X}} [1-\eta_N(x)]K_f^{(3)}(h)(x)\,d\mu(x) =: \mathrm{I}_N + \mathrm{II}_N.$$

By $\int_{\mathcal{X}} h(y)\,d\mu(y) = 0$, we conclude that

$$\begin{aligned}
\mathrm{II}_N &= \int_{\mathcal{X}} [1-\eta_N(x)] \int_{\mathcal{X}} K_f^{(3)}(x,y)h(y)\,d\mu(y)d\mu(x) \\
&= \int_{\mathcal{X}} \int_{B(x_0,r_0)} K_f^{(3)}(x,y)[1-\eta_N(x)]h(y)\,d\mu(y)d\mu(x) \\
&= \int_{\mathcal{X}} \int_{B(x_0,r_0)} \left[K_f^{(3)}(x,y) - K_f^{(3)}(x,x_0)\right][1-\eta_N(x)]h(y)\,d\mu(y)d\mu(x).
\end{aligned}$$

We observe that if $d(y,x_0) < r_0$ and $d(x,x_0) \ge 2A_0Nr_0$, then $d(x,x_0) \ge 2A_0Nr_0 \ge 2A_0d(y,x_0)$, which implies that

$$\begin{aligned}
|\mathrm{II}_N| &\le \int_{\mathcal{X}\setminus B(x_0,2A_0Nr_0)} \int_{B(x_0,r_0)} \left|K_f^{(3)}(x,y) - K_f^{(3)}(x,x_0)\right| |1-\eta_N(x)||h(y)|\,d\mu(y)d\mu(x) \\
&\lesssim \|h\|_{L^\infty(\mathcal{X})} \int_{\mathcal{X}\setminus B(x_0,2A_0Nr_0)} \int_{B(x_0,r_0)} \left[\frac{d(y,x_0)}{d(x,x_0)}\right]^\eta \frac{1}{V(x,x_0)}\,d\mu(y)d\mu(x) \\
&\lesssim \|h\|_{L^\infty(\mathcal{X})} r_0^\eta V(x_0,r_0) \int_{\mathcal{X}\setminus B(x_0,2A_0Nr_0)} \left[\frac{1}{d(x,x_0)}\right]^\eta \frac{1}{V(x,x_0)}\,d\mu(x) \to 0, \quad \text{as} \quad N\to\infty.
\end{aligned}$$

Then we need to prove $\lim_{N\to\infty} \mathrm{I}_N = \langle f,h\rangle$. Indeed, for any $h \in \mathring{C}_b^\eta(\mathcal{X})$, observe that, by ([29], Corollary 3.14) and the boundedness of Q_j on $L^2(\mathcal{X})$, we know that $Q_j(f)\eta_N \in L^2(\mathcal{X})$, which, combined with the assumptions (a) and (b) in Theorem 3, the fact that h is a multiple of a $(1,2)$-atom, and the Lebesgue dominated convergence theorem, further implies that

$$\begin{aligned}
\lim_{N\to\infty} \left\langle \left(K_f^{(3)}\right)^*(\eta_N),h\right\rangle &= \lim_{N\to\infty} \sum_{j\in\mathbb{Z}} \left\langle Q_j^*(Q_j(f)\eta_N),h\right\rangle = \lim_{N\to\infty} \sum_{j\in\mathbb{Z}} \langle Q_j(f)\eta_N, Q_j(h)\rangle \\
&= \lim_{N\to\infty} \sum_{j\in\mathbb{Z}} \langle \eta_N, Q_j(f)Q_j(h)\rangle = \lim_{N\to\infty} \left\langle \eta_N, \sum_{j\in\mathbb{Z}} Q_j(f)Q_j(h)\right\rangle
\end{aligned}$$

$$= \left\langle 1, \sum_{j\in\mathbb{Z}} Q_j(f)Q_j(h) \right\rangle = \sum_{j\in\mathbb{Z}} \langle Q_j(f), Q_j(h) \rangle$$
$$= \sum_{j\in\mathbb{Z}} \left\langle f, Q_j^* Q_j(h) \right\rangle = \left\langle f, \sum_{j\in\mathbb{Z}} Q_j(h) \right\rangle = \langle f, h \rangle,$$

where in the fifth inequality of this equation, we need to show that the series $\sum_{j\in\mathbb{Z}} Q_j(f)Q_j(h)$ absolutely converges in $L^1(\mathcal{X})$. Indeed, from (13), ([62], (2.4)) and the fact that $C_b^\eta(\mathcal{X}) \subset \mathcal{G}(x_1, r, \eta, \vartheta)$ for any given $x_1 \in \mathcal{X}$ and $r, \vartheta \in (0, \infty)$ (see ([33], p. 19)), it follows that

$$\begin{aligned} \left\| \sum_{j\in\mathbb{Z}} |Q_j(f)Q_j(h)| \right\|_{L^1(\mathcal{X})} &\le \sum_{j\in\mathbb{Z}} \|Q_j(f)Q_j(h)\|_{L^1(\mathcal{X})} \le \sum_{j\in\mathbb{Z}} \|Q_j(f)\|_{L^\infty(\mathcal{X})} \|Q_j(h)\|_{L^1(\mathcal{X})} \\ &\lesssim \|f\|_{\mathrm{BMO}(\mathcal{X})} \sum_{j\in\mathbb{Z}} \|Q_j(h)\|_{L^1(\mathcal{X})} \lesssim \|f\|_{\mathrm{BMO}(\mathcal{X})} \|h\|_{\mathcal{G}(x_1,r,\eta,\vartheta)} < \infty, \end{aligned}$$

which proves the desired result. This shows $\lim_{N\to\infty} \mathrm{I}_N = \langle f, h \rangle$, which, together with the estimate of II_N, implies that $(K_f^{(3)})^*(1) = f$ on $(C_b^\eta(\mathcal{X}))'$ and hence $(K_f^{(3)})^*(1) = f \in \mathrm{BMO}(\mathcal{X})$.

Moreover, from the $T(1)$ theorem (see Lemma 3) ([32], Theorem 12.2), we deduce that $K_f^{(3)}$ is bounded on $L^2(\mathcal{X})$. Then, by the boundedness of the Calderón–Zygmund operator (see, for example, ([27], Theorem 2.4 in Chapter III), ([28], p. 599), ([35], Theorem 1.12), and ([58], Theorem 3.4)), we find that (i)–(iv) of Theorem 3 hold true.

Now we begin to show (v) and (vi) of Theorem 3. To this end, we temporarily fix a $g \in L^\infty(\mathcal{X})$. From the fact that $g \in L^\infty(\mathcal{X}) \subset \mathrm{BMO}(\mathcal{X})$, and the arguments used in the proof of (i)–(iv) of Theorem 3, it follows that the kernel of the operator $K_g^{(3)}(\cdot) := \Pi(\cdot, g)$, defined by setting, for any $(x, y) \in \mathcal{X} \times \mathcal{X}$,

$$K_g^{(3)}(x, y) := \sum_{j\in\mathbb{Z}} Q_j(x, y) Q_j(g)(y)$$

satisfies (3) through (5) and WBP(η) with $\|f\|_{\mathrm{BMO}(\mathcal{X})}$ replaced by $\|g\|_{L^\infty(\mathcal{X})}$, $K_g^{(3)}(1) = 0 \in \mathrm{BMO}(\mathcal{X})$, $(K_g^{(3)})^*(1) = g \in L^\infty(\mathcal{X}) \subset \mathrm{BMO}(\mathcal{X})$ and $K_g^{(3)}$ is bounded on $L^2(\mathcal{X})$.

Thus, $K_g^{(3)}$ is an η-Calderón–Zygmund operator, which, combined with the fact that $(K_g^{(3)})^*(1) = g \in L^\infty(\mathcal{X}) \subset \mathrm{BMO}(\mathcal{X})$ and the $T(1)$ theorem (see Lemma 3) and ([27], Theorem 2.4 in Chapter III), further completes the proof of (v) and (vi) of Theorem 3 and hence of Theorem 3. □

Proof of Theorem 4. Similar to the proof of Theorem 3, without loss of generality, we may assume that the sum $\sum_{j\in\mathbb{Z}}$ in $\Pi_1(f, g)$ is a finite sum $\sum_{j=-N}^{N}$ for any fixed $N \in \mathbb{N}$.

We first prove (i) through (iv) of Theorem 4. Fix $f \in L^\infty(\mathcal{X})$, we consider the operator $K_f^{(1)}$ and its kernel, which is still denoted by $K_f^{(1)}$, defined by setting, for any $x \in \mathcal{X}$,

$$K_f^{(1)}(g)(x) := \Pi_1(f, g)(x) = \sum_{j\in\mathbb{Z}} P_j(f)(x) Q_j(g)(x) = \int_{\mathcal{X}} K_f^{(1)}(x, y) g(y)\, d\mu(y),$$

where $K_f^{(1)}(x, y) = \sum_{j\in\mathbb{Z}} Q_j(x, y) P_j(f)(x)$.

Now we show that $K_f^{(1)}(\cdot,\cdot)$ satisfies (3) through (5). To this end, we first prove that $K_f^{(1)}(\cdot,\cdot)$ satisfies (3). From Remark 2(ii), it follows that there exists a positive constant C such that, for any $f \in L^\infty(\mathcal{X})$,

$$\sup_{j\in\mathbb{Z}} \|P_j(f)\|_{L^\infty(\mathcal{X})} \le C\|f\|_{L^\infty(\mathcal{X})}, \tag{23}$$

which, together with (13) and some arguments used in the proof of (18), further implies that, for any $x,\ y \in \mathcal{X}$,

$$\left|K_f^{(1)}(x,y)\right| \le \sum_{j\in\mathbb{Z}} |Q_j(x,y)||P_j(f)(x)| \lesssim \|f\|_{L^\infty(\mathcal{X})} \sum_{j\in\mathbb{Z}} |Q_j(x,y)| \lesssim \|f\|_{L^\infty(\mathcal{X})} \frac{1}{V(x,y)},$$

which completes the proof of (3) for $K_f^{(1)}(\cdot,\cdot)$.

Then we prove that $K_f^{(1)}(\cdot,\cdot)$ satisfies (4). Let $x,\ \tilde{x}, y \in \mathcal{X}$, $d(x,\tilde{x}) \le \frac{1}{2A_0} d(x,y)$ with $x \neq y$. We write

$$\begin{aligned}\left|K_f^{(1)}(x,y) - K_f^{(1)}(\tilde{x},y)\right| &\le \sum_{j\in\mathbb{Z}} |Q_j(x,y) - Q_j(\tilde{x},y)||P_j(f)(x)| \\ &\quad + \sum_{j\in\mathbb{Z}} |Q_j(\tilde{x},y)||P_j(f)(x) - P_j(f)(\tilde{x})| \\ &=: \mathrm{A}_1 + \mathrm{A}_2.\end{aligned}$$

From (20) and (23), we deduce that

$$\mathrm{A}_1 \lesssim \|f\|_{L^\infty(\mathcal{X})} \sum_{j\in\mathbb{Z}} |Q_j(x,y) - Q_j(\tilde{x},y)| \lesssim \|f\|_{L^\infty(\mathcal{X})} \left[\frac{d(x,\tilde{x})}{d(x,y)}\right]^\eta \frac{1}{V(x,y)}.$$

Moreover, by the fact that $L^\infty(\mathcal{X}) \subset \mathrm{BMO}\,(\mathcal{X})$ and some arguments similar to those used in the proof of (22), we know that

$$\begin{aligned}\mathrm{A}_2 &\lesssim \sum_{j\in\mathbb{Z}} |Q_j(\tilde{x},y)| \int_{\mathcal{X}} |P_j(x,z) - P_j(\tilde{x},z)| \left|f(z) - m_{B(x,\delta^j)}(f)\right| d\mu(z) \\ &\lesssim \|f\|_{\mathrm{BMO}\,(\mathcal{X})} \left[\frac{d(x,\tilde{x})}{d(y,x)}\right]^\eta \frac{1}{V(x,y)} \lesssim \|f\|_{L^\infty(\mathcal{X})} \left[\frac{d(x,\tilde{x})}{d(y,x)}\right]^\eta \frac{1}{V(x,y)},\end{aligned}$$

which completes the proof of (4) for $K_f^{(1)}(\cdot,\cdot)$.

Now we prove that $K_f^{(1)}(\cdot,\cdot)$ satisfies (5). Let $x,\ \tilde{x}, y \in \mathcal{X}$, $d(x,\tilde{x}) \le \frac{1}{2A_0} d(x,y)$ with $x \neq y$. From (18) and (21), it follows that

$$\begin{aligned}\left|K_f^{(1)}(y,x) - K_f^{(1)}(y,\tilde{x})\right| &\le \sum_{j\in\mathbb{Z}} |Q_j(y,x) - Q_j(y,\tilde{x})||P_j(f)(y)| \\ &\lesssim \|f\|_{L^\infty(\mathcal{X})} \sum_{j\in\mathbb{Z}} |Q_j(y,x) - Q_j(y,\tilde{x})| \\ &\lesssim \|f\|_{L^\infty(\mathcal{X})} \left[\frac{d(x,\tilde{x})}{d(y,x)}\right]^\eta \frac{1}{V(x,y)},\end{aligned}$$

which completes the proof of (5) for $K_f^{(1)}(\cdot,\cdot)$. This completes the proof of (3) through (5) for $K_f^{(1)}(\cdot,\cdot)$.

Then we claim that $K_f^{(1)}$ has WBP(η) and hence maps from $C_b^\eta(\mathcal{X})$ into $(C_b^\eta(\mathcal{X}))'$. Indeed, let g, $h \in C_b^\eta(\mathcal{X})$, supported on some ball $B(x_0, r_0)$ with $x_0 \in \mathcal{X}$ and $r_0 \in (0, \infty)$, normalized by

$$\|g\|_{L^\infty(\mathcal{X})} + r_0^\eta \|g\|_{\dot{C}_b^\eta(\mathcal{X})} \le 1 \quad \text{and} \quad \|h\|_{L^\infty(\mathcal{X})} + r_0^\eta \|h\|_{\dot{C}_b^\eta(\mathcal{X})} \le 1.$$

Then, by (23), the Hölder inequality and (14), we conclude that

$$\begin{aligned}
\left|\left\langle K_f^{(1)}(g), h\right\rangle\right| &= |\langle \Pi_1(f,g), h\rangle| = \left|\int_{\mathcal{X}} \Pi_1(f,g)(x)h(x)\,d\mu(x)\right| \\
&\le \sum_{j\in\mathbb{Z}} \int_{\mathcal{X}} |P_j(f)(x)||Q_j(g)(x)||h(x)|\,d\mu(x) \\
&\lesssim \|f\|_{L^\infty(\mathcal{X})} \sum_{j\in\mathbb{Z}} \int_{\mathcal{X}} |Q_j(g)(x)||h(x)|\,d\mu(x) \\
&\lesssim \|f\|_{L^\infty(\mathcal{X})} \sum_{j\in\mathbb{Z}} \|Q_j(g)\|_{L^2(\mathcal{X})} \|h\|_{L^2(\mathcal{X})} \\
&\lesssim \|f\|_{L^\infty(\mathcal{X})} [V(x_0, r_0)]^{1/2} \sum_{j\in\mathbb{Z}} \|Q_j(g)\|_{L^2(\mathcal{X})} \lesssim \|f\|_{L^\infty(\mathcal{X})} V(x_0, r_0).
\end{aligned}$$

Next we show that $K_f^{(1)}(1)$, $(K_f^{(1)})^*(1) \in \mathrm{BMO}(\mathcal{X})$. Obviously, $K_f^{(1)}(1) = 0 \in \mathrm{BMO}(\mathcal{X})$. Now we prove that $(K_f^{(1)})^*(1) = 0 \in \mathrm{BMO}(\mathcal{X})$. It is easy to see that

$$\left(K_f^{(1)}\right)^*(x,y) = K_f^{(1)}(y,x) = \sum_{j\in\mathbb{Z}} Q_j(y,x) P_j(f)(y).$$

For any $h \in \mathring{C}_b^\eta(\mathcal{X})$ with $\operatorname{supp} h \subset B(x_0, r_0)$ and any $N \in \mathbb{N}$, choose $\eta_N \in C_b^\eta(\mathcal{X})$ with $\eta_N \equiv 1$ on $B(x_0, 2A_0Nr_0)$, $\operatorname{supp}(\eta_N) \subset B(x_0, 4A_0Nr_0)$ and $0 \le \eta_N \le 1$. We write

$$\left\langle \left(K_f^{(1)}\right)^*(1), h\right\rangle = \left\langle \left(K_f^{(1)}\right)^*(\eta_N), h\right\rangle + \int_{\mathcal{X}} [1 - \eta_N(x)] K_f^{(1)}(h)\,d\mu(x) =: \mathrm{I}_N + \mathrm{II}_N.$$

By the same arguments used in the proof of $(K_f^{(3)})^*(1) \in \mathrm{BMO}(\mathcal{X})$, we conclude that $\lim_{N\to\infty} \mathrm{II}_N = 0$. Then we show that $\lim_{N\to\infty} \mathrm{I}_N = 0$. Indeed, for any $h \in \mathring{C}_b^\eta(\mathcal{X})$, observe that, by ([29], Corollary 3.14) and the boundedness of P_j on $L^2(\mathcal{X})$, we know that $P_j(f)\eta_N \in L^2(\mathcal{X})$, which, combined with the assumption (10), and the Lebesgue dominated convergence theorem, further implies that

$$\begin{aligned}
\lim_{N\to\infty} \left\langle \left(K_f^{(1)}\right)^*(\eta_N), h\right\rangle &= \lim_{N\to\infty} \sum_{j\in\mathbb{Z}} \left\langle Q_j^*(P_j(f)\eta_N), h\right\rangle = \lim_{N\to\infty} \sum_{j\in\mathbb{Z}} \langle P_j(f)\eta_N, Q_j(h)\rangle \\
&= \lim_{N\to\infty} \sum_{j\in\mathbb{Z}} \langle \eta_N, P_j(f)Q_j(h)\rangle = \lim_{N\to\infty} \left\langle \eta_N, \sum_{j\in\mathbb{Z}} P_j(f)Q_j(h)\right\rangle \\
&= \left\langle 1, \sum_{j\in\mathbb{Z}} P_j(f)Q_j(h)\right\rangle = \langle 1, \Pi_1(f,h)\rangle = 0,
\end{aligned}$$

where in the third to the last inequality of this equation, we have used the fact that the series $\sum_{j\in\mathbb{Z}} P_j(f)Q_j(h)$ absolutely converges in $L^1(\mathcal{X})$, which is similar to that of $\sum_{j\in\mathbb{Z}} Q_j(f)Q_j(h)$.

This shows $\lim_{N\to\infty} \mathrm{I}_N = 0$, which, together with the estimate of II_N, implies that $(K_f^{(1)})^*(1) = 0$ on $(C_b^\eta(\mathcal{X}))'$ and hence $(K_f^{(1)})^*(1) = 0 \in \mathrm{BMO}(\mathcal{X})$.

Moreover, from the $T(1)$ theorem (see Lemma 3), we deduce that $K_f^{(1)}$ is bounded on $L^2(\mathcal{X})$. Then, by the boundedness of the Calderón–Zygmund operator (see, for instance, ([27], Theorem 2.4 in Chapter III)), we find that (i) through (iv) of Theorem 4 hold true.

Now we begin to prove (v) and (vi) of Theorem 4. To this end, we temporarily fix a $g \in L^\infty(\mathcal{X})$. By the fact that $L^\infty(\mathcal{X}) \subset \mathrm{BMO}\,(\mathcal{X})$ and checking the proofs of Theorem 3 and (i) and (ii) of Theorem 4 carefully, we conclude that there exists a positive constant C such that, for any $f \in L^\infty(\mathcal{X})$, $g \in L^2(\mathcal{X})$, and $N \in \mathbb{N}$,

$$\left\|\Pi_1^{(N)}(f,g)\right\|_{L^2(\mathcal{X})} + \left\|\Pi_3^{(N)}(f,g)\right\|_{L^2(\mathcal{X})} \le C\|f\|_{L^\infty(\mathcal{X})}\|g\|_{L^2(\mathcal{X})},$$

which, further implies that, for any $g \in L^\infty(\mathcal{X})$ and $f \in L^2(\mathcal{X})$,

$$\begin{aligned}\left\|\Pi_2^{(N)}(f,g)\right\|_{L^2(\mathcal{X})} + \left\|\Pi_3^{(N)}(f,g)\right\|_{L^2(\mathcal{X})} &= \left\|\Pi_1^{(N)}(g,f)\right\|_{L^2(\mathcal{X})} + \left\|\Pi_3^{(N)}(g,f)\right\|_{L^2(\mathcal{X})}\\ &\lesssim \|g\|_{L^\infty(\mathcal{X})}\|f\|_{L^2(\mathcal{X})}.\end{aligned}$$

From this and Theorem 1, we deduce that $\Pi_1(f,g)$ is bounded from $L^2(\mathcal{X}) \times L^\infty(\mathcal{X})$ into $L^2(\mathcal{X})$, which, combined with the fact that $L^\infty(\mathcal{X}) \subset \mathrm{BMO}\,(\mathcal{X})$ and some arguments used in the proof of (i) through (iv) of Theorem 3, implies that the kernel of the operator $K_g^{(1)}(\cdot) := \Pi_1(\cdot, g)$, defined by setting, for any $(x,y) \in \mathcal{X} \times \mathcal{X}$,

$$K_g^{(1)}(x,y) := \sum_{j\in\mathbb{Z}} P_j(x,y)Q_j(g)(x)$$

satisfies (3) through (5), and hence $K_g^{(1)}$ is an η-Calderón–Zygmund operator which is bounded on $L^2(\mathcal{X})$. By these and the boundedness of Calderón–Zygmund operators (see, for example, ([27], Theorem 2.4 in Chapter III)), we finish the proof of (v) and (vi) of Theorem 4 and hence of Theorem 4. □

Remark 9. *We observe that the proofs of Theorems 3 and 4 do not use the second difference regularity condition of $\{Q_j\}_{j\in\mathbb{Z}}$ in Definition 4. Thus, the results in Theorems 3 and 4 hold true for more general approximations of identity.*

Author Contributions: Conceptualization, X.F.; methodology, X.F.; software, X.F.; validation, X.F.; formal analysis, X.F.; investigation, X.F.; resources, X.F.; data curation, X.F.; writing—original draft preparation, X.F.; writing—review and editing, X.F.; visualization, X.F.; supervision, X.F.; project administration, X.F.; funding acquisition, X.F. The author has read and agreed to the published version of the manuscript.

Funding: This research was funded by the National Natural Science Foundation of China (Grant Nos. 11701160 and 11871100).

Institutional Review Board Statement: Not applicable.

Informed Consent Statement: Not applicable.

Data Availability Statement: Not applicable.

Acknowledgments: We would like to express our deep thanks to the anonymous referees for their valuable comments, which improve the presentation of this paper.

Conflicts of Interest: The author declares no conflict of interest.

References

1. Fujita, H.; Kato, T. On the Navier-Stokes initial value problem. I. *Arch. Rational Mech. Anal.* **1964**, *16*, 269–315. [CrossRef]
2. Kato, T. Strong L^p-solutions of the Navier–Stokes equation in $\mathbb{R}^m$, with applications to weak solutions. *Math. Z.* **1984**, *187*, 471–480. [CrossRef]

3. Meyer, Y.; Coifman, R.R. Opérateurs pseudo-différentiels et théorème de Calderón. In *(French) Séminaire d'Analyse Harmonique (1976–1977)*; Publ. Math. Orsay, No. 77-77; Dépt. Math., Univ. Paris-Sud: Orsay, France, 1977; pp. 28–40.
4. Meyer, Y.; Coifman, R.R. *Au Delà des Opérateurs Pseudo-différentiels, (French) [Beyond Pseudodifferential Operators] With an English Summary, Astérisque 57*; Société Mathématique de France: Paris, France, 1978.
5. Meyer, Y.; Coifman, R.R. Wavelets. In *Calderón–Zygmund and Multilinear Operators, Translated from the 1990 and 1991 French Originals by David Salinger, Cambridge Studies in Advanced Mathematics 48*; Cambridge University Press: Cambridge, UK, 1997.
6. Bony, J.-M. Calcul symbolique et propagation des singularités pour les équations aux dérivées partielles non linéaires, (French) [Symbolic calculus and propagation of singularities for nonlinear partial differential equations]. *Ann. Sci. École Norm. Sup.* **1981**, *14*, 209–246. [CrossRef]
7. Bényi, Á.; Maldonado, D.; Nahmod, A.R.; Torres, R.H. Bilinear paraproducts revisited. *Math. Nachr.* **2010**, *283*, 1257–1276. [CrossRef]
8. Bernicot, F. Uniform estimates for paraproducts and related multilinear multipliers. *Rev. Mat. Iberoam.* **2009**, *25*, 1055–1088. [CrossRef]
9. Bernicot, F. A $T(1)$-theorem in relation to a semigroup of operators and applications to new paraproducts. *Trans. Am. Math. Soc.* **2012**, *364*, 6071–6108. [CrossRef]
10. Bernicot, F. Fiber-wise Calderón–Zygmund decomposition and application to a bi-dimen-sional paraproduct. *Ill. J. Math.* **2012**, *56*, 415–422.
11. David, G.; Journé, J.-L. A boundedness criterion for generalized Calderón–Zygmund operators. *Ann. Math.* **1984**, *120*, 371–397. [CrossRef]
12. Gilbert, J.E.; Nahmod, A.R. Bilinear operators with non-smooth symbol. I. *J. Fourier Anal. Appl.* **2001**, *7*, 435–467. [CrossRef]
13. Gilbert, J.E.; Nahmod, A.R. L^p-boundedness for time-frequency paraproducts. II. *J. Fourier Anal. Appl.* **2002**, *8*, 109–172. [CrossRef]
14. Grafakos, L.; Kalton, N.J. The Marcinkiewicz multiplier condition for bilinear operators. *Studia Math.* **2001**, *146*, 115–156. [CrossRef]
15. Muscalu, C.; Pipher, J.; Tao, T.; Thiele, C. Bi-parameter paraproducts. *Acta Math.* **2004**, *193*, 269–296. [CrossRef]
16. Muscalu, C.; Tao, T.; Thiele, C. Uniform estimates on multi-linear operators with modulation symmetry. Dedicated to the memory of Tom Wolff. *J. Anal. Math.* **2002**, *88*, 255–309. [CrossRef]
17. Germain, P.; Masmoudi, N.; Shatah, J. Global solutions for the gravity water waves equation in dimension 3. *Ann. Math.* **2012**, *175*, 691–754. [CrossRef]
18. Germain, P.; Masmoudi, N.; Shatah, J. Global solutions for 2D quadratic Schrödinger equations. *J. Math. Pures Appl.* **2012**, *97*, 505–543. [CrossRef]
19. Grafakos, L.; Torres, R.H. Discrete decompositions for bilinear operators and almost diagonal conditions. *Trans. Am. Math. Soc.* **2002**, *354*, 1153–1176. [CrossRef]
20. Bonami, A.; Grellier, S.; Ky, L.D. Paraproducts and products of functions in $\mathrm{BMO}(\mathbb{R}^n)$ and $H^1(\mathbb{R}^n)$ through wavelets. *J. Math. Pures Appl.* **2012**, *97*, 230–241. [CrossRef]
21. Ky, L.D. Bilinear decompositions for the product space $H^1_L \times \mathrm{BMO}_L$. *Math. Nachr.* **2014**, *287*, 1288–1297. [CrossRef]
22. Ky, L.D. Bilinear decompositions and commutators of singular integral operators. *Trans. Am. Math. Soc.* **2013**, *365*, 2931–2958. [CrossRef]
23. Ky, L.D. Endpoint estimates for commutators of singular integrals related to Schrödinger operators. *Rev. Mat. Iberoam.* **2015**, *31*, 1333–1373.
24. Bényi, Á.; Maldonado, D.; Naibo, V. What is ... a paraproduct? *Notices Am. Math. Soc.* **2010**, *57*, 858–860.
25. Muscalu, C.; Schlag, W. *Classical and Multilinear Harmonic Analysis, Volume II*; Cambridge Studies in Advanced Mathematics 138; Cambridge University Press: Cambridge, UK, 2013.
26. Yang, D.; Liang, Y.; Ky, L.D. *Real-Variable Theory of Musielak—Orlicz Hardy Spaces*; Lecture Notes in Mathematics 2182; Springer: Cham, Switzerland, 2017.
27. Coifman, R.R.; Weiss, G. *Analyse Harmonique Non-Commutative sur Certains Espaces Homogènes, (French) Étude de Certaines Intégrales Singulières*; Lecture Notes in Mathematics 242; Springer: Berlin, Germany; New York, NY, USA, 1971.
28. Coifman, R.R.; Weiss, G. Extensions of Hardy spaces and their use in analysis. *Bull. Am. Math. Soc.* **1977**, *83*, 569–645. [CrossRef]
29. Fu, X.; Yang, D.; Yang, S. Endpoint boundedness of linear commutators on local Hardy spaces over metric measure spaces of homogeneous type. *J. Geom. Anal.* **2020**, *31*, 4092–4164. [CrossRef]
30. Chang, D.-C.; Fu, X.; Yang, D. Boundedness of paraproducts on spaces of homogeneous type I. *Appl. Anal.* **2020**. [CrossRef]
31. Nakai, E.; Yabuta, K. Pointwise multipliers for functions of weighted bounded mean oscillation on spaces of homogeneous type. *Math. Japon.* **1997**, *46*, 15–28.
32. Auscher, P.; Hytönen, T. Orthonormal bases of regular wavelets in spaces of homogeneous type. *Appl. Comput. Harmon. Anal.* **2013**, *34*, 266–296. [CrossRef]
33. Han, Y.; Müller, D.; Yang, D. A theory of Besov and Triebel—Lizorkin spaces on metric measure spaces modeled on Carnot—Carathéodory spaces. *Abstr. Appl. Anal.* **2008**, *2008*, 893409. [CrossRef]
34. Han, Y.; Müller, D.; Yang, D. Littlewood—Paley characterizations for Hardy spaces on spaces of homogeneous type. *Math. Nachr.* **2006**, *279*, 1505–1537. [CrossRef]

35. Deng, D.; Han, Y. *Harmonic Analysis on Spaces of Homogeneous Type*; Lecture Notes in Mathematics 1966; Springer: Berlin, Germany, 2009.
36. Macías, R.A.; Segovia, C. A decomposition into atoms of distributions on spaces of homogeneous type. *Adv. Math.* **1979**, *33*, 271–309. [CrossRef]
37. Bui, H.-Q.; Bui, T.A.; Duong, X.T. Weighted Besov and Triebel—Lizorkin spaces associated to operators and applications. *Forum Math. Sigma* **2020**, *8*, e11. [CrossRef]
38. Bui, T.A.; Duong, X.T. Sharp weighted estimates for square functions associated to operators on spaces of homogeneous type. *J. Geom. Anal.* **2020**, *30*, 874–900. [CrossRef]
39. Bui, T.A.; Duong, X.T.; Ky, L.D. Hardy spaces associated to critical functions and applications to $T1$ theorems. *J. Fourier Anal. Appl.* **2020**, *26*, 27. [CrossRef]
40. Bui, T.A.; Duong, X.T.; Ly, F.K. Maximal function characterizations for new local Hardy type spaces on spaces of homogeneous type. *Trans. Am. Math. Soc.* **2018**, *370*, 7229–7292. [CrossRef]
41. Bui, T.A.; Duong, X.T.; Ly, F.K. Maximal function characterizations for Hardy spaces on spaces of homogeneous type with finite measure and applications. *J. Funct. Anal.* **2020**, *278*, 108423. [CrossRef]
42. Grafakos, L.; Liu, L.; Maldonado, D.; Yang, D. Multilinear analysis on metric spaces. *Diss. Math.* **2014**, *497*, 1–121. [CrossRef]
43. Grafakos, L.; Liu, L.; Yang, D. Boundedness of paraproduct operators on RD-spaces. *Sci. China Math.* **2010**, *53*, 2097–2114. [CrossRef]
44. Hu, G.; Yang, D.; Zhou, Y. Boundedness of singular integrals in Hardy spaces on spaces of homogeneous type. *Taiwan. J. Math.* **2009**, *13*, 91–135. [CrossRef]
45. Yang, D.; Zhou, Y. Radial maximal function characterizations of Hardy spaces on RD-spaces and their applications. *Math. Ann.* **2010**, *346*, 307–333. [CrossRef]
46. Yang, D.; Zhou, Y. New properties of Besov and Triebel—Lizorkin spaces on RD-spaces. *Manuscripta Math.* **2011**, *134*, 59–90. [CrossRef]
47. Fu, X.; Ma, T.; Yang, D. Real-variable characterizations of Musielak—Orlicz Hardy spaces on spaces of homogeneous type. *Ann. Acad. Sci. Fenn. Math.* **2020**, *45*, 343–410. [CrossRef]
48. Fu, X.; Yang, D.; Liang, Y. Products of functions in BMO$(\mathcal{X})$ and $H^1_{\mathrm{at}}(\mathcal{X})$ via wavelets over spaces of homogeneous type. *J. Fourier Anal. Appl.* **2017**, *23*, 919–990. [CrossRef]
49. Han, Y.; Han, Y.; He, Z.; Li, J.; Pereyra, C. Geometric characteriztions of embedding theorems—For Sobolev, Besov and Triebel—Lizorkin spaces on spaces of homogeneous type—Via orthonormal wavelets. *J. Geom. Anal.* **2021**, *31*, 8947–8978. [CrossRef]
50. Han, Y.; Han, Y.; Li, J. Criterion of the boundedness of singular integrals on spaces of homogeneous type. *J. Funct. Anal.* **2016**, *271*, 3423–3464. [CrossRef]
51. Han, Y.; Han, Y.; Li, J. Geometry and Hardy spaces on spaces of homogeneous type in the sense of Coifman and Weiss. *Sci. China Math.* **2017**, *60*, 2199–2218. [CrossRef]
52. Han, Y.; Li, J.; Ward, L.A. Hardy space theory on spaces of homogeneous type via orthonormal wavelet bases. *Appl. Comput. Harmon. Anal.* **2018**, *45*, 120–169. [CrossRef]
53. He, Z.; Han, Y.; Li, J.; Liu, L.; Yang, D.; Yuan, W. A complete real-variable theory of Hardy spaces on spaces of homogeneous type. *J. Fourier Anal. Appl.* **2019**, *25*, 2197–2267. [CrossRef]
54. He, Z.; Liu, L.; Yang, D.; Yuan, W. New Calderón reproducing formulae with exponential decay on spaces of homogeneous type. *Sci. China Math.* **2019**, *62*, 283–350. [CrossRef]
55. He, Z.; Wang, F.; Yang, D.; Yuan, W. Wavelet characterizations of Besov and Triebel—Lizorkin spaces on spaces of homogeneous type and their applications. *Appl. Comput. Harmon. Anal.* **2021**, *54*, 176–226. [CrossRef]
56. He, Z.; Yang, D.; Yuan, W. Real-variable characterizations of local Hardy spaces on spaces of homogeneous type. *Math. Nachr.* **2021**, *294*, 900–955. [CrossRef]
57. Liu, L.; Chang, D.-C.; Fu, X.; Yang, D. Endpoint boundedness of commutators on spaces of homogeneous type. *Appl. Anal.* **2017**, *96*, 2408–2433. [CrossRef]
58. Liu, L.; Chang, D.-C.; Fu, X.; Yang, D. Endpoint estimates of linear commutators on Hardy spaces over spaces of homogeneous type. *Math. Meth. Appl. Sci.* **2018**, *41*, 5951–5984. [CrossRef]
59. Liu, L.; Yang, D.; Yuan, W. Bilinear decompositions for products of Hardy and Lipschitz spaces on spaces of homogeneous type. *Diss. Math. (Rozpr. Mat.)* **2018**, *533*, 1–93. [CrossRef]
60. Wang, F.; Han, Y.; He, Z.; Yang, D. Besov spaces and Triebel—Lizorkin spaces on spaces of homogeneous type with their applications to a boundedness of Calderón–Zygmund operators. *arXiv* **2020**, arXiv:2012.13035.
61. Zhou, X.; He, Z.; Yang, D. Real-variable characterizations of Hardy—Lorentz spaces on spaces of homogeneous type with applications to real interpolation and boundedness of Calderón–Zygmund operators. *Anal. Geom. Metr. Spaces* **2020**, *8*, 182–260. [CrossRef]
62. Chang, D.-C.; Fu, X.; Yang, D. Boundedness of paraproducts on spaces of homogeneous type II. *Appl. Anal.* **2020**. [CrossRef]
63. Fu, X.; Chang, D.-C.; Yang, D. Recent progress in bilinear decompositions. *Appl. Anal. Optim.* **2017**, *1*, 153–210.
64. Hytönen, T.; Kairema, A. Systems of dyadic cubes in a doubling metric space. *Colloq. Math.* **2012**, *126*, 1–33. [CrossRef]
65. Fu, X.; Yang, D. Wavelet characterizations of the atomic Hardy space H^1 on spaces of homogeneous type. *Appl. Comput. Harmon. Anal.* **2018**, *44*, 1–37. [CrossRef]

66. Stein, E.M. *Harmonic Analysis: Real-Variable Methods, Orthogonality and Oscillatory Integrals;* Princeton University Press: Princeton, NJ, USA, 1993.
67. Grafakos, L. *Modern Fourier Analysis*, 3rd ed.; Graduate Texts in Mathematics 250; Springer: New York, NY, USA, 2014.

Σ mathematics

MDPI

Article

Lebesgue Points of Besov and Triebel–Lizorkin Spaces with Generalized Smoothness

Ziwei Li, Dachun Yang and Wen Yuan *

Laboratory of Mathematics and Complex Systems (Ministry of Education of China), School of Mathematical Sciences, Beijing Normal University, Beijing 100875, China; zwli@mail.bnu.edu.cn (Z.L.); dcyang@bnu.edu.cn (D.Y.)
* Correspondence: wenyuan@bnu.edu.cn

Abstract: In this article, the authors study the Lebesgue point of functions from Hajłasz–Sobolev, Besov, and Triebel–Lizorkin spaces with generalized smoothness on doubling metric measure spaces and prove that the exceptional sets of their Lebesgue points have zero capacity via the capacities related to these spaces. In case these functions are not locally integrable, the authors also consider their generalized Lebesgue points defined via the γ-medians instead of the classical ball integral averages and establish the corresponding zero-capacity property of the exceptional sets.

Keywords: Hajłasz–Sobolev space; Hajłasz–Besov space; Hajłasz–Triebel–Lizorkin space; generalized smoothness; Lebesgue point; capacity

Citation: Li, Z.; Yang, D.; Yuan, W. Lebesgue Points of Besov and Triebel–Lizorkin Spaces with Generalized Smoothness. *Mathematics* **2021**, *9*, 2724. https://doi.org/10.3390/math9212724

Academic Editor: Andrea Scapellato

Received: 22 September 2021
Accepted: 22 October 2021
Published: 27 October 2021

1. Introduction

The study of function spaces on the Euclidean space $\mathbb{R}^n$ and its subsets with generalized smoothness started from the middle of the 1970s (see, for instance, [1–4]), and has found various applications in interpolations, embedding properties of function spaces [5–8], fractal analysis ([9], Chapters 18–23), and many other fields such as probability theory and stochastic processes [10,11]. Recall that, in [11], Farkas and Leopold studied the generalized Besov spaces $B^{(\sigma,N)}_{p,q}(\mathbb{R}^n)$ and Triebel–Lizorkin spaces $F^{(\sigma,N)}_{p,q}(\mathbb{R}^n)$ for the full range of parameters, in which the smoothness, instead of the classical smoothness sequence $\{2^{js}\}_{j\geq 0}$, was given via a weight sequence $\sigma := \{\sigma_j\}_{j\geq 0}$ of positive numbers. Intensive investigations on generalized Besov and Triebel–Lizorkin spaces also exist in which smoothness is described by a parameter function; see, for instance [6,12–16]. In recent years, a lot of attention has been paid to Besov and Triebel–Lizorkin spaces on $\mathbb{R}^n$ with logarithmic smoothness; see, for instance [17–27].

Recently, using Hajłasz gradient sequences, the authors [28] introduced Hajłasz–Besov and Hajłasz–Triebel–Lizorkin spaces with generalized smoothness on a given metric space $\mathcal{X}$ with a doubling measure and, when $\mathcal{X} = \mathbb{R}^n$, proved their coincidence with the classical Besov and Triebel–Lizorkin spaces with generalized smoothness. Recall that the Hajłasz gradients were originally introduced by Hajłasz [29] and have been an important tool used to develop Sobolev spaces on metric measure spaces (see, for instance [30–34]). The fractional Hajłasz gradients were introduced independently by Hu [35] and Yang [36] in 2003. In 2011, Koskela et al. [37] introduced the notion of sequences of Hajłasz gradients and characterized Besov and Triebel–Lizorkin spaces via some pointwise inequalities involving these Hajłasz gradient sequences; as an application, this pointwise characterization has been used in [37] to show the invariance of quasi-conformal mappings on some Triebel–Lizorkin spaces.

It is well known, by the Lebesgue differentiation theorem, that almost every point is a Lebesgue point of a locally integrable function. Then, it is very natural to expect a smaller exceptional set when the function has higher regularity. In [38], Kinnunen and Latvala considered the Lebesgue point of functions in the Hajłasz–Sobolev space $M^{1,p}(\mathcal{X})$

on a given metric measure space $\mathcal{X}$ and proved that, when the measure doubles and $p \in (1, Q]$, a Hajłasz–Sobolev function has Lebesgue points outside a set of zero Hajłasz–Sobolev capacity, where Q represents the doubling dimension of $\mathcal{X}$. This result leads to a series of related work on many other function spaces such as fractional Hajłasz–Sobolev spaces [39], Orlicz–Sobolev spaces [40], as well as Hajłasz–Besov and Hajłasz–Triebel–Lizorkin spaces [41]. We also refer the reader to [42,43] for a related study on variable function spaces.

Inspired by these works, in this article, we study the Lebesgue point of functions from the Hajłasz–Sobolev space $M^{\phi,p}(\mathcal{X})$, the Hajłasz–Besov space $N^{\phi}_{p,q}(\mathcal{X})$, and the Hajłasz–Triebel–Lizorkin space $M^{\phi}_{p,q}(\mathcal{X})$ with generalized smoothness on a given doubling measure space $\mathcal{X}$, via measuring the related exceptional sets of Lebeguse points. Note that functions in the Hajłasz–Besov or Hajłasz–Triebel–Lizorkin spaces with generalized smoothness might fail to be locally integrable when their index p or q is close to zero. To overcome this obstacle, similar to [41,44,45], we also consider a class of generalized Lebesgue points, which are defined via the γ-medians introduced in [46,47], instead of the classical integrals. As the main results of this article, we prove that the exceptional sets of (generalized) Lebesgue points of functions from the above spaces have zero capacity, where those capacities are defined by related spaces. These results can apply to a wide class of function spaces due to the generality of the smoothness factor ϕ. In particular, the logarithmic Hajłasz–Sobolev space is an admissible function space for our main results.

The structure of this article is as follows.

In Section 2, we state some basic notions and assumptions on the smoothness function ϕ. We also introduce the inhomogeneous Hajłasz–Sobolev space $M^{\phi,p}(\mathcal{X})$, the inhomogeneous Hajłasz–Besov space $N^{\phi}_{p,q}(\mathcal{X})$, and the inhomogeneous Hajłasz–Triebel–Lizorkin space $M^{\phi}_{p,q}(\mathcal{X})$ with generalized smoothness and establish their coincidence with those classical Besov and Triebel–Lizorkin spaces with generalized smoothness when $\mathcal{X} = \mathbb{R}^n$.

Section 3 is devoted to studying the Lebesgue point of functions from $N^{\phi}_{p,q}(\mathcal{X})$ and $M^{\phi}_{p,q}(\mathcal{X})$ and, in particular, $M^{\phi,p}(\mathcal{X}) = M^{\phi}_{p,\infty}(\mathcal{X})$, via the capacities $\mathrm{Cap}_{N^{\phi}_{p,q}(\mathcal{X})}$ and $\mathrm{Cap}_{M^{\phi}_{p,q}(\mathcal{X})}$ related to the spaces $N^{\phi}_{p,q}(\mathcal{X})$ and $M^{\phi}_{p,q}(\mathcal{X})$, respectively. To this end, via establishing some Poincaré-type inequalities and estimates related to Hajłasz-type spaces with generalized smoothness, we first prove the convergence of discrete convolution approximations in $N^{\phi}_{p,q}(\mathcal{X})$ and $M^{\phi}_{p,q}(\mathcal{X})$ when $p,\ q < \infty$, and a dense subset in $M^{\phi,p}(\mathcal{X}) = M^{\phi}_{p,\infty}(\mathcal{X})$ exists when $p < \infty$, which consists of continuous functions. Recall that, when $s \in (0, 1]$ and $p \in (0, \infty)$, the class of all s-Hölder continuous functions is dense in the classical Hajłasz–Sobolev space $M^{s,p}(\mathcal{X})$ (see, for instance, ([48], Theorem 5.19)), which was proved via an extension argument together with the inequality

$$[d(x,y)]^s \leq [d(x,z)]^s + [d(z,y)]^s$$

for any $x,\ y,\ z \in \mathcal{X}$. However, this inequality may not be true if one replaces $[d(\cdot,\cdot)]^s$ by $\phi(d(\cdot,\cdot))$ due to the generality of ϕ. To overcome the difficulties caused by this, we borrow the notion of the modulus of continuity and, for certain ϕ that satisfies such assumptions, find a dense subset of $M^{\phi,p}(\mathcal{X})$ consisting of generalized Lipschitz functions. Applying these dense properties, we obtain the boundedness of discrete maximal operators on these Hajłasz-type spaces and then a weak-type capacitary estimate for restricted maximal functions, which is further used to prove that the exceptional sets of Lebesgue points of functions from $M^{\phi,p}(\mathcal{X})$, $N^{\phi}_{p,q}(\mathcal{X})$, and $M^{\phi}_{p,q}(\mathcal{X})$ have zero $\mathrm{Cap}_{M^{\phi,p}(\mathcal{X})}$, $\mathrm{Cap}_{N^{\phi}_{p,q}(\mathcal{X})}$, and $\mathrm{Cap}_{M^{\phi}_{p,q}(\mathcal{X})}$ capacities, respectively.

In Section 4, we deal with the generalized Lebesgue point of functions from the spaces $M^{\phi,p}(\mathcal{X})$, $N^{\phi}_{p,q}(\mathcal{X})$, and $M^{\phi}_{p,q}(\mathcal{X})$, which are defined via the γ-medians instead of the classical ball integral averages. Following a procedure similar to that of Section 3, we also prove

that the exceptional sets of generalized Lebesgue points of functions from $\mathcal{F}$ have zero $\mathrm{Cap}_{\mathcal{F}}$-capacity with
$$\mathcal{F} \in \left\{N^{\phi}_{p,q}(\mathcal{X}), M^{\phi}_{p,q}(\mathcal{X}), M^{\phi,p}(\mathcal{X})\right\}.$$

Finally, we compare the capacity $\mathrm{Cap}_{\mathcal{F}}$ with some Netrusov–Hausdorff contents and prove that they have the same null sets. This enables us to also use some Netrusov–Hausdorff contents to measure the exceptional set of Lebesgue points of functions from these Hajłasz-type spaces.

2. Hajłasz–Besov and Hajłasz–Triebel–Lizorkin Spaces with Generalized Smoothness

In this section, we recall some basic notation and notions as well as the definitions of the function spaces used in this article. Let $\mathbb{Z}$ be the collection of all integers, $\mathbb{N}$ be the collection of all positive integers, and $\mathbb{Z}_+ := \mathbb{N} \cup \{0\}$. We write $A \lesssim B$ if there exists a positive constant C that is independent of the main parameters such that $A \leq CB$ and write $A \sim B$ if $A \lesssim B \lesssim A$. We also denote by $C_{(a_1,a_2,\ldots)}$ a positive constant depending on the parameters $a_1, a_2, \ldots$.

A triple $(\mathcal{X}, d, \mu)$ is called a *metric measure space* if $\mathcal{X}$ is a non-empty set, d is a metric on $\mathcal{X}$, and μ is a regular Borel measure on $\mathcal{X}$ such that all of the balls defined by d have finite and positive measures. Recall that (see [48], [Convention 1.4]) a measure μ on $\mathcal{X}$ is called a *regular Borel measure* if open sets are μ-measurable and every set is contained in a Borel set with the same measure. Additionally, the measure μ is said to *double* if there exists a positive constant $C_\mu \in [1, \infty)$ such that, for any ball $B \subset \mathcal{X}$,
$$\mu(2B) \leq C_\mu \mu(B).$$

Here and thereafter, for any $\lambda \in (0, \infty)$, λB denotes the ball with the same center as B but λ-times radius of B. The doubling property of μ implies that, for any ball $B \subset \mathcal{X}$ and any $\lambda \in [1, \infty)$,
$$\mu(\lambda B) \leq C_\mu \lambda^D \mu(B), \tag{1}$$
where $D := \log_2 C_\mu$. Here and thereafter, we assume that C_μ is the smallest positive constant such that (1) holds true. Clearly, when $\mathcal{X} = \mathbb{R}^n$, $D = n$. Throughout this article, we always let$(\mathcal{X}, d, \mu)$ be a *metric space with a doubling measure* (for short, a *doubling metric measure space*). For any subset $E \subset \mathcal{X}$, we denote by $\mathbf{1}_E$ the characteristic function of E.

Let $L^0(\mathcal{X})$ be the collection of all measurable functions on $\mathcal{X}$ that are finite almost everywhere and $L^1_{\mathrm{loc}}(\mathcal{X})$ be the collection of all measurable functions on $\mathcal{X}$ satisfying that, for any $x_0 \in \mathcal{X}$, there exists an $r_0 \in (0, \infty)$ such that $f\mathbf{1}_{B(x_0,r_0)} \in L^1(\mathcal{X})$. For any $p,\ q \in (0, \infty]$, let $L^p(\mathcal{X}, l^q)$ and $l^q(\mathcal{X}, L^p)$ be, respectively, the collections of all sequences $\{u_k\}_{k\in\mathbb{Z}} \subset L^0(\mathcal{X})$ such that
$$\|\{u_k\}_{k\in\mathbb{Z}}\|_{L^p(\mathcal{X},l^q)} := \left\|\left(\sum_{k\in\mathbb{Z}} |u_k|^q\right)^{1/q}\right\|_{L^p(\mathcal{X})} < \infty$$
and
$$\|\{u_k\}_{k\in\mathbb{Z}}\|_{l^q(\mathcal{X},L^p)} := \left[\sum_{k\in\mathbb{Z}} \|u_k\|^q_{L^p(\mathcal{X})}\right]^{1/q} < \infty$$
with the usual modifications made when $p = \infty$ or $q = \infty$.

For any $u \in L^0(\mathcal{X})$ and $E \subset \mathcal{X}$ with $\mu(E) \in (0, \infty)$, let
$$u_E := \fint_E u\, d\mu := \frac{1}{\mu(E)} \int_E u\, d\mu := \frac{1}{\mu(E)} \int_E u(x)\, d\mu(x). \tag{2}$$

For any $L \in (0, \infty)$, a function f is said to be *L-Lipschitz* if it satisfies
$$|f(x) - f(y)| \leq L\, d(x, y), \qquad \forall\, x,\ y \in \mathcal{X}.$$

For a Lipschitz function f, the smallest constant L satisfying the above inequality is called the *Lipschitz constant* of f and denoted by $\operatorname{Lip} f$.

We also frequently use the following inequality: if $q \in (0,1]$, then, for any $\{a_i\}_{i\in\mathbb{Z}} \subset \mathbb{C}$,

$$\left(\sum_{i\in\mathbb{Z}} |a_i|\right)^q \le \sum_{i\in\mathbb{Z}} |a_i|^q. \tag{3}$$

We now recall the definition and some basic properties of weight functions used to describe the smoothness of function spaces under consideration. We begin with a classical notion of admissible sequences; see, for instance [11,49].

Definition 1. *Let $E \in \{\mathbb{Z}, \mathbb{Z}_+\}$. A sequence of positive numbers, $\{\sigma_j\}_{j\in E}$, is said to be* admissible *if there exist two positive constants d_0 and d_1 such that, for any $j \in E$, $d_0\sigma_j \le \sigma_{j+1} \le d_1\sigma_j$.*

Several examples of admissible sequences can be found in [11], which illustrate the flexibility of this assumption.

Definition 2. *A continuous function $\phi : [0,\infty) \to [0,\infty)$ is said to be* of admissible growth *if $\{\phi(2^j)\}_{j\in\mathbb{Z}}$ is an admissible sequence and $\phi(t) \sim \phi(2^k)$ for any $k \in \mathbb{Z}$ and $t \in [2^k, 2^{k+1})$ with the positive equivalence constants independent of both t and k.*

We point out that, for any given admissible sequence $\sigma := \{\sigma_j\}_{j\in\mathbb{Z}}$, there exists a continuous function ϕ of admissible growth such that, for any $j \in \mathbb{Z}$, $\phi(2^{-j}) = 1/\sigma_j$. Indeed, the function

$$\phi_\sigma(t) := 2^{j+1}\left(\frac{1}{\sigma_j} - \frac{1}{\sigma_{j+1}}\right)(t - 2^{-j-1}) + \frac{1}{\sigma_{j+1}}, \quad \forall\, t \in [2^{-j-1}, 2^{-j}),\ \forall\, j \in \mathbb{Z}, \tag{4}$$

suits this job; see ([28] [Proposition 2.4]) or ([14] [Example 2.3]). Throughout this article, for any given admissible sequence $\sigma := \{\sigma_j\}_{j\in\mathbb{Z}}$, we *always* let ϕ_σ be as in (4).

For any given sequence $\sigma := \{\sigma_k\}_{k\in\mathbb{Z}}$ of positive numbers or any given function $\phi : [0,\infty) \to [0,\infty)$, let

$$\alpha_\sigma := \max\{\alpha_\sigma^-, \alpha_\sigma^+\} := \max\left\{\limsup_{k\to-\infty} \frac{\sigma_k}{\sigma_{k+1}},\ \limsup_{k\to\infty} \frac{\sigma_k}{\sigma_{k+1}}\right\},$$

$$\beta_\sigma := \max\{\beta_\sigma^-, \beta_\sigma^+\} := \max\left\{\limsup_{k\to-\infty} \frac{\sigma_{k+1}}{\sigma_k},\ \limsup_{k\to\infty} \frac{\sigma_{k+1}}{\sigma_k}\right\},$$

$$\alpha_\phi := \max\{\alpha_\phi^-, \alpha_\phi^+\} := \max\left\{\limsup_{k\to-\infty} \frac{\phi(2^k)}{\phi(2^{k+1})},\ \limsup_{k\to\infty} \frac{\phi(2^k)}{\phi(2^{k+1})}\right\},$$

and

$$\beta_\phi := \max\{\beta_\phi^-, \beta_\phi^+\} := \max\left\{\limsup_{k\to-\infty} \frac{\phi(2^{k+1})}{\phi(2^k)},\ \limsup_{k\to\infty} \frac{\phi(2^{k+1})}{\phi(2^k)}\right\}.$$

Since, for any $j \in \mathbb{Z}$, $\phi_\sigma(2^{-j}) = 1/\sigma_j$, then $\alpha_{\phi_\sigma}^- = \alpha_\sigma^+$, $\alpha_{\phi_\sigma}^+ = \alpha_\sigma^-$, $\beta_{\phi_\sigma}^- = \beta_\sigma^+$, and $\beta_{\phi_\sigma}^+ = \beta_\sigma^-$, which means that $\alpha_{\phi_\sigma} = \alpha_\sigma$ and $\beta_{\phi_\sigma} = \beta_\sigma$. By an obvious observation that $1/\alpha_\sigma^- \le \beta_\sigma^-$ and $1/\alpha_\sigma^+ \le \beta_\sigma^+$, it is also easy to show that $1/\alpha_\sigma \le \beta_\sigma$; furthermore, $\alpha_\phi < 1$ implies $\beta_\phi > 1$, and $\beta_\phi < 2$ implies $\alpha_\phi > 1/2$.

Observe that, if $\alpha_\phi^- \in (0,1)$ (resp., $\alpha_\phi^+ \in (0,1)$), then there exists a $\delta_1 \in (0,\infty)$ such that $\alpha_\phi^- + \delta_1 < 1$ (resp., $\alpha_\phi^+ + \delta_1 < 1$). Let K_0 be a given integer. By the definition of α_ϕ^- (resp., α_ϕ^+), we find that there exists an integer K_1 (resp., K_2) such that, for any $k \in (-\infty, \min\{K_1, K_0\}]$ (resp., $k \in [\max\{K_2, K_0\}, \infty)$),

$$\frac{\phi(2^k)}{\phi(2^{k+1})} < \alpha_\phi^- + \delta_1 \qquad \left(\text{resp., } \frac{\phi(2^k)}{\phi(2^{k+1})} < \alpha_\phi^- + \delta_1\right)$$

and hence, for any $i,\ j \in (-\infty, \min\{K_1, K_0\}]$ (resp., $i,\ j \in [\max\{K_2, K_0\}, \infty)$) with $i \le j$,

$$\frac{\phi(2^i)}{\phi(2^j)} < (\alpha_\phi^- + \delta_1)^{j-i} \quad \left(\text{resp., } \frac{\phi(2^i)}{\phi(2^j)} < (\alpha_\phi^+ + \delta_1)^{j-i}\right). \tag{5}$$

Since $\phi(2^k)/\phi(2^{k+1})$ is bounded on $[\min\{K_1, K_0\}, K_0]$ (resp., $k \in [K_0, \max\{K_2, K_0\}]$), then, from (5), we deduce that there exists a positive constant C, depending only on K_0, ϕ, and δ_1, such that, for any $i,\ j \in (-\infty, K_0] \cap \mathbb{Z}$ (resp., $i,\ j \in [K_0, \infty) \cap \mathbb{Z}$) with $i \le j$,

$$\frac{\phi(2^i)}{\phi(2^j)} \le C(\alpha_\phi^- + \delta_1)^{j-i} \quad \left(\text{resp., } \frac{\phi(2^i)}{\phi(2^j)} \le C(\alpha_\phi^+ + \delta_1)^{j-i}\right). \tag{6}$$

By this, we further obtain, for any $k_0 \in (-\infty, K_0] \cap \mathbb{Z}$ (resp., $k_0 \in [K_0, \infty) \cap \mathbb{Z}$) and $r \in (0, \infty]$,

$$\begin{aligned}\left\{\sum_{k\le k_0}\left[\phi(2^k)\right]^r\right\}^{1/r} &= \phi(2^{k_0})\left\{\sum_{k\le k_0}\left[\frac{\phi(2^k)}{\phi(2^{k_0})}\right]^r\right\}^{1/r}\\ &\lesssim \phi(2^{k_0})\left\{\sum_{k\le k_0}(\alpha_\phi^- + \delta_1)^{(k_0-k)r}\right\}^{1/r} \lesssim \phi(2^{k_0})\end{aligned}$$

$$\left(\text{resp., } \left\{\sum_{k\ge k_0}\left[\frac{1}{\phi(2^k)}\right]^r\right\}^{1/r} \le \frac{1}{\phi(2^{k_0})}\left\{\sum_{k\ge k_0}(\alpha_\phi^+ + \delta_1)^{(k-k_0)r}\right\}^{1/r} \lesssim \frac{1}{\phi(2^{k_0})}\right),$$

where the implicit positive constants depend only on K_0, ϕ, and δ_1.

If $\beta_\phi^- \in (0,2)$ (resp., $\beta_\phi^+ \in (0,2)$), by an argument similar to the above, we conclude that there exist a $\delta_2 \in (0,\infty)$ such that $\beta_\phi^- + \delta_2 < 2$ (resp., $\beta_\phi^+ + \delta_2 < 2$) and a positive constant C, depending only on K_0, ϕ, and δ_2, such that, for any $i,\ j \in (-\infty, K_0] \cap \mathbb{Z}$ (resp., $i,\ j \in [K_0, \infty) \cap \mathbb{Z}$) with $i \le j$,

$$2^{i-j}\frac{\phi(2^j)}{\phi(2^i)} \le C\left(\frac{\beta_\phi^- + \delta_2}{2}\right)^{j-i} \quad \left(\text{resp., } 2^{i-j}\frac{\phi(2^j)}{\phi(2^i)} \le C\left(\frac{\beta_\phi^+ + \delta_2}{2}\right)^{j-i}\right). \tag{7}$$

Furthermore, for any $k_0 \in (-\infty, K_0] \cap \mathbb{Z}$ (resp., $k_0 \in [K_0, \infty) \cap \mathbb{Z}$) and $r \in (0, \infty]$, we have

$$\begin{aligned}\left\{\sum_{k\le k_0}\left[\frac{2^k}{\phi(2^k)}\right]^r\right\}^{1/r} &= \frac{2^{k_0}}{\phi(2^{k_0})}\left\{\sum_{k\le k_0}\left[2^{k-k_0}\frac{\phi(2^{k_0})}{\phi(2^k)}\right]^r\right\}^{1/r}\\ &\lesssim \frac{2^{k_0}}{\phi(2^{k_0})}\left\{\sum_{k\le k_0}\left(\frac{\beta_\phi^- + \delta_2}{2}\right)^{(k_0-k)r}\right\}^{1/r} \lesssim \frac{2^{k_0}}{\phi(2^{k_0})}\end{aligned} \tag{8}$$

$$\left(\text{resp., } \left\{\sum_{k\ge k_0}\left[2^{-k}\phi(2^k)\right]^r\right\}^{1/r} \lesssim 2^{-k_0}\phi(2^{k_0})\left\{\sum_{k\ge k_0}\left(\frac{\beta_\phi^+ + \delta_2}{2}\right)^{(k-k_0)r}\right\}^{1/r} \lesssim 2^{-k_0}\phi(2^{k_0})\right).$$

If $\alpha_\phi \in (0,1)$ (resp.,$\beta_\phi \in (0,2)$), then $\alpha_\phi^- \in (0,1)$ and $\alpha_\phi^+ \in (0,1)$ (resp.,$\beta_\phi^- \in (0,2)$ and $\beta_\phi^+ \in (0,2)$). Thus, by (6) and (7), we obtain, for any $i,\ j \in \mathbb{Z}$ with $i \le j$,

$$\frac{\phi(2^i)}{\phi(2^j)} \lesssim (\alpha_\phi + \delta_1)^{j-i} \quad \left(\text{resp., } 2^{i-j}\frac{\phi(2^j)}{\phi(2^i)} \lesssim \left(\frac{\beta_\phi + \delta_2}{2}\right)^{j-i}\right),$$

where δ_1 (resp., δ_2) is any given positive constant such that $\alpha_\phi + \delta_1 < 1$ (resp., $\beta_\phi + \delta_2 < 2$), and the implicit positive constants depend only on ϕ and δ_1 (resp., δ_2). By this, we conclude that, for any $r \in (0,\infty]$ and $k_0 \in \mathbb{Z}$,

$$\left\{\sum_{k\le k_0}[\phi(2^k)]^r\right\}^{1/r} \lesssim \phi(2^{k_0}) \quad \text{and} \quad \left\{\sum_{k\ge k_0}\left[\frac{1}{\phi(2^k)}\right]^r\right\}^{1/r} \lesssim \frac{1}{\phi(2^{k_0})} \tag{9}$$

$$\left(\text{resp., } \left\{\sum_{k\le k_0}\left[\frac{2^k}{\phi(2^k)}\right]^r\right\}^{1/r} \lesssim \frac{2^{k_0}}{\phi(2^{k_0})} \quad \text{and} \quad \left\{\sum_{k\ge k_0}\left[2^{-k}\phi(2^k)\right]\right\}^{1/r} \lesssim 2^{-k_0}\phi(2^{k_0})\right). \tag{10}$$

Here, the implicit positive constants depend only on ϕ.

The following lemma is just ([28] [Lemma 2.5]).

Lemma 1. *Let $\phi : [0,\infty) \to [0,\infty)$ satisfy $\alpha_\phi \in (0,1)$, $\varepsilon \in (0, -\log_2 \alpha_\phi)$, and $\delta \in (\log_2 \beta_\phi, \infty)$. Then,*

(i) *there exist positive constants C_1 and C_2, depending on ϕ, such that, for any $k \in \mathbb{Z}$,*

$$\sum_{j\ge k}\frac{2^{j\varepsilon}}{\phi(2^j)} \le C_1\frac{2^{k\varepsilon}}{\phi(2^k)} \quad \textit{and} \quad \sum_{j\le k}2^{-j\varepsilon}\phi(2^j) \le C_2\, 2^{-k\varepsilon}\phi(2^k);$$

(ii) *there exist positive constants c_1 and c_2, depending on ϕ, such that, for any $i,\ j \in \mathbb{Z}$ with $i \le j$,*

$$2^{(j-i)\varepsilon}\frac{\phi(2^i)}{\phi(2^j)} \le c_1 \quad \textit{and} \quad 2^{(i-j)\delta}\frac{\phi(2^j)}{\phi(2^i)} \le c_2.$$

We recall another widely used notion (see, for instance, [50], Section 2.2.1) to describe the smoothness function as follows.

Definition 3. *A function $f : [0,\infty) \to [0,\infty)$ is said to be* almost increasing *(resp.,* decreasing*) if there exists a positive constant $C \in [1,\infty)$ such that, for any $t_1,\ t_2 \in [0,\infty)$ with $t_1 \le t_2$ (resp., $t_1 \ge t_2$), $f(t_1) \le C f(t_2)$.*

Throughout this article, for simplicity, we *always denote* by $\mathcal{A}$ the class of all continuous and almost increasing functions $\phi : [0,\infty) \to [0,\infty)$ satisfying that $\phi(0) = 0$, $\phi(1) = 1$, and $\{\phi(2^j)\}_{j\in\mathbb{Z}}$ is admissible.

Let $\mathcal{A}_\infty$ be the set of all functions $\phi \in \mathcal{A}$ satisfying that the function $\widetilde{\phi}$, defined by setting, for any $t \in [0,\infty)$, $\widetilde{\phi}(t) := \phi(t)/t$, almost decreases.

For any $r \in (0,\infty)$, let $\mathcal{A}_r$ be the set of all functions $\phi \in \mathcal{A}_\infty$ satisfying that ϕ is of admissible growth and that there exist a $k_0 \in \mathbb{Z}$ and two positive constants X_{k_0} and Y_{k_0}, depending on k_0 and r, such that

$$\left\{\sum_{j\ge k_0}\left[\phi(2^j)\right]^{-r}\right\}^{1/r} \le X_{k_0} \quad \text{and} \quad \left\{\sum_{j\ge k_0}2^{-jr}\left[\phi(2^{-j})\right]^{-r}\right\}^{1/r} \le Y_{k_0}. \tag{11}$$

We claim that if, for some $k_0 \in \mathbb{Z}$, there exist positive constants X_{k_0} and Y_{k_0} such that (11) holds true, then, for any $k \in \mathbb{Z}$, there exist positive constants X_k and Y_k, depending on k and r, such that (11) holds true with k_0 replaced by k. Indeed, this claim is trivial when $k \geq k_0$, while when $k < k_0$, it easily follows from the fact that $\sum_{j=k}^{k_0-1}[\phi(2^j)]^{-r}$ and $\sum_{j=k}^{k_0-1} 2^{-jr}[\phi(2^{-j})]^{-r}$ are always finite. This proves the above claim.

Clearly, by (3), $\mathcal{A}_{r_1} \subset \mathcal{A}_{r_2} \subset \mathcal{A}_\infty$ for any $r_1,\ r_2 \in (0,\infty)$ with $r_1 \leq r_2$. For instance, for any $b \in (0,\infty)$ and $r \in (1/b,\infty]$, the function

$$\phi(t) := \begin{cases} [\log_2(1+t)]^b, & t \in (0,1), \\ (1+\log_2 t)^b, & t \in [1,\infty) \end{cases} \tag{12}$$

belongs to $\mathcal{A}_r$.

If ϕ is of admissible growth, then $\alpha_\phi \in (0,1)$ implies $\phi \in \mathcal{A}$; furthermore, $\alpha_\phi \in (0,1)$, together with $\beta_\phi^- \in (0,2)$, implies that, for any $r \in (0,\infty]$, $\phi \in \mathcal{A}_r$. In view of these, we let $\mathcal{A}_0$ be the class of all functions ϕ satisfying that $\alpha_\phi \in (0,1)$, $\beta_\phi^- \in (0,2)$, and ϕ is of admissible growth.

Now, we state the notions of generalized Hajłasz gradients and the related Hajłasz-type spaces with respect to the smoothness function $\phi \in \mathcal{A}$.

Definition 4. *Let $\phi \in \mathcal{A}$ and $u \in L^0(\mathcal{X})$.*

(i) *A nonnegative measurable function g is called a ϕ-*Hajłasz gradient *of u if there exists a set $E \subset \mathcal{X}$ with $\mu(E) = 0$ such that, for any $x,\ y \in \mathcal{X} \setminus E$,*

$$|u(x) - u(y)| \leq \phi(d(x,y))[g(x) + g(y)]. \tag{13}$$

Denote by $\mathcal{D}^\phi(u)$ the collection of all ϕ-Hajłasz gradients of u.

(ii) *A sequence of nonnegative measurable functions, $\vec{g} := \{g_k\}_{k\in\mathbb{Z}}$, is called a ϕ-*Hajłasz gradient sequence *of u if, for any $k \in \mathbb{Z}$, there exists a set $E_k \subset \mathcal{X}$ with $\mu(E_k) = 0$ such that, for any $x,\ y \in \mathcal{X} \setminus E_k$ with $2^{-k-1} \leq d(x,y) < 2^{-k}$,*

$$|u(x) - u(y)| \leq \phi(d(x,y))[g_k(x) + g_k(y)].$$

Denote by $\mathbb{D}^\phi(u)$ the collection of all ϕ-Hajłasz gradient sequences of u.

The following are basic properties of these generalized gradients, which can be proved by an argument similar to those about classical Hajłasz gradients (see, for instance, ([51] [Lemma 2.4]), ([38] [Lemma 2.6]), ([41] [Lemmas 2.3 and 2.4]), and ([45][Lemmas 4 and 5])); we omit the details.

Lemma 2. (i) *Let $u,\ v \in L^0(\mathcal{X})$, $\{g_k\}_{k\in\mathbb{Z}} \in \mathbb{D}^\phi(u)$, and $\{h_k\}_{k\in\mathbb{Z}} \in \mathbb{D}^\phi(v)$. Then,*

$$\{\max(g_k,h_k)\}_{k\in\mathbb{Z}} \in \mathbb{D}^\phi(\max\{u,v\}) \quad \text{and} \quad \{\max(g_k,h_k)\}_{k\in\mathbb{Z}} \in \mathbb{D}^\phi(\min\{u,v\}).$$

(ii) *Let $\{u_i\}_{i\in\mathbb{N}} \subset L^0(\mathcal{X})$ and, for any $i \in \mathbb{N}$, let $\{g_k^{(i)}\}_{k\in\mathbb{Z}} \in \mathbb{D}^\phi(u_i)$. Let $u := \sup_{i\in\mathbb{N}} u_i$ and $\{g_k\}_{k\in\mathbb{Z}} := \{\sup_{i\in\mathbb{N}} g_k^{(i)}\}_{k\in\mathbb{Z}}$. If $u \in L^0(\mathcal{X})$, then $\{g_k\}_{k\in\mathbb{Z}} \in \mathbb{D}^\phi(u)$.*

Using these generalized gradients, we introduced the following homogeneous ϕ-Hajłasz–Triebel–Lizorkin and ϕ-Hajłasz–Besov spaces in [28].

Definition 5. *Let $\phi \in \mathcal{A}$ and $p,\ q \in (0,\infty]$.*

(i) *The homogeneous ϕ-Hajłasz–Triebel–Lizorkin space $\dot{M}^\phi_{p,q}(\mathcal{X})$ is defined to be the set of all $u \in L^0(\mathcal{X})$ such that*

$$\|u\|_{\dot{M}^\phi_{p,q}(\mathcal{X})} := \inf_{\vec{g}\in\mathbb{D}^\phi(u)} \|\vec{g}\|_{L^p(\mathcal{X},l^q)} < \infty$$

when $p \in (0, \infty)$ *and* $q \in (0, \infty]$, *or* $p = q = \infty$, *and*

$$\|u\|_{\dot{M}^{\phi}_{\infty,q}(\mathcal{X})} := \inf_{\vec{g}\in\mathbb{D}^{\phi}(u)} \sup_{k\in\mathbb{Z}} \sup_{x\in\mathcal{X}} \left\{ \sum_{j\geq k} \fint_{B(x,2^{-k})} [g_j(y)]^q \, d\mu(y) \right\}^{\frac{1}{q}} < \infty$$

when $p = \infty$ *and* $q \in (0, \infty)$.

(ii) *The homogeneous* ϕ*-Hajłasz–Besov space* $\dot{N}^{\phi}_{p,q}(\mathcal{X})$ *is defined to be the set of all* $u \in L^0(\mathcal{X})$ *such that*

$$\|u\|_{\dot{N}^{\phi}_{p,q}(\mathcal{X})} := \inf_{\vec{g}\in\mathbb{D}^{\phi}(u)} \|\vec{g}\|_{l^q(\mathcal{X},L^p)} < \infty.$$

In [28], we proved that, when $\mathcal{X} = \mathbb{R}^n$, for any given admissible sequence $\sigma := \{\sigma_j\}_{j\in\mathbb{Z}}$ with $\alpha_\sigma \in (0,1)$ and $\beta_\sigma \in (0,2)$, $\dot{M}^{\phi_\sigma}_{p,q}(\mathbb{R}^n) = \dot{F}^{\sigma}_{p,q}(\mathbb{R}^n)$ for any given p, $q \in (n/[n - \log_2 \alpha_\sigma], \infty]$, and $\dot{N}^{\phi_\sigma}_{p,q}(\mathbb{R}^n) = \dot{B}^{\sigma}_{p,q}(\mathbb{R}^n)$ for any given $p \in (n/[n - \log_2 \alpha_\sigma], \infty]$ and $q \in (0, \infty]$, where $\dot{B}^{\sigma}_{p,q}(\mathbb{R}^n)$ and $\dot{F}^{\sigma}_{p,q}(\mathbb{R}^n)$ are, respectively, the classical generalized Besov and Triebel–Lizorkin spaces in which smoothness is described by an admissible sequence σ (see Definition 7 below). In this sense, the spaces $\dot{M}^{\phi}_{p,q}(\mathcal{X})$ and $\dot{N}^{\phi}_{p,q}(\mathcal{X})$ serve as natural generalizations of classical Besov and Triebel–Lizorkin spaces with generalized smoothness on metric measure spaces.

In this article, we also consider the inhomogeneous version of the above spaces.

Definition 6. *Let* $\phi \in \mathcal{A}$ *and* p, $q \in (0, \infty]$.

(i) *The inhomogeneous* ϕ*-Hajłasz–Triebel-Lizorkin space* $M^{\phi}_{p,q}(\mathcal{X})$ *is defined as the set* $L^p(\mathcal{X}) \cap \dot{M}^{\phi}_{p,q}(\mathcal{X})$. *Moreover, for any* $u \in M^{\phi}_{p,q}(\mathcal{X})$, *let*

$$\|u\|_{M^{\phi}_{p,q}(\mathcal{X})} := \|u\|_{L^p(\mathcal{X})} + \|u\|_{\dot{M}^{\phi}_{p,q}(\mathcal{X})}.$$

(ii) *The inhomogeneous* ϕ*-Hajłasz–Besov space* $N^{\phi}_{p,q}(\mathcal{X})$ *is defined as the set* $L^p(\mathcal{X}) \cap \dot{N}^{\phi}_{p,q}(\mathcal{X})$. *Moreover, for any* $u \in M^{\phi}_{p,q}(\mathcal{X})$, *let*

$$\|u\|_{M^{\phi}_{p,q}(\mathcal{X})} := \|u\|_{L^p(\mathcal{X})} + \|u\|_{\dot{N}^{\phi}_{p,q}(\mathcal{X})}.$$

Remark 1. (i) *Recall that, for any given* $p \in (0, \infty]$, $\dot{M}^{\phi}_{p,\infty}(\mathcal{X}) = \dot{M}^{\phi,p}(\mathcal{X})$ *(see [28], [Remark 3.4(i)]), where* $\dot{M}^{\phi,p}(\mathcal{X})$ *denotes the homogeneous Hajłasz–Sobolev space with respect to* ϕ, *which consists of all* $u \in L^0(\mathcal{X})$ *such that*

$$\|u\|_{\dot{M}^{\phi,p}(\mathcal{X})} := \inf_{g\in\mathcal{D}^{\phi}(u)} \|g\|_{L^p(\mathcal{X})} < \infty.$$

Consequently, if the inhomogeneous Hajłasz–Sobolev space $M^{\phi,p}(\mathcal{X})$ *is defined as the set* $L^p(\mathcal{X}) \cap \dot{M}^{\phi,p}(\mathcal{X})$, *then* $M^{\phi}_{p,\infty}(\mathcal{X}) = M^{\phi,p}(\mathcal{X})$. *In particular, when* ϕ *is as in* (12), *the related spaces are called the* logarithmic Hajłasz–Sobolev spaces.

(ii) *Let* $\phi \in \mathcal{A}$, $k_0 \in \mathbb{Z}$, *and* $u \in L^0(\mathcal{X})$. *Let* $\mathbb{D}^{\phi}_{k_0}(u)$ *be the set of all sequences* $\vec{h} := \{h_k\}_{k\in\mathbb{Z}}$, *defined by setting* $h_k := \widetilde{h}_k$ *when* $k \geq k_0$ *and* $h_k \equiv 0$ *when* $k < k_0$, *where* $\widetilde{h} := \{\widetilde{h}_k\}_{k\in\mathbb{Z}}$ *is a* ϕ*-Hajłasz gradient sequence of* u. *Naturally,* $\mathcal{D}^{\phi}_{k_0}(u)$ *denotes the set of all functions* g *such that, for almost every* x, $y \in \mathcal{X}$ *with* $d(x,y) < 2^{-k_0}$, (13) *holds true. Then, for any given* $p \in (0, \infty]$, $q = \infty$, *and* $\phi \in \mathcal{A}$ *or for any given* $p \in (0, \infty]$, $q \in (0, \infty)$, *and* $\phi \in \mathcal{A}$ *with* $\alpha^+_\phi \in (0,1)$,

$$|||u|||_{N^{\phi}_{p,q}(\mathcal{X})} := \|u\|_{L^p(\mathcal{X})} + \inf_{\vec{h}\in\mathbb{D}^{\phi}_{k_0}(u)} \|\vec{h}\|_{l^q(\mathcal{X},L^p)}, \quad \forall\, u \in N^{\phi}_{p,q}(\mathcal{X}),$$

is an equivalent quasi-norm of $N^{\phi}_{p,q}(\mathcal{X})$ *with the positive equivalence constants depending on* k_0. *Indeed, for any* $u \in L^0(\mathcal{X})$, $|||u|||_{N^{\phi}_{p,q}(\mathcal{X})} \le \|u\|_{N^{\phi}_{p,q}(\mathcal{X})}$ *obviously holds true. Conversely, let* $q \in (0,\infty)$ *and* $u \in L^0(\mathcal{X})$. *Notice that, for any* $k \in \mathbb{Z}$ *and* $x, y \in \mathcal{X}$,

$$|u(x) - u(y)| \le \phi(2^{-k})\left[\frac{|u(x)|}{\phi(2^{-k})} + \frac{|u(y)|}{\phi(2^{-k})}\right].$$

Then, $\{\frac{|u|}{\phi(2^{-k})}\}_{k\in\mathbb{Z}}$ *is a* ϕ*-Hajłasz gradient sequence of u modulo some uniform constant, which implies that, for any* $\vec{h} := \{h_k\}_{k\in\mathbb{Z}} \in \mathbb{D}^{\phi}_{k_0}(u)$, *the sequence* $\vec{g} := \{g_k\}_{k\in\mathbb{Z}}$, *defined by setting, for any* $k \ge k_0$, $g_k := h_k$ *and, for any* $k < k_0$, $g_k := \frac{|u|}{\phi(2^{-k})}$ *is an element of* $\mathbb{D}^{\phi}_{k_0}(u)$. *By* $\alpha^+_\phi \in (0,1)$, *we can choose a* $\delta_1 \in (0,\infty)$ *such that* $\alpha^+_\phi + \delta_1 < 1$. *Then, there exists a* $K \in \mathbb{Z}$ *such that, for any integer* $k \le K$, $\phi(2^{-k})/\phi(2^{-k+1}) < \alpha^+_\phi + \delta_1$. *Notice that* $\phi(2^{-k})/\phi(2^{-k+1})$ *is bounded when* $k \in [K, k_0]$. *We then have*

$$\begin{aligned}\sum_{k\le k_0}[\phi(2^{-k})]^{-q} &= [\phi(2^{-k_0})]^{-q}\sum_{k\le k_0}\left[\frac{\phi(2^{-k-1})}{\phi(2^{-k})}\frac{\phi(2^{-k-2})}{\phi(2^{-k-1})}\cdots\frac{\phi(2^{-k_0})}{\phi(2^{-k_0+1})}\right]^q\\ &\lesssim \sum_{k\le k_0}(\alpha^+_\phi + \delta_1)^{(k_0-k)q} \lesssim 1,\end{aligned}$$

where the implicit positive constants depend only on ϕ, q, *and* k_0. *This implies that*

$$\|u\|_{N^{\phi}_{p,q}(\mathcal{X})} \le \|\vec{g}\|_{l^q(\mathcal{X},L^p)} + \|u\|_{L^p(\mathcal{X})} \lesssim \|\vec{h}\|_{l^q(\mathcal{X},L^p)} + \|u\|_{L^p(\mathcal{X})} \lesssim |||u|||_{N^{\phi}_{p,q}(\mathcal{X})}.$$

The proof for the case $q = \infty$ *is similar, and we omit the details here.*

Similarly, for any $\phi \in \mathcal{A}$ *with* $\alpha^+_\phi \in (0,1)$, $p \in (0,\infty]$, *and* $q \in (0,\infty)$ *or any* $\phi \in \mathcal{A}$ *with* $p \in (0,\infty]$ *and* $q = \infty$, $|||u|||_{M^{\phi}_{p,q}(\mathcal{X})}$, *defined by replacing* $\vec{g} \in \mathbb{D}^{\phi}(u)$ *in* $\|u\|_{M^{\phi}_{p,q}(\mathcal{X})}$ *by* $\vec{h} \in \mathbb{D}^{\phi}_{k_0}(u)$, *is also an equivalent quasi-norm of* $M^{\phi}_{p,q}(\mathcal{X})$.

As was mentioned above, the spaces $\dot{M}^{\phi_\sigma}_{p,q}(\mathbb{R}^n)$ and $\dot{N}^{\phi_\sigma}_{p,q}(\mathbb{R}^n)$ coincide, respectively, with the Triebel–Lizorkin space $\dot{F}^{\sigma}_{p,q}(\mathbb{R}^n)$ and the Besov space $\dot{B}^{\sigma}_{p,q}(\mathbb{R}^n)$ with generalized smoothness; see [28]. It is natural to expect to obtain their inhomogeneous counterparts. To this end, we let $\mathcal{S}(\mathbb{R}^n)$ be the collection of all Schwartz functions on $\mathbb{R}^n$, in which the topology is determined by a family of norms, $\{\|\cdot\|_{\mathcal{S}_{k,m}(\mathbb{R}^n)}\}_{k,m\in\mathbb{Z}_+}$, where, for any $k, m \in \mathbb{Z}_+$ and any $\varphi \in \mathcal{S}(\mathbb{R}^n)$,

$$\|\varphi\|_{\mathcal{S}_{k,m}(\mathbb{R}^n)} := \sup_{\alpha\in\mathbb{Z}^n_+,|\alpha|\le k}\ \sup_{x\in\mathbb{R}^n}(1+|x|)^m|\partial^\alpha\varphi(x)|$$

with $\alpha := (\alpha_1,\ldots,\alpha_n) \in \mathbb{Z}^n_+$, $|\alpha| := \alpha_1 + \cdots + \alpha_n$, and $\partial^\alpha := (\frac{\partial}{\partial x_1})^{\alpha_1}\cdots(\frac{\partial}{\partial x_n})^{\alpha_n}$. Additionally, let $\mathcal{S}'(\mathbb{R}^n)$ be the space of all tempered distributions on $\mathbb{R}^n$ equipped with the weak-$*$ topology. Define

$$\mathcal{S}_\infty(\mathbb{R}^n) := \left\{\varphi \in \mathcal{S}(\mathbb{R}^n) : \int_{\mathbb{R}^n}\varphi(x)\,x^\gamma\,dx = 0 \text{ for all multi-indices } \gamma \in \mathbb{Z}^n_+\right\},$$

and let $\mathcal{S}'_\infty(\mathbb{R}^n)$ be the topological dual of $\mathcal{S}'_\infty(\mathbb{R}^n)$ equipped with the weak-$*$ topology. For any $f \in \mathcal{S}'_\infty(\mathbb{R}^n)$, we use $\widehat{f}$ to denote its Fourier transform in the sense of $\mathcal{S}'_\infty(\mathbb{R}^n)$; in particular, for any $f \in L^1(\mathbb{R}^n)$ and $\xi \in \mathbb{R}^n$, $\widehat{f}(\xi) := \int_{\mathbb{R}^n} f(x)\,e^{-2\pi i x\cdot\xi}\,dx$. For any $t \in (0,\infty)$ and $x \in \mathbb{R}^n$, let $\varphi_t(x) := t^{-n}\varphi(x/t)$.

Definition 7. *Let $\sigma := \{\sigma_j\}_{j\in\mathbb{Z}}$ be an admissible sequence. Let p, $q \in (0,\infty]$ and φ, $\Phi \in \mathcal{S}(\mathbb{R}^n)$ be such that*

$$\operatorname{supp}\widehat{\varphi} \subset \{\xi \in \mathbb{R}^n : 1/2 \le |\xi| \le 2\}, \text{ and } |\widehat{\varphi}(\xi)| \ge C_1 \text{ if } 3/5 \le |\xi| \le 5/3$$

and

$$\operatorname{supp}\widehat{\Phi} \subset \{\xi \in \mathbb{R}^n : |\xi| \le 2\}, \text{ and } |\widehat{\Phi}(\xi)| \ge C_2 \text{ if } |\xi| \le 5/3,$$

where C_1, C_2 are two positive constants.

(i) *The* homogeneous Triebel–Lizorkin space $\dot{F}^\sigma_{p,q}(\mathbb{R}^n)$ with generalized smoothness *is defined as the set of all $u \in \mathcal{S}'_\infty(\mathbb{R}^n)$ such that $\|u\|_{\dot{F}^\sigma_{p,q}(\mathbb{R}^n)} < \infty$, where, when $p < \infty$,*

$$\|u\|_{\dot{F}^\sigma_{p,q}(\mathbb{R}^n)} := \|\{\sigma_k\,\varphi_{2^{-k}} * u\}_{k\in\mathbb{Z}}\|_{L^p(\mathbb{R}^n,l^q)} := \left\|\left(\sum_{k\in\mathbb{Z}} \sigma_k^q \left|\varphi_{2^{-k}} * u\right|^q\right)^{1/q}\right\|_{L^p(\mathbb{R}^n)}$$

with the usual modification made if $q = \infty$ and, when $p = \infty$,

$$\|u\|_{\dot{F}^\sigma_{\infty,q}(\mathbb{R}^n)} := \sup_{x\in\mathbb{R}^n} \sup_{l\in\mathbb{Z}} \left\{ \fint_{B(x,2^{-l})} \sum_{k\ge l} \sigma_k^q \left|\varphi_{2^{-k}} * u(y)\right|^q dy \right\}^{1/q}$$

with the usual modification made if $q = \infty$.

(ii) *The* homogeneous Besov space $\dot{B}^\sigma_{p,q}(\mathbb{R}^n)$ with generalized smoothness *is defined as the set of all $u \in \mathcal{S}'_\infty(\mathbb{R}^n)$ such that*

$$\|u\|_{\dot{B}^\sigma_{p,q}(\mathbb{R}^n)} := \|\{\sigma_k\,\varphi_{2^{-k}} * u\}_{k\in\mathbb{Z}}\|_{l^q(\mathbb{R}^n,L^p)} := \left[\sum_{k\in\mathbb{Z}} \sigma_k^q \left\|\varphi_{2^{-k}} * u\right\|^q_{L^p(\mathbb{R}^n)}\right]^{1/q} < \infty$$

with the usual modification made if $q = \infty$.

(iii) *The* inhomogeneous Triebel–Lizorkin space $F^\sigma_{p,q}(\mathbb{R}^n)$ with generalized smoothness *is defined as the set of all $u \in \mathcal{S}'(\mathbb{R}^n)$ such that $\|u\|_{F^\sigma_{p,q}(\mathbb{R}^n)}$ is finite, where $\|u\|_{F^\sigma_{p,q}(\mathbb{R}^n)}$ is defined as $\|u\|_{\dot{F}^\sigma_{p,q}(\mathbb{R}^n)}$ with $\{\sigma_k\,\varphi_{2^{-k}} * u\}_{k\in\mathbb{Z}}$ and φ_1 replaced, respectively, by $\{\sigma_k\,\varphi_{2^{-k}} * u\}_{k\in\mathbb{Z}_+}$ and Φ.*

(iv) *The* inhomogeneous Besov space $B^\sigma_{p,q}(\mathbb{R}^n)$ with generalized smoothness *is defined as the set of all $u \in \mathcal{S}'(\mathbb{R}^n)$ such that $\|u\|_{B^\sigma_{p,q}(\mathbb{R}^n)}$ is finite, where $\|u\|_{B^\sigma_{p,q}(\mathbb{R}^n)}$ is defined as $\|u\|_{\dot{B}^\sigma_{p,q}(\mathbb{R}^n)}$ with $\{\sigma_k\,\varphi_{2^{-k}} * u\}_{k\in\mathbb{Z}}$ and φ_1 replaced, respectively, by $\{\sigma_k\,\varphi_{2^{-k}} * u\}_{k\in\mathbb{Z}_+}$ and Φ.*

We then have the following relation between homogeneous and inhomogeneous spaces.

Proposition 1. *Let $p \in [1,\infty]$, $q \in (0,\infty]$, and $\sigma := \{\sigma_j\}_{j\in\mathbb{Z}_+}$ be admissible sequences with $\alpha^+_\sigma \in (0,1)$. Then, for $A \in \{B, F\}$, $A^\sigma_{p,q}(\mathbb{R}^n) = [L^p(\mathbb{R}^n) \cap \dot{A}^{\widetilde{\sigma}}_{p,q}(\mathbb{R}^n)]$, where $\widetilde{\sigma} := \{\widetilde{\sigma}_j\}_{j\in\mathbb{Z}}$ is any given admissible sequence satisfying that, for any $j \in \mathbb{Z}_+$ and $\alpha^-_{\widetilde{\sigma}} \in (0,1)$, $\widetilde{\sigma}_j = \sigma_j$.*

Proof. By similarity, we only consider the Triebel–Lizorkin case.

First, we show $F^\sigma_{p,q}(\mathbb{R}^n) \subset [L^p(\mathbb{R}^n) \cap \dot{F}^{\widetilde{\sigma}}_{p,q}(\mathbb{R}^n)]$. From $p \in [1,\infty]$, $\alpha^+_\sigma < 1$, ([14] [Corollary 3.18]), or ([52] [Theorem 4.1]), we deduce that $B^\sigma_{p,\max\{p,q\}}(\mathbb{R}^n) \subset L^p(\mathbb{R}^n)$, which, together with the trivial embedding $F^\sigma_{p,q}(\mathbb{R}^n) \subset B^\sigma_{p,\max\{p,q\}}(\mathbb{R}^n)$, implies that $F^\sigma_{p,q}(\mathbb{R}^n) \subset L^p(\mathbb{R}^n)$ and, for any $u \in F^\sigma_{p,q}(\mathbb{R}^n)$, $\|u\|_{L^p(\mathbb{R}^n)} \lesssim \|u\|_{F^\sigma_{p,q}(\mathbb{R}^n)}$. Moreover, if $p \in [1,\infty)$, applying (3) when $p/q \le 1$, the Minkowski inequality when $p/q > 1$, or the Minkowski integral inequality, we conclude that, for any $u \in F^\sigma_{p,q}(\mathbb{R}^n)$,

$$\left\|\left(\sum_{k\le 0} \widetilde{\sigma}_k^q \left|\varphi_{2^{-k}} * u\right|^q\right)^{1/q}\right\|_{L^p(\mathbb{R}^n)} \le \left[\sum_{k\le 0} \widetilde{\sigma}_k^{\min\{p,q\}} \|\varphi_{2^{-k}} * u\|^{\min\{p,q\}}_{L^p(\mathbb{R}^n)}\right]^{1/\min\{p,q\}}$$

$$\lesssim \left[\sum_{k\le 0}\widetilde{\sigma}_k^{\min\{p,q\}}\right]^{1/\min\{p,q\}}\|u\|_{L^p(\mathbb{R}^n)}.$$

By $\alpha_{\widetilde{\sigma}}^- \in (0,1)$, we know that there exists a $\delta_1 \in (0,\infty)$ small enough such that $\alpha_{\widetilde{\sigma}}^- + \delta_1 < 1$. Then we have, for any $k \le 0$ and $r \in (0,\infty]$,

$$\sum_{k\le 0}\widetilde{\sigma}_k^r \lesssim (\alpha_{\widetilde{\sigma}}^- + \delta_1)^{kr},$$

where the implicit positive constant only depends on $\widetilde{\sigma}$ and δ_1. Therefore, we obtain

$$\left\|\left(\sum_{k\le 0}\widetilde{\sigma}_k^q\left|\varphi_{2^{-k}} * u\right|^q\right)^{1/q}\right\|_{L^p(\mathbb{R}^n)} \lesssim \left[\sum_{k\le 0}(\alpha_{\widetilde{\sigma}}^- + \delta_1)^{k\min\{p,q\}}\right]^{1/\min\{p,q\}}\|u\|_{L^p(\mathbb{R}^n)} \lesssim \|u\|_{L^p(\mathbb{R}^n)},$$

which implies that $\|u\|_{\dot{F}^{\widetilde{\sigma}}_{p,q}(\mathbb{R}^n)} \lesssim \|u\|_{L^p(\mathbb{R}^n)} + \|u\|_{F^{\widetilde{\sigma}}_{p,q}(\mathbb{R}^n)}$. Similar estimates also holds true for the case $p = \infty$. Altogether, we obtain the embedding $F^{\sigma}_{p,q}(\mathbb{R}^n) \subset [L^p(\mathbb{R}^n) \cap \dot{F}^{\widetilde{\sigma}}_{p,q}(\mathbb{R}^n)]$.

Conversely, let $u \in [L^p(\mathbb{R}^n) \cap \dot{F}^{\widetilde{\sigma}}_{p,q}(\mathbb{R}^n)]$. By the Minkowski integral inequality, we know that, for any given $p \in [1,\infty]$, $\|\Phi * u\|_{L^p(\mathbb{R}^n)} \lesssim \|u\|_{L^p(\mathbb{R}^n)}$. This, combined with the obvious fact that $\|\{\sigma_k|\varphi_{2^{-k}} * u|\}_{k\ge 1}\|_{L^p(\mathbb{R}^n,l^q)} \le \|u\|_{\dot{F}^{\widetilde{\sigma}}_{p,q}(\mathbb{R}^n)}$, implies the embedding $[L^p(\mathbb{R}^n) \cap \dot{F}^{\widetilde{\sigma}}_{p,q}(\mathbb{R}^n)] \subset F^{\sigma}_{p,q}(\mathbb{R}^n)$. This finishes the proof of Proposition 1. □

As an application of Proposition 1 and ([28], Theorem 3.10), we immediately obtain the following conclusion; we omit the details.

Corollary 1. *Let $p \in [1,\infty]$, and $\sigma := \{\sigma_j\}_{j\in\mathbb{Z}_+}$ be an admissible sequence with $\alpha_{\sigma}^+ \in (0,1)$ and $\beta_{\sigma}^+ \in (0,2)$. Then, $F^{\sigma}_{p,q}(\mathbb{R}^n) = M^{\phi_{\widetilde{\sigma}}}_{p,q}(\mathbb{R}^n)$ for any $q \in (n/[n-\log_2\alpha_{\sigma}^+],\infty]$ and $B^{\sigma}_{p,q}(\mathbb{R}^n) = N^{\phi_{\widetilde{\sigma}}}_{p,q}(\mathbb{R}^n)$ for any $q \in (0,\infty]$, where $\widetilde{\sigma} := \{\widetilde{\sigma}_j\}_{j\in\mathbb{Z}}$ is any given admissible sequence satisfying $\widetilde{\sigma}_j = \sigma_j$ for any $j \in \mathbb{Z}_+$, $\alpha_{\widetilde{\sigma}}^- \in (0,1)$, and $\beta_{\widetilde{\sigma}}^- \in (0,2)$.*

3. Lebesgue Points of ϕ-Hajłasz-Type Functions

Let u be a function on the metric measure space $(\mathcal{X},d,\mu)$. A point $x \in \mathcal{X}$ is called a *Lebesgue point* of u if it satisfies

$$\lim_{r\to 0^+}\fint_{B(x,r)}\left|u(y) - u(x)\right| d\mu(y) = 0.$$

For such an x,

$$u(x) = \lim_{r\to 0^+}\fint_{B(x,r)} u(y)\, d\mu(y).$$

Here and thereafter, $t \to 0^+$ means $t \in (0,\infty)$ and $t \to 0$. The classical Lebesgue differentiation theorem states that almost every point is a Lebesgue point of a locally integrable function on $\mathbb{R}^n$. If the function has higher regularity, one could expect a smaller exceptional set. In 2002, Kinnunen and Latvala [38] studied the Lebesgue point of functions of Hajłasz–Sobolev spaces on doubling metric measure spaces, which has led to a lot of related works; see, for instance [39–44].

In this section, we study the Lebesgue point of ϕ-Hajłasz–Besov and ϕ-Hajłasz–Triebel–Lizorkin functions on a given doubling metric measure space $(\mathcal{X},d,\mu)$. To this end, one key tool is the maximal operators. Let $R \in (0,\infty]$. The *restricted maximal operator* $\mathcal{M}_R$ is defined by setting, for any $u \in L^0(\mathcal{X})$ and $x \in \mathcal{X}$,

$$\mathcal{M}_R u(x) := \sup_{B_r\ni x,\, r\in(0,R)}\fint_{B_r}|u|\, d\mu, \tag{14}$$

where the supremum is taken over all balls B_r in $\mathcal{X}$ containing x with the radius $r \in (0, R)$. Obviously, $\mathcal{M} := \mathcal{M}_\infty$ is just the classical Hardy–Littlewood maximal operator, which is known to be bounded on $L^p(\mathcal{X})$ for any given $p \in (1, \infty]$ when $\mathcal{X}$ is a doubling measure space; see, for instance ([53], Theorem 14.13). We also need the discrete Hardy–Littlewood-type maximal operator defined via discrete convolutions (see, for instance [38,11,54]). To recall this, we first need the notion of the *partition of unity*.

Definition 8. *Let $r \in (0, \infty)$, $\mathcal{J} \subset \mathbb{N}$ be an index set, and balls $\{B_j\}_{j\in\mathcal{J}}$ be a covering of $\mathcal{X}$ with the radius r such that $\sum_{j\in\mathcal{J}} \mathbf{1}_{2B_j} \lesssim 1$, where the implicit positive constant is some positive absolute constant. A sequence $\{\varphi_j\}_{j\in\mathcal{J}}$ of functions is called a* partition of unity *with respect to the above ball covering $\{B_j\}_{j\in\mathcal{J}}$ if, for any $j \in \mathcal{J}$, φ_j is a Lipschitz function with the Lipschitz constant cr^{-1}, $\varphi_j \geq C > 0$ on B_j, $\operatorname{supp}\varphi_j \subset \overline{2B_j}$, $0 \leq \varphi_j \leq 1$, and $\sum_{j\in\mathcal{J}} \varphi_j \equiv 1$, where c and C are two positive constants depending only on the doubling constant.*

The existence of the partition of unity in Definition 8 with respect to any given ball covering of $\mathcal{X}$ can be seen, for instance, in ([38], p. 690).

Definition 9. (i) *Let $u \in L^0(\mathcal{X})$. The* discrete convolution of u at the scale $r \in (0, \infty)$ *is defined by setting*

$$u_r := \sum_{j\in\mathcal{J}} u_{B_j} \varphi_j,$$

where $\{B_j\}_{j\in\mathcal{J}}$ is a ball covering of $\mathcal{X}$ with the radius r and $\{\varphi_j\}_{j\in\mathcal{J}}$ a partition of unity with respect to $\{B_j\}_{j\in\mathcal{J}}$ as in Definition 8.

(ii) *The* discrete maximal operator $\mathcal{M}^*$ *is defined by setting, for any $u \in L^0(\mathcal{X})$,*

$$\mathcal{M}^* u := \sup_{k\in\mathbb{Z}} |u|_{2^{-k}},$$

where $|u|_{2^{-k}}$ is the discrete convolution of $|u|$ at the scale 2^{-k}.

(iii) *Let $R \in (0, \infty]$. The* restricted discrete maximal operator $\mathcal{M}_R^*$ *is defined by setting, for any $u \in L^0(\mathcal{X})$,*

$$\mathcal{M}_R^* u := \sup_{\{k\in\mathbb{Z}:\, 2^{-k}<R\}} |u|_{2^{-k}},$$

where $|u|_{2^{-k}}$ is the discrete convolution of $|u|$ at the scale 2^{-k}.

Obviously, $\mathcal{M}_\infty^* = \mathcal{M}^*$. Now, we present two Poincaré-type inequalities with respect to ϕ as below. The first one is easy to prove using the definition of Hajłasz gradients, and the other is provided in ([28], Lemma 3.7).

Lemma 3. *Let $\phi \in \mathcal{A}$. Then, there exists a positive constant $C = C_{(\phi, C_\mu)}$ such that, for any $x \in \mathcal{X}$, $k \in \mathbb{Z}$, $u \in L^0(B(x, 2^{-k}))$, and $g \in \mathcal{D}^\phi(u)$,*

$$\inf_{c\in\mathbb{R}} \fint_{B(x,2^{-k})} |u(y) - c|\, d\mu(y) \leq C\,\phi(2^{-k}) \fint_{B(x,2^{-k})} g(y)\, d\mu(y),$$

where C_μ is as in (1).

Proof. Let $x \in \mathcal{X}$, $k \in \mathbb{Z}$, $u \in L^0(B(x, 2^{-k}))$ and $g \in \mathcal{D}^\phi(u)$. Then,

$$\begin{aligned}
\inf_{c\in\mathbb{R}} \fint_{B(x,2^{-k})} |u(y) - c|\, d\mu(y) &\leq \fint_{B(x,2^{-k})} |u(y) - u_{B(x,2^{-k})}|\, d\mu(y) \\
&\leq \fint_{B(x,2^{-k})} \fint_{B(x,2^{-k})} |u(y) - u(z)|\, d\mu(z)\, d\mu(y) \\
&\leq \fint_{B(x,2^{-k})} \fint_{B(x,2^{-k})} \phi(2^{-k+1})[g(y) + g(z)]\, d\mu(z)\, d\mu(y)
\end{aligned}$$

$$\lesssim \phi(2^{-k}) \fint_{B(x,2^{-k})} g(y)\, d\mu(y).$$

This finishes the proof of Lemma 3. □

Lemma 4. *Let $\phi \in \mathcal{A}$ with $\alpha_\phi \in (0,1)$. Then, for any ε, $\varepsilon' \in (0, -\log_2 \alpha_\phi)$ with $\varepsilon < \varepsilon'$ and $p \in (0, D/\varepsilon)$, there exists a positive constant $C = C_{(\phi,p,\varepsilon',C_\mu)}$ such that, for any $x \in \mathcal{X}$, $k \in \mathbb{Z}$, $u \in L^0(B(x, 2^{-k+1}))$ and $\vec{g} := \{g_j\}_{j\in\mathbb{Z}} \in \mathbb{D}^\phi(u)$,*

$$\begin{aligned} &\inf_{c\in\mathbb{R}} \left[\fint_{B(x,2^{-k})} |u(y) - c|^{\frac{Dp}{D-\varepsilon p}}\, d\mu(y) \right]^{\frac{D-\varepsilon p}{np}} \\ &\quad \le C\, 2^{-k\varepsilon'} \sum_{j \ge k-2} 2^{j\varepsilon'} \phi(2^{-j}) \left\{ \fint_{B(x,2^{-k+1})} [g_j(y)]^p\, d\mu(y) \right\}^{1/p}, \end{aligned} \tag{15}$$

where D and C_μ are as in (1).

Remark 2. *Let D and C_μ be as in (1).*

(i) *Let ϕ, ε, and p be as in Lemma 4. By taking, for any $k \in \mathbb{Z}$, $x \in \mathcal{X}$, $u \in L^0(B(x, 2^{-k+1}))$, and $g \in \mathcal{D}^\phi(u)$, $\varepsilon' := (\varepsilon - \log_2 \alpha_\phi)/2$ and $\vec{g} := \{g_j := g\}_{j\in\mathbb{Z}}$ in (15), we obtain*

$$\inf_{c\in\mathbb{R}} \left[\fint_{B(x,2^{-k})} |u(y) - c|^{\frac{Dp}{D-\varepsilon p}}\, d\mu(y) \right]^{\frac{D-\varepsilon p}{Dp}} \lesssim \phi(2^{-k}) \left\{ \fint_{B(x,2^{-k+1})} [g(y)]^p\, d\mu(y) \right\}^{1/p}, \tag{16}$$

where the implicit positive constant depends only on ϕ, p, ε, and C_μ.

(ii) *Notice that, if $Dp/(D - \varepsilon p) = 1$, then $p = D/(D + \varepsilon)$. In this case, (15) and (16) become, respectively,*

$$\begin{aligned} &\inf_{c\in\mathbb{R}} \fint_{B(x,2^{-k})} |u(y) - c|\, d\mu(y) \\ &\quad \le C_{(\phi,p,\varepsilon',C_\mu)}\, 2^{-k\varepsilon'} \sum_{j\ge k-2} 2^{j\varepsilon'} \phi(2^{-j}) \left\{ \fint_{B(x,2^{-k+1})} [g_j(y)]^{\frac{D}{D+\varepsilon}}\, d\mu(y) \right\}^{\frac{D+\varepsilon}{D}} \end{aligned} \tag{17}$$

and

$$\begin{aligned} &\inf_{c\in\mathbb{R}} \fint_{B(x,2^{-k})} |u(y) - c|\, d\mu(y) \\ &\quad \le C_{(\phi,p,\varepsilon,C_\mu)}\, \phi(2^{-k}) \left\{ \fint_{B(x,2^{-k+1})} [g(y)]^{\frac{D}{D+\varepsilon}}\, d\mu(y) \right\}^{\frac{D+\varepsilon}{D}}. \end{aligned} \tag{18}$$

Applying these Poincaré-type inequalities, we obtain the following estimates.

Lemma 5. *Let $\phi \in \mathcal{A}$, D, and C_μ be as in (1) and $\mathcal{M}$ be the Hardy–Littlewood maximal operator.*

(i) *Then, there exists a positive constant $C = C_{(\phi,C_\mu)}$ such that, for any $u \in L^1_{\mathrm{loc}}(\mathcal{X})$, $g \in \mathcal{D}^\phi(u)$, $i \in \mathbb{Z}$, $y \in \mathcal{X}$ with $u_{B(y,2^{-i})} < \infty$, and almost every $x \in B(y, 2^{-i+1})$,*

$$\left| u(x) - u_{B(y,2^{-i})} \right| \le C\, \phi(2^{-i}) \mathcal{M}(g)(x).$$

(ii) *Let $\alpha_\phi \in (0,1)$. Then, for any $\lambda \in (D/[D - \log_2 \alpha_\phi], \infty)$, there exists a positive constant $C = C_{(\phi,\lambda,C_\mu)}$ such that, for any $u \in L^1_{\mathrm{loc}}(\mathcal{X})$, $g \in \mathcal{D}^\phi(u)$, $i \in \mathbb{Z}$, $y \in \mathcal{X}$ with $u_{B(y,2^{-i})} < \infty$, and almost every $x \in B(y, 2^{-i+1})$,*

$$\left| u(x) - u_{B(y,2^{-i})} \right| \le C\, \phi(2^{-i}) \left[\mathcal{M}\left(g^\lambda\right)(x) \right]^{1/\lambda}. \tag{19}$$

(iii) *Let* $\alpha_\phi \in (0,1)$. *Then, for any* $\lambda \in (D/[D - \log_2 \alpha_\phi], \infty)$, *there exist an* $\epsilon \in (0, -\log_2 \alpha_\phi)$ *depending on* λ, *and a positive constant* $C = C_{(\phi,\lambda,C_\mu)}$ *such that, for any* $u \in L^1_{\rm loc}(\mathcal{X})$, $\vec{g} := \{g_l\}_{l\in\mathbb{Z}} \in \mathbb{D}^\phi(u)$, $i \in \mathbb{Z}$, $y \in \mathcal{X}$ *with* $u_{B(y,2^{-i})} < \infty$, *and almost every* $x \in B(y, 2^{-i+1})$,

$$\left|u(x) - u_{B(y,2^{-i})}\right| \le C \sum_{l\ge i-4} 2^{(l-i)\epsilon}\phi(2^{-l})\left[\mathcal{M}\left(g_l^\lambda\right)(x)\right]^{1/\lambda}. \tag{20}$$

Proof. Let u, g, i, y, and x be as in the present lemma. By the definition of Hajłasz gradients, the doubling property of μ, the geometrical observation that, for any $x \in B(y, 2^{-i+1})$, $B(y, 2^{-i+1}) \subset B(x, 2^{-i+2})$ and, for almost every $x \in \mathcal{X}$, $g(x) \le \mathcal{M}(g)(x)$, we have, for almost every $x \in B(y, 2^{-i+1})$,

$$\begin{aligned}\left|u(x) - u_{B(y,2^{-i})}\right| &\le \fint_{B(y,2^{-i})} |u(x) - u(z)|\, d\mu(z)\\ &\lesssim \phi(2^{-i}) \fint_{B(y,2^{-i})} [g(x) + g(z)]\, d\mu(z)\\ &\lesssim \phi(2^{-i})\left[g(x) + \fint_{B(x,2^{-i+2})} g(z)\, d\mu(z)\right]\\ &\lesssim \phi(2^{-i})\mathcal{M}(g)(x),\end{aligned}$$

which proves (i) of the present lemma.

To complete the proof of the present lemma, we observe that, for any $i \in \mathbb{Z}$, $y \in \mathcal{X}$ and $x \in B(y, 2^{-i+1})$, $B(y, 2^{-i}) \subset B(x, 2^{-i+2})$. Thus, by the Lebesgue differentiation theorem and the doubling property of μ, we find that, for almost every $x \in B(y, 2^{-i+1})$,

$$\begin{aligned}\left|u(x) - u_{B(y,2^{-i})}\right| &\le \left|u(x) - u_{B(x,2^{-i+2})}\right| + \left|u_{B(x,2^{-i+2})} - u_{B(y,2^{-i})}\right|\\ &\lesssim \sum_{k\ge i-2} \fint_{B(x,2^{-k})} \left|u(z) - u_{B(x,2^{-k})}\right| d\mu(z)\\ &\quad + \fint_{B(x,2^{-i+2})} \left|u(z) - u_{B(x,2^{-i+2})}\right| d\mu(z)\\ &\lesssim \sum_{k\ge i-2} \fint_{B(x,2^{-k})} \left|u(z) - u_{B(x,2^{-k})}\right| d\mu(z)\\ &\lesssim \sum_{k\ge i-2} \inf_{c\in\mathbb{R}} \fint_{B(x,2^{-k})} \left|u(z) - c\right| d\mu(z).\end{aligned} \tag{21}$$

If $\lambda \in (D/[D - \log_2 \alpha_\phi], 1)$, choose $\omega \in (0, -\log_2 \alpha_\phi)$ such that $\lambda = D/(D + \omega)$. By $\alpha_\phi \in (0,1)$, (21), and the definition of $\mathcal{M}$, we conclude that (19) and (20) follow from (18) and (17) with $\varepsilon = \omega$ therein, respectively.

If $\lambda \in [1, \infty)$, then, for any $\epsilon \in (0, -\log_2 \alpha_\phi)$, by the Hölder inequality, we also obtain the same estimate as the case $\lambda \in (D/[D - \log_2 \alpha_\phi], 1)$. This finishes the proof of Lemma 5. □

Remark 3. (i) *Let* $\phi \in \mathcal{A}$ *with* $\alpha_\phi \in (0,1)$. *Recall that, for any* $p \in (D/(D - \log_2 \alpha_\phi), \infty]$, $q \in (0, \infty]$, *and* $u \in [\dot{M}^\phi_{p,q}(\mathcal{X}) \cup \dot{N}^\phi_{p,q}(\mathcal{X})]$, *the integral of* u *on any ball in* $\mathcal{X}$ *is finite (see [28], Remark 3.8), where* D *is as in* (1).

(ii) *Let* $\phi \in \mathcal{A}$. *For any* $u \in \dot{\mathcal{F}}$, *the integral of* $|u|^p$ *on any ball* $B := B(x, 2^{-k})$ *in* $\mathcal{X}$ *with* $k \in \mathbb{Z}$ *is also finite, where*

$$\begin{aligned}\dot{\mathcal{F}} \in &\left\{\dot{M}^{\phi,p}(\mathcal{X}) : \ p \in [1,\infty)\right\} \cup \left\{\dot{M}^{\phi,p}(\mathcal{X}) : \ p \in (0,1),\ \alpha_\phi \in (0,1)\right\}\\ &\cup \left\{\dot{M}^\phi_{p,q}(\mathcal{X}),\ \dot{N}^\phi_{p,q}(\mathcal{X}) : \ p,\ q \in (0,\infty],\ \alpha_\phi \in (0,1)\right\}.\end{aligned}$$

To see this, by similarity, we only prove the case $\dot{\mathcal{F}} = \dot{M}^{\phi}_{p,q}(\mathcal{X})$ *with* $p,\ q \in (0,\infty]$ *and* $\alpha_\phi \in (0,1)$. *Indeed, by (15), the Hölder inequality, Lemma 1(i), and the definition of* $\mathcal{A}$, *we find that*

$$\begin{aligned}\inf_{c\in\mathbb{R}}\left[\fint_B |u(y)-c|^p\,d\mu(y)\right]^{1/p} &\lesssim 2^{-k\varepsilon'}\sum_{j\ge k-2} 2^{j\varepsilon'}\phi(2^{-j})\left\{\fint_{2B}[g_j(y)]^p\,d\mu(y)\right\}^{1/p}\\ &\lesssim 2^{-k\varepsilon'}\sum_{j\ge k-2} 2^{j\varepsilon'}\phi(2^{-j})[\mu(2B)]^{-1/p}\left\|\{g_j\}_{j\in\mathbb{Z}}\right\|_{L^p(2B,l^q)}\\ &\lesssim \phi(2^{-k})[\mu(2B)]^{-1/p}\left\|\{g_j\}_{j\in\mathbb{Z}}\right\|_{L^p(\mathcal{X},l^q)}<\infty,\end{aligned}$$

where $\varepsilon' \in (0, -\log_2\alpha_\phi)$ *and* $\{g_j\}_{j\in\mathbb{Z}} \in \mathbb{D}^\phi(u)\cap L^p(\mathcal{X},l^q)$. *Let* $c_0 \in \mathbb{R}$ *be such that*

$$\fint_B |u(y)-c_0|^p\,d\mu(y) < \infty.$$

Then,

$$\int_B |u(y)|^p\,d\mu(y) \lesssim \mu(B)\fint_B |u(y)-c_0|^p\,d\mu(y) + \mu(B)c_0^p < \infty.$$

Thus, the above claim holds true.

Due to Remark 3(i), the classical Lebesgue differentiation theorem implies that almost every point is a Lebesgue point of u. As u has certain regularity, one would expect a smaller exceptional set than that of usual locally integrable functions. Inspired by [41,45], we introduce capacities related, respectively, to $M^{\phi}_{p,q}(\mathcal{X})$ and $N^{\phi}_{p,q}(\mathcal{X})$ to measure such exceptional sets.

Below, for simplicity, we use $\mathcal{F}$ to denote either $M^{\phi}_{p,q}(\mathcal{X})$ or $N^{\phi}_{p,q}(\mathcal{X})$, or $\dot{\mathcal{F}}$ to denote either $\dot{M}^{\phi}_{p,q}(\mathcal{X})$ or $\dot{N}^{\phi}_{p,q}(\mathcal{X})$.

Definition 10. *Let E be a subset of $\mathcal{X}$. Recall that a set U is called a* neighborhood *of E if it is open and $E \subset U$. Let* $\mathcal{F} \in \{M^{\phi}_{p,q}(\mathcal{X}), N^{\phi}_{p,q}(\mathcal{X})\}$ *with* $\phi \in \mathcal{A}$ *and* $p,\ q \in (0,\infty]$, *and*

$$\mathcal{G}_{\mathcal{F}}(E) := \{u \in \mathcal{F} :\ u \ge 1 \text{ on a neighborhood of } E\}.$$

The $\mathcal{F}$-capacity $\mathrm{Cap}_{\mathcal{F}}(E)$ *of E is defined by setting*

$$\mathrm{Cap}_{\mathcal{F}}(E) := \inf\left\{\|u\|^p_{\mathcal{F}} :\ u \in \mathcal{G}_{\mathcal{F}}(E)\right\}.$$

Remark 4. *Let* $E,\ E_1,\ E_2 \subset \mathcal{X}$ *and* $\mathcal{F} \in \{M^{\phi}_{p,q}(\mathcal{X}), N^{\phi}_{p,q}(\mathcal{X})\}$ *with* $\phi \in \mathcal{A}$ *and* $p,\ q \in (0,\infty]$.

(i) *Let* $\mathcal{G}'_{\mathcal{F}}(E) := \{u \in \mathcal{G}_{\mathcal{F}}(E) :\ 0 \le u \le 1\}$. *By Lemma 2(i),* $\|\max\{\min\{u,1\},0\}\|_{\mathcal{F}} \le \|u\|_{\mathcal{F}}$, *and an argument similar to that used in ([55], Remark 3.2), we have*

$$\mathrm{Cap}_{\mathcal{F}}(E) = \inf\left\{\|u\|^p_{\mathcal{F}} :\ u \in \mathcal{G}'_{\mathcal{F}}(E)\right\}.$$

(ii) *If* $\mathrm{Cap}_{\mathcal{F}}(E) = 0$ *with* $p \in (0,\infty)$, *then* $\mu(E) = 0$. *Indeed, for any* $\epsilon \in (0,\infty)$, *there always exists a neighborhood* U_ϵ *of E such that* $\|\mathbf{1}_{U_\epsilon}\|_{\mathcal{F}} < \epsilon$, *which implies that*

$$[\mu(E)]^{1/p} = \|\mathbf{1}_E\|_{L^p(\mathcal{X})} \le \|\mathbf{1}_E\|_{\mathcal{F}} \le \epsilon.$$

Letting $\epsilon \to 0^+$, *we obtain* $\mu(E) = 0$.

(iii) *If* $E_1 \subset E_2$, *then* $\mathcal{G}_{\mathcal{F}}(E_2) \subset \mathcal{G}_{\mathcal{F}}(E_1)$, *which means that* $\mathrm{Cap}_{\mathcal{F}}(E_1) \le \mathrm{Cap}_{\mathcal{F}}(E_2)$.

The following lemma provides a basic property of the capacity which is a slight generalization of ([41], Lemma 6.4); we omit the details.

Lemma 6. *Let* $\mathcal{F} \in \{M^{\phi}_{p,q}(\mathcal{X}), N^{\phi}_{p,q}(\mathcal{X})\}$ *with* $\phi \in \mathcal{A}$ *and* $p \in (0,\infty)$ *and* $q \in (0,\infty]$. *Let* $\theta := \min\{1, q/p\}$. *Then, there exists a positive constant* $C = C_{(p,q)} \in [1,\infty)$ *such that, for any sequence* $\{E_i\}_{i\in\mathbb{N}}$ *of subsets of* $\mathcal{X}$,

$$\left[\mathrm{Cap}_{\mathcal{F}}\left(\bigcup_{i\in\mathbb{N}} E_i\right)\right]^{\theta} \le C\sum_{i\in\mathbb{N}}[\mathrm{Cap}_{\mathcal{F}}(E_i)]^{\theta}.$$

Via $\mathcal{F}$-capacities, we introduce the $\mathcal{F}$-quasi-continuity as follows.

Definition 11. *Let* $\mathcal{F} \in \{M^{\phi}_{p,q}(\mathcal{X}), N^{\phi}_{p,q}(\mathcal{X})\}$ *with* $\phi \in \mathcal{A}$ *and* $p,\ q \in (0,\infty]$. *A function u is said to be* $\mathcal{F}$*-quasi-continuous if, for any* $\varepsilon \in (0,\infty)$, *there exists a set* U_ε *such that* $\mathrm{Cap}_{\mathcal{F}}(U_\varepsilon) < \varepsilon$ *and the restriction* $u|_{\mathcal{X}\setminus U_\varepsilon}$ *of u on* $\mathcal{X}\setminus U_\varepsilon$ *is continuous.*

The following theorem shows the convergence of discrete convolution approximations in $\mathcal{F}$, which generalizes ([41], Theorem 5.1).

Theorem 1. *Let* $\phi \in \mathcal{A}_0$, $p \in (D/(D-\log_2\alpha_\phi),\infty)$, $\mathcal{F} = M^{\phi}_{p,q}(\mathcal{X})$ *[resp.,* $\dot{\mathcal{F}} = \dot{M}^{\phi}_{p,q}(\mathcal{X})$*] with* $q \in (D/(D-\log_2\alpha_\phi),\infty)$, *or* $\mathcal{F} = N^{\phi}_{p,q}(\mathcal{X})$ *[resp.,* $\dot{\mathcal{F}} = \dot{N}^{\phi}_{p,q}(\mathcal{X})$*] with* $q \in (0,\infty)$, *and* $u \in \dot{\mathcal{F}}$. *Then,* $\|u - u_{2^{-i}}\|_{\mathcal{F}} \to 0$ *as* $i \to \infty$, *where* $\{u_{2^{-i}}\}_{i\in\mathbb{Z}_+}$ *are the discrete convolutions as in Definition 9(i).*

To prove Theorem 1, we need the following lemma, which generalizes ([41], Lemma 3.1) (see also [47], Lemma 3.10).

Lemma 7. *Let* $E \subset \mathcal{X}$ *be a measurable set,* $L \in (0,\infty)$, φ *be a bounded L-Lipschitz function supported in E,* $u \in L^0(\mathcal{X})$, *and* $\phi \in \mathcal{A}_\infty$.

(i) *If* $\{g_k\}_{k\in\mathbb{Z}} \in \mathbb{D}^{\phi}(u)$, *then, for any* $i \in \mathbb{Z}$, *the sequence* $\{h_k\}_{k\in\mathbb{Z}}$, *defined by setting*

$$h_k := \begin{cases} \left\{2^{-k}\left[\phi(2^{-k})\right]^{-1} L|u| + \|\varphi\|_{L^\infty(\mathcal{X})}\, g_k\right\}\mathbf{1}_E, & k > i, \\ \|\varphi\|_{L^\infty(\mathcal{X})}\left[\phi(2^{-k})\right]^{-1}|u|\,\mathbf{1}_E, & k \le i, \end{cases} \tag{22}$$

is an element of $\mathbb{D}^{\phi}(u\varphi)$ *modulo a positive constant that is independent of i and L.*

(ii) *If* $g \in \mathcal{D}^{\phi}(u)$, *then*

$$h := \left\{\|\varphi\|_{L^\infty(\mathcal{X})}\, g + \left[\|\varphi\|_{L^\infty(\mathcal{X})} + 1\right]\left[\phi(L^{-1})\right]^{-1}|u|\right\}\mathbf{1}_E$$

is an element of $\mathcal{D}^{\phi}(u\varphi)$ *modulo a positive constant that is independent of L.*

Proof. We first prove (i). Let φ be a bounded L-Lipschitz function supported in E, $u \in L^0(\mathcal{X})$, and $\{g_k\}_{k\in\mathbb{Z}} \in \mathbb{D}^{\phi}(u)$. For any $k \in \mathbb{Z}$ and $x,\ y \in \mathcal{X}$ with $d(x,y) \in [2^{-k-1}, 2^{-k})$, we have

$$d(x,y)/\phi(d(x,y)) \lesssim 2^{-k}/\phi(2^{-k}) \quad \text{and} \quad [\phi(d(x,y))]^{-1} \lesssim [\phi(2^{-k})]^{-1}.$$

Then, from the Lipschitz continuity of φ and the definition of $\mathbb{D}^{\phi}(u)$, it follows that, for any $k \in \mathbb{Z}$ and almost every $x,\ y \in E$ with $d(x,y) \in [2^{-k-1}, 2^{-k})$,

$$\begin{aligned} \left|u(x)\varphi(x) - u(y)\varphi(y)\right| &\le |u(x)|\left|\varphi(x) - \varphi(y)\right| + \|\varphi\|_{L^\infty}\left|u(x) - u(y)\right| \\ &\le \phi(d(x,y))\left\{\frac{Ld(x,y)|u(x)|}{\phi(d(x,y))} + \|\varphi\|_{L^\infty(\mathcal{X})}[g_k(x) + g_k(y)]\right\} \\ &\lesssim \phi(d(x,y))\left\{\frac{L2^{-k}|u(x)|}{\phi(2^{-k})} + \|\varphi\|_{L^\infty(\mathcal{X})}[g_k(x) + g_k(y)]\right\} \end{aligned}$$

and

$$\left|u(x)\varphi(x) - u(y)\varphi(y)\right| \lesssim |u(x)|\|\varphi\|_{L^\infty(\mathcal{X})} + \|\varphi\|_{L^\infty}(|u(x)| + |u(y)|)$$
$$\lesssim \phi(d(x,y))\frac{\|\varphi\|_{L^\infty(\mathcal{X})}(|u(x)| + |u(y)|)}{\phi(d(x,y))}$$
$$\lesssim \phi(d(x,y))\frac{\|\varphi\|_{L^\infty(\mathcal{X})}(|u(x)| + |u(y)|)}{\phi(2^{-k})}.$$

For any $k \in \mathbb{Z}$ and almost every $x \in E$ and $y \in \mathcal{X} \setminus E$ with $d(x,y) \in [2^{-k-1}, 2^{-k})$, we have

$$\left|u(x)\varphi(x) - u(y)\varphi(y)\right| \le |u(x)||\varphi(x) - \varphi(y)|$$
$$\le \phi(d(x,y))\frac{Ld(x,y)|u(x)|}{\phi(d(x,y))} \lesssim \phi(d(x,y))\frac{L2^{-k}|u(x)|}{\phi(2^{-k})}$$

and

$$\left|u(x)\varphi(x) - u(y)\varphi(y)\right| \le \|\varphi\|_{L^\infty(\mathcal{X})}|u(x)| \lesssim \phi(d(x,y))\frac{\|\varphi\|_{L^\infty(\mathcal{X})}|u(x)|}{\phi(2^{-k})}.$$

Similarly, for any $k \in \mathbb{Z}$ and almost every $x \in E$ and $y \in \mathcal{X} \setminus E$ with $d(x,y) \in [2^{-k-1}, 2^{-k})$, we have

$$\left|u(x)\varphi(x) - u(y)\varphi(y)\right| \lesssim \phi(d(x,y))\frac{L2^{-k}|u(y)|}{\phi(2^{-k})}$$

and

$$\left|u(x)\varphi(x) - u(y)\varphi(y)\right| \lesssim \phi(d(x,y))\frac{\|\varphi\|_{L^\infty(\mathcal{X})}|u(y)|}{\phi(2^{-k})}.$$

From these estimates, we deduce that $\{h_k\}_{k\in\mathbb{Z}}$ as in (22) is a positive constant multiple of an element in $\mathcal{D}^\phi(u\varphi)$, with the positive constant independent of i and L. This proves (i).

The item (ii) is easy to show using the result in (i) and choosing $h := \sup_{k\in\mathbb{Z}} h_k$ and $i \in \mathbb{Z}$ such that $L \in [2^i, 2^{i+1})$. This finishes the proof of Lemma 7. □

We now state some corollaries of Lemma 7 as follows.

Corollary 2. *Let $E \subset \mathcal{X}$ be a measurable set, $L \in [1/2, \infty)$, φ be a bounded L-Lipschitz function supported in E and $p \in (0,\infty)$. Let $\mathcal{F} \in \{M^\phi_{p,q}(\mathcal{X}), N^\phi_{p,q}(\mathcal{X})\}$ with $q \in (0,\infty)$ and $\phi \in \mathcal{A}_q$, or $\mathcal{F} \in \{M^\phi_{p,\infty}(\mathcal{X}) = M^{\phi,p}(\mathcal{X}), N^\phi_{p,\infty}(\mathcal{X})\}$ with $\phi \in \mathcal{A}_\infty$. Then, for any $u \in \mathcal{F}$, $u\varphi \in \mathcal{F}$ with $\|u\varphi\|_{\mathcal{F}} \lesssim \|u\|_{\mathcal{F}}$, where the implicit positive constant is independent of u.*

Proof. By similarity, we only consider $\mathcal{F} = M^\phi_{p,q}(\mathcal{X})$ with $p,\ q \in (0,\infty)$ and $\phi \in \mathcal{A}_q$. Let $i \in \mathbb{Z}_+$ be such that $2^{i-1} \le L < 2^i$, $u \in L^0(\mathcal{X})$, $\{g_k\}_{k\in\mathbb{Z}} \in \mathbb{D}^\phi(u)$ satisfy $\|\{g_k\}_{k\in\mathbb{Z}}\|_{L^p(\mathcal{X},l^q)} \lesssim \|u\|_{\dot{M}^\phi_{p,q}(\mathcal{X})}$, and $\{h_k\}_{k\in\mathbb{Z}}$ be as in (22). By the definition of $\mathcal{A}_q$, we have

$$\sum_{k\le i}\frac{1}{[\phi(2^{-k})]^q} \lesssim X_L^q \quad\text{and}\quad \sum_{k>i}\frac{2^{-kq}}{[\phi(2^{-k})]^q} \lesssim Y_L^q,$$

where X_L and Y_L are two positive constants independent of ϕ. From this, we deduce that

$$\|\{h_k\}_{k\in\mathbb{Z}}\|_{L^p(\mathcal{X},l^q)} \lesssim \left\{\sum_{k>i}\left(2^{-k}\left[\phi(2^{-k})\right]^{-1}\right)^q\right\}^{1/q} L\|u\mathbf{1}_E\|_{L^p(\mathcal{X})}$$

$$
\begin{aligned}
&+\|\varphi\|_{L^\infty(\mathcal{X})}\|\{g_k\}_{k\in\mathbb{Z}}\|_{L^p(E,l^q)} \\
&+\left\{\sum_{k\le i}\left(\left[\phi(2^{-k})\right]^{-1}\right)^q\right\}^{1/q}\|\varphi\|_{L^\infty(\mathcal{X})}\|u\mathbf{1}_E\|_{L^p(\mathcal{X})} \\
&\lesssim \|\varphi\|_{L^\infty(\mathcal{X})}\|\{g_k\}_{k\in\mathbb{Z}}\|_{L^p(E,l^q)} \\
&+\left[X_L\|\varphi\|_{L^\infty(\mathcal{X})}+Y_L L\right]\|u\|_{L^p(E)},
\end{aligned} \tag{23}
$$

which, combined with Lemma 7 and $\|u\varphi\|_{L^p(\mathcal{X})}\le\|u\|_{L^p(\mathcal{X})}\|\varphi\|_{L^\infty(\mathcal{X})}$, implies that

$$
\begin{aligned}
\|u\varphi\|_{M^\phi_{p,q}(\mathcal{X})} &\lesssim \|u\varphi\|_{L^p(\mathcal{X})}+\|\{h_k\}_{k\in\mathbb{Z}}\|_{L^p(\mathcal{X},l^q)} \\
&\lesssim \left[(X_L+1)\|\varphi\|_{L^\infty(\mathcal{X})}+Y_L L\right]\|u\|_{M^\phi_{p,q}(\mathcal{X})},
\end{aligned}
$$

where the implicit positive constants are independent of L, φ, and u. This finishes the proof of Corollary 2. □

Corollary 3. *With the same assumptions as in Corollary 2, if the set E is bounded, then, for any $u\in\dot{\mathcal{F}}$, $u\varphi\in\mathcal{F}$.*

Proof. Again, by similarity, we only consider $\mathcal{F}=M^\phi_{p,q}(\mathcal{X})$ with $p,\ q\in(0,\infty)$ and $\phi\in\mathcal{A}_q$. Let $i\in\mathbb{Z}_+$ be such that $2^{i-1}\le L<2^i$, $u\in L^0(\mathcal{X})$, and $\{g_k\}_{k\in\mathbb{Z}}\in\mathbb{D}^\phi(u)$ be such that $\|\{g_k\}_{k\in\mathbb{Z}}\|_{L^p(\mathcal{X},l^q)}\lesssim\|u\|_{\dot{M}^\phi_{p,q}(\mathcal{X})}$. Since E is bounded, we can find a ball B containing E. Then, by Remark 3(ii), we conclude that $\|u\|_{L^p(E)}\le\|u\|_{L^p(B)}<\infty$. Let $\{h_k\}_{k\in\mathbb{Z}}$ be as in (22). Then, from (23), we deduce that $\|\{h_k\}_{k\in\mathbb{Z}}\|_{L^p(\mathcal{X},l^q)}<\infty$, which, combined with Lemma 7, implies that $\|u\varphi\|_{\dot{M}^\phi_{p,q}(\mathcal{X})}<\infty$. Notice that $\|u\varphi\|_{L^p(\mathcal{X})}=\|u\|_{L^p(E)}\|\varphi\|_{L^\infty(\mathcal{X})}<\infty$. We then obtain $\|u\varphi\|_{M^\phi_{p,q}(\mathcal{X})}<\infty$, which completes the proof of Corollary 3. □

Corollary 4. *Let $E\subset\mathcal{X}$ be a measurable set with $\mu(E)\in(0,\infty)$; $L\in(0,\infty)$; φ be a bounded L-Lipschitz function supported in E; and $\mathcal{F}\in\{M^\phi_{p,q}(\mathcal{X}),N^\phi_{p,q}(\mathcal{X})\}$ with $p,\ q\in(0,\infty)$, $\alpha_\phi\in(0,1)$, and $\beta_\phi\in(0,2)$ or $\mathcal{F}\in\{M^\phi_{p,\infty}(\mathcal{X})=M^{\phi,p}(\mathcal{X}),N^\phi_{p,\infty}(\mathcal{X})\}$ with $p\in(0,\infty)$, $\phi\in\mathcal{A}_\infty$, and $u\in L^0(\mathcal{X})$. Then,*

$$
\|\varphi\|_{\mathcal{F}}\lesssim\left[1+\|\varphi\|_{L^\infty(\mathcal{X})}\right]\left\{1+\left[\phi(L^{-1})\right]^{-1}\right\}[\mu(E)]^{1/p} \tag{24}
$$

with the implicit positive constant independent of L, φ, and E.

Proof. We first consider $\mathcal{F}=M^\phi_{p,q}(\mathcal{X})$ with $p,\ q\in(0,\infty)$, $\alpha_\phi\in(0,1)$, and $\beta_\phi\in(0,2)$. Let $\{h_k\}_{k\in\mathbb{Z}}$ be as in (22). From Lemma 7(i) and choosing $u\equiv1$, $g_k\equiv0$ for any $k\in\mathbb{Z}$, and $i\in\mathbb{Z}$ such that $2^i\le L<2^{i+1}$ in (22), we deduce that

$$
\begin{aligned}
\|\varphi\|_{\dot{M}^\phi_{p,q}(\mathcal{X})} &\lesssim \|\{h_k\}_{k\in\mathbb{Z}}\|_{L^p(\mathcal{X},l^q)} \\
&\lesssim \left[\left\{\sum_{k>i}\left(2^{-k}\left[\phi(2^{-k})\right]^{-1}\right)^q\right\}^{1/q}L\right. \\
&\quad\left.+\left\{\sum_{k\le i}\left(\left[\phi(2^{-k})\right]^{-1}\right)^q\right\}^{1/q}\|\varphi\|_{L^\infty(\mathcal{X})}\right]\|\mathbf{1}_E\|_{L^p(\mathcal{X})} \\
&\lesssim \left\{\left[\phi(L^{-1})\right]^{-1}+\left[\phi(L^{-1})\right]^{-1}\|\varphi\|_{L^\infty(\mathcal{X})}\right\}[\mu(E)]^{1/p},
\end{aligned}
$$

where, in the last inequality, we used (9) and (8). This, combined with the fact that

$$
\|\varphi\|_{L^p(\mathcal{X})}\le\|\varphi\|_{L^\infty(\mathcal{X})}[\mu(E)]^{1/p},
$$

implies (24) with $\mathcal{F}=M^\phi_{p,q}(\mathcal{X})$.

By choosing $u \equiv 1$ and $g \equiv 0$ in Lemma 7(ii), the case

$$\mathcal{F} \in \left\{M^{\phi}_{p,\infty}(\mathcal{X}) = M^{\phi,p}(\mathcal{X}), N^{\phi}_{p,\infty}(\mathcal{X})\right\}$$

with $p \in (0, \infty)$ and $\phi \in \mathcal{A}_\infty$ can be similarly proved. This finishes the proof of Corollary 4. □

Now, we prove Theorem 1.

Proof of Theorem 1. By similarity, we only consider the case $\mathcal{F} = M^{\phi}_{p,q}(\mathcal{X})$. Let p, q, and ϕ be as in the present theorem; C_μ be as in (1); $i \in \mathbb{Z}_+$; $u \in \dot{M}^{\phi}_{p,q}(\mathcal{X})$; and $\{g_k\}_{k\in\mathbb{Z}} \in \mathbb{D}^{\phi}(u) \cap L^p(\mathcal{X}, l^q)$. Let $\{B_j\}_{j\in\mathcal{J}}$ be any given ball covering of $\mathcal{X}$ with the radius 2^{-i} such that $\sum_{j\in\mathcal{J}} \mathbf{1}_{2B_j} \lesssim 1$ and $\{\varphi_j\}_{j\in\mathcal{J}}$, consisting of a sequence of $c2^i$-Lipschitz functions, be a partition of unity with respect to $\{B_j\}_{j\in\mathcal{J}}$ as in Definition 8, where c is a positive constant depending only on C_μ. For any $j \in \mathcal{J}$, let u_{B_j} be as in (2). By ([28], Remark 3.8), we have, for any $j \in \mathcal{J}$, $|u_{B_j}| < \infty$. Let $u_{2^{-i}}$ be as in Definition 9(i). Thus, by the properties of $\{\varphi_j\}_{j\in\mathcal{J}}$, we obtain

$$u - u_{2^{-i}} = \sum_{j\in\mathcal{J}} \left(u - u_{B_j}\right)\varphi_j. \tag{25}$$

Noticing that φ_j is a $c2^i$-Lipschitz function and $\|\varphi_j\|_{L^\infty(\mathcal{X})} \le 1$, from Lemma 7 with u and L replaced, respectively, by $u - u_{B_j}$ and $c2^i$, we deduce that, for any $j \in \mathcal{J}$, $\vec{h}_j := \{h_{k,j}\}_{k\in\mathbb{Z}}$, defined by setting, for any $k \in \mathbb{Z}$,

$$h_{k,j} := \begin{cases} \left\{2^{i-k}\left[\phi(2^{-k})\right]^{-1}|u - u_{B_j}| + g_k\right\}\mathbf{1}_{2B_j}, & k > i, \\ \left[\phi(2^{-k})\right]^{-1}\left|u - u_{B_j}\right|\mathbf{1}_{2B_j}, & k \le i, \end{cases}$$

is a positive constant multiple of an element of $\mathbb{D}^{\phi}([u - u_{B_j}]\varphi_j)$. By this, (25), and $\sum_{j\in\mathcal{J}} \mathbf{1}_{2B_j} \lesssim 1$, we conclude that, for almost every x, $y \in \mathcal{X}$ with $d(x, y) \in [2^{-k-1}, 2^{-k})$,

$$\begin{aligned} &\left|(u - u_{2^{-i}})(x) - (u - u_{2^{-i}})(y)\right| \\ &\quad = \sum_{j\in\mathcal{J}} \left(u(x) - u_{B_j}\right)\varphi_j(x) - \sum_{j\in\mathcal{J}} \left(u(y) - u_{B_j}\right)\varphi_j(y) \\ &\quad \le \sum_{j\in\mathcal{J},\, 2B_j\cap\{x,y\}\neq\emptyset} \left|\left(u(x) - u_{B_j}\right)\varphi_j(x) - \left(u(y) - u_{B_j}\right)\varphi_j(y)\right| \\ &\quad \lesssim \phi(d(x, y)) \sum_{j\in\mathcal{J},\, 2B_j\cap\{x,y\}\neq\emptyset} [h_{k,j}(x) + h_{k,j}(y)]. \end{aligned} \tag{26}$$

For any given $\epsilon \in (0, -\log_2 \alpha_\phi)$ and $\lambda \in (n/[n - \log_2 \alpha_\phi], \infty)$, by Lemma 5(iii), we obtain, for any $j \in \mathcal{J}$ and almost every $x \in 2B_j$,

$$\left|u(x) - u_{B_j}\right| \lesssim \sum_{l\ge i-4} 2^{(l-i)\epsilon}\phi(2^{-l})\left[\mathcal{M}\left(g_l^{\lambda}\right)(x)\right]^{1/\lambda}.$$

Then,

$$\begin{aligned} h_{k,j} &\le \begin{cases} \left\{2^{i-k}\left[\phi(2^{-k})\right]^{-1} \sum_{l\ge i-4} 2^{(l-i)\epsilon}\phi(2^{-l})\left[\mathcal{M}\left(g_l^{\lambda}\right)\right]^{1/\lambda} + g_k\right\}\mathbf{1}_{2B_j}, & k > i, \\ \left[\phi(2^{-k})\right]^{-1} \sum_{l\ge i-4} 2^{(l-i)\epsilon}\phi(2^{-l})\left[\mathcal{M}\left(g_l^{\lambda}\right)\right]^{1/\lambda}\mathbf{1}_{2B_j}, & k \le i \end{cases} \\ &=: \widetilde{h_{k,j}}. \end{aligned} \tag{27}$$

Define the sequence $\{h_k\}_{k\in\mathbb{Z}}$ by setting, for any $k \in \mathbb{Z}$,

$$h_k := \begin{cases} 2^{i-k}2^{-i\epsilon}\left[\phi(2^{-k})\right]^{-1}\sum_{l\ge i-4}2^{l\epsilon}\phi(2^{-l})\left[\mathcal{M}\left(g_l^{\lambda}\right)\right]^{1/\lambda}+g_k, & k>i, \\ 2^{-i\epsilon}\left[\phi(2^{-k})\right]^{-1}\sum_{l\ge i-4}2^{l\epsilon}\phi(2^{-l})\left[\mathcal{M}\left(g_l^{\lambda}\right)\right]^{1/\lambda}, & k\le i. \end{cases} \tag{28}$$

Then, by (26), (27), and $\sum_{j\in\mathcal{J}}\mathbf{1}_{2B_j}\lesssim 1$, we conclude that, for almost every $x, y \in \mathcal{X}$,

$$\begin{aligned}\left|(u-u_{2^{-i}})(x)-(u-u_{2^{-i}})(y)\right| &\lesssim \phi(d(x,y))\sum_{j\in\mathcal{J},\,2B_j\cap\{x,y\}\neq\emptyset}\left[\widetilde{h_{k,j}}(x)+\widetilde{h_{k,j}}(y)\right] \\ &\lesssim \phi(d(x,y))\left[h_k(x)+h_k(y)\right],\end{aligned}$$

which implies that $\{h_k\}_{k\in\mathbb{Z}}$ is a positive constant multiple of an element in $\mathbb{D}^{\phi}(u-u_{2^{-i}})$.

Let $\lambda \in (n/[n-\log_2\alpha_\phi], \min\{p,q\})$. Using the Hölder inequality, the fact that $\alpha_\phi < 2^{-\epsilon}$, and Lemma 1, we have

$$\sum_{l\ge i-4}2^{l\epsilon}\phi(2^{-l})\left[\mathcal{M}\left(g_l^{\lambda}\right)\right]^{1/\lambda}\lesssim\left[2^{i\epsilon}\phi(2^{-i})\right]^{(q-1)/q}\left\{\sum_{l\ge i-4}2^{l\epsilon}\phi(2^{-l})\left[\mathcal{M}\left(g_l^{\lambda}\right)\right]^{q/\lambda}\right\}^{1/q} \tag{29}$$

with the implicit positive constant independent of i. Notice that, by (10) and $\beta_\phi^- < 2$,

$$\begin{aligned}\left\{\sum_{k>i}\left\{2^{i-k}2^{-i\epsilon}\left[\phi(2^{-k})\right]^{-1}\right\}^q\right\}^{1/q} &= 2^{-i(\epsilon-1)}\left\{\sum_{k>i}\left[\frac{2^{-k}}{\phi(2^{-k})}\right]^q\right\}^{1/q} \\ &\lesssim \frac{2^{-i\epsilon}}{\phi(2^{-i})}\end{aligned} \tag{30}$$

and, by (9) and $\alpha_\phi < 1$,

$$\left\{\sum_{k\le i}\left\{2^{-i\epsilon}\left[\phi(2^{-k})\right]^{-1}\right\}^q\right\}^{1/q}\lesssim\frac{2^{-i\epsilon}}{\phi(2^{-i})}. \tag{31}$$

Thus, by (29)–(31), Lemma 1, and the Fefferman–Stein vector-valued maximal inequality in $L^{p/\lambda}(\mathcal{X}, l^{q/\lambda})$ (see ([56], Theorem 1.2) or ([57], Theorem 1.3)), we obtain

$$\begin{aligned}\|\{h_k\}_{k\in\mathbb{Z}}\|_{L^p(\mathcal{X},l^q)} &\lesssim \left\|\left\{\sum_{l\ge i-4}2^{(l-i)\epsilon}\frac{\phi(2^{-l})}{\phi(2^{-i})}\left[\mathcal{M}(g_l^{\lambda})\right]^{q/\lambda}\right\}^{1/q}\right\|_{L^p(\mathcal{X})}+\left\|\left(\sum_{k>i}g_k^q\right)^{1/q}\right\|_{L^p(\mathcal{X})} \\ &\lesssim \left\|\left\{\sum_{l\ge i-4}\left[\mathcal{M}(g_l^{\lambda})\right]^{q/\lambda}\right\}^{1/q}\right\|_{L^p(\mathcal{X})}+\left\|\left(\sum_{k>i}g_k^q\right)^{1/q}\right\|_{L^p(\mathcal{X})} \\ &\lesssim \left\|\left(\sum_{l\ge i-4}g_l^q\right)^{1/q}\right\|_{L^p(\mathcal{X})}+\left\|\left(\sum_{k>i}g_k^q\right)^{1/q}\right\|_{L^p(\mathcal{X})} \\ &\lesssim \left\|\left(\sum_{k\ge i-4}g_k^q\right)^{1/q}\right\|_{L^p(\mathcal{X})},\end{aligned} \tag{32}$$

which, combined with $\|\{g_k\}_{k\in\mathbb{Z}}\|_{L^p(\mathcal{X},l^q)} < \infty$, implies that

$$\left\|u-u_{2^{-i}}\right\|_{\dot{M}^{\phi}_{p,q}(\mathcal{X})}\lesssim\left\|\left(\sum_{k\ge i-4}g_k^q\right)^{1/q}\right\|_{L^p(\mathcal{X})}\to 0 \quad \text{as } i\to\infty.$$

On the other hand, from (25), Lemmas 5(iii), and 1(ii) with $\epsilon \in (0, -\log_2 \alpha_\phi)$, the properties of $\{\varphi_j\}_{j\in\mathcal{J}}$, the Fefferman–Stein vector-valued maximal inequality, and $\phi(0) = 0$, it follows that

$$\begin{aligned}\|u - u_{2^{-i}}\|_{L^p(\mathcal{X})} &= \left\| \sum_{j\in\mathcal{J}} (u - u_{B_j})\varphi_j \right\|_{L^p(\mathcal{X})} \\ &\lesssim \left\| \sum_{j\in\mathcal{J}} \left\{ \sum_{l\geq i-4} 2^{(l-i)\epsilon} \phi(2^{-l}) \left[\mathcal{M}\left(g_l^\lambda\right)\right]^{1/\lambda} \right\} \varphi_j \right\|_{L^p(\mathcal{X})} \\ &\lesssim \phi(2^{-i}) \left\| \sum_{l\geq i-4} 2^{(l-i)\epsilon} \frac{\phi(2^{-l})}{\phi(2^{-i})} \left[\mathcal{M}(g_l^\lambda)\right]^{1/\lambda} \right\|_{L^p(\mathcal{X})} \\ &\lesssim \phi(2^{-i}) \left\| \sum_{l\geq i-4} \left[\mathcal{M}(g_l^\lambda)\right]^{1/\lambda} \right\|_{L^p(\mathcal{X})} \\ &\lesssim \phi(2^{-i}) \left\| \left(\sum_{l\geq i-4} g_l^q \right)^{1/q} \right\|_{L^p(\mathcal{X})} \to 0 \quad \text{as } i \to \infty. \end{aligned} \tag{33}$$

This finishes the proof of Theorem 1. □

Recall that, when $q = \infty$, $M^\phi_{p,\infty}(\mathcal{X}) = M^{\phi,p}(\mathcal{X})$ (see Remark 1(i)). Even in the classical case $\phi(t) := t$ for any $t \in [0,\infty)$, Theorem 1 is not true for $q = \infty$; we refer the reader to ([41], Example 3.5) with $m_u^\gamma(B_j)$ therein replaced by u_{B_j} for any $j \in \mathbb{N}$ for a counterexample. For any given Hajłasz–Sobolev function, to find a convergent sequence consisting of continuous functions to this given Hajłasz–Sobolev function in Hajłasz–Sobolev spaces, instead of Theorem 1, we turn to find a dense subspace of $M^\phi_{p,\infty}(\mathcal{X})$, which consists of some generalized Lipschitz continuous functions.

Definition 12. *Let $\phi \in \mathcal{A}$. A function u on $\mathcal{X}$ is said to be in the ϕ-*Lipschitz class $\mathrm{Lip}_\phi(\mathcal{X})$ *if there exists a positive constant C such that, for any $x,\ y \in \mathcal{X}$,*

$$|u(x) - u(y)| \leq C\phi(d(x,y)).$$

Observe that $\mathrm{Lip}_\phi(\mathcal{X})$ is just the classical Hölder space of order $s \in (0,1]$ when $\phi(t) := t^s$ for any $t \in [0,\infty)$.

Recall that a function $\phi : [0,\infty) \to [0,\infty)$ is called a *modulus of continuity* if it is increasing, the function $\widetilde{\phi}$, defined by setting, for any $t \in [0,\infty)$, $\widetilde{\phi}(t) := \phi(t)/t$, is decreasing, $\phi(0) = 0$, and, for any $t \in (0,\infty)$, $\phi(t) > 0$; see [58]. Obviously, the collection of all moduli of continuity is contained in $\mathcal{A}_\infty$. It is well known that, if ϕ is a modulus of continuity, then, for any $x,\ y \in [0,\infty)$,

$$\phi(x+y) \leq \phi(x) + \phi(y).$$

Borrowing some ideas similar to that used in the proof of ([48], Theorem 5.19) (see also ([59], Proposition 4.5)), we can prove the following conclusion.

Theorem 2. *Let ϕ be a modulus of continuity, and $p \in (0,\infty)$. Then* $\mathrm{Lip}_\phi(\mathcal{X}) \cap M^{\phi,p}(\mathcal{X})$ *is a dense subspace of $M^{\phi,p}(\mathcal{X})$.*

Proof. Let $p \in (0,\infty)$, $u \in M^{\phi,p}(\mathcal{X})$, $g \in \mathcal{D}^\phi(u) \cap L^p(\mathcal{X})$, and E be the exceptional zero-measure set such that (13) holds true. For any $\lambda \in (0,\infty)$, let

$$E_\lambda := \{x \in \mathcal{X} \setminus E : \ g(x) \leq \lambda,\ |u(x)| \leq \lambda\}. \tag{34}$$

Then, the facts that $u \in L^p(\mathcal{X})$ and $g \in L^p(\mathcal{X})$ imply that, for any $\lambda \in (0, \infty)$,

$$\mu(\mathcal{X} \setminus E_\lambda) < \infty. \tag{35}$$

Moreover, by the definitions of $\mathcal{D}^\phi(u)$ and E_λ, we know that, for any $x,\ y \in E_\lambda$,

$$|u(x) - u(y)| \le \phi(d(x,y))[g(x) + g(y)] \le 2\lambda\phi(d(x,y)).$$

Thus, $u|_{E_\lambda}$ is ϕ-Lipschitz continuous on E_λ. By ([60], Theorem 2) with the function ω therein replaced by $2\lambda\phi$, we find that u_λ, defined by setting, for any $x \in \mathcal{X}$,

$$u_\lambda(x) := \sup\{u(y) - 2\lambda\phi(d(x,y)) :\ y \in E_\lambda\},$$

is a *ϕ-Lipschitz continuous extension* of $u|_{E_\lambda}$ from E_λ to $\mathcal{X}$ and, furthermore, for any $x_1,\ x_2 \in \mathcal{X}$,

$$\left|u_\lambda(x_1) - u_\lambda(x_2)\right| \le 2\lambda\phi(d(x_1,x_2)). \tag{36}$$

Define $v_\lambda :=\ \operatorname{sgn}(u_\lambda)\min\{|u_\lambda|, \lambda\}$. By $u_\lambda|_{E_\lambda} = u|_{E_\lambda}$, (34), and the definition of v_λ, we find that

$$v_\lambda|_{E_\lambda} = u_\lambda|_{E_\lambda} = u|_{E_\lambda}. \tag{37}$$

By the definition of v_λ and (36), we find that, for any $x,\ y \in \mathcal{X}$,

$$\left|v_\lambda(x) - v_\lambda(y)\right| \le |u_\lambda(x) - u_\lambda(y)| \le 2\lambda\phi(d(x,y)), \tag{38}$$

which means that v_λ is still ϕ-Lipschitz continuous on $\mathcal{X}$.

We now show $v_\lambda \in M^{\phi,p}(\mathcal{X})$. If $x,\ y \in E_\lambda$, then, by (37) and the definition of $\mathcal{D}^\phi(u)$, we have

$$\begin{aligned}|v_\lambda(x) - v_\lambda(y)| &= |u(x) - u(y)| \\ &\le \phi(d(x,y))[g(x) + g(y)].\end{aligned} \tag{39}$$

Otherwise, if at least one of x and y lies in $\mathcal{X} \setminus E_\lambda$, then, by (38), we find that

$$\left|v_\lambda(x) - v_\lambda(y)\right| \le 2\lambda\phi(d(x,y)),$$

which, combined with (39) and the definition of $\mathcal{D}^\phi(v_\lambda)$, implies that

$$g_\lambda := g\mathbf{1}_{E_\lambda} + 2\lambda\mathbf{1}_{\mathcal{X}\setminus E_\lambda} \in \mathcal{D}^\phi(v_\lambda).$$

By the definitions of v_λ and g_λ, (37), $|v_\lambda| \le \lambda$, and (35), we conclude that

$$\begin{aligned}\|v_\lambda\|_{L^p(\mathcal{X})} &\lesssim \|v_\lambda\mathbf{1}_{E_\lambda}\|_{L^p(\mathcal{X})} + \|v_\lambda\mathbf{1}_{\mathcal{X}\setminus E_\lambda}\|_{L^p(\mathcal{X})} \\ &\lesssim \|u\|_{L^p(\mathcal{X})} + \lambda[\mu(\mathcal{X} \setminus E_\lambda)]^{1/p} < \infty\end{aligned}$$

and

$$\|g_\lambda\|_{L^p(\mathcal{X})} \lesssim \|g\|_{L^p(\mathcal{X})} + 2\lambda[\mu(\mathcal{X} \setminus E_\lambda)]^{1/p} < \infty,$$

which, combined with the definition of $\|\cdot\|_{M^{\phi,p}(\mathcal{X})}$, implies that $v_\lambda \in M^{\phi,p}(\mathcal{X})$.

Now, we consider $v_\lambda - u$. Let $x,\ y \in \mathcal{X} \setminus E$. If $x,\ y \in E_\lambda$, then, by (37), it is obvious that

$$\left|(v_\lambda - u)(x) - (v_\lambda - u)(y)\right| = 0.$$

If $x,\ y \in \mathcal{X} \setminus (E_\lambda \cup E)$, then, by (38) and the definition of $\mathcal{D}^\phi(u)$, we obtain

$$\begin{aligned}\left|(v_\lambda - u)(x) - (v_\lambda - u)(y)\right| &\le \left|v_\lambda(x) - v_\lambda(y)\right| + \left|u(x) - u(y)\right| \\ &\le \phi(d(x,y))[2\lambda + g(x) + g(y)].\end{aligned}$$

If $x \in E_\lambda$ and $y \in \mathcal{X} \setminus (E_\lambda \cup E)$, then, by (38) and the definitions of $\mathcal{D}^\phi(u)$ and E_λ, we conclude that

$$\begin{aligned}\left|(v_\lambda - u)(x) - (v_\lambda - u)(y)\right| &\le \left|v_\lambda(x) - v_\lambda(y)\right| + \left|u(x) - u(y)\right| \\ &\le \phi(d(x,y))[2\lambda + g(x) + g(y)] \\ &\le \phi(d(x,y))[3\lambda + g(y)]\end{aligned}$$

and, similarly, if $x \in \mathcal{X} \setminus (E_\lambda \cup E)$ and $y \in E_\lambda$, by (38) and the definitions of $\mathcal{D}^\phi(u)$ and E_λ again, we find that

$$\left|(v_\lambda - u)(x) - (v_\lambda - u)(y)\right| \le \phi(d(x,y))[3\lambda + g(x)].$$

Altogether, from the definition of $\mathcal{D}^\phi(v_\lambda - u)$ and $\mu(E) = 0$, we deduce that

$$\widetilde{g}_\lambda := (3\lambda + g)\,\mathbf{1}_{\mathcal{X}\setminus E_\lambda} \in \mathcal{D}^\phi(v_\lambda - u).$$

Moreover, by $|v_\lambda| \le \lambda$ and the definitions of $\widetilde{g}_\lambda$ and E_λ, we have

$$\begin{aligned}\left\|(v_\lambda - u)\mathbf{1}_{\mathcal{X}\setminus E_\lambda}\right\|_{L^p(\mathcal{X})} &\lesssim \left\|(g+u)\mathbf{1}_{\mathcal{X}\setminus E_\lambda}\right\|_{L^p(\mathcal{X})} \\ &\lesssim \|g\|_{L^p(\mathcal{X})} + \|u\|_{L^p(\mathcal{X})} < \infty\end{aligned}$$

and

$$\begin{aligned}\left\|\widetilde{g}_\lambda \mathbf{1}_{\mathcal{X}\setminus E_\lambda}\right\|_{L^p(\mathcal{X})} &\lesssim \left\|(3\lambda + g)\mathbf{1}_{\mathcal{X}\setminus E_\lambda}\right\|_{L^p(\mathcal{X})} \\ &\lesssim \|u\|_{L^p(\mathcal{X})} + \|g\|_{L^p(\mathcal{X})} < \infty.\end{aligned}$$

Then, using this, (37), the dominated convergence theorem with respect to μ, and $\mu(\mathcal{X} \setminus E_\lambda) \to 0$ as $\lambda \to \infty$, we conclude that

$$\lim_{\lambda\to\infty} \|v_\lambda - u\|_{L^p(\mathcal{X})} = \lim_{\lambda\to\infty} \left\|(v_\lambda - u)\mathbf{1}_{\mathcal{X}\setminus E_\lambda}\right\|_{L^p(\mathcal{X})} = 0$$

and

$$\lim_{\lambda\to\infty} \|\widetilde{g}_\lambda\|_{L^p(\mathcal{X})} = \lim_{\lambda\to\infty} \left\|\widetilde{g}_\lambda \mathbf{1}_{\mathcal{X}\setminus E_\lambda}\right\|_{L^p(\mathcal{X})} = 0,$$

which imply $\lim_{\lambda\to\infty} \|v_\lambda - u\|_{M^{\phi,p}(\mathcal{X})} = 0$. This finishes the proof of Theorem 2. □

Now, we state the main result of this section, which generalizes ([41], Theorem 8.1) from fractional Hajłasz-type spaces to those with generalized smoothness.

Theorem 3. *Let $\phi \in \mathcal{A}$ and $\mathcal{F}$ be one of the following cases:*

(i) $\mathcal{F} = M^\phi_{p,\infty}(\mathcal{X}) = M^{\phi,p}(\mathcal{X})$ *with ϕ being a modulus of continuity and $p \in (1,\infty)$;*

(ii) $\mathcal{F} = M^\phi_{p,\infty}(\mathcal{X}) = M^{\phi,p}(\mathcal{X})$ *with ϕ being a modulus of continuity, $\alpha_\phi \in (0,1)$, and $p \in (D/(D - \log_2 \alpha_\phi), 1]$;*

(iii) $\mathcal{F} = M^\phi_{p,q}(\mathcal{X})$ *with $\alpha_\phi \in (0,1)$, $\beta_\phi \in (0,2)$, and $p,\ q \in (D/(D - \log_2 \alpha_\phi), \infty)$;*

(iv) $\mathcal{F} = N^\phi_{p,q}(\mathcal{X})$ *with $\alpha_\phi \in (0,1)$, $\beta_\phi \in (0,2)$, $p \in (D/(D - \log_2 \alpha_\phi), \infty)$, and $q \in (0,\infty)$,*

where D is as in (1). If $u \in \dot{\mathcal{F}}$, then there exist a set E with $\mathrm{Cap}_{\mathcal{F}}(E) = 0$ and an $\mathcal{F}$-quasi-continuous function u^* on $\mathcal{X}$ such that, for any $x \in \mathcal{X} \setminus E$,

$$u^*(x) = \lim_{r\to 0^+} u_{B(x,r)}. \tag{40}$$

To prove Theorem 3, we need a weak-type estimate of the $\mathcal{F}$-capacity. To this end, we need several technical lemmas. The first one is on the Hajłasz gradient of $\mathcal{M}^*u$ for any u in which the integral on any ball is finite. Recall that, for any $u \in L^1_{\mathrm{loc}}(\mathcal{X})$, either $\mathcal{M}^*u \equiv \infty$ or

$\mathcal{M}^*u < \infty$ almost everywhere (see ([54], (3.1) and Lemma 4.8) or ([61], Remark 2.2)), where $\mathcal{M}^*$ is as in Definition 9(ii).

Lemma 8. (i) *Let $\phi \in \mathcal{A}_\infty$. Then, for any $u \in L^1_{\mathrm{loc}}(\mathcal{X})$ satisfying that its integral on any ball of $\mathcal{X}$ is finite and $\mathcal{M}^*u \not\equiv \infty$ and for any $g \in \mathcal{D}^\phi(u)$, $\mathcal{M}(g)$ is an element of $\mathcal{D}^\phi(\mathcal{M}^*u)$ modulo a positive constant independent of u and g, where $\mathcal{M}$ is the classical Hardy–Littlewood maximal operator and $\mathcal{M}^*$ as in Definition 9(ii).*

(ii) *Let $\phi \in \mathcal{A}_\infty$ with $\alpha_\phi \in (0,1)$. Then, for any $\lambda \in (D/[D-\log_2 \alpha_\phi], \infty)$, any $u \in L^1_{\mathrm{loc}}(\mathcal{X})$ satisfying that its integral on any ball of $\mathcal{X}$ is finite and $\mathcal{M}^*u \not\equiv \infty$, and for any $g \in \mathcal{D}^\phi(u)$, $[\mathcal{M}(g^\lambda)]^{1/\lambda}$ is an element of $\mathcal{D}^\phi(\mathcal{M}^*u)$ modulo a positive constant independent of u and g.*

Proof. Due to similarity, we only prove (ii). For any given $r \in (0,\infty)$, let $\{B_j\}_{j\in\mathcal{J}}$ be any given sequence of balls as in the definition of $\mathcal{M}^*$ with the radius r, and $\{\varphi_j\}_{j\in\mathcal{J}}$ be a partition of unity with respect to $\{B_j\}_{j\in\mathcal{J}}$ as in Definition 8, where $\mathcal{J}$ is an index set. Let u and g be as in the present lemma. From the definition of $\mathcal{M}^*$ and the observation that $\mathcal{D}^\phi(u) \subset \mathcal{D}^\phi(|u|)$, without loss of generality, we may assume that $u \geq 0$.

Let u_r be as in Definition 9(i). By $\sum_{j\in\mathcal{J}} \varphi_j \equiv 1$, we have

$$u_r = u + \sum_{j\in\mathcal{J}} \left(u_{B_j} - u\right)\varphi_j. \tag{41}$$

Therefore, for any $j \in \mathcal{J}$, using Lemma 7(ii) with u, E, and L^{-1} therein replaced, respectively, by $u - u_{B_j}$, $2B_j$, and r, and the properties of φ_j, we find that, for any $j \in \mathcal{J}$,

$$\widetilde{g^{(j)}} := \left\{g + [\phi(r)]^{-1}\left|u - u_{B_j}\right|\right\}\mathbf{1}_{2B_j}$$

is a positive constant multiple of an element in $\mathcal{D}^\phi([u-u_{B_j}]\varphi_j)$, where the positive constant is independent of r, u, and g. Let $\lambda \in (D/[D-\log_2\alpha_\phi],\infty)$. Notice that, for any $j \in \mathcal{J}$, by Lemma 5(ii) with $B(y, 2^{-i})$ and 2^{-i} therein replaced, respectively, by B_j and r, we have, for any $x \in 2B_j$,

$$\left|u(x) - u_{B_j}\right| \lesssim \phi(r)\left[\mathcal{M}\left(g^\lambda\right)(x)\right]^{1/\lambda}$$

with the implicit positive constant independent of u, g, x, j, and r. From this; the proven conclusion that, for any $j \in \mathcal{J}$, $\widetilde{g^{(j)}}$ is a positive constant multiple of an element in $\mathcal{D}^\phi([u - u_{B_j}]\varphi_j)$; the definition of $\widetilde{g^{(j)}}$, $\sum_{j\in\mathcal{J}} \mathbf{1}_{2B_j} \lesssim 1$; and $g \leq [\mathcal{M}(g^\lambda)]^{1/\lambda}$, we deduce that, for almost every $x, y \in \mathcal{X}$,

$$\begin{aligned}
&\left|\sum_{j\in\mathcal{J}}\left[u_{B_j} - u(x)\right]\varphi_j(x) - \sum_{j\in\mathcal{J}}\left[u_{B_j} - u(y)\right]\varphi_j(y)\right| \\
&\quad \lesssim \phi(d(x,y)) \sum_{j\in\mathcal{J}}\left[\widetilde{g^{(j)}}(x) + \widetilde{g^{(j)}}(y)\right] \\
&\quad \lesssim \phi(d(x,y)) \sum_{j\in\mathcal{J}}\left[\left\{g(x) + \left[\mathcal{M}\left(g^\lambda\right)(x)\right]^{1/\lambda}\right\}\mathbf{1}_{2B_j}(x) + \left\{g(y) + \left[\mathcal{M}\left(g^\lambda\right)(y)\right]^{1/\lambda}\right\}\mathbf{1}_{2B_j}(y)\right] \\
&\quad \lesssim \phi(d(x,y))\left\{g(x) + \left[\mathcal{M}\left(g^\lambda\right)(x)\right]^{1/\lambda} + g(y) + \left[\mathcal{M}\left(g^\lambda\right)(y)\right]^{1/\lambda}\right\} \\
&\quad \lesssim \phi(d(x,y))\left\{\left[\mathcal{M}\left(g^\lambda\right)(x)\right]^{1/\lambda} + \left[\mathcal{M}\left(g^\lambda\right)(y)\right]^{1/\lambda}\right\},
\end{aligned}$$

which implies that $[\mathcal{M}(g^\lambda)]^{1/\lambda}$ is a positive constant multiple of an element of $\mathcal{D}^\phi(\sum_{j\in\mathcal{J}}[u_{B_j} - u]\varphi_j)$. By this, (41), the definition of $\mathcal{D}^\phi(u_r)$, $g \in \mathcal{D}^\phi(u)$, and $g \leq [\mathcal{M}(g^\lambda)]^{1/\lambda}$, we further conclude that $[\mathcal{M}(g^\lambda)]^{1/\lambda}$ is a positive constant multiple of an element in $\mathcal{D}^\phi(u_r)$ with the positive constant independent of u, g, and r. Moreover, if $\mathcal{M}^*u \not\equiv \infty$, then by the definition of $\mathcal{M}^*$ and Lemma 2(ii), we conclude that $[\mathcal{M}(g^\lambda)]^{1/\lambda}$ is an element of $\mathcal{D}^\phi(\mathcal{M}^*u)$ modulo a positive constant independent of u and g. This finishes the proof of (ii) and hence of Lemma 8. □

Borrowing some ideas from the proof of ([41], Lemma 7.1), we can prove the following lemma on the Hajłasz gradient sequence of $\mathcal{M}^*u$ for any $u \in L^1_{\mathrm{loc}}(\mathcal{X})$ with its integral on any ball being finite.

Lemma 9. *Let $\phi \in \mathcal{A}_0$ with $\beta^+_\phi \in (0,2)$, $\epsilon \in (0, -\log_2 \alpha_\phi)$, and*

$$\delta \in \left(0, \min\left\{1 - \log_2 \beta_\phi, -\log_2 \alpha_\phi - \epsilon\right\}\right).$$

Then, for any $\lambda \in (D/[D+\epsilon], \infty)$, any $u \in L^1_{\mathrm{loc}}(\mathcal{X})$ such that its integral on any ball in $\mathcal{X}$ is finite and $\mathcal{M}^*u \not\equiv \infty$, and any $\vec{g} := \{g_k\}_{k\in\mathbb{Z}} \in \mathbb{D}^\phi(u)$, the sequence $\{\widetilde{g_k}\}_{k\in\mathbb{Z}}$ of functions, defined by setting, for any $k \in \mathbb{Z}$,

$$\widetilde{g_k} := \sum_{l\in\mathbb{Z}} 2^{-|l-k|\delta} \left[\mathcal{M}\left(g_l^\lambda\right)\right]^{1/\lambda}, \tag{42}$$

is a positive constant multiple of an element in $\mathbb{D}^\phi(\mathcal{M}^*u)$, where the positive constant is independent of u and $\vec{g}$, D as in (1), and $\mathcal{M}^*$ as in Definition 9(ii).

Proof. Let all of the symbols be as in the present lemma. By the definition of $\mathcal{M}^*u$ and the observation that $\mathbb{D}^\phi(u) \subset \mathbb{D}^\phi(|u|)$, without loss of generality, we may assume that $u \geq 0$. Moreover, by Lemma 2 and the definition of $\mathcal{M}^*$, to prove the present lemma, it suffices to show that, for any $i \in \mathbb{Z}$, $\{\widetilde{g_k}\}_{k\in\mathbb{Z}}$ is a positive constant multiple of an element in $\mathbb{D}^\phi(u_{2^{-i}})$ with the positive constant independent of i, where $u_{2^{-i}}$ is as in Definition 9(i).

To this end, we first recall that, in the proof of Theorem 1, we have shown that, for any $i \in \mathbb{Z}$, $\{h_k\}_{k\in\mathbb{Z}}$, defined as in (28), is a positive constant multiple of an element in $\mathbb{D}^\phi(u - u_{2^{-i}})$. From this, $\vec{g} \in \mathbb{D}^\phi(u)$, the definitions of $\mathbb{D}^\phi(u)$ and $\mathbb{D}^\phi(u - u_{2^{-i}})$, and, for any x, $y \in \mathcal{X}$,

$$\begin{aligned}&\left|u_{2^{-i}}(x) - u_{2^{-i}}(y)\right| \\ &\quad \leq |u(x) - u(y)| + \left|(u - u_{2^{-i}})(x) - (u - u_{2^{-i}})(y)\right|,\end{aligned}$$

it follows that, for any $i \in \mathbb{Z}$, $\{g_k + h_k\}_{k\in\mathbb{Z}}$ is a positive constant multiple of an element in $\mathbb{D}^\phi(u_{2^{-i}})$, where the positive constant is independent of i, u, and $\vec{g}$. Thus, to prove that $\{\widetilde{g_k}\}_{k\in\mathbb{Z}}$ is a positive constant multiple of an element in $\mathbb{D}^\phi(u_{2^{-i}})$ for any $i \in \mathbb{Z}$, it suffices to show that

$$g_k + h_k \lesssim \widetilde{g_k}, \quad \forall\, k \in \mathbb{Z}. \tag{43}$$

Indeed, by the definition of $\widetilde{g_k}$ and the fact that, for almost every $x \in \mathcal{X}$, $g_k(x) \leq [\mathcal{M}(g_k^\lambda)(x)]^{1/\lambda}$, we have $g_k \leq \widetilde{g_k}$ for any $k \in \mathbb{Z}$ almost everywhere. Then, to show (43), it suffices to prove that, for any $k \in \mathbb{Z}$, $h_k \lesssim \widetilde{g_k}$ almost everywhere. Let ϵ and δ be as in the present lemma. By $1-\delta > \log_2 \beta_\phi$, $\epsilon + \delta < -\log_2 \alpha_\phi$, and Lemma 1(ii) with δ and ϵ therein replaced, respectively, by $1-\delta$ and $\epsilon+\delta$, we find that, for any k, $l \in \mathbb{Z}$ with $l \leq k$,

$$2^{l-k}\phi(2^{-l})\left[\phi(2^{-k})\right]^{-1} \lesssim 2^{(l-k)\delta} \tag{44}$$

and, for any k, $l \in \mathbb{Z}$ with $l \geq k-4$,

$$2^{(l-k)\epsilon}\phi(2^{-l})\left[\phi(2^{-k})\right]^{-1} \lesssim 2^{-(l-k)\delta}. \tag{45}$$

Let $i \in \mathbb{Z}$. Observe that, for any $l \geq i-4$, $2^{(i-l)(1-\epsilon)} \lesssim 1$ and, for any $k \leq i$, $2^{(k-i)\epsilon} \lesssim 1$. By this, (44) and (45), we obtain, for any $x \in \mathcal{X}$ and $k > i$,

$$\frac{2^{i-k}2^{-i\epsilon}}{\phi(2^{-k})} \sum_{i-4\leq l\leq k} 2^{l\epsilon}\phi(2^{-l})\left[\mathcal{M}\left([g_l(x)]^\lambda\right)\right]^{1/\lambda}$$

$$\begin{aligned}&\lesssim \sum_{i-4\le l\le k} 2^{(i-l)(1-\epsilon)}2^{(l-k)\delta}\Big[\mathcal{M}\Big([g_l(x)]^{\lambda}\Big)\Big]^{1/\lambda}\\&\lesssim \sum_{i-4\le l\le k} 2^{(l-k)\delta}\Big[\mathcal{M}\Big([g_l(x)]^{\lambda}\Big)\Big]^{1/\lambda}\end{aligned}$$

and

$$\begin{aligned}&\frac{2^{i-k}2^{-i\epsilon}}{\phi(2^{-k})}\sum_{l>k}2^{l\epsilon}\phi(2^{-l})\Big[\mathcal{M}\Big([g_l(x)]^{\lambda}\Big)\Big]^{1/\lambda}\\&\quad\lesssim\sum_{l>k}2^{(i-k)(1-\epsilon)}2^{-(l-k)\delta}\Big[\mathcal{M}\Big([g_l(x)]^{\lambda}\Big)\Big]^{1/\lambda}\\&\quad\lesssim\sum_{l>k}2^{-(l-k)\delta}\Big[\mathcal{M}\Big([g_l(x)]^{\lambda}\Big)\Big]^{1/\lambda}\end{aligned}$$

and, for any $x \in \mathcal{X}$ and $k \le i$,

$$\begin{aligned}&\frac{2^{-i\epsilon}}{\phi(2^{-k})}\sum_{l\ge i-4}2^{l\epsilon}\phi(2^{-l})\Big[\mathcal{M}\Big([g_l(x)]^{\lambda}\Big)\Big]^{1/\lambda}\\&\quad\lesssim\sum_{l\ge i-4}2^{(k-i)\epsilon}2^{-(l-k)\delta}\Big[\mathcal{M}\Big([g_l(x)]^{\lambda}\Big)\Big]^{1/\lambda}\\&\quad\lesssim\sum_{l\ge i-4}2^{-(l-k)\delta}\Big[\mathcal{M}\Big([g_l(x)]^{\lambda}\Big)\Big]^{1/\lambda},\end{aligned}$$

which, combined with the proved conclusion that, for any $k \in \mathbb{Z}$, $g_k \le \widetilde{g_k}$ almost everywhere, implies that, for any $k \in \mathbb{Z}$, $h_k \lesssim \widetilde{g_k}$ almost everywhere. Thus, for any $k \in \mathbb{Z}$, $g_k + h_k \lesssim \widetilde{g_k}$ almost everywhere. Furthermore, noticing that $\{g_k + h_k\}_{k\in\mathbb{Z}}$ is a positive constant multiple of an element in $\mathbb{D}^{\phi}(u_{2^{-i}})$, from the definition of $\mathbb{D}^{\phi}(u_{2^{-i}})$, we deduce that $\{\widetilde{g_k}\}_{k\in\mathbb{Z}}$ is also a positive constant multiple of an element in $\mathbb{D}^{\phi}(u_{2^{-i}})$, where the positive constant is independent of u, $\vec{g}$, and i. Thus, by Lemma 2 and the definition of $\mathcal{M}^*$, we conclude that $\{\widetilde{g_k}\}_{k\in\mathbb{Z}}$ is a positive constant multiple of an element in $\mathbb{D}^{\phi}(\mathcal{M}^*u)$, which completes the proof of Lemma 9. □

The next two lemmas are used to show the boundedness of the discrete maximal operator $\mathcal{M}^*$ on ϕ-Hajłasz-type spaces, which is a generalization of ([61], Theorem 4.7) and ([41], Lemma 8.3), respectively.

Lemma 10. *With the assumptions same as in Theorem 3, there exists a positive constant C, independent of u, such that, for any $u \in \dot{\mathcal{F}}$ with $\mathcal{M}^*u \not\equiv \infty$,*

$$\|\mathcal{M}^*u\|_{\dot{\mathcal{F}}} \le C\|u\|_{\dot{\mathcal{F}}}, \tag{46}$$

where $\mathcal{M}^*$ is as in Definition 9(ii).

Proof. If $\mathcal{F}$ belongs to the case (i) of Theorem 3, then (46) follows from Lemma 8(i) and the boundedness of the Hardy–Littlewood maximal operator on $L^p(\mathcal{X})$.

If $\mathcal{F}$ belongs to the case (ii) of Theorem 3, then (46) follows from Lemma 8(ii) and the boundedness of the classical Hardy–Littlewood maximal operator $\mathcal{M}$ on $L^{p/\lambda}(\mathcal{X})$ with $\lambda \in (D/[D - \log_2 \alpha_\phi], p)$.

Now, let $\mathcal{F}$ belong to the case (iii) of Theorem 3. Let $u \in \dot{M}^{\phi}_{p,q}(\mathcal{X})$, $\{g_k\}_{k\in\mathbb{Z}} \in \mathbb{D}^{\phi}(u)$ with $\|\{g_k\}_{k\in\mathbb{Z}}\|_{L^p(\mathcal{X},l^q)} \lesssim \|u\|_{\dot{M}^{\phi}_{p,q}(\mathcal{X})}$, and $r := \min\{p,q\}$. Let D be as in (1), $\lambda \in (D/[D - \log_2 \alpha_\phi], r)$, and $\epsilon' \in (0, -\log_2 \alpha_\phi)$ be such that $\lambda = D/(D + \epsilon')$. We also choose $\epsilon := (\epsilon' - \log_2 \alpha_\phi)/2$. From $\alpha_\phi < 1$, it follows that $0 < \epsilon' < \epsilon < -\log_2 \alpha_\phi$ and hence $\lambda \in (D/[D + \epsilon], r)$.

Let $\{\widetilde{g_k}\}_{k\in\mathbb{Z}}$ be as in (42) with $\delta \in (0, \min\{1 - \log_2 \beta_\phi, -\log_2 \alpha_\phi - \epsilon\})$. Then, by the definition of $\{\widetilde{g_k}\}_{k\in\mathbb{Z}}$ and the Fefferman–Stein vector-valued maximal inequality on $L^{\frac{p}{\lambda}}(\mathcal{X}, l^{\frac{q}{\lambda}})$ (see ([56], Theorem 1.2) or ([57], Theorem 1.3)), we have

$$\left\|\{\widetilde{g_k}\}_{k\in\mathbb{Z}}\right\|_{L^p(\mathcal{X},l^q)} \leq \left\|\left\{\left[\mathcal{M}\left(g_k^\lambda\right)\right]^{1/\lambda}\right\}_{k\in\mathbb{Z}}\right\|_{L^p(\mathcal{X},l^q)} \lesssim \left\|\{g_k\}_{k\in\mathbb{Z}}\right\|_{L^p(\mathcal{X},l^q)}.$$

Thus, using this, the definition of $\|\cdot\|_{\dot{M}^\phi_{p,q}(\mathcal{X})}$, Lemma 9, and $\|\{g_k\}_{k\in\mathbb{Z}}\|_{L^p(\mathcal{X},l^q)} \lesssim \|u\|_{\dot{M}^\phi_{p,q}(\mathcal{X})}$, we obtain

$$\left\|\mathcal{M}^* u\right\|_{\dot{M}^\phi_{p,q}(\mathcal{X})} \lesssim \left\|\{\widetilde{g_k}\}_{k\in\mathbb{Z}}\right\|_{L^p(\mathcal{X},l^q)} \lesssim \left\|\{g_k\}_{k\in\mathbb{Z}}\right\|_{L^p(\mathcal{X},l^q)} \lesssim \|u\|_{\dot{M}^\phi_{p,q}(\mathcal{X})}.$$

This finishes the proof of Lemma 10. □

Lemma 11. *Let $x_0 \in \mathcal{X}$, $r \in (0, \infty)$, $B_0 := B(x_0, r)$, $\phi \in \mathcal{A}$, and*

$$\mathcal{F} \in \left\{M^\phi_{p,q}(\mathcal{X}),\ N^\phi_{p,q}(\mathcal{X}) :\ p,\ q \in (0, \infty]\right\}.$$

If $\{y \in \mathcal{X} :\ d(x_0, y) = \tau r\}$ for some $\tau \in (2, \infty)$ is not empty, then there exists a positive constant C, depending only on τ, ϕ, and C_μ, such that, for any $u \in \dot{\mathcal{F}}$ supported in B_0,

$$\|u\|_{\mathcal{F}} \leq C[1 + \phi(r)]\|u\|_{\dot{\mathcal{F}}},$$

where C_μ is as in (1).

Proof. By similarity, we only prove the case $\mathcal{F} = M^\phi_{p,q}(\mathcal{X})$. Let $B_0 := B(x_0, r)$; τ, C_μ, and u be as in the present lemma; E be the exceptional zero-measure set such that (13) holds true; and $\{g_k\}_{k\in\mathbb{Z}} \in \mathbb{D}^\phi(u)$ with $\|\{g_k\}_{k\in\mathbb{Z}}\|_{L^p(\mathcal{X},l^q)} \lesssim \|u\|_{\dot{M}^\phi_{p,q}(\mathcal{X})}$. Notice that, for any $x \in B_0$ and $y \in 2\tau B_0 \setminus 2B_0$, we have $d(x, y) \in (r, [1 + 2\tau]r)$. From this, the fact that $u|_{2\tau B_0\setminus 2B_0} \equiv 0$, and the definitions of both $\mathcal{A}$ and $\mathbb{D}^\phi(u)$, we deduce that, for any $x \in B_0 \setminus E$,

$$\begin{aligned} \left|u(x)\right| &= \inf_{y\in 2\tau B_0\setminus(2B_0\cup E)} \left|u(x) - u(y)\right| \\ &\lesssim \phi([1 + 2\tau]r)\left[g(x) + \inf_{y\in 2\tau B_0\setminus(2B_0\cup E)} g(y)\right], \end{aligned} \tag{47}$$

where $g := \sup_{\{k:\, r\leq 2^{-k}\leq(1+2\tau)r\}} g_k$ and $g \geq 0$.

Let $z \in \mathcal{X}$ be such that $d(x_0, z) = \tau r$. Then, by a geometrical observation, we have

$$B_0 \subset B(z, [1 + \tau]r) \quad \text{and} \quad B(z, [\tau - 2]r/2) \subset (2\tau B_0 \setminus 2B_0),$$

which, together with the doubling property of μ, implies that

$$\mu(B_0) \leq \mu(B(z, [1 + \tau]r)) \lesssim \mu(B(z, [\tau - 2]r/2)) \lesssim \mu(2\tau B_0 \setminus 2B_0), \tag{48}$$

where the implicit positive constants depend only on τ and C_μ. Thus, from $u_{\mathcal{X}\setminus B_0} \equiv 0$, $\mu(E) = 0$, (47), (48), and the definitions of g and $\mathcal{A}$, we deduce that

$$\begin{aligned} \|u\|_{L^p(\mathcal{X})} &= \|u\|_{L^p(B_0)} \\ &\lesssim \phi([1 + 2\tau]r)\left\{\|g\|_{L^p(B_0)} + [\mu(B_0)]^{1/p} \inf_{y\in 2\tau B_0\setminus(2B_0\cup E)} g(y)\right\} \\ &\lesssim \phi([1 + 2\tau]r)\left\{\|g\|_{L^p(B_0)} + [\mu(2\tau B_0 \setminus 2B_0)]^{1/p} \inf_{y\in 2\tau B_0\setminus(2B_0\cup E)} g(y)\right\} \\ &\lesssim \phi([1 + 2\tau]r)\|g\|_{L^p(2\tau B_0)} \lesssim \phi(r)\|\{g_k\}\|_{L^p(\mathcal{X},l^q)} \end{aligned}$$

with the usual modification made when $p = \infty$, which, together with the assumption of $\{g_k\}_{k\in\mathbb{Z}}$, implies that

$$\begin{aligned}\|u\|_{M^{\phi}_{p,q}(\mathcal{X})} &\leq \|u\|_{L^p(\mathcal{X})} + \|\{g_k\}_{k\in\mathbb{Z}}\|_{L^p(\mathcal{X},l^q)} \\ &\lesssim \phi(r)\|\{g_k\}\|_{L^p(\mathcal{X},l^q)} + \|u\|_{\dot{M}^{\phi}_{p,q}(\mathcal{X})} \\ &\lesssim [1+\phi(r)]\|u\|_{\dot{M}^{\phi}_{p,q}(\mathcal{X})}.\end{aligned}$$

This finishes the proof of Lemma 11. □

Based on the above lemmas, we can obtain the following localized weak-type capacitary estimate for the restricted maximal operator $\mathcal{M}_R$, where $R \in (0,\infty]$. Recall that there exists a positive constant c, depending only on C_μ, such that, for any $u \in L^0(\mathcal{X})$,

$$c^{-1}\mathcal{M}_{R/c}u \leq \mathcal{M}^*_R u \leq c\mathcal{M}_{cR}u \tag{49}$$

(see, for instance, [41], [(8.1)]), where $\mathcal{M}_R$ is as in (14), $\mathcal{M}^*_R$ as in Definition 9(iii), and C_μ as in (1).

Lemma 12. *With the same assumptions as in Theorem 3, let $x_0 \in \mathcal{X}$, $R \in (0,\infty)$, and $B := B(x_0, R)$. If $\tau B \setminus 10B$ for some $\tau \in (10,\infty)$ is not empty, then there exist positive constants $c = c_{(C_\mu)}$ and $C = C_{(\mathcal{F},R,\tau,C_\mu)}$ such that, for any $u \in \mathcal{F}$ and $\kappa \in (0,\infty)$,*

$$\mathrm{Cap}_{\mathcal{F}}\left(\left\{x \in B: \ \mathcal{M}_{R/c}u(x) > \kappa\right\}\right) \leq C\kappa^{-p}\|u\|^p_{\mathcal{F}},$$

where $\mathcal{M}_R$ is as in (14) and C_μ as in (1).

Proof. Let all of the symbols be as in the present lemma, $\mathcal{M}^*_R$ as in Definition 9(iii), and $u \in \mathcal{F}$. Let φ be a Lipschitz function supported in $4B$ such that $0 \leq \varphi \leq 1$ and $\varphi \equiv 1$ on $3B$. By the definition of $\mathcal{M}^*_R$ and the assumption of φ, we have $\mathcal{M}^*_R u = \mathcal{M}^*_R(u\varphi)$ on B and $\mathcal{M}^*_R(u\varphi) \equiv 0$ on $\mathcal{X} \setminus 5B$. Then, from (49), we deduce that

$$\begin{aligned}\left\{x \in B: \ \mathcal{M}_{R/c}u(x) > \kappa\right\} &\subset \left\{x \in B: \ c\mathcal{M}^*_R u(x) > \kappa\right\} \\ &\subset \left\{x \in \mathcal{X}: \ c\mathcal{M}^*_R(u\varphi)(x) > \kappa\right\} \\ &= \left\{x \in \mathcal{X}: \ c\kappa^{-1}\mathcal{M}^*_R(u\varphi)(x) > 1\right\} =: Q,\end{aligned} \tag{50}$$

where $c = c_{(C_\mu)}$ is just the positive constant as in (49).

By the lower semi-continuity of $\mathcal{M}^*_R(u\varphi)$ (see [54], p. 376), we conclude that, for any $x \in Q$, there exists a $\delta_x \in (0,1)$ such that, for any $y \in B(x,\delta_x)$, $c\kappa^{-1}\mathcal{M}^*_R(u\varphi)(y) > 1$. Thus, $Q' := \bigcup_{x\in Q} B(x,\delta_x)$ is a neighborhood of Q and $c\kappa^{-1}\mathcal{M}^*_R(u\varphi) > 1$ on Q'. By this; (50); Remark 4(iii); Definition 10; Lemma 11 with u and B_0 therein replaced, respectively, by $\mathcal{M}^*_R(u\varphi)$ and $5B$; Lemma 10; and Corollary 2, we obtain

$$\begin{aligned}\mathrm{Cap}_{\mathcal{F}}\left(\left\{x \in B: \ \mathcal{M}_{R/c}u(x) > \kappa\right\}\right) &\leq \mathrm{Cap}_{\mathcal{F}}(Q) \\ &\leq \left\|c\kappa^{-1}\mathcal{M}^*_R(u\varphi)\right\|^p_{\mathcal{F}} \lesssim \kappa^{-p}\left\|\mathcal{M}^*_R(u\varphi)\right\|^p_{\dot{\mathcal{F}}} \\ &\lesssim \kappa^{-p}\left\|\mathcal{M}^*(u\varphi)\right\|^p_{\dot{\mathcal{F}}} \lesssim \kappa^{-p}\|u\varphi\|^p_{\dot{\mathcal{F}}} \\ &\lesssim \kappa^{-p}\|u\|^p_{\mathcal{F}},\end{aligned}$$

where the implicit positive constants depend on $\mathcal{F}$, R, τ, and C_μ. This finishes the proof of Lemma 12. □

Now, we show Theorem 3.

Proof of Theorem 3. Again, by similarity, we only consider the case $\mathcal{F} = M^{\phi}_{p,q}(\mathcal{X})$. Without loss of generality, we may assume that $\mathcal{X}$ contains at least two points. By this, we easily know that there exist balls $\{B(x_l, r_l)\}_{l\in\mathcal{I}}$ with $\mathcal{I} \subset \mathbb{N}$ being an index set such that $\mathcal{X} \subset \bigcup_{l\in\mathcal{I}} B(x_l, r_l)$ and, for any $l \in \mathcal{I}$, $5B(x_l, r_l) \setminus 4B(x_l, r_l)$ is not empty.

Let $\mathcal{F}$ be any given function space as in (i), (ii), or (iii) of the present theorem, and $u \in \mathcal{F}$. Then, from Theorem 2 when $\mathcal{F}$ is as in either (i) or (ii), or from Theorem 1 when $\mathcal{F}$ is as in case (iii), we deduce that there exists a sequence $\{v_i\}_{i\in\mathbb{N}}$ of continuous functions such that, for any $i \in \mathbb{N}$,

$$\|u - v_i\|^p_{\mathcal{F}} < 2^{-i(1+p)}. \tag{51}$$

Let $\{B(x_l, r_l)\}_{l\in\mathcal{I}}$ be a ball covering of $\mathcal{X}$ as above and $c = c(C_\mu)$ the positive constant as in Lemma 12. For any $l \in \mathcal{I}$, any $i, j \in \mathbb{N}$, and any $u \in \mathcal{F}$, let

$$A_{l,i} := \left\{x \in B(x_l, r_l) : \mathcal{M}_{r_l/c}(u - v_i)(x) > 2^{-i}\right\}$$

and

$$B_{l,j} := \bigcup_{i\geq j} A_{l,i}.$$

Then, by Lemma 12 and (51), we have

$$\mathrm{Cap}_{\mathcal{F}}(A_{l,i}) \lesssim 2^{ip}\|u - v_i\|^p_{\mathcal{F}} \lesssim 2^{-i}$$

and, furthermore, by Lemma 6, we obtain

$$\mathrm{Cap}_{\mathcal{F}}(B_{l,j}) \lesssim \left\{\sum_{i\geq j}[\mathrm{Cap}_{\mathcal{F}}(A_{l,i})]^\theta\right\}^{1/\theta} \lesssim 2^{-j},$$

where $\theta := \min\{1, q/p\}$. Thus, the set $F_l := \bigcap_{j\in\mathbb{N}} B_{l,j}$ is of zero $\mathcal{F}$-capacity.

Let $l \in \mathcal{I}$. For any $i \in \mathbb{N}$, using the continuity of v_i and the Lebesgue differentiation theorem, we conclude that, for any $x \in \mathcal{X}$,

$$\lim_{r\to 0^+} \fint_{B(x,r)} |v_i(y) - v_i(x)|\, d\mu(y) = 0. \tag{52}$$

Since u is locally integrable (see Remark 3(i)), then, for any $i \in \mathbb{N}$, from (52) and the definition of $A_{l,i}$, we deduce that, for any $x \in B(x_l, r_l) \setminus A_{l,i}$,

$$\begin{aligned}
\limsup_{r\to 0^+}\left|v_i(x) - u_{B(x,r)}\right| &\leq \limsup_{r\to 0^+} \fint_{B(x,r)} |v_i(x) - u(y)|\, d\mu(y) \\
&\leq \limsup_{r\to 0^+} \fint_{B(x,r)} |v_i(y) - u(y)|\, d\mu(y) \\
&\leq \mathcal{M}_{r_l/c}(u - v_i)(x) \leq 2^{-i}.
\end{aligned} \tag{53}$$

Therefore, by (53), we find that, for any $j \in \mathbb{N}$, i_1, $i_2 \in \mathbb{N}$ with i_1, $i_2 \geq j$ and $x \in B(x_l, r_l) \setminus B_{l,j} = \bigcap_{i\geq j}[B(x_l, r_l) \setminus A_{l,i}]$,

$$\begin{aligned}
\left|v_{i_1}(x) - v_{i_2}(x)\right| &\leq \limsup_{r\to 0^+}\left|v_{i_1}(x) - u_{B(x,r)}\right| + \limsup_{r\to 0^+}\left|v_{i_2}(x) - u_{B(x,r)}\right| \\
&\leq 2^{-i_1} + 2^{-i_2},
\end{aligned}$$

which means that, for any given $j \in \mathbb{N}$, $\{v_i|_{B(x_l,r_l)\setminus B_{l,j}}\}_{i\in\mathbb{N}}$ is a Cauchy sequence uniformly in $B(x_l, r_l) \setminus B_{l,j}$. Thus, for any $j \in \mathbb{N}$, $\{v_i|_{B(x_l,r_l)\setminus B_{l,j}}\}_{i\in\mathbb{N}}$ converge to some continuous function

$v_{l,j}$ uniformly in $B(x_l, r_l) \setminus B_{l,j}$ as $i \to \infty$. Due to the observation that $B(x_l, r_l) \setminus B_{l,j}$ increases on j and the uniqueness of the limit, we conclude that, for any $j_1,\ j_2 \in \mathbb{N}$ with $j_1 \le j_2$,

$$v_{l,j_2}|_{B(x_l,r_l)\setminus B_{l,j_1}} = v_{l,j_1}.$$

Therefore, the function v_l^*, defined by setting, for any $x \in B(x_l, r_l) \setminus F_l$,

$$v_l^*(x) := \lim_{j\to\infty} v_{l,j}(x),$$

exists and, for any given $j \in \mathbb{N}$, $v_l^*|_{B(x_l,r_l)\setminus B_{l,j}} = v_{l,j}$. Since $v_{l,j}$ is continuous in $B(x_l, r_l) \setminus B_{l,j}$, we deduce that, for any given $j \in \mathbb{N}$, v_l^* is continuous in $B(x_l, r_l) \setminus B_{l,j}$. By the definitions of v_l^* and $v_{l,j}$, and (53) with $i \to \infty$, we conclude that, for any $x \in B(x_l, r_l) \setminus F_l = \bigcup_{j\in\mathbb{N}}[B(x_l, r_l) \setminus B_{l,j}] = \bigcup_{j\in\mathbb{N}} \bigcap_{i\ge j}[B(x_l, r_l) \setminus A_{l,i}]$,

$$v_l^*(x) = \lim_{j\to\infty} v_{l,j}(x) = \lim_{j\to\infty}\lim_{i\to\infty} v_i|_{B(x_l,r_l)\setminus B_{l,j}}(x) = \lim_{r\to 0^+} u_{B(x,r)}.$$

Altogether, we find a function v_l^* and a set F_l with $\mathrm{Cap}_{\mathcal{F}}(F_l) = 0$ such that

$$v_l^*(\cdot) = \lim_{r\to 0^+} u_{B(\cdot,r)}$$

in $B(x_l, r_l) \setminus F_l$ and, for any $\epsilon \in (0, \infty)$, there exist a $j \in \mathbb{N}$ and a set $B_{l,j}$ with $\mathrm{Cap}_{\mathcal{F}}(B_{l,j}) < \epsilon$ such that v_l^* is continuous in $B(x_l, r_l) \setminus B_{l,j}$.

Next, let $u \in \dot{\mathcal{F}}$. For any given $\widetilde{x} \in \mathcal{X}$ and $k \in \mathbb{N}$, let φ_k be a Lipschitz function such that $\varphi_k \mathbf{1}_{B(\widetilde{x},2k)} = 1$ and $\varphi_k \mathbf{1}_{\mathcal{X}\setminus B(\widetilde{x},3k)} = 0$. By the boundedness of the support of φ_k and Corollary 3, we find that $u\varphi_k \in \mathcal{F}$. Thus, from the conclusion proved in the above paragraph, we deduce that, for any $k \in \mathbb{N}$, there exist a set $E_{l,k}$ with $\mathrm{Cap}_{\mathcal{F}}(E_{l,k}) = 0$ and a function $u_{l,k}$ defined on $B(x_l, r_l) \setminus E_{l,k}$ such that, for any $x \in B(x_l, r_l) \setminus E_{l,k}$,

$$u_{l,k}(x) = \lim_{r\to 0^+} (u\varphi_k)_{B(x,r)}$$

and, for any $\epsilon \in (0, \infty)$, there exists a set $U_{l,k}$ with $\mathrm{Cap}_{\mathcal{F}}(U_{l,k}) < 2^{-k-l}\epsilon$ such that $u_{l,k}$ is continuous in $B(x_l, r_l) \setminus U_{l,k}$.

Define $E_l := \bigcup_{k\in\mathbb{N}} E_{l,k}$ and $U_l := \bigcup_{k\in\mathbb{N}} U_{l,k}$. Then, by Lemma 6, we have $\mathrm{Cap}_{\mathcal{F}}(E_l) = 0$ and, for the above given $\epsilon \in (0, \infty)$, $\mathrm{Cap}_{\mathcal{F}}(U_l) \le 2^{-l}\epsilon$ and, moreover, $\mathrm{Cap}_{\mathcal{F}}(E_l \cup U_l) \le 2^{-l}\epsilon$. For any $x \in B(x_l, r_l) \setminus E_l = \bigcap_{k\in\mathbb{N}} B(x_l, r_l) \setminus E_{l,k}$ and any $k_x \in \mathbb{N}$ big enough such that $x \in B(\widetilde{x}, k_x)$, since, for any $r \in (0, k_x]$, we have $B(x, r) \subset B(\widetilde{x}, 2k_x)$, then, from the fact that $\varphi_{k_x}\mathbf{1}_{B(\widetilde{x},2k_x)} = 1$, we deduce that

$$\begin{aligned}\lim_{r\to 0^+} (u\varphi_{k_x})_{B(x,r)} &= \lim_{r\to 0^+,\, r\in(0,k_x]} \fint_{B(x,r)} u\varphi_{k_x}\, d\mu \\ &= \lim_{r\to 0^+,\, r\in(0,k_x]} \fint_{B(x,r)} u\, d\mu \\ &= \lim_{r\to 0^+} u_{B(x,r)}.\end{aligned} \tag{54}$$

Define u_l by setting, for any $x \in B(x_l, r_l) \setminus E_l$, $u_l(x) := \lim_{r\to 0^+} u_{B(x,r)}$. Then, by (54) and the definition of $u_{l,k}$, we conclude that, for any $k \in \mathbb{N}$, $u_l = u_{l,k}$ in $[B(x_l, r_l) \cap B(\widetilde{x}, k)] \setminus E_l$. From this, the fact that $u_{l,k}$ is continuous in $B(x_l, r_l) \setminus U_{l,k}$, and the definition of U_l, we deduce that, for any $k \in \mathbb{N}$, u_l is continuous in $[B(x_l, r_l) \cap B(\widetilde{x}, k)] \setminus (E_l \cup U_l)$. Therefore, u_l is continuous in $B(x_l, r_l) \setminus (E_l \cup U_l)$.

Finally, we turn to the whole space $\mathcal{X}$ using the covering $\mathcal{X} \subset \bigcup_{l\in\mathcal{I}} B(x_l, r_l)$. Let $u \in \dot{\mathcal{F}}$. On the one hand, we have shown that, for any $l \in \mathcal{I}$, there exists a set E_l with $\mathrm{Cap}_{\mathcal{F}}(E_l) = 0$ such that $u_l(\cdot) := \lim_{r\to 0^+} u_{B(\cdot,r)}$ exists on $B(x_l, r_l) \setminus E_l$. Define $E := \bigcup_{l\in\mathcal{I}} E_l$ and, for any $x \in \mathcal{X} \setminus E$, $\widetilde{u}(x) := \lim_{r\to 0^+} u_{B(x,r)}$. Then, for any $l \in \mathcal{I}$, $\widetilde{u} = u_l$ in $B(x_l, r_l) \setminus E$.

On the other hand, by the above proof, we conclude that, for any given $\epsilon \in (0, \infty)$ and any $l \in \mathcal{I}$, there exists a set $\widetilde{U_l}$ with $\mathrm{Cap}_{\mathcal{F}}(\widetilde{U_l}) \le 2^{-l}\epsilon$ such that u_l is continuous in $B(x_l, r_l) \setminus \widetilde{U_l}$. Define $U := \bigcup_{l\in\mathcal{I}} \widetilde{U_l}$. Then, for any $l \in \mathcal{I}$, u_l is continuous in $B(x_l, r_l) \setminus U$. From this and the fact that, for any $l \in \mathcal{I}, \widetilde{u} = u_l$ in $B(x_l, r_l) \setminus E$, we deduce that $\widetilde{u}$ is continuous in $B(x_l, r_l) \setminus (E \cup U)$ for any $l \in \mathcal{I}$ and hence in $\mathcal{X} \setminus (E \cup U)$.

By Lemma 6, we have $\mathrm{Cap}_{\mathcal{F}}(E) = 0$ and $\mathrm{Cap}_{\mathcal{F}}(U) \le \epsilon$ and, furthermore,

$$\mathrm{Cap}_{\mathcal{F}}(E \cup U) \le \epsilon.$$

Let u^* be any function defined in $\mathcal{X}$ such that $u^* = \widetilde{u}$ in $\mathcal{X} \setminus E$. Then, u^* is continuous in $\mathcal{X} \setminus (E \cup U)$. Thus, u^* is one of the desired $\mathcal{F}$-quasi-continuous functions on $\mathcal{X}$, which completes the proof of Theorem 3. □

Remark 5. *With the same assumptions as in Theorem 3, by (40), the local integrability of u (see [28], Remark 3.8), Remark 4(ii), and the Lebesgue differentiation theorem, we have the following two obvious observations:*

(i) *$u^* = u$ almost everywhere;*
(ii) *every point outside E is a Lebesgue point of u^*.*

In this sense, u^* is called an $\mathcal{F}$-quasi-continuous representative *of u. Furthermore, from the conclusion in (ii) of the present remark and ([45], Lemma 17), we deduce that, for any given $\mathcal{F}$-quasi-continuous function u in $\mathcal{F}$, there exists a set of zero $\mathcal{F}$-capacity such that all the outside points are Lebesgue points of u. Observe that, by Remark 4(ii), any set of zero $\mathcal{F}$-capacity is of zero measure. This implies that, for any $\mathcal{F}$-quasi-continuous function, compared with only locally integrable functions, there exist more Lebesgue points.*

4. Generalized Lebesgue Points of ϕ-Hajłasz-Type Functions

If a function fails to be locally integrable, which may happen, for instance, when the index p of the ϕ-Hajłasz-type space is close to zero, the γ-median serves as a reasonable substitute of the integral average (see, for instance [41,45,46]). That is because the γ-median is defined, instead of integrals, only by the distribution sets of functions and their measures, which removes the necessity for the local integrability of functions. Due to the similarity between the behavior of the γ-median and that of the integral average, the Lebesgue point can naturally be generalized to the γ-median case; see (56). In this section, we still use the capacity to measure the set of such generalized Lebesgue points of ϕ-Hajłasz-type functions. We first recall the notion of the γ-median and some of its basic properties; see ([41], Section 2.4) (see also ([46], Section 1) for a different definition).

Definition 13. *Let $u \in L^0(\mathcal{X})$ and $\gamma \in (0, 1/2]$. The* γ-median *$m_u^\gamma(E)$ of u over a set $E \subset \mathcal{X}$ of finite measure is defined by setting*

$$m_u^\gamma(E) := \inf\{\lambda \in \mathbb{R} : \mu(\{x \in E : u(x) > \lambda\}) < \gamma\mu(E)\}.$$

Observe that, if $E \subset \mathcal{X}$, $\mu(E) \in (0, \infty)$ and $u \in L^0(E)$, then $m_u^\gamma(E)$ is finite.

Lemma 13. *Let E, E_1, $E_2 \subset \mathcal{X}$ be sets of finite measure, γ, γ_1, $\gamma_2 \in (0, 1/2]$, and u, $v \in L^0(\mathcal{X})$. The following statements hold true:*

(i) *If $\gamma_1 \le \gamma_2$, then $m_u^{\gamma_1}(E) \ge m_u^{\gamma_2}(E)$.*
(ii) *If $u \le v$ almost everywhere, then $m_u^\gamma(E) \le m_v^\gamma(E)$.*
(iii) *If $E_1 \subset E_2$ and, for some positive constant c, $\mu(E_2) \le c\mu(E_1)$, then*

$$m_u^\gamma(E_1) \le m_u^{\gamma/c}(E_2).$$

(iv) *For any $c \in \mathbb{R}$, $m_u^\gamma(E) + c = m_{u+c}^\gamma(E)$.*
(v) *For any $c \in (0, \infty)$, $m_{cu}^\gamma(E) = cm_u^\gamma(E)$.*

(vi) $\left|m_u^\gamma(E)\right| \le m_{|u|}^\gamma(E)$.

(vii) $m_{u+v}^\gamma(E) \le m_u^{\gamma/2}(E) + m_v^{\gamma/2}(E)$.

(viii) *For any* $t \in (0,\infty)$,

$$m_{|u|}^\gamma(E) \le \left(\gamma^{-1} \fint_E |u|^t \, d\mu\right)^{1/t}. \tag{55}$$

The following lemma (see, for instance, ([46], Theorem 2.1)) implies that the γ-median over small balls can behave similar to the classical integral average of locally integrable functions at Lebesgue points and becomes a reasonable substitute of the classical Lebesgue differentiation theorem when the function fails to be locally integrable.

Lemma 14. *Let* $u \in L^0(\mathcal{X})$*. Then, there exists a set* $E \subset \mathcal{X}$ *with* $\mu(E) = 0$ *such that, for any* $\gamma \in (0, 1/2]$ *and* $x \in \mathcal{X} \setminus E$,

$$\lim_{r \to 0^+} m_u^\gamma(B(x,r)) = u(x). \tag{56}$$

In particular, (56) holds true at every continuous point x of u.

Let $u \in L^0(\mathcal{X})$. Recall that a point $x \in \mathcal{X}$ is called a *generalized Lebesgue point* of u if (56) holds true for x and any $\gamma \in (0, 1/2]$; see, for instance [41,44,45]. If u is locally integrable, as was pointed by ([46], p. 231), any Lebesgue point of u is a generalized Lebesgue point of u. This means that the generalized Lebesgue point is a more extensive notion than the Lebesgue point.

Next, we recall the variants of both $\mathcal{M}$ and $\mathcal{M}^*$ in the γ-median version (see, for instance [41,45]), where $\mathcal{M} = \mathcal{M}_\infty$ is as in (14), and $\mathcal{M}^*$ as in Definition 9(ii).

Definition 14. *Let* $\gamma \in (0, 1/2]$ *and* $u \in L^0(\mathcal{X})$*. The* γ-median maximal function $\mathcal{M}^\gamma(u)$ *of u is defined by setting, for any* $x \in \mathcal{X}$,

$$\mathcal{M}^\gamma(u)(x) := \sup_{r \in (0,\infty)} m_{|u|}^\gamma(B(x,r)).$$

Definition 15. *Let* $\gamma \in (0, 1/2]$ *and* $u \in L^0(\mathcal{X})$.

(i) *The* discrete γ-median convolution u_r^γ *of u at scale* $r \in (0,\infty)$ *is defined by setting, for any* $x \in \mathcal{X}$,

$$u_r^\gamma(x) := \sum_{j \in \mathcal{J}} m_u^\gamma(B_j)\, \varphi_j(x),$$

where $\mathcal{J}$ *is an index set,* $\{B_j\}_{j \in \mathcal{J}}$ *is a ball covering of* $\mathcal{X}$ *with the radius r such that* $\sum_{j \in \mathcal{J}} \mathbf{1}_{2B_j} \lesssim 1$*, and* $\{\varphi_j\}_{j \in \mathcal{J}}$ *is a partition of unity with respect to* $\{B_j\}_{j \in \mathcal{J}}$ *as in Definition 8.*

(ii) *The* discrete γ-median maximal function $\mathcal{M}^{\gamma,*}u$ *of u is defined by setting, for any* $x \in \mathcal{X}$,

$$\mathcal{M}^{\gamma,*}u(x) := \sup_{k \in \mathbb{Z}} |u|_{2^{-k}}^\gamma(x),$$

where $|u|_{2^{-k}}^\gamma$ *is as in (i) with u and r replaced, respectively, by* $|u|$ *and* 2^{-k}.

Remark 6. *Let* $\mathcal{M}^\gamma$ *and* $\mathcal{M}^{\gamma,*}$ *be as in Definitions 14 and 15. Recall that there exists a positive constant c such that, for any* $u \in L^0(\mathcal{X})$,

$$\mathcal{M}^\gamma u \le c\mathcal{M}^{\gamma/c,*}u \le c^2 \mathcal{M}^{\gamma/c^2}u; \tag{57}$$

see ([41], (2.10)). Additionally, recall that either $\mathcal{M}^\gamma u \equiv \infty$ *or* $\mathcal{M}^\gamma u < \infty$ *almost everywhere in* $\mathcal{X}$ *and either* $\mathcal{M}^{\gamma,*}u \equiv \infty$ *or* $\mathcal{M}^{\gamma,*}u < \infty$ *almost everywhere in* $\mathcal{X}$*; see ([41], (2.10)) and ([41], Remark 2.11).*

The following two lemmas are the variants of Poincaré-type inequalities, respectively, in Lemma 3, (18), and (17), where the second lemma is a generalization of ([41], Lemma 3.2).

Lemma 15. *Let $\gamma \in (0, 1/2]$, $\phi \in \mathcal{A}$, and C_μ be as in (1).*

(i) *Then, there exists a positive constant $C = C_{(\phi, C_\mu)}$ such that, for any $k \in \mathbb{Z}$, $u \in L^0(\mathcal{X})$, $g \in \mathcal{D}^\phi(u)$, and $x \in \mathcal{X}$,*

$$\inf_{c\in\mathbb{R}} m^\gamma_{|u-c|}(B(x, 2^{-k})) \le C\gamma^{-1}\phi(2^{-k}) \fint_{B(x,2^{-k})} g(y)\, d\mu(y).$$

(ii) *If $\alpha_\phi \in (0, 1)$, then, for any given $\lambda \in (0, \infty)$, there exists a positive constant $C = C_{(\gamma,\lambda,\phi,C_\mu)}$ such that, for any $k \in \mathbb{Z}$, $u \in L^0(\mathcal{X})$, $g \in \mathcal{D}^\phi(u)$, and $x \in \mathcal{X}$,*

$$\inf_{c\in\mathbb{R}} m^\gamma_{|u-c|}(B(x, 2^{-k})) \le C\phi(2^{-k})\left\{\fint_{B(x,2^{-k+1})} [g(y)]^\lambda\, d\mu(y)\right\}^{1/\lambda}.$$

Proof. We first prove (i). For any $k \in \mathbb{Z}$, $u \in L^0(\mathcal{X})$, $g \in \mathcal{D}^\phi(u)$, $x \in \mathcal{X}$, and $c \in \mathbb{R}$, from (55) with $t = 1$, and E and u therein replaced, respectively, by $B(x, 2^{-k})$ and $u - c$, we deduce that

$$m^\gamma_{|u-c|}(B(x, 2^{-k})) \le \gamma^{-1} \fint_{B(x,2^{-k})} |u(y) - c|\, d\mu(y). \tag{58}$$

Taking the infimum of $c \in \mathbb{R}$ in (58), and using Lemma 3, we obtain (i) of the present lemma.

Now we prove (ii). By $\alpha_\phi < 1$, we choose $\varepsilon := -(\log_2 \alpha_\phi)/2 > 0$. For any $k \in \mathbb{Z}$, $\lambda \in (0, D/\varepsilon)$, $u \in L^0(\mathcal{X})$, $g \in \mathcal{D}^\phi(u)$, $x \in \mathcal{X}$, and $c \in \mathbb{R}$, applying (55) with $t = (D\lambda)/(D - \varepsilon\lambda) \in (0, \infty)$, and E and u therein replaced, respectively, by $B(x, 2^{-k})$ and $u - c$, we conclude that

$$\begin{aligned} &m^\gamma_{|u-c|}(B(x, 2^{-k})) \\ &\quad\le \left\{\gamma^{-1} \fint_{B(x,2^{-k})} [u(y) - c]^{(D\lambda)/(D-\varepsilon\lambda)}\, d\mu(y)\right\}^{(D-\varepsilon\lambda)/(D\lambda)}. \end{aligned} \tag{59}$$

Taking the infimum of $c \in \mathbb{R}$ in (59) and using (16) with $p = \lambda$, we obtain the conclusion of (ii) when $\lambda \in (0, D/\varepsilon)$. From this and the Hölder inequality, we deduce that the conclusion of (ii) also holds true when $\lambda \in [D/\varepsilon, \infty)$, which completes the proof of Lemma 15. □

Lemma 16. *Let $\gamma \in (0, 1/2]$, $\phi \in \mathcal{A}$ with $\alpha_\phi \in (0, 1)$, and C_μ be as in (1). Then, for any given $\lambda \in (0, \infty)$ and $\epsilon \in (0, -\log_2 \alpha_\phi)$, there exists a positive constant $C = C_{(\gamma,\phi,\epsilon,\lambda,C_\mu)}$ such that, for any $k \in \mathbb{Z}$, $u \in L^0(\mathcal{X})$, $\{g_k\}_{k\in\mathbb{Z}} \in \mathbb{D}^\phi(u)$, and $x \in \mathcal{X}$,*

$$\begin{aligned} &\inf_{c\in\mathbb{R}} m^\gamma_{|u-c|}(B(x, 2^{-k})) \\ &\quad\le C2^{-k\epsilon} \sum_{l\ge k-2} 2^{l\epsilon}\phi(2^{-l}) \left\{\fint_{B(x,2^{-k+1})} [g_l(y)]^\lambda\, d\mu(y)\right\}^{1/\lambda}. \end{aligned} \tag{60}$$

Proof. Let $\lambda \in (0, \infty)$ and $\nu \in (0, \epsilon)$, where ϵ is given as in Lemma 16. When $\lambda \in (0, D/\nu)$, (60) follows from (55) with $t = (D\lambda)/(D - \nu\lambda) \in (0, \infty)$, E and u therein replaced, respectively, by $B(x, 2^{-k})$ and $u - c$ for arbitrary $c \in \mathbb{R}$, and from Lemma 4 with p and ε' therein replaced, respectively, by λ and ϵ. This, combined with the Hölder inequality, further implies (60) when $\lambda \in [D/\nu, \infty)$. This finishes the proof of Lemma 16. □

The following lemma is a variant of Lemma 5 in the γ-median version.

Lemma 17. *Let $\gamma \in (0, 1/2]$, $\phi \in \mathcal{A}$, C_μ be as in (1), and $\mathcal{M}$ the classical Hardy–Littlewood maximal operator.*

(i) *Then, there exists a positive constant $C = C_{(\phi,C_\mu)}$ such that, for any $k \in \mathbb{Z}$, $u \in L^0(\mathcal{X})$, $g \in \mathcal{D}^\phi(u)$, $y \in \mathcal{X}$, and almost every $x \in B(y, 2^{-k+1})$,*

$$\left|u(x) - m_u^\gamma(B(y, 2^{-k}))\right| \le C\gamma^{-1}\phi(2^{-k})\,\mathcal{M}(g)(x).$$

(ii) *Let $\alpha_\phi \in (0, 1)$. Then, for any given $\lambda \in (0, \infty)$, there exists a positive constant $C = C_{(\gamma,\phi,\lambda,C_\mu)}$ such that, for any $k \in \mathbb{Z}$, $u \in L^0(\mathcal{X})$, $g \in \mathcal{D}^\phi(u)$, $y \in \mathcal{X}$, and any generalized Lebesgue point $x \in B(y, 2^{-k+1})$,*

$$\left|u(x) - m_u^\gamma(B(y, 2^{-k}))\right| \le C\phi(2^{-k})\left[\mathcal{M}\left(g^\lambda\right)(x)\right]^{1/\lambda}.$$

(iii) *Let $\alpha_\phi \in (0, 1)$. Then, for any given $\lambda \in (0, \infty)$ and $\epsilon \in (0, -\log_2 \alpha_\phi)$, there exists a positive constant $C = C_{(\gamma,\phi,\lambda,\epsilon,C_\mu)}$ such that, for any $k \in \mathbb{Z}$, $u \in L^0(\mathcal{X})$, $\{g_l\}_{l\in\mathbb{Z}} \in \mathbb{D}^\phi(u)$, $y \in \mathcal{X}$, and any generalized Lebesgue point $x \in B(y, 2^{-k+1})$,*

$$\left|u(x) - m_u^\gamma(B(y, 2^{-k}))\right| \le C2^{-k\epsilon} \sum_{l\ge k-4} 2^{l\epsilon}\phi(2^{-l})\left[\mathcal{M}\left(g_l^\lambda\right)(x)\right]^{1/\lambda}.$$

Proof. Let all of the symbols be as in the present lemma. We first prove (i). For any $k \in \mathbb{Z}$, $y \in \mathcal{X}$ and almost every $x \in B(y, 2^{-k+1})$, by (iv) and (vi) of Lemma 13; (55) with $t = 1$; and E and u therein replaced, respectively, by $B(y, 2^{-k})$ and $u - u(x)$; the geometric observation that, for any $x \in B(y, 2^{-k+1})$, $B(y, 2^{-k}) \subset B(x, 2^{-k+2})$; the doubling property of μ; the definitions of $\mathcal{D}^\phi(u)$ and $\mathcal{A}$; and $g \le \mathcal{M}(g)$ almost everywhere, we have, for almost every $x \in B(y, 2^{-k+1}) \setminus E$,

$$\begin{aligned}
&\left|u(x) - m_u^\gamma(B(y, 2^{-k}))\right| \\
&= \left|m_{u-u(x)}^\gamma(B(y, 2^{-k}))\right| \le m_{|u-u(x)|}^\gamma(B(y, 2^{-k})) \\
&\le \gamma^{-1} \fint_{B(y,2^{-k})} |u(z) - u(x)|\,d\mu(z) \lesssim \gamma^{-1} \fint_{B(x,2^{-k+2})} |u(z) - u(x)|\,d\mu(z) \\
&\lesssim \gamma^{-1}\phi(2^{-k})\left[g(x) + \fint_{B(x,2^{-k+2})} g(z)\,d\mu(z)\right] \lesssim \gamma^{-1}\phi(2^{-k})\mathcal{M}(g)(x),
\end{aligned}$$

which completes the proof of (i).

Now, we prove (ii) and (iii). Let λ and ϵ be as in (ii) and (iii) of the present lemma. Similar to ([41] (3.3)), by (ii), (iv), and (vi) of Lemma 13, we have, for any γ, $\gamma' \in (0, 1/2]$ and any ball B,

$$\begin{aligned}
m_{|u-m_u^\gamma(B)|}^{\gamma'}(B) &\le \inf_{c\in\mathbb{R}} m_{|u-c|+|c-m_u^\gamma(B)|}^{\gamma'}(B) \\
&= \inf_{c\in\mathbb{R}}\left[m_{|u-c|}^{\gamma'}(B) + |c - m_u^\gamma(B)|\right] \\
&\le \inf_{c\in\mathbb{R}}\left[m_{|u-c|}^{\gamma'}(B) + m_{|u-c|}^{\gamma}(B)\right].
\end{aligned} \tag{61}$$

Moreover, by the geometrical observation that, for any $k \in \mathbb{Z}$, $y \in \mathcal{X}$, and $x \in B(y, 2^{-k+1})$; $B(x, 2^{-k}) \subset B(y, 2^{-k+2})$; and the doubling property of μ, we obtain

$$\mu\left(B(x, 2^{-k+2})\right) \le C_\mu^2\mu\left(B(x, 2^{-k})\right) \le C_\mu^2\mu\left(B(y, 2^{-k+2})\right) \le C_\mu^4\mu\left(B(y, 2^{-k})\right).$$

Therefore, from this, the definition of generalized Lebesgue points; the doubling property of μ; (i), (iii), (iv), and (vi) of Lemma 13; $C_\mu \in [1,\infty)$; and (61) with $\gamma' = \gamma/C_\mu^4$ and B replaced by $B(x, 2^{-j})$, we deduce that, for any generalized Lebesgue point $x \in B(y, 2^{-k+1})$,

$$\begin{aligned}
&\left|u(x) - m_u^\gamma(B(y,2^{-k}))\right| \\
&\quad\le \left|u(x) - m_u^\gamma(B(x,2^{-k+2}))\right| + \left|m_u^\gamma(B(x,2^{-k+2})) - m_u^\gamma(B(y,2^{-k}))\right| \\
&\quad\le \sum_{j\ge k-2} \left|m_u^\gamma(B(x,2^{-j-1})) - m_u^\gamma(B(x,2^{-j}))\right| + \left|m_u^\gamma(B(y,2^{-k})) - m_u^\gamma(B(x,2^{-k+2}))\right| \\
&\quad\le \sum_{j\ge k-2} m^\gamma_{|u-m_u^\gamma(B(x,2^{-j}))|}(B(x,2^{-j-1})) + m^\gamma_{|u-m_u^\gamma(B(x,2^{-k+2}))|}(B(y,2^{-k})) \\
&\quad\le \sum_{j\ge k-2} m^{\gamma/C_\mu}_{|u-m_u^\gamma(B(x,2^{-j}))|}(B(x,2^{-j})) + m^{\gamma/C_\mu^4}_{|u-m_u^\gamma(B(x,2^{-k+2}))|}(B(x,2^{-k+2})) \\
&\quad\lesssim \sum_{j\ge k-2} m^{\gamma/C_\mu^4}_{|u-m_u^\gamma(B(x,2^{-j}))|}(B(x,2^{-j})) \\
&\quad\lesssim \sum_{j\ge k-2} \inf_{c\in\mathbb{R}}\left[m^{\gamma/C_\mu^4}_{|u-c|}(B(x,2^{-j})) + m^\gamma_{|u-c|}(B(x,2^{-j}))\right].
\end{aligned} \tag{62}$$

On the one hand, (62), combined with Lemma 15(ii) with k therein replaced by j, (9) with k and k_0 therein replaced, respectively, by $-j$ and $-k+2$ and the definitions of $\mathcal{M}$ and $\mathcal{A}$, implies (ii) of the present lemma. On the other hand, (62), combined with Lemma 16, the definition of $\mathcal{M}$, and $\sum_{j\ge k-2} 2^{-j\epsilon} \lesssim 2^{-k\epsilon}$, implies (iii) of the present lemma, which completes the proof of Lemma 17. □

We now establish the convergence of approximations by discrete γ-median convolutions as below, which is a generalization of ([41], Theorem 1.1) from fractional Hajłasz-type spaces to those with generalized smoothness.

Theorem 4. *Let $\gamma \in (0,1/2]$, $\mathcal{F} \in \{M^\phi_{p,q}(\mathcal{X}), N^\phi_{p,q}(\mathcal{X})\}$ with $\phi \in \mathcal{A}_0$ and $p,\ q \in (0,\infty)$, and $u \in \dot{\mathcal{F}}$. Then, $\|u - u^\gamma_{2^{-i}}\|_{\mathcal{F}} \to 0$ as $i \to \infty$, where $\{u^\gamma_{2^{-i}}\}_{i\ge 0}$ are the discrete γ-median convolutions as in Definition 15(i).*

Proof. By similarity, we only consider the case $\mathcal{F} = M^\phi_{p,q}(\mathcal{X})$. Let $\gamma \in (0,1/2]$, $i \in \mathbb{Z}_+$, $u^\gamma_{2^{-i}}$ be as in Definition 15(i), $u \in \dot{M}^\phi_{p,q}(\mathcal{X})$, and $\{g_k\}_{k\in\mathbb{Z}} \in \mathbb{D}^\phi(u) \cap L^p(\mathcal{X}, l^q)$.

Let $\lambda \in (0, \min(p,q))$, $\epsilon \in (0, -\log_2 \alpha_\phi)$, $\{B_j\}_{j\in\mathcal{J}}$ be any given ball covering of $\mathcal{X}$ with the radius 2^{-i} such that $\sum_{j\in\mathcal{J}} \mathbf{1}_{2B_j} \lesssim 1$, and $\{\varphi_j\}_{j\in\mathcal{J}}$, consisting of a sequence of $c2^i$-Lipschitz functions, a partition of unity with respect to $\{B_j\}_{j\in\mathcal{J}}$ as in Definition 8. For any $j \in \mathcal{J}$, let $m_u^\gamma(B_j)$ be as in Definition 13. Then, by the properties of $\{\varphi_j\}_{j\in\mathcal{J}}$, we have

$$u - u^\gamma_{2^{-i}} = \sum_{j\in\mathcal{J}} \left(u - m_u^\gamma(B_j)\right)\varphi_j. \tag{63}$$

Using Lemma 7 with u and L^{-1} therein replaced, respectively, by $u - m_u^\gamma(B_j)$ and $c2^i$, we conclude that, for any $j \in \mathcal{J}$, $\{h^*_{k,j}\}_{k\in\mathbb{Z}}$, defined by setting, for any $k \in \mathbb{Z}$,

$$h^*_{k,j} := \begin{cases} \left\{2^{i-k}\left[\phi(2^{-k})\right]^{-1}\left|u - m_u^\gamma(B_j)\right| + g_k\right\}\mathbf{1}_{2B_j}, & k > i, \\ \left[\phi(2^{-k})\right]^{-1}\left|u - m_u^\gamma(B_j)\right|\mathbf{1}_{2B_j}, & k \le i, \end{cases}$$

is a positive constant multiple of an element of $\mathbb{D}^\phi([u - m_u^\gamma(B_j)]\varphi_j)$. From this, (63), an argument similar to that used in the estimation of (26) with $u_{2^{-i}}$, u_{B_j}, and $h_{k,j}$ therein replaced, respectively, by $u^\gamma_{2^{-i}}$, $m_u^\gamma(B_j)$, and $h^*_{k,j}$, Lemma 17(iii), (27), and $\sum_{j\in\mathcal{J}} \mathbf{1}_{2B_j} \lesssim 1$, we deduce that $\{h_k\}_{k\in\mathbb{Z}}$, defined as in (28) with the above λ and ϵ, is also a positive constant multiple of an element in $\mathbb{D}^{\phi}(u-u^{\gamma}_{2^{-i}})$. By this, (32),

$$\left\|\left(\sum_{k\geq i-4} g_k^q\right)^{1/q}\right\|_{L^p(\mathcal{X})} \leq \|\{g_k\}_{k\in\mathbb{Z}}\|_{L^p(\mathcal{X},l^q)} < \infty$$

with $i\in\mathbb{Z}_+$, and the dominated convergence theorem with respect to μ, we obtain

$$\left\|u-u^{\gamma}_{2^{-i}}\right\|_{\dot{M}^{\phi}_{p,q}(\mathcal{X})} \lesssim \|h_k\|_{L^p(\mathcal{X},l^q)} \lesssim \left\|\left(\sum_{k\geq i-4} g_k^q\right)^{1/q}\right\|_{L^p(\mathcal{X})} \to 0$$

as $i\to\infty$. Then, using (63), Lemma 17(iii) instead of Lemma 5(iii), the properties of $\{\varphi_j\}_{j\in\mathcal{J}}$, Lemma 1(ii) with $\epsilon\in(0,-\log_2\alpha_\phi)$, the Fefferman–Stein vector-valued maximal inequality on $L^{p/\lambda}(\mathcal{X},l^{q/\lambda})$ (see ([56], Theorem 1.2) or ([57], Theorem 1.3)), $\phi(0)=0$, and an argument similar to that used in the estimation of (33), we conclude that

$$\left\|u-u^{\gamma}_{2^{-i}}\right\|_{L^p(\mathcal{X})} \lesssim \phi(2^{-i})\left\|\left(\sum_{l\geq i-4} g_l^q\right)^{1/q}\right\|_{L^p(\mathcal{X})} \to 0$$

as $i\to\infty$. This finishes the proof of Theorem 4. □

Now, we state the following variant of Theorem 3 for γ-medians.

Theorem 5. *Let $\gamma\in(0,1/2]$, $\phi\in\mathcal{A}$, and $\mathcal{F}$ be one of the following cases:*

(i) $\mathcal{F}=M^{\phi}_{p,\infty}(\mathcal{X})=M^{\phi,p}(\mathcal{X})$ *with ϕ being a modulus of continuity and $p\in(1,\infty)$;*

(ii) $\mathcal{F}=M^{\phi}_{p,\infty}(\mathcal{X})=M^{\phi,p}(\mathcal{X})$ *with ϕ being a modulus of continuity, $\alpha_\phi\in(0,1)$, and $p\in(0,1]$;*

(iii) $\mathcal{F}\in\{M^{\phi}_{p,q}(\mathcal{X}),N^{\phi}_{p,q}(\mathcal{X})\}$ *with $\alpha_\phi\in(0,1)$, $\beta_\phi\in(0,2)$, and $p,\ q\in(0,\infty)$.*

Then, for any $u\in\dot{\mathcal{F}}$, there exists a set E with $\mathrm{Cap}_{\mathcal{F}}(E)=0$ satisfying that, for any $\gamma\in(0,1/2]$, there exists an $\mathcal{F}$-quasi-continuous function u^* on $\mathcal{X}$ such that, for any $x\in\mathcal{X}\setminus E$,

$$u^*(x)=\lim_{r\to 0} m^{\gamma}_u(B(x,r)). \tag{64}$$

To show Theorem 5, similar to the proof of Theorem 3, we need a weak-type capacitary estimate with respect to $\mathcal{M}^{\gamma}$. To this end, we first prove an auxiliary lemma as below, which is about the boundedness of $\mathcal{M}^{\gamma,*}$ in ϕ-Hajłasz-type spaces and generalizes ([41], Theorem 7.6). Here and thereafter, $\mathcal{M}^{\gamma}$ and $\mathcal{M}^{\gamma,*}$ are as in Definitions 14 and 15(ii), respectively.

Lemma 18. *With the same assumptions as in Theorem 5, there exists a positive constant $C=C_{(\mathcal{F},\gamma,C_\mu)}$ such that, for any $u\in\mathcal{F}$,*

$$\left\|\mathcal{M}^{\gamma,*}u\right\|_{\mathcal{F}} \leq C\|u\|_{\mathcal{F}}, \tag{65}$$

where $\mathcal{M}^{\gamma,*}$ is as in Definition 15.

Proof. Let all of the symbols be as in the present lemma. Without loss of generality, by the definition of $\mathcal{M}^{\gamma,*}$, $\mathcal{D}^{\phi}(u)\subset\mathcal{D}^{\phi}(|u|)$, and $\mathbb{D}^{\phi}(u)\subset\mathbb{D}^{\phi}(|u|)$, we may assume that $u\geq 0$.

Let $i\in\mathbb{Z}$, $\{B_j\}_{j\in\mathcal{J}}$ be any given ball covering of $\mathcal{X}$ with the radius 2^{-i} such that

$$\sum_{j\in\mathcal{J}} \mathbf{1}_{2B_j} \lesssim 1,$$

$\{\varphi_j\}_{j\in\mathcal{J}}$ be a partition of unity with respect to $\{B_j\}_{j\in\mathcal{J}}$ as in Definition 8, $u^{\gamma}_{2^{-i}}$ be as in Definition 15, and $\mathcal{M}^{\gamma}$ be as in Definition 14. Then, by (57) and ([41], (2.7)), we have, for any given $p \in (0, \infty)$,

$$\left\|\mathcal{M}^{\gamma,*}u\right\|_{L^p(\mathcal{X})} \lesssim \left\|\mathcal{M}^{\gamma/c}u\right\|_{L^p(\mathcal{X})} \lesssim \|u\|_{L^p(\mathcal{X})} < \infty, \tag{66}$$

where c is the same positive constant as in (57). From this and Remark 6, we deduce that $\mathcal{M}^{\gamma,*}u < \infty$ almost everywhere.

Let $\mathcal{F} = M^{\phi,p}(\mathcal{X})$ and $g \in \mathcal{D}^{\phi}(u)$. Using (i) and (ii) of Lemma 17 instead of (i) and (ii) of Lemma 5, and $\mathcal{M}^{\gamma,*}u < \infty$ almost everywhere, from an argument similar to that used in the proof of Lemma 8 with $\{u_{B_j}\}_{j\in\mathcal{J}}$ and $u_{2^{-i}}$ therein replaced, respectively, by $\{m^{\gamma}_u(B_j)\}_{j\in\mathcal{J}}$ and $u^{\gamma}_{2^{-i}}$, we deduce that $\mathcal{M}(g)$ is a positive constant multiple of an element in $\mathcal{D}^{\phi}(\mathcal{M}^{\gamma,*}u)$ and, if $\alpha_\phi \in (0, 1)$, then for any $\lambda \in (0, \infty)$, $[\mathcal{M}(g^{\lambda})]^{1/\lambda}$ is a positive constant multiple of an element in $\mathcal{D}^{\phi}(\mathcal{M}^{\gamma,*}u)$, where both of the positive constants are independent of u and g. Below, we let $\lambda \in (0, \min(p, q))$. Thus, by the boundedness of $\mathcal{M}$ on $L^p(\mathcal{X})$ when $p \in (1, \infty)$, and on $L^{p/\lambda}(\mathcal{X})$ with $\lambda \in (0, p)$ when $p \in (0, 1]$, we obtain, when $p \in (1, \infty)$,

$$\left\|\mathcal{M}^{\gamma,*}u\right\|_{\dot{M}^{\phi,p}(\mathcal{X})} \lesssim \left\|\mathcal{M}(g)\right\|_{L^p(\mathcal{X})} \lesssim \|g\|_{L^p(\mathcal{X})}$$

and, when $\alpha_\phi \in (0, 1)$ and $p \in (0, 1]$,

$$\left\|\mathcal{M}^{\gamma,*}u\right\|_{\dot{M}^{\phi,p}(\mathcal{X})} \lesssim \left\|[\mathcal{M}(g^{\lambda})]^{1/\lambda}\right\|_{L^p(\mathcal{X})} \lesssim \|g\|_{L^p(\mathcal{X})}.$$

This, combined with (66), proves (65) when $\mathcal{F}$ belongs to either (i) or (ii) of the assumptions of Theorem 5.

Next, we prove (65) when $\mathcal{F}$ belongs to the case (iii) of Theorem 5. By similarity, we only consider the case $\mathcal{F} = M^{\phi}_{p,q}(\mathcal{X})$ with $\alpha_\phi \in (0, 1)$, $\beta_\phi \in (0, 2)$, and $p,\ q \in (0, \infty)$. To prove (65), by (66), it suffices to show

$$\left\|\mathcal{M}^{\gamma,*}u\right\|_{\dot{M}^{\phi}_{p,q}(\mathcal{X})} \lesssim \|u\|_{\dot{M}^{\phi}_{p,q}(\mathcal{X})}.$$

Let $\{g_k\}_{k\in\mathbb{Z}} \in \mathbb{D}^{\phi}(u)$ be such that $\|\{g_k\}_{k\in\mathbb{Z}}\|_{L^p(\mathcal{X},l^q)} \lesssim \|u\|_{\dot{M}^{\phi}_{p,q}(\mathcal{X})}$, and $\epsilon \in (0, -\log_2 \alpha_\phi)$. Recall that we have proved in the proof of Theorem 4 that $\{h_k\}_{k\in\mathbb{Z}}$, defined as in (28) with the above λ and ϵ, is a positive constant multiple of an element in $\mathbb{D}^{\phi}(u - u^{\gamma}_{2^{-i}})$. Thus, by $\{g_k\}_{k\in\mathbb{Z}} \in \mathbb{D}^{\phi}(u)$, we conclude that $\{g_k + h_k\}_{k\in\mathbb{Z}}$ is a positive constant multiple of an element in $\mathbb{D}^{\phi}(u^{\gamma}_{2^{-i}})$.

Let $\delta \in (0, \min\{1 - \log_2 \beta_\phi, -\log_2 \alpha_\phi - \epsilon\})$ and $\{\widetilde{g_k}\}_{k\in\mathbb{Z}}$ be as in (42) with the above λ and δ. Similar to the proof of Lemma 9, we know that, for any $k \in \mathbb{Z}$, $g_k + h_k \lesssim \widetilde{g_k}$ almost everywhere. By this and the proved conclusion that $\{g_k + h_k\}_{k\in\mathbb{Z}}$ is a positive constant multiple of an element in $\mathbb{D}^{\phi}(u^{\gamma}_{2^{-i}})$, we conclude that $\{\widetilde{g_k}\}_{k\in\mathbb{Z}}$ is also a positive constant multiple of an element in $\mathbb{D}^{\phi}(u^{\gamma}_{2^{-i}})$ with the positive constant independent of i. Furthermore, using the fact that $\mathcal{M}^{\gamma,*}u < \infty$ almost everywhere and Lemma 2(ii), we find that $\{\widetilde{g_k}\}_{k\in\mathbb{Z}}$ is a positive constant multiple of an element in $\mathbb{D}^{\phi}(\mathcal{M}^{\gamma,*}u)$. From this, the Fefferman–Stein vector-valued maximal inequality on $L^{p/\lambda}(\mathcal{X}, l^{q/\lambda})$ (see ([56], Theorem 1.2) or ([57], Theorem 1.3)), and the choice of $\{g_k\}_{k\in\mathbb{Z}}$, we deduce that

$$\begin{aligned}\left\|\mathcal{M}^{\gamma,*}u\right\|_{\dot{M}^{\phi}_{p,q}(\mathcal{X})} &\lesssim \|\{\widetilde{g_k}\}_{k\in\mathbb{Z}}\|_{L^p(\mathcal{X},l^q)} \\ &\lesssim \left\|\left\{\sum_{l\in\mathbb{Z}}\left[\mathcal{M}(g_l^{\lambda})\right]^{q/\lambda}\right\}^{\lambda/q}\right\|^{1/\lambda}_{L^{p/\lambda}(\mathcal{X})} \\ &\lesssim \|\{g_l\}_{l\in\mathbb{Z}}\|_{L^p(\mathcal{X},l^q)} \lesssim \|u\|_{\dot{M}^{\phi}_{p,q}(\mathcal{X})}.\end{aligned}$$

Thus, by (66), we conclude that (65) holds true for $\mathcal{F} = M^{\phi}_{p,q}(\mathcal{X})$ with $\alpha_\phi \in (0,1)$, $\beta_\phi \in (0,2)$, and $p,\ q \in (0,\infty)$. This finishes the proof of Lemma 18. □

The following weak-type capacitary estimate plays a crucial role in the proof of Theorem 5. Since it is just a generalization of ([41], Theorem 7.7), and a straight corollary of both Lemma 18 and the lower semi-continuity of $\mathcal{M}^{\gamma,*}u$ for any $u \in L^0(\mathcal{X})$, we omit its proof.

Lemma 19. *With the assumptions same as in Theorem 5, there exists a positive constant C, depending only on $\mathcal{F}$, γ, and C_μ, such that, for any $u \in \mathcal{F}$ and $\kappa \in (0,\infty)$,*

$$\mathrm{Cap}_{\mathcal{F}}(\{x \in \mathcal{X}:\ \mathcal{M}^{\gamma}u(x) > \kappa\}) \leq C\kappa^{-p}\|u\|^p_{\mathcal{F}},$$

where $\mathcal{M}^{\gamma}$ is as in Definition 14 and C_μ as in (1).

Now, we turn to prove Theorem 5. Since the proof of Theorem 5 is quite similar to that of Theorem 3, we only sketch the main steps.

Proof of Theorem 5. Let $\mathcal{F}$ be any given function space as in (i), (ii), or (iii) of the present theorem, and $p \in (0,\infty)$. We first let $u \in \mathcal{F}$. By Theorems 2 and 4, we find that, in any case as above, there always exists a sequence $\{u_i\}_{i\in\mathbb{N}}$ of continuous functions such that, for any $i \in \mathbb{N}$,

$$\|u - u_i\|^p_{\mathcal{F}} < 2^{-i(1+p)}.$$

For any $k,\ i \in \mathbb{N}$, define

$$A_{k,i} := \left\{x \in \mathcal{X}:\ \mathcal{M}^{1/(2k)}(u-u_i)(x) > 2^{-i}\right\}$$

and

$$E := \bigcup_{k\geq 2} E_k := \bigcup_{k\geq 2}\bigcap_{j\in\mathbb{N}} B_{k,j} := \bigcup_{k\geq 2}\bigcap_{j\in\mathbb{N}}\bigcup_{i\geq j} A_{k,i}.$$

Then, by Lemma 19, we have, for any given $k \in \mathbb{N}$, $\mathrm{Cap}_{\mathcal{F}}(A_{k,i}) \lesssim 2^{-i}$ and, by Lemma 6, for any $j \in \mathbb{N}$, $\mathrm{Cap}_{\mathcal{F}}(B_{k,j}) \lesssim 2^{-j}$, which implies that, for any given $k \in \mathbb{N}$, $\mathrm{Cap}_{\mathcal{F}}(E_k) = 0$ and hence $\mathrm{Cap}_{\mathcal{F}}(E) = 0$.

For any given $k \in \mathbb{N}\setminus\{1\}$ and any $i \in \mathbb{N}$, by the continuity of u_i and (55) with $t = 1$, we find that, for any $x \in \mathcal{X}$,

$$\begin{aligned}\limsup_{r\to 0^+} m^{1/k}_{|u_i-u_i(x)|}(B(x,r)) &\leq k \lim_{r\to 0^+} \fint_{B(x,r)} \left|u(y)-u(x)\right|dy \\ &= 0.\end{aligned}$$

From this, (i), (iv), (vi), and (vii) of Lemma 13 and the definitions of $\mathcal{M}^{1/(2k)}$ and $A_{k,i}$, we deduce that, for any given $k \in \mathbb{N}\setminus\{1\}$, any $\gamma \in [1/k, 1/2]$, $i \in \mathbb{N}$, and $x \in \mathcal{X}\setminus A_{k,i}$,

$$\begin{aligned}\limsup_{r\to 0^+}\left|u_i(x) - m^{\gamma}_u(B(x,r))\right| &\leq \limsup_{r\to 0^+} m^{\gamma}_{|u-u_i(x)|}(B(x,r)) \\ &\leq \limsup_{r\to 0^+}\left[m^{1/(2k)}_{|u-u_i|}(B(x,r)) + m^{1/(2k)}_{|u_i-u_i(x)|}(B(x,r))\right] \\ &\leq \mathcal{M}^{1/(2k)}(u-u_i)(x) \leq 2^{-i}.\end{aligned} \tag{67}$$

By an argument similar to that used in the proof of Theorem 3, with (53) replaced by (67), we conclude that, for any given $k \in \mathbb{N}\setminus\{1\}$, there exists a function v_k on $\mathcal{X}\setminus E_k$ such that, for any $\gamma \in [1/k, 1/2]$ and $x \in \mathcal{X}\setminus E_k$,

$$v_k(x) = \lim_{r\to 0^+} m^{\gamma}_u(B(x,r))$$

and, moreover, for any $j \in \mathbb{N}$, v_k is continuous on $\mathcal{X} \setminus B_{k,j}$.

For any given $\gamma \in (0, 1/2]$, define v_γ^* by setting, for any $x \in \mathcal{X} \setminus E$,

$$v_\gamma^*(x) := \lim_{r \to 0^+} m_u^\gamma(B(x,r)).$$

Then, for any $k \in \mathbb{N}$ with $k \geq 2$, $v_\gamma^* = v_k$ in $\mathcal{X} \setminus E$ and hence v_γ^* is continuous in $\mathcal{X} \setminus (E \cup B_{k,j})$ for any $j \in \mathbb{N}$. Notice that, by Lemma 6, for any $j \in \mathbb{N}$,

$$\mathrm{Cap}_{\mathcal{F}}(E \cup B_{k,j}) \lesssim 2^{-j}.$$

By choosing j big enough, we conclude that any function u^* satisfying $u^* = v_\gamma^*$ in $\mathcal{X} \setminus E$ is $\mathcal{F}$-quasi-continuous in $\mathcal{X}$ and hence the desired function in the present theorem.

Similar to the proof of Theorem 3, by Corollary 3, the proved conclusion for the case $u \in \mathcal{F}$, and Lemma 6, via choosing a sequence of Lipschitz continuous functions supported in balls, we obtain the desired conclusion of the present theorem when $u \in \dot{\mathcal{F}}$. This finishes the proof of Theorem 5. □

Remark 7. *With the same assumptions as in Theorem 5, by Lemma 14, (64), and Remark 4(ii), we have the following two observations:*

(i) *$u^* = u$ almost everywhere;*

(ii) *every point outside E is a generalized Lebesgue point of u.*

From (ii) and ([45], Lemma 17), we further deduce that, if $u \in \mathcal{F}$ is $\mathcal{F}$-quasi-continuous, then there exists a set E with $\mathrm{Cap}_{\mathcal{F}}(E) = 0$ such that every point outside E is a generalized Lebesgue point of u. This means that $\mathcal{F}$-quasi-continuous functions may have more Lebesgue points, compared with the functions that are only locally integrable.

In the following, we consider another technical tool, the generalized Hausdorff measure, which can also be applied to measure the exceptional set of (generalized) Lebesgue points. To see this, we study the comparison between the capacity and the above generalized Hausdorff measure. We refer the reader to [55,62,63] for more studies on the comparison between the capacity and the generalized Hausdorff measure, and to [64] for a study on measuring the exceptional set of Lebesgue points via the generalized Hausdorff measure straightly.

Let $h \in \mathcal{A}$, $\theta \in (0, 1]$, and $R \in (0, \infty]$. The *Netrusov–Hausdorff cocontent* $\mathcal{H}_R^{h,\theta}$, related to h, θ, and R, is defined by setting, for any $E \subset \mathcal{X}$,

$$\mathcal{H}_R^{h,\theta}(E) := \inf\left\{ \left[\sum_{i \in I} \left\{ \frac{\mu(B(x_i, r_i))}{h(r_i)} \right\}^\theta \right]^{1/\theta} : E \subset \bigcup_{i \in I} B(x_i, r_i),\ r_i \leq R \right\}, \tag{68}$$

where the infimum is taken over all coverings $\{B(x_i, r_i)\}_{i \in I}$ of E, and $I \subset \mathbb{N}$ an index set. Then, the *generalized Hausdorff measure* $\mathcal{H}^{h,\theta}(E)$, related to h and θ, is defined by setting, for any $E \subset \mathcal{X}$,

$$\mathcal{H}^{h,\theta}(E) := \limsup_{R \to 0^+} \mathcal{H}_R^{h,\theta}(E). \tag{69}$$

Recall that the Netrusov–Hausdorff content on $\mathbb{R}^n$ defined via the powers of the radius was first considered by Netrusov [65] and generalized to metric spaces via an increasing function h by Nuutinen ([55], Definition 5.1).

Observe that some lower bound and upper bound estimates for the $N_{p,q}^s$-capacity and the $M_{p,q}^s$-capacity with $p,\ q \in (0, \infty)$, in terms of the related Netrusov–Hausdorff contents, have been established, respectively, in ([55], Theorems 5.4 and 5.5) and ([63], Theorems 3.6 and 3.7) where $N_{p,q}^s$ and $M_{p,q}^s$ denote the classical fractional Hajłasz–Besov and Hajłasz–Triebel–Lizorkin spaces, respectively. By some arguments similar to those used in the proofs of ([55], Theorems 5.4 and 5.5) and ([63], Theorems 3.6 and 3.7), we have the following conclusions (Theorems 6 and 7) on the generalized spaces $M^{\phi}_{p,q}(\mathcal{X})$ and $N^{\phi}_{p,q}(\mathcal{X})$; we omit the details of their proofs.

Theorem 6. *Let $\phi \in \mathcal{A}_0$, $p \in (0,\infty)$, $q \in (0,\infty]$, $\theta := \min\{1, q/p\}$, $\mathcal{F} \in \{M^{\phi}_{p,q}(\mathcal{X}), N^{\phi}_{p,q}(\mathcal{X})\}$, and C_μ be as in (1). Then, there exists a positive constant $C = C_{(\mathcal{F})}$ such that, for any $E \subset \mathcal{X}$ and $R \in (0,\infty)$,*

$$\mathrm{Cap}_{\mathcal{F}}(E) \leq C\mathcal{H}^{h,\theta}_{R}(E), \tag{70}$$

where $\mathcal{H}^{h,\theta}_{R}$ is as in (68).

Remark 8. *Let $\phi(r) := r^s$ with $s \in (0,1)$ for any $r \in [0,\infty)$. In this case, (70) with $\mathcal{F} = N^{\phi}_{p,q}(\mathcal{X})$ becomes*

$$\mathrm{Cap}_{N^{\phi}_{p,q}(\mathcal{X})}(E) \lesssim \mathcal{H}^{h,\theta}_{R}(E)$$

with the implicit positive constant independent of E, which is just ([55] Theorem 5.4); moreover, taking $\mathcal{F} = M^{\phi}_{p,q}(\mathcal{X})$ and letting $R \to 0^+$ in (70), we obtain

$$\mathrm{Cap}_{M^{\phi}_{p,q}(\mathcal{X})}(E) \lesssim \mathcal{H}^{h,\theta}(E)$$

with the implicit positive constant independent of E, which is just ([63] Theorem 3.6), where $\mathcal{H}^{h,\theta}$ is as in (69).

Theorem 7. *Let $\phi \in \mathcal{A}_0$, p, $q \in (0,\infty)$, $\mathcal{F} \in \{M^{\phi}_{p,q}(\mathcal{X}), N^{\phi}_{p,q}(\mathcal{X})\}$, ω be any given function of admissible growth such that, for any $L \in \mathbb{Z}_+$,*

$$\sum_{k\geq L} \frac{1}{\omega(2^{-k})} < \infty;$$

and C_μ as in (1). Let $x_0 \in \mathcal{X}$, $R \in (0,1)$, and $B_0 := B(x_0, R)$. If there exist two positive constants $\kappa_1 \in (2,\infty)$ and $\kappa_2 \in (\kappa_1,\infty)$ such that $\kappa_2 B_0 \setminus \kappa_1 B_0 \neq \emptyset$, then there exist two positive constants τ and $C = C_{(\kappa_1,\kappa_2,R,\omega,\mathcal{F},C_\mu)}$ such that, for any compact set $E \subset B_0$,

$$\mathcal{H}^{h_\omega,1}_{\tau R}(E) \leq C\,\mathrm{Cap}_{\mathcal{F}}(E),$$

where, for any $r \in (0,R]$, $h_\omega(r) := [\phi(r)\omega(r)]^p$, and $\mathcal{H}^{h_\epsilon,1}_{\tau R}(E)$ is as in (68) with $\theta = 1$.

Remark 9. *Let $\phi(r) := r^s$ with $s \in (0,1)$ for any $r \in [0,\infty)$. When $\mathcal{F} = N^{\phi}_{p,q}(\mathcal{X})$, if h_ω, as in Theorem 7, satisfies that, for any $N \in \mathbb{Z}$,*

$$\int_0^{2^N} [h_\omega(t)]^{-1/p} t^{s-1}\, dt < \infty$$

(which is just the assumption in ([55], Theorem 5.5)), then, for any $L \in \mathbb{Z}_+$,

$$\begin{aligned}\sum_{k\geq L} \frac{1}{\omega(2^{-k})} &\sim \sum_{k\geq L} \int_{2^{-k-1}}^{2^{-k}} [\omega(t)]^{-1} t^{-1}\, dt \\ &\sim \int_0^{2^{-L}} [\omega(t)]^{-1} t^{-1}\, dt \\ &\sim \int_0^{2^{-L}} [h_\omega(t)]^{-1/p} t^{s-1}\, dt < \infty.\end{aligned}$$

Thus, Theorem 7 implies ([55], Theorem 5.5) with $\kappa_1 = 4$ and $\kappa_2 = 8$.

When $\mathcal{F} = N^{\phi}_{p,q}(\mathcal{X})$, *for any given* $\varepsilon \in (0,\infty)$, *let* $\omega(r) := [\log(1/r)]^{-1-\varepsilon/p}$ *for any* $r \in [0,\infty)$. *Obviously, we have*

$$\sum_{k\geq L}[\log(2^k)]^{-1-\varepsilon/p} < \infty.$$

Moreover, if $\mathrm{Cap}_{\mathcal{F}}(E) = 0$, *then, by* (71), *we obtain* $\mathcal{H}^{h_\omega,1}_{\infty}(E) = 0$, *which implies* ([63], *Theorem 3.7) with* $\kappa_1 = 4$ *and* $\kappa_2 = 8$, *where* $\mathcal{H}^{h_\omega,1}_{\infty}(E)$ *is as in* (68) *with* $R = \infty$.

Finally, we concentrate on the space $M^{\phi,p}(\mathcal{X})$ with $\alpha_\phi \in (0,1)$ and $p \in (D/(-\log_2 \alpha_\phi),\infty)$, where D is as in (1). We point out that, similarly to ([55], Theorems 5.4 and 5.5) and ([63], Theorems 3.6 and 3.7), the proofs of Theorems 6 and 7 rely on some equivalent characterizations of the related capacities $\mathrm{Cap}_{N^{\phi}_{p,q}(\mathcal{X})}$ and $\mathrm{Cap}_{M^{\phi}_{p,q}(\mathcal{X})}$, in which the counterpart for the capacity $\mathrm{Cap}_{M^{\phi,p}(\mathcal{X})}$ is unknown. Instead, we use Lemma 14 and the doubling property of the measure to obtain the following result.

Theorem 8. *Let* $\mathcal{F} = M^{\phi}_{p,\infty}(\mathcal{X}) = M^{\phi,p}(\mathcal{X})$ *with* $\phi \in \mathcal{A}_\infty$, $\alpha_\phi \in (0,1)$, $p \in (D/(-\log_2 \alpha_\phi),\infty)$, *and* D *and* C_μ *be as in* (1). *Let* B_0 *be a ball with the radius* $\widetilde{R_0} \in (0,\infty)$. *If there exist an* $R_0 \in (0,\widetilde{R_0}]$ *and a* $\tau \in (2,\infty)$ *such that, for any ball* $B \subset 2B_0$ *with the radius no more than* R_0, $\tau B \setminus 2B \neq \emptyset$, *then, for any compact set* $E \subset B_0$,

$$\mathrm{Cap}_{\mathcal{F}}(E) = 0 \iff \mathcal{H}^{h,1}(E) = 0, \tag{71}$$

where, for any $r \in (0,R_0]$, $h(r) := [\phi(r)]^p$, *and* $\mathcal{H}^{h,1}(E)$ *is as in* (69) *with* $\theta = 1$.

Proof. Let all the symbols be as in the present theorem and $L \in \mathbb{Z}$ such that $R_0 \in (2^{L-1},2^L]$.

We first prove $\mathcal{H}^{h,1}(E) = 0 \Longrightarrow \mathrm{Cap}_{\mathcal{F}}(E) = 0$. To this end, let $R \in (0,\min\{1,R_0\}]$ and $\{B(x_i,r_i) : r_i \leq R\}_{i\in\mathcal{I}}$ be a ball covering of E, where $\mathcal{I}$ is an index set. For any $i \in \mathcal{I}$, we let φ_i be an r_i^{-1}-Lipschitz function supported in $2B(x_i,r_i)$ such that $0 \leq \varphi_i \leq 1$ and $\varphi_i|_{B(x_i,r_i)} \equiv 1$. The existence of such $\{\varphi_i\}_{i\in\mathcal{I}}$ can be found in the proof of ([63], Theorem 3.6). For any $i \in \mathcal{I}$, by Definition 10; the continuity of φ_i; Corollary 4 with L^{-1} and E therein replaced, respectively, by r_i and $2B(x_i,r_i)$; the doubling property of μ; the definition of $\mathcal{A}$; and $r_i \leq 1$, we have

$$\begin{aligned}\mathrm{Cap}_{\mathcal{F}}(B(x_i,r_i)) &\leq \|2\varphi\|^p_{\mathcal{F}}\\ &\lesssim \left\{1+[\phi(r_i)]^{-1}\right\}^p \mu(2B(x_i,r_i))\\ &\lesssim [\phi(r_i)]^{-p}\mu(B(x_i,r_i))\end{aligned}$$

with the implicit positive constants independent of x_i and r_i. From this, Remark 4(iii), and Lemma 6 with $\theta = 1$ and E_i replaced by $B(x_i,r_i)$, we deduce that

$$\mathrm{Cap}_{\mathcal{F}}(E) \leq \mathrm{Cap}_{\mathcal{F}}\left(\bigcup_{i\in\mathcal{I}} B(x_i,r_i)\right) \lesssim \sum_{i\in\mathcal{I}} \frac{\mu(B(x_i,r_i))}{[\phi(r_i)]^p} \sim \sum_{i\in\mathcal{I}} \frac{\mu(B(x_i,r_i))}{h(r_i)},$$

which, combined with (68) with $\theta = 1$, implies that $\mathrm{Cap}_{\mathcal{F}}(E) \lesssim \mathcal{H}^{h,1}_R(E)$ with the implicit positive constant independent of R and E. Letting $R \to 0^+$, we obtain $\mathrm{Cap}_{\mathcal{F}}(E) \lesssim \mathcal{H}^{h,1}(E)$, which implies that, if $\mathcal{H}^{h,1}(E) = 0$, then $\mathrm{Cap}_{\mathcal{F}}(E) = 0$.

Conversely, if $\mathrm{Cap}_{\mathcal{F}}(E) = 0$, then by the definition of $\mathrm{Cap}_{\mathcal{F}}(E)$, we find that, for any given $\varepsilon \in (0,\infty)$, there exists a function v such that $v \geq 1$ in a neighborhood of E and

$$\|v\|^p_{M^{\phi,p}(\mathcal{X})} < \mathrm{Cap}_{\mathcal{F}}(E) + \varepsilon = \varepsilon. \tag{72}$$

For any given generalized Lebesgue point $x \in E$ and any given $k \in \mathbb{Z}$ with $k \geq -L+1$, take $B := B(x,2^{-k})$. Then $B \subset 2B_0$, which together with the assumption of the present theorem, means that $\tau B \setminus 2B \neq \emptyset$. Let φ be a Lipschitz function such that $\varphi|_B \equiv 1$ and

$\varphi|_{\mathcal{X}\setminus 2B} \equiv 0$. Define $u := v\varphi$. Then, by Lemma 7(ii) with E and u therein replaced, respectively, by $2B$ and v, we conclude that there exists a $g \in \mathcal{D}^{\phi}(u)$, supported in $2B$, such that

$$\|g\|_{L^p(\mathcal{X})} \lesssim \|v\|_{M^{\phi,p}(2B)}, \tag{73}$$

where the implicit positive constant depends only on ϕ, p, and K.

Since $\tau B \setminus 2B \neq \emptyset$, it follows that there always exists a point $z \in \tau B \setminus 2B$. Observe that, for any $y \in 2B$, we have $d(y,z) < (\tau+2)2^{-k}$, $u(z) = 0$, and $g(z) = 0$. Then, by the definition of $\mathcal{D}^{\phi}(u)$ and $\phi \in \mathcal{A}$, we conclude that, for almost every $y \in 2B$,

$$|u(y)| = \inf_{z\in\tau B\setminus 2B} |u(y) - u(z)| \le \inf_{z\in\tau B\setminus 2B} \phi(d(y,z))[g(y) + g(z)] \lesssim \phi(2^{-k})g(y),$$

which combined with (ii), (v), and (vi) of Lemma 13 and the doubling property of μ, implies that

$$|m_u^{\gamma}(B)| \le m_{|u|}^{\gamma}(B) \lesssim \phi(2^{-k})\, m_g^{\gamma}(B).$$

From this; the definition of the generalized Lebesgue point; the doubling property of μ; (iii), (iv), and (vi) of Lemma 13; (61); Lemma 15(ii) with $\lambda = p$; and (55) with $t = p$ and $E = B$, we deduce that, for the above given x,

$$\begin{aligned}
|u(x)| &\le |u(x) - m_u^{\gamma}(B)| + |m_u^{\gamma}(B)| \\
&\le \sum_{j\ge k-2} |m_u^{\gamma}(B(x,2^{-j-1})) - m_u^{\gamma}(B(x,2^{-j}))| + |m_u^{\gamma}(B)| \\
&\le \sum_{j\ge k-2} m^{\gamma/C_\mu}_{|u-m_u^{\gamma}(B(x,2^{-j}))|}(B(x,2^{-j})) + |m_u^{\gamma}(B)| \\
&\lesssim \sum_{j\ge k-2} \inf_{c\in\mathbb{R}}\Big[m^{\gamma/C_\mu}_{|u-c|}(B(x,2^{-j})) + m^{\gamma}_{|u-c|}(B(x,2^{-j}))\Big] + \phi(2^{-K})\, m_g^{\gamma}(B) \\
&\lesssim \sum_{j\ge k-2} \phi(2^{-j})\left\{\fint_{B(x,2^{-j+1})} [g(y)]^p\, d\mu(y)\right\}^{1/p} \\
&\quad + \phi(2^{-K})\left\{\fint_B [g(y)]^p\, d\mu(y)\right\}^{1/p} \\
&\lesssim \sum_{j\ge k-2} \phi(2^{-j})\left\{\fint_{B(x,2^{-j+1})} [g(y)]^p\, d\mu(y)\right\}^{1/p} \\
&\sim \sum_{j\ge k-2} \phi(2^{-j})\big[\mu(B(x,2^{-j+1}))\big]^{-1/p} \|g\|_{L^p(B(x,2^{-j+1}))}.
\end{aligned}$$

Using this and $u|_{E\cap B} \ge 1$, we conclude that, for this x,

$$1 \lesssim \sum_{j\ge k-2} \phi(2^{-j})\big[\mu(B(x,2^{-j+1}))\big]^{-1/p} \|g\|_{L^p(B(x,2^{-j+1}))}, \tag{74}$$

where the implicit positive constant is independent of x and k. Moreover, by the doubling property of μ, we find that, for any $j \ge k-2$,

$$\big[\mu(B(x,2^{-j+1}))\big]^{-1} \lesssim 2^{(j-k)D}\big[\mu(B(x,2^{-k+2}))\big]^{-1},$$

where the implicit positive constant depends only on C_μ. From this, (74), the fact that g is supported in $2B$, the doubling property of μ, $p \in (D/(-\log_2 \alpha_\phi), \infty)$, and Lemma 1(i) with $\varepsilon = D/p$, it follows that, for x and k as above,

$$1 \lesssim \sum_{j\ge k-2} 2^{(j-k)D/p}\phi(2^{-j})\big[\mu(B(x,2^{-k+2}))\big]^{-1/p} \|g\|_{L^p(B(x,2^{-k+1}))}$$

$$\lesssim \phi(2^{-k})\left[\mu(B(x,2^{-k}))\right]^{-1/p}\|g\|_{L^p(B(x,2^{-k+1}))}, \quad (75)$$

where the implicit positive constant is independent of x and k. By (73), (75), and the definition of h, we conclude that, for any given $k \in \mathbb{Z}$ with $k \geq -L+1$ and any generalized Lebesgue point $x \in E$,

$$\begin{aligned}\frac{\mu(B(x,2^{-k}))}{h(2^{-k})} &\lesssim \frac{[\phi(2^{-k})]^p}{h(2^{-k})}\|g\|^p_{L^p(B(x,2^{-k+1}))} \\ &\lesssim \frac{[\phi(2^{-k})]^p}{h(2^{-k})}\|v\|^p_{M^{\phi,p}(B(x,2^{-k+1}))} \\ &\sim \|v\|^p_{M^{\phi,p}(B(x,2^{-k+1}))}, \end{aligned} \quad (76)$$

where the implicit positive constants depend only on k, γ, ϕ, p, and C_μ.

Recall that, for any ball B' with the radius $r \in (0,\infty)$, $\mu(B') \in (0,\infty)$. Then, by Lemma 14, we have that, for any $k \in \mathbb{Z}$ with $k \geq -L+1$ and $x' \in E$, there always exists a generalized Lebesgue point y in $B(x', 2^{-k})$. Thus, $B(x', 2^{-k}) \subset B(y, 2^{-k+1})$ and $B(y, 2^{-k}) \subset B(x', 2^{-k+1})$. Using this, (76) with x therein replaced by y, the definition of h, $\phi \in \mathcal{A}$, and the doubling property of μ, we further conclude that, for any given $k \in \mathbb{Z}$ with $k \geq -L+1$ and any $x' \in E$,

$$\begin{aligned}\frac{\mu(B(x',2^{-k}))}{h(2^{-k})} &\leq \frac{\mu(B(y,2^{-k+1}))}{h(2^{-k})} \\ &\lesssim \frac{\mu(B(y,2^{-k-1}))}{h(2^{-k-1})} \lesssim \|v\|^p_{M^{\phi,p}(B(y,2^{-k}))} \lesssim \|v\|^p_{M^{\phi,p}(B(x',2^{-k+1}))}. \end{aligned} \quad (77)$$

For any given $R \in (0, 2^{L-1}]$, let $k_0 \in \mathbb{Z}$ be such that $2^{-k_0} \leq R < 2^{-k_0+1}$. Obviously, $\{B(x, 2^{-k_0}) : x \in E\}$ is a covering, consisting of balls with uniformly bounded diameter, of E. Thus, by a covering lemma for doubling metric spaces (see, for instance, ([66], Theorem 3.1.3) and ([67], Lemma 2.9)), we obtain a countable subfamily $\{B(x_i, 2^{-k_0}) : x_i \in E,\ i \in \mathcal{I}\}$ of disjoint balls with the radius no more than R such that

$$E \subset \bigcup_{i\in\mathcal{I}} 5B(x_i, 2^{-k_0}),$$

where $\mathcal{I}$ is an index set. From this, (68) with $\theta = 1$ and R replaced by $5R$, the doubling property of μ, $\phi \in \mathcal{A}$, (77) with $k = k_0+1$, the property of $\{B(x_i, 2^{-k_0}) : x_i \in E,\ i \in \mathcal{I}\}$, and (72), we deduce that

$$\begin{aligned}\mathcal{H}^{h,1}_{5R}(E) &\leq \sum_{i\in\mathcal{I}} \frac{\mu(5B(x_i,2^{-k_0}))}{h(5\cdot 2^{-k_0})} \\ &\lesssim \sum_{i\in\mathcal{I}} \frac{\mu(B(x_i,2^{-k_0-1}))}{h(2^{-k_0-1})} \lesssim \sum_{i\in\mathcal{I}} \|v\|^p_{M^{\phi,p}(B(x_i,2^{-k_0}))} \lesssim \|v\|^p_{M^{\phi,p}(\mathcal{X})} \lesssim \varepsilon, \end{aligned}$$

where the implicit positive constants depend only on ϕ, p, C_μ, and R. Letting $\varepsilon \to 0^+$, we then conclude that, for any $R \in (0, 2^{L-1}]$, $\mathcal{H}^{h,1}_{5R}(E) = 0$, which further implies that

$$\mathcal{H}^{h,1}(E) = \limsup_{R\to 0^+} \mathcal{H}^{h,1}_{5R}(E) = 0.$$

This finishes the proof of Theorem 8. □

Remark 10. *Let* $\mathcal{F}$ *and* h *be as in Theorem 8, and* D *and* C_μ *be as in (1). We point out that, by the proof of Theorem 8, the implication*

$$\mathrm{Cap}_{\mathcal{F}}(E) = 0 \Longrightarrow \mathcal{H}^{h,1}(E) = 0$$

holds true for any set $E \subset \mathcal{X}$.

Author Contributions: Conceptualization, Z.L., D.Y. and W.Y.; methodology, Z.L., D.Y. and W.Y.; software, Z.L., D.Y. and W.Y.; validation, Z.L., D.Y. and W.Y.; formal analysis, Z.L., D.Y. and W.Y.; investigation, Z.L., D.Y. and W.Y.; resources, Z.L., D.Y. and W.Y.; data curation, Z.L., D.Y. and W.Y.; writing—original draft preparation, Z.L., D.Y. and W.Y.; writing—review and editing, Z.L., D.Y. and W.Y.; visualization, Z.L., D.Y. and W.Y.; supervision, Z.L., D.Y. and W.Y.; project administration, Z.L., D.Y. and W.Y.; funding acquisition, Z.L., D.Y. and W.Y. All authors have read and agreed to the published version of the manuscript.

Funding: This research was funded by the National Key Research and Development Program of China (grant No. 2020YFA0712900) and the National Natural Science Foundation of China (grant Nos. 11971058, 12071197, 12122102, and 11871100).

Institutional Review Board Statement: Not applicable.

Informed Consent Statement: Not applicable.

Data Availability Statement: Not applicable.

Conflicts of Interest: The authors declare no conflict of interest.

References

1. Goldman, M.L. A description of the trace space for functions of a generalized Hölder class. *Dokl. Akad. Nauk. SSSR* **1976**, *231*, 525–528.
2. Goldman, M.L. The method of coverings for description of general spaces of Besov type. *Trudy Mat. Inst. Steklov.* **1980**, *156*, 47–81.
3. Kaljabin, G.A. Imbedding theorems for generalized Besov and Liouville spaces. *Dokl. Akad. Nauk SSSR* **1977**, *232*, 1245–1248.
4. Kaljabin, G.A. Descriptions of functions from classes of Besov–Lizorkin–Triebel type. *Trudy Mat. Inst. Steklov.* **1980**, *156*, 82–109.
5. Merucci, C. Applications of interpolation with a function parameter to Lorentz, Sobolev and Besov spaces. In *Interpolation Spaces and Allied Topics in Analysis (Lund, 1983)*; Lecture Notes in Math., 1070; Springer: Berlin, Germany, 1984; pp. 183–201.
6. Cobos, F.; Fernandez, D.L. Hardy-Sobolev spaces and Besov spaces with a function parameter. In *Function Spaces and Applications (Lund, 1986)*; Lecture Notes in Math., 1302; Springer: Berlin, Germany, 1988; pp. 158–170.
7. Leopold, H.-G. Embeddings and entropy numbers in Besov spaces of generalized smoothness. In *Function Spaces (Poznań, 1998)*; Lecture Notes in Pure and Appl. Math., 213; Dekker: New York, NY, USA, 2000; pp. 323–336.
8. Moura, S.D. Function spaces of generalised smoothness. *Dissertationes Math.* **2001**, *398*, 88. [CrossRef]
9. Triebel, H. *The Structure of Functions*; Monographs in Math., 97; Birkhäuser: Basel, Switzerland, 2001.
10. Farkas, W.; Jacob, N.; Schilling, R.L. Function spaces related to continuous negative definite functions: ψ-Bessel potential spaces. *Dissertationes Math.* **2001**, *393*, 62. [CrossRef]
11. Farkas, W.; Leopold, H.G. Characterisations of function spaces of generalized smoothness. *Ann. Mat. Pura Appl.* **2006**, *185*, 1–62. [CrossRef]
12. Caetano, A.M.; Moura, S.D. Local growth envelopes of spaces of generalized smoothness: The subcritical case. *Math. Nachr.* **2004**, *273*, 43–57. [CrossRef]
13. Caetano, A.M.; Moura, S.D. Local growth envelopes of spaces of generalized smoothness: The critical case. *Math. Inequal. Appl.* **2004**, *7*, 573–606. [CrossRef]
14. Caetano, A.M.; Farkas, W. Local growth envelopes of Besov spaces of generalized smoothness. *Z. Anal. Anwend.* **2006**, *25*, 265–298.
15. Caetano, A.M.; Leopold, H.G. Local growth envelopes of Triebel–Lizorkin spaces of generalized smoothness. *J. Fourier Anal. Appl.* **2006**, *12*, 427–445. [CrossRef]
16. Cobos, F.; Kühn, T. Approximation and entropy numbers in Besov spaces of generalized smoothness. *J. Approx. Theory* **2009**, *160*, 56–70. [CrossRef]
17. Opic, B.; Trebels, W. Sharp embeddings of Bessel potential spaces with logarithmic smoothness. *Math. Proc. Camb. Philos. Soc.* **2003**, *134*, 347–384. [CrossRef]
18. Gurka, P.; Opic, B. Sharp embeddings of Besov spaces with logarithmic smoothness. *Rev. Mat. Complut.* **2005**, *18*, 81–110. [CrossRef]
19. Caetano, A.M.; Gogatishvili, A.; Opic, B. Sharp embeddings of Besov spaces involving only logarithmic smoothness. *J. Approx. Theory* **2008**, *152*, 188–214. [CrossRef]

20. Cobos, F.; Domínguez, Ó. Embeddings of Besov spaces of logarithmic smoothness. *Studia Math.* **2014**, *223*, 193–204. [CrossRef]
21. Cobos, F.; Domínguez, Ó. On Besov spaces of logarithmic smoothness and Lipschitz spaces. *J. Math. Anal. Appl.* **2015**, *425*, 71–84. [CrossRef]
22. Cobos, F.; Domínguez, Ó. On the relationship between two kinds of Besov spaces with smoothness near zero and some other applications of limiting interpolation. *J. Fourier Anal. Appl.* **2016**, *22*, 1174–1191. [CrossRef]
23. Cobos, F.; Domínguez, Ó.; Triebel, H. Characterizations of logarithmic Besov spaces in terms of differences, Fourier-analytical decompositions, wavelets and semi-groups. *J. Funct. Anal.* **2016**, *270*, 4386–4425. [CrossRef]
24. Domínguez, Ó. Sharp embeddings of Besov spaces with logarithmic smoothness in sub-critical cases. *Anal. Math.* **2017**, *43*, 219–240. [CrossRef]
25. Besoy, B.; Cobos, F. Duality for logarithmic interpolation spaces when $0 < q < 1$ and applications. *J. Math. Anal. Appl.* **2018**, *466*, 373–399.
26. Domínguez, Ó.; Haroske, D.D.; Tikhonov, S. Embeddings and characterizations of Lipschitz spaces. *J. Math. Pures Appl.* **2020**, *144*, 69–105. [CrossRef]
27. Domínguez, Ó.; Tikhonov, S. Function Spaces of Logarithmic Smoothness: Embeddings and Characterizations. *arXiv* **2018**, arXiv: 1811.06399.
28. Li, Z.; Yang, D.; Yuan, W. Pointwise characterization of Besov and Triebel–Lizorkin spaces with generalized smoothness and their applications. *Acta Math. Sin. (Engl. Ser.)* **2021**, in press.
29. Hajłasz, P. Sobolev spaces on an arbitrary metric space. *Potential Anal.* **1996**, *5*, 403–415.
30. Franchi, B.; Hajłasz, P.; Koskela, P. Definitions of Sobolev classes on metric spaces. *Ann. Inst. Fourier (Grenoble)* **1999**, *49*, 1903–1924. [CrossRef]
31. Hajłasz, P. A new characterization of the Sobolev space. *Studia Math.* **2003**, *159*, 263–275. [CrossRef]
32. Hajłasz, P. Sobolev spaces on metric-measure spaces. In *Heat Kernels and Analysis on Manifolds, Graphs, and Metric Spaces (Paris, 2002)*; Contemp. Math., 338; American Mathematical Society: Providence, RI, USA, 2003; pp. 173–218.
33. Koskela, P.; Saksman, E. Pointwise characterizations of Hardy–Sobolev functions. *Math. Res. Lett.* **2008**, *15*, 727–744. [CrossRef]
34. Koskela, P.; Yang, D.; Zhou, Y. A characterization of Hajłasz–Sobolev and Triebel–Lizorkin spaces via grand Littlewood–Paley functions. *J. Funct. Anal.* **2010**, *258*, 2637–2661. [CrossRef]
35. Hu, J. A note on Hajłasz–Sobolev spaces on fractals. *J. Math. Anal. Appl.* **2003**, *280*, 91–101. [CrossRef]
36. Yang, D. New characterizations of Hajłasz–Sobolev spaces on metric spaces. *Sci. China Ser. A* **2003**, *46*, 675–689. [CrossRef]
37. Koskela, P.; Yang, D.; Zhou, Y. Pointwise characterizations of Besov and Triebel–Lizorkin spaces and quasiconformal mappings. *Adv. Math.* **2011**, *226*, 3579–3621. [CrossRef]
38. Kinnunen, J.; Latvala, V. Lebesgue points for Sobolev functions on metric spaces. *Rev. Mat. Iberoam.* **2002**, *18*, 685–700. [CrossRef]
39. Prokhorovich, M.A. Capacities and Lebesgue points for Hajłasz–Sobolev fractional classes on metric measure spaces. *Vestsī Nats. Akad. Navuk Belarusī Ser. Fīz. Mat. Navuk* **2006**, *124*, 19–23.
40. Mocanu, M. Lebesgue points for Orlicz–Sobolev functions on metric measure spaces. *An. Stiint. Univ. Al. I. Cuza Iasi. Mat. (N.S.)* **2011**, *57*, 175–186. [CrossRef]
41. Heikkinen, T.; Koskela, P.; Tuominen, H. Approximation and quasicontinuity of Besov and Triebel-Lizorkin functions. *Trans. Am. Math. Soc.* **2017**, *369*, 3547–3573. [CrossRef]
42. Harjulehto, P.; Hästö, P. Lebesgue points in variable exponent spaces. *Ann. Acad. Sci. Fenn. Math.* **2004**, *29*, 295–306.
43. Hakkarainen, H.; Nuortio, M. The variable exponent Sobolev capacity and quasi-fine properties of Sobolev functions in the case $p^- = 1$. *J. Math. Anal. Appl.* **2014**, *412*, 168–180. [CrossRef]
44. Karak, N. Generalized Lebesgue points for Sobolev functions. *Czechoslov. Math. J.* **2017**, *67*, 143–150. [CrossRef]
45. Heikkinen, T. Generalized Lebesgue points for Hajłasz functions. *J. Funct. Spaces* **2018**, *2018*, 5637042. [CrossRef]
46. Poelhuis, J.; Torchinsky, A. Medians, continuity, and vanishing oscillation. *Studia Math.* **2012**, *213*, 227–242. [CrossRef]
47. Heikkinen, T.; Ihnatsyeva, L.; Tuominen, H. Measure density and extension of Besov and Triebel-Lizorkin functions. *J. Fourier Anal. Appl.* **2016**, *22*, 334–382. [CrossRef]
48. Heinonen, J. *Lectures on Analysis on Metric Spaces*; Universitext.; Springer: New York, NY, USA, 2001; p. 140.
49. Kaljabin, G.A.; Lizorkin, P.I. Spaces of functions of generalized smoothness. *Math. Nachr.* **1987**, *133*, 7–32. [CrossRef]
50. Kokilashvili, V.; Meskhi, A.; Rafeiro, H.; Samko, S. Integral Operators in Non-standard Function Spaces. Vol. 1. Variable Exponent Lebesgue and Amalgam Spaces. In *Operator Theory: Advances and Applications, 248*; Birkhäuser/Springer: Cham, Switzerland, 2016; p. 567.
51. Kinnunen, J.; Martio, O. The Sobolev capacity on metric spaces. *Ann. Acad. Sci. Fenn. Math.* **1996**, *21*, 367–382.
52. Moura, S.D. On some characterizations of Besov spaces of generalized smoothness. *Math. Nachr.* **2007**, *280*, 1190–1199. [CrossRef]
53. Hajłasz, P.; Koskela, P. Sobolev Met Poincaré. *Mem. Am. Math. Soc.* **2000**, *145*, 101. [CrossRef]
54. Aalto, D.; Kinnunen, J. The discrete maximal operator in metric spaces. *J. Anal. Math.* **2010**, *111*, 369–390. [CrossRef]
55. Nuutinen, J. The Besov capacity in metric spaces. *Ann. Polon. Math.* **2016**, *117*, 59–78. [CrossRef]
56. Grafakos, L.; Liu, L.; Yang, D. Vector-valued singular integrals and maximal functions on spaces of homogeneous type. *Math. Scand.* **2009**, *104*, 296–310. [CrossRef]
57. Sawano, Y. Sharp estimates of the modified Hardy–Littlewood maximal operator on the nonhomogeneous space via covering lemmas. *Hokkaido Math. J.* **2005**, *34*, 435–458. [CrossRef]

58. Lappalainen, V. Lip$_h$-extension domains. *Ann. Acad. Sci. Fenn. Ser. A I Math. Dissertationes* **1985**, *56*, 52.
59. Shanmugalingam, N.; Yang, D.; Yuan, W. Newton–Besov spaces and Newton–Triebel–Lizorkin spaces on metric measure spaces. *Positivity* **2015**, *19*, 177–220. [CrossRef]
60. McShane, E.J. Extension of range of functions. *Bull. Am. Math. Soc.* **1934**, *40*, 837–842. [CrossRef]
61. Heikkinen, T.; Tuominen, H. Smoothing properties of the discrete fractional maximal operator on Besov and Triebel–Lizorkin spaces. *Publ. Mat.* **2014**, *58*, 379–399. [CrossRef]
62. Karak, N.; Koskela, P. Capacities and Hausdorff measures on metric spaces. *Rev. Mat. Complut.* **2015**, *28*, 733–740. [CrossRef]
63. Karak, N. Triebel–Lizorkin capacity and Hausdorff measure in metric spaces. *Math. Slovaca* **2020**, *70*, 617–624. [CrossRef]
64. Karak, N.; Koskela, P. Lebesgue points via the Poincaré inequality. *Sci. China Math.* **2015**, *58*, 1697–1706. [CrossRef]
65. Netrusov, Y.V. Estimates of capacities associated with Besov spaces. (Russian) *Zap. Nauchn. Sem. S. Peterburg. Otdel. Mat. Inst. Steklov. (POMI) 201 (1992), Issled. po Lineĭn. Oper. Teor. Funktsiĭ. 20*, 124–156. 191. *J. Math. Sci.* **1996**, *78*, 199–217. [CrossRef]
66. Coifman, R.R.; Weiss, G. *Analyse Harmonique Non-Commutative sur Certains Espaces Homogènes. (French) Étude de Certaines Intégrales Singulières*; Lecture Notes in Math., 242; Springer: Berlin, Germany; New York, NY, USA, 1971; p. 160.
67. Macías, R.A.; Segovia, C. A decomposition into atoms of distributions on spaces of homogeneous type. *Adv. Math.* **1979**, *33*, 271–309. [CrossRef]

mathematics

MDPI

Article

Spaces of Pointwise Multipliers on Morrey Spaces and Weak Morrey Spaces

Eiichi Nakai [1] and Yoshihiro Sawano [2,*]

[1] Department of Mathematics, Ibaraki University, Mito 310-8512, Ibaraki, Japan; eiichi.nakai.math@vc.ibaraki.ac.jp
[2] Department of Mathematics, Chuo University, 1-13-27, Kasuga 112-8551, Bunkyo, Japan
* Correspondence: yoshihiro-sawano@celery.ocn.ne.jp

Abstract: The spaces of pointwise multipliers on Morrey spaces are described in terms of Morrey spaces, their preduals, and vector-valued Morrey spaces introduced by Ho. This paper covers weak Morrey spaces as well. The result in the present paper completes the characterization of the earlier works of the first author's papers written in 1997 and 2000, as well as Lemarié-Rieusset's 2013 paper. As a corollary, the main result in the present paper shows that different quasi-Banach lattices can create the same vector-valued Morrey spaces. The goal of the present paper is to provide a complete picture of the pointwise multiplier spaces.

Keywords: pointwise multipliers; Morrey spaces; block spaces; convexification

MSC: 42B35; 26A33

Citation: Nakai, E.; Sawano, Y. Spaces of Pointwise Multipliers on Morrey Spaces and Weak Morrey Spaces. *Mathematics* **2021**, *9*, 2754. https://doi.org/10.3390/math9212754

Academic Editor: Maria C. Mariani

Received: 5 September 2021
Accepted: 24 October 2021
Published: 29 October 2021

1. Introduction

The aim of this note is to consider spaces of pointwise multipliers on Morrey spaces and weak Morrey spaces. Our results supplement the ones in [1–4]. We state our main results in Section 2. Section 1 is devoted to the formulation of the results.

We denote by $L^0(\mathbb{R}^n)$ the space of all measurable functions from $\mathbb{R}^n$ to $\mathbb{R}$ or $\mathbb{C}$. Let $E_1, E_2 \subset L^0(\mathbb{R}^n)$ be linear subspaces. We say that a function $g \in L^0(\mathbb{R}^n)$ is a pointwise multiplier from E_1 to E_2, if the pointwise multiplication $f \cdot g$ is in E_2 for any $f \in E_1$. We denote by $\mathrm{PWM}(E_1, E_2)$ the set of all pointwise multipliers from E_1 to E_2. We abbreviate this as $\mathrm{PWM}(E, E)$ to $\mathrm{PWM}(E)$.

For $p \in (0, \infty]$, $L^p(\mathbb{R}^n)$ denotes the usual Lebesgue space equipped with the norm $\|\cdot\|_{L^p}$. It is well known by Hölder's inequality that:

$$\|f \cdot g\|_{L^{p_2}} \le \|f\|_{L^{p_1}} \|g\|_{L^{p_3}} \quad (f \in L^{p_1}(\mathbb{R}^n), g \in L^{p_3}(\mathbb{R}^n))$$

for $1/p_2 = 1/p_1 + 1/p_3$ with $p_j \in (0, \infty]$, $j = 1, 2, 3$, so that $p_1 \ge p_2$. This shows that:

$$\mathrm{PWM}(L^{p_1}(\mathbb{R}^n), L^{p_2}(\mathbb{R}^n)) \supset L^{p_3}(\mathbb{R}^n).$$

Conversely, we can show the reverse inclusion by using the uniform boundedness theorem or the closed graph theorem, that is,

$$\mathrm{PWM}(L^{p_1}(\mathbb{R}^n), L^{p_2}(\mathbb{R}^n)) = L^{p_3}(\mathbb{R}^n). \tag{1}$$

In particular, if $p_1 = p_2 = p$, then:

$$\mathrm{PWM}(L^p(\mathbb{R}^n)) = L^\infty(\mathbb{R}^n). \tag{2}$$

Meanwhile, if $p_1 < p_2$, then:

$$\mathrm{PWM}(L^{p_1}(\mathbb{R}^n), L^{p_2}(\mathbb{R}^n)) = \{0\} \tag{3}$$

since $L^{p_1}_{\mathrm{loc}}(\mathbb{R}^n)$ is not included in $L^{p_2}_{\mathrm{loc}}(\mathbb{R}^n)$. Proofs of (1) and (2) can be found in the work of Maligranda and Persson [5], Proposition 3 and Theorem 1. See also [4]. We do not prove (3) directly in this paper, but we mention that (3) is a direct consequence in Section 2. The goal of this note is to generalize this observation to Morrey spaces motivated by the works [2–4,6]. For $p \in (0, \infty)$ and $\lambda \in [0, n]$, the (classical/strong) Morrey space $L^{p,\lambda}(\mathbb{R}^n)$ is defined as the space of $f \in L^0(\mathbb{R}^n)$ such that:

$$\|f\|_{L^{p,\lambda}} = \sup_{Q \in \mathcal{Q}} \left(\frac{1}{|Q|^{\frac{\lambda}{n}}} \int_Q |f(y)|^p \, dy \right)^{1/p} < \infty, \tag{4}$$

where $\mathcal{Q}$ stands for the set of all cubes in $\mathbb{R}^n$ whose edges are parallel to the coordinate axes. The parameter p serves to describe the local integrability of functions, while λ describes the growth of $\int_Q |f(y)|^p dy$ in comparison with $|Q|$. It is easy to see that $L^{p,\lambda}(\mathbb{R}^n)$ is a quasi-Banach space, which is subject to the scaling law $\|f(t\cdot)\|_{L^{p,\lambda}} = t^{-\frac{n-\lambda}{p}} \|f\|_{L^{p,\lambda}}$ for all $f \in L^{p,\lambda}(\mathbb{R}^n)$ and $t > 0$. The notation $L^{p,\lambda}(\mathbb{R}^n)$ was used, for instance, by Peetre [7]. The weak Morrey space $\mathrm{w}L^{p,\lambda}(\mathbb{R}^n)$ is defined by a routine procedure: The weak Morrey space $\mathrm{w}L^{p,\lambda}(\mathbb{R}^n)$ is the set of all measurable functions $f \in L^0(\mathbb{R}^n)$ for which $\|f\|_{\mathrm{w}L^{p,\lambda}} = \sup_{\lambda > 0} \lambda \|\chi_{(\lambda,\infty]}(|f|)\|_{L^{p,\lambda}}$ is finite, where χ_A stands for the characteristic function of the set A.

To describe various properties of functions in $L^{p,\lambda}(\mathbb{R}^n)$, it is sometimes convenient to use the notation $\mathcal{M}^p_q(\mathbb{R}^n)$. Let $0 < q \le p \le \infty$. Recall that for an $L^q_{\mathrm{loc}}(\mathbb{R}^n)$-function f, its Morrey norm $\|f\|_{\mathcal{M}^p_q}$ is defined by:

$$\|f\|_{\mathcal{M}^p_q} \equiv \sup_{Q \in \mathcal{Q}} |Q|^{\frac{1}{p} - \frac{1}{q}} \left(\int_Q |f(y)|^q \mathrm{d}y \right)^{\frac{1}{q}}. \tag{5}$$

The Morrey space $\mathcal{M}^p_q(\mathbb{R}^n)$ is the set of all $L^q(\mathbb{R}^n)$-locally integrable functions f for which the norm $\|f\|_{\mathcal{M}^p_q}$ is finite. Once again, by the routine procedure, we define the weak Morrey space $\mathrm{w}\mathcal{M}^p_q(\mathbb{R}^n)$ as the set of all measurable functions $f \in L^0(\mathbb{R}^n)$ for which $\|f\|_{\mathrm{w}\mathcal{M}^p_q} = \sup_{\lambda > 0} \lambda \|\chi_{(\lambda,\infty]}(|f|)\|_{\mathcal{M}^p_q}$ is finite. The parameter q describes the local integrability of functions. As is seen from the scaling law $\|f(t\cdot)\|_{\mathcal{M}^p_q} = t^{-\frac{n}{p}} \|f\|_{\mathcal{M}^p_q}$ for all $f \in \mathcal{M}^p_q(\mathbb{R}^n)$ and $t > 0$, the parameter p in the Morrey space $\mathcal{M}^p_q(\mathbb{R}^n)$ describes the global integrability. We remark that some authors swap the role of p and q; see [6] for example.

By (4) and (5), we have:

$$L^{q,\lambda}(\mathbb{R}^n) = \mathcal{M}^p_q(\mathbb{R}^n), \quad \text{if} \quad \lambda = n(1 - q/p) \quad \text{or equivalently} \quad p = \frac{qn}{n - \lambda}.$$

Let $0 < p < \infty$. It is noteworthy that $L^{p,0}(\mathbb{R}^n) = \mathcal{M}^p_p(\mathbb{R}^n) = L^p(\mathbb{R}^n)$ and that $L^{p,n}(\mathbb{R}^n) = \mathcal{M}^\infty_p(\mathbb{R}^n) = L^\infty(\mathbb{R}^n)$, so that Morrey spaces generalize Lebesgue spaces.

Let $0 < q_i \le p_i < \infty$, $i = 1, 2$. We consider the space of pointwise multipliers from $\mathcal{M}^{p_1}_{q_1}(\mathbb{R}^n)$ to $\mathcal{M}^{p_2}_{q_2}(\mathbb{R}^n)$. A direct consequence of the closed graph theorem is that there exists a constant $M > 0$ such that, for $f \in \mathcal{M}^{p_1}_{q_1}(\mathbb{R}^n)$ and $g \in \mathrm{PWM}(\mathcal{M}^{p_1}_{q_1}(\mathbb{R}^n), \mathcal{M}^{p_2}_{q_2}(\mathbb{R}^n))$,

$$\|f \cdot g\|_{\mathcal{M}^{p_2}_{q_2}} \le M \|f\|_{\mathcal{M}^{p_1}_{q_1}}. \tag{6}$$

One naturally defines a norm on $\mathrm{PWM}(\mathcal{M}^{p_1}_{q_1}(\mathbb{R}^n), \mathcal{M}^{p_2}_{q_2}(\mathbb{R}^n))$ by:

$$\|g\|_{\mathrm{PWM}(\mathcal{M}^{p_1}_{q_1}, \mathcal{M}^{p_2}_{q_2})} \equiv \inf\{M > 0 : (6) \text{ holds for all } f \in \mathcal{M}^{p_1}_{q_1}(\mathbb{R}^n)\}$$

for $g \in \mathrm{PWM}(\mathcal{M}^{p_1}_{q_1}(\mathbb{R}^n), \mathcal{M}^{p_2}_{q_2}(\mathbb{R}^n))$. In the following, unless otherwise stated, the equality:

$$\mathrm{PWM}(E_1, E_2) = E_3$$

tacitly means the norm equivalence, that is a function $g \in L^0(\mathbb{R}^n)$ belongs to E_3 if and only if $g \in \mathrm{PWM}(E_1, E_2)$, and in this case:

$$\|g\|_{\mathrm{PWM}(E_1,E_2)} \sim \|g\|_{E_3},$$

where the implicit constants in $\sim$ do not depend on g. It follows from the scaling law of Morrey spaces that:

$$\|g(t\cdot)\|_{\mathrm{PWM}(\mathcal{M}^{p_1}_{q_1},\mathcal{M}^{p_2}_{q_2})} = t^{-\frac{n}{p_2}+\frac{n}{p_1}} \|g\|_{\mathrm{PWM}(\mathcal{M}^{p_1}_{q_1},\mathcal{M}^{p_2}_{q_2})}$$

for all $g \in \mathrm{PWM}(\mathcal{M}^{p_1}_{q_1}(\mathbb{R}^n), \mathcal{M}^{p_2}_{q_2}(\mathbb{R}^n))$.

An easy consequence of Hölder's inequality is that:

$$\|f \cdot g\|_{L^{p_2,\lambda_2}} \le \|f\|_{L^{p_1,\lambda_1}} \|g\|_{L^{p_3,\lambda_3}},$$

if $p_j \in (0,\infty)$ and $\lambda_j \in [0,n]$, $j = 1,2,3$ satisfy $1/p_2 = 1/p_1 + 1/p_3$ and $\lambda_2/p_2 = \lambda_1/p_1 + \lambda_3/p_3$. This shows that:

$$\mathrm{PWM}(L^{p_1,\lambda_1}(\mathbb{R}^n), L^{p_2,\lambda_2}(\mathbb{R}^n)) \supset L^{p_3,\lambda_3}(\mathbb{R}^n). \quad (7)$$

Therefore, the aim of this note is to investigate the difference between the two spaces above. It is important to note that the scaling laws considered above force the parameters p_1, p_2, p_3 to satisfy $\lambda_2/p_2 = \lambda_1/p_1 + \lambda_3/p_3$.

In this paper, we describe $\mathrm{PWM}(L^{p_1,\lambda_1}(\mathbb{R}^n), L^{p_2,\lambda_2}(\mathbb{R}^n))$ for all parameters $p_j \in (0,\infty)$ and $\lambda_j \in [0,n)$, $j = 1,2$. Of interest is the case where $\lambda_2 < \lambda_1$, since we already specified $\mathrm{PWM}(L^{p_1,\lambda_1}(\mathbb{R}^n), L^{p_2,\lambda_2}(\mathbb{R}^n))$ in the case $\lambda_1 \le \lambda_2$ in our earlier paper [3].

Theorem 1 ([3], Corollary 2.4). *Let $p_i \in (0,\infty)$ and $\lambda_i \in [0,n)$, $i = 1,2$. Then:*

$$\mathrm{PWM}(L^{p_1,\lambda_1}(\mathbb{R}^n),\ L^{p_2,\lambda_2}(\mathbb{R}^n)) \begin{cases} = \{0\}, & p_1 < p_2 \text{ or } n + (\lambda_1 - n)\frac{p_2}{p_1} < \lambda_2, \\ = L^\infty(\mathbb{R}^n), & p_2 \le p_1 \text{ and } \lambda_2 = n + (\lambda_1 - n)\frac{p_2}{p_1}, \\ = L^{p_3,\lambda_3}(\mathbb{R}^n), & p_2 < p_1 \text{ and } \lambda_1 \le \lambda_2 < n + (\lambda_1 - n)\frac{p_2}{p_1}, \\ \supsetneq L^{p_3,\lambda_3}(\mathbb{R}^n), & p_2 < p_1 \text{ and } 0 < \lambda_1\frac{p_2}{p_1} \le \lambda_2 < \lambda_1, \\ \supsetneq \{0\}, & p_2 \le p_1 \text{ and } 0 \le \lambda_2 < \lambda_1\frac{p_2}{p_1}, \end{cases}$$

where $p_3 = p_1p_2/(p_1 - p_2)$ and $\lambda_3 = (p_1\lambda_2 - p_2\lambda_1)/(p_1 - p_2)$.

Let $p_i \in (0,\infty)$, $i = 1,2$. As the endpoint cases of $\lambda_1 = n$ or/and $\lambda_2 = n$, we have:

$$\mathrm{PWM}(L^{p_1,\lambda_1}(\mathbb{R}^n),\ L^{p_2,\lambda_2}(\mathbb{R}^n)) \begin{cases} = \{0\}, & 0 \le \lambda_1 < \lambda_2 = n, \\ = L^{p_2,\lambda_2}(\mathbb{R}^n), & 0 \le \lambda_2 \le \lambda_1 = n. \end{cases}$$

We rephrase Theorem 1 as follows:

Theorem 2. *Let* $0 < q_i \le p_i < \infty$, $i = 1,2$. *Then:*

$$\mathrm{PWM}(\mathcal{M}^{p_1}_{q_1}(\mathbb{R}^n), \mathcal{M}^{p_2}_{q_2}(\mathbb{R}^n)) \begin{cases} = \{0\}, & q_1 < q_2 \text{ or } p_1 < p_2, \\ = L^\infty(\mathbb{R}^n), & q_2 \le q_1 \text{ and } p_1 = p_2, \\ = \mathcal{M}^{p_3}_{q_3}(\mathbb{R}^n), & q_2 < q_1 \text{ and } p_1 q_2/q_1 \le p_2 < p_1, \\ \supsetneq \mathcal{M}^{p_3}_{q_3}(\mathbb{R}^n), & q_2 < q_1 < p_1 \text{ and } 1/(1/p_1 + 1/q_2 - 1/q_1) \le p_2 < p_1 q_2/q_1, \\ \supsetneq \{0\}, & q_2 \le q_1 < p_1 \text{ and } p_2 < 1/(1/p_1 + 1/q_2 - 1/q_1), \end{cases}$$

where $q_3 = q_1 q_2/(q_1 - q_2)$ *and* $p_3 = p_1 p_2/(p_1 - p_2)$.

We have notation for the scale $L^{p,\lambda}(\mathbb{R}^n)$ analogous to the scale $\mathcal{M}^p_q(\mathbb{R}^n)$. We may also replace $\mathcal{M}^{p_1}_{q_1}(\mathbb{R}^n)$ and/or $\mathcal{M}^{p_2}_{q_2}(\mathbb{R}^n)$ by $\mathrm{w}\mathcal{M}^{p_1}_{q_1}(\mathbb{R}^n)$ and/or $\mathrm{w}\mathcal{M}^{p_2}_{q_2}(\mathbb{R}^n)$ to define the corresponding multiplier spaces. According to [1], we have a counterpart of Theorem 1 to weak Morrey spaces: we can replace $L^{p_i,\lambda_i}(\mathbb{R}^n)$ by $\mathrm{w}L^{p_i,\lambda_i}(\mathbb{R}^n)$ in Theorem 1 and $\mathcal{M}^{p_i}_{q_i}(\mathbb{R}^n)$ by $\mathrm{w}\mathcal{M}^{p_i}_{q_i}(\mathbb{R}^n)$ in Theorem 2. As for weak Morrey spaces, the following results were obtained in [1].

Theorem 3 ([8], Corollary 3). *The same conclusion as Theorem 1 remains valid if we replace* $L^{p_k,\lambda_k}(\mathbb{R}^n)$ *by* $\mathrm{w}L^{p_k,\lambda_k}(\mathbb{R}^n)$ *for* $k = 1,2,3$. *As a result, the same conclusion as Theorem 2 remains valid if we replace* $\mathcal{M}^{p_k}_{q_k}(\mathbb{R}^n)$ *by* $\mathrm{w}\mathcal{M}^{p_k}_{q_k}(\mathbb{R}^n)$ *for* $k = 1,2,3$.

It is interesting to compare these results with the following endpoint cases:

$$\mathrm{PWM}(L^\infty(\mathbb{R}^n), \mathcal{M}^p_q(\mathbb{R}^n)) = \mathcal{M}^p_q(\mathbb{R}^n),$$
$$\mathrm{PWM}(L^\infty(\mathbb{R}^n), \mathrm{w}\mathcal{M}^p_q(\mathbb{R}^n)) = \mathrm{w}\mathcal{M}^p_q(\mathbb{R}^n),$$
$$\mathrm{PWM}(\mathcal{M}^p_q(\mathbb{R}^n), L^\infty(\mathbb{R}^n)) = \mathrm{PWM}(\mathrm{w}\mathcal{M}^p_q(\mathbb{R}^n), L^\infty(\mathbb{R}^n)) = \{0\}$$

for all $0 < q \le p < \infty$.

The goal of this note is to give complete characterizations of:

$$\mathrm{PWM}(\mathcal{M}^{p_1}_{q_1}(\mathbb{R}^n), \mathcal{M}^{p_2}_{q_2}(\mathbb{R}^n))$$

including

$$\mathrm{PWM}(\mathrm{w}\mathcal{M}^{p_1}_{q_1}(\mathbb{R}^n), \mathcal{M}^{p_2}_{q_2}(\mathbb{R}^n)),$$
$$\mathrm{PWM}(\mathcal{M}^{p_1}_{q_1}(\mathbb{R}^n), \mathrm{w}\mathcal{M}^{p_2}_{q_2}(\mathbb{R}^n))$$

and:

$$\mathrm{PWM}(\mathrm{w}\mathcal{M}^{p_1}_{q_1}(\mathbb{R}^n), \mathrm{w}\mathcal{M}^{p_2}_{q_2}(\mathbb{R}^n)).$$

Here are tables of the characterization of these spaces. For example, in Table 1, we deal with the case of $p_1 > p_2$ and $q_1 > q_2$ in Theorem 4 to follow.

Table 1. $\mathrm{PWM}(\mathcal{M}^{p_1}_{q_1}(\mathbb{R}^n), \mathcal{M}^{p_2}_{q_2}(\mathbb{R}^n))$.

	$p_1 < p_2$	$p_1 = p_2$	$p_1 > p_2$
$q_1 < q_2$	Theorem 2	Theorem 2	Theorem 2
$q_1 = q_2$	Theorem 2	Theorem 2	Theorem 4
$q_1 > q_2$	Theorem 2	Theorem 2	Theorem 4

The remaining part of this paper is organized as follows: In Section 2, we present our main results summarized as Tables 1–4. Section 3 deals with preliminary and general facts of the multiplier spaces. Section 4 is devoted to the proof of the results summarized in the tables above.

Table 2. $\mathrm{PWM}(\mathrm{w}\mathcal{M}^{p_1}_{q_1}(\mathbb{R}^n), \mathcal{M}^{p_2}_{q_2}(\mathbb{R}^n))$.

	$p_1 < p_2$	$p_1 = p_2$	$p_1 > p_2$
$q_1 < q_2$	Proposition 1, 1 and 2.	Proposition 1, 2.	Proposition 1, 2.
$q_1 = q_2$	Proposition 1, 1.	Propositions 1, 3.	Theorem 6
$q_1 > q_2$	Proposition 1, 1.	Proposition 2	Theorem 6

Table 3. $\mathrm{PWM}(\mathcal{M}^{p_1}_{q_1}(\mathbb{R}^n), \mathrm{w}\mathcal{M}^{p_2}_{q_2}(\mathbb{R}^n))$.

	$p_1 < p_2$	$p_1 = p_2$	$p_1 > p_2$
$q_1 < q_2$	Proposition 3, 1.	Proposition 3, 1.	Proposition 3, 1.
$q_1 = q_2$	Proposition 3, 1.	Proposition 3, 2.	Theorem 7
$q_1 > q_2$	Proposition 3, 1.	Proposition 3, 2.	Theorem 7

Table 4. $\mathrm{PWM}(\mathrm{w}\mathcal{M}^{p_1}_{q_1}(\mathbb{R}^n), \mathrm{w}\mathcal{M}^{p_2}_{q_2}(\mathbb{R}^n))$.

	$p_1 < p_2$	$p_1 = p_2$	$p_1 > p_2$
$q_1 < q_2$	Theorem 3	Theorem 3	Theorem 3
$q_1 = q_2$	Theorem 3	Theorem 3	Theorem 8
$q_1 > q_2$	Theorem 3	Theorem 3	Theorem 8

2. Main Results

2.1. *Characterization of* $\mathrm{PWM}(\mathcal{M}^{p_1}_{q_1}(\mathbb{R}^n), \mathcal{M}^{p_2}_{q_2}(\mathbb{R}^n))$

To characterize the pointwise multiplier space $\mathrm{PWM}(\mathcal{M}^{p_1}_{q_1}(\mathbb{R}^n), \mathcal{M}^{p_2}_{q_2}(\mathbb{R}^n))$, we recall a couple of notions in [9,10].

A quasi-Banach (resp. Banach) lattice on $\mathbb{R}^n$ is a nonzero quasi-Banach (resp. Banach) space $(E, \|\cdot\|)$ contained in $L^0(\mathbb{R}^n)$ such that $\|f\|_E \le \|g\|_E$ holds for all $f, g \in E$ such that $|f| \le |g|$. Let $u \in (0, \infty)$. For a quasi-Banach lattice $E \subset L^0(\mathbb{R}^n)$, we define its u-convexification E^u by:

$$E^u \equiv \{f : |f|^u \in E\}, \quad \|f\|_{E^u} \equiv (\||f|^u\|_E)^{1/u}.$$

For example, $(L^1(\mathbb{R}^n))^p = L^p(\mathbb{R}^n)$.

We next recall the notion of block spaces introduced by Long [10].

Definition 1. *Let $1 \le q \le p < \infty$. A function $A \in L^0(\mathbb{R}^n)$ is a (p,q)-block if there exists a cube Q that supports A and:*

$$\|A\|_{L^{q'}} \le |Q|^{\frac{1}{q'} - \frac{1}{p'}}, \tag{8}$$

where p' and q' stand for the conjugate exponent of p and q, respectively. If we need to specify Q, then we say that b is a (p,q)-block supported on Q. Let $1 \le q \le p < \infty$, and define the block space $\mathcal{H}^{p'}_{q'}(\mathbb{R}^n)$ as the set of all $f \in L^{p'}(\mathbb{R}^n)$ for which f is realized as the sum $f = \sum_{j=1}^{\infty} \tau_j A_j$ with some $\{\tau_j\}_{j=1}^{\infty} \in \ell^1(\mathbb{N})$ and some sequence $\{A_j\}_{j=1}^{\infty}$ of (p,q)-blocks. Define the norm $\|f\|_{\mathcal{H}^{p'}_{q'}}$ for $f \in \mathcal{H}^{p'}_{q'}(\mathbb{R}^n)$ as:

$$\|f\|_{\mathcal{H}^{p'}_{q'}} \equiv \inf_{\tau} \|\tau\|_{\ell^1}, \tag{9}$$

where $\tau = \{\tau_j\}_{j=1}^{\infty}$ runs over all admissible expressions as above.

Finally, to state our result, we recall the definition of vector-valued Morrey spaces proposed by Ho [9]. Let $E(\mathbb{R}^n) \subset L^0(\mathbb{R}^n)$ be a quasi-Banach lattice, and let $p > 0$. Then, the E-based vector-valued Morrey space $\mathcal{M}^p_E(\mathbb{R}^n)$ is the set of all $f \in L^0(\mathbb{R}^n)$ for which:

$$\|f\|_{\mathcal{M}^p_E} \equiv \sup_{Q\in\mathcal{Q}} |Q|^{\frac{1}{p}} \frac{\|\chi_Q f\|_E}{\|\chi_Q\|_E}$$

is finite.

Recall that a quasi-Banach lattice E enjoys the Fatou property if $\sup_{j\in\mathbb{N}} f_j \in E$ and $\lim_{j\to\infty} \|f_j\|_E = \left\|\lim_{j\in\mathbb{N}} f_j\right\|$ for any sequence $\{f_j\}_{j=1}^{\infty}$ in E satisfying $0 \le f_1 \le f_2 \le \cdots$. We make a brief remark on the relation among these notions introduced above.

Remark 1. *If $\|\chi_Q\|_E = |Q|^{\frac{1}{p}}$ for all cubes Q and if E has the Fatou property, then a simple observation shows $\mathcal{M}^p_E(\mathbb{R}^n) = E(\mathbb{R}^n)$ with the equivalence of norms. In particular, If $E(\mathbb{R}^n) = \mathcal{H}^{p'}_{q'}(\mathbb{R}^n)$, then $\mathcal{M}^{p'}_E(\mathbb{R}^n) = E(\mathbb{R}^n)$.*

We provide a complete picture of the description of $\mathrm{PWM}(\mathcal{M}^{p_1}_{q_1}(\mathbb{R}^n), \mathcal{M}^{p_2}_{q_2}(\mathbb{R}^n))$.

Theorem 4. *Let $0 < q_i \le p_i < \infty$, $i = 1, 2$:*

1. *If $q_1 < q_2$ or $p_1 < p_2$, then $\mathrm{PWM}(\mathcal{M}^{p_1}_{q_1}(\mathbb{R}^n), \mathcal{M}^{p_2}_{q_2}(\mathbb{R}^n)) = \{0\}$;*
2. *If $q_1 \ge q_2$ and $p_1 = p_2$, then $\mathrm{PWM}(\mathcal{M}^{p_1}_{q_1}(\mathbb{R}^n), \mathcal{M}^{p_2}_{q_2}(\mathbb{R}^n)) = L^\infty(\mathbb{R}^n)$;*
3. *If $q_1 \ge q_2$ and $p_1 > p_2$, then $\mathrm{PWM}(\mathcal{M}^{p_1}_{q_1}(\mathbb{R}^n), \mathcal{M}^{p_2}_{q_2}(\mathbb{R}^n)) = \mathcal{M}^{p_3}_{X^{q_2}}(\mathbb{R}^n)$, where p_3 and X are given by:*

$$p_3 = \frac{p_1 p_2}{p_1 - p_2}, \quad X = \mathcal{H}^{\left(\frac{p_1}{q_2}\right)'}_{\left(\frac{q_1}{q_2}\right)'}(\mathbb{R}^n).$$

In particular,

$$\mathrm{PWM}(\mathcal{M}^{p_1}_{q_1}(\mathbb{R}^n), \mathcal{M}^{p_2}_{1}(\mathbb{R}^n)) = \mathcal{M}^{p_3}_{\mathcal{H}^{p_1'}_{q_1'}}(\mathbb{R}^n),$$

where p_3 is defined by $p_3 = \frac{p_1 p_2}{p_1 - p_2}$.

It is significant that Theorem 4 does not require $\frac{q_1}{p_1} \ge \frac{q_2}{p_2}$, unlike Theorem 2. We give an equivalent form using the scale $L^{p,\lambda}(\mathbb{R}^n)$.

Theorem 5. *Let $p_i \in (0, \infty)$ and $\lambda_i \in [0, n)$, $i = 1, 2$:*

1. *If $p_1 < p_2$ or $\frac{p_1}{n-\lambda_1} < \frac{p_2}{n-\lambda_2}$, then $\mathrm{PWM}(L^{p_1,\lambda_1}(\mathbb{R}^n), L^{p_2,\lambda_2}(\mathbb{R}^n)) = \{0\}$;*
2. *If $p_1 \ge p_2$ and $\frac{p_1}{n-\lambda_1} = \frac{p_2}{n-\lambda_2}$, then $\mathrm{PWM}(L^{p_1,\lambda_1}(\mathbb{R}^n), L^{p_2,\lambda_2}(\mathbb{R}^n)) = L^\infty(\mathbb{R}^n)$;*
3. *If $p_1 \ge p_2$ and $\frac{p_1}{n-\lambda_1} > \frac{p_2}{n-\lambda_2}$, then $\mathrm{PWM}(L^{p_1,\lambda_1}(\mathbb{R}^n), L^{p_2,\lambda_2}(\mathbb{R}^n)) = \mathcal{M}^{v_3}_{X^{p_2}}(\mathbb{R}^n)$, where v_3 and X are given by:*

$$\frac{n-\lambda_1}{p_1 n} + \frac{1}{v_3} = \frac{n-\lambda_2}{p_2 n}, \quad X(\mathbb{R}^n) = \mathcal{H}^{\left(\frac{p_1 n}{p_2(n-\lambda_1)}\right)'}_{\left(\frac{p_1}{p_2}\right)'}(\mathbb{R}^n).$$

We prove Theorem 4 in Section 4.1.

We combine Theorems 2 and 4 to have a nontrivial coincidence of function spaces.

Corollary 1. *Let $q_i \in (0, \infty)$ and $p_i \in [q_i, \infty)$, $i = 1, 2$. Assume that $q_1 > q_2$ and $p_1 > p_2$. Write $q_3 = \frac{q_1 q_2}{q_1 - q_2}$, $p_3 = \frac{p_1 p_2}{p_1 - p_2}$ and $X(\mathbb{R}^n) = \mathcal{H}^{\left(\frac{p_1}{q_2}\right)'}_{\left(\frac{q_1}{q_2}\right)'}(\mathbb{R}^n)$. If $\frac{q_1}{p_1} \ge \frac{q_2}{p_2}$, then $\mathcal{M}^{p_3}_{q_3}(\mathbb{R}^n) = \mathcal{M}^{p_3}_{X^{q_2}}(\mathbb{R}^n)$.*

A remark about Corollary 1 may be in order.

Remark 2. *Let $X(\mathbb{R}^n)$ be as in Corollary 1, and let $Y(\mathbb{R}^n) = L^{q_3}(\mathbb{R}^n)$. Corollary 1 reveals that $\mathcal{M}^{p_3}_Y(\mathbb{R}^n) = \mathcal{M}^{p_3}_{X^{q_2}}(\mathbb{R}^n)$, although $X(\mathbb{R}^n)^{q_2} \neq Y(\mathbb{R}^n)$.*

2.2. Characterization of $\mathrm{PWM}(\mathrm{w}\mathcal{M}^{p_1}_{q_1}(\mathbb{R}^n), \mathcal{M}^{p_2}_{q_2}(\mathbb{R}^n))$

Once we prove Theorem 4, we can pass the results above from $\mathcal{M}^{p_1}_{q_1}(\mathbb{R}^n)$ to $\mathrm{w}\mathcal{M}^{p_1}_{q_1}(\mathbb{R}^n)$ with ease if $0 < q_2 < q_1 < \infty$. To describe the multiplier space $\mathrm{PWM}(\mathrm{w}\mathcal{M}^{p_1}_{q_1}(\mathbb{R}^n), \mathcal{M}^{p_2}_{q_2}(\mathbb{R}^n))$, we will recall the definition given in [11,12]:

Definition 2. *1.* ([11], Definition 1.4.1) *Let $f : \mathbb{R}^n \to \mathbb{C}$ be a measurable function. Then, define its decreasing rearrangement f^* by:*

$$f^*(t) = \inf\{\lambda > 0 : |\{|f| > \lambda\}| \le t\};$$

2. ([11], Definition 1.4.6) *Let $1 < p, q < \infty$. The Lorentz space $L^{p,q}$ is the set of all measurable functions $f : \mathbb{R}^n \to \mathbb{C}$ for which:*

$$\|f\|_{L^{p,q}} = \left(\int_0^\infty (t^{\frac{1}{p}} f^*(t))^q \frac{dt}{t}\right)^{\frac{1}{q}}$$

is finite;

3. ([12], Definition 2.3) *Let $1 < q \le p < \infty$. A measurable function b is said to be a $(p'; q', 1)$-block if there exists a cube Q such that:*

$$\mathrm{supp}(b) \subset Q, \quad \|b\|_{L^{q',1}} \le |Q|^{\frac{1}{p} - \frac{1}{q}}; \tag{10}$$

4. ([12], Definition 2.3) *Let $1 < q \le p < \infty$. The space $\mathcal{H}^{p'}_{q',1}(\mathbb{R}^n)$ is the set of all $L^p(\mathbb{R}^n)$-functions f for which there exist a sequence $\{\lambda_j\}_{j=1}^\infty \in \ell^1(\mathbb{N})$ and a sequence $\{b_j\}_{j=1}^\infty$ of $(p'; q', 1)$-blocks for which:*

$$f = \sum_{j=1}^\infty \lambda_j b_j \tag{11}$$

in $L^p(\mathbb{R}^n)$. For $f \in \mathcal{H}^{p'}_{q',1}(\mathbb{R}^n)$, one defines:

$$\|f\|_{\mathcal{H}^{p'}_{q',1}} = \inf \sum_{j=1}^\infty |\lambda_j|,$$

where inf *is over all possible decompositions in (11).*

Concerning Lorentz spaces, a couple of remarks may be in order:

Remark 3. *Let $0 < p, p_1, p_2, q, q_1, q_2 \le \infty$:*

1. *Let G be a measurable set in $\mathbb{R}^n$. Then:*

$$\|\chi_G\|_{L^{p,q}} = \left(\frac{p}{q}\right)^{\frac{1}{q}} |G|^{\frac{1}{p}},$$

where we understand $(p/q)^{1/q} = 1$ for $q = \infty$. See [11], Example 1.4.8;

2. *Assume that:*

$$\frac{1}{p} = \frac{1}{p_1} + \frac{1}{p_2}, \quad \frac{1}{q} = \frac{1}{q_1} + \frac{1}{q_2}.$$

Then:

$$\|f \cdot g\|_{L^{p,q}} \le e^{\frac{1}{p}} \|f\|_{L^{p_1,q_1}} \|g\|_{L^{p_2,q_2}}$$

for all $f \in L^{p_1,q_1}(\mathbb{R}^n)$ *and* $g \in L^{p_2,q_2}(\mathbb{R}^n)$*, or equivalently:*

$$\|g\|_{\mathrm{PWM}(L^{p_1,q_1},L^{p,q})} \le e^{\frac{1}{p}} \|g\|_{L^{p_2,q_2}}$$

for all $g \in L^{p_2,q_2}(\mathbb{R}^n)$*. See [8], p. 6, Corollary 3, for the precise constant;*

3. *We have an equivalent expression if* $p > 1$*: For all* $f \in L^0(\mathbb{R}^n)$*,*

$$\begin{aligned} \|f\|_{\mathrm{w}L^p} &= \|f\|_{L^{p,\infty}} \\ &\sim \sup\left\{ |E|^{\frac{1}{p}-1}\|f\|_{L^1(E)} \, : \, E \text{ is a measurable set with } |E| \in (0,\infty) \right\}. \end{aligned} \tag{12}$$

See [11], Exercise 1.1.12.

Theorem 6. *Let* $0 < q_i \le p_i < \infty$*,* $i = 1, 2$*. If* $p_1 > p_2$ *and* $q_1 \ge q_2$*, then:*

$$\mathrm{PWM}(\mathrm{w}\mathcal{M}^{p_1}_{q_1}(\mathbb{R}^n), \mathcal{M}^{p_2}_{q_2}(\mathbb{R}^n)) = \mathcal{M}^{p_3}_{Xq_2}(\mathbb{R}^n),$$

where p_3 *and* X *are given by:*

$$p_3 = \frac{p_1 p_2}{p_1 - p_2}, \quad X = \mathcal{H}^{\left(\frac{p_1}{q_2}\right)'}_{\left(\frac{q_1}{q_2}\right)',1}(\mathbb{R}^n).$$

We prove Theorem 6 in Section 4.2.

The special case of $p_1 = q_1 > p_2 = q_2$ deserves attention.

Corollary 2. *In addition to the assumption in Theorem 6, we let* $p_1 = q_1 > p_2 = q_2$*. Then:*

$$\mathrm{PWM}(\mathrm{w}L^{p_1}(\mathbb{R}^n), L^{p_2}(\mathbb{R}^n)) = \mathcal{M}^{p_3}_{Xp_2}(\mathbb{R}^n),$$

where p_3 *and* X *are given by:*

$$p_3 = \frac{p_1 p_2}{p_1 - p_2}, \quad X = \mathcal{H}^{\left(\frac{p_1}{p_2}\right)'}_{\left(\frac{p_1}{p_2}\right)',1}(\mathbb{R}^n).$$

We complement Corollary 2.

Proposition 1. *Let* $0 < q_i \le p_i < \infty$*,* $i = 1, 2$*. If either one of the following conditions holds, then:*

$$\mathrm{PWM}(\mathrm{w}\mathcal{M}^{p_1}_{q_1}(\mathbb{R}^n), \mathcal{M}^{p_2}_{q_2}(\mathbb{R}^n)) = \{0\} :$$

1. $p_1 < p_2$*;*
2. $q_1 < q_2$*;*
3. $p_1 = p_2$ *and* $q_1 = q_2$*.*

We prove Proposition 1 in Section 4.3.

If $p_1 = p_2$ and $q_1 > q_2$, then we have something similar to the case of classical Morrey spaces.

Proposition 2. *Let* $0 < q_i \le p_i < \infty$*,* $i = 1, 2$*. Assume* $p_1 = p_2$ *and* $q_1 > q_2$*. Then:*

$$\mathrm{PWM}(\mathrm{w}\mathcal{M}^{p_1}_{q_1}(\mathbb{R}^n), \mathcal{M}^{p_2}_{q_2}(\mathbb{R}^n)) = L^\infty(\mathbb{R}^n).$$

We prove Proposition 2 in Section 4.4.

2.3. Characterization of $\mathrm{PWM}(\mathcal{M}^{p_1}_{q_1}(\mathbb{R}^n), \mathrm{w}\mathcal{M}^{p_2}_{q_2}(\mathbb{R}^n))$

Next, we pass from $\mathcal{M}^{p_2}_{q_2}(\mathbb{R}^n)$ to $\mathrm{w}\mathcal{M}^{p_2}_{q_2}(\mathbb{R}^n)$.

Theorem 4 allows us to characterize $\mathrm{PWM}(\mathcal{M}^{p_1}_{q_1}(\mathbb{R}^n), \mathrm{w}\mathcal{M}^{p_2}_{q_2}(\mathbb{R}^n))$.

Theorem 7. *Let $0 < q_i \le p_i < \infty$, $i = 1, 2$, satisfy $p_1 > p_2$ and $q_1 \ge q_2$. Define p_3 by:*

$$\frac{1}{p_3} = \frac{1}{p_2} + \frac{1}{q_2'} - \frac{1}{p_1}.$$

Then. a function $h \in L^0(\mathbb{R}^n)$ belongs to $\mathrm{PWM}(\mathcal{M}^{p_1}_{q_1}(\mathbb{R}^n), \mathrm{w}\mathcal{M}^{p_2}_{q_2}(\mathbb{R}^n))$ if and only if $\chi_E h \in \mathcal{M}^{p_3}_{\mathcal{H}^{p_1'}_{q_1'}}(\mathbb{R}^n)$ for all measurable sets E with $|E| \in (0,\infty)$ and:

$$\sup\{|E|^{\frac{1}{q_2}-1} \|\chi_E h\|_{\mathcal{M}^{p_3}_{\mathcal{H}^{p_1'}_{q_1'}}} : E \text{ is a measurable set with } |E| \in (0,\infty)\} < \infty.$$

In this case,

$$\|h\|_{\mathrm{PWM}(\mathcal{M}^{p_1}_{q_1}, \mathrm{w}\mathcal{M}^{p_2}_{q_2})} \sim \sup\{|E|^{\frac{1}{q_2}-1} \|\chi_E h\|_{\mathcal{M}^{p_3}_{\mathcal{H}^{p_1'}_{q_1'}}} : E \text{ is a measurable set with } |E| \in (0,\infty)\}.$$

We prove Theorem 7 in Section 4.5.

We supplement Theorem 7 by considering the case of $p_1 \le p_2$.

Proposition 3. *Let $0 < q_i \le p_i < \infty$, $i = 1, 2$:*

1. *Assume $p_1 < p_2$ or $q_1 < q_2$. Then:*

$$\mathrm{PWM}(\mathcal{M}^{p_1}_{q_1}(\mathbb{R}^n), \mathrm{w}\mathcal{M}^{p_2}_{q_2}(\mathbb{R}^n)) = \{0\};$$

2. *Assume $p_1 = p_2$ and $q_1 \ge q_2$. Then:*

$$\mathrm{PWM}(\mathcal{M}^{p_1}_{q_1}(\mathbb{R}^n), \mathrm{w}\mathcal{M}^{p_2}_{q_2}(\mathbb{R}^n)) = L^\infty(\mathbb{R}^n).$$

We prove Proposition 3 in Section 4.6.

2.4. Characterization of $\mathrm{PWM}(\mathrm{w}\mathcal{M}^{p_1}_{q_1}(\mathbb{R}^n), \mathrm{w}\mathcal{M}^{p_2}_{q_2}(\mathbb{R}^n))$

Finally, we pass both $\mathcal{M}^{p_1}_{q_1}(\mathbb{R}^n)$ and $\mathcal{M}^{p_2}_{q_2}(\mathbb{R}^n)$ to $\mathrm{w}\mathcal{M}^{p_1}_{q_1}(\mathbb{R}^n)$ and $\mathrm{w}\mathcal{M}^{p_2}_{q_2}(\mathbb{R}^n)$, respectively. The proof is a mere combination of Theorems 6 and 7. Therefore, we omit the detail again.

Theorem 8. *Let $0 < q_i \le p_i < \infty$, $i = 1, 2$, satisfy $p_1 > p_2$ and $q_1 \ge q_2$. Define p_3 by:*

$$\frac{1}{p_3} = \frac{1}{p_2} + \frac{1}{q_2'} - \frac{1}{p_1}.$$

Then $h \in L^0(\mathbb{R}^n)$ belongs to $\mathrm{PWM}(\mathrm{w}\mathcal{M}^{p_1}_{q_1}(\mathbb{R}^n), \mathrm{w}\mathcal{M}^{p_2}_{q_2}(\mathbb{R}^n))$ if and only if $\chi_E h \in \mathcal{M}^{p_3}_{\mathcal{H}^{p_1'}_{q_1',1}}(\mathbb{R}^n)$ for all measurable sets E with $|E| \in (0,\infty)$ and:

$$\sup\{|E|^{\frac{1}{q_2}-1} \|\chi_E h\|_{\mathcal{M}^{p_3}_{\mathcal{H}^{p_1'}_{q_1',1}}} : E \text{ is a measurable set with } |E| \in (0,\infty)\} < \infty$$

and in this case:

$$\|h\|_{\mathrm{PWM}(\mathrm{w}\mathcal{M}^{p_1}_{q_1},\mathrm{w}\mathcal{M}^{p_2}_{q_2})} \sim \sup\{|E|^{\frac{1}{q_2}-1}\|\chi_E h\|_{\mathcal{M}^{p_3}_{\mathcal{H}^{p_1'}_{q_1',1}}} : E \text{ is a measurable set with } |E| \in (0,\infty)\}.$$

In particular, $h \in L^0(\mathbb{R}^n)$ belongs to $\mathrm{PWM}(\mathrm{w}L^{p_1}(\mathbb{R}^n), \mathrm{w}L^{p_2}(\mathbb{R}^n))$ if and only if, for all measurable sets E with $|E| \in (0,\infty)$, $\chi_E h \in \mathcal{M}^{p_1'}_{\mathcal{H}^{p_1'}_{p_1',1}}(\mathbb{R}^n)$ and

$$\sup\{|E|^{\frac{1}{p_2}-1}\|\chi_E h\|_{\mathcal{M}^{p_1'}_{\mathcal{H}^{p_1'}_{p_1',1}}} : E \text{ is a measurable set with } |E| \in (0,\infty)\} < \infty$$

and in this case:

$$\|h\|_{\mathrm{PWM}(\mathrm{w}L^{p_1},\mathrm{w}L^{p_2})} \sim \sup\{|E|^{\frac{1}{p_2}-1}\|\chi_E h\|_{\mathcal{M}^{p_1'}_{\mathcal{H}^{p_1'}_{p_1',1}}} : E \text{ is a measurable set with } |E| \in (0,\infty)\}.$$

In the above, the implicit constants do not depend on h.

In Theorem 8, the case of $p_1 \le p_2$ is covered in Theorem 3.

It seems to make sense to compare Theorems 7 and 8 with an existing result. Let $p_1 = q_1$ and $p_2 = q_2$ in Theorems 7 and 8.

Corollary 3. *Let $0 < p_2 < p_1 < \infty$. Then:*

$$\mathrm{PWM}(L^{p_1}(\mathbb{R}^n), \mathrm{w}L^{p_2}(\mathbb{R}^n)) = \mathrm{PWM}(\mathrm{w}L^{p_1}(\mathbb{R}^n), \mathrm{w}L^{p_2}(\mathbb{R}^n)) = \mathrm{w}L^{\frac{p_1p_2}{p_1-p_2}}(\mathbb{R}^n). \tag{13}$$

In [8], Corollary 3, the first author showed the second equality in (13). We reprove Corollary 3 by the use of Theorems 7 and 8 in Section 4.7.

3. Preliminaries

For the proof of the theorems in the present paper, we use a scaling property. Arithmetic shows that the following scaling property holds:

Lemma 1. ([5], (g) p. 326) *Let E_1 and E_2 be quasi-Banach lattices, and let $u > 0$. Then:*

$$\mathrm{PWM}(E_1^u, E_2^u) = \mathrm{PWM}(E_1, E_2)^u.$$

We move on to the convexification of E-based Morrey spaces. Actually, as the next lemma shows, E-based Morrey spaces are closed under the convexification of quasi-Banach lattices.

Lemma 2. *Let $E \subset L^0(\mathbb{R}^n)$ be a quasi-Banach lattice and $p, u > 0$. Then: $(\mathcal{M}^{\frac{p}{u}}_E(\mathbb{R}^n))^u = \mathcal{M}^p_{E^u}(\mathbb{R}^n)$.*

Proof. For $f \in L^0(\mathbb{R}^n)$, a direct computation shows:

$$\|f\|_{\mathcal{M}^p_{E^u}} = \sup_{Q\in\mathcal{Q}} |Q|^{\frac{1}{p}} \frac{\|\chi_Q f\|_{E^u}}{\|\chi_Q\|_{E^u}} = \sup_{Q\in\mathcal{Q}} \left(|Q|^{\frac{u}{p}} \frac{\|\chi_Q |f|^u\|_E}{\|\chi_Q\|_E}\right)^{\frac{1}{u}} = \left(\||f|^u\|_{\mathcal{M}^{\frac{p}{u}}_E}\right)^{\frac{1}{u}} = \|f\|_{(\mathcal{M}^{\frac{p}{u}}_E)^u}.$$

□

We also investigate how $\mathcal{M}^p_E(\mathbb{R}^n)$ inherits the dilation property from E.

Lemma 3. *We have* $\|f(t\cdot)\|_{\mathcal{M}^p_E} = t^{-\frac{n}{p}}\|f\|_{\mathcal{M}^p_E}$ *for all* $f \in \mathcal{M}^p_E(\mathbb{R}^n)$ *and* $t > 0$ *as long as* E *is subject to the scaling law* $\|g(t\cdot)\|_E = t^{-\frac{n}{u}}\|g\|_E$ *for some* $u > 0$ *and for all* $g \in E$ *and* $t > 0$.

Proof. The proof is straightforward, and we omit the detail. □

Remark that Lemma 3 is not used for the proof of the main results in the present paper. However, Lemma 3 allows us to compare the scaling laws in the function spaces in question.

In Section 2, we introduced block spaces together with some of their variants. We recall that these spaces can be identified with the Köthe dual of Morrey spaces.

If E is a Banach lattice, then recall that its "Köthe dual" E' is defined in $L^0(\mathbb{R}^n)$ by the set of all $g \in L^0(\mathbb{R}^n)$ such that:

$$\|g\|_{E'} \equiv \sup\left\{\|f \cdot g\|_{L^1} : f \in L^0(\mathbb{R}^n), \|f\|_E \le 1\right\} < \infty. \tag{14}$$

We can specify the Köthe dual of Morrey spaces as follows:

Lemma 4. *1.* *Let* $1 \le q \le p < \infty$. *Then, the Köthe dual of* $\mathcal{M}^p_q(\mathbb{R}^n)$ *is* $\mathcal{H}^{p'}_{q'}(\mathbb{R}^n)$ *with the coincidence of norms;*

2. *Let* $1 < q \le p < \infty$. *Then, the Köthe dual of* $\mathrm{w}\mathcal{M}^p_q(\mathbb{R}^n)$ *is isomorphic to* $\mathcal{H}^{p'}_{q',1}(\mathbb{R}^n)$ *with the equivalence of norms.*

Lemma 4 is a culmination of what we proved in various papers. See [13], Theorem 3.1, for 1. with $q = 1$, and see [14], Theorem 4.1, for example, for 1. with $1 < q < \infty$, while 2. was proven in [12], Theorem 2.7.

A direct consequence of Lemma 4 is that we have:

$$\|\chi_Q\|_{\mathcal{H}^{p'}_{q'}} = |Q|^{\frac{1}{p'}} \tag{15}$$

for all cubes Q.

When E_1 and E_2 are both homogeneous in the sense that the translation operator induces isomorphism, we can mollify $\mathrm{PWM}(E_1, E_2)$. Furthermore, in this case, by the next lemma, we see that the functions in $\mathrm{PWM}(E_1, E_2)$ do not increase the local integrability of the functions.

Lemma 5. *Let* E_1, E_2 *be Banach lattices, which are translation invariant in the sense that* $\|h(\cdot - y)\|_{E_j} = \|h\|_{E_j}$ *for all* $h \in E_j$, $j = 1, 2$. *Assume that* E_1 *and* E_2 *enjoy the Fatou property and that* $E_2 \subset L^u_{\mathrm{loc}}(\mathbb{R}^n)$ *for some* $u \in (0, \infty)$:

1. $\chi_{[0,1]^n} \in E_1 \cap E_2$.
2. *The space* $\mathrm{PWM}(E_1, E_2)$ *is a translation-invariant Banach lattice, and any element in* $\mathrm{PWM}(E_1, E_2)$ *is almost everywhere finite;*
3. *If* $f \in L^1(\mathbb{R}^n)$ *and* $g \in \mathrm{PWM}(E_1, E_2)$, *then* $f * g \in \mathrm{PWM}(E_1, E_2)$ *and:*

$$\|f * g\|_{\mathrm{PWM}(E_1,E_2)} \le \|f\|_{L^1}\|g\|_{\mathrm{PWM}(E_1,E_2)}. \tag{16}$$

In particular, for almost all $x \in \mathbb{R}^n$,

$$\int_{\mathbb{R}^n} |g(x-y)f(y)|dy < \infty. \tag{17}$$

4. *If* $\mathrm{PWM}(E_1, E_2) \neq \{0\}$, *then* $\chi_{[-1,1]^n} \in \mathrm{PWM}(E_1, E_2)$.
5. $\mathrm{PWM}(E_1, E_2) \subset L^u_{\mathrm{loc}}(\mathbb{R}^n)$.
6. *If there exists a function* $f \in E_1 \setminus L^u_{\mathrm{loc}}(\mathbb{R}^n)$, *then* $\mathrm{PWM}(E_1, E_2) = \{0\}$.

Proof. 1. We concentrate on E_1; E_2 can be dealt with similarly. Let $f \in E_1$ be a nonzero function. By truncation, the linearity of E_1, and the lattice property of E_1, we may assume that $f = \chi_F$ for some bounded measurable set F. Notice that:

$$g_N := \frac{1}{N^n} \sum_{k_1=1}^{N} \sum_{k_2=1}^{N} \cdots \sum_{k_n=1}^{N} f\left(\cdot - \frac{(k_1, k_2, \ldots, k_n)}{N}\right) \in E_1$$

satisfies $\|g_N\|_{E_1} \le \|f\|_{E_1}$ due to the translation invariance and the triangle inequality. Since $g_N \to \chi_{[0,1]^n} * f$ in the topology of $L^1(\mathbb{R}^n)$ as $N \to \infty$, by the Fatou property of E_1, $\chi_{[0,1]^n} * f \in E_1$. Since $\|\chi_{[0,1]^n} * f\|_{L^1} = |F| > 0$, it follows that $\chi_{[0,1]^n} * f$ is a nonzero continuous function. By the translation invariance and the lattice property of E_1, it follows that $\chi_{[0,1]^n} \in E_1$;

2. Let $g \in \mathrm{PWM}(E_1, E_2)$ and $y \in \mathbb{R}^n$. Then:

$$\|g(\cdot - y)f\|_{E_2} = \|f(\cdot + y)g\|_{E_2} \le \|g\|_{\mathrm{PWM}(E_1,E_2)} \|f(\cdot + y)\|_{E_1} = \|g\|_{\mathrm{PWM}(E_1,E_2)} \|f\|_{E_1}$$

for all $f \in E_1$. Thus, we see that $g(\cdot - y) \in \mathrm{PWM}(E_1, E_2)$ and that:

$$\|g(\cdot - y)\|_{\mathrm{PWM}(E_1,E_2)} \le \|g\|_{\mathrm{PWM}(E_1,E_2)}.$$

Likewise, if we swap the role of g and $g(\cdot - y)$, then we have:

$$\|g\|_{\mathrm{PWM}(E_1,E_2)} \le \|g(\cdot - y)\|_{\mathrm{PWM}(E_1,E_2)}.$$

Thus, $\mathrm{PWM}(E_1, E_2)$ is translation invariant. Since E_2 is a Banach lattice, we see that $\mathrm{PWM}(E_1, E_2)$ is a Banach lattice. To check that any element $g \in \mathrm{PWM}(E_1, E_2)$ is finite almost everywhere, we only need to show that $g\chi_{[-1,1]^n}$ is finite almost everywhere. Assume otherwise; $F := \{x \in [-1,1]^n : |g(x)| = \infty\}$ has a positive measure. Then, $\chi_F \in \mathrm{PWM}(E_1, E_2)$ since $\chi_F \le |g| \in \mathrm{PWM}(E_1, E_2)$. Thus, $\chi_F = \chi_F \cdot \chi_{[-1,1]^n} \in E_2$. This implies that $\|\chi_F\|_{E_2} \in (0, \infty)$. However, this is a contradiction since $\infty > \|g\chi_{[-1,1]^n}\|_{E_2} \ge \|\infty \chi_F\|_{E_2} = \infty$;

3. We prove:

$$|f| * |g| \in \mathrm{PWM}(E_1, E_2) \text{ and } \| \, |f| * |g| \, \|_{\mathrm{PWM}(E_1,E_2)} \le \|f\|_{L^1} \|g\|_{\mathrm{PWM}(E_1,E_2)},$$

which is slightly stronger than (16). For $h \in E_1$, we have:

$$\begin{aligned}
\|h \cdot |f| * |g|\|_{E_2} &\le \int_{\mathbb{R}^n} \|h(\cdot)g(\cdot - y)f(y)\|_{E_2} dy \\
&= \int_{\mathbb{R}^n} \|h(\cdot + y)g(\cdot)f(y)\|_{E_2} dy \\
&\le \|g\|_{\mathrm{PWM}(E_1,E_2)} \int_{\mathbb{R}^n} \|h(\cdot + y)f(y)\|_{E_1} dy \\
&= \|g\|_{\mathrm{PWM}(E_1,E_2)} \int_{\mathbb{R}^n} \|h(\cdot + y)\|_{E_1} |f(y)| dy \\
&= \|g\|_{\mathrm{PWM}(E_1,E_2)} \int_{\mathbb{R}^n} \|h\|_{E_1} |f(y)| dy \\
&= \|g\|_{\mathrm{PWM}(E_1,E_2)} \|f\|_{L^1} \|h\|_{E_1}.
\end{aligned}$$

Finally, (17) is a consequence of 2. and the fact that:

$$\int_{\mathbb{R}^n} |f(\cdot - y)g(y)| dy = |f| * |g| \in \mathrm{PWM}(E_1, E_2);$$

4. If $\mathrm{PWM}(E_1, E_2) \neq \{0\}$, then by the lattice property of $\mathrm{PWM}(E_1, E_2)$, there exists a nonzero and non-negative function $g \in \mathrm{PWM}(E_1, E_2)$. By 1., $\chi_{[-R,R]^n} * g \in \mathrm{PWM}(E_1, E_2) \setminus \{0\}$. If we choose $R \geq 1$ large enough, then $\chi_{[-R,R]^n} * g \geq \kappa\chi_{[-1,1]^n}$ for some $\kappa > 0$. Due to the lattice property of $\mathrm{PWM}(E_1, E_2)$, we obtain $\chi_{[-1,1]^n} \in \mathrm{PWM}(E_1, E_2)$;
5. By 2., the lattice property, and the translation invariance of E_1, $\chi_K \in E_1$ for all compact sets K. Thus, if $f \in \mathrm{PWM}(E_1, E_2)$, then $\chi_K f \in E_2 \subset L^u_{\mathrm{loc}}(\mathbb{R}^n)$;
6. Assume $\mathrm{PWM}(E_1, E_2) \neq \{0\}$. By translation, we may assume $f\chi_{[-1,1]^n} \notin L^u_{\mathrm{loc}}(\mathbb{R}^n)$. Meanwhile, by 3., $|f\chi_{[-1,1]^n}| \in E_2 \subset L^u_{\mathrm{loc}}(\mathbb{R}^n)$. This is a contradiction.
□

4. Proof of the Main Results

4.1. Proof of Theorem 4

The proof of Theorem 4 is not so long. Furthermore, the statements in Theorem 4, 1. and 2. are already included in Theorem 2. Therefore, we consider 3. solely. First, assume that $p_2 = q_2 = 1$. In this case, we need to find a description of $\mathrm{PWM}(\mathcal{M}^{p_1}_{q_1}(\mathbb{R}^n), L^1(\mathbb{R}^n))$. According to [5], this is nothing but the Köthe dual of $\mathcal{M}^{p_1}_{q_1}(\mathbb{R}^n)$. In this case, it remains to note that $\mathcal{H}^{p_1'}_{q_1'}(\mathbb{R}^n) = \mathcal{M}^{p_1'}_{\mathcal{H}^{p_1'}_{q_1'}}(\mathbb{R}^n)$ thanks to Remark 1 and that $p_1' = \frac{p_3}{q_2} = p_3$.

Next, we assume that $p_2 > q_2 = 1$. Then by the definition of $\mathcal{M}^{p_2}_1(\mathbb{R}^n)$, $g \in L^0(\mathbb{R}^n)$ belongs to $\mathrm{PWM}(\mathcal{M}^{p_1}_{q_1}(\mathbb{R}^n), \mathcal{M}^{p_2}_{q_2}(\mathbb{R}^n)) = \mathrm{PWM}(\mathcal{M}^{p_1}_{q_1}(\mathbb{R}^n), \mathcal{M}^{p_2}_1(\mathbb{R}^n))$ if and only if $|Q|^{\frac{1}{p_2}-1}\chi_Q g \in \mathrm{PWM}(\mathcal{M}^{p_1}_{q_1}(\mathbb{R}^n), L^1(\mathbb{R}^n))$ for each $Q \in \mathcal{Q}$ and fulfills:

$$\sup_{Q \in \mathcal{Q}} |Q|^{\frac{1}{p_2}-1} \|\chi_Q g\|_{\mathrm{PWM}(\mathcal{M}^{p_1}_{q_1}, L^1)} < \infty.$$

According to the previous paragraph, this is equivalent to $|Q|^{\frac{1}{p_2}-1}\chi_Q g \in \mathcal{H}^{p_1'}_{q_1'}(\mathbb{R}^n)$ for each $Q \in \mathcal{Q}$ and $\sup_{Q\in\mathcal{Q}} |Q|^{\frac{1}{p_2}-1}\|\chi_Q g\|_{\mathcal{H}^{p_1'}_{q_1'}} = \sup_{Q\in\mathcal{Q}} |Q|^{\frac{1}{p_3}} \frac{\|\chi_Q g\|_{\mathcal{H}^{p_1'}_{q_1'}}}{\|\chi_Q\|_{\mathcal{H}^{p_1'}_{q_1'}}} < \infty$, i.e., $g \in \mathcal{M}^{p_3}_{\mathcal{H}^{p_1'}_{q_1'}}(\mathbb{R}^n)$.

We handle the general case. Let $L > 0$. According to Lemma 1,

$$g \in \mathrm{PWM}(\mathcal{M}^{Lp_1}_{Lq_1}(\mathbb{R}^n), \mathcal{M}^{Lp_2}_{Lq_2}(\mathbb{R}^n))$$

if and only if $|g|^L \in \mathrm{PWM}(\mathcal{M}^{p_1}_{q_1}(\mathbb{R}^n), \mathcal{M}^{p_2}_{q_2}(\mathbb{R}^n))$. Therefore, from Lemma 2 and what we proved in the previous paragraph, we deduce:

$$\begin{aligned}
\mathrm{PWM}(\mathcal{M}^{p_1}_{q_1}(\mathbb{R}^n), \mathcal{M}^{p_2}_{q_2}(\mathbb{R}^n)) &= \left(\mathrm{PWM}(\mathcal{M}^{\frac{p_1}{q_2}}_{\frac{q_1}{q_2}}(\mathbb{R}^n), \mathcal{M}^{\frac{p_2}{q_2}}_1(\mathbb{R}^n))\right)^{q_2} \\
&= \left(\mathcal{M}^{\frac{p_3}{q_2}}_X(\mathbb{R}^n)\right)^{q_2} \\
&= \mathcal{M}^{p_3}_{X^{q_2}}(\mathbb{R}^n).
\end{aligned}$$

The proof is therefore complete.

4.2. Proof of Theorem 6

In the proof of Theorem 4, we may replace $\mathcal{M}^{p_1}_{q_1}(\mathbb{R}^n)$ by $\mathrm{w}\mathcal{M}^{p_1}_{q_1}(\mathbb{R}^n)$. Then, accordingly, we have to replace $\mathcal{H}^{\left(\frac{p_1}{q_2}\right)'}_{\left(\frac{q_1}{q_2}\right)'}(\mathbb{R}^n)$ by $\mathcal{H}^{\left(\frac{p_1}{q_2}\right)'}_{\left(\frac{q_1}{q_2}\right)',1}(\mathbb{R}^n)$. Thus, the proof is similar to Theorem 4.

4.3. *Proof of Proposition 1*

We may assume $q_1, q_2 > 1$ by the scaling argument by Lemma 1:

1. Since $q_1, q_2 > 1$, we may regard $\mathrm{w}\mathcal{M}^{p_1}_{q_1}(\mathbb{R}^n)$ and $\mathcal{M}^{p_2}_{q_2}(\mathbb{R}^n)$ as Banach spaces as in (12). Assume:
$$\mathrm{PWM}(\mathrm{w}\mathcal{M}^{p_1}_{q_1}(\mathbb{R}^n), \mathcal{M}^{p_2}_{q_2}(\mathbb{R}^n)) \neq \{0\}.$$
Then, $\chi_{[-1,1]^n} \in \mathrm{PWM}(\mathrm{w}\mathcal{M}^{p_1}_{q_1}(\mathbb{R}^n), \mathcal{M}^{p_2}_{q_2}(\mathbb{R}^n))$ by virtue of Lemma 5, 4.. This implies $\|f \cdot \chi_{[-1,1]^n}\|_{\mathcal{M}^{p_2}_{q_2}} \leq \|g\|_{\mathrm{PWM}(\mathrm{w}\mathcal{M}^{p_1}_{q_1}, \mathcal{M}^{p_2}_{q_2})} \|f\|_{\mathrm{w}\mathcal{M}^{p_1}_{q_1}}$ for all $f \in \mathrm{w}\mathcal{M}^{p_1}_{q_1}(\mathbb{R}^n)$. If we substitute $f(t\cdot)$ instead of f into this condition, we obtain $\|f \cdot \chi_{[-t,t]^n}\|_{\mathcal{M}^{p_2}_{q_2}} \leq t^{\frac{n}{p_2}-\frac{n}{p_1}} \|g\|_{\mathrm{PWM}(\mathrm{w}\mathcal{M}^{p_1}_{q_1}, \mathcal{M}^{p_2}_{q_2})} \|f\|_{\mathrm{w}\mathcal{M}^{p_1}_{q_1}}$. Since $p_2 > p_1$, if we let $t \to \infty$, then we have $\|f\|_{\mathcal{M}^{p_2}_{q_2}} = 0$ for all $f \in \mathrm{w}\mathcal{M}^{p_1}_{q_1}(\mathbb{R}^n)$. This is a contradiction;
2. Let $r \in (q_1, q_2)$. According to [15], p. 67 (see also [3], Theorem 2.2 and Remark 2.3, and [16], Theorem 4.9), there exists $f \in \mathcal{M}^{p_1}_{r}(\mathbb{R}^n) \setminus L^{q_2}(\mathbb{R}^n)$ such that $\mathrm{supp}(f) \subset [0,1]^n$. Thus, we are in the position of using Lemma 5, 6. to have the conclusion;
3. By virtue of Lemma 5, 4., if:
$$\mathrm{PWM}(\mathrm{w}\mathcal{M}^{p_1}_{q_1}(\mathbb{R}^n), \mathcal{M}^{p_2}_{q_2}(\mathbb{R}^n)) = \mathrm{PWM}(\mathrm{w}\mathcal{M}^{p_1}_{q_1}(\mathbb{R}^n), \mathcal{M}^{p_1}_{q_1}(\mathbb{R}^n)) \neq \{0\},$$
then $\chi_{[-1,1]^n} \in \mathrm{PWM}(\mathrm{w}\mathcal{M}^{p_1}_{q_1}(\mathbb{R}^n), \mathcal{M}^{p_1}_{q_1}(\mathbb{R}^n))$. Then, for $f \in \mathrm{w}\mathcal{M}^{p_1}_{q_1}(\mathbb{R}^n)$ and $r > 0$,
$$\begin{aligned}\|\chi_{[-r,r]^n} f\|_{\mathcal{M}^{p_1}_{q_1}} &= r^{\frac{n}{p_1}} \|\chi_{[-1,1]^n} f(r\cdot)\|_{\mathcal{M}^{p_1}_{q_1}} \\ &\leq \|\chi_{[-1,1]^n}\|_{\mathrm{PWM}(\mathrm{w}\mathcal{M}^{p_1}_{q_1}, \mathcal{M}^{p_1}_{q_1})} r^{\frac{n}{p_1}} \|f(r\cdot)\|_{\mathrm{w}\mathcal{M}^{p_1}_{q_1}} \\ &= \|\chi_{[-1,1]^n}\|_{\mathrm{PWM}(\mathrm{w}\mathcal{M}^{p_1}_{q_1}, \mathcal{M}^{p_1}_{q_1})} \|f\|_{\mathrm{w}\mathcal{M}^{p_1}_{q_1}}.\end{aligned}$$
Letting $r \to \infty$, we obtain:
$$\|f\|_{\mathcal{M}^{p_1}_{q_1}} \leq \|\chi_{[-1,1]^n}\|_{\mathrm{PWM}(\mathrm{w}\mathcal{M}^{p_1}_{q_1}, \mathcal{M}^{p_1}_{q_1})} \|f\|_{\mathrm{w}\mathcal{M}^{p_1}_{q_1}}.$$
This implies $\mathrm{w}\mathcal{M}^{p_1}_{q_1}(\mathbb{R}^n) \subset \mathcal{M}^{p_1}_{q_1}(\mathbb{R}^n)$. This is impossible; see [16,17] as well as [18], Section 4.

4.4. *Proof of Proposition 2*

Thanks to Theorem 2 and the embedding:
$$\mathcal{M}^{p_1}_{q_1}(\mathbb{R}^n) \subset \mathrm{w}\mathcal{M}^{p_1}_{q_1}(\mathbb{R}^n) \subset \mathcal{M}^{p_1}_{q_2}(\mathbb{R}^n),$$
we have:
$$\begin{aligned}L^{\infty}(\mathbb{R}^n) &= \mathrm{PWM}(\mathcal{M}^{p_1}_{q_2}(\mathbb{R}^n), \mathcal{M}^{p_2}_{q_2}(\mathbb{R}^n)) \\ &\subset \mathrm{PWM}(\mathrm{w}\mathcal{M}^{p_1}_{q_1}(\mathbb{R}^n), \mathcal{M}^{p_2}_{q_2}(\mathbb{R}^n)) \\ &\subset \mathrm{PWM}(\mathcal{M}^{p_1}_{q_1}(\mathbb{R}^n), \mathcal{M}^{p_2}_{q_2}(\mathbb{R}^n)) \\ &= L^{\infty}(\mathbb{R}^n).\end{aligned}$$

4.5. *Proof of Theorem 7*

We may assume $q_1, q_2 > 1$ by Lemma 1. The proof of Theorem 7 is a direct combination of Theorem 4 and Lemma 6 below.

Lemma 6. *Let $1 < q_i \le p_i < \infty$, $i = 1, 2$. Assume $p_1 > p_2$ and $q_1 \ge q_2$. Define $r > 0$ by:*

$$\frac{1}{r} = \frac{1}{p_2} - \frac{1}{q_2} + 1.$$

Then, $f \in L^0(\mathbb{R}^n)$ belongs to $\mathrm{PWM}(\mathcal{M}^{p_1}_{q_1}(\mathbb{R}^n), \mathrm{w}\mathcal{M}^{p_2}_{q_2}(\mathbb{R}^n))$ if and only if, for all measurable sets E with $|E| \in (0,\infty)$, $\chi_E f \in \mathrm{PWM}(\mathcal{M}^{p_1}_{q_1}(\mathbb{R}^n), \mathcal{M}^r_1(\mathbb{R}^n))$ and:

$$\sup\{|E|^{\frac{1}{q_2}-1}\|\chi_E f\|_{\mathrm{PWM}(\mathcal{M}^{p_1}_{q_1}, \mathcal{M}^r_1)} : E \text{ is a measurable set with } |E| \in (0,\infty)\} < \infty.$$

In this case,

$$\|f\|_{\mathrm{PWM}(\mathcal{M}^{p_1}_{q_1}, \mathrm{w}\mathcal{M}^{p_2}_{q_2})} \sim \sup\{|E|^{\frac{1}{q_2}-1}\|\chi_E f\|_{\mathrm{PWM}(\mathcal{M}^{p_1}_{q_1}, \mathcal{M}^r_1)} : E \text{ is a measurable set with } |E| \in (0,\infty)\}.$$

Once Lemma 6 is established, Theorem 4 immediately gives the proof of Theorem 7. Therefore, we concentrate on Lemma 6.

Proof of Lemma 6. Let $h \in L^0(\mathbb{R}^n)$. Thanks to (12), $h \in \mathrm{w}\mathcal{M}^{p_2}_{q_2}(\mathbb{R}^n)$ if and only if $\chi_E h \in \mathcal{M}^{p_2}_1(\mathbb{R}^n)$ for all measurable sets E with $|E| \in (0,\infty)$ and:

$$\sup_Q \sup_E |Q|^{\frac{1}{p_2}-\frac{1}{q_2}}|E|^{\frac{1}{q_2}-1}\|\chi_{E\cap Q} h\|_{L^1} = \sup_E |E|^{\frac{1}{q_2}-1}\|\chi_E h\|_{\mathcal{M}^r_1} < \infty,$$

where E moves over all measurable sets with $0 < |E| < \infty$. Therefore, supposing that E moves over all measurable sets with $|E| \in (0,\infty)$, we obtain:

$$\begin{aligned}
\|f\|_{\mathrm{PWM}(\mathcal{M}^{p_1}_{q_1}, \mathrm{w}\mathcal{M}^{p_2}_{q_2})} &= \sup\{\|f\cdot g\|_{\mathrm{w}\mathcal{M}^{p_2}_{q_2}} : g \in \mathcal{M}^{p_1}_{q_1}(\mathbb{R}^n), \|g\|_{\mathcal{M}^{p_1}_{q_1}} = 1\} \\
&\sim \sup_E \sup\{|E|^{\frac{1}{q_2}-1}\|f\cdot g\cdot\chi_E\|_{\mathcal{M}^r_1} : g \in \mathcal{M}^{p_1}_{q_1}(\mathbb{R}^n), \|g\|_{\mathcal{M}^{p_1}_{q_1}} = 1\} \\
&= \sup_E |E|^{\frac{1}{q_2}-1}\|f\chi_E\|_{\mathrm{PWM}(\mathcal{M}^{p_1}_{q_1}, \mathcal{M}^r_1)},
\end{aligned}$$

as required. □

4.6. *Proof of Proposition 3*

1. Suppose $p_1 < p_2$. We can go through the same argument as Proposition 1, 2. to conclude that by using the function:

$$\chi_{B(x_0,r)}|\cdot - x_0|^{-\frac{n(p_1+p_2)}{2p_1p_2}} \in \mathcal{M}^{p_1}_{q_1}(\mathbb{R}^n) \setminus \mathrm{w}\mathcal{M}^{p_2}_{q_2}(\mathbb{R}^n)$$

for some $x_0 \in \mathbb{R}^n$ and $r > 0$. If $q_1 < q_2$, then we take r_1, r_2 so that $q_1 < r_1 < r_2 < q_2$. Then, we have:

$$\mathrm{PWM}(\mathcal{M}^{p_1}_{q_1}(\mathbb{R}^n), \mathrm{w}\mathcal{M}^{p_2}_{q_2}(\mathbb{R}^n)) \subset \mathrm{PWM}(\mathrm{w}\mathcal{M}^{p_1}_{r_1}(\mathbb{R}^n), \mathcal{M}^{p_2}_{r_2}(\mathbb{R}^n)) = \{0\}$$

thanks to Proposition 1, 2.;

2. It is clear that:

$$L^\infty(\mathbb{R}^n) \subset \mathrm{PWM}(\mathcal{M}^{p_1}_{q_1}(\mathbb{R}^n), \mathrm{w}\mathcal{M}^{p_2}_{q_2}(\mathbb{R}^n)).$$

Thus, it suffices to show that:

$$\mathrm{PWM}(\mathcal{M}^{p_1}_{q_1}(\mathbb{R}^n), \mathrm{w}\mathcal{M}^{p_2}_{q_2}(\mathbb{R}^n)) \subset L^\infty(\mathbb{R}^n).$$

To this end, let $g \in \mathrm{PWM}(\mathcal{M}^{p_1}_{q_1}(\mathbb{R}^n), \mathrm{w}\mathcal{M}^{p_2}_{q_2}(\mathbb{R}^n))$. Then:

$$|Q|^{\frac{1}{p_2}-\frac{1}{q_2}}\|g\chi_Q\|_{\mathrm{w}L^{q_2}} \le \|g\|_{\mathrm{PWM}(\mathcal{M}^{p_1}_{q_1},\mathrm{w}\mathcal{M}^{p_2}_{q_2})}|Q|^{\frac{1}{p_1}} = \|g\|_{\mathrm{PWM}(\mathcal{M}^{p_1}_{q_1},\mathrm{w}\mathcal{M}^{p_2}_{q_2})}|Q|^{\frac{1}{p_2}}.$$

Thus, by the Lebesgue differentiation theorem, we obtain:

$$\|g\|_{L^\infty} \le \|g\|_{\mathrm{PWM}(\mathcal{M}^{p_1}_{q_1},\mathrm{w}\mathcal{M}^{p_2}_{q_2})},$$

as required.

4.7. *Proof of Corollary 3*

Theorems 7 and 8 can be shown to recover this result as follows:

1. Thanks to the fact that $\|\chi_Q\|_{\mathcal{H}^{p_1'}_{p_1'}} = \|\chi_Q\|_{L^{p_1'}}$ and the Fatou property of $\mathcal{H}^{p_1'}_{p_1'}(\mathbb{R}^n)$ established in [14], $\mathcal{M}^{p_1'}_{\mathcal{H}^{p_1'}_{p_1'}}(\mathbb{R}^n)$ coincides with $L^{p_1'}(\mathbb{R}^n)$. Thus, according to [11], Exercise 1.4.14, we see that $\mathrm{PWM}(L^{p_1}(\mathbb{R}^n), \mathrm{w}L^{p_2}(\mathbb{R}^n)) = \mathrm{w}L^{\frac{p_1p_2}{p_1-p_2}}(\mathbb{R}^n)$;
2. Using Lemma 4, we deduce:

$$\begin{aligned}
&\sup\{|E|^{\frac{1}{p_2}-1}\|\chi_E h\|_{\mathcal{H}^{p_1'}_{p_1',1}} : E \text{ is a measurable set with } |E| \in (0,\infty)\} \\
&\gtrsim \sup\{|E|^{\frac{1}{p_2}-1}\|\chi_E h\|_{L^{p_1'}} : E \text{ is a measurable set with } |E| \in (0,\infty)\} \\
&\sim \|h\|_{\mathrm{w}L^{\frac{p_1p_2}{p_1-p_2}}}.
\end{aligned}$$

Let r_1 be a number slightly less than p_1, so that r_1' is slightly larger than p_1'. Define v_1 by:

$$\frac{1}{p_1'} = \frac{1}{v_1} + \frac{1}{r_1'}.$$

Thanks to Remark 3,

$$\|\chi_E h\|_{L^{p_1',1}} \lesssim \|\chi_E\|_{L^{v_1,1}}\|h\|_{\mathrm{w}L^{r_1'}} \sim |E|^{\frac{1}{v_1}}\|h\|_{\mathrm{w}L^{r_1'}}$$

for all $h \in \mathrm{w}L^{p_1}(\mathbb{R}^n)$. Using this estimate and Remark 1, we have:

$$\begin{aligned}
\|\chi_E h\|_{\mathcal{M}^{p_1'}_{\mathcal{H}^{p_1'}_{p_1',1}}} &\sim \|\chi_E h\|_{\mathcal{H}^{p_1'}_{p_1',1}} \\
&\sim \sup\{\|\chi_E g h\|_{L^1} : g \in \mathrm{w}\mathcal{M}^{p_1}_{p_1}(\mathbb{R}^n), \|g\|_{\mathrm{w}\mathcal{M}^{p_1}_{p_1}} \le 1\} \\
&= \sup\{\|\chi_E g h\|_{L^1} : g \in \mathrm{w}L^{p_1}(\mathbb{R}^n), \|g\|_{\mathrm{w}L^{p_1}} \le 1\} \\
&\lesssim \|\chi_E h\|_{\mathrm{w}L^{p_1'}} \\
&\lesssim |E|^{\frac{1}{v_1}}\|\chi_E h\|_{\mathrm{w}L^{r_1'}}.
\end{aligned}$$

Thus, it follows from the embedding $L^{r_1'}(\mathbb{R}^n) \hookrightarrow \mathrm{w}L^{r_1'}(\mathbb{R}^n)$ that:

$$\sup\{|E|^{\frac{1}{p_2}-1}\|\chi_E h\|_{\mathcal{H}^{p_1'}_{p_1',1}} : E \text{ is a measurable set with } |E| \in (0,\infty)\}$$

$$\lesssim \sup\{|E|^{\frac{1}{p_2}+\frac{1}{r_1}-1}\|\chi_E h\|_{L^{r_1'}} : E \text{ is a measurable set with } |E| \in (0,\infty)\}$$

$$= \sup\{|E|^{\frac{p_1-p_2}{p_1p_2}-\frac{1}{r_1'}}\|\chi_E h\|_{L^{r_1'}} : E \text{ is a measurable set with } |E| \in (0,\infty)\}.$$

Invoking [11], Exercise 1.4.14, once again, one obtains:

$$\sup\{|E|^{\frac{1}{p_2}-1}\|\chi_E h\|_{\mathcal{H}^{p_1'}_{p_1',1}} : E \text{ is a measurable set with } |E| \in (0,\infty)\} \lesssim \|h\|_{\mathrm{w}L^{\frac{p_1p_2}{p_1-p_2}}}.$$

Thus, Theorems 7 and 8 can recover the result in [8].

Author Contributions: Conceptualization, E.N. and Y.S.; methodology, E.N. and Y.S.; software, E.N. and Y.S.; validation, E.N. and Y.S.; formal analysis, E.N. and Y.S.; investigation, E.N. and Y.S.; resources, E.N. and Y.S.; data curation, E.N. and Y.S.; writing—original draft preparation, E.N. and Y.S.; writing—review and editing, E.N. and Y.S.; visualization, E.N. and Y.S.; supervision, E.N. and Y.S.; project administration, E.N. and Y.S.; funding acquisition, E.N. and Y.S. All authors have read and agreed to the published version of the manuscript.

Funding: The first author was supported by Grant-in-Aid for Scientific Research (B), No. 15H03621, and, Research (C), No. 21K03304, Japan Society for the Promotion of Science. The second author was supported by Grant-in-Aid for Scientific Research (C) No. 16K05209, the Japan Society for the Promotion of Science.

Institutional Review Board Statement: Not applicable.

Informed Consent Statement: Not applicable.

Data Availability Statement: Data sharing not applicable to this article as no datasets were generated or analyzed during the current study.

Conflicts of Interest: The authors declare no conflict of interest.

References

1. Kawasumi, R.; Nakai, E. Pointwise multipliers on weak Morrey spaces. *Anal. Geom. Metr. Spaces* **2020**, *8*, 363–381. [CrossRef]
2. Nakai, E. Pointwise multipliers on the Morrey spaces. *Mem. Osaka Kyouiku Univ. III Natur. Sci. Appl. Sci.* **1997**, *46*, 1–11.
3. Nakai, E. A characterization of pointwise multipliers on the Morrey spaces, *Sci. Math.* **2000**, *3*, 445–454.
4. Nakai, E. Pointwise multipliers on several function spaces—A survey. *Linear Nonlinear Anal.* **2017**, *3*, 27–59.
5. Maligranda, L.; Persson, L.E. Generalized duality of some Banach function spaces. *Indag. Math. (Proc.)* **1989**, *92*, 323–338. [CrossRef]
6. Lemarié-Rieusset, P.G. Multipliers and Morrey spaces. *Potential Anal.* **2013**, *38*, 741–752. [CrossRef]
7. Peetre, J. On convolution operators leaving $L^{p,\lambda}$ spaces invariant. *Ann. Mat. Pura Appl.* **1966**, *72*, 295–304. [CrossRef]
8. Nakai, E. Pointwise Multipliers on the Lorentz Spaces. *Mem. Osaka Kyouiku Univ. III Natur. Sci. Appl. Sci.* **1996**, *45*, 1–7.
9. Ho, K.-P. Vector-valued operators with singular kernel and Triebel–Lizorkin block spaces with variable exponents. *Kyoto J. Math.* **2016**, *56*, 97–124. [CrossRef]
10. Long, R.L. The spaces generated by blocks. *Sci. Sin. Ser. A* **1984**, *27*, 16–26.
11. Grafakos, L. Classical Fourier Analysis. In *Graduate Texts in Mathematics*, 2nd ed.; Springer: New York, NY, USA, 2014; Volume 249.
12. Sawano, Y.; El-Shabrawy, S.R. Weak Morrey spaces with applications. *Math. Nachr.* **2018**, *291*, 178–186. [CrossRef]
13. Mastyło, M.; Sawano, Y.; Tanaka, H. Morrey type space and its Köthe dual space. *Bull. Malays. Math. Soc.* **2018**, *41*, 1181–1198. [CrossRef]
14. Sawano, Y.; Tanaka, H. The Fatou property of block spaces. *J. Math. Sci. Univ. Tokyo* **2015**, *22*, 663–683.
15. Giaquinta, M. *Multiple Integrals in the Calculus of Variations and Nonlinear Elliptic Systems*; Annals of Mathematics Studies; Princeton Universiy Press: Princeton, NJ, USA, 1983; Volume 105.
16. Nakai, E. Orlicz–Morrey spaces and the Hardy–Littlewood maximal function. *Studia Math.* **2008**, *188*, 193–221. [CrossRef]

17. Gunawan, H.; Hakim, D.I.; Nakai, E.; Sawano, Y. On inclusion relation between weak Morrey spaces and Morrey spaces. *Nonlinear Anal.* **2018**, *168*, 27–31. [CrossRef]
18. Mastyło, M.; Sawano, Y. Complex interpolation and Calderón–Mityagin couples of Morrey spaces. *Anal. PDE* **2019**, *12*, 1711–1740. [CrossRef]

mathematics

MDPI

Article

Calderón Operator on Local Morrey Spaces with Variable Exponents

Kwok-Pun Ho

Department of Mathematics and Information Technology, The Education University of Hong Kong, 10 Lo Ping Road, Tai Po, Hong Kong, China; vkpho@eduhk.hk

Abstract: In this paper, we establish the boundedness of the Calderón operator on local Morrey spaces with variable exponents. We obtain our result by extending the extrapolation theory of Rubio de Francia to the local Morrey spaces with variable exponents. The exponent functions of the local Morrey spaces with the exponent functions are only required to satisfy the log-Hölder continuity assumption at the origin and infinity only. As special cases of the main result, we have Hardy's inequalities, the Hilbert inequalities and the boundedness of the Riemann–Liouville and Weyl averaging operators on local Morrey spaces with variable exponents.

Keywords: Calderón operator; Hardy's inequality; variable Lebesgue space; local Morrey space; local block space; extrapolation

Citation: Ho, K.-P. Calderón Operator on Local Morrey Spaces with Variable Exponents. *Mathematics* **2021**, *9*, 2977. https://doi.org/10.3390/math9222977

Academic Editor: Palle E. T. Jorgensen

Received: 20 October 2021
Accepted: 18 November 2021
Published: 22 November 2021

1. Introduction

The main theme of this paper is the boundedness of the Calderón operator on local Morrey spaces with variable exponents.

The Calderón operator is one of the important operators in harmonic analysis and theory of function spaces. The Calderón operator is related with the Hardy' inequality, the Stieltjes transformation, the Riemann–Liouville and Weyl averaging operators. It also gives an estimate for the maximal Hilbert transform ([1], Chapter 3, Theorem 4.7). Moreover, the boundedness of the Calderón operator is also related with the convergence of Fourier series on rearrangement-invariant Banach function spaces ([1], Chapter 3, Theorem 6.10).

The boundedness of the Calderón operator on Lebesgue spaces is a well known result [2]. Recently, the boundedness property has been extended to the weighted Lebesgue spaces [3] and the weighted Lebesgue spaces with variable exponents [4]. In this paper, we further extend the boundedness of the Calderón operator to local Morrey spaces with variable exponents.

The local Morrey spaces with variable exponents are extensions of the classical Morrey spaces introduced and studied by Morrey [5] and the Lebesgue spaces with variable exponents [6,7]. The mapping properties of singular integral operators, the fractional integral operators, the geometric maximal operators and the spherical maximal functions were obtained in [8–14].

In this paper, we obtain our main results by extending the techniques from the extrapolation theory introduced by Rubio de Francia [15–17] to local Morrey spaces with variable exponents. An extrapolation theory for local Morrey spaces with variable exponents was obtained in [14], while the extrapolation theory given in [14] is based on the Hardy–Littlewood maximal function. In this paper, we use another maximal function from [3] which is defined via the basis $\{(0,r) : r > 0\}$. Similar to the results in [4], by using this maximal function, the exponent functions for the local Morrey spaces with variable exponents is not required to be globally log-Hölder continuous function. The exponent function is just required to be log-Hölder continuous at origin and infinity.

This paper is organized as follows. The definition and the boundedness of the Calderón operator on weighted Lebesgue spaces were presented in Section 2. The definitions of local

Morrey spaces with variable exponents and local block spaces with variable exponents are given in Section 3. The local block spaces with variable exponents are pre-duals of local Morrey spaces with variable exponents, and the boundedness of the maximal function associated with the the basis $\{(0,r) : r > 0\}$ on the local block spaces with variable exponents is obtained in Section 3. This boundedness result is one of the crucial results for the boundedness of the Calderón operator obtained in Section 4. As applications of our main results, we obtain the Hardy' inequalities, the boundedness of the Stieltjes transformation, the Riemann–Liouville and Weyl averaging operators on local Morrey spaces with variable exponents.

2. Definitions and Preliminaries

Let $\mathcal{M}$ be the class of Lebesgue measurable functions on $(0,\infty)$.

For any non-negative function f on $(0,\infty)$, the Calderón operator is defined as

$$Sf(x) = \frac{1}{x}\int_0^x f(y)dy + \int_x^\infty \frac{f(y)}{y}dy, \quad x \in (0,\infty).$$

For any non-negative function f on $(0,\infty)$, the Hardy operator is defined as

$$\mathcal{H}f(x) = \frac{1}{x}\int_0^x f(y)dy, \quad x \in (0,\infty).$$

We see that the adjoint operator of $\mathcal{H}$ is given by

$$\mathcal{H}^*f(x) = \int_x^\infty \frac{f(y)}{y}dy, \quad x \in (0,\infty).$$

The boundedness of $\mathcal{H}$ and $\mathcal{H}^*$ on Lebesgue spaces is called the Hardy's inequalities. We see that $S = \mathcal{H} + \mathcal{H}^*$. Thus, the boundedness of the Calderón operator on Lebesgue spaces follow from the Hardy's inequalities. The reader is referred to [2,18,19] for the studies of Hardy's inequalities.

Let $\alpha \geq 0$; the Stieltjes transformation, the Riemann–Liouville and Weyl averaging operators are defined as

$$\begin{aligned}
Hf(x) &= \int_0^\infty \frac{f(y)}{x+y}dy \\
I_\alpha f(x) &= \frac{\alpha+1}{x^{\alpha+1}}\int_0^x (x-y)^\alpha f(y)dy, \\
J_\alpha f(x) &= (\alpha+1)\int_x^\infty \frac{(y-x)^\alpha}{y^{\alpha+1}} f(y)dy.
\end{aligned}$$

For any non-negative function f, we have $Hf(x) \leq Sf(x)$, $I_\alpha f(x) \leq Sf(x)$ and $J_\alpha f(x) \leq Sf(x)$. The reader is referred to [20–22] for the studies of the Stieltjes transformation and its application on the Hilbert's double series.

We recall the following maximal operator and the Muckenhoupt type classes of weight functions for S. They were introduced in [3]. For any locally integrable function f, define

$$Nf(x) = \sup_{b>x} \frac{1}{b}\int_0^b |f(y)|dy, \quad x > 0.$$

The operator N is the maximal operator on $(0,\infty)$ with the basis $\{(0,r) : r > 0\}$. Notice that for any non-negative function f, we have $Nf \leq Sf$.

We recall the following class of weighted functions from ([3], (1.2)).

Definition 1. *Let $p \in (1, \infty)$. We say that a Lebesgue measurable function $\omega : (0, \infty) \to [0, \infty)$ belongs to $A_{p,0}$ if*

$$\sup_{b>0} \left(\frac{1}{b} \int_0^b \omega(x) dx \right) \left(\frac{1}{b} \int_0^b \omega(x)^{1-p'} dx \right)^{p-1} < \infty$$

where p' is the conjugate of p.

The class $A_{1,0}$ consists of all Lebesgue measurable function $\omega : (0, \infty) \to [0, \infty)$ satisfying

$$N\omega(x) \le C\omega(x), \quad x \in (0, \infty)$$

and $[w]_{A_{1,0}}$ denotes the smallest constant for which the above inequality holds.

In view of ([3], Theorem 1.1), we have the following weighted norm inequalities for N.

Theorem 1. *Let $p \in (1, \infty)$. We have a constant $C > 0$ such that*

$$\int_0^\infty |Nf(x)|^p \omega(x) dx \le C \int_0^\infty |f(x)|^p \omega(x) dx$$

if and only if $\omega \in A_{p,0}$.

When $p \in (1, \infty)$, the class $A_{p,0}$ coincides with the class $\mathcal{C}_p$ introduced in [23]; see ([3], Theorem 1.2). In addition, as a special case of ([3], Theorem 1.2), we have the weighted norm inequalities for the Calderón operator.

Theorem 2. *Let $p \in (1, \infty)$. We have a constant $C > 0$ such that*

$$\int_0^\infty |Sf(x)|^p \omega(x) dx \le C \int_0^\infty |f(x)|^p \omega(x) dx \tag{1}$$

if and only if $\omega \in A_{p,0}$.

3. Local Morrey Spaces with Variable Exponents

In this section, we recall the definition of local Morrey space with variable exponent and study a pre-dual of this space, namely, the local block space with variable exponent. As a crucial supporting result for our main result, we obtain the boundedness of the maximal function N on local block spaces with variable exponents at the end of this section.

We recall the definition of Lebesgue spaces with variable exponents.

Definition 2. *Let $p(\cdot) : (0, \infty) \to [1, \infty)$ be a Lebesgue measurable function. The Lebesgue space with variable exponent $L^{p(\cdot)}$ consists of all Lebesgue measurable functions $f : (0, \infty) \to \mathbb{C}$ satisfying*

$$\|f\|_{L^{p(\cdot)}} = \inf\left\{ \lambda > 0 : \rho_{p(\cdot)}(f/\lambda) \le 1 \right\} < \infty$$

where

$$\rho_{p(\cdot)}(f) = \int_0^\infty |f(x)|^{p(x)} dx.$$

We call $p(x)$ the exponent function of $L^{p(\cdot)}$.

Let $p'(x)$ be the conjugate function of $p(x)$. That is, they satisfy $\frac{1}{p(x)} + \frac{1}{p'(x)} = 1$, $x \in (0, \infty)$. Let $p_- = \operatorname{ess\,inf}_{x \in (0,\infty)} p(x)$ and $p_+ = \operatorname{ess\,sup}_{x \in (0,\infty)} p(x)$.

Definition 3. *A continuous function g on $(0,\infty)$ is log-Hölder continuous at the origin if there exist $c_{\log} > 0$ and g_0 such that*

$$|g(x) - g_0| \leq \frac{c_{\log}}{-\log(x)}, \quad \forall x \in (0,1/2). \tag{2}$$

A continuous function is log-Hölder continuous at infinity if there exist $g_\infty \in \mathbb{R}$ and $c_\infty > 0$ so that

$$|g(x) - g_\infty| \leq \frac{c_\infty}{\log(e+|x|)}, \quad \forall x \in (0,\infty). \tag{3}$$

We write $g \in C^{\log}$ if g is log-Hölder continuous at origin and log-Hölder continuous at infinity.

The above classes of log-Hölder continuous functions are used in [24–26] for the studies of Herz spaces with variable exponents.

We have the boundedness of the maximal operator N on $L^{p(\cdot)}$ whenever $p(\cdot) \in C^{\log}$ with $1 < p_- \leq p_+ < \infty$.

Theorem 3. *Let $p(\cdot) \in C^{\log}$. If $1 < p_- \leq p_+ < \infty$, then there exists a constant $C > 0$ such that*

$$\|Nf\|_{L^{p(\cdot)}} \leq C\|f\|_{L^{p(\cdot)}}.$$

For the proof of the above theorem, the reader is referred to ([4], Theorem 1.6 and Section 3).

We now give the definitions of local Morrey spaces with variable exponents from [14].

Definition 4. *Let $p(\cdot) : (0,\infty) \to (1,\infty)$ and $u : (0,\infty) \to (0,\infty)$ be Lebesgue measurable functions. The local Morrey space with variable exponent $LM_u^{p(\cdot)}$ consists of all $f \in \mathcal{M}$ satisfying*

$$\|f\|_{LM_u^{p(\cdot)}} = \sup_{r>0} \frac{1}{u(r)} \|\chi_{(0,r)} f\|_{L^{p(\cdot)}} < \infty.$$

When $p(\cdot) = p$, $1 \leq p < \infty$, the local Morrey space with variable exponent becomes the local Morrey space LM_u^p. For the studies of local Morrey spaces, the reader is referred to [9–13]. For the mapping properties of the Carleson operator, the local sharp maximal functions, the geometrical maximal functions and the rough maximal functions on $LM_u^{p(\cdot)}$, see [14,27].

The local Morrey spaces with variable exponents are ball Banach function spaces defined and studied in [28,29]; see the discussion after ([27], Theorem 2.3).

We recall a class of weight functions for the studies of the local Morrey spaces with variable exponents defined in ([14], Definition 2.5).

Definition 5. *Let $q_0 \in (0,\infty)$, $p(\cdot) : (0,\infty) \to [1,\infty]$. We say that a Lebesgue measurable function, $u(r) : (0,\infty) \to (0,\infty)$, belongs to $\mathbb{LW}_{p(\cdot)}^{q_0}$ if there exists a constant $C > 0$ such that for any $r > 0$, u fulfills*

$$C \leq u(r), \quad \forall r \geq 1, \tag{4}$$

$$\|\chi_{(0,r)}\|_{L^{p(\cdot)}} \leq Cu(r), \quad \forall r < 1, \tag{5}$$

$$\sum_{j=0}^{\infty} \frac{\|\chi_{(0,r)}\|_{L^{p(\cdot)/q_0}}}{\|\chi_{(0,2^{j+1}r)}\|_{L^{p(\cdot)/q_0}}} (u(2^{j+1}r))^{q_0} < C(u(r))^{q_0} \tag{6}$$

for all $r > 0$.

When $q_0 = 1$, we write $\mathbb{LW}_{p(\cdot)} = \mathbb{LW}^1_{p(\cdot)}$. Let $0 \le \theta < 1$ and $u_\theta(r) = \|\chi_{B(0,r)}\|^\theta_{L^{p(\cdot)}}$. The discussion at the end of ([30], Section 2) shows that $u_\theta \in \mathbb{LW}_{p(\cdot)}$. Particularly, $u \equiv 1$ is a member of $\mathbb{LW}_{p(\cdot)}$.

Next, we recall a pre-dual of the local Morrey space with variable exponent from ([14], Definition 3.1).

Definition 6. *Let $p(\cdot) : (0,\infty) \to (0,\infty)$ and $u(r) : (0,\infty) \to (0,\infty)$ be Lebesgue measurable functions. A $b \in \mathcal{M}$ is a local $(u, L^{p(\cdot)})$-block if it is supported in $(0,r)$, $r > 0$ and*

$$\|b\|_{L^{p(\cdot)}} \le \frac{1}{u(r)}. \tag{7}$$

We write $b \in \mathfrak{lb}_{u,L^{p(\cdot)}}$ if b is a local $(u, L^{p(\cdot)})$-block.
Define $\mathfrak{LB}_{u,p(\cdot)}$ by

$$\mathfrak{LB}_{u,p(\cdot)} = \left\{ \sum_{k=1}^{\infty} \lambda_k b_k : \sum_{k=1}^{\infty} |\lambda_k| < \infty \text{ and } b_k \text{ is a local } (u, L^{p(\cdot)})\text{-block} \right\}. \tag{8}$$

The space $\mathfrak{LB}_{u,p(\cdot)}$ is endowed with the norm

$$\|f\|_{\mathfrak{LB}_{u,p(\cdot)}} = \inf\left\{ \sum_{k=1}^{\infty} |\lambda_k| \text{ such that } f = \sum_{k=1}^{\infty} \lambda_k b_k \text{ a.e.} \right\}. \tag{9}$$

We call $\mathfrak{LB}_{u,p(\cdot)}$ the local block space with variable exponent.

In view of ([14], Theorem 3.3), $\mathfrak{LB}_{u,p(\cdot)}$ is a Banach space and $\mathfrak{LB}_{u,p(\cdot)} \subset L^1_{\text{loc}}$. In addition, whenever $f, g \in \mathcal{M}$ satisfying $|f| \le |g|$ and $g \in \mathfrak{LB}_{u,p(\cdot)}$, we have $f \in \mathfrak{LB}_{u,p(\cdot)}$ ([14], Proposition 3.2).

We present the following results for the block spaces with variable exponent from ([14], Section 3). Notice that the results in [14] are for local Morrey spaces with variable exponents on $\mathbb{R}^n$, while with some simple modifications, the results and the proofs in [14] can be extended to local Morrey spaces with variable exponents on $(0,\infty)$.

Theorem 4. *Let $p(\cdot) : (0,\infty) \to (1,\infty)$ and $u : (0,\infty) \to (0,\infty)$ be Lebesgue measurable functions. We have*

$$\mathfrak{LB}^*_{u,p(\cdot)} = LM^{p'(\cdot)}_u$$

where $\mathfrak{LB}^*_{u,p(\cdot)}$ denotes the dual space of $\mathfrak{LB}_{u,p(\cdot)}$.

The reader is referred to ([14], Theorem 3.1) for the proof of the above results. Furthermore, the proof of ([14], Theorem 3.1) gives the Hölder inequalities for $f \in LM^{p'(\cdot)}_u$ and $g \in \mathfrak{LB}_{u,p(\cdot)}$

$$\int_0^\infty |f(x)g(x)|dx \le C\|g\|_{\mathfrak{LB}_{u,p(\cdot)}} \|f\|_{LM^{p'(\cdot)}_u} \tag{10}$$

for some $C > 0$.

Moreover, in the proof of ([14], Theorem 3.1), we also have the norm conjugate formula

$$C_0\|f\|_{LM^{p(\cdot)}_u} \le \sup_{h \in \mathfrak{lb}_{u,p(\cdot)}} \int_0^\infty |f(x)h(x)|dx \le C_1\|f\|_{LM^{p(\cdot)}_u} \tag{11}$$

for some $C_0, C_1 > 0$.

Proposition 1. *Let $p(\cdot) : (0,\infty) \to (1,\infty)$, $u : (0,\infty) \to (0,\infty)$ be Lebesgue measurable functions and $f \in \mathfrak{LB}_{u,p(\cdot)}$. If $g \in \mathcal{M}$ satisfying $|g| \le |f|$, then $g \in \mathfrak{LB}_{u,p(\cdot)}$.*

The proof of the preceding proposition is given in ([14], Proposition 3.2.). We establish a supporting lemma in the following paragraphs.

Lemma 1. *Let $p(\cdot) \in C^{\log}$ with $1 < p_- \leq p_+ < \infty$. We have constants $C_0, C_1 > 0$ such that for any $r > 0$, we have*

$$C_0 r \leq \|\chi_{(0,r)}\|_{L^{p'(\cdot)}} \|\chi_{(0,r)}\|_{L^{p(\cdot)}} \leq C_1 r. \tag{12}$$

Proof. The first inequality in (12) follows from the Hölder inequality for Lebesgue spaces with variable exponents.

For any $r > 0$ and locally integrable function f, define

$$P_r f(y) = \left(\frac{1}{r}\int_0^r f(x)dx\right)\chi_{(0,r)}(y).$$

The definition of N guarantees that $|P_r f| \leq Nf$. Therefore, we have $\|P_r\|_{L^{p(\cdot)} \to L^{p(\cdot)}} \leq \|N\|_{L^{p(\cdot)} \to L^{p(\cdot)}}$. According to ([7], Corollary 3.2.14), we have

$$\|\chi_{(0,r)}\|_{L^{p'(\cdot)}} \|\chi_{(0,r)}\|_{L^{p(\cdot)}} = \sup\left\{\left|\int_0^r g(x)dx\right| \|\chi_{(0,r)}\|_{L^{p(\cdot)}} : \|g\|_{L^{p(\cdot)}} \leq 1\right\}.$$

Theorem 3 yields a constant $C_1 > 0$ such that for any $r > 0$, we have

$$\begin{aligned}\|\chi_{(0,r)}\|_{L^{p'(\cdot)}} \|\chi_{(0,r)}\|_{L^{p(\cdot)}} &\leq \sup\{r\|P_r g\|_{L^{p(\cdot)}} : \|g\|_{L^{p(\cdot)}} \leq 1\} \\ &\leq \sup\{r\|Ng\|_{L^{p(\cdot)}} : \|g\|_{L^{p(\cdot)}} \leq 1\} \leq C_1 r.\end{aligned}$$

Therefore, the second inequality in (12) holds. □

We are now ready to obtain the boundedness of the maximal function N on $\mathfrak{LB}_{u,p(\cdot)}$.

Theorem 5. *Let $p(\cdot) : (0,\infty) \to (1,\infty)$ and $u : (0,\infty) \to (0,\infty)$ be Lebesgue measurable functions. If $p(\cdot) \in C^{\log}$ with $1 < p_- \leq p_+ < \infty$ and $u \in \mathbb{LW}_{p'(\cdot)}$, then the maximal operator N is bounded on $\mathfrak{LB}_{u,p(\cdot)}$.*

Proof. In view of ([14], Theorem 3.3), we have $\mathfrak{LB}_{u,p(\cdot)} \subset L^1_{\text{loc}}$; therefore, the maximal operator N is well defined on $\mathfrak{LB}_{u,p(\cdot)}$.

Let $b \in \mathfrak{lb}_{u,L^{p(\cdot)}}$ with support $(0,r)$, $r > 0$. For any $k \in \mathbb{N}$, write $B_k = (0, 2^k r)$. Define $n_k = \chi_{B_{k+1} \backslash B_k} Nb$, $k \in \mathbb{N}\backslash\{0\}$ and $n_0 = \chi_{(0,2r)} Nb$. We have $\operatorname{supp} n_k \subseteq B_{k+1}\backslash B_k$ and $Nb = \sum_{k=0}^{\infty} n_k$.

As $p(\cdot) \in C^{\log}$ with $1 < p_- \leq p_+ < \infty$, Theorem 3 guarantees that

$$\|n_0\|_{L^{p(\cdot)}} \leq C\|Nb\|_{L^{p(\cdot)}} \leq \frac{C}{u(r)} \leq \frac{C}{u(2r)}$$

for some constant $C > 0$ independent r. The last inequality holds since (6) asserts that $\frac{\|\chi_{B(0,r)}\|_{L^{p(\cdot)}}}{\|\chi_{B(0,2r)}\|_{L^{p(\cdot)}}} u(2r) \leq Cu(r)$ and ([4], Lemma 2.3) yields $\|\chi_{B(0,2r)}\|_{L^{p(\cdot)}} \leq C\|\chi_{B(0,r)}\|_{L^{p(\cdot)}}$ for some $C > 0$ independent of $r > 0$. As a result of the above inequalities, n_0 is a constant-multiple of a local $(u, L^{p(\cdot)})$-block.

The Hölder inequality for $L^{p(\cdot)}$ yields

$$\begin{aligned} n_k = \chi_{B_{k+1}\backslash B_k} Nb &\leq \frac{\chi_{B_{k+1}\backslash B_k}}{2^k r} \int_0^r |b(x)|dx \\ &\leq C\chi_{B_{k+1}\backslash B_k} \frac{1}{2^k r} \|b\|_{L^{p(\cdot)}} \|\chi_{(0,r)}\|_{L^{p'(\cdot)}} \end{aligned}$$

for some $C > 0$ independent of k.

Consequently, (12) gives

$$\|n_k\|_{L^{p(\cdot)}} \le \frac{\|\chi_{B_{k+1}\setminus B_k}\|_{L^{p(\cdot)}}}{2^k r}\|b\|_{L^{p(\cdot)}}\|\chi_{B(0,r)}\|_{L^{p'(\cdot)}}$$
$$\le C\frac{\|\chi_{(0,r)}\|_{L^{p'(\cdot)}}}{\|\chi_{B_{k+1}}\|_{L^{p'(\cdot)}}}\frac{u(2^{k+1}r)}{u(r)}\frac{1}{u(2^{k+1}r)}.$$

Write $n_k = \sigma_k d_k$, where

$$\sigma_k = \frac{\|\chi_{(0,r)}\|_{L^{p'(\cdot)}}}{\|\chi_{B_{k+1}}\|_{L^{p'(\cdot)}}}\frac{u(2^{k+1}r)}{u(r)}.$$

We find that d_k is a constant-multiple of a local $(u, L^{p(\cdot)})$-block, and this constant does not depend on k. As $u \in \mathbb{LW}_{p'(\cdot)}$, we have

$$\sum_{j=0}^{\infty}\frac{\|\chi_{(0,r)}\|_{L^{p'(\cdot)}}}{\|\chi_{(0,2^{j+1}r)}\|_{L^{p'(\cdot)}}}u(2^{j+1}r) \le Cu(r).$$

We have $\sum_{k=0}^{\infty}\sigma_k < C$ for some $C > 0$. Hence, $Nb \in \mathfrak{LB}_{u,p(\cdot)}$. Moreover, there exists a constant $C_0 > 0$ so that for any local $(u, L^{p(\cdot)})$-block b,

$$\|Nb\|_{\mathfrak{LB}_{u,p(\cdot)}} < C_0.$$

Let $f \in \mathfrak{LB}_{u,p(\cdot)}$. The definition of $\mathfrak{LB}_{u,p(\cdot)}$ yields a family of local $(u, L^{p(\cdot)})$-blocks $\{c_k\}_{k=1}^{\infty}$ and a sequence $\Lambda = \{\lambda_k\}_{k=1}^{\infty} \in l^1$ such that $f = \sum_{k=1}^{\infty}\lambda_k c_k$ with $\|\Lambda\|_{l^1} \le 2\|f\|_{\mathfrak{LB}_{u,p(\cdot)}}$. Since N is sublinear, we find that

$$\left\|\sum_{k=1}^{\infty}\lambda_k N c_k\right\|_{\mathfrak{LB}_{u,p(\cdot)}} \le \sum_{k=1}^{\infty}|\lambda_k|\|Nc_k\|_{\mathfrak{LB}_{u,p(\cdot)}}$$
$$\le C_0\sum_{k=1}^{\infty}|\lambda_k| \le 2C_0\|f\|_{\mathfrak{LB}_{u,p(\cdot)}}.$$

As $Nf \le \sum_{k=1}^{\infty}|\lambda_k|Nc_k$, Proposition 1 guarantees that $Nf \in \mathfrak{LB}_{u,p(\cdot)}$ and $\|Nf\|_{\mathfrak{LB}_{u,p(\cdot)}} \le C\|f\|_{\mathfrak{LB}_{u,p(\cdot)}}$ for some $C > 0$. □

4. Calderón Operator

The boundedness of the Calderón operator on local Morrey spaces with variable exponents is established in this section. As applications of our main result, we obtain the Hardy's inequalities and the Hilbert inequalities on local Morrey spaces with variable exponents.

We use the techniques from the extrapolation theory. We first recall an operator from the Rubio de Francia algorithm. Let $p_0 \in (0,\infty)$ and $p(\cdot) \in C^{\log}$ with $p_0 < p_- \le p_+ < \infty$. The operator $\mathcal{R}$ is defined by

$$\mathcal{R}h = \sum_{k=0}^{\infty}\frac{N^k h}{2^k\|N^k\|_{\mathfrak{LB}_{u^{p_0},(p(\cdot)/p_0)'}\to\mathfrak{LB}_{u^{p_0},(p(\cdot)/p_0)'}}}, \quad h \in L^1_{loc},$$

where N^k is the k iterations of the operator N and $N^0 h = |h|$. The following are the boundedness of N and $\mathcal{R}$ on the local block spaces with variable exponents.

Proposition 2. *Let $p_0 \in (0,\infty)$ and $p(\cdot) \in C^{\log}$ with $p_0 < p_- \leq p_+ < \infty$. If $u \in \mathbb{LW}^{p_0}_{p(\cdot)}$, then the operator $\mathcal{R}$ is well defined on $\mathfrak{LB}_{u^{p_0},(p(\cdot)/p_0)'}$ and there is a constant $C > 0$ such that for any $h \in \mathfrak{LB}_{u^{p_0},(p(\cdot)/p_0)'}$,*

$$|h(x)| \leq \mathcal{R}h(x) \tag{13}$$
$$\|\mathcal{R}h\|_{\mathfrak{LB}_{u^{p_0},(p(\cdot)/p_0)'}} \leq 2\|h\|_{\mathfrak{LB}_{u^{p_0},(p(\cdot)/p_0)'}} \tag{14}$$
$$[\mathcal{R}h]_{A_{1,0}} \leq C\|N\|_{\mathfrak{LB}_{u^{p_0},(p(\cdot)/p_0)'}\to\mathfrak{LB}_{u^{p_0},(p(\cdot)/p_0)'}}. \tag{15}$$

Proof. As $u \in \mathbb{LW}^{p_0}_{p(\cdot)}$ implies $u^{p_0} \in \mathbb{LW}_{p(\cdot)/p_0}$, Theorem 5 guarantees that the maximal operator N is bounded on $\mathfrak{LB}_{u^{p_0},(p(\cdot)/p_0)'}$. Consequently, the operator $\mathcal{R}$ is well defined in $\mathfrak{LB}_{u^{p_0},(p(\cdot)/p_0)'}$, and the definition of $\mathcal{R}$ yields (13) and (14). In addition, since N is a sublinear operator, for any $h \in \mathfrak{LB}_{u^{p_0},(p(\cdot)/p_0)'}$, we obtain

$$\begin{aligned} N\mathcal{R}h &\leq \sum_{k=0}^{\infty} \frac{N^{k+1}h}{2^k\|N^k\|_{\mathfrak{LB}_{u^{p_0},(p(\cdot)/p_0)'}\to\mathfrak{LB}_{u^{p_0},(p(\cdot)/p_0)'}}} \\ &\leq 2\|N\|_{\mathfrak{LB}_{u^{p_0},(p(\cdot)/p_0)'}\to\mathfrak{LB}_{u^{p_0},(p(\cdot)/p_0)'}}\mathcal{R}h. \end{aligned}$$

According to Definition 1, $\mathcal{R}h \in A_{1,0}$, and hence, (15) holds. □

Theorem 6. *Let $p(\cdot) \in C^{\log}$ with $1 < p_- \leq p_+ < \infty$. If there exists a $p_0 \in (0, p_-)$ such that $u \in \mathbb{LW}^{p_0}_{p(\cdot)}$, then the Calderón operator S is bounded on $LM^{p(\cdot)}_u$.*

Proof. Let $f \in LM^{p(\cdot)}_u$. For any $h \in \mathfrak{LB}_{u^{p_0},(p(\cdot)/p_0)'}$, (10) and (14) yield

$$\begin{aligned} \int_0^\infty |f(x)|^{p_0}\mathcal{R}h(x)dx &\leq C\||f|^{p_0}\|_{LM^{p(\cdot)/p_0}_{u^{p_0}}}\|\mathcal{R}h\|_{\mathfrak{LB}_{u^{p_0},(p(\cdot)/p_0)'}} \\ &\leq \|f\|_{LM^{p(\cdot)}_u}\|h\|_{\mathfrak{LB}_{u^{p_0},(p(\cdot)/p_0)'}}. \end{aligned}$$

Thus, we have

$$LM^{p(\cdot)}_u \hookrightarrow \bigcap_{h\in\mathfrak{LB}_{u^{p_0},(p(\cdot)/p_0)'}} L^{p_0}(\mathcal{R}h). \tag{16}$$

Theorem 4 guarantees

$$\begin{aligned} \|Sf\|^{p_0}_{LM^{p(\cdot)}_u} &= \||Sf|^{p_0}\|_{LM^{p(\cdot)/p_0}_{u^{p_0}}} \\ &\leq C\sup\left\{\int_0^\infty |Sf(x)|^{p_0}|h(x)|dx : \|h\|_{\mathfrak{LB}_{u^{p_0},(p(\cdot)/p_0)'}} \leq 1\right\} \end{aligned} \tag{17}$$

for some $C > 0$.

In view of (15), $\mathcal{R}h \in A_{1,0}$. Furthermore, the embedding (16) guarantees that (1) holds for all $f \in LM^{p(\cdot)}_u$. Consequently, by applying $\omega = \mathcal{R}h$ on (1) and using (13), we find that

$$\begin{aligned} \int_0^\infty |Sf(x)|^{p_0}h(x)dx &\leq \int_0^\infty |Sf(x)|^{p_0}\mathcal{R}h(x)dx \\ &\leq C\int_0^\infty |f(x)|^{p_0}\mathcal{R}h(x)dx. \end{aligned}$$

Consequently, (10) and (14) give

$$\int_0^\infty |Sf(x)|^{p_0}h(x)dx \le C\||f|^{p_0}\|_{LM^{p(\cdot)/p_0}_{u^{p_0}}}\|\mathcal{R}h\|_{\mathfrak{LB}_{u^{p_0},(p(\cdot)/p_0)'}}$$
$$\le C\|f\|_{LM^{p(\cdot)}_u}\|h\|_{\mathfrak{LB}_{u^{p_0},(p(\cdot)/p_0)'}} \le C\|f\|^{p_0}_{LM^{p(\cdot)}_u}. \quad (18)$$

By taking supremum over all $h \in \mathfrak{LB}_{u^{p_0},(p(\cdot)/p_0)'}$ with $\|h\|_{\mathfrak{LB}_{u^{p_0},(p(\cdot)/p_0)'}} \le 1$, Theorem 4, (17) and (18) yield the boundedness of the Calderón operator S on $LM^{p(\cdot)}_u$. □

We also use the technique from the extrapolation theory to study the mapping properties of the local sharp maximal functions, the geometrical maximal functions and the rough maximal functions on local Morrey spaces with variable exponents in [14]. The results in [14] rely on the boundedness of the Hardy–Littlewood maximal operator. Therefore, the results obtained in [14] are valid for local Morrey spaces with variable exponents with the exponent functions being globally log-Hölder continuous. Our results use the maximal function N. Therefore, in view of Theorems 1 and 3, we just require $p(\cdot)$ to be log-Hölder continuous at origin and infinity for the boundedness of the Calderón operator on $LM^{p(\cdot)}_u$.

We give a concrete example for the weight function u that satisfies the conditions in Theorem 6. Let $p(\cdot) \in C^{\log}$ with $1 < p_- \le p_+ < \infty$. Let $0 \le \theta < 1$ and $u_\theta(r) = \|\chi_{B(0,r)}\|^\theta_{L^{p(\cdot)}}$. The discussion at the end of ([30], Section 2) shows that $u_\theta \in \mathbb{LW}_{p(\cdot)}$. For any $p_0 \in (1, p_-)$, we have

$$u_\theta(r)^{p_0} = \|\chi_{B(0,r)}\|^{p_0\theta}_{L^{p(\cdot)}} = \|\chi_{B(0,r)}\|^\theta_{L^{p(\cdot)/p_0}}.$$

The discussion at the end of ([30], Section 2) asserts that $u_\theta(r)^{p_0} \in \mathbb{LW}_{p(\cdot)/p_0}$. Therefore, the conditions in Theorem 6 are fulfilled, and the Calderón operator S is bounded on $LM^{p(\cdot)}_{u_\theta}$.

As $|\mathcal{H}f| \le \mathcal{H}|f| \le S|f|$ and $|\mathcal{H}^*f| \le \mathcal{H}^*|f| \le S|f|$, Theorem 6 yields the Hardy's inequalities on $LM^{p(\cdot)}_u$.

Theorem 7. *Let $p(\cdot) \in C^{\log}$ with $1 < p_- \le p_+ < \infty$. If there exists a $p_0 \in (0, p_-)$ such that $u \in \mathbb{LW}^{p_0}_{p(\cdot)}$, then there exists a constant $C > 0$ such that for any $f \in LM^{p(\cdot)}_u$*

$$\|\mathcal{H}f\|_{LM^{p(\cdot)}_u} \le C\|f\|_{LM^{p(\cdot)}_u},$$
$$\|\mathcal{H}^*f\|_{LM^{p(\cdot)}_u} \le C\|f\|_{LM^{p(\cdot)}_u}.$$

In particular, when $p(\cdot) = p$, $1 < p < \infty$ is a constant function, we have the Hardy's inequality on the local Morrey space LM^p_u. In addition, when $u \equiv 1$, the above results become the Hardy's inequalities on Lebesgue spaces with variable exponents, which recover the results in [31].

The reader is referred to [2,18,19] for the history and applications of the Hardy' inequalities. For the Hardy's inequalities on the Hardy type spaces, the Lebesgue spaces with variable exponents and the Herz–Morrey spaces, the reader may consult [31–37].

Theorem 6 also yields the boundedness of the Stieltjes transformation, the Riemann–Liouville and Weyl averaging operators on $LM^{p(\cdot)}_u$.

Theorem 8. *Let $p(\cdot) \in \mathcal{C}^{\log}$ with $1 < p_- \leq p_+ < \infty$. If there exists a $p_0 \in (0, p_-)$ such that $u \in \mathbb{LW}^{p_0}_{p(\cdot)}$, then there exists a constant $C > 0$ such that for any $f \in LM^{p(\cdot)}_u$*

$$\|Hf\|_{LM^{p(\cdot)}_u} \leq C\|f\|_{LM^{p(\cdot)}_u},$$
$$\|I_\alpha f\|_{LM^{p(\cdot)}_u} \leq C\|f\|_{LM^{p(\cdot)}_u},$$
$$\|J_\alpha f\|_{LM^{p(\cdot)}_u} \leq C\|f\|_{LM^{p(\cdot)}_u}.$$

The boundedness of the Stieltjes transformation on Lebesgue space is called as the Hilbert inequality. Therefore, as special cases of the preceding theorem, we also have the Hilbert inequality and the boundedness of the Riemann–Liouville and Weyl averaging operators on the local Morrey spaces LM^p_u and the Lebesgue spaces with variable exponents $L^{p(\cdot)}$.

5. Discussion

We establish the boundedness of the Calderón operator on local Morrey spaces with variable exponents by extending the extrapolation theory. The exponent functions used in the local Morrey spaces with variable exponents are required to be log-Hölder continuous at the origin and infinity only. We need to refine the extrapolation theory for the maximal operator N and the class of weight functions $A_{p,0}$. In addition, in order to get rid of the approximation argument, we need to establish the embedding (16).

As applications of the main result, we have Hardy's inequalities, the Hilbert inequalities and the boundedness of the Riemann–Liouville and Weyl averaging operators on local Morrey spaces with variable exponents.

Moreover, we see that whenever we can establish the weighted norm inequalities with the class of weight function $A_{p,0}$ for an operator T, even if T is nonlinear, we can apply our extrapolation theory to obtain the boundedness of T on the local Morrey spaces with variable exponents where the exponent function is log-Hölder continuous at 0 and infinity.

6. Conclusions

We extend the extrapolation theory to the local Morrey spaces with variable exponents with the exponent functions being log-Hölder continuous at the origin and infinity only. With this refined extrapolation theory, we obtain Hardy's inequalities and the Hilbert inequalities on the local Morrey spaces with variable exponents. Furthermore, the boundedness of the Calderón operator, the Riemann–Liouville operators and the Weyl averaging operators has been extended to the local Morrey spaces with variable exponents.

In particular, we have the Hardy's inequalities, the Hilbert inequalities on local Morrey spaces and the boundedness of the Calderón operator, the Riemann–Liouville averaging operators and the Weyl averaging operators on local Morrey spaces.

In conclusion, the results obtained in this paper generalize the existing results on the studies of local Morrey spaces with variable exponent, the Hardy's inequalities, the Hilbert inequalities on local Morrey spaces and the boundedness of the Calderón operator, the Riemann–Liouville averaging operators and the Weyl averaging operators.

Funding: This research was funded by The Education University of Hong Kong, Additional Reserach Fund R6391.

Institutional Review Board Statement: Not applicable.

Informed Consent Statement: Not applicable.

Conflicts of Interest: The author declares no conflict of interest.

References

1. Bennett, C.; Sharpley, R. *Interpolations of Operators*; Academic Press: Cambridge, MA, USA, 1988.
2. Hardy, G.; Littlewood, J.; Pólya, G. *Inequalities*; Cambridge University Press: Cambridge, UK, 1952.
3. Duoandikoetxea, J.; Martin-Reyes, F.; Ombrosi, S. Calderón weights as Muckenhoupt weights. *Indiana Univ. Math. J.* **2013**, *62*, 891–910. [CrossRef]
4. Cruz-Uribe, D.; Dalmasso, E.; Martin-Reyes, F.; Ortega Salvador, P. The Calderón operator and the Stieltjes transform on variable Lebesgue spaces with weights. *Collect. Math.* **2020**, *71*, 443–469. [CrossRef]
5. Morrey, C.B. On the solutions of quasi-linear elliptic partial differential equations. *Trans. Am. Math. Soc.* **1938**, *43*, 126–166. [CrossRef]
6. Cruz-Uribe, D.; Fiorenza, A. *Variable Lebesgue Spaces*; Birkhäuser: Basel, Switzerland, 2013.
7. Diening, L.; Harjulehto, P.; Hästö, P.; Ružička, M. *Lebesgue and Sobolev Spaces with Variable Exponents*; Springer: Berlin/Heidelberg, Germany, 2011.
8. Burenkov, V.I.; Guliyev, H.V. Necessary and sufficient conditions for boundedness of the maximal operator in local Morrey-type spaces. *Studia Math.* **2004**, *163*, 157–176. [CrossRef]
9. Burenkov, V.I.; Guliyev, H.V.; Guliyev, V.S. Necessary and sufficient conditions for boundedness of the fractional maximal operator in the local Morrey-type spaces. *J. Comp. Appl. Math.* **2007**, 208, 280–301. [CrossRef]
10. Burenkov, V.I.; Guliyev, V.S.; Tararykova, T.V.; Serbetci, A. Necessary and sufficient conditions for the boundedness of genuine singular integral operators in Local Morrey-type spaces. *Dokl. Akad. Nauk* **2008**, *422*, 11–14. [CrossRef]
11. Burenkov, V.; Gogatishvili, A.; Guliyev, V.S.; Mustafayev, R. Boundedness of the fractional maximal operator in local Morrey-type spaces. *Compl. Variabl. Ellipt. Equat.* **2010**, *55*, 739–758. [CrossRef]
12. Burenkov, V.; Nursultanov, E. Description of interpolation spaces for local Morrey-type spaces. *Proc. Steklov Inst. Math.* **2010**, *269*, 46–56. [CrossRef]
13. Guliyev, V.S. Generalized local Morrey spaces and fractional integral operators with rough kernel. *J. Math. Sci.* **2013**, *193*, 211–227. [CrossRef]
14. Yee, T.-L.; Cheung, K.L.; Ho, K.-P.; Suen, C.K. Local sharp maximal functions, geometrical maximal functions and rough maximal functions on local Morrey spaces with variable exponents. *Math. Inequal. Appl.* **2020**, *23*, 1509–1528. [CrossRef]
15. Rubio de Francia, J. Factorization and extrapolation of weights. *Bull. Am. Math. Soc.* **1982**, *7*, 393–395. [CrossRef]
16. Rubio de Francia, J. A new technique in the theory of A_p weights. In *Topics in Modern Harmonic Analysis, Vol. I, II (Turin/Milan, (1982)*; Istituto Nazionale di Alta Matematica Francesco Severi: Rome, Italy, 1983; p. 571579.
17. Rubio de Francia, J. Factorization theory and A_p weights. *Am. J. Math.* **1984**, *106*, 533–547. [CrossRef]
18. Kufner, A.; Maligranda, L.; Persson, L.-E. The prehistory of the Hardy inequality. *Am. Math. Mon.* **2006**, *113*, 715–732. [CrossRef]
19. Kufner, A.; Persson, L.-E.; Samko, N. *Weighted Inequalities of Hardy Type*; World Scientific Publishing Company: Singapore, 2017.
20. Andersen, K. Weighted inequalities for the Stieltjes transformation and Hubert's double series. *Proc. R. Soc. Edinb. Sect. A* **1980**, *86*, 75–84. [CrossRef]
21. Gogatishvili, A.; Kufner, A.; Persson, L.-E. The weighted Stieltjes inequality and applications. *Math. Nachr.* **2013**, 286, 659–668. [CrossRef]
22. Gogatishvili, A.; Persson, L.-E.; Stepanov, V.; Wall, P. Some scales of equivalent conditions to characterize the Stieltjes inequality: The case $q < p$. *Math. Nachr.* **2014**, *287*, 242–253.
23. Bastero, J.; Milman, M.; Ruiz, F. On the connection between weighted norm inequalities, commutators and real interpolation. In *Memoirs of the American Mathematical Society*; American Mathematical Society: Providence, RI, USA, 2001; Volume 154.
24. Almeida, A.; Drihem, D. Maximal, potential and singular type operators on Herz spaces with variable exponents. *J. Math. Anal. Appl.* **2012**, *394*, 781–795. [CrossRef]
25. Ho, K.-P. Extrapolation to Herz spaces with variable exponents and applications. *Rev. Mat. Complut.* **2020**, *33*, 437–463. [CrossRef]
26. Ho, K.-P. Spherical maximal function, maximal Bochner-Riesz mean and geometrical maximal function on Herz spaces with variable exponents. *Rend. Circ. Mat. Palermo Ser. II* **2021**, *70*, 559–574 . [CrossRef]
27. Ho, K.-P. Singular integral operators and sublinear operators on Hardy local Morrey spaces with variable exponents. *Bull. Sci. Math.* **2021**, *171*, 103033.
28. Sawano, Y.; Ho, K.-P.; Yang, D.; Yang, S. Hardy spaces for ball quasi-Banach function spaces. *Diss. Math.* **2017**, *525*, 1–102. [CrossRef]
29. Tao, J.; Yang, D.; Yuan, W.; Zhang, Y. Compactness Characterizations of Commutators on Ball Banach Function Spaces. *Potential Anal.* **2021**. [CrossRef]
30. Ho, K.-P. Definability of singular integral operators on Morrey-Banach spaces. *J. Math. Soc. Jpn.* **2020**, *72*, 155–170. [CrossRef]
31. Diening, L.; Samko, S. Hardy inequality in variable exponent Lebesgue spaces. *Fract. Calc. Appl. Anal.* **2007**, *10*, 1–18.
32. Andersen, K.; Sawyer, E. Weighted norm Inequalities for the Riemann-Liouville and Weyl fractional integral operators. *Trans. Am. Math. Soc.* **1988**, 308, 547–558. [CrossRef]
33. Cruz-Uribe, D.; Mamedov, F. On a general weighted Hardy type inequality in the variable exponent Lebesgue spaces. *Rev. Mat. Complut.* **2012**, *25*, 335–367. [CrossRef]
34. Ho, K.-P. Hardy's inequalities, Hilbert inequalities and fractional integrals on function spaces of q-integral p-variation. *Ann. Polon. Math.* **2021**, *126*, 251–263. [CrossRef]

35. Mamedov, F.; Harman, A. On a Hardy type general weighted inequality in spaces $L^{p(\cdot)}$. *Integral Equ. Oper. Theory* **2010**, *66*, 565–592. [CrossRef]
36. Mamedov, F.; Zeren, Y. On equivalent conditions for the general weighted Hardy type inequality in space $L^{p(\cdot)}$. *Z. Anal. Anwend.* **2012**, *31*, 55–74. [CrossRef]
37. Yee, T.-L.; Ho, K.-P. Hardy's inequalities and integral operators on Herz-Morrey spaces. *Open Math.* **2020**, *18*, 106–121. [CrossRef]

mathematics

Article

Maximal Function Characterizations of Hardy Spaces on $\mathbb{R}^n$ with Pointwise Variable Anisotropy

Aiting Wang [1,2,†], Wenhua Wang [1,†] and Baode Li [1,*,†]

1 College of Mathematics and System Science, Xinjiang University, Urumqi 830017, China; aitingwangmath@163.com (A.W.); wangwhmath@163.com (W.W.)
2 College of Mathematics and Statistics, Qinghai Minzu University, Xining 810007, China
* Correspondence: baodeli@xju.edu.cn
† These authors contributed equally to this work.

Abstract: In 2011, Dekel et al. developed highly geometric Hardy spaces $H^p(\Theta)$, for the full range $0 < p \leq 1$, which were constructed by continuous multi-level ellipsoid covers Θ of $\mathbb{R}^n$ with high anisotropy in the sense that the ellipsoids can rapidly change shape from point to point and from level to level. In this article, when the ellipsoids in Θ rapidly change shape from level to level, the authors further obtain some real-variable characterizations of $H^p(\Theta)$ in terms of the radial, the non-tangential, and the tangential maximal functions, which generalize the known results on the anisotropic Hardy spaces of Bownik.

Keywords: anisotropy; Hardy space; continuous ellipsoid cover; maximal function

Citation: Wang, A.; Wang, W.; Li, B. Maximal Function Characterizations of Hardy Spaces on $\mathbb{R}^n$ with Pointwise Variable Anisotropy. *Mathematics* **2021**, *9*, 3246. https://doi.org/10.3390/math9243246

Academic Editor: Juan Benigno Seoane-Sepúlveda

Received: 21 November 2021
Accepted: 9 December 2021
Published: 15 December 2021

1. Introduction

As a generalization of the classical isotropic Hardy spaces $H^p(\mathbb{R}^n)$ [1], anisotropic Hardy spaces $H^p_A(\mathbb{R}^n)$ were introduced and investigated by Bownik [2] in 2003. These spaces were defined on $\mathbb{R}^n$, associated with a fixed expansive matrix, which acts on an ellipsoid instead of Euclidean balls. In [3–8], many authors also studied Bownik's anisotropic Hardy spaces. In 2011, Dekel et al. [9] further generalized Bownik's spaces by constructing Hardy spaces with pointwise variable anisotropy $H^p(\Theta)$, $0 < p \leq 1$, associated with an ellipsoid cover Θ. The anisotropy in Bownik's Hardy spaces is the same one at each point in $\mathbb{R}^n$, while the anisotropy in $H^p(\Theta)$ can change rapidly from point to point and from level to level. Moreover, the ellipsoid cover Θ is a very general setting that includes the classical isotropic setting, non-isotropic setting of Calderón and Torchinsky [10], and the anisotropic setting of Bownik [2] as special cases; see more details in ([2], pp. 2–3) and ([11], p. 157).

On the other hand, maximal function characterizations are very fundamental characterizations of Hardy spaces, and they are crucial to conveniently apply the real-variable theory of Hardy spaces $H^p(\mathbb{R}^n)$ with $p \in (0, 1]$. Maximal function characterizations were first shown for the classical isotropic Hardy spaces $H^p(\mathbb{R}^n)$ by Fefferman and Stein in their fundamental work [1], ([12], Chapter III). Analogous results were shown by Calderón and Torchinsky [10,13] for parabolic H^p spaces and Uchiyama [14] for H^p on a homogeneous-type space. In 2003, Bownik ([2], p. 42) obtained the maximal function characterizations of the anisotropic Hardy space $H^p_A(\mathbb{R}^n)$. This was further extended to anisotropic Hardy spaces of the Musielak–Orlicz type in [15], to anisotropic Hardy–Lorentz spaces in [16], to variable anisotropic Hardy spaces in [17], and to anisotropic mixed-norm Hardy spaces in [18].

Motivated by the abovementioned facts, a natural question arises: Do the maximal function characterizations still hold for Hardy spaces $H^p(\Theta)$ with variable anisotropy? In this article, we answer this question affirmatively in the sense that the ellipsoids in Θ

can change shape rapidly from level to level, which is a variable anisotropic extension of Bownik's [2].

This article is organized as follows.

In Section 2, we recall some notation and definitions concerning anisotropic continuous ellipsoid cover Θ, several maximal functions, and anisotropic Hardy spaces $H^p(\Theta)$ defined via the grand radial maximal function. We also give some propositions about $H^p(\Theta)$, several classes of variable anisotropic maximal functions, and Schwartz functions since they provide tools for further work. In Section 3, we first state the main result: if the ellipsoids in Θ can rapidly change shape from level to level (see Definition 1), denoted as Θ_t, we may obtain some real-variable characterizations of $H^p(\Theta_t)$ in terms of the radial, the non-tangential, and the tangential maximal functions (see Theorem 1). Then, we present several lemmas that are isotropic extensions in the setting of variable anisotropy, and finally, we show the proof for the main result.

In the process of proving the main result, we used the methods from Stein [1] and Bownik [2]. However, it is worth pointing out that these ellipsoids of Bownik were images of the unit ball by powers of a fixed expansive matrix, whereas in our case, the ellipsoids of Dekel are images of the unit ball by powers of a group of matrices satisfying some "shape condition". This makes the proof complicated and needs many subtle estimates such as Propositions 5 and 6, and Lemma 1.

However, this article left an open question: if the maximal function characterizations of $H^p(\Theta)$ still hold true in the sense that the ellipsoids of Θ change rapidly from level to level and from point to point?

Finally, we note some conventions on notation. Let $\mathbb{N}_0 := \{0, 1, 2, \ldots\}$ and $\lceil t \rceil$ be the smallest integer no less than t. For any $\alpha := (\alpha_1, \ldots, \alpha_n) \in \mathbb{N}_0^n$, $|\alpha| := \alpha_1 + \cdots + \alpha_n$ and $\partial^\alpha := (\frac{\partial}{\partial x_1})^{\alpha_1} \cdots (\frac{\partial}{\partial x_n})^{\alpha_n}$. Throughout the whole paper, we denote by C a positive constant that is independent on the main parameters but may vary from line to line. For any sets $E, F \subset \mathbb{R}^n$, we use $E^\complement$ to denote the set $\mathbb{R}^n \setminus E$. If there are no special instructions, any space $\mathcal{X}(\mathbb{R}^n)$ is denoted simply by $\mathcal{X}$. Denote by $\mathcal{S}$ the space of all Schwartz functions and $\mathcal{S}'$ the space of all tempered distributions.

2. Preliminary and Some Basic Propositions

In this section, we first recall the notion of continuous ellipsoid covers Θ and we introduce the pointwise continuity for Θ. An *ellipsoid* ξ in $\mathbb{R}^n$ is an image of the Euclidean unit ball $\mathbb{B}^n := \{x \in \mathbb{R}^n : |x| < 1\}$ under an affine transform, i.e.,

$$\xi := M_\xi(\mathbb{B}^n) + c_\xi,$$

where M_ξ is a non-singular matrix and $c_\xi \in \mathbb{R}^n$ is the center.

Let us begin with the definition of continuous ellipsoid covers, which was introduced in ([11], Definition 2.4).

Definition 1. *We say that*

$$\Theta := \{\theta(x, t) : x \in \mathbb{R}^n, t \in \mathbb{R}\}$$

is a continuous ellipsoid cover of $\mathbb{R}^n$ or, in short, an ellipsoid cover if there exist positive constants $p(\Theta) := \{a_1, \ldots, a_6\}$ such that

(i) For every $x \in \mathbb{R}^n$ and $t \in \mathbb{R}$, there exists an ellipsoid $\theta(x, t) := M_{x,t}(\mathbb{B}^n) + x$ satisfying

$$a_1 2^{-t} \le |\theta(x, t)| \le a_2 2^{-t}. \tag{1}$$

(ii) Intersecting ellipsoids from Θ satisfy a "shape condition", i.e., for any $x, y \in \mathbb{R}^n$, $t \in \mathbb{R}$ and $s \ge 0$, if $\theta(x, t) \cap \theta(y, t+s) \neq \emptyset$, then

$$a_3 2^{-a_4 s} \le \frac{1}{\|(M_{y,t+s})^{-1} M_{x,t}\|} \le \|(M_{x,t})^{-1} M_{y,t+s}\| \le a_5 2^{-a_6 s}. \tag{2}$$

where $\|\cdot\|$ *is the matrix norm given by* $\|M\| := \max_{|x|=1} |Mx|$ *for an* $n \times n$ *real matrix* M.

Particularly, for any $\theta(x,t) \in \Theta$, *when the related matrix function* $M_{x,t}$ *of* $x \in \mathbb{R}^n$ *and* $t \in \mathbb{R}$ *is reduced to the matrix function* M_t *of* $t \in \mathbb{R}$, *we call a cover* Θ *a t-continuous ellipsoid cover, denoted as* Θ_t.

The word continuous refers to the fact that ellipsoids $\theta x, t$ *are defined for all values of* $x \in \mathbb{R}^n$ *and* $t \in \mathbb{R}$, *and we say that a continuous ellipsoid cover* Θ *is pointwise continuous if, for every* $t \in \mathbb{R}$, *the matrix valued function* $x \mapsto M_{x,t}$ *is continuous:*

$$\|M_{x',t} - M_{x,t}\| \to 0 \text{ as } x' \to x. \tag{3}$$

Remark 1. *By ([19], Theorem 2.2), we know that the pointwise continuous assumption is not necessary since it is always possible to construct an equivalent ellipsoid cover*

$$\Xi := \{\zeta x, t : x \in \mathbb{R}^n, t \in \mathbb{R}\}$$

such that Ξ *is pointwise continuous and* Ξ *is equivalent to* Θ. *Here, we say that two ellipsoid covers* Θ *and* Ξ *are equivalent if there exists a constant* $C > 0$ *such that, for any* $x \in \mathbb{R}^n$ *and* $t \in \mathbb{R}$, *we have*

$$\frac{1}{C}\zeta x, t \subset \theta x, t \subset C\zeta x, t.$$

Taking $M_{y,t+s} = M_{x,t}$ in (2), we have

$$a_3 \leq 1 \text{ and } a_5 \geq 1. \tag{4}$$

For more properties about ellipsoid covers, see [9,11].

For any N, $\widetilde{N} \in \mathbb{N}_0$ with $N \leq \widetilde{N}$, let

$$\mathcal{S}_{N,\widetilde{N}} := \left\{ \psi \in \mathcal{S} :\ \|\psi\|_{\mathcal{S}_{N,\widetilde{N}}} := \max_{\alpha \in \mathbb{N}_0^n, |\alpha| \leq N} \sup_{y \in \mathbb{R}^n} (1+|y|)^{\widetilde{N}} |\partial^\alpha \psi(y)| \leq 1 \right\}.$$

For any $\varphi \in \mathcal{S}$, $x \in \mathbb{R}^n$, $t \in \mathbb{R}$ and $\theta(x,t) = M_{x,t}(\mathbb{B}^n) + x$, denote

$$\varphi_{x,t}(y) := \left|\det(M_{x,t}^{-1})\right| \varphi(M_{x,t}^{-1} y), \quad y \in \mathbb{R}^n.$$

Particularly, when the matrix $M_{x,t}$ is reduced to M_t, $\varphi_{x,t}(y)$ is simply denoted as $\varphi_t(y)$.

Now, we give the notions of anisotropic variants of the non-tangential, the grand non-tangential, the radial, the grand radial, and the tangential maximal functions.

Definition 2. *Let* $f \in \mathcal{S}'$, $\varphi \in \mathcal{S}$ *and* N, $\widetilde{N} \in \mathbb{N}_0$ *with* $N \leq \widetilde{N}$. *We define the non-tangential, the grand non-tangential, the radial, the rand radial, and the tangential maximal functions, respectively as*

$$M_\varphi f(x) := \sup_{t \in \mathbb{R}} \sup_{y \in \theta(x,t)} |f * \varphi_{x,t}(y)|, \quad x \in \mathbb{R}^n,$$

$$M_{N,\widetilde{N}} f(x) := \sup_{\varphi \in \mathcal{S}_{N,\widetilde{N}}} M_\varphi f(x), \quad x \in \mathbb{R}^n,$$

$$M_\varphi^0 f(x) := \sup_{t \in \mathbb{R}} |f * \varphi_{x,t}(x)|, \quad x \in \mathbb{R}^n,$$

$$M_{N,\widetilde{N}}^0 f(x) := \sup_{\varphi \in \mathcal{S}_{N,\widetilde{N}}} M_\varphi^\circ f(x), \quad x \in \mathbb{R}^n,$$

$$T_\varphi^N f(x) := \sup_{t \in \mathbb{R}} \sup_{y \in \mathbb{R}^n} |f * \varphi_{x,t}(y)| \left(1 + \left|M_{x,t}^{-1}(x-y)\right|\right)^{-N}, \quad x \in \mathbb{R}^n.$$

Here and hereafter, the symbol "$*$" always represents a convolution.

Remark 2. *We immediately have the following pointwise estimate among the radial, the non-tangential, and the tangential maximal functions:*

$$M_\varphi^0 f(x) \le M_\varphi f(x) \le 2^N T_\varphi^N f(x), \quad x \in \mathbb{R}^n.$$

Next, we recall the definition of Hardy spaces with pointwise variable anisotropy ([9], Definition 3.6) via the grand radial maximal function.

Let Θ be an ellipsoid cover of $\mathbb{R}^n$ with parameters $p(\Theta) = \{a_1, \cdots, a_6\}$ and $0 < p \le 1$. We define $N_p(\Theta)$ as the minimal integer satisfying

$$N_p := N_p(\Theta) > \frac{\max(1, a_4)n + 1}{a_6 p}, \tag{5}$$

and then $\widetilde{N}_p(\Theta)$ as the minimal integer satisfying

$$\widetilde{N}_p := \widetilde{N}_p(\Theta) > \frac{a_4 N_p(\Theta) + 1}{a_6}. \tag{6}$$

Definition 3. *Let Θ be a continuous ellipsoid cover and $0 < p \le 1$. Define $M^0 := M^0_{N_p, \widetilde{N}_p}$, and the anisotropic Hardy space is defined as*

$$H^p_{N_p, \widetilde{N}_p}(\Theta) := \{f \in \mathcal{S}' : M^0 f \in L^p\}$$

with the (quasi-)norm $\|f\|_{H^p(\Theta)} := \|M^0 f\|_{L^p}$.

Remark 3. *By Remark 1, we know that, for every continuous ellipsoid cover Θ, there exists an equivalent pointwise continuous ellipsoid cover Ξ. This implies that their corresponding (quasi-)norms $\rho_\Theta(\cdot,\cdot)$ and $\rho_\Xi(\cdot,\cdot)$ are also equivalent, and hence, the corresponding Hardy spaces $H^p(\Theta) = H^p(\Xi)(0 < p \le 1)$ with equivalent (quasi-)norms (see ([9], Theorem 5.8)). Therefore, here and hereafter, we always consider Θ of $H^p(\Theta)$ to be a pointwise continuous ellipsoid cover.*

Proposition 1. *Let Θ be an ellipsoid cover, $0 < p \le 1 \le q \le \infty$, $p < q$ and $l \ge N_p$ with N_p as in (5). If $N \ge N_p$ and $\widetilde{N} \ge (a_4 N + 1)/a_6$, then*

$$H^p_{N_p, \widetilde{N}_p}(\Theta) = H^p_{q,l}(\Theta) = H^p_{N, \widetilde{N}}(\Theta)$$

with equivalent (quasi-)norms, where $H^p_{q,l}(\Theta)$ denotes the atomic Hardy space with pointwise variable anisotropy; see ([9], Definition 4.2).

Proof. This proposition is a corollary of ([9], Theorems 4.4 and 4.19). Indeed, by Definition 3, we obtain that, for any $N \ge N_p$ and $\widetilde{N} \ge (a_4 N + 1)/a_6$,

$$H^p_{N_p, \widetilde{N}_p}(\Theta) \subseteq H^p_{N, \widetilde{N}}(\Theta).$$

Combining this and $H^p_{q,l}(\Theta) \subseteq H^p_{N_p, \widetilde{N}_p}(\Theta)$ (see ([9], Theorem 4.4)), we obtain

$$H^p_{q,l}(\Theta) \subseteq H^p_{N, \widetilde{N}}(\Theta). \tag{7}$$

By checking the definition of anisotropic (p, q, l)-atom (see ([9], Definition 4.1)), we know that every (p, ∞, l)-atom is also a (p, q, l)-atom and hence

$$H^p_{\infty,l}(\Theta) \subseteq H^p_{q,l}(\Theta).$$

Let $l' \geq \max(l, N)$. By a similar argument to the proof of ([9], Theorem 4.19), we obtain

$$H^p_{N,\widetilde{N}}(\Theta) \subseteq H^p_{\infty,l'}(\Theta),$$

where $N \geq N_p$ and $\widetilde{N} \geq (a_4 N + 1)/a_6$. Thus,

$$H^p_{N,\widetilde{N}}(\Theta) \subseteq H^p_{\infty,l'}(\Theta) \subseteq H^p_{\infty,l}(\Theta) \subseteq H^p_{q,l}(\Theta). \tag{8}$$

Combining (7) and (8), we conclude that

$$H^p_{N_p,\widetilde{N}_p}(\Theta) = H^p_{q,l}(\Theta) = H^p_{N,\widetilde{N}}(\Theta)$$

with equivalent (quasi-)norms. □

Remark 4. *From Proposition 1, we deduce that, for any integers $N \geq N_p$ and $\widetilde{N} \geq (a_4 N + 1)/a_6$, the definition of $H^p_{N,\widetilde{N}}(\Theta)$ is independent of N and $\widetilde{N}$. Therefore, from now on, we denote $H^p_{N,\widetilde{N}}(\Theta)$ with $N \geq N_p$ and $\widetilde{N} \geq (a_4 N + 1)/a_6$ simply by $H^p(\Theta)$.*

Proposition 2 ([9], Lemma 2.3)**.** *Let Θ be an ellipsoid cover. Then, there exists a constant $J := J(p(\Theta)) \geq 1$ such that, for any $x \in \mathbb{R}^n$ and $t \in \mathbb{R}$,*

$$2M_{x,t}(\mathbb{B}) + x \subset \theta(x, t - J).$$

Here and hereafter, let J always be as in Proposition 2.

Definition 4 ([9], Definition 3.1)**.** *Let Θ be an ellipsoid cover. For any locally integrable function f, the maximal function of the Hardy–Littlewood type of f is defined by*

$$M_\Theta f(x) := \sup_{t \in \mathbb{R}} \frac{1}{|\theta(x,t)|} \int_{\theta(x,t)} |f(y)|\, dy, \quad x \in \mathbb{R}^n.$$

Proposition 3 ([9], Theorem 3.3)**.** *Let Θ be an ellipsoid cover. Then,*

(i) There exists a constant C depending only on $p(\Theta)$ and n such that for all $f \in L^1$ and $\alpha > 0$,

$$|\{x : M_\Theta f(x) > \alpha\}| \leq C\alpha^{-1} \|f\|_{L^1}; \tag{9}$$

(ii) For $1 < p < \infty$, there exists a constant C_p depending only on C and p such that, for all $f \in L^p$,

$$\|M_\Theta f\|_{L^p} \leq C_p \|f\|_{L^p}. \tag{10}$$

We give some useful results about variable anisotropic maximal functions with different apertures. They also play important roles in obtaining the maximal function characterizations of $H^p(\Theta)$. For any given $x \in \mathbb{R}^n$, suppose that $F : \mathbb{R}^n \times \mathbb{R} \to (0, \infty)$ is a Lebesgue measurable function. Let Θ be an ellipsoid cover. For fixed $l \in \mathbb{Z}$ and $t_0 < 0$, define the maximal function of F with aperture l as

$$F_l^{*\,t_0}(x) := \sup_{t \geq t_0} \sup_{y \in \theta(x, t - lJ)} F(y,t). \tag{11}$$

Proposition 4. *For any $l \in \mathbb{Z}$ and $t_0 < 0$, let $F_l^{*\,t_0}$ be as in (11). If the ellipsoid cover Θ is pointwise continuous, then $F_l^{*\,t_0} : \mathbb{R}^n \to (0, \infty]$ is lower semi-continuous, i.e.,*

$$\{x \in \mathbb{R}^n : F_l^{*\,t_0}(x) > \lambda\} \text{ is open for any } \lambda > 0.$$

Proof. If $F_l^{*\,t_0}(x) > \lambda$ for some $x \in \mathbb{R}^n$, then there exist $t \geq t_0$ and $y \in \theta(x, t - lJ)$ such that $F(y,t) > \lambda$. Since $\theta(x,t)$ is continuous for variable x (see Remark 1), there exists $\delta_1 > 0$ such that, for any $x' \in U(x,\delta) := \{z \in \mathbb{R}^n : |z - x| < \delta\}$, $y \in \theta(x', t - lJ)$ and hence $F_l^{*\,t_0}(x') > \lambda$. □

By Proposition 4, we obtain that $\{x \in \mathbb{R}^n : F_l^{*\,t_0}(x) > \lambda\}$ is Lebesgue measurable. Based on this and inspired by ([2], Lemma 7.2), the following Proposition 5 shows some estimates for maximal function $F_l^{*\,t_0}$.

Proposition 5. *Let Θ be an ellipsoid cover, $F_l^{*\,t_0}$ and $F_{l'}^{*\,t_0}$ as in (11) with integers $l > l'$ and $t_0 < 0$. Then, there exists a constant $C > 0$ that depends on parameters $p(\Theta)$ such that, for any functions $F_l^{*\,t_0}$, $F_{l'}^{*\,t_0}$ and $\lambda > 0$, we have*

$$\left|\left\{x \in \mathbb{R}^n : F_l^{*\,t_0}(x) > \lambda\right\}\right| \leq C 2^{(l-l')J} \left|\left\{x \in \mathbb{R}^n : F_{l'}^{*\,t_0}(x) > \lambda\right\}\right| \tag{12}$$

and

$$\int_{\mathbb{R}^n} F_l^{*\,t_0}(x)\, dx \leq C 2^{(l-l')J} \int_{\mathbb{R}^n} F_{l'}^{*\,t_0}(x)\, dx. \tag{13}$$

Proof. Let $\Omega := \{x \in \mathbb{R}^n : F_{l'}^{*\,t_0}(x) > \lambda\}$. We claim that

$$\left\{x \in \mathbb{R}^n : F_l^{*\,t_0}(x) > \lambda\right\} \subset \left\{x \in \mathbb{R}^n : M_\Theta(\chi_\Omega)(x) \geq C_1 2^{(l'-l)J}\right\}, \tag{14}$$

where C_1 is a positive constant to be fixed later. Assuming that the claim holds for the moment, from this and a weak type (1,1) of M_Θ (see (9)), we deduce

$$\begin{aligned}\left|\left\{x \in \mathbb{R}^n : F_l^{*\,t_0}(x) > \lambda\right\}\right| &\leq \left|\left\{x \in \mathbb{R}^n : M_\Theta(\chi_\Omega)(x) \geq C_1 2^{(l'-l)J}\right\}\right| \\ &\leq C_1^{-1} 2^{(l-l')J} \|\chi_\Omega\|_{L^1} \leq C 2^{(l-l')J} |\Omega|\end{aligned}$$

and hence (12) holds true, where $C := 1/C_1$. Furthermore, integrating (12) on $(0, \infty)$ with respect to λ yields (13). Therefore, (14) remains to be shown.

Suppose $F_l^{*\,t_0}(x) > \lambda$ for some $x \in \mathbb{R}^n$. Then, there exist t with $t \geq t_0$ and $y \in \theta(x, t - lJ)$ such that $F(y, t) > \lambda$. For any $l, l' \in \mathbb{Z}$ and $l \geq l'$, we first prove that the following holds true:

$$a_5{}^{-1}\theta(y, t - l'J) \subseteq \theta(x, t - (l+1)J) \cap \Omega. \tag{15}$$

For any $z \in a_5{}^{-1}\theta(y, t - l'J)$, by (4), we have $z \in \theta(y, t - l'J)$ and hence

$$\theta(z, t - l'J) \cap \theta(y, t - l'J) \neq \emptyset.$$

Thus, by (2), we have

$$\left\| M_{z,t-l'J}^{-1} M_{y,t-l'J} \right\| \leq a_5.$$

From this, it follows that

$$a_5{}^{-1} M_{z,t-l'J}^{-1} M_{y,t-l'J}(\mathbb{B}^n) \subseteq \mathbb{B}^n$$

and hence

$$a_5{}^{-1} M_{y,t-l'J}(\mathbb{B}^n) \subseteq M_{z,t-l'J}(\mathbb{B}^n).$$

By this and $y \in a_5^{-1} M_{y,t-l'J}(\mathbb{B}^n) + z$, we obtain $y \in \theta(z, t - l'J)$. From this and $F(y,t) > \lambda$ with $t \geq t_0$, we deduce that $F_{l'}^{*t_0}(z) > \lambda$, and hence, $z \in \Omega$, which implies

$$a_5^{-1}\theta(y, t - l'J) \subseteq \Omega. \tag{16}$$

Moreover, by $y \in \theta(x, t - lJ)$, (2), and $l \geq l'$, we have

$$\left\| M_{x,t-lJ}^{-1} M_{y,t-l'J} \right\| \leq a_5 2^{-a_6(l-l')J} \leq a_5.$$

From this, it follows that

$$a_5{}^{-1} M_{x,t-lJ}^{-1} M_{y,t-l'J}(\mathbb{B}^n) \subseteq \mathbb{B}^n$$

and hence

$$a_5{}^{-1} M_{y,t-l'J}(\mathbb{B}^n) \subseteq M_{x,t-lJ}(\mathbb{B}^n).$$

By this, (4), $y \in \theta(x, t - lJ)$, and Proposition 2, we obtain

$$a_5{}^{-1} M_{y,t-l'J}(\mathbb{B}^n) + y \subseteq 2M_{x,t-lJ}(\mathbb{B}^n) + x \subseteq \theta(x, t - (l+1)J).$$

From this and (16), we deduce that (15) holds true.

Next, let us prove (14). By (15) and (1), we obtain

$$\begin{aligned} |\theta(x, t - (l+1)J) \cap \Omega| &\geq (a_5)^{-n} |\theta(y, t - l'J)| \\ &\geq \frac{a_1}{(a_5)^n} 2^{l'J-t}. \end{aligned} \tag{17}$$

Taking $b_0 := t - (l+1)J$, by (1) and (17), we have

$$\frac{1}{|\theta(x, b_0)|} \int_{\theta(x,b_0)} |\chi_\Omega(y)| dy \geq a_2{}^{-1} 2^{b_0} |\theta(x, b_0) \cap \Omega| \geq \frac{a_1}{(a_5)^n a_2} 2^{(l'-l-1)J},$$

which implies $M_\Theta(\chi_\Omega)(x) \geq C_1 2^{(l'-l)J}$ and hence (14) holds true, where $C_1 := 2^{-J} a_1 / [(a_5)^n a_2]$. □

The following result enables us to pass from one function in $\mathcal{S}$ to the sum of dilates of another function in $\mathcal{S}$ with nonzero mean, which is a variable anisotropic extension of ([12], p. 93, Lemma 2) of Stein and ([2], Lemma 7.3) of Bownik.

Proposition 6. *Let Θ be an ellipsoid cover of $\mathbb{R}^n$ and $\varphi \in \mathcal{S}$, with $\int_{\mathbb{R}^n} \varphi(x)\, dx \neq 0$. Then, for any $\psi \in \mathcal{S}$, $x \in \mathbb{R}^n$, and $t \in \mathbb{R}$, there exists a sequence $\{\eta^k\}_{k=0}^{\infty}$ and $\eta^k \in \mathcal{S}$, such that*

$$\psi = \sum_{k=0}^{\infty} \eta^k * \varphi^k \tag{18}$$

converges in $\mathcal{S}$, where

$$\varphi^k := |\det(M_{x,t+kJ}^{-1} M_{x,t})| \varphi(M_{x,t+kJ}^{-1} M_{x,t} \cdot), \quad k > 0,$$

where $J > 0$ is as in Proposition 2.

Furthermore, for any positive integers N, $\widetilde{N}$ and L, there exists a constant $C > 0$ depending on φ, L, N, $\widetilde{N}$, and $p(\Theta)$ but not ψ, such that

$$\|\eta^k\|_{\mathcal{S}_{N,\widetilde{N}}} \leq C 2^{-kL} \|\psi\|_{\mathcal{S}_{N+n+1+\lceil L/(a_6 J) \rceil, \widetilde{N}+n+1}}. \tag{19}$$

Proof. The following simplified proof is accomplished by Dekel. By scaling φ, we can assume that $\int_{\mathbb{R}^n}\varphi(x)dx = 1$ and $|\widehat{\varphi}(\xi)| \geq 1/2$, for $|\xi| \leq 2$. This assumption only impacts the constant in (19). Let $\zeta \in \mathcal{S}$ such that $0 \leq \zeta \leq 1$ on $\mathbb{B}^n$ and supp $(\zeta) \subset 2\mathbb{B}^n$. We fix $x \in \mathbb{R}^n$ and $t \in \mathbb{R}$, denote $M_k := M_{x,t+kJ}$, and define the sequence of functions $\{\zeta_k\}_{k=0}^{\infty}$, where $\zeta_0 := \zeta$, and

$$\zeta_k := \zeta\Big(\big(M_{x,t}^{-1}M_k\big)^T\cdot\Big) - \zeta\Big(\big(M_{x,t}^{-1}M_{k-1}\big)^T\cdot\Big), \quad k \geq 1,$$

where M^T denotes the transpose of a matrix M. We claim that

$$\operatorname{supp}(\zeta_k) \subset \Big\{\xi \in \mathbb{R}^n : a_5^{-1}2^{-a_6 J}2^{a_6 kJ} \leq |\xi| \leq 2a_3^{-1}2^{a_4 kJ}\Big\}. \tag{20}$$

Indeed, by the properties of ζ, Proposition 2 and (2),

$$\begin{aligned}\xi \in \operatorname{supp}(\zeta_k) &\Rightarrow \big(M_{x,t}^{-1}M_k\big)^T(\xi) \in 2\mathbb{B}^n \vee \big(M_{x,t}^{-1}M_{k-1}\big)^T(\xi) \in 2\mathbb{B}^n \\ &\Rightarrow \xi \in 2\big(M_k^{-1}M_{x,t}\big)^T(\mathbb{B}^n) \vee \xi \in 2\big(M_{k-1}^{-1}M_{x,t}\big)^T(\mathbb{B}^n) \\ &\Rightarrow \xi \in 2a_3^{-1}2^{a_4 kJ}\mathbb{B}^n.\end{aligned}$$

In the other direction, Proposition 2 and the properties of ζ yield

$$\begin{aligned}\xi \in \big(M_{k-1}^{-1}M_{x,t}\big)^T(\mathbb{B}^n) &\Rightarrow \big(M_{x,t}^{-1}M_k\big)^T(\xi) \in \mathbb{B}^n, \big(M_{x,t}^{-1}M_{k-1}\big)^T(\xi) \in \mathbb{B}^n \\ &\Rightarrow \zeta_k(\xi) = 0.\end{aligned}$$

Applying (2), we have

$$\xi \notin \big(M_{k-1}^{-1}M_{x,t}\big)^T(\mathbb{B}^n) \Rightarrow |\xi| \geq 2a_5^{-1}2^{a_6(k-1)J}.$$

This proves (20). Additionally, by (2), for any $\xi \in \mathbb{R}^n$,

$$\Big|\big(M_{x,t}^{-1}M_k\big)^T\xi\Big| \leq \Big\|M_{x,t}^{-1}M_k\Big\||\xi| \leq a_5 2^{-a_6 kJ}|\xi| \to 0, \quad k \to \infty.$$

From this, we deduce that, for any $\xi \in \mathbb{R}^n$, for a large enough k, $(M_{x,t}^{-1}M_k)^T\xi \in \mathbb{B}^n$. This implies that

$$\sum_{k=0}^{\infty}\zeta_k(\xi) = 1, \quad \forall\xi \in \mathbb{R}^n.$$

Thus, formally, a Fourier transform of (18) is given by

$$\widehat{\psi} = \sum_{k=0}^{\infty}\widehat{\eta^k}\widehat{\varphi}\Big(\big(M_{x,t}^{-1}M_k\big)^T\cdot\Big), \quad \widehat{\eta^k} := \frac{\zeta_k}{\widehat{\varphi}\Big(\big(M_{x,t}^{-1}M_k\big)^T\cdot\Big)}\widehat{\psi}.$$

Observe that η^k is well defined and in $\mathcal{S}$. Indeed, $\widehat{\eta^k}$ is well defined with $0/0 := 0$, since by our assumption on φ,

$$\begin{aligned}\xi \in supp(\zeta_k) &\Rightarrow \xi \in 2\big(M_k^{-1}M_{x,t}\big)^T(\mathbb{B}^n) \\ &\Rightarrow \Big|\big(M_{x,t}^{-1}M_k\big)^T\xi\Big| \leq 2 \\ &\Rightarrow \widehat{\varphi}\Big(\big(M_{x,t}^{-1}M_k\big)^T\xi\Big) \geq \frac{1}{2}.\end{aligned}$$

From this, it is obvious that $\widehat{\eta^k} \in \mathcal{S}$, and therefore, $\eta^k \in \mathcal{S}$. We now proceed to prove (19). First, observe that, for any $\eta \in \mathcal{S}$, N, $\widetilde{N} \in \mathbb{N}$,

$$\|\eta\|_{\mathcal{S}_{N,\widetilde{N}}} \leq C(N, \widetilde{N}, n)\|\widehat{\eta}\|_{\mathcal{S}_{\widetilde{N},N+n+1}}. \tag{21}$$

Next, we claim that, for any $K \in \mathbb{N}$,

$$\max_{|\alpha|\leq K}\left\|\partial^\alpha\left(\zeta_k/\widehat{\varphi}\left(\left(M_k^{-1}M_{x,t}\right)^T\cdot\right)\right)\right\|_\infty \leq C(K, n, \varphi). \tag{22}$$

Indeed, on its support, any partial derivative of $\zeta_k/\widehat{\varphi}((M_{x,t}^{-1}M_k)^T\cdot)$ has a denominator with its absolute value bounded from below and a numerator that is a superposition of compositions of partial derivatives of η and φ with contractive matrices of the type $(M_{x,t}^{-1}M_k)^T$. Using (20)–(22), we obtain

$$\begin{aligned}
\left\|\eta^k\right\|_{\mathcal{S}_{N,\widetilde{N}}} &\leq C\left\|\widehat{\eta^k}\right\|_{\mathcal{S}_{\widetilde{N},N+n+1}} \\
&\leq C \sup_{|\xi|\geq a_5^{-1}2^{-a_6J}2^{a_6kJ}} \max_{|\alpha|\leq\widetilde{N}}\left|\partial^\alpha\widehat{\eta^k}(\xi)\right|(1+|\xi|)^{N+n+1} \\
&\leq C \sup_{|\xi|\geq a_5^{-1}2^{-a_6J}2^{a_6kJ}} \max_{|\alpha|\leq\widetilde{N}}\left|\partial^\alpha\widehat{\psi}(\xi)\right|(1+|\xi|)^{N+n+1} \\
&\leq C \sup_{|\xi|\geq a_5^{-1}2^{-a_6J}2^{a_6kJ}} \max_{|\alpha|\leq\widetilde{N}}\left|\partial^\alpha\widehat{\psi}(\xi)\right|(1+|\xi|)^{N+n+1+\lceil L/(a_6J)\rceil} \\
&\quad\times(1+|\xi|)^{-\lceil L/(a_6J)\rceil} \\
&\leq C2^{-kL}\|\widehat{\psi}\|_{\mathcal{S}_{\widetilde{N},N+n+1+\lceil L/(a_6J)\rceil}} \\
&\leq C2^{-kL}\|\psi\|_{\mathcal{S}_{N+n+1+\lceil L/(a_6J)\rceil,\widetilde{N}+n+1}}.
\end{aligned}$$

□

3. Maximal Function Characterizations of $H^p(\Theta_t)$

In this section, we show the maximal function characterizations of $H^p(\Theta_t)$ using the radial, the non-tangential, and the tangential maximal functions of a single test function $\varphi \in \mathcal{S}$.

Theorem 1. *Let Θ_t be a t-continuous ellipsoid cover, $0 < p \leq 1$, and $\varphi \in \mathcal{S}$ satisfy $\int_{\mathbb{R}^n}\varphi(x)\,dx \neq 0$. Then, for any $f \in \mathcal{S}'$, the following are mutually equivalent:*

$$f \in H^p(\Theta_t); \tag{23}$$

$$M_\varphi f \in L^p; \tag{24}$$

$$M_\varphi^0 f \in L^p; \tag{25}$$

$$T_\varphi^N f \in L^p, \quad N > \frac{1}{a_6p}. \tag{26}$$

In this case,

$$\|f\|_{H^p(\Theta_t)} = \left\|M^0f\right\|_{L^p} \leq C_1\left\|T_\varphi^N f\right\|_{L^p} \leq C_2\|M_\varphi f\|_{L^p} \leq C_3\left\|M_\varphi^0 f\right\|_{L^p} \leq C_4\|f\|_{H^p(\Theta_t)},$$

where the positive constants C_1, C_2, C_3 and C_4 are independent of f.

The framework to prove Theorem 1 is motivated by Fefferman and Stein [1], ([12], Chapter III), and Bownik ([2], p. 42, Theorem 7.1).

Inspired by Fefferman and Stein ([12], p. 97), and Bownik ([2], p. 47), we now start with maximal functions obtained from truncation with an additional extra decay term. Namely, for $t_0 < 0$ representing the truncation level and real number $L \geq 0$ representing the decay level, we define the *radial*, the *non-tangential*, the *tangential*, the *grand radial*, and the *grand non-tangential maximal functions*, respectively, as

$$M_\varphi^{0\,(t_0,L)} f(x) := \sup_{t\geq t_0} |(f*\varphi_{x,t})(x)| \left(1+\left|M_{x,t_0}^{-1}x\right|\right)^{-L} (1+2^{t+t_0})^{-L},$$

$$M_\varphi^{(t_0,L)} f(x) := \sup_{t\geq t_0} \sup_{y\in\theta(x,t)} |(f*\varphi_{x,t})(y)| \left(1+\left|M_{x,t_0}^{-1}y\right|\right)^{-L} (1+2^{t+t_0})^{-L},$$

$$T_\varphi^{N\,(t_0,L)} f(x) := \sup_{t\geq t_0} \sup_{y\in\mathbb{R}^n} \frac{|(f*\varphi_{x,t})(y)|}{\left[1+\left|M_{x,t}^{-1}(x-y)\right|\right]^N} \frac{1}{(1+2^{t+t_0})^L\left(1+\left|M_{x,t_0}^{-1}y\right|\right)^L},$$

$$M_{N,\tilde{N}}^{0\,(t_0,L)} f(x) := \sup_{\varphi\in\mathcal{S}_{N,\tilde{N}}} M_\varphi^{0\,(t_0,L)} f(x)$$

and

$$M_{N,\tilde{N}}^{(t_0,L)} f(x) := \sup_{\varphi\in\mathcal{S}_{N,\tilde{N}}} M_\varphi^{(t_0,L)} f(x).$$

The following Lemma 1 guarantees control of the tangential by the non-tangential maximal function in $L^p(\mathbb{R}^n)$ independent of t_0 and L.

Lemma 1. *Let Θ_t be a t-continuous ellipsoid cover. Suppose $p > 0$, $N > 1/(a_6 p)$, and $\varphi \in \mathcal{S}$. Then, there exists a positive constant C such that, for any $t_0 < 0$, $L \geq 0$ and $f \in \mathcal{S}'$,*

$$\left\|T_\varphi^{N\,(t_0,L)} f\right\|_{L^p} \leq C\left\|M_\varphi^{(t_0,L)} f\right\|_{L^p}.$$

Proof. Consider the function $F : \mathbb{R}^n \times \mathbb{R} \longrightarrow [0,\infty)$ given by

$$F(y,t) := |(f*\varphi_t)(y)|^p \left(1+\left|M_{t_0}^{-1}y\right|\right)^{-pL} (1+2^{t+t_0})^{-pL}.$$

Let $F_l^{*\,t_0}$ be as in (11) with $l = 0$. When $y \in \theta(x,t)$, we have $M_t^{-1}(x-y) \in \mathbb{B}^n$ and hence $|M_t^{-1}(x-y)| < 1$. If $t \geq t_0$, then

$$F(y,t)\left[1+\left|M_t^{-1}(x-y)\right|\right]^{-pN} \leq F_0^{*\,t_0}(x).$$

When $y \in \theta(x,t-kJ)\backslash\theta(x,t-(k-1)J)$ for some $k \geq 1$, we have

$$M_t^{-1}(x-y) \notin M_t^{-1}M_{t-(k-1)J}(\mathbb{B}^n). \tag{27}$$

By (2), we obtain

$$\left\|M_{t-(k-1)J}^{-1}M_t\right\| \leq a_5 2^{-a_6(k-1)J}$$

and hence,

$$M_{t-(k-1)J}^{-1}M_t(\mathbb{B}^n) \subseteq a_5 2^{-a_6(k-1)J}\mathbb{B}^n,$$

which implies

$$(2^{a_6(k-1)J}/a_5)\mathbb{B}^n \subseteq M_t^{-1}M_{t-(k-1)J}(\mathbb{B}^n).$$

From this and (27), it follows that $|M_t^{-1}(x-y)| \geq 2^{a_6(k-1)J}/a_5$. Thus, for any $t \geq t_0$, we have

$$F(y,t)\left[1+\left|M_t^{-1}(x-y)\right|\right]^{-pN} \leq a_5{}^{pN} 2^{-pNa_6(k-1)J} F_k^{*\,t_0}(x).$$

By taking the supremum over all $y \in \mathbb{R}^n$ and $t \geq t_0$, we know that

$$\left[T_\varphi^{N\,(t_0,L)} f(x)\right]^p \leq a_5{}^{pN} \sum_{k=0}^{\infty} 2^{-pNa_6(k-1)J} F_k^{*\,t_0}(x).$$

Therefore, using this and Proposition 5, we obtain

$$\begin{aligned}\left\|T_\varphi^{N\,(t_0,L)} f\right\|_{L^p(\mathbb{R}^n)}^p &\leq a_5{}^{pN} \sum_{k=0}^{\infty} 2^{-pNa_6(k-1)J} \int_{\mathbb{R}^n} F_k^{*\,t_0}(x)dx \\ &\leq Ca_5{}^{pN} \sum_{k=0}^{\infty} 2^{-pNa_6(k-1)J} 2^{kJ} \int_{\mathbb{R}^n} F_0^{*\,t_0}(x)dx \\ &= C'\left\|M_\varphi^{(t_0,L)} f\right\|_{L^p(\mathbb{R}^n)}^p,\end{aligned}$$

where $C' := Ca_5{}^{pN} 2^{pNa_6 J} \sum_{k=0}^{\infty} 2^{(1-pNa_6)kJ} = Ca_5{}^{pN} 2^J/(1-2^{(1-pNa_6)J})$. □

The following Lemma 2 gives the pointwise majorization of the grand radial maximal function by the tangential one, which is a variable anisotropic extension of ([2], Lemma 7.5).

Lemma 2. *Let Θ be an ellipsoid cover of $\mathbb{R}^n$, $\varphi \in \mathcal{S}$, $\int_{\mathbb{R}^n} \varphi(x)\,dx \neq 0$, and $f \in \mathcal{S}'$. For any given positive integers N and L, there exist integers $0 < U \leq \widetilde{U}$, $U \geq N_p$, and $\widetilde{U} \geq \widetilde{N_p}$ that are large enough and constant $C > 0$ such that, for any $t_0 < 0$,*

$$M_{U,\widetilde{U}}^{0\,(t_0,L)} f(x) \leq C T_\varphi^{N\,(t_0,L)} f(x), \quad \forall x \in \mathbb{R}^n.$$

Proof. The simplified proof of this final version is from Dekel (Lemma 6.20). By Proposition 6, for any $\psi \in \mathcal{S}$, $x \in \mathbb{R}^n$, $t \in \mathbb{R}$, there exists a sequence $\{\eta^k\}_{k=0}^{\infty}$, $\eta^k \in \mathcal{S}$ that satisfies

$$\psi = \sum_{k=0}^{\infty} \eta^k * \varphi^k$$

converging in $\mathcal{S}$, where

$$\varphi^k := |\det(M_{x,t+kJ}^{-1} M_{x,t})| \varphi(M_{x,t+kJ}^{-1} M_{x,t}\cdot), \quad k \geq 0.$$

Furthermore, for any positive integers U, $\widetilde{U}$ and V,

$$\|\eta^k\|_{\mathcal{S}_{U,\widetilde{U}}} \leq C2^{-kV} \|\psi\|_{\mathcal{S}_{U+n+1+\lceil V/(a_6 J)\rceil,\widetilde{U}+n+1}}. \tag{28}$$

where the constant depends on φ, U, $\widetilde{U}$, V, $p(\Theta)$ but not ψ. Denoting $M_k := M_{x,t+kJ}$, for $t \geq t_0$, implies

$$
\begin{aligned}
|(f*\psi_{x,t})(x)| &= \left|\left[f*\sum_{k=0}^{\infty}\left(\eta^k*\varphi^k\right)_{x,t}\right](x)\right| \\
&\leq C\left|\left[f*\sum_{k=0}^{\infty}\left|\det\left(M_k^{-1}\right)\right|\int_{\mathbb{R}^n}\eta^k(y)\,\varphi\left(M_k^{-1}(\cdot-M_{x,t}y)\right)dy\right](x)\right| \\
&= C\left|\left[f*\sum_{k=0}^{\infty}\left|\det\left(M_k^{-1}M_{x,t}^{-1}\right)\right|\int_{\mathbb{R}^n}\eta^k\left(M_{x,t}^{-1}y\right)\varphi\left(M_k^{-1}(\cdot-y)\right)dy\right](x)\right| \\
&\leq C\sum_{k=0}^{\infty}\left|\left[f*\left(\eta^k\right)_{x,t}*\varphi_{x,t+kJ}\right](x)\right| \\
&\leq C\sum_{k=0}^{\infty}\int_{\mathbb{R}^n}|f*\varphi_{x,t+kJ}(x-y)|\left|\left(\eta^k\right)_{x,t}(y)\right|dy \\
&\leq CT_\varphi^{N\,(t_0,L)}f(x)\sum_{k=0}^{\infty}\int_{\mathbb{R}^n}\left(1+\left|M_k^{-1}y\right|\right)^N \\
&\quad\times\left(1+\left|M_{x,t_0}^{-1}(x-y)\right|\right)^L\left(1+2^{t+t_0+kJ}\right)^L\left|\left(\eta^k\right)_{x,t}(y)\right|dy.
\end{aligned}
$$

Therefore,

$$
\begin{aligned}
M_\psi^{0\,(t_0,L)}f(x) &\leq T_\varphi^{N\,(t_0,L)}f(x)\sup_{t\geq t_0}\sum_{k=0}^{\infty}\int_{\mathbb{R}^n}\left(1+\left|M_k^{-1}y\right|\right)^N \\
&\quad\times\frac{\left(1+\left|M_{x,t_0}^{-1}(x-y)\right|\right)^L\left(1+2^{t+t_0+kJ}\right)^L}{\left(1+\left|M_{x,t_0}^{-1}x\right|\right)^L(1+2^{t+t_0})^L}\left|\left(\eta^k\right)_{x,t}(y)\right|dy \\
&=: T_\varphi^{N\,(t_0,L)}f(x)\sup_{t\geq t_0}\sum_{k=0}^{\infty}\mathrm{I}_{t,k}.
\end{aligned}
\tag{29}
$$

Let us now estimate $\mathrm{I}_{t,k}$ for $t\geq t_0$, $k\geq 0$. We begin with the simple observations that

$$
\frac{1+2^{t+t_0+kJ}}{1+2^{t+t_0}}=\frac{2^{kJ}(2^{-kJ}+2^{t+t_0})}{1+2^{t+t_0}}\leq C2^{kJ}
$$

and

$$
1+|x+y|\leq 1+|x|+|y|\leq(1+|x|)(1+|y|),\quad x,\,y\in\mathbb{R}^n. \tag{30}
$$

Therefore, we may obtain

$$
\begin{aligned}
\mathrm{I}_{t,k} &\leq C2^{t+kJL}\int_{\mathbb{R}^n}\left(1+\left|M_k^{-1}y\right|\right)^N\left(1+\left|M_{x,t_0}^{-1}y\right|\right)^L\left|\eta^k\left(M_{x,t}^{-1}y\right)\right|dy \\
&\leq C2^{kJL}\int_{\mathbb{R}^n}\left(1+\left\|M_k^{-1}M_{x,t}\right\||y|\right)^N\left(1+\left\|M_{x,t_0}^{-1}M_{x,t}\right\||y|\right)^L\left|\eta^k(y)\right|dy,
\end{aligned}
$$

which, together with

$$
\|M_k^{-1}M_{x,t}\|\leq a_3 2^{a_4 kJ}\ \text{ and }\ \|M_{x,t_0}^{-1}M_{x,t}\|\leq a_5 2^{-a_6(t-t_0)}\leq a_5\ (\text{by } t\geq t_0 \text{ and (2)}),
$$

further implies that

$$\begin{aligned} \mathrm{I}_{t,k} &\le C2^{kJ(L+a_4N)} \int_{\mathbb{R}^n} (1+|y|)^{N+L} \left|\eta^k(y)\right| dy \\ &\le C2^{kJ(L+a_4N)} \left\|\eta^k\right\|_{\mathcal{S}_{0,\widetilde{N}+n+L}}. \end{aligned} \tag{31}$$

We now apply (28) with $V := \lceil J(L + a_4N)\rceil + 1$, which gives

$$\mathrm{I}_{t,k} \le C2^{-kV} \|\psi\|_{\mathcal{S}_{n+1+\lceil V/(a_6J)\rceil,\, N+L+2n+2}}. \tag{32}$$

This yields for any $\psi \in \mathcal{S}_{U,\widetilde{U}}$, $U := \max(N_p, n+1+\lceil V/(a_6J)\rceil)$, $\widetilde{U} := \max(\widetilde{N}_p, N+L+2n+2)$

$$M^{0\,(t_0,L)}_{U,\widetilde{U}} f(x) = \sup_{\psi\in\mathcal{S}_{U,\widetilde{U}}} M^{0\,(t_0,L)}_{\psi} f(x) \le C\, T^{N\,(t_0,L)}_{\varphi} f(x).$$

This finishes the proof of Lemma 2. □

The following Lemma 3 shows that the radial and the grand non-tangential maximal functions are pointwise equivalent, which is a variable anisotropic extension of ([2], Proposition 3.10).

Lemma 3 ([19], Theorem 3.4). *For any N, $\widetilde{N} \in \mathbb{N}$ with $N \le \widetilde{N}$, there exists a positive constant $C := C(\widetilde{N})$ such that, for any $f \in \mathcal{S}'$,*

$$M^0_{N,\widetilde{N}} f(x) \le M_{N,\widetilde{N}} f(x) \le C M^0_{N,\widetilde{N}} f(x), \quad x \in \mathbb{R}^n.$$

The following Lemma 4 is a variable anisotropic extension of ([2], p. 46, Lemma 7.6).

Lemma 4. *Let Θ_t be a t-continuous ellipsoid cover, $\varphi \in \mathcal{S}$, and $f \in \mathcal{S}'$. Then, for every $M > 0$ and $t_0 < 0$, there exist $L > 0$ and $N' > 0$ large enough such that*

$$M^{(t_0,L)}_{\varphi} f(x) \le C2^{-t_0(2a_4N'+2L+a_4L)}(1+|x|)^{-M}, \quad x \in \mathbb{R}^n, \tag{33}$$

where C is a positive constant dependent on $p(\Theta)$, N', f, and φ.

Proof. For any $\varphi \in \mathcal{S}$, there exist an integer $N > 0$ and positive constant $C := C(\varphi)$ such that, for any $N' \ge N$ and $y \in \mathbb{R}^n$,

$$|f * \varphi(y)| \le C\|\varphi\|_{\mathcal{S}_{N,N'}}(1+|y|)^{N'}. \tag{34}$$

Therefore, for any $t_0 < 0$, $t \ge t_0$ and $x \in \mathbb{R}^n$, by (34), we have

$$\begin{aligned} &|(f*\varphi_t)(y)| \left(1+\left|M_{t_0}^{-1}y\right|\right)^{-L} \left(1+2^{t+t_0}\right)^{-L} \\ &\le C2^{-L(t+t_0)} \|\varphi_t\|_{\mathcal{S}_{N,N'}} (1+|y|)^{N'} \left(1+\left|M_{t_0}^{-1}y\right|\right)^{-L}. \end{aligned} \tag{35}$$

Let us first estimate $\|\varphi_t\|_{\mathcal{S}_{N,N'}}$. By the chain rule and (1), we have

$$\begin{aligned}\|\varphi_t\|_{\mathcal{S}_{N,N'}} &= |\det M_t^{-1}| \sup_{z\in\mathbb{R}^n} \sup_{|\alpha|\le N} (1+|z|)^{N'} \left|\partial^\alpha\left(\varphi\left(M_t^{-1}\cdot\right)\right)(z)\right| \\ &\le C2^t \sup_{z\in\mathbb{R}^n} \sup_{|\alpha|\le N} (1+|z|)^{N'} \left\|M_t^{-1}\right\|^{|\alpha|} \left|(\partial^\alpha\varphi)\left(M_t^{-1}z\right)\right| \\ &\le C2^t \sup_{z\in\mathbb{R}^n} \sup_{|\alpha|\le N} (1+|M_t z|)^{N'} \left\|M_t^{-1}\right\|^{|\alpha|} |\partial^\alpha\varphi(z)|. \end{aligned} \tag{36}$$

Now, let us further estimate (36) in the following two cases.
Case 1: $t \ge 0$. By (2), we have

$$\left\|M_t^{-1}\right\| = \left\|M_t^{-1}M_0M_0^{-1}\right\| \le \left\|M_t^{-1}M_0\right\|\left\|M_0^{-1}\right\| \le \left\|M_0^{-1}\right\| a_3^{-1}2^{a_4 t} = C2^{a_4 t}$$

and

$$\begin{aligned}|M_t z| &= \left|M_0M_0^{-1}M_t z\right| \le \|M_0\|\left|M_0^{-1}M_t z\right| \le \|M_0\|\left\|M_0^{-1}M_t\right\||z| \\ &\le \|M_0\| a_5 2^{-a_6 t}|z| \le C|z|.\end{aligned}$$

Inserting the above two estimates into (36) with $t \ge 0$, we know that

$$\begin{aligned}\|\varphi_t\|_{\mathcal{S}_{N,N'}} &\le C2^t \sup_{z\in\mathbb{R}^n} \sup_{|\alpha|\le N} (1+|M_t z|)^{N'} \left\|M_t^{-1}\right\|^{|\alpha|} |\partial^\alpha\varphi(z)| \\ &\le C2^t 2^{a_4 tN}\|\varphi\|_{\mathcal{S}_{N,N'}}.\end{aligned} \tag{37}$$

Case 2: $t_0 \le t < 0$. By (2), we have

$$\left\|M_t^{-1}\right\| = \left\|M_t^{-1}M_0M_0^{-1}\right\| \le \left\|M_t^{-1}M_0\right\|\left\|M_0^{-1}\right\| \le \left\|M_0^{-1}\right\| a_5 2^{a_6 t} \le C$$

and

$$\begin{aligned}|M_t z| &= \left|M_0M_0^{-1}M_t z\right| \le \|M_0\|\left|M_0^{-1}M_t z\right| \le \|M_0\|\left\|M_0^{-1}M_t\right\||z| \\ &\le \|M_0\| a_3^{-1}2^{-a_4 t}|z| = C2^{-a_4 t_0}|z|.\end{aligned}$$

Inserting the above two estimates into (36) with $t_0 \le t < 0$, we know that

$$\begin{aligned}\|\varphi_t\|_{\mathcal{S}_{N,N'}} &\le C2^t \sup_{z\in\mathbb{R}^n} \sup_{|\alpha|\le N} (1+|M_t z|)^{N'} \left\|M_t^{-1}\right\|^{|\alpha|} |\partial^\alpha\varphi(z)| \\ &\le C2^{-a_4 t_0 N'}\|\varphi\|_{\mathcal{S}_{N,N'}}.\end{aligned} \tag{38}$$

For any $M > 0$, let $L := M + N'$. For any $t_0 < 0$, $t \ge t_0$ and taking some integer $N' > 0$ large enough, by (37) and (38), we obtain

$$2^{-L(t+t_0)}\|\varphi_t\|_{\mathcal{S}_{N,N'}} \le C2^{-t_0(a_4N'+2L)}\|\varphi\|_{\mathcal{S}_{N,N'}}. \tag{39}$$

Inserting (39) into (35), we further obtain

$$\begin{aligned}&|(f*\varphi_t)(y)|\left(1+\left|M_{t_0}^{-1}y\right|\right)^{-L}\left(1+2^{t+t_0}\right)^{-L} \\ &\le C2^{-t_0(a_4N'+2L)}\|\varphi\|_{\mathcal{S}_{N,N'}}(1+|y|)^{N'}\left(1+\left|M_{t_0}^{-1}y\right|\right)^{-L}.\end{aligned} \tag{40}$$

For any $y \in \theta(x, t)$, there exists $z \in \mathbb{B}^n$ such that $y = x + M_t z$. By (30), we have

$$1 + |y| = 1 + |x + M_t z| \leq (1 + |x|)(1 + |M_t z|). \tag{41}$$

If $t \geq 0$, by (2), then

$$\begin{aligned} |M_t z| &= \left|M_0 M_0^{-1} M_t z\right| \leq \|M_0\| \left|M_0^{-1} M_t z\right| \leq \|M_0\| \left\|M_0^{-1} M_t\right\| |z| \\ &\leq \|M_0\| a_5 2^{-a_6 t} |z| \leq C. \end{aligned}$$

If $t_0 \leq t < 0$, by (2), then

$$\begin{aligned} |M_t z| &= \left|M_0 M_0^{-1} M_t z\right| \leq \|M_0\| \left|M_0^{-1} M_t z\right| \leq \|M_0\| \left\|M_0^{-1} M_t\right\| |z| \\ &\leq \|M_0\| a_3^{-1} 2^{-a_4 t} |z| = C 2^{-a_4 t_0}. \end{aligned}$$

Therefore, for any $t \geq t_0$, by using the above two estimates, we have

$$|M_t z| \leq C 2^{-a_4 t_0}.$$

From this and (41), it follows that

$$(1 + |y|) \leq C 2^{-a_4 t_0} (1 + |x|). \tag{42}$$

Moreover, for any $t_0 < 0$, by (2), we have

$$1 + |x| \leq 1 + \|M_0\| \left\|M_0^{-1} M_{t_0}\right\| \left|M_{t_0}^{-1} x\right| \leq C 2^{-a_4 t_0} \left(1 + \left|M_{t_0}^{-1} x\right|\right).$$

Furthermore, for any $y \in \theta(x, t)$, we have $x \in M_t(\mathbb{B}^n) + y$. Thus, there exists $z \in \mathbb{B}^n$ such that $x = M_t z + y$. Hence, for any $t \geq t_0$, by (30) and (2), we obtain

$$\begin{aligned} \left(1 + \left|M_{t_0}^{-1} x\right|\right) &= \left(1 + \left|M_{t_0}^{-1}(y + M_t z)\right|\right) \leq \left(1 + \left|M_{t_0}^{-1} y\right|\right) \left(1 + \left\|M_{t_0}^{-1} M_t\right\| |z|\right) \\ &\leq \left(1 + \left|M_{t_0}^{-1} y\right|\right) \left(1 + a_5 2^{-a_6(t - t_0)} |z|\right) \leq C\left(1 + \left|M_{t_0}^{-1} y\right|\right). \end{aligned}$$

Combining with the above two inequalities, we have

$$(1 + |M_{t_0}^{-1} y|) \geq C 2^{a_4 t_0} (1 + |x|). \tag{43}$$

Thus, for any $t \geq t_0$ and $y \in \theta(x,t)$, inserting (42) and (43) into (40) with $L = M + N'$, we obtain

$$|(f * \varphi_t)(y)| \left(1 + \left|M_{t_0}^{-1} y\right|\right)^{-L} \left(1 + 2^{t + t_0}\right)^{-L} \leq C 2^{-t_0(2a_4 N' + 2L + a_4 L)} (1 + |x|)^{-M},$$

which implies that (33) holds true and hence completes the proof of Lemma 4. □

Note that the above argument gives the same estimate for the truncated grand maximal function $M_{N,\tilde{N}}^{0\,(t_0,\,L)} f(x)$. As a consequence of Lemma 4, we obtain that, for any choice of $t_0 < 0$ and any $f \in \mathcal{S}'$, we can find an appropriate $L > 0$ so that the maximal function, say $M_\varphi^{(t_0,\,L)} f$, is bounded and belongs to $L^p(\mathbb{R}^n)$. This becomes crucial in the proof of Theorem 1, where we work with truncated maximal functions, The complexity of the preceding argument stems from the fact that, a priori, we do not know whether $M_\varphi^0 f \in L^p$ implies $M_\varphi f \in L^p$. Instead, we must work with variants of maximal functions for which this is satisfied.

Proof of Theorem 1. Suppose that Θ_t is a t-continuous ellipsoid cover and $\varphi \in \mathcal{S}$ satisfying $\int_{\mathbb{R}^n} \varphi(x)\,dx \neq 0$. From Remark 2 and the definition of the grand radial maximal function, it follows that

$$(26) \Rightarrow (24) \Rightarrow (25)$$

and

$$(23) \Rightarrow (25).$$

By Lemma 1 applied for $L = 0$, we have

$$\left\| T_\varphi^{N(t_0,0)} f \right\|_{L^p} \leq C \left\| M_\varphi^{(t_0,0)} f \right\|_{L^p} \quad \text{for any } f \in \mathcal{S}' \text{ and } t_0 < 0.$$

As $t_0 \to -\infty$, by the monotone convergence theorem, we obtain

$$\left\| T_\varphi^N f \right\|_{L^p} \leq C \|M_\varphi f\|_{L^p},$$

which shows (24) $\Rightarrow$ (26).

Combining Lemma 2 applied for $N > 1/(a_6\,p)$ and $L = 0$ and Lemma 1 applied for $L = 0$, we conclude that there exist integers $0 < U \leq \widetilde{U}$, $U > N_p$, $\widetilde{U} \geq \widetilde{N_p}$ that are large enough and a positive constant C such that

$$\left\| M_{U,\widetilde{U}}^{0(t_0,0)} f \right\|_{L^p} \leq C \left\| M_\varphi^{(t_0,0)} f \right\|_{L^p} \quad \text{for any } f \in \mathcal{S}' \text{ and } t_0 < 0.$$

As $t_0 \to -\infty$, by the monotone convergence theorem, we obtain

$$\left\| M_{U,\widetilde{U}}^0 f \right\|_{L^p} \leq C \|M_\varphi f\|_{L^p}.$$

From this and Proposition 1, we deduce that

$$\|f\|_{H^p(\Theta_t)} = \left\| M_{N_p,\widetilde{N}_p}^0 f \right\|_{L^p} \leq C \left\| M_{U,\widetilde{U}}^0 f \right\|_{L^p} \leq C \|M_\varphi f\|_{L^p}$$

and hence (24) $\Rightarrow$ (23). (25) $\Rightarrow$ (24) remain to be shown.

Suppose now $M_\varphi^\circ f \in L^p$. By Lemma 4, we can find a $L > 0$ large enough such that (33) holds true, which implies $M_\varphi^{(t_0,L)} f \in L^p$ for all $t_0 < 0$. Combining Lemmas 1 and 2, we obtain that there exist $0 < U \leq \widetilde{U}$, $U > N_p$, and $\widetilde{U} \geq \widetilde{N_p}$ large enough such that

$$\left\| M_{U,\widetilde{U}}^{0(t_0,L)} f \right\|_p \leq C_1 \left\| M_\varphi^{(t_0,L)} f \right\|_p, \tag{44}$$

where constant C_1 is independent of $t_0 < 0$. For a given $t_0 < 0$, let

$$\Omega_{t_0} := \left\{ x \in \mathbb{R}^n : M_{U,\widetilde{U}}^{0(t_0,L)} f(x) \leq C_2 M_\varphi^{(t_0,L)} f(x) \right\}, \tag{45}$$

where $C_2 := 2^{1/p} C_1$. We claim that

$$\int_{\mathbb{R}^n} \left[M_\varphi^{(t_0,L)} f(x) \right]^p dx \leq 2 \int_{\Omega_{t_0}} \left[M_\varphi^{(t_0,L)} f(x) \right]^p dx. \tag{46}$$

Indeed, this follows from (44), $M_\varphi^{(t_0,L)} f \in L^p$ and

$$\int_{\Omega_{t_0}^c} \left[M_\varphi^{(t_0,L)} f(x) \right]^p dx \leq C_2^{-p} \int_{\Omega_{t_0}^c} \left[M_{U,\widetilde{U}}^{0(t_0,L)} f(x) \right]^p dx \leq (C_1/C_2)^p \int_{\mathbb{R}^n} \left[M_\varphi^{(t_0,L)} f(x) \right]^p dx,$$

where $(C_1/C_2)^p = 1/2$.

We also claim that, for $0 < q < p$, there exists a constant $C_3 > 0$ such that, for any $t_0 < 0$,

$$M_{\varphi}^{(t_0,L)}f(x) \le C_3\left[M_{\Theta}\left(M_{\varphi}^{0\,(t_0,L)}f\right)^q(x)\right]^{1/q}, \tag{47}$$

where M_{Θ} is as in Definition 4. Indeed, let $t \ge t_0$, $y \in \theta(x,t)$ and

$$F(y,\,t) := |(f * \varphi_t)(y)|\,(1 + |M_{t_0}^{-1}y|)^{-L}(1 + 2^{t+t_0})^{-L}.$$

Suppose that $x \in \Omega_{t_0}$ and let $F_l^{*\,t_0}(x)$ be as in (11) with $l = 0$. Then, there exist $t' \in \mathbb{R}$ with $t' \ge t_0$ and $y' \in \theta(x,\,t')$ such that

$$F(y',\,t') \ge F_0^{*\,t_0}(x)/2 = M_{\varphi}^{(t_0,L)}f(x)/2. \tag{48}$$

Consider $x' \in y' + M_{t'+lJ}(\mathbb{B}^n)$ for some integer $l \ge 1$ to be specified later. Let $\Phi(z) := \varphi\left(z + M_{t'}^{-1}(x' - y')\right) - \varphi(z)$. Obviously, we have

$$f * \varphi_{t'}(x') - f * \varphi_{t'}(y') = f * \Phi_{t'}(y'). \tag{49}$$

Let us first estimate $\|\Phi\|_{\mathcal{S}_{U,\tilde{U}}}$. From $x' \in y' + M_{t'+lJ}(\mathbb{B}^n)$, we deduce that

$$M_{t'}^{-1}(x' - y') \in M_{t'}^{-1}M_{t'+lJ}(\mathbb{B}^n).$$

By this and the mean value theorem, we obtain

$$\begin{aligned}\|\Phi\|_{\mathcal{S}_{U,\tilde{U}}} &\le \sup_{h\in M_{t'}^{-1}M_{t'+lJ}(\mathbb{B}^n)} \|\varphi(\cdot + h) - \varphi(\cdot)\|_{\mathcal{S}_{U,\tilde{U}}} \\ &= \sup_{h\in M_{t'}^{-1}M_{t'+lJ}(\mathbb{B}^n)} \sup_{z\in\mathbb{R}^n} \sup_{|\alpha|\le U} (1+|z|)^{\tilde{U}}|(\partial^\alpha\varphi)(z+h) - \partial^\alpha\varphi(z)| \\ &\le C \sup_{h\in M_{t'}^{-1}M_{t'+lJ}(\mathbb{B}^n)} \sup_{z\in\mathbb{R}^n} \sup_{|\alpha|\le U+1} (1+|z|)^{\tilde{U}}|(\partial^\alpha\varphi)(z+h)| \\ &\quad\times \sup_{h\in M_{t'}^{-1}M_{t'+lJ}(\mathbb{B}^n)} |h|.\end{aligned} \tag{50}$$

From (2), we deduce

$$\|M_{t'}^{-1}M_{t'+lJ}\| \le a_5 2^{-a_6 lJ},$$

which implies

$$M_{t'}^{-1}M_{t'+lJ}(\mathbb{B}^n) \subset a_5 2^{-a_6 lJ}\mathbb{B}^n.$$

By this and $h \in M_{t'}^{-1}M_{t'+lJ}(\mathbb{B}^n)$, we have $|h| \le a_5 2^{-a_6 lJ}$. From this and (30), we deduce that

$$1 + |z| \le (1 + |z+h|)(1 + |h|) \le C(1 + |z+h|),\ \ z \in \mathbb{R}^n.$$

Applying this and $|h| \le a_5 2^{-a_6 lJ}$ in (50), we obtain

$$\begin{aligned}\|\Phi\|_{\mathcal{S}_{U,\tilde{U}}} &\le C \sup_{h\in M_{t'}^{-1}M_{t'+lJ}(\mathbb{B}^n)} \sup_{z\in\mathbb{R}^n} \sup_{|\alpha|\le U+1} (1+|z+h|)^{\tilde{U}}|(\partial^\alpha\varphi)(z+h)| \\ &\quad\times \sup_{h\in M_{t'}^{-1}M_{t'+lJ}(\mathbb{B}^n)} |h| \le C\|\varphi\|_{\mathcal{S}_{U+1,\tilde{U}}} a_5 2^{-a_6 lJ} \le C_4 2^{-a_6 lJ},\end{aligned} \tag{51}$$

where a positive constant C_4 does not depend on L.

Moreover, notice that, for any $x' \in M_{t'+lJ}(\mathbb{B}^n) + y'$, there exists $z \in \mathbb{B}^n$ such that $x' = M_{t'+lJ}z + y'$. By (30), (2), and $t' \geq t_0$, we have

$$\begin{aligned}\left(1 + \left|M_{t_0}^{-1}x'\right|\right) &\leq \left(1 + \left|M_{t_0}^{-1}y'\right|\right)\left(1 + \left\|M_{t_0}^{-1}M_{t'+lJ}\right\||z|\right) \\ &\leq \left(1 + \left|M_{t_0}^{-1}y'\right|\right)\left(1 + a_5 2^{-a_6(t'-t_0+lJ)}|z|\right) \leq 2a_5\left(1 + \left|M_{t_0}^{-1}y'\right|\right).\end{aligned} \tag{52}$$

Thus, for any $x \in \Omega_{t_0}$, from (49), (52), (48), (51), Lemma 3, and (45), it follows that

$$\begin{aligned}2^L a_5^L F(x', t') &= 2^L a_5^L\left[|(f * \varphi_{t'})(x')|\,(1 + |M_{t_0}^{-1}x'|)^{-L}(1 + 2^{t'+t_0})^{-L}\right] \\ &\geq [|f * \varphi_{t'}(y')| - |f * \Phi_{t'}(y')|]\left(1 + \left|M_{t_0}^{-1}y'\right|\right)^{-L}\left(1 + 2^{t'+t_0}\right)^{-L} \\ &\geq F(y', t') - M_{U,\tilde{U}}^{(t_0,L)}f(x)\|\Phi\|_{S_{U,\tilde{U}}} \\ &\geq M_\varphi^{(t_0,L)}f(x)/2 - C_4 2^{-a_6 lJ} C M_{U,\tilde{U}}^{0\,(t_0,L)}f(x) \\ &\geq M_\varphi^{(t_0,L)}f(x)/2 - C_4C_2C2^{-a_6 lJ}M_\varphi^{(t_0,L)}f(x).\end{aligned}$$

We choose an integer $l \geq 1$ large enough such that $C_4C_2C2^{-a_6 lJ} \leq 1/4$. Therefore, for any $x \in \Omega_{t_0}$ and $x' \in M_{t'+lJ}(\mathbb{B}^n) + y'$, we further have

$$2^L a_5^L F(x', t') \geq M_\varphi^{(t_0,L)}f(x)/2 - C_4C_2C2^{-a_6 lJ}M_\varphi^{(t_0,L)}f(x) \geq M_\varphi^{(t_0,L)}f(x)/4. \tag{53}$$

Moreover, by $y' \in \theta(x, t')$ and Proposition 2, we have

$$\begin{aligned}M_{t'+lJ}(\mathbb{B}^n) + y' &\subseteq M_{t'+lJ}(\mathbb{B}^n) + M_{t'}(\mathbb{B}^n) + x \\ &\subseteq 2M_{t'}(\mathbb{B}^n) + x \subseteq \theta(x, t' - J).\end{aligned} \tag{54}$$

Thus, for any $x \in \Omega_{t_0}$ and $t \geq t_0$, by (53) and (54), we obtain

$$\begin{aligned}\left[M_\varphi^{(t_0,L)}f(x)\right]^q &\leq \frac{4^q 2^{Lq} a_5^{Lq}}{|M_{t'+lJ}(\mathbb{B}^n)|}\int_{y'+M_{t'+lJ}(\mathbb{B}^n)}[F(z,t')]^q dz \\ &\leq C4^q 2^{Lq} a_5^{Lq}\frac{2^{(l+1)J}}{|\theta(x, t'-J)|}\int_{\theta(x,t'-J)}\left[M_\varphi^{0\,(t_0,L)}f(z)\right]^q dz \\ &\leq C_3 M_\Theta\left(\left(M_\varphi^{0\,(t_0,L)}f\right)^q\right)(x),\end{aligned}$$

which shows the above claim (47).

Consequently, by (46), (47), and Proposition 3 with $p/q > 1$, we have

$$\begin{aligned}\int_{\mathbb{R}^n}\left[M_\varphi^{(t_0,L)}f(x)\right]^p dx &\leq 2\int_{\Omega_{t_0}}\left[M_\varphi^{(t_0,L)}f(x)\right]^p dx \\ &\leq 2C_3^p\int_{\Omega_{t_0}}\left[M_\Theta\left(\left(M_\varphi^{0(t_0,L)}f\right)^q\right)(x)\right]^{p/q} dx \\ &\leq C_5\int_{\mathbb{R}^n}\left[M_\varphi^{0(t_0,L)}f(x)\right]^p dx,\end{aligned} \tag{55}$$

where the constant C_5 depends on $p/q > 1$, $L \geq 0$ and $p(\Theta)$ but is independent of $t_0 < 0$. This inequality is crucial as it gives a bound of the non-tangential by the radial maximal function in L^p. The rest of the proof is immediate.

For any $x \in \mathbb{R}^n$, $y \in \mathbb{R}^n$ and $t < 0$, by (2), we obtain

$$\left|M_t^{-1}y\right| = \left|M_t^{-1}M_0M_0^{-1}y\right| \leq \left\|M_t^{-1}M_0\right\|\left\|M_0^{-1}\right\||y|$$
$$\leq a_5 2^{a_6 t}\left\|M_0^{-1}\right\||y| \to 0 \text{ as } t \to -\infty.$$

Hence, we obtain that $M_\varphi^{(t_0,L)}f(x)$ converges pointwise and monotonically to $M_\varphi f(x)$ for all $x \in \mathbb{R}^n$ as $t_0 \to -\infty$, which together with (55) and the monotone convergence theorem, further implies that $M_\varphi f \in L^p$. Therefore, we can now choose $L = 0$, and again, by (55) and the monotone convergence theorem, we have $\|M_\varphi f\|_p^p \leq C_5\|M_\varphi^0 f\|_p^p$, where C_5 corresponds to $L = 0$ and is independent of $f \in \mathcal{S}'$. This finishes the proof of Theorem 1. □

Author Contributions: Formal analysis, W.W.; Writing—original draft, A.W. and B.L. All authors have read and agreed to the published version of the manuscript.

Funding: This research is supported by the National Natural Science Foundation of China (No. 11861062) and the Xinjiang Training of Innovative Personnel of China (No. 2020D01C048).

Institutional Review Board Statement: Not applicable.

Informed Consent Statement: Not applicable.

Conflicts of Interest: The authors declare no conflict of interest.

References

1. Fefferman, C.; Stein, E. H^p spaces of several variables. *Acta Math.* **1972**, *129*, 137–193. [CrossRef]
2. Bownik, M. Anisotropic Hardy spaces and wavelets. *Mem. Am. Math. Soc.* **2003**, *164*, 1–122. [CrossRef]
3. Barrios, B.; Betancor, J. Anisotropic weak Hardy spaces and wavelets. *J. Funct. Spaces Appl.* **2012**, *17*, 809121. [CrossRef]
4. Betancor, J.; Damián, W. Anisotropic local Hardy spaces. *J. Fourier Anal. Appl.* **2010**, *16*, 658–675. [CrossRef]
5. Bownik, M.; Li, B.; Yang, D.; Zhou, Y. Weighted anisotropic Hardy spaces and their applications in boundedness of sublinear operators. *Indiana Univ. Math. J.* **2008**, *57*, 3065–3100.
6. Hu, G. Littlewood-Paley characterization of weighted anisotropic Hardy spaces. *Taiwan. J. Math.* **2013**, *17*, 675–700. [CrossRef]
7. Wang, L.-A. *Multiplier Theorems on Anisotropic Hardy Spaces*; ProQuest LLC: Ann Arbor, MI, USA, 2012.
8. Zhao, K.; Li, L. Molecular decomposition of weighted anisotropic Hardy spaces. *Taiwan. J. Math.* **2013**, *17*, 583–599. [CrossRef]
9. Dekel, S.; Petrushev, P.; Weissblat, T. Hardy spaces on $\mathbb{R}^n$ with pointwise variable anisotropy. *J. Fourier Anal. Appl.* **2011**, *17*, 1066–1107. [CrossRef]
10. Calderón, A.-P.; Torchinsky, A. Parabolic maximal functions associated with a distribution. *Adv. Math.* **1975**, *16*, 1–64. [CrossRef]
11. Dahmen, W.; Dekel, S.; Petrushev, P. Two-level-split decomposition of anisotropic Besov spaces. *Constr. Approx.* **2010**, *31*, 149–194. [CrossRef]
12. Stein, E. *Harmonic Analysis: Real-Variable Methods, Orthogonality, and Oscillatintegrals*; Princeton Mathematical Series, no. 43; Princeton University Press: Princeton, NJ, USA, 1993.
13. Calderón, A.-P.; Torchinsky, A. Parabolic maximal functions associated with a distribution II. *Adv. Math.* **1977**, *25*, 216–225. [CrossRef]
14. Uchiyama, A. A maximal function characterization of H^p on the space of homogeneous type. *Trans. Am. Math. Soc.* **1980**, *262*, 579–592.
15. Li, B.; Yang, D.; Yuan, W. Anisotropic Hardy spaces of Musielak-Orlicz type with applications to boundedness of sublinear operators. *Sci. World J.* **2014**, *2014*, 306214. [CrossRef] [PubMed]
16. Liu, J.; Yang, D.; Yuan, W. Anisotropic Hardy-Lorentz spaces and their applications. *Sci. China Math.* **2016**, *59*, 1669–1720. [CrossRef]
17. Liu, J.; Weisz, F.; Yang, D.; Yuan W. Variable anisotropic Hardy spaces and their applications. *Taiwan. J. Math.* **2018**, *22*, 1173–1216. [CrossRef]
18. Huang, L.; Liu, J.; Yang, D.; Yuan, W. Real-variable characterizations of new anisotropic mixed-norm Hardy spaces. *Comm. Pure Appl. Anal.* **2020**, *19*, 3033–3082. [CrossRef]
19. Bownik, M.; Li, B.; Li, J. Variable anisotropic singular integral operators. *arXiv* **2004**, arXiv:2004.09707.

mathematics

Article

An Optimal Estimate for the Anisotropic Logarithmic Potential

Shaoxiong Hou

Key Laboratory of Computational Mathematics and Applications of Hebei Province, College of Mathematical Science, Hebei Normal University, Shijiazhuang 050024, China; sxhou@mail.hebtu.edu.cn

Abstract: This paper introduces the new annulus body to establish the optimal lower bound for the anisotropic logarithmic potential as the complement to the theory of its upper bound estimate which has already been investigated. The connections with convex geometry analysis and some metric properties are also established. For the application, a polynomial dual log-mixed volume difference law is deduced from the optimal estimate.

Keywords: anisotropic log-potential; optimal polynomial inequality; annulus body; dual log-mixed volume

1. Backgrounds

The Riesz potential $I_\alpha(\alpha > 0)$ operator is defined by

$$I_\alpha f(x) = \int_{\mathbb{R}^n} \frac{f(y)}{|x-y|^\alpha} dy,$$

where f is a measurable function. It has been widely developed in harmonic analysis including function spaces, mathematical physics and partial differential equations (see [1–4]).

For the endpoint case $\alpha = 0$, it is trivial to study the limitation

$$\lim_{\alpha \to 0} |x-y|^{-\alpha} = 1 \quad as \quad x \neq y.$$

Instead, the convolution kernel is usually changed in such a derivative way

$$\frac{\partial}{\partial \alpha}|x-y|^{-\alpha}\Big|_{\alpha=0} = \frac{\log|x-y|^{-1}}{|x-y|^\alpha}\Big|_{\alpha=0} = \log|x-y|^{-1} \quad as \quad x \neq y.$$

This logarithmic kernel produces a corresponding logarithmic potential operator, which represents a the better complement for the endpoint case of Riesz potential operator by virtue of effective properties and applications. For example, $|x|^{2-n}(n \geq 3)$ is harmonic on $\mathbb{R}^n \setminus o$, while for teh lower dimension $n = 2$, $\log|x|$ is studied since it is harmonic on $\mathbb{R}^n \setminus o$ (see [5,6]).

Recently, both Riesz potential and logarithmic potential have been studied in an anisotropic way, which is closely related with convex geometry analysis and mathematical physics (see [7–11]). Here we first recall some basic concepts and results in convex geometry.

If the intersection of each line through the origin with a set $K \subsetneq \mathbb{R}^n$ is a compact line segment, K is called star-shaped with respect to the origin. Let

$$\rho_K(x) = \max\{\lambda \geq 0 : \lambda x \in K\} \quad \text{for} \quad x \in \mathbb{R}^n \setminus o,$$

where o is the origin, be the radial function of the star-shaped set K. K is called a star body with respect to the origin, if ρ_K is positive and continuous. We assume that K is a star body

Citation: Hou, S. . *Mathematics* **2022**, *10*, 261. https://doi.org/10.3390/math10020261

Academic Editor: Juan Benigno Seoane-Sepúlveda

Received: 28 December 2021
Accepted: 7 January 2022
Published: 15 January 2022

with respect to the origin and E is a bounded measurable set in this paper. Note that the radial function ρ_K is positively homogeneous with degree -1, i.e.,

$$\rho_A(sx) = s^{-1}\rho_A(x) \text{ for all } s > 0.$$

Let $V(E)$ and E^c denote, respectively, the n-dimensional volume of E and the complement of E. We assume $V(E) \neq 0$ in this paper, since when $V(E) = 0$, some trivial result follows directly. Let $dS(\cdot)$ denote the natural spherical measure on the boundary $\mathbb{S}^{n-1}$ of the unit ball B_2^n centered at the origin. Then

$$V(K) = \frac{1}{n}\int_{\mathbb{S}^{n-1}} \rho_K^n(u)\, dS(u).$$

Let $\|\cdot\|_K$ denote by the Minkowski functional of K:

$$\|x\|_K = \inf\{s > 0 : x \in sK\} \text{ for all } x \in \mathbb{R}^n \tag{1}$$

where

$$sK = \{sy : y \in K\}.$$

Note that $\rho_K^{-1}(x) = \|x\|_K$ and $\|\cdot\|_{B_2^n} = |\cdot|$, where $|\cdot|$ denotes the Euclidean norm. We refer to [12,13] for more information on convex geometry.

Let $y \in \mathbb{R}^n$, $a > 1$ and denote by

$$R_a^K(y) = \{x \in \mathbb{R}^n : \frac{1}{a} \leq \|x - y\|_K \leq a\}$$

the K-annulus body centered at y with outer radius a and inner radius $\frac{1}{a}$. Then, by the definition of the Minkowski functional, it follows that

$$V(R_a^K(y)) = \left(a^n - \left(\frac{1}{a}\right)^n\right)V(K).$$

Several anisotropic Riesz potentials are introduced and their optimal extreme values estimates are systematically studied in [10]. We omit the details here for the brevity of this paper. Let

$$P_{\log,m}(K, E; y) = \int_E \left(\log \frac{1}{\|x - y\|_K}\right)^m dx$$

be the anisotropic m-log-potential of measurable set E at $y \in \mathbb{R}^n$ with respect to K, and

$$V_{\log,m}(K, E) = \sup_{y \in \mathbb{R}^n} P_{\log,m}(K, E; y)$$

be the mixed volume of K and E. We refer to [11] for these definitions and [14,15] for their relations with engineering and mathematical physics.

Note that $V_{\log,m}(K, E)$ is obviously an extreme value of the anisotropic m-log-potential. It is also closely related to convex geometry analysis. In [11], when m is an odd number, the optimal estimate for $V_{\log,m}(K, E)$ is established as follows:

$$V_{\log,m}(K, E) \leq \begin{cases} \frac{V(E)}{n^m}\sum_{i=0}^{m} \frac{m!}{(m-i)!}\left(\log \frac{V(K)}{V(E)}\right)^{m-i} & for\ V(E) > 0, \\ 0 & for\ V(E) = 0. \end{cases} \tag{2}$$

When $V(E) > 0$, the equality in (2) holds if and only if E is a K-ball introduced in [11] up to the difference of a measure zero set.

For the application of the sharp estimate in (2), the dual polynomial log-Minkowski inequality is established in [11]:

$$\sum_{i=0}^{m} \frac{n^{m-i}m!}{(m-i)!}\int_{\mathbb{S}^{n-1}} \left(\log\frac{\rho_K(u)}{\rho_L(u)}\right)^{m-i} dV_L(u) \le \sum_{i=0}^{m} \frac{m!}{(m-i)!}\left(\log\frac{V(K)}{V(L)}\right)^{m-i} \tag{3}$$

where m is an odd number, K, L are two star bodies and $dV_L(\cdot)$ is the normalized cone-volume measure

$$dV_L(\cdot) = \left(\frac{\rho_L^n(\cdot)}{nV(L)}\right)dS(\cdot). \tag{4}$$

The equality in (3) holds if and only if there exists $s > 0$ such that $K = sL$.

Note that (3) generalizes the dual log-Minkowski inequality for a mixed volume of two star bodies (see [12,16]) and produces the polynomial dual for the conjectured log-Minkowski inequality (see [17]).

In this paper, we study the other extreme value of the anisotropic m-log-potential:

Definition 1. *For $m \in \mathbb{N}$, define*

$$W_{\log,m}(K,E) = \inf_{y\in\mathbb{R}^n} P_{\log,m}(K,E;y).$$

Note that because $\log\|x-y\|_K^{-1}$ may be negative, $W_{\log,m}(K,E)$ is defined for integer m.

In Section 2, some fundamental properties of $W_{\log,m}(K,E)$ are established. Then, in Section 3, we are able to introduce the new annulus body to solve the problem of optimal estimate for $W_{\log,m}(K,E)$ in a precise analytic way. For the application, a polynomial dual log-mixed volume difference law is induced from the optimal estimate.

2. Fundamental Properties

First we recall a metric property in [11] for the Minkowski functional of a star body with respect to the origin.

Proposition 1. *Let B_2^n be the unit ball and*

$$\begin{cases} I_K = \sup\{\tilde{r} \ge 0 : \tilde{r}B_2^n \subseteq K\}, \\ O_K = \inf\{\tilde{r} \ge 0 : K \subseteq \tilde{r}B_2^n\}. \end{cases} \tag{5}$$

Then

$$O_K^{-1}|x| \le \|x\|_K \le I_K^{-1}|x| \quad \text{for all} \quad x \in \mathbb{R}^n, \tag{6}$$

and a quasi-triangle inequality holds for $\|\cdot\|_K$

$$\|x+y\|_K \le I_K^{-1}O_K(\|x\|_K + \|y\|_K) \quad \text{for all} \quad x,y \in \mathbb{R}^n.$$

If m is an even number, the supremum of the anisotropic m-log-potential $V_{\log,m}(K,E) \equiv +\infty$ (see [11]). For the infimum of the anisotropic m-log-potential $W_{\log,m}(K,E)$, it follows

Proposition 2. *$W_{\log,m}(K,E) \equiv -\infty$ for m as an odd number.*

Proof. Note that K is a star body with respect to the origin and E is a bounded measurable set. Then $\sup_{x\in E}|x| < +\infty$. For all $C > 0$, let $C_1 = e^{\left(\frac{C}{V(E)}\right)^{\frac{1}{m}}} > 1$, $|y| > \max\{2O_K C_1, 2\sup_{x\in E}|x|\}$, where O_K is defined in (5). Hence, for all $x \in E$,

$$\|x-y\|_K \ge {O_K}^{-1}|x-y| \ge {O_K}^{-1}(|y|-|x|) > {O_K}^{-1}\frac{|y|}{2} > C_1 > 1.$$

Since m is odd, it follows that

$$\begin{aligned}P_{\log,m}(K,E;y) &= \int_E \left(\log \frac{1}{\|x-y\|_K}\right)^m dx \\ &< \int_E \left(\log C_1^{-1}\right)^m dx \\ &= \int_E \left(\log e^{\left(\frac{-C}{V(E)}\right)^{\frac{1}{m}}}\right)^m dx \\ &= -C,\end{aligned}$$

which implies

$$W_{\log,m}(K,E) = -\infty \quad \text{via} \quad W = \inf_{y\in\mathbb{R}^n} P_{\log,m}(K,E;y).$$

□

$W_{\log,m}(K,E)$ has the following metric properties for the nontrivial case (m is an even number).

Proposition 3. *Let m be an even number.*

(i) Monotonicity: let E_1 and E_2 are bounded measurable sets and $E_1 \subseteq E_2$. Then $W_{\log,m}(K,E_1) \leq W_{\log,m}(K,E_2)$.

(ii) Translation-invariance: for all $z \in \mathbb{R}^n$, let $z+E = \{z+y : y \in E\}$. Then $W_{\log,m}(K,z+E) = W_{\log,m}(K,E)$.

(iii) Homogeneity: for all $s > 0$, $W_{\log,m}(sK,sE) = s^n W_{\log,m}(K,E)$.

Proof. (i) Since $E_1 \subseteq E_2$, then for all $y \in \mathbb{R}^n$,

$$\int_{E_1} \left(\log \frac{1}{\|x-y\|_K}\right)^m dx \leqslant \int_{E_2} \left(\log \frac{1}{\|x-y\|_K}\right)^m dx.$$

Hence,

$$\begin{aligned}W_{\log,m}(K,E_1) &= \inf_{y\in\mathbb{R}^n} \int_{E_1} \left(\log \frac{1}{\|x-y\|_K}\right)^m dx \\ &\leqslant \inf_{y\in\mathbb{R}^n} \int_{E_2} \left(\log \frac{1}{\|x-y\|_K}\right)^m dx = W_{\log,m}(K,E_2).\end{aligned}$$

(ii) For all $z \in \mathbb{R}^n$, by changing the variables $x = z + x_1$ and $y = z + y_1$, it follows

$$\begin{aligned}W_{\log,m}(K,z+E) &= \inf_{y\in\mathbb{R}^n} \int_{z+E} \left(\log \frac{1}{\|x-y\|_K}\right)^m dx \\ &= \inf_{y\in\mathbb{R}^n} \int_E \left(\log \frac{1}{\|x_1+z-y\|_K}\right)^m dx_1 \\ &= \inf_{y_1\in\mathbb{R}^n} \int_E \left(\log \frac{1}{\|x_1-y_1\|_K}\right)^m dx_1 \\ &= W_{\log,m}(K,E).\end{aligned}$$

(iii) For all $\forall\, s > 0$, by changing the variables $x = s\tilde{x}$ and $y = s\tilde{y}$ and the definition of Minkowski functional in (1), it follows that

$$\begin{aligned} W_{\log,m}(sK, sE) &= \inf_{y\in\mathbb{R}^n} \int_{sE} \left(\log \frac{1}{\|x-y\|_{sK}}\right)^m dx \\ &= \inf_{s\tilde{y}\in\mathbb{R}^n} \int_{E} \left(\log \frac{1}{\|s\tilde{x}-s\tilde{y}\|_{sK}}\right)^m ds\tilde{x} \\ &= \inf_{\tilde{y}\in\mathbb{R}^n} \int_{E} \left(\log \frac{1}{\|\tilde{x}-\tilde{y}\|_{K}}\right)^m ds\tilde{x} \\ &= s^n W_{\log,m}(K, E). \end{aligned}$$

□

The continuity of the anisotropic m-log-potential $P_{\log,m}(K, E; \cdot)$ has already been proven in [11]. From this, it follows that

Lemma 1. *Let m be an even number. The infimum in*

$$W_{\log,m}(K, E) = \inf_{y\in\mathbb{R}^n} P_{\log,m}(K, E; y)$$

is achieved at some $y \in \mathbb{R}^n$.

Proof. We first conclude that

$$\lim_{|y|\to+\infty} P_{\log,m}(K, E; y) = +\infty. \tag{7}$$

Actually, note that E is a bounded measurable set, then $\sup_{x\in E}|x| < +\infty$. For all $M_1 > 0$, let

$$|y| \geq \max\left\{2\sup_{x\in E}|x|, 2O_K e^{\left(\frac{M_1}{V(E)}\right)^{\frac{1}{m}}}\right\},$$

where O_K is defined in (5). It follows from m being an even number and (6) that

$$\begin{aligned} P_{\log,m}(K, E; y) &= \int_E \left(\log \frac{1}{\|x-y\|_K}\right)^m dx \\ &= \int_E (\log\|x-y\|_K)^m dx \\ &\geq \int_E \left(\log|O_K|^{-1}|x-y|\right)^m dx \\ &\geq \int_E \left(\log|O_K|^{-1}(|y|-|x|)\right)^m dx \\ &\geq \int_E \left(\log(2|O_K|)^{-1}|y|\right)^m dx \\ &\geq \int_E \left(\log e^{\left(\frac{M_1}{V(E)}\right)^{\frac{1}{m}}}\right)^m dx \\ &\geq M_1, \end{aligned}$$

which implies that (7) holds.

In the following, we will show that $P_{\log,m}(K, E; \cdot) \not\equiv +\infty$. As a matter of fact, for $z \in \mathbb{R}^n$ and $|z| \geq \sup_{x\in E}|x|$,

$$\begin{aligned} P_{\log,m}(K,E;z) &= \int_E \left(\log \frac{1}{\|x-z\|_K}\right)^m dx \\ &= \int_E (\log \|x-z\|_K)^m dx \\ &\leq \int_E \left(\log I_K^{-1}|x-z|\right)^m dx \\ &\leq \int_E \left(\log I_K^{-1}(|z|+|x|)\right)^m dx \\ &\leq \int_E \left(\log 2I_K^{-1}|z|\right)^m dx \\ &= \left(\log 2I_K^{-1}|z|\right)^m V(E) \\ &< +\infty, \end{aligned}$$

where I_K is in (5). Let $M_2 = \left(\log 2I_K^{-1}|z|\right)^m V(E)$. Because of (7), there exists $D_1 \geq 0$ such that for all $y \in \{y \in \mathbb{R}^n : |y| > D_1\}$, $P_{\log,m}(K,E;y) > M_2$, which implies that

$$z \in D = \{y \in \mathbb{R}^n : |y| \leq D_1\}.$$

Since $P_{\log,m}(K,E;\cdot)$ is continuous and D is compact, it can attain its minimum at a point y_0. Then

$$P_{\log,m}(K,E;y_0) = \inf_{y \in D} P_{\log,m}(K,E;y) \leq P_{\log,m}(K,E;z) \leq M_2 \leq \inf_{y \in D^c} P_{\log,m}(K,E;y),$$

which implies

$$P_{\log,m}(K,E;y_0) = \inf_{y \in \mathbb{R}^n} P_{\log,m}(K,E;y).$$

□

3. Optimal Estimate and Application

Now we are ready to establish the optimal estimate for the infimum of the anisotropic m-log-potential.

Theorem 1. *Let m be an even number. Then*

$$\begin{aligned} W_{\log,m}(K,E) \geq \frac{m!V(K)}{n^m} \sum_{i=0}^{m} \frac{1}{(m-i)!} \left[\log\left(\left(\left(\frac{V(E)}{2V(K)}\right)^2+1\right)^{\frac{1}{2}} + \frac{V(E)}{2V(K)}\right)\right]^{m-i} \\ \times \left[\left((-1)^i-1\right)\left(\left(\frac{V(E)}{2V(K)}\right)^2+1\right)^{\frac{1}{2}} + \left((-1)^i+1\right)\frac{V(E)}{2V(K)}\right], \end{aligned} \tag{8}$$

where the equality holds if and only if E is a K-annulus body with outer radius a and inner radius $\frac{1}{a}$ up to a difference of a measure zero set, namely there exists $y \in \mathbb{R}^n$ such that

$$V\left(E \cap (R_a^K(y))^c\right) = V\left(R_a^K(y) \cap E^c\right) = 0$$

where $a = \left(\left(\left(\frac{V(E)}{2V(K)}\right)^2+1\right)^{\frac{1}{2}} + \frac{V(E)}{2V(K)}\right)^{\frac{1}{n}}$.

Proof. Let $y \in \mathbb{R}^n$ be fixed, and note that $a = \left(\left(\left(\frac{V(E)}{2V(K)}\right)^2 + 1\right)^{\frac{1}{2}} + \frac{V(E)}{2V(K)}\right)^{\frac{1}{n}} > 1$ and

$$0 < \frac{1}{a} = \left(\left(\left(\frac{V(E)}{2V(K)}\right)^2 + 1\right)^{\frac{1}{2}} - \frac{V(E)}{2V(K)}\right)^{\frac{1}{n}} < 1,$$

which imply

$$V\left(R_a^K(y)\right) = \left[a^n - \left(\frac{1}{a}\right)^n\right] V(K) = V(E).$$

Note that

$$\begin{aligned} V\left(E \cap (R_a^K(y))^c\right) &= V\left(E \setminus R_a^K(y)\right) \\ &= V(E) - V\left(R_a^K(y) \cap E\right) \\ &= V\left(R_a^K(y)\right) - V\left(R_a^K(y) \cap E\right) \\ &= V\left(R_a^K(y) \setminus E\right) \\ &= V\left(R_a^K(y) \cap E^c\right), \end{aligned}$$

which, together with the following elementary computations

$$\begin{cases} \|x-y\|_K > a (or < \frac{1}{a}) \text{ and } (\log a)^m < (\log \|x-y\|_K)^m \text{ for all } x \in E \cap (R_a^K(y))^c, \\ \frac{1}{a} \leqslant \|x-y\|_K \leq a \text{ and } 0 \leqslant (\log \|x-y\|_K)^m \leqslant (\log a)^m \text{ for all } x \in R_a^K(y) \cap E^c, \end{cases}$$

implies

$$\begin{aligned} \int_{R_a^K(y) \cap E^c} (\log \|x-y\|_K)^m \, dx &\leqslant (\log a)^m V\left(R_a^K(y) \cap E^c\right) \\ &= (\log a)^m V\left(E \cap (R_a^K(y))^c\right) \\ &\leqslant \int_{E \cap \left(R_a^K(y)\right)^c} (\log \|x-y\|_K)^m \, dx. \end{aligned} \tag{9}$$

Note that m is an even number, then

$$\begin{aligned} &P_{\log,m}(K, E; y) \\ &= \int_E \left(\log \frac{1}{\|x-y\|_K}\right)^m dx \\ &= \int_E (\log \|x-y\|_K)^m \, dx \\ &= \int_{\left(R_a^K(y)\right)^c \cap E} (\log \|x-y\|_K)^m \, dx + \int_{R_a^K(y) \cap E} (\log \|x-y\|_K)^m \, dx \\ &\geq \int_{R_a^K(y) \cap E^c} (\log \|x-y\|_K)^m \, dx + \int_{R_a^K(y) \cap E} (\log \|x-y\|_K)^m \, dx \\ &= \int_{R_a^K(y)} (\log \|x-y\|_K)^m \, dx. \\ &= m \int_{\{x: \frac{1}{a} \leq \|x-y\|_K \leq a\}} \int_1^{\|x-y\|_K} s^{-1} (\log s)^{m-1} \, ds \, dx \\ &= m \int_{\{x: 1 \leq \|x-y\|_K \leq a\}} \int_1^{\|x-y\|_K} s^{-1} (\log s)^{m-1} \, ds \, dx \\ &\quad - m \int_{\{x: \frac{1}{a} \leq \|x-y\|_K \leq 1\}} \int_{\|x-y\|_K}^1 s^{-1} (\log s)^{m-1} \, ds \, dx \\ &:= I_1 + I_2. \end{aligned} \tag{10}$$

By Fubini's theorem, it follows

$$
\begin{aligned}
I_1 &= m\int_1^a s^{-1}(\log s)^{m-1}\int_{\{x:s\le\|x-y\|_K\le a\}} dx\,ds \\
&= m\int_1^a s^{-1}(\log s)^{m-1}(a^n - s^n)V(K)\,ds \\
&= mV(K)a^n\int_1^a s^{-1}(\log s)^{m-1}\,ds - mV(K)\int_1^a s^{n-1}(\log s)^{m-1}\,ds,
\end{aligned}
$$

and

$$
\begin{aligned}
I_2 &= -m\int_{\frac{1}{a}}^1 s^{-1}(\log s)^{m-1}\int_{\{x:\frac{1}{a}\le\|x-y\|_K\le s\}} dx\,ds \\
&= -m\int_{\frac{1}{a}}^1 s^{-1}(\log s)^{m-1}\left(s^n - \frac{1}{a^n}\right)V(K)\,ds \\
&= -mV(K)\int_{\frac{1}{a}}^1 s^{n-1}(\log s)^{m-1}\,ds + \frac{mV(K)}{a^n}\int_{\frac{1}{a}}^1 s^{-1}(\log s)^{m-1}\,ds.
\end{aligned}
$$

Then, by integration by parts, it follows

$$
\begin{aligned}
&I_1 + I_2 \qquad (11)\\
&= mV(K)\left[\frac{1}{a^n}\int_{\frac{1}{a}}^1 s^{-1}(\log s)^{m-1}\,ds + a^n\int_1^a s^{-1}(\log s)^{m-1}\,ds\right.\\
&\left.-\int_{\frac{1}{a}}^a s^{n-1}(\log s)^{m-1}\,ds\right]\\
&= mV(K)\left[\frac{1}{ma^n}(\log s)^m\Big|_{\frac{1}{a}}^1 + \frac{a^n}{m}(\log s)^m\Big|_1^a\right.\\
&\left.-(m-1)!s^n\sum_{i=1}^m \frac{(-1)^{i-1}(\log s)^{m-i}}{n^i(m-i)!}\Bigg|_{\frac{1}{a}}^a\right]\\
&= m!V(K)\sum_{i=0}^m \frac{1}{n^i(m-i)!}(\log a)^{m-i}\left[-\left(\frac{1}{a}\right)^n - (-1)^{i-1}a^n\right]\\
&= \frac{m!V(K)}{n^m}\sum_{i=0}^m \frac{1}{(m-i)!}\left[\log\left(\left(\left(\frac{V(E)}{2V(K)}\right)^2+1\right)^{\frac{1}{2}} + \frac{V(E)}{2V(K)}\right)\right]^{m-i}\\
&\times\left[\left((-1)^i - 1\right)\left(\left(\frac{V(E)}{2V(K)}\right)^2+1\right)^{\frac{1}{2}} + \left((-1)^i+1\right)\frac{V(E)}{2V(K)}\right].
\end{aligned}
$$

Hence, by (10) and (11), it follows that

$$\begin{aligned}
W_{\log,m}(K,E) &= \inf_{y\in\mathbb{R}^n} P_{\log,m}(K,E;y)\\
&= \inf_{y\in\mathbb{R}^n} \int_E \left(\log\frac{1}{\|x-y\|_K}\right)^m dx\\
&= \inf_{y\in\mathbb{R}^n} \int_E (\log\|x-y\|_K)^m\, dx\\
&\geq \inf_{y\in\mathbb{R}^n} \int_{R_a^K(y)} (\log\|x-y\|_K)^m\, dx\\
&= \frac{m!V(K)}{n^m}\sum_{i=0}^{m}\frac{1}{(m-i)!}\left[\log\left(\left(\left(\frac{V(E)}{2V(K)}\right)^2+1\right)^{\frac{1}{2}}+\frac{V(E)}{2V(K)}\right)\right]^{m-i}\\
&\quad\times\left[\left((-1)^i-1\right)\left(\left(\frac{V(E)}{2V(K)}\right)^2+1\right)^{\frac{1}{2}}+\left((-1)^i+1\right)\frac{V(E)}{2V(K)}\right].
\end{aligned}$$

To prove the equality in (8) , if E is almost a K-annulus body up to a difference of a measure zero set, which means there exists $z_1\in\mathbb{R}^n$ and a such that

$$V\left(E\cap(R_a^K(z_1))^c\right)=V\left(R_a^K(z_1)\cap E^c\right)=0,$$

which, together with (9), implies

$$\int_{R_a^K(z_1)\cap E^c}(\log\|x-z_1\|_K)^m\,dx=\int_{E\cap\left(R_a^K(z_1)\right)^c}(\log\|x-z_1\|_K)^m\,dx=0,$$

and hence

$$\int_E\left(\log\frac{1}{\|x-z_1\|_K}\right)^m dx=\int_{R_a^K(z_1)}\left(\log\frac{1}{\|x-z_1\|_K}\right)^m dx, \tag{12}$$

from (10).

By (10)–(12), it follows

$$\begin{aligned}
P_{\log,m}(K,E;z_1) &= \int_{R_a^K(z_1)}\left(\log\frac{1}{\|x-z_1\|_K}\right)^m dx\\
&= \frac{m!V(K)}{n^m}\sum_{i=0}^{m}\frac{1}{(m-i)!}\left[\log\left(\left(\left(\frac{V(E)}{2V(K)}\right)^2+1\right)^{\frac{1}{2}}+\frac{V(E)}{2V(K)}\right)\right]^{m-i}\\
&\quad\times\left[\left((-1)^i-1\right)\left(\left(\frac{V(E)}{2V(K)}\right)^2+1\right)^{\frac{1}{2}}+\left((-1)^i+1\right)\frac{V(E)}{2V(K)}\right],
\end{aligned}$$

which means the equality in (8) holds.

On the other hand, by Lemma 1, there exists $z_2\in\mathbb{R}^n$, $W_{\log,m}(K,E)=P_{\log,m}(K,E;z_2)$. If E is not a K-annulus body up to a difference of a measure zero set, it follows

$$V\left(E\cap R_a^K(z_2)^c\right)\neq 0 \text{ and } V\left(R_a^K(z_2)\cap E^c\right)\neq 0.$$

Then the following strict inequality holds from (9):

$$\int_{R_a^K(z_2)\cap E^c}(\log\|x-z_2\|_K)^m\,dx<\int_{E\cap\left(R_a^K(z_2)\right)^c}(\log\|x-z_2\|_K)^m\,dx,$$

which implies the inequality in (10) is also strict, and hence

$$\begin{aligned}
&W_{\log,m}(K,E) \\
&= P_{\log,m}(K,E;z_2) \\
&= \int_E \left(\log \frac{1}{\|x-z_2\|_K}\right)^m dx \\
&= \int_E (\log \|x-z_2\|_K)^m \, dx \\
&> \int_{R_a^K(z_2)} (\log \|x-z_2\|_K)^m \, dx \\
&= \frac{m!V(K)}{n^m} \sum_{i=0}^{m} \frac{1}{(m-i)!} \left[\log\left(\left(\left(\frac{V(E)}{2V(K)}\right)^2+1\right)^{\frac{1}{2}} + \frac{V(E)}{2V(K)}\right)\right]^{m-i} \\
&\times \left[\left((-1)^i-1\right)\left(\left(\frac{V(E)}{2V(K)}\right)^2+1\right)^{\frac{1}{2}} + \left((-1)^i+1\right)\frac{V(E)}{2V(K)}\right],
\end{aligned}$$

which means, if the equality in (8) holds, E must be almost a K-annulus body up to a difference of a measure zero set. □

Remark 1. *We claim that there is no such upper bound for $W_{\log,m}(K,E)$ by using $V(K)$ and $V(E)$ as in Theorem 1 when m is an even number.*

Proof. Actually, let $V(E)$ be fixed. For all $M>0$, let $E = E_1 \bigcup E_2$, where $V(E_1) = V(E_2) = 2^{-1}V(E)$ and

$$\mathrm{dist}\{E_1,E_2\} = \inf\{|x_1-x_2||x_1\in E_1, x_2\in E_2\} > 2O_K e^{\left(\frac{2M}{V(E)}\right)^{\frac{1}{m}}}.$$

Then, for all $y\in\mathbb{R}^n$, $\mathrm{dist}\{\{y\},E_1\} > O_K e^{\left(\frac{2M}{V(E)}\right)^{\frac{1}{m}}}$ or $\mathrm{dist}\{\{y\},E_2\} > O_K e^{\left(\frac{2M}{V(E)}\right)^{\frac{1}{m}}}$. Without loss of generality, suppose $\mathrm{dist}\{\{y\},E_1\} > O_K e^{\left(\frac{2M}{V(E)}\right)^{\frac{1}{m}}}$, then, by (6), it follows

$$\begin{aligned}
P_{\log,m}(K,E;y) &= \int_E \left(\log \frac{1}{\|x-y\|_K}\right)^m dx, \\
&= \int_E (\log \|x-y\|_K)^m \, dx \\
&\geq \int_E \left(\log O_K^{-1}|x-y|\right)^m dx \\
&> \int_{E_1} \left(\log O_K^{-1}|x-y|\right)^m dx \\
&> M,
\end{aligned}$$

which implies

$$W_{\log,m}(K,E) = \inf_{y\in\mathbb{R}^n} P_{\log,m}(K,E;y) \geq M.$$

This completes the proof of the remark. □

The infimum of the anisotropic m-log-potential is closely related with the convex geometry analysis. For this, a polynomial dual log-mixed volume difference law can be deduced from the optimal estimate for $W_{\log,m}(K,E)$ in Theorem 1.

Theorem 2. *Let m be an even number, L_1, L_2, K be star bodies with respect to the origin, $L_1 \subseteq L_2$, and $dV_{L_1}(u)$, $dV_{L_2}(u)$ be the normalized cone-volume measures defined in (4), then*

$$V(L_2)\int_{\mathbb{S}^{n-1}}\sum_{i=0}^{m}\frac{m!}{n^i(m-i)!}\log\left(\frac{\rho_K(u)}{\rho_{L_2}(u)}\right)^{m-i}dV_{L_2}(u) \tag{13}$$
$$-V(L_1)\int_{\mathbb{S}^{n-1}}\sum_{i=0}^{m}\frac{m!}{n^i(m-i)!}\log\left(\frac{\rho_K(u)}{\rho_{L_1}(u)}\right)^{m-i}dV_{L_1}(u)\geq$$
$$\frac{m!V(K)}{n^m}\sum_{i=0}^{m}\frac{1}{(m-i)!}\left[\log\left(\left(\left(\frac{V(L_2)-V(L_1)}{2V(K)}\right)^2+1\right)^{\frac{1}{2}}+\frac{V(L_2)-V(L_1)}{2V(K)}\right)\right]^{m-i}$$
$$\times\left[\left((-1)^i-1\right)\left(\left(\frac{V(L_2)-V(L_1)}{2V(K)}\right)^2+1\right)^{\frac{1}{2}}+\left((-1)^i+1\right)\frac{V(L_2)-V(L_1)}{2V(K)}\right],$$

where the equality holds if and only if $L_2 \setminus L_1$ is a K-annulus body centered at origin with outer radius a and inner radius $\frac{1}{a}(a>0)$ up to a difference of a measure zero set.

Proof. Note that $\rho_K^{-1}(\cdot) = \|\cdot\|_K$, then, by changing to the polar coordinates and integration by parts, it follows that

$$
\begin{aligned}
&P_{\log,m}(K, L_2 \setminus L_1; 0) && (14)\\
&= \int_{L_2\setminus L_1}\left(\log\frac{1}{\|x\|_K}\right)^m dx\\
&= \int_{L_2}\left(\log\frac{1}{\|x\|_K}\right)^m dx - \int_{L_1}\left(\log\frac{1}{\|x\|_K}\right)^m dx\\
&= \int_{L_2}(\log\rho_K(x))^m dx - \int_{L_1}(\log\rho_K(x))^m dx\\
&= \int_{\mathbb{S}^{n-1}}\int_0^{\rho_{L_2}(u)} s^{n-1}(\log\rho_K(su))^m dsdu\\
&- \int_{\mathbb{S}^{n-1}}\int_0^{\rho_{L_1}(u)} s^{n-1}(\log\rho_K(su))^m dsdu\\
&= n^{-1}\int_{\mathbb{S}^{n-1}}\int_0^{\rho_{L_2}(u)}\left(\log\left(s^{-1}\rho_K(u)\right)\right)^m ds^n du\\
&- n^{-1}\int_{\mathbb{S}^{n-1}}\int_0^{\rho_{L_1}(u)}\left(\log\left(s^{-1}\rho_K(u)\right)\right)^m ds^n du\\
&= n^{-1}\int_{\mathbb{S}^{n-1}}\rho_{L_2}(u)^n\left(\log\frac{\rho_K(u)}{\rho_L(u)}\right)^m du\\
&+ n^{-1}m\int_{\mathbb{S}^{n-1}}\int_0^{\rho_{L_2}(u)} s^{n-1}\left(\log\left(s^{-1}\rho_K(u)\right)\right)^{m-1} dsdu\\
&- n^{-1}\int_{\mathbb{S}^{n-1}}\rho_{L_1}(u)^n\left(\log\frac{\rho_K(u)}{\rho_L(u)}\right)^m du\\
&- n^{-1}m\int_{\mathbb{S}^{n-1}}\int_0^{\rho_{L_1}(u)} s^{n-1}\left(\log\left(s^{-1}\rho_K(u)\right)\right)^{m-1} dsdu\\
&\vdots\\
&= V(L_2)\int_{\mathbb{S}^{n-1}}\sum_{i=0}^{m}\frac{m!}{n^i(m-i)!}\log\left(\frac{\rho_K(u)}{\rho_{L_2}(u)}\right)^{m-i}dV_{L_2}(u)\\
&- V(L_1)\int_{\mathbb{S}^{n-1}}\sum_{i=0}^{m}\frac{m!}{n^i(m-i)!}\log\left(\frac{\rho_K(u)}{\rho_{L_1}(u)}\right)^{m-i}dV_{L_1}(u),
\end{aligned}
$$

where dV_{L_1} and dV_{L_2} are defined as in (4).

By Theorem 1, it follows that

$$\begin{aligned}
&P_{\log,m}(K, L_2 \setminus L_1; 0) \\
&= \int_{L_2\setminus L_1} \left(\log \frac{1}{\|x\|_K}\right)^m dx \\
&\geq \inf_{y\in\mathbb{R}^n} \int_{L_2\setminus L_1} \left(\log \frac{1}{\|x-y\|_K}\right)^m dx \\
&\geq \frac{m!V(K)}{n^m} \sum_{i=0}^{m} \frac{1}{(m-i)!} \left[\log\left(\left(\left(\frac{V(L_2\setminus L_1)}{2V(K)}\right)^2+1\right)^{\frac{1}{2}} + \frac{V(L_2\setminus L_1)}{2V(K)}\right)\right]^{m-i} \\
&\times \left[\left((-1)^i-1\right)\left(\left(\frac{V(L_2\setminus L_1)}{2V(K)}\right)^2+1\right)^{\frac{1}{2}} + \left((-1)^i+1\right)\frac{V(L_2\setminus L_1)}{2V(K)}\right] \\
&= \\
&\frac{m!V(K)}{n^m} \sum_{i=0}^{m} \frac{1}{(m-i)!} \left[\log\left(\left(\left(\frac{V(L_2)-V(L_1)}{2V(K)}\right)^2+1\right)^{\frac{1}{2}} + \frac{V(L_2)-V(L_1)}{2V(K)}\right)\right]^{m-i} \\
&\times \left[\left((-1)^i-1\right)\left(\left(\frac{V(L_2)-V(L_1)}{2V(K)}\right)^2+1\right)^{\frac{1}{2}} + \left((-1)^i+1\right)\frac{V(L_2)-V(L_1)}{2V(K)}\right],
\end{aligned}$$

which, together with (14), implies (13) holds with the equality holds if and only if $L_2 \setminus L_1$ is a K-annulus body centered at origin with outer radius a and inner radius $\frac{1}{a}(a > 0)$ up to a difference of a measure zero set. □

4. Conclusions

Theorem 1 and its Remark 1 complete the systematic study of the optimal upper and lower bounds of the extreme value of the anisotropic m-log-potential on a bounded measurable set (for the part of its supremum, we refer to [11]). Note that the anisotropic m-log-potential extends the classical logarithmic potential two-fold in anisotropic and higher order of m ways. By virtue of the wide development of Riesz potential with its better complement logarithmic potential for the end point case in harmonic analysis including function spaces, mathematical physics and partial differential equations (see [1–6]), these optimal estimates can be further applied to these related topics.

On the other hand, Brunn–Minkowski inequality and Minkowski inequality including their dual versions and generalizations are main topics in convex geometry analysis (see [12,13,16,17] and their references). The dual log-Minkowski inequality deals with the optimal estimate for mixed volume of two star bodies (see [12,16]), which exists as the dual version for the conjectured log-Minkowski inequality (see [17]). The polynomial dual log-mixed volume difference law in Theorem 2 deduced from the optimal estimate in Theorem 1, deals with the optimal estimate for the difference of mixed volumes of two star bodies, which is totally new and contributes to these theories.

Author Contributions: Conceptualization, S.H.; methodology, S.H.; software, S.H.; validation, S.H.; formal analysis, S.H.; investigation, S.H.; resources, S.H.; data curation, S.H.; writing—original draft preparation, S.H.; writing—review and editing, S.H.; visualization, S.H.; supervision, S.H.; project administration, S.H.; funding acquisition, S.H. All authors have read and agreed to the published version of the manuscript.

Funding: This research was funded by National Natural Science Foundation of China (No.12001157 and 11871191) and Natural Science Foundation of Hebei (No.A2021205013).

Institutional Review Board Statement: Not applicable.

Informed Consent Statement: Not applicable.

Data Availability Statement: Not applicable.

Acknowledgments: We will like to express our deep thanks to the anonymous referees for their valuable comments.

Conflicts of Interest: The author declares no conflict of interest.

References

1. Stein, E. *Singular Integrals and Differentiability Properties of Functions*; Princeton University Press: Princeton, NJ, USA, 1970.
2. Sawano, Y.; Sugano, S.; Tanaka, H. Olsen's inequality and its applications to Schrödinger equations. *RIMS Kôkyûroku Bessatsu* **2011**, *B26*, 51–80.
3. Liu, L.; Wu, S.; Yang, D.; Yuan, W. New characterizations of Morrey spaces and their preduals with applications to fractional Laplace equations. *J. Differ. Equ.* **2018**, *266*, 5118–5167. [CrossRef]
4. Rozenblum, G.; Ruzhansky, M.; Suragan, D. Isoperimetric inequalities for Schatten norms of Riesz potentials. *J. Funct. Anal.* **2016**, *271*, 224–239. [CrossRef]
5. Bent, F. The logarithmic potential in higher dimensions. *Mat. Fys. Medd. Dan. Cid. Selsk.* **1960**, *33*, 1–14.
6. Hou, S.; Xiao, J. Convex bodies via gravitational potentials. *Expo. Math.* **2017**, *35*, 478–482. [CrossRef]
7. Ludwig, M. Anisotropic fractional perimeters. *J. Differ. Geom.* **2014**, *96*, 77–93. [CrossRef]
8. Ludwig, M. Anisotropic fractional Sobolev norms. *Adv. Math.* **2014**, *252*, 150–157. [CrossRef]
9. Xiao, J.; Ye, D. Anisotropic Sobolev capacity with fractional order. *Can. J. Math.* **2017**, *69*, 873–889. [CrossRef]
10. Hou, S.; Xiao, J.; Ye, D. A mixed volume from the anisotropic Riesz-potential. *Trans. Lond. Math. Soc.* **2018**, *5*, 71–96. [CrossRef]
11. Hou, S.; Xiao, J. A mixed volumetry for the anisotropic logarithmic potential. *J. Geom. Anal.* **2018**, *28*, 2028–2049. [CrossRef]
12. Gardner, R.; Hug, D.; Weil, W.; Ye, D. The dual Orlicz–Brunn–Minkowski theory. *J. Math. Anal. Appl.* **2015**, *430*, 810–829. [CrossRef]
13. Schneider, R. *Convex Bodies: The Brunn-Minkowski Theory*, 2nd ed.; Cambridge University Press: Cambridge, UK, 2014.
14. Fairweather, G.; Johnston, R. The method of fundamental solutions for problems in potential theory. In *Treatment of Integral Equations by Numerical Methods*; Baker, C., Miller, G., Eds.; Academic Press Inc. (London) Ltd.: London, UK, 1982.
15. Jaswon, M.; Symm, G. *Integral Equation Methods in Potential Theory and Elastostatics*; Academic Press: London, UK, 1977.
16. Wang, W.; Liu, L. The dual log-Brunn-Minkowski inequalities. *Taiwan J. Math.* **2016**, *20*, 909–919. [CrossRef]
17. Böröczky, K.; Lutwak, E.; Yang, D.; Zhang, G. The log-Brunn-Minkowski inequality. *Adv. Math.* **2012**, *231*, 1974–1997. [CrossRef]

mathematics

MDPI

Article

Sobolev-Slobodeckij Spaces on Compact Manifolds, Revisited

Ali Behzadan [1,*] and Michael Holst [2]

[1] Department of Mathematics and Statistics, California State University Sacramento, Sacramento, CA 95819, USA
[2] Department of Mathematics, University of California San Diego, La Jolla, San Diego, CA 92093, USA; mholst@math.ucsd.edu
* Correspondence: a.behzadan@csus.edu

Abstract: In this manuscript, we present a coherent rigorous overview of the main properties of Sobolev-Slobodeckij spaces of sections of vector bundles on compact manifolds; results of this type are scattered through the literature and can be difficult to find. A special emphasis has been put on spaces with noninteger smoothness order, and a special attention has been paid to the peculiar fact that for a general nonsmooth domain Ω in $\mathbb{R}^n$, $0 < t < 1$, and $1 < p < \infty$, it is not necessarily true that $W^{1,p}(\Omega) \hookrightarrow W^{t,p}(\Omega)$. This has dire consequences in the multiplication properties of Sobolev-Slobodeckij spaces and subsequently in the study of Sobolev spaces on manifolds. We focus on establishing certain fundamental properties of Sobolev-Slobodeckij spaces that are particularly useful in better understanding the behavior of elliptic differential operators on compact manifolds. In particular, by introducing notions such as "geometrically Lipschitz atlases" we build a general framework for developing multiplication theorems, embedding results, etc. for Sobolev-Slobodeckij spaces on compact manifolds. To the authors' knowledge, some of the proofs, especially those that are pertinent to the properties of Sobolev-Slobodeckij spaces of sections of general vector bundles, cannot be found in the literature in the generality appearing here.

Keywords: Sobolev spaces; compact manifolds; tensor bundles; differential operators

Citation: Behzadan, A.; Holst, M. Sobolev-Slobodeckij Spaces on Compact Manifolds, Revisited. *Mathematics* **2022**, *10*, 522. https://doi.org/10.3390/math10030522

Academic Editors: Dachun Yang and Wen Yuan

Received: 7 January 2022
Accepted: 30 January 2022
Published: 7 February 2022

1. Introduction

Suppose $s \in \mathbb{R}$ and $p \in (1, \infty)$. With each nonempty open set Ω in $\mathbb{R}^n$ we can associate a complete normed function space denoted by $W^{s,p}(\Omega)$ called the Sobolev-Slobodeckij space with smoothness degree s and integrability degree p. Similarly, given a compact smooth manifold M and a vector bundle E over M, there are several ways to define the normed spaces $W^{s,p}(M)$ and more generally $W^{s,p}(E)$. The main goal of this manuscript is to review these various definitions and rigorously study the key properties of these spaces. Some of the properties that we are interested in are as follows:

- Density of smooth functions
- Completeness, separability, reflexivity
- Embedding properties
- Behavior under differentiation
- Being closed under multiplication by smooth functions:

$$u \in W^{s,p}, \quad \varphi \text{ is smooth} \overset{?}{\Longrightarrow} \varphi u \in W^{s,p}$$

- Invariance under change of coordinates:

$$u \in W^{s,p}, \quad T \text{ is a diffeomorphism} \overset{?}{\Longrightarrow} u \circ T \in W^{s,p}$$

- Invariance under composition by a smooth function:

$$u \in W^{s,p}, \quad F \text{ is smooth} \overset{?}{\Longrightarrow} F(u) \in W^{s,p}$$

As we shall see, there are several ways to define $W^{s,p}(E)$. In particular, $\|u\|_{W^{s,p}(E)}$ can be defined using the components of the local representations of u with respect to a fixed augmented total trivialization atlas Λ, or it can be defined using the notion of connection in E. Here are some of the questions that we have studied in this paper regarding this issue:

- Are the different characterizations that exist in the literature equivalent? If not, what is the relationship between the various characterizations of Sobolev-Slobodeckij spaces on M?
- In particular, does the corresponding space depend on the chosen atlas (more precisely the chosen augmented total trivialization atlas) used in the definition?
- Suppose $f \in W^{s,p}(M)$. Does this imply that the local representation of f with respect to each chart $(U_\alpha, \varphi_\alpha)$ is in $W^{s,p}(\varphi_\alpha(U_\alpha))$? If g is a metric on M and $g \in W^{s,p}$, can we conclude that $g_{ij} \circ \varphi_\alpha^{-1} \in W^{s,p}(\varphi_\alpha(U_\alpha))$?
- Suppose that $P : C^\infty(M) \to C^\infty(M)$ is a linear differential operator. Is it possible to gain information about the mapping properties of P by studying the mapping properties of its local representations with respect to charts in a given atlas? For example, suppose that the local representations of P with respect to each chart $(U_\alpha, \varphi_\alpha)$ in an atlas is continuous from $W^{s,p}(\varphi_\alpha(U_\alpha))$ to $W^{\tilde{s},\tilde{p}}(\varphi_\alpha(U_\alpha))$. Is it possible to extend P to a continuous linear map from $W^{s,p}(M)$ to $W^{\tilde{s},\tilde{p}}(M)$?

To further motivate the questions that are studied in this paper and the study of the key properties mentioned above, let us consider a concrete example. For any two sets A and B, let $\mathrm{Func}(A, B)$ denote the collection of all functions from A to B. Consider the differential operator

$$\mathrm{div}_g : C^\infty(TM) \to \mathrm{Func}(M, \mathbb{R}), \qquad \mathrm{div}_g X = (\mathrm{tr} \circ \mathrm{sharp}_g \circ \nabla \circ \mathrm{flat}_g)X,$$

on a compact Riemannian manifold (M, g) with $g \in W^{s,p}$. Let $\{(U_\alpha, \varphi_\alpha)\}$ be a smooth atlas for M. It can be shown that for each α

$$(\mathrm{div}_g X) \circ \varphi_\alpha^{-1} = \sum_{j=1}^{n} \frac{1}{\sqrt{\det g_\alpha}} \frac{\partial}{\partial x^j} \left[(\sqrt{\det g_\alpha})(X^j \circ \varphi_\alpha^{-1}) \right],$$

where $g_\alpha(x)$ is the matrix whose (i,j)-entry is $(g_{ij} \circ \varphi_\alpha^{-1})(x)$. As it will be discussed in detail in Section 10, we call $Q^\alpha : C^\infty(\varphi_\alpha(U_\alpha), \mathbb{R}^n) \to \mathrm{Func}(\varphi_\alpha(U_\alpha), \mathbb{R})$ defined by

$$Q^\alpha(Y) = \sum_{j=1}^{n} \underbrace{\frac{1}{\sqrt{\det g_\alpha}} \frac{\partial}{\partial x^j} \left[(\sqrt{\det g_\alpha})(Y^j) \right]}_{Q_j^\alpha(Y^j)}$$

the *local representation* of div_g with respect to the local chart $(U_\alpha, \varphi_\alpha)$. Let us say we can prove that for each α and j, Q_j^α maps $W_0^{e,q}(\varphi_\alpha(U_\alpha))$ to $W^{e-1,q}(\varphi_\alpha(U_\alpha))$. Can we conclude that div_g maps $W^{e,q}(TM)$ to $W^{e-1,q}(M)$? Furthermore, how can we find exponents e and q such that

$$Q_j^\alpha : W_0^{e,q}(\varphi_\alpha(U_\alpha)) \to W^{e-1,q}(\varphi_\alpha(U_\alpha))$$

is a well-defined continuous map? We will see how the properties we mentioned above play a key role in answering these questions.

Since $W^{0,p}(\Omega) = L^p(\Omega)$, Sobolev-Slobodeckij spaces can be viewed as a generalization of classical Lebesgue spaces. Of course, unlike Lebesgue spaces, some of the key properties of $W^{s,p}(\Omega)$ (for $s \neq 0$) depend on the geometry of the boundary of Ω. Indeed, to thoroughly study the properties of $W^{s,p}(\Omega)$ one should consider the following cases independently:

(1) $\Omega = \mathbb{R}^n$

(2) Ω is an arbitrary open subset of $\mathbb{R}^n$ $\begin{cases} 2a) \text{ bounded} \\ 2b) \text{ unbounded} \end{cases}$

(3) Ω is an open subset of $\mathbb{R}^n$ with smooth boundary $\begin{cases} 3a) \text{ bounded} \\ 3b) \text{ unbounded} \end{cases}$

Let us mention here four facts to highlight the dependence on domain and some atypical behaviors of certain fractional Sobolev spaces. Let $s \in (0, \infty)$ and $p \in (1, \infty)$.

- **Fact 1:**

$$\forall j \qquad \frac{\partial}{\partial x^j} : W^{s,p}(\mathbb{R}^n) \to W^{s-1,p}(\mathbb{R}^n)$$

is a well-defined bounded linear operator.

- **Fact 2:** If we further assume that $s \neq \frac{1}{p}$ and Ω has smooth boundary then

$$\forall j \qquad \frac{\partial}{\partial x^j} : W^{s,p}(\Omega) \to W^{s-1,p}(\Omega)$$

is a well-defined bounded linear operator.

- **Fact 3:** If $\tilde{s} \leq s$, then

$$W^{s,p}(\mathbb{R}^n) \hookrightarrow W^{\tilde{s},p}(\mathbb{R}^n).$$

- **Fact 4:** If Ω does NOT have Lipschitz boundary, then it is NOT necessarily true that

$$W^{1,p}(\Omega) \hookrightarrow W^{\tilde{s},p}(\Omega)$$

for $0 < \tilde{s} < 1$.

Let M be an n-dimensional compact smooth manifold and let $\{(U_\alpha, \varphi_\alpha)\}$ be a smooth atlas for M. As we will see, the properties of Sobolev-Slobodeckij spaces of sections of vector bundles on M are closely related to the properties of spaces of locally Sobolev-Slobodeckij functions on domains in $\mathbb{R}^n$. Primarily we will be interested in the properties of $W^{s,p}(\varphi_\alpha(U_\alpha))$ and $W^{s,p}_{loc}(\varphi_\alpha(U_\alpha))$. Furthermore, when we want to patch things together consistently and move from "local" to "global", we will need to consider spaces $W^{s,p}(\varphi_\alpha(U_\alpha \cap U_\beta))$ and $W^{s,p}(\varphi_\beta(U_\alpha \cap U_\beta))$. However, as we pointed out earlier, some of the properties of $W^{s,p}(\Omega)$ depend heavily on the geometry of the boundary of Ω. Considering that the intersection of two Lipschitz domains is not necessarily a Lipschitz domain, we need to consider the following question:

- Is it possible to find an atlas such that the image of each coordinate domain in the atlas (and the image of the intersection of any two coordinate domains in the atlas) under the corresponding coordinate map is either the entire $\mathbb{R}^n$ or a nonempty bounded set with smooth boundary? Furthermore, if we define the Sobolev spaces using such an atlas, will the results be independent of the chosen atlas?

This manuscript is an attempt to collect some results concerning these questions and certain other fundamental questions similar to the ones stated above, and we pay special attention to spaces with noninteger smoothness order and to general sections of vector bundles. There are a number of standard sources for properties of integer order Sobolev spaces of functions and related elliptic operators on domains in $\mathbb{R}^n$ (cf. [1–3]), real order Sobolev spaces of functions [4–8], Sobolev spaces of functions on manifolds [9–12], and Sobolev spaces of sections of vector bundles on manifolds [13,14]. However, most of these works focus on spaces of functions rather than general sections, and in many cases the focus is on integer order spaces. This paper should be viewed as a part of our efforts to build a more complete foundation for the study and use of Sobolev-Slobodeckij spaces on manifolds through a sequence of related manuscripts [15–18].

Outline of Paper. In Section 2, we summarize some of the basic notation and conventions used throughout the paper. In Section 3, we will review a number of basic constructions in linear algebra that are essential in the study of function spaces of generalized sections of vector bundles. In Section 4 we will recall some useful tools from analysis and topology. In particular, a concise overview of some of the main properties of topological vector spaces is presented in this section. Section 5 deals with reviewing some results we need from differential geometry. The main purpose of this section is to set the notation, definitions, and conventions straight. This section also includes some less well-known facts about topics such as higher order covariant derivatives in the context of vector bundles. In Section 6 we collect the results that we need from the theory of generalized functions on Euclidean spaces and vector bundles. Section 7 is concerned with various definitions and properties of Sobolev spaces that are needed for developing a coherent theory of such spaces on the vector bundles. In Sections 8 and 9 we introduce Lebesgue spaces and Sobolev–Slobodeckij spaces of sections of vector bundles and we present a rigorous account of their various properties. Finally in Section 10 we study the continuity of certain differential operators between Sobolev spaces of sections of vector bundles. Although the purpose of Section 3 through Section 7 is to give a quick overview of the prerequisites that are needed to understand the proofs of the results in later sections and set the notation straight, as it was pointed out earlier, several theorems and proofs that appear in these sections cannot be found elsewhere in the generality that are stated here.

2. Notation and Conventions

Throughout this paper, $\mathbb{R}$ denotes the set of real numbers, $\mathbb{N}$ denotes the set of positive integers, and $\mathbb{N}_0$ denotes the set of nonnegative integers. For any nonnegative real number s, the integer part of s is denoted by $\lfloor s \rfloor$. The letter n is a positive integer and stands for the dimension of the space.

Ω is a nonempty open set in $\mathbb{R}^n$. The collection of all compact subsets of Ω will be denoted by $\mathcal{K}(\Omega)$. Lipschitz domain in $\mathbb{R}^n$ refers to a nonempty bounded open set in $\mathbb{R}^n$ with Lipschitz continuous boundary.

Each element of $\mathbb{N}_0^n$ is called a multi-index. For a multi-index $\alpha = (\alpha_1, \dots, \alpha_n) \in \mathbb{N}_0^n$, we let

- $|\alpha| := \alpha_1 + \dots + \alpha_n$;
- $\alpha! := \alpha_1! \dots \alpha_n!$.

If $\alpha, \beta \in \mathbb{N}_0^n$, we say $\beta \leq \alpha$ provided that $\beta_i \leq \alpha_i$ for all $1 \leq i \leq n$. If $\beta \leq \alpha$, we let

$$\binom{\alpha}{\beta} := \frac{\alpha!}{\beta!(\alpha-\beta)!} = \binom{\alpha_1}{\beta_1} \cdots \binom{\alpha_n}{\beta_n}.$$

Suppose that $\alpha \in \mathbb{N}_0^n$. For sufficiently smooth functions $u : \Omega \to \mathbb{R}$ (or for any distribution u) we define the αth order partial derivative of u as follows:

$$\partial^\alpha u := \frac{\partial^{|\alpha|} u}{\partial x_1^{\alpha_1} \dots \partial x_n^{\alpha_n}}.$$

We use the notation $A \preceq B$ to mean $A \leq cB$, where c is a positive constant that does not depend on the non-fixed parameters appearing in A and B. We write $A \simeq B$ if $A \preceq B$ and $B \preceq A$.

For any nonempty set X and $r \in \mathbb{N}$, $X^{\times r}$ stands for $\underbrace{X \times \dots \times X}_{r \text{ times}}$.

For any two nonempty sets X and Y, $\mathrm{Func}(X, Y)$ denotes the collection of all functions from X to Y.

We write $L(X, Y)$ for the space of all *continuous* linear maps from the normed space X to the normed space Y. $L(X, \mathbb{R})$ is called the (topological) dual of X and is denoted by X^*. We use the notation $X \hookrightarrow Y$ to mean $X \subseteq Y$ and the inclusion map is continuous.

$\mathrm{GL}(n, \mathbb{R})$ is the set of all $n \times n$ invertible matrices with real entries. Note that $\mathrm{GL}(n, \mathbb{R})$ can be identified with an open subset of $\mathbb{R}^{n^2}$ and so it can be viewed as a smooth manifold (more precisely, $\mathrm{GL}(n, \mathbb{R})$ is a Lie group).

Throughout this manuscript, all manifolds are assumed to be smooth, Hausdorff, and second-countable.

Let M be an n-dimensional compact smooth manifold. The tangent space of the manifold M at point $p \in M$ is denoted by T_pM, and the cotangent space by T_p^*M. If $(U, \varphi = (x^i))$ is a local coordinate chart and $p \in U$, we denote the corresponding coordinate basis for T_pM by $\partial_i|_p$ while $\frac{\partial}{\partial x^i}|_x$ denotes the basis for the tangent space to $\mathbb{R}^n$ at $x = \varphi(p) \in \mathbb{R}^n$; that is,

$$\varphi_* \partial_i = \frac{\partial}{\partial x^i} .$$

Note that for any smooth function $f : M \to \mathbb{R}$ we have

$$(\partial_i f) \circ \varphi^{-1} = \frac{\partial}{\partial x^i}(f \circ \varphi^{-1}) .$$

The vector space of all k-covariant, l-contravariant tensors on T_pM is denoted by $T_l^k(T_pM)$. So, each element of $T_l^k(T_pM)$ is a multilinear map of the form

$$F : \underbrace{T_p^*M \times \cdots \times T_p^*M}_{l \text{ copies}} \times \underbrace{T_pM \times \cdots \times T_pM}_{k \text{ copies}} \to \mathbb{R} .$$

We are primarily interested in the vector bundle of $\binom{k}{l}$-tensors on M whose total space is

$$T_l^k(M) = \bigsqcup_{p \in M} T_l^k(T_pM) .$$

A section of this bundle is called a $\binom{k}{l}$-tensor field. We set $T^kM := T_0^k(M)$. TM denotes the tangent bundle of M and T^*M is the cotangent bundle of M. We set

$$\tau_l^k(M) = C^\infty(M, T_l^k(M)) = \text{collection of smooth } \tbinom{k}{l}\text{-tensor fields on } M$$

and

$$\chi(M) = C^\infty(M, TM) = \text{the collection of smooth vector fields on } M .$$

A symmetric positive definite section of T^2M is called a Riemannian metric on M. If M is equipped with a Riemannian metric g, the combination (M, g) will be referred to as a Riemannian manifold. If there is no possibility of confusion, we may write $\langle X, Y \rangle$ instead of $g(X, Y)$. The norm induced by g on each tangent space will be denoted by $\|.\|_g$. We say that g is smooth (or the Riemannian manifold is smooth) if $g \in C^\infty(M, T^2M)$.

d denotes the exterior derivative and $\mathrm{grad} : C^\infty(M) \to C^\infty(M, TM)$ denotes the gradient operator which is defined by $g(\mathrm{grad}\, f, X) = df(X)$ for all $f \in C^\infty(M)$ and $X \in C^\infty(M, TM)$.

Given a metric g on M, one can define the musical isomorphisms as follows:

$$\begin{aligned} \mathrm{flat}_g : T_pM &\to T_p^*M \\ X &\mapsto X^\flat := g(X, \cdot\,) , \\ \mathrm{sharp}_g : T_p^*M &\to T_pM \\ \psi &\mapsto \psi^\sharp := \mathrm{flat}_g^{-1}(\psi) . \end{aligned}$$

Using sharp_g we can define the $\binom{0}{2}$-tensor field g^{-1} (which is called the **inverse metric tensor**) as follows

$$g^{-1}(\psi_1, \psi_2) := g(\mathrm{sharp}_g(\psi_1), \mathrm{sharp}_g(\psi_2)) .$$

Let $\{E_i\}$ be a local frame on an open subset $U \subset M$ and $\{\eta^i\}$ be the corresponding dual coframe. So we can write $X = X^i E_i$ and $\psi = \psi_i \eta^i$. It is standard practice to denote the ith component of $\text{flat}_g X$ by X_i and the ith component of $\text{sharp}_g(\psi)$ by ψ^i:

$$\text{flat}_g X = X_i \eta^i, \quad \text{sharp}_g \psi = \psi^i E_i .$$

It is easy to show that

$$X_i = g_{ij} X^j, \quad \psi^i = g^{ij} \psi_j,$$

where $g_{ij} = g(E_i, E_j)$ and $g^{ij} = g^{-1}(\eta^i, \eta^j)$. It is said that $\text{flat}_g X$ is obtained from X by lowering an index and $\text{sharp}_g \psi$ is obtained from ψ by raising an index.

3. Review of Some Results from Linear Algebra

In this section, we summarize a collection of definitions and results from linear algebra that play an important role in our study of function spaces and differential operators on manifolds.

There are several ways to construct new vector spaces from old ones: subspaces, products, direct sums, quotients, etc. The ones that are particularly important for the study of Sobolev spaces of sections of vector bundles are the vector space of linear maps between two given vector spaces, the tensor product of vector spaces, and the vector space of all densities on a given vector space which we briefly review here in order to set the notation straight.

- Let V and W be two vector spaces. The collection of all linear maps from V to W is a new vector space which we denote by $\text{Hom}(V, W)$. In particular, $\text{Hom}(V, \mathbb{R})$ is the (algebraic) dual of V. If V and W are finite-dimensional, then $\text{Hom}(V, W)$ is a vector space whose dimension is equal to the product of dimensions of V and W. Indeed, if we choose a basis for V and a basis for W, then $\text{Hom}(V, W)$ is isomorphic with the space of matrices with $\dim W$ rows and $\dim V$ columns.
- Let U and V be two vector spaces. Roughly speaking, the tensor product of U and V (denoted by $U \otimes V$) is the unique vector space (up to isomorphism of vector spaces) such that for any vector space W, $\text{Hom}(U \otimes V, W)$ is isomorphic to the collection of bilinear maps from $U \times V$ to W. Informally, $U \otimes V$ consists of finite linear combinations of symbols $u \otimes v$, where $u \in U$ and $v \in V$. It is assumed that these symbols satisfy the following identities:

$$\begin{aligned} &(u_1 + u_2) \otimes v - u_1 \otimes v - u_2 \otimes v = 0, \\ &u \otimes (v_1 + v_2) - u \otimes v_1 - u \otimes v_2 = 0, \\ &\alpha(u \otimes v) - (\alpha u) \otimes v = 0, \\ &\alpha(u \otimes v) - u \otimes (\alpha v) = 0, \end{aligned}$$

for all $u, u_1, u_2 \in U$, $v, v_1, v_2 \in V$ and $\alpha \in \mathbb{R}$. These identities simply say that the map

$$\otimes : U \times V \to U \otimes V, \quad (u, v) \mapsto u \otimes v,$$

is a bilinear map. The image of this map spans $U \otimes V$.

Definition 1. *Let U and V be two vector spaces. Tensor product is a vector space $U \otimes V$ together with a bilinear map $\otimes : U \times V \to U \otimes V$, $(u, v) \mapsto u \otimes v$ such that given any vector space W and any* ***bilinear map*** *$b : U \times V \to W$, there is a unique* ***linear map*** *$\bar{b} : U \otimes V \to W$ with $\bar{b}(u \otimes v) = b(u, v)$. That is, the following diagram commutes:*

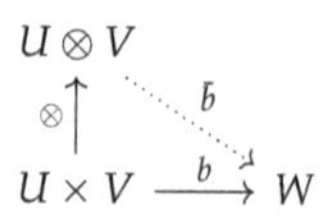

For us, the most useful property of the tensor product of finite dimensional vector spaces is the following property:

$$\text{Hom}(V,W) \cong V^* \otimes W\,.$$

Indeed, the following map is an isomorphism of vector spaces:

$$F : V^* \otimes W \to \text{Hom}(V,W), \qquad \underbrace{F(v^* \otimes w)}_{\text{an element of Hom}(V,W)} \quad (v) = \underbrace{[v^*(v)]}_{\text{a real number}} \; w\,.$$

It is useful to obtain an expression for the inverse of F too. That is, given $T \in \text{Hom}(V,W)$, we want to find an expression for the corresponding element of $V^* \otimes W$. To this end, let $\{e_i\}_{1\le i\le n}$ be a basis for V and $\{e^i\}_{1\le i\le n}$ denote the corresponding dual basis. Let $\{s_a\}_{1\le a\le r}$ be a basis for W. Then $\{e^i \otimes s_b\}$ is a basis for $V^* \otimes W$. Suppose $\sum_{i,a} R_i^a e^i \otimes s_a$ is the element of $V^* \otimes W$ that corresponds to T. We have

$$F(\sum_{i,a} R_i^a e^i \otimes s_a) = T \Longrightarrow \forall u \in V \quad \sum_{i,a} R_i^a F[e^i \otimes s_a](u) = T(u)$$
$$\Longrightarrow \forall u \in V \quad \sum_{i,a} R_i^a e^i(u) s_a = T(u)\,.$$

In particular, for all $1 \le j \le n$,

$$T(e_j) = \sum_{i,a} R_i^a \underbrace{e^i(e_j)}_{\delta^i_j} s_a = \sum_a R_j^a s_a\,.$$

That is, R_i^a is the entry in the ath row and ith column of the matrix of the linear transformation T.

- Let V be an n-dimensional vector space. A density on V is a function $\mu : \underbrace{V \times \ldots \times V}_{n \text{ copies}} \to \mathbb{R}$ with the property that

$$\mu(Tv_1,\ldots,Tv_n) = |\text{det}T|\mu(v_1,\ldots,v_n)\,,$$

for all $T \in \text{Hom}(V,V)$.
We denote the collection of all densities on V by $\mathcal{D}(V)$. It can be shown that $\mathcal{D}(V)$ is a one dimensional vector space under the obvious vector space operations. Indeed, if $(e_1,\ldots,e_n)$ is a basis for V, then each element $\mu \in \mathcal{D}(V)$ is uniquely determined by its value at $(e_1,\ldots,e_n)$ because for any $(v_1,\ldots,v_n) \in V^{\times n}$, we have $\mu(v_1,\ldots,v_n) = |\text{det}T|\mu(e_1,\ldots,e_n)$ where $T : V \to V$ is the linear transformation defined by $T(e_i) = v_i$ for all $1 \le i \le n$. Thus

$$F : \mathcal{D}(V) \to \mathbb{R}, \qquad F(\mu) = \mu(e_1,\ldots,e_n)\,,$$

will be an isomorphism of vector spaces.
Moreover, if $\omega \in \Lambda^n(V)$ where $\Lambda^n(V)$ is the collection of all alternating covariant n-tensors, then $|\omega|$ belongs to $\mathcal{D}(V)$. Thus, if ω is any nonzero element of $\Lambda^n(V)$, then $\{|\omega|\}$ will be a basis for $\mathcal{D}(V)$ ([19], p. 428).

4. Review of Some Results from Analysis and Topology

4.1. Euclidean Space

Let Ω be a nonempty open set in $\mathbb{R}^n$ and $m \in \mathbb{N}_0$. Here is a list of several useful function spaces on Ω:

$$
\begin{aligned}
&C(\Omega) = \{f : \Omega \to \mathbb{R} : f \text{ is continuous}\} \\
&C^m(\Omega) = \{f : \Omega \to \mathbb{R} : \forall\,|\alpha| \leq m \quad \partial^\alpha f \in C(\Omega)\} \qquad (C^0(\Omega) = C(\Omega)) \\
&BC(\Omega) = \{f : \Omega \to \mathbb{R} : f \text{ is continuous and bounded on } \Omega\} \\
&BC^m(\Omega) = \{f \in C^m(\Omega) : \forall\,|\alpha| \leq m \quad \partial^\alpha f \text{ is bounded on } \Omega\} \\
&BC(\bar{\Omega}) = \{f : \Omega \to \mathbb{R} : f \in BC(\Omega) \text{ and } f \text{ is uniformly continuous on } \Omega\} \\
&BC^m(\bar{\Omega}) = \{f : \Omega \to \mathbb{R} : f \in BC^m(\Omega), \forall\,|\alpha| \leq m \quad \partial^\alpha f \text{ is uniformly continuous on } \Omega\} \\
&C^\infty(\Omega) = \bigcap_{m\in\mathbb{N}_0} C^m(\Omega), \quad BC^\infty(\Omega) = \bigcap_{m\in\mathbb{N}_0} BC^m(\Omega), \quad BC^\infty(\bar{\Omega}) = \bigcap_{m\in\mathbb{N}_0} BC^m(\bar{\Omega})
\end{aligned}
$$

Remark 1 ([1]). *If $g : \Omega \to \mathbb{R}$ is in $BC(\bar{\Omega})$, then it possesses a unique, bounded, continuous extension to the closure $\bar{\Omega}$ of Ω.*

Notation: Let Ω be a nonempty open set in $\mathbb{R}^n$. The collection of all compact sets in Ω is denoted by $\mathcal{K}(\Omega)$. If $f : \Omega \to \mathbb{R}$ is a function, the support of f is denoted by $\operatorname{supp} f$. Notice that, in some references $\operatorname{supp} f$ is defined as the closure of $\{x \in \Omega : f(x) \neq 0\}$ in Ω, while in certain other references it is defined as the closure of $\{x \in \Omega : f(x) \neq 0\}$ in $\mathbb{R}^n$. Of course, if we are concerned with functions whose support is inside an element of $\mathcal{K}(\Omega)$, then the two definitions agree. For the sake of definiteness, in this manuscript we always use the former interpretation of support. Furthermore, support of a distribution will be discussed in Section 6.

Remark 2. *If $\mathcal{F}(\Omega)$ is any function space on Ω and $K \in \mathcal{K}(\Omega)$, then $\mathcal{F}_K(\Omega)$ denotes the collection of elements in $\mathcal{F}(\Omega)$ whose support is inside K. Furthermore,*

$$\mathcal{F}_c(\Omega) = \mathcal{F}_{comp}(\Omega) = \bigcup_{K\in\mathcal{K}(\Omega)} \mathcal{F}_K(\Omega)\,.$$

Let $0 < \lambda \leq 1$. A function $F : \Omega \subseteq \mathbb{R}^n \to \mathbb{R}^k$ is called *λ-Holder continuous* if there exists a constant L such that

$$|F(x) - F(y)| \leq L|x - y|^\lambda \quad \forall x, y \in \Omega\,.$$

Clearly, a λ-Holder continuous function on Ω is uniformly continuous on Ω. 1-Holder continuous functions are also called *Lipschitz continuous* functions or simply Lipschitz functions. We define

$$
\begin{aligned}
BC^{m,\lambda}(\Omega) &= \{f : \Omega \to \mathbb{R} : \forall\,|\alpha| \leq m \quad \partial^\alpha f \text{ is } \lambda\text{-Holder continuous and bounded}\} \\
&= \{f \in BC^m(\Omega) : \forall\,|\alpha| \leq m \quad \partial^\alpha f \text{ is } \lambda\text{-Holder continuous}\} \\
&= \{f \in BC^m(\bar{\Omega}) : \forall\,|\alpha| \leq m \quad \partial^\alpha f \text{ is } \lambda\text{-Holder continuous}\}
\end{aligned}
$$

and $BC^{\infty,\lambda}(\Omega) := \bigcap_{m\in\mathbb{N}_0} BC^{m,\lambda}(\Omega)$.

Remark 3. *Let $F : \Omega \subseteq \mathbb{R}^n \to \mathbb{R}^k$ ($F = (F^1, \cdots, F^k)$). Then*

$$F \text{ is Lipschitz} \iff \forall\, 1 \leq i \leq k \quad F^i \text{ is Lipschitz}\,.$$

Indeed, for each i

$$|F^i(x) - F^i(y)| \leq \sqrt{\sum_{j=1}^{k} |F^j(x) - F^j(y)|^2} = |F(x) - F(y)| \leq L|x - y|\,,$$

which shows that if F is Lipschitz so will be its components. Furthermore, if for each i, there exists L_i such that

$$|F^i(x) - F^i(y)| \leq L_i|x-y|,$$

then

$$\sum_{j=1}^{k} |F^j(x) - F^j(y)|^2 \leq nL^2|x-y|^2,$$

where $L = \max\{L_1, \cdots, L_k\}$. *This proves that if each component of F is Lipschitz so is F itself.*

Theorem 1 ([20]). *Let Ω be a nonempty open set in $\mathbb{R}^n$ and let $K \in \mathcal{K}(\Omega)$. There is a function $\psi \in C_c^\infty(\Omega)$ taking values in $[0,1]$ such that $\psi = 1$ on a neighborhood of K.*

Theorem 2 (Exhaustion by Compact Sets [20]). *Let Ω be a nonempty open subset of $\mathbb{R}^n$. There exists a sequence of compact subsets $(K_j)_{j\in\mathbb{N}}$ such that $\cup_{j\in\mathbb{N}}\mathring{K}_j = \Omega$ and*

$$K_1 \subseteq \mathring{K}_2 \subseteq K_2 \subseteq \cdots \subseteq \mathring{K}_j \subseteq K_j \subseteq \cdots.$$

Moreover, as a direct consequence, if K is any compact subset of the open set Ω, then there exists an open set V such that $K \subseteq V \subseteq \bar{V} \subseteq \Omega$.

Theorem 3 ([20]). *Let Ω be a nonempty open subset of $\mathbb{R}^n$. Let $\{K_j\}_{j\in\mathbb{N}}$ be an exhaustion of Ω by compact sets. Define*

$$V_0 = \mathring{K}_4, \qquad \forall j \in \mathbb{N} \quad V_j = \mathring{K}_{j+4} \setminus K_j.$$

Then

(1) *Each V_j is an open bounded set and $\Omega = \cup_j V_j$;*
(2) *The cover $\{V_j\}_{j\in\mathbb{N}_0}$ is* ***locally finite*** *in Ω, that is, each compact subset of Ω has nonempty intersection with only a finite number of the V_j's;*
(3) *There is a family of functions $\psi_j \in C_c^\infty(\Omega)$ taking values in $[0,1]$ such that supp $\psi_j \subseteq V_j$ and*

$$\sum_{j\in\mathbb{N}_0} \psi_j(x) = 1 \quad \text{for all } x \in \Omega.$$

Theorem 4 ([21], p. 74). *Suppose Ω is an open set in $\mathbb{R}^n$ and $G : \Omega \to G(\Omega) \subseteq \mathbb{R}^n$ is a C^1-diffeomorphism (i.e., G and G^{-1} are both C^1 maps). If f is a Lebesgue measurable function on $G(\Omega)$, then $f \circ G$ is Lebesgue measurable on Ω. If $f \geq 0$ or $f \in L^1(G(\Omega))$, then*

$$\int_{G(\Omega)} f(x)dx = \int_\Omega f \circ G(x)|detG'(x)|dx.$$

Theorem 5 ([21], p. 79). *If f is a nonnegative measurable function on $\mathbb{R}^n$ such that $f(x) = g(|x|)$ for some function g on $(0,\infty)$, then*

$$\int f(x)dx = \sigma(S^{n-1})\int_0^\infty g(r)r^{n-1}dr,$$

where $\sigma(S^{n-1})$ is the surface area of $(n-1)$-sphere.

Theorem 6 ([22], Section 12.11). *Suppose U is an open set in $\mathbb{R}^n$ and $f : U \to \mathbb{R}$ is differentiable. Let x and y be two points in U and suppose the line segment joining x and y is contained in U. Then there exists a point z on the line joining x to y such that*

$$f(y) - f(x) = \nabla f(z).(y-x).$$

As a consequence, if U is ***convex*** *and all first order partial derivatives of f are bounded, then f is Lipschitz on U.*

Warning: Suppose $f \in BC^{\infty}(U)$. By the above item, if U is convex, then f is Lipschitz. However, if U is not convex, then f is not necessarily Lipschitz. For example, let $U = \cup_{n=0}^{\infty}(n, n+1)$ and define

$$f : U \to \mathbb{R}, \qquad f(x) = (-1)^n, \forall x \in (n, n+1).$$

Clearly, all derivatives of U are equal to zero, so $f \in BC^{\infty}(U)$. However, f is not uniformly continuous and thus it is not Lipschitz. Indeed, for any $1 > \delta > 0$, we can let $x = 2 - \delta/4$ and $y = 2 + \delta/4$. Clearly $|x - y| < \delta$, however, $|f(x) - f(y)| = 2$.

Of course, if $f \in C_c^1(U)$, then f can be extended by zero to a function in $C_c^1(\mathbb{R}^n)$. Since $\mathbb{R}^n$ is convex, we may conclude that the extension by zero of f is Lipschitz which implies that $f : U \to \mathbb{R}$ is Lipschitz. As a consequence, $C_c^1(U) \subseteq BC^{0,1}(U)$ and $C_c^{\infty}(U) \subseteq BC^{\infty,1}(U)$. Furthermore, Theorem 60 and the following theorem provide useful information regarding this issue.

Theorem 7. *Let $U \subseteq \mathbb{R}^n$ and $V \subseteq \mathbb{R}^k$ be two nonempty open sets and let $T : U \to V$ $(T = (T^1, \ldots, T^k))$ be a C^1 map (that is, for each $1 \le i \le k$, $T^i \in C^1(U)$). Suppose $B \subseteq U$ is a bounded set such that $B \subseteq \bar{B} \subseteq U$. Then $T : B \to V$ is Lipschitz.*

Proof. By Remark 3 it is enough to show that each T^i is Lipschitz on B. Fix a function $\varphi \in C_c^{\infty}(\mathbb{R}^n)$ such that $\varphi = 1$ on $\bar{B}$ and $\varphi = 0$ on $\mathbb{R}^n \setminus U$. Then φT^i can be viewed as an element of $C_c^1(\mathbb{R}^n)$. Therefore, it is Lipschitz ($\mathbb{R}^n$ is convex) and there exists a constant L, which may depend on φ, B and T^i, such that

$$|\varphi T^i(x) - \varphi T^i(y)| \le L|x - y| \quad \forall x, y \in \mathbb{R}^n.$$

Since $\varphi = 1$ on $\bar{B}$, it follows that

$$|T^i(x) - T^i(y)| \le L|x - y| \quad \forall x, y \in B.$$

□

4.2. Normed Spaces

Theorem 8. *Let X and Y be normed spaces. Let A be a dense subspace of X and B be a dense subspace of Y. Then*

- *$A \times B$ is dense in $X \times Y$;*
- *If $T : A \times B \to \mathbb{R}$ is a continuous bilinear map, then T has a unique extension to a continuous bilinear operator $\tilde{T} : X \times Y \to \mathbb{R}$.*

Theorem 9 ([1]). *Let X be a normed space and let M be a closed vector subspace of X.*

(1) *If X is reflexive, then X is a Banach space.*
(2) *X is reflexive if and only if X^* is reflexive.*
(3) *If X^* is separable, then X is separable.*
(4) *If X is reflexive and separable, then so is X^*.*
(5) *If X is a reflexive Banach space, then so is M.*
(6) *If X is a separable Banach space, then so is M.*

Moreover, if $X_1, \ldots, X_r$ are reflexive Banach spaces, then $X_1 \times \ldots \times X_r$ equipped with the norm

$$\|(x_1, \ldots, x_r)\| = \|x_1\|_{X_1} + \ldots + \|x_r\|_{X_r}$$

is also a reflexive Banach space.

4.3. Topological Vector Spaces

There are different, generally nonequivalent, ways to define topological vector spaces. The conventions in this section mainly follow Rudin's functional analysis [23]. Statements in this section are either taken from Rudin's functional analysis, Grubb's distributions and operators [20], excellent presentation of Reus [24], and Treves' topological vector spaces [25] or are direct consequences of statements in the aforementioned references. Therefore we will not give the proofs.

Definition 2. *A topological vector space is a vector space X together with a topology τ with the following properties:*

(i) *For all $x \in X$, the singleton $\{x\}$ is a closed set.*
(ii) *The maps*

$$(x, y) \mapsto x + y \qquad (\text{from } X \times X \text{ into } X),$$
$$(\lambda, x) \mapsto \lambda x \qquad (\text{from } \mathbb{R} \times X \text{ into } X),$$

are continuous where $X \times X$ and $\mathbb{R} \times X$ are equipped with the product topology.

Definition 3. *Suppose (X, τ) is a topological vector space and $Y \subseteq X$.*

- *Y is said to be **convex** if for all $y_1, y_2 \in Y$ and $t \in (0,1)$ it is true that $ty_1 + (1-t)y_2 \in Y$.*
- *Y is said to be **balanced** if for all $y \in Y$ and $|\lambda| \leq 1$ it holds that $\lambda y \in Y$. In particular, any balanced set contains the origin.*
- *We say Y is **bounded** if for any neighborhood U of the origin (i.e., any open set containing the origin), there exits $t > 0$ such that $Y \subseteq tU$.*

Theorem 10 (Important Properties of Topological Vector Spaces)**.**

- *Every topological vector space is Hausdorff.*
- *If (X, τ) is a topological vector space, then*
  (1) *For all $a \in X$: $E \in \tau \Longleftrightarrow a + E \in \tau$ (that is, τ is **translation invariant**);*
  (2) *For all $\lambda \in \mathbb{R} \setminus \{0\}$: $E \in \tau \Longleftrightarrow \lambda E \in \tau$ (that is, τ is **scale invariant**);*
  (3) *If $A \subseteq X$ is convex and $x \in X$, then so is $A + x$;*
  (4) *If $\{A_i\}_{i \in I}$ is a family of convex subsets of X, then $\cap_{i \in I} A_i$ is convex.*

Note: Some authors do not include condition (i) in the definition of topological vector spaces. In that case, a topological vector space will not necessarily be Hausdorff.

Definition 4. *Let (X, τ) be a topological space.*

- *A collection $\mathcal{B} \subseteq \tau$ is said to be a **basis** for τ, if every element of τ is a union of elements in $\mathcal{B}$.*
- *Let $p \in X$. If $\gamma \subseteq \tau$ is such that each element of γ contains p and every neighborhood of p (i.e., every open set containing p) contains at least one element of γ, then we say γ is a **local base at** p. If X is a vector space, then the local base γ is said to be convex if each element of γ is a convex set.*
- *(X, τ) is called **first-countable** if each point has a countable local base.*
- *(X, τ) is called **second-countable** if there is a countable basis for τ.*

Theorem 11. *Let (X, τ) be a topological space and suppose for all $x \in X$, γ_x is a local base at x. Then $\mathcal{B} = \cup_{x \in X} \gamma_x$ is a basis for τ.*

Theorem 12. *Let X be a vector space and suppose τ is a translation invariant topology on X. Then for all $x_1, x_2 \in X$, the collection γ_{x_1} is a local base at x_1 if and only if the collection $\{A + (x_2 - x_1)\}_{A \in \gamma_{x_1}}$ is a local base at x_2.*

Remark 4. *Let X be a vector space and suppose τ is a translation invariant topology on X. As a direct consequence of the previous theorems the topology τ is uniquely determined by giving a local base γ_{x_0} at some point $x_0 \in X$.*

Definition 5. *Let (X, τ) be a topological vector space. X is said to be* ***metrizable*** *if there exists a metric $d : X \times X \to [0, \infty)$ whose induced topology is τ. In this case we say that the metric d is compatible with the topology τ.*

Theorem 13. *Let (X, τ) be a topological vector space.*

- *X is metrizable $\Longleftrightarrow$ there exists a metric d on X such that for all $x \in X$, $\{B(x, \frac{1}{n})\}_{n \in \mathbb{N}}$ is a local base at x.*
- *A metric d on X is compatible with $\tau \Longleftrightarrow$ for all $x \in X$, $\{B(x, \frac{1}{n})\}_{n \in \mathbb{N}}$ is a local base at x. ($B(x, \frac{1}{n})$ is the open ball of radius $\frac{1}{n}$ centered at x).*

Definition 6. *Let X be a vector space and d be a metric on X. d is said to be translation invariant provided that*

$$\forall\, x, y, a \in X \qquad d(x + a, y + a) = d(x, y)\,.$$

Remark 5. *Let (X, τ) be a topological vector space and suppose d is a translation invariant metric on X. Then the following statements are equivalent:*

(1) *For all $x \in X$, $\{B(x, \frac{1}{n})\}_{n \in \mathbb{N}}$ is a local base at x.*
(2) *There exists $x_0 \in X$ such that $\{B(x_0, \frac{1}{n})\}_{n \in \mathbb{N}}$ is a local base at x_0.*

Therefore, d is compatible with τ if and only if $\{B(0, \frac{1}{n})\}_{n \in \mathbb{N}}$ is a local base at the origin.

Theorem 14. *Let (X, τ) be a topological vector space. Then (X, τ) is metrizable if and only if it has a countable local base at the origin. Moreover, if (X, τ) is metrizable, then one can find a translation invariant metric that is compatible with τ.*

Definition 7. *Let (X, τ) be a topological vector space and let $\{x_n\}$ be a sequence in X.*

- *We say that $\{x_n\}$ converges to a point $x \in X$ provided that*

$$\forall\, U \in \tau,\ x \in U \quad \exists N \quad \forall\, n \geq N \quad x_n \in U\,.$$

- *We say that $\{x_n\}$ is a Cauchy sequence provided that*

$$\forall\, U \in \tau,\ 0 \in U \quad \exists N \quad \forall\, m, n \geq N \quad x_n - x_m \in U\,.$$

Theorem 15. *Let (X, τ) be a topological vector space, $\{x_n\}$ be a sequence in X, and $x, y \in X$. Additionally, suppose γ is a local base at the origin. The following statements are equivalent:*

(1) $x_n \to x$;
(2) $(x_n - x) \to 0$;
(3) $x_n + y \to x + y$;
(4) $\forall\, V \in \gamma \quad \exists N \quad \forall\, n \geq N \quad x_n - x \in V$.

Moreover, $\{x_n\}$ is a Cauchy sequence if and only if

$$\forall\, V \in \gamma \quad \exists N \quad \forall\, n, m \geq N \quad x_n - x_m \in V\,.$$

Remark 6. *In contrast with properties like continuity of a function and convergence of a sequence which depend only on the topology of the space, the property of being a Cauchy sequence is not a topological property. Indeed, it is easy to construct examples of two metrics d_1 and d_2 on a vector space X that induce the same topology (i.e., the metrics are equivalent) but have different collection of Cauchy sequences. However, it can be shown that if d_1 and d_2 are two translation invariant* *metrics that induce the same topology on X, then the Cauchy sequences of (X, d_1) will be exactly the same as the Cauchy sequences of (X, d_2).*

Theorem 16. *Let (X, τ) be a metrizable topological vector space and d be a translation invariant metric on X that is compatible with τ. Let $\{x_n\}$ be a sequence in X. The following statements are equivalent:*

(1) $\{x_n\}$ is a Cauchy sequence in the topological vector space (X, τ).
(2) $\{x_n\}$ is a Cauchy sequence in the metric space (X, d).

Definition 8. *Let (X, τ) be a topological vector space. We say (X, τ) is* **locally convex** *if it has a convex local base at the origin.*

Note that, as a consequence of Theorems 10 and 12, the following statements are equivalent:

(1) (X, τ) is a locally convex topological vector space.
(2) There exists $p \in X$ with a convex local base at p.
(3) For every $p \in X$ there exists a convex local base at p.

Definition 9. *Let (X, τ) be a metrizable locally convex topological vector space. Let d be a translation invariant metric on X that is compatible with τ. We say that X is* **complete** *if and only if the metric space (X, d) is a complete metric space. A complete metrizable locally convex topological vector space is called a* **Frechet space***.*

Remark 7. *Our previous remark about Cauchy sequences shows that the above definition of completeness is independent of the chosen translation invariant metric d. Indeed one can show that the locally convex topological vector space (X, τ) is complete in the above sense if and only if every Cauchy net in (X, τ) is convergent.*

Theorem 17 ([26], p. 63). *A linear continuous bijective mapping of a Frechet space X onto a Frechet space Y has a continuous linear inverse.*

Definition 10. *A* ***seminorm*** *on a vector space X is a real-valued function $p : X \to \mathbb{R}$ such that*

(i) $\forall\, x, y \in X \quad p(x + y) \le p(x) + p(y)$
(ii) $\forall\, x \in X\ \forall\, \alpha \in \mathbb{R} \quad p(\alpha x) = |\alpha| p(x)$

If $\mathcal{P}$ is a family of seminorms on X, then we say $\mathcal{P}$ is **separating** *provided that for all $x \neq 0$ there exists at least one $p \in \mathcal{P}$ such that $p(x) \neq 0$ (that is, if $p(x) = 0$ for all $p \in \mathcal{P}$, then $x = 0$).*

Remark 8. *It follows from conditions (i) and (ii) that if $p : X \to \mathbb{R}$ is a seminorm, then $p(x) \ge 0$ for all $x \in X$.*

Theorem 18. *Suppose $\mathcal{P}$ is a separating family of seminorms on a vector space X. For all $p \in \mathcal{P}$ and $n \in \mathbb{N}$ let*

$$V(p, n) := \{x \in X : p(x) < \frac{1}{n}\}.$$

Furthermore, let γ be the collection of all finite intersections of $V(p, n)$'s. That is,

$$A \in \gamma \Longleftrightarrow \exists k \in \mathbb{N}, \exists p_1, \ldots, p_k \in \mathcal{P}, \exists n_1, \ldots, n_k \in \mathbb{N} \text{ such that } A = \cap_{i=1}^{k} V(p_i, n_i)$$

Then each element of γ is a convex balanced subset of X. Moreover, there exists a unique topology τ on X that satisfies both of the following properties:

(1) τ is translation invariant (that is, if $U \in \tau$ and $a \in X$, then $a + U \in \tau$).
(2) γ is a local base at the origin for τ.

This unique topology is called the **natural topology** *induced by the family of seminorms $\mathcal{P}$. Furthermore, if X is equipped with the natural topology τ, then*

(i) (X, τ) *is a locally convex topological vector space,*
(ii) *every* $p \in \mathcal{P}$ *is a continuous function from* X *to* $\mathbb{R}$.

Theorem 19. *Suppose* $\mathcal{P}$ *is a separating family of seminorms on a vector space* X. *Let* τ *be the natural topology induced by* $\mathcal{P}$. *Then*

(1) τ *is the smallest topology on* X *that is translation invariant and with respect to which every* $p \in \mathcal{P}$ *is continuous,*
(2) τ *is the smallest topology on* X *with respect to which addition is continuous and every* $p \in \mathcal{P}$ *is continuous.*

Theorem 20. *Let* X *and* Y *be two vector spaces and suppose* $\mathcal{P}$ *and* $\mathcal{Q}$ *are two separating families of seminorms on* X *and* Y, *respectively. Equip* X *and* Y *with the corresponding natural topologies.*

(1) *A sequence* x_n *converges to* x *in* X *if and only if for all* $p \in \mathcal{P}$, $p(x_n - x) \to 0$.
(2) *A linear operator* $T : X \to Y$ *is continuous if and only if*

$$\forall q \in \mathcal{Q} \quad \exists c > 0,\, k \in \mathbb{N},\, p_1, \ldots, p_k \in \mathcal{P} \quad \text{such that} \quad \forall x \in X \quad |q \circ T(x)| \le c \max_{1 \le i \le k} p_i(x)\,.$$

(3) *A linear operator* $T : X \to \mathbb{R}$ *is continuous if and only if*

$$\exists c > 0,\, k \in \mathbb{N},\, p_1, \ldots, p_k \in \mathcal{P} \quad \text{such that} \quad \forall x \in X \quad |T(x)| \le c \max_{1 \le i \le k} p_i(x)\,.$$

Theorem 21. *Let* X *be a Frechet space and let* Y *be a topological vector space. When* T *is a linear map of* X *into* Y, *the following two properties are equivalent:*

(1) T *is continuous.*
(2) $x_n \to 0$ *in* $X \Longrightarrow Tx_n \to 0$ *in* Y.

Theorem 22. *Let* $\mathcal{P} = \{p_k\}_{k \in \mathbb{N}}$ *be a* ***countable*** *separating family of seminorms on a vector space* X. *Let* τ *be the corresponding natural topology. Then the locally convex topological vector space* (X, τ) *is metrizable and the following translation invariant metric on* X *is compatible with* τ:

$$d(x, y) = \sum_{k=1}^{\infty} \frac{1}{2^k} \frac{p_k(x-y)}{1 + p_k(x-y)}\,.$$

Let (X, τ) be a locally convex topological vector space. Consider the topological dual of X,

$$X^* := \{f : X \to \mathbb{R} : f \text{ is linear and continuous}\}\,.$$

There are several ways to topologize X^*: the weak* topology, the topology of convex compact convergence, the topology of compact convergence, and the strong topology (see [25], Chapter 19). Here we describe the weak* topology and the strong topology on X^*.

Definition 11. *Let* (X, τ) *be a locally convex topological vector space.*

- *The* ***weak* topology*** *on* X^* *is the natural topology induced by the separating family of seminorms* $\{p_x\}_{x \in X}$ *where*

$$\forall x \in X \qquad p_x : X^* \to \mathbb{R}, \quad p_x(f) := |f(x)|\,.$$

A sequence $\{f_m\}$ *converges to* f *in* X^* *with respect to the weak* topology if and only if* $f_m(x) \to f(x)$ *in* $\mathbb{R}$ *for all* $x \in X$.

- *The* ***strong topology*** *on* X^* *is the natural topology induced by the separating family of seminorms* $\{p_B\}_{B \subseteq X bounded}$ *where for any bounded subset* B *of* X

$$p_B : X^* \to \mathbb{R} \qquad p_B(f) := \sup\{|f(x)| : x \in B\}\,.$$

(It can be shown that for any bounded subset B of X and $f \in X^*$, $f(B)$ is a bounded subset of $\mathbb{R}$.)

Remark 9.

(1) *If X is a normed space, then the topology induced by the norm*

$$\forall f \in X^* \qquad \|f\|_{op} = \sup_{\|x\|_X=1} |f(x)|$$

on X^* is the same as the strong topology on X^* ([25], p. 198).

(2) *In this manuscript, we always consider the topological dual of a locally convex topological vector space with the strong topology. Of course, it is worth mentioning that for many of the spaces that we will consider (including $X = \mathcal{E}(\Omega)$ or $X = D(\Omega)$ where Ω is an open subset of $\mathbb{R}^n$) a sequence in X^* converges with respect to the weak* topology if and only if it converges with respect to the strong topology (for more details on this see the definition and properties of* ***Montel spaces*** *in Section 34.4, page 356 of [25]).*

The following theorem, which is easy to prove, will later be used in the proof of completeness of Sobolev spaces of sections of vector bundles.

Theorem 23 ([24], p. 160)**.** *If X and Y are topological vector spaces and $I : X \to Y$ and $P : Y \to X$ are continuous linear maps such that $P \circ I = id_X$, then $I : X \to I(X) \subseteq Y$ is a linear topological isomorphism and $I(X)$ is closed in Y.*

Now we briefly review the relationship between the dual of a product of topological vector spaces and the product of the dual spaces. This will play an important role in our discussion of local representations of distributions in vector bundles in later sections.

Let $X_1, \ldots, X_r$ be topological vector spaces. Recall that the product topology on $X_1 \times \ldots \times X_r$ is the smallest topology such that the projection maps

$$\pi_k : X_1 \times \ldots \times X_r \to X_k, \qquad \pi_k(x_1, \ldots, x_r) = x_k\,,$$

are continuous for all $1 \leq k \leq r$. It can be shown that if each X_k is a locally convex topological vector space whose topology is induced by a family of seminorms $\mathcal{P}_k$, then $X_1 \times \ldots \times X_r$ equipped with the product topology is a locally convex topological vector space whose topology is induced by the following family of seminorms

$$\{p_1 \circ \pi_1 + \ldots + p_r \circ \pi_r : p_k \in \mathcal{P}_k \ \forall 1 \leq k \leq r\}\,.$$

Theorem 24 ([24], p. 164)**.** *Let $X_1, \ldots, X_r$ be locally convex topological vector spaces. Equip $X_1 \times \ldots \times X_r$ and $X_1^* \times \ldots \times X_r^*$ with the product topology. The mapping $\tilde{L} : X_1^* \times \ldots \times X_r^* \to (X_1 \times \ldots \times X_r)^*$ defined by*

$$\tilde{L}(u_1, \ldots, u_r) = u_1 \circ \pi_1 + \ldots + u_r \circ \pi_r$$

is a linear topological isomorphism. Its inverse is

$$L(v) = (v \circ i_1, \ldots, v \circ i_r)\,,$$

where for all $1 \leq k \leq r$, $i_k : X_k \to X_1 \times \ldots \times X_r$ is defined by

$$i_k(z) = (0, \ldots, 0, \underbrace{z}_{k^{th} \text{ position}}, 0, \ldots, 0)\,.$$

The notion of adjoint operator, which frequently appears in the future sections, is introduced in the following theorem.

Theorem 25 ([24], p. 163). *Let X and Y be locally convex topological vector spaces and suppose $T : X \to Y$ is a continuous linear map. Then*

(1) *The map*

$$T^* : Y^* \to X^* \qquad \langle T^* y, x\rangle_{X^* \times X} = \langle y, Tx\rangle_{Y^* \times Y},$$

*is well-defined, linear, and continuous. (T^* is called the **adjoint** of T).*

(2) *If $T(X)$ is dense in Y, then $T^* : Y^* \to X^*$ is injective.*

Remark 10. *In the subsequent sections we will focus heavily on certain function spaces on domains Ω in the Euclidean space. For approximation purposes, it is always desirable to have $D(\Omega)(= C_c^\infty(\Omega))$ as a dense subspace of our function spaces. However, there is another, may be more profound, reason for being interested in having $D(\Omega)$ as a dense subspace. It is important to note that we would like to use the term "function spaces" for topological vector spaces that can be continuously embedded in $D'(\Omega)$ (see Section 6 for the definition of $D'(\Omega)$) so that concepts such as differentiation will be meaningful for the elements of our function spaces. Given a function space $A(\Omega)$ it is usually helpful to consider its dual too. In order to be able to view the dual of $A(\Omega)$ as a function space we need to ensure that $[A(\Omega)]^*$ can be viewed as a subspace of $D'(\Omega)$. To this end, according to the above theorem, it is enough to ensure that the identity map from $D(\Omega)$ to $A(\Omega)$ is continuous with dense image in $A(\Omega)$.*

Let us consider more closely two special cases of Theorem 25.

(1) Suppose Y is a normed space and H is a dense subspace of Y. Clearly, the identity map $i : H \to Y$ is continuous with dense image. Therefore, $i^* : Y^* \to H^*$ ($F \mapsto F|_H$) is continuous and injective. Furthermore, by the Hahn–Banach theorem for all $\varphi \in H^*$ there exists $F \in Y^*$ such that $F|_H = \varphi$ and $\|F\|_{Y^*} = \|\varphi\|_{H^*}$. So the above map is indeed bijective and Y^* and H^* are isometrically isomorphic. As an important example, let Ω be a nonempty open set in $\mathbb{R}^n$, $s \geq 0$, and $1 < p < \infty$. Consider the space $W_0^{s,p}(\Omega)$ (see Section 7 for the definition of $W_0^{s,p}(\Omega)$). $C_c^\infty(\Omega)$ is a dense subspace of $W_0^{s,p}(\Omega)$. Therefore, $W^{-s,p'}(\Omega) := [W_0^{s,p}(\Omega)]^*$ is isometrically isomorphic to $[(C_c^\infty(\Omega), \|.\|_{s,p})]^*$. In particular, if $F \in W^{-s,p'}(\Omega)$, then

$$\|F\|_{W^{-s,p'}(\Omega)} = \sup_{0 \not\equiv \psi \in C_c^\infty(\Omega)} \frac{|F(\psi)|}{\|\psi\|_{s,p}}.$$

(2) Suppose $(Y, \|.\|_Y)$ is a normed space, (X, τ) is a locally convex topological vector space, $X \subseteq Y$, and the identity map $i : (X, \tau) \to (Y, \|.\|_Y)$ is continuous with dense image. So $i^* : Y^* \to X^*$ ($F \mapsto F|_X$) is continuous and injective and can be used to identify Y^* with a subspace of X^*.

- **Question:** Exactly what elements of X^* are in the image of i^*? That is, which elements of X^* "belong to" Y^*?
- **Answer:** $\varphi \in X^*$ belongs to the image of i^* if and only if $\varphi : (X, \|.\|_Y) \to \mathbb{R}$ is continuous, that is, $\varphi \in X^*$ belongs to the image of i^* if and only if $\sup_{x \in X \setminus \{0\}} \frac{|\varphi(x)|}{\|x\|_Y} < \infty$.

So, an element $\varphi \in X^*$ can be considered as an element of Y^* if and only if

$$\sup_{x \in X \setminus \{0\}} \frac{|\varphi(x)|}{\|x\|_Y} < \infty.$$

Furthermore, if we denote the unique corresponding element in Y^* by $\tilde{\varphi}$ (normally we identify φ and $\tilde{\varphi}$ and we use the same notation for both) then since X is dense in Y

$$\|\tilde{\varphi}\|_{Y^*} = \sup_{y \in Y \setminus \{0\}} \frac{|\tilde{\varphi}(y)|}{\|y\|_Y} = \sup_{x \in X \setminus \{0\}} \frac{|\varphi(x)|}{\|x\|_Y} < \infty.$$

Remark 11. *To sum up, given an element $\varphi \in X^*$ in order to show that φ can be considered as an element of Y^* we just need to show that $\sup_{x \in X \setminus \{0\}} \frac{|\varphi(x)|}{\|x\|_Y} < \infty$ and in that case, norm of φ as an element of Y^* is $\sup_{x \in X \setminus \{0\}} \frac{|\varphi(x)|}{\|x\|_Y}$. However, it is important to notice that if $F : Y \to \mathbb{R}$ is a linear map, X is a dense subspace of Y, and $F|_X : (X, \|.\|_Y) \to \mathbb{R}$ is bounded, that does NOT imply that $F \in Y^*$. It just shows that there exists $G \in Y^*$ such that $G|_X = F|_X$.*

We conclude this section by a quick review of the inductive limit topology.

Definition 12. *Let X be a vector space and let $\{X_\alpha\}_{\alpha \in I}$ be a family of vector subspaces of X with the property that*

- *For each $\alpha \in I$, X_α is equipped with a topology that makes it a locally convex topological vector space, and*
- $\bigcup_{\alpha \in I} X_\alpha = X$.

*The **inductive limit topology** on X with respect to the family $\{X_\alpha\}_{\alpha \in I}$ is defined to be the largest topology with respect to which*

(1) *X is a locally convex topological vector space;*
(2) *All the inclusions $X_\alpha \subseteq X$ are continuous.*

Theorem 26 ([24], p. 161)**.** *Let X be a vector space equipped with the inductive limit topology with respect to $\{X_\alpha\}$ as described above. If Y is a locally convex vector space, then a linear map $T : X \to Y$ is continuous if and only if $T|_{X_\alpha} : X_\alpha \to Y$ is continuous for all $\alpha \in I$.*

Theorem 27 ([24], p. 162)**.** *Let X be a vector space equipped with the inductive limit topology with respect to $\{X_\alpha\}$ as described above. A convex subset W of X is a neighborhood of the origin (i.e., an open set containing the origin) in X if and only if for all α, the set $W \cap X_\alpha$ is a neighborhood of the origin in X_α.*

Theorem 28 ([24], p. 165)**.** *Let X be a vector space and let $\{X_j\}_{j \in \mathbb{N}_0}$ be a nested family of vector subspaces of X:*

$$X_0 \subsetneq X_1 \subsetneq \ldots \subsetneq X_j \subsetneq \ldots .$$

Suppose each X_j is equipped with a topology that makes it a locally convex topological vector space. Equip X with the inductive limit topology with respect to $\{X_j\}$. Then the following topologies on $X^{\times r}$ are equivalent (=they are the same):

(1) *The product topology;*
(2) *The inductive limit topology with respect to the family $\{X_j^{\times r}\}$ (For each j, $X_j^{\times r}$ is equipped with the product topology).*

As a consequence, if Y is a locally convex vector space, then a linear map $T : X^{\times r} \to Y$ is continuous if and only if $T|_{X_j^{\times r}} : X_j^{\times r} \to Y$ is continuous for all $j \in \mathbb{N}_0$.

5. Review of Some Results from Differential Geometry

The main purpose of this section is to set the notation and terminology straight. To this end we cite the definitions of several basic terms and a number of basic properties that we will frequently use. The main reference for the majority of the definitions is one of the invaluable books by John M. Lee [19].

5.1. Smooth Manifolds

Suppose M is a topological space. We say that M is a topological manifold of dimension n if it is Hausdorff, second-countable, and locally Euclidean in the sense that each point of M has a neighborhood that is homeomorphic to an open subset of $\mathbb{R}^n$. It is easy to see that the following statements are equivalent ([19], p. 3):

(1) Each point of M has a neighborhood that is homeomorphic to an open subset of $\mathbb{R}^n$.
(2) Each point of M has a neighborhood that is homeomorphic to an open ball in $\mathbb{R}^n$.
(3) Each point of M has a neighborhood that is homeomorphic to $\mathbb{R}^n$.

By a **coordinate chart** (or just **chart**) on M we mean a pair (U, φ), where U is an open subset of M and $\varphi : U \to \hat{U}$ is a homeomorphism from U to an open subset $\hat{U} = \varphi(U) \subseteq \mathbb{R}^n$. U is called a **coordinate domain** or a **coordinate neighborhood** of each of its points and φ is called a **coordinate map**. An **atlas for** M is a collection of charts whose domains cover M. Two charts (U, φ) and (V, ψ) are said to be **smoothly compatible** if either $U \cap V = \emptyset$ or the transition map $\psi \circ \varphi^{-1}$ is a C^∞-diffeomorphism. An atlas $\mathcal{A}$ is called a **smooth atlas** if any two charts in $\mathcal{A}$ are smoothly compatible with each other. A smooth atlas $\mathcal{A}$ on M is **maximal** if it is not properly contained in any larger smooth atlas. A **smooth structure** on M is a maximal smooth atlas. A **smooth manifold** is a pair $(M, \mathcal{A})$, where M is a topological manifold and $\mathcal{A}$ is a smooth structure on M. Any chart (U, φ) contained in the given maximal smooth atlas is called a **smooth chart**. If M and N are two smooth manifolds, a map $F : M \to N$ is said to be a **smooth** (C^∞) **map** if for every $p \in M$, there exist smooth charts (U, φ) containing p and (V, ψ) containing $F(p)$ such that $F(U) \subseteq V$ and $\psi \circ F \circ \varphi^{-1} \in C^\infty(\varphi(U))$. It can be shown that if F is smooth, then its restriction to every open subset of M is smooth. Furthermore, if every $p \in M$ has a neighborhood U such that $F|_U$ is smooth, then F is smooth.

Remark 12.

- *Sometimes we use the shorthand notation M^n to indicate that M is n-dimensional.*
- *Clearly, if (U, φ) is a chart in a maximal smooth atlas and V is an open subset of U, then (V, ψ) where $\psi = \varphi|_V$ is also a smooth chart (i.e., it belongs to the same maximal atlas).*
- *Every smooth atlas $\mathcal{A}$ for M is contained in a unique maximal smooth atlas, called the* ***smooth structure determined by*** $\mathcal{A}$.
- *If M is a compact smooth manifold, then there exists a smooth atlas with finitely many elements that determines the smooth structure of M (this is immediate from the definition of compactness).*

Definition 13.

- *We say that a smooth atlas for a smooth manifold M is a* ***geometrically Lipschitz (GL)*** *smooth atlas if the image of each coordinate domain in the atlas under the corresponding coordinate map is a nonempty bounded open set with Lipschitz boundary.*
- *We say that a smooth atlas for a smooth manifold M^n is a* ***generalized geometrically Lipschitz (GGL)*** *smooth atlas if the image of each coordinate domain in the atlas under the corresponding coordinate map is the entire $\mathbb{R}^n$ or a nonempty bounded open set with Lipschitz boundary.*
- *We say that a smooth atlas for a smooth manifold M^n is a* ***nice*** *smooth atlas if the image of each coordinate domain in the atlas under the corresponding coordinate map is a ball in $\mathbb{R}^n$.*
- *We say that a smooth atlas for a smooth manifold M^n is a* ***super nice*** *smooth atlas if the image of each coordinate domain in the atlas under the corresponding coordinate map is the entire $\mathbb{R}^n$.*
- *We say that two smooth atlases $\{(U_\alpha, \varphi_\alpha)\}_{\alpha \in I}$ and $\{(\tilde{U}_\beta, \tilde{\varphi}_\beta)\}_{\beta \in J}$ for a smooth manifold M^n are* ***geometrically Lipschitz compatible (GLC)*** *smooth atlases provided that each atlas is GGL and moreover for all $\alpha \in I$ and $\beta \in J$ with $U_\alpha \cap \tilde{U}_\beta \neq \emptyset$, $\varphi_\alpha(U_\alpha \cap \tilde{U}_\beta)$ and $\tilde{\varphi}_\beta(U_\alpha \cap \tilde{U}_\beta)$ are nonempty bounded open sets with Lipschitz boundary or the entire $\mathbb{R}^n$.*

Clearly, every super nice smooth atlas is also a GGL smooth atlas; every nice smooth atlas is also a GL smooth atlas, and every GL smooth atlas is also a GGL smooth atlas. Furthermore, note that two arbitrary GL smooth atlases are not necessarily GLC smooth atlases because the intersection of two Lipschitz domains is not necessarily Lipschitz (see, e.g., [27], pp. 115–117).

Given a smooth atlas $\{(U_\alpha, \varphi_\alpha)\}$ for a compact smooth manifold M, it is not necessarily possible to construct a new atlas $\{(U_\alpha, \tilde{\varphi}_\alpha)\}$ such that this new atlas is nice; for instance if

U_α is not connected we cannot find $\tilde{\varphi}_\alpha$ such that $\tilde{\varphi}_\alpha(U_\alpha) = \mathbb{R}^n$ (or any ball in $\mathbb{R}^n$). However, as the following lemma states, it is always possible to find a refinement that is nice.

Lemma 1. *Suppose $\{(U_\alpha, \varphi_\alpha)\}_{1\leq\alpha\leq N}$ is a smooth atlas for a compact smooth manifold M. Then there exists a finite open cover $\{V_\beta\}_{1\leq\beta\leq L}$ of M such that*

$$\forall\,\beta \qquad \exists 1 \leq \alpha(\beta) \leq N \text{ s.t.} \quad V_\beta \subseteq U_{\alpha(\beta)}, \quad \varphi_{\alpha(\beta)}(V_\beta) \text{ is a ball in } \mathbb{R}^n\,.$$

Therefore, $\{(V_\beta, \varphi_{\alpha(\beta)}|_{V_\beta})\}_{1\leq\beta\leq L}$ is a nice smooth atlas.

Proof. For each $1 \leq \alpha \leq N$ and $p \in U_\alpha$, there exists $r_{\alpha p} > 0$ such that $B_{r_{\alpha p}}(\varphi_\alpha(p)) \subseteq \varphi_\alpha(U_\alpha)$. Let $V_{\alpha p} := \varphi_\alpha^{-1}(B_{r_{\alpha p}}(\varphi_\alpha(p)))$. $\bigcup_{1\leq\alpha\leq N}\bigcup_{p\in U_\alpha} V_{\alpha p}$ is an open cover of M and so it has a finite subcover $\{V_{\alpha_1 p_1}, \ldots, V_{\alpha_L p_L}\}$. Let $V_\beta = V_{\alpha_\beta p_\beta}$. Clearly, $V_\beta \subseteq U_{\alpha_\beta}$ and $\varphi_{\alpha_\beta}(V_\beta)$ is a ball in $\mathbb{R}^n$. □

Remark 13. *Every open ball in $\mathbb{R}^n$ is C^∞-diffeomorphic to $\mathbb{R}^n$. Furthermore, compositions of diffeomorphisms is a diffeomorphism. Therefore, existence of a finite nice smooth atlas on a compact smooth manifold, which is guaranteed by the above lemma, implies the existence of a finite super nice smooth atlas.*

Lemma 2. *Let M be a compact smooth manifold. Let $\{U_\alpha\}_{1\leq\alpha\leq N}$ be an open cover of M. Suppose C is a closed set in M (so C is compact) which is contained in U_β for some $1 \leq \beta \leq N$. Then there exists an open cover $\{A_\alpha\}_{1\leq\alpha\leq N}$ of M such that $C \subseteq A_\beta \subseteq \bar{A}_\beta \subseteq U_\beta$ and $A_\alpha \subseteq \bar{A}_\alpha \subseteq U_\alpha$ for all $\alpha \neq \beta$.*

Proof. Without loss of generality we may assume that $\beta = 1$. For each $1 \leq \alpha \leq N$ and $p \in U_\alpha$, there exists $r_{\alpha p} > 0$ such that $B_{2r_{\alpha p}}(\varphi_\alpha(p)) \subseteq \varphi_\alpha(U_\alpha)$. Let $V_{\alpha p} := \varphi_\alpha^{-1}(B_{r_{\alpha p}}(\varphi_\alpha(p)))$. Clearly, $p \in V_{\alpha p} \subseteq \bar{V}_{\alpha p} \subseteq U_\alpha$. Since M is compact, the open cover $\bigcup_{1\leq\alpha\leq N}\bigcup_{p\in U_\alpha} V_{\alpha p}$ of M has a finite subcover $\mathcal{A}$. For each $1 \leq \alpha \leq N$ let $E_\alpha = \{p \in U_\alpha : V_{\alpha p} \in \mathcal{A}\}$ and

$$I_1 = \{\alpha : E_\alpha \neq \emptyset\}\,.$$

If $\alpha \in I_1$, we let $W_\alpha = \bigcup_{p\in E_\alpha} V_{\alpha p}$. For $\alpha \notin I_1$ choose one point $p \in U_\alpha$ and let $W_\alpha = V_{\alpha p}$. C is compact so $\varphi_1(C)$ is a compact set inside the open set $\varphi_1(U_1)$. Therefore, there exists an open set B such that

$$\varphi_1(C) \subseteq B \subseteq \bar{B} \subseteq \varphi_1(U_1)\,.$$

Let $W = \varphi_1^{-1}(B)$. Clearly, $C \subseteq W \subseteq \bar{W} \subseteq U_\alpha$. Now Let

$$A_1 = W \bigcup W_1\,,$$
$$A_\alpha = W_\alpha \quad \forall \alpha > 1\,.$$

Clearly, A_1 contains W which contains C. Furthermore, union of A_α's contains $\bigcup_{\alpha=1}^N \bigcup_{p\in E_\alpha} V_{\alpha p}$ which is equal to M. Closure of a union of sets is a subset of the union of closures of those sets. Therefore, for each α, $\bar{A}_\alpha \subseteq U_\alpha$. □

Theorem 29 (Exhaustion by Compact Sets for Manifolds). *Let M be a smooth manifold. There exists a sequence of compact subsets $(K_j)_{j\in\mathbb{N}}$ such that $\bigcup_{j\in\mathbb{N}} \mathring{K}_j = M$, $\mathring{K}_{j+1} \setminus K_j \neq \emptyset$ for all j and*

$$K_1 \subseteq \mathring{K}_2 \subseteq K_2 \subseteq \ldots \subseteq \mathring{K}_j \subseteq K_j \subseteq \ldots\,.$$

Definition 14. *A C^∞ partition of unity on a smooth manifold is a collection of nonnegative C^∞ functions $\{\psi_\alpha : M \to \mathbb{R}\}_{\alpha\in A}$ such that*

(i) The collection of supports, $\{\text{supp}\,\psi_\alpha\}_{\alpha\in A}$ is locally finite in the sense that every point in M has a neighborhood that intersects only finitely many of the sets in $\{\text{supp}\,\psi_\alpha\}_{\alpha\in A}$.

(ii) $\sum_{\alpha \in A} \psi_\alpha = 1$.

Given an open cover $\{U_\alpha\}_{\alpha \in A}$ *of* M, *we say that a partition of unity* $\{\psi_\alpha\}_{\alpha \in A}$ *is subordinate to the open cover* $\{U_\alpha\}_{\alpha \in A}$ *if supp* $\psi_\alpha \subseteq U_\alpha$ *for every* $\alpha \in A$.

Theorem 30 ([28], p. 146)**.** *Let* M *be a* ***compact*** *smooth manifold and* $\{U_\alpha\}_{\alpha \in A}$ *an open cover of* M. *There exists a* C^∞ *partition of unity* $\{\psi_\alpha\}_{\alpha \in A}$ *subordinate to* $\{U_\alpha\}_{\alpha \in A}$ *(notice that the index sets are the same).*

Theorem 31 ([28], p. 347)**.** *Let* $\{U_\alpha\}_{\alpha \in A}$ *be an open cover of a smooth manifold* M.

(i) *There is a* C^∞ *partition of unity* $\{\varphi_k\}_{k=1}^\infty$ *with every* φ_k ***having compact support*** *such that for each* k, *supp* $\varphi_k \subseteq U_\alpha$ *for some* $\alpha \in A$.

(ii) *If we do not require compact support, then there is a* C^∞ *partition of unity* $\{\psi_\alpha\}_{\alpha \in A}$ *subordinate to* $\{U_\alpha\}_{\alpha \in A}$.

Remark 14. *Let* M *be a compact smooth manifold. Suppose* $\{U_\alpha\}_{\alpha \in A}$ *is an open cover of* M *and* $\{\psi_\alpha\}_{\alpha \in A}$ *is a partition of unity subordiante to* $\{U_\alpha\}_{\alpha \in A}$.

- *For all* $m \in \mathbb{N}$, $\{\tilde{\psi}_\alpha = \frac{\psi_\alpha^m}{\sum_{\alpha \in A} \psi_\alpha^m}\}$ *is another partition of unity subordinate to* $\{U_\alpha\}_{\alpha \in A}$.
- *If* $\{V_\beta\}_{\beta \in B}$ *is an open cover of* M *and* $\{\xi_\beta\}$ *is a partition of unity subordinate to* $\{V_\beta\}_{\beta \in B}$, *then* $\{\psi_\alpha \xi_\beta\}_{(\alpha,\beta) \in A \times B}$ *is a partition of unity subordinate to the open cover* $\{U_\alpha \cap V_\beta\}_{(\alpha,\beta) \in A \times B}$.

Lemma 3. *Let* M *be a compact smooth manifold. Suppose* $\{U_\alpha\}_{1 \le \alpha \le N}$ *is an open cover of* M. *Suppose* C *is a closed set in* M *(so* C *is compact) which is contained in* U_β *for some* $1 \le \beta \le N$. *Then there exists a partition of unity* $\{\psi_\alpha\}_{1 \le \alpha \le N}$ *subordinate to* $\{U_\alpha\}_{1 \le \alpha \le N}$ *such that* $\psi_\beta = 1$ *on* C.

Proof. We follow the argument in [29]. Without loss of generality we may assume $\beta = 1$. We can construct a partition of unity with the desired property as follows: Let A_α be a collection of open sets that covers M and such that $C \subseteq A_1 \subseteq \bar{A}_1 \subseteq U_1$ and for $\alpha > 1$, $A_\alpha \subseteq \bar{A}_\alpha \subseteq U_\alpha$ (see Lemma 2). Let $\eta_\alpha \in C_c^\infty(U_\alpha)$ be such that $0 \le \eta_\alpha \le 1$ and $\eta_\alpha = 1$ on a neighborhood of $\bar{A}_\alpha$. Of course $\sum_{\alpha=1}^N \eta_\alpha$ is not necessarily equal to 1 for all $x \in M$. However, if we define $\psi_1 = \eta_1$ and for $\alpha > 1$

$$\psi_\alpha = \eta_\alpha(1 - \eta_1)\dots(1 - \eta_{\alpha-1}),$$

by induction one can easily show that for $1 \le l \le N$

$$1 - \sum_{\alpha=1}^{l} \psi_\alpha = (1 - \eta_1)\dots(1 - \eta_l).$$

In particular,

$$1 - \sum_{\alpha=1}^{N} \psi_\alpha = (1 - \eta_1)\dots(1 - \eta_N) = 0,$$

since for each $x \in M$ there exists α such that $x \in A_\alpha$ and so $\eta_\alpha(x) = 1$. Consequently, $\sum_{\alpha=1}^N \psi_\alpha = 1$. □

5.2. Vector Bundles, Basic Definitions

Let M be a smooth manifold. A (smooth real) **vector bundle** of rank r over M is a smooth manifold E together with a surjective smooth map $\pi : E \to M$ such that

(1) For each $x \in M$, $E_x = \pi^{-1}(x)$ is an r-dimensional (real) vector space;

(2) For each $x \in M$, there exists a neighborhood U of x in M and a smooth map $\rho = (\rho^1, \dots, \rho^r)$ from $E|_U := \pi^{-1}(U)$ onto $\mathbb{R}^r$ such that

- For every $x \in U$, $\rho|_{E_x} : E_x \to \mathbb{R}^r$ is an isomorphism of vector spaces,
- $\Phi = (\pi|_{E_U}, \rho) : E_U \to U \times \mathbb{R}^r$ is a diffeomorphism.

We denote the projection onto the last r components by π'. So $\pi' \circ \Phi = \rho$. The expressions "E is a vector bundle over M", or "$E \to M$ is a vector bundle", or "$\pi : E \to M$ is a vector bundle" are all considered to be equivalent in this manuscript.

If $\pi : E \to M$ is a vector bundle of rank r, U is an open set in M, $\rho : E_U = \pi^{-1}(U) \to \mathbb{R}^r$ and $\Phi = (\pi|_{E_U}, \rho) : E_U \to U \times \mathbb{R}^r$ satisfy the properties stated in item (2), then we refer to both $\Phi : E_U \to U \times \mathbb{R}^r$ and $\rho : E_U \to \mathbb{R}^r$ as a (smooth) **local trivialization** of E over U (it will be clear from the context which one we are referring to). We say that $E|_U$ is trivial. The pair (U, ρ) (or (U, Φ)) is sometimes called a **vector bundle chart**. It is easy to see that if (U, ρ) is a vector bundle chart and $\emptyset \neq V \subseteq U$ is open, then $(V, \rho|_{E_V})$ is also a vector bundle chart for E. Moreover, if V is any nonempty open subset of M, then E_V is a vector bundle over the manifold V. We say that a triple (U, φ, ρ) is a **total trivialization triple** of the vector bundle $\pi : E \to M$ provided that (U, φ) is a smooth coordinate chart and $\rho = (\rho^1, \cdots, \rho^r) : E_U \to \mathbb{R}^r$ is a trivialization of E over U. A collection $\{(U_\alpha, \varphi_\alpha, \rho_\alpha)\}$ is called a **total trivialization atlas** for the vector bundle $E \to M$ provided that for each α, $(U_\alpha, \varphi_\alpha, \rho_\alpha)$ is a total trivialization triple and $\{(U_\alpha, \varphi_\alpha)\}$ is a smooth atlas for M.

Lemma 4 ([19], p. 252). *Let $\pi : E \to M$ be a smooth vector bundle of rank r over M. Suppose $\Phi : \pi^{-1}(U) \to U \times \mathbb{R}^r$ and $\Psi : \pi^{-1}(V) \to V \times \mathbb{R}^r$ are two smooth local trivializations of E with $U \cap V \neq \emptyset$. There exists a smooth map $\tau : U \cap V \to GL(r, \mathbb{R})$ such that the composition*

$$\Phi \circ \Psi^{-1} : (U \cap V) \times \mathbb{R}^r \to (U \cap V) \times \mathbb{R}^r$$

has the form

$$\Phi \circ \Psi^{-1}(p, v) = (p, \tau(p)v).$$

Remark 15. *Let E be a vector bundle over an n-dimensional smooth manifold M. Suppose $\{(U_\alpha, \varphi_\alpha, \rho_\alpha)\}_{\alpha \in I}$ is a total trivialization atlas for the vector bundle $\pi : E \to M$. Then for each $\alpha \in I$, the mapping*

$$E_{U_\alpha} = \pi^{-1}(U_\alpha) \to \varphi_\alpha(U_\alpha) \times \mathbb{R}^r \subseteq \mathbb{R}^{n+r}, \qquad s \mapsto (\varphi_\alpha(\pi(s)), \rho_\alpha(s))$$

will be a coordinate map for the manifold E over the coordinate domain E_{U_α}. The collection $\{(E_{U_\alpha}, (\varphi_\alpha \circ \pi, \rho_\alpha))\}_{\alpha \in I}$ will be a smooth atlas for the manifold E.

The following statements show that any vector bundle has a total trivialization atlas.

Lemma 5 ([30], p. 77). *Let E be a vector bundle over an n-dimensional smooth manifold M (M does not need to be compact). Then M can be covered by $n+1$ open sets $V_0, \ldots, V_n$ where the restriction $E|_{V_i}$ is trivial.*

Theorem 32. *Let E be a vector bundle of rank r over an n-dimensional smooth manifold M. Then $E \to M$ has a total trivialization atlas. In particular, if M is compact, then it has a total trivialization atlas that consists of only finitely many total trivialization triples.*

Proof. Let $V_0, \ldots, V_n$ be an open cover of M such that E is trivial over V_β with the mapping $\rho_\beta : E_{V_\beta} \to \mathbb{R}^r$. Let $\{(U_\alpha, \varphi_\alpha)\}_{\alpha \in I}$ be a smooth atlas for M (if M is compact, the index set I can be chosen to be finite). For all $\alpha \in I$ and $0 \leq \beta \leq n$ let $W_{\alpha\beta} = U_\alpha \cap V_\beta$. Let $J = \{(\alpha, \beta) : W_{\alpha\beta} \neq \emptyset\}$. Clearly, $\{(W_{\alpha\beta}, \varphi_{\alpha\beta}, \rho_{\alpha\beta})\}_{(\alpha,\beta) \in J}$ where $\varphi_{\alpha\beta} = \varphi_\alpha|_{W_{\alpha\beta}}$ and $\rho_{\alpha\beta} = \rho_\beta|_{\pi^{-1}(W_{\alpha\beta})}$ is a total trivialization atlas for $E \to M$. $\square$

Definition 15.

- *We say that a total trivialization triple (U,φ,ρ) is* ***geometrically Lipschitz (GL)*** *provided that $\varphi(U)$ is a nonempty bounded open set with Lipschitz boundary. A total trivialization atlas is called* ***geometrically Lipschitz*** *if each of its total trivialization triples is GL.*
- *We say that a total trivialization triple (U,φ,ρ) is* ***nice*** *provided that $\varphi(U)$ is equal to a ball in $\mathbb{R}^n$. A total trivialization atlas is called nice if each of its total trivialization triples is nice.*
- *We say that a total trivialization triple (U,φ,ρ) is* ***super nice*** *provided that $\varphi(U)$ is equal to $\mathbb{R}^n$. A total trivialization atlas is called super nice if each of its total trivialization triples is super nice.*
- *A total trivialization atlas is called* ***generalized geometrically Lipschitz (GGL)*** *if each of its total trivialization triples is GL or super nice.*
- *We say that two total trivialization atlases $\{(U_\alpha,\varphi_\alpha,\rho_\alpha)\}_{\alpha\in I}$ and $\{(\tilde{U}_\beta,\tilde{\varphi}_\beta,\tilde{\rho}_\beta)\}_{\beta\in J}$ are* ***geometrically Lipschitz compatible (GLC)*** *if the corresponding atlases $\{(U_\alpha,\varphi_\alpha)\}_{\alpha\in I}$ and $\{(\tilde{U}_\beta,\tilde{\varphi}_\beta)\}_{\beta\in J}$ are GLC.*

Theorem 33. *Let E be a vector bundle of rank r over an n-dimensional compact smooth manifold M. Then E has a nice total trivialization atlas (and a super nice total trivialization atlas) that consists of only finitely many total trivialization triples.*

Proof. By Theorem 32, $E\to M$ has a finite total trivialization atlas $\{(U_\alpha,\varphi_\alpha,\rho_\alpha)\}$. By Lemma 1 (and Remark 13) there exists a finite open cover $\{V_\beta\}_{1\le\beta\le L}$ of M such that

$$\begin{aligned}&\forall\beta \quad \exists 1\le\alpha(\beta)\le N \text{ s.t. } \quad V_\beta\subseteq U_{\alpha(\beta)},\quad \varphi_{\alpha(\beta)}(V_\beta)\text{ is a ball in }\mathbb{R}^n\\ &(\text{or }\forall\beta \quad \exists 1\le\alpha(\beta)\le N \text{ s.t. } \quad V_\beta\subseteq U_{\alpha(\beta)},\quad \varphi_{\alpha(\beta)}(V_\beta)=\mathbb{R}^n),\end{aligned}$$

and thus $\{(V_\beta,\varphi_{\alpha(\beta)}|_{V_\beta})\}_{1\le\beta\le L}$ is a nice (resp. super nice) smooth atlas. Now, clearly, $\{(V_\beta,\varphi_{\alpha(\beta)}|_{V_\beta},\rho_{\alpha(\beta)}|_{E_{V_\beta}})\}_{1\le\beta\le L}$ is a nice (resp. super nice) total trivialization atlas. □

Theorem 34. *Let E be a vector bundle of rank r over an n-dimensional compact smooth manifold M. Then E admits a finite total trivialization atlas that is GL compatible with itself. In fact, there exists a total trivialization atlas $\{(U_\alpha,\varphi_\alpha,\rho_\alpha)\}_{1\le\alpha\le N}$ such that*

- *For all $1\le\alpha\le N$, $\varphi_\alpha(U_\alpha)$ is bounded with Lipschitz continuous boundary;*
- *For all $1\le\alpha,\beta\le N$, $U_\alpha\cap U_\beta$ is either empty or else $\varphi_\alpha(U_\alpha\cap U_\beta)$ and $\varphi_\beta(U_\alpha\cap U_\beta)$ are bounded with Lipschitz continuous boundary.*

Proof. The proof of this theorem is based on the argument presented in the proof of Lemma 3.1 in [31]. Equip M with a smooth Riemannian metric g. Let r_{inj} denote the injectivity radius of M which is strictly positive because M is compact. Let $V_0,\dots,V_n$ be an open cover of M such that E is trivial over V_β with the mapping $\rho_\beta: E_{V_\beta}\to\mathbb{R}^r$. For every $x\in M$ choose $0\le i(x)\le n$ such that $x\in V_{i(x)}$. For all $x\in M$ let r_x be a positive number less than $\frac{r_{inj}}{2}$ such that $\exp_x(B_{r_x})\subseteq V_{i(x)}$ where B_{r_x} denotes the open ball in T_xM of radius r_x (with respect to the inner product induced by the Riemannian metric g) and $\exp_x: T_xM\to M$ denotes the exponential map at x. For every $x\in M$ define the normal coordinate chart centered at x, (U_x,φ_x), as follows:

$$U_x=\exp_x(B_{r_x}),\quad \varphi_x:=\lambda_x^{-1}\circ\exp_x^{-1}: U_x\to\mathbb{R}^n,$$

where $\lambda_x:\mathbb{R}^n\to T_xM$ is an isomorphism defined by $\lambda_x(y^1,\dots,y^n)=y^iE_{ix}$; Here $\{E_{ix}\}_{i=1}^n$ is a an arbitrary but fixed orthonormal basis for T_xM. It is well-known that (see, e.g., [32])

- $\varphi_x(x)=(0,\dots,0)$;
- $g_{ij}(x)=\delta_{ij}$ where g_{ij} denotes the components of the metric with respect to the normal coordinate chart (U_x,φ_x);
- $E_{ix}=\partial_i|_x$ where $\{\partial_i\}_{1\le i\le n}$ is the coordinate basis induced by (U_x,φ_x).

As a consequence of the previous items, it is easy to show that if $X \in T_xM$ ($X = X^i\partial_i|_x$), then the Euclidean norm of X will be equal to the norm of X with respect to the metric g, that is, $|X|_{\bar{g}} = |X|_g$ where

$$|X|_{\bar{g}} = \sqrt{(X^1)^2 + \ldots + (X^n)^2} \quad |X|_g = \sqrt{g(X,X)}\,.$$

Consequently, for every $x \in M$, $\varphi_x(U_x)$ will be a ball in the Euclidean space, in particular, $\{(U_x, \varphi_x)\}_{x \in M}$ is a GL atlas. The proof of Lemma 3.1 in [31] in part shows that the atlas $\{(U_x, \varphi_x)\}_{x \in M}$ is GL compatible with itself. Since M is compact there exists $x_1, \ldots, x_N \in M$ such that $\{U_{x_j}\}_{1 \le j \le N}$ also covers M.
Now, clearly, $\{(U_{x_j}, \varphi_{x_j}, \rho_{i(x_j)}|_{U_{x_j}})\}_{1 \le j \le N}$ is a total trivialization atlas for E that is GL compatible with itself. □

Corollary 1. *Let E be a vector bundle of rank r over an n-dimensional compact smooth manifold M. Then E admits a finite super nice total trivialization atlas that is GL compatible with itself.*

Proof. Let $\{(U_\alpha, \varphi_\alpha, \rho_\alpha)\}_{1 \le \alpha \le N}$ be the total trivialization atlas that was constructed above. For each α, $\varphi_\alpha(U_\alpha)$ is a ball in the Euclidean space and so it is diffeomorphic to $\mathbb{R}^n$; let $\xi_\alpha : \varphi_\alpha(U_\alpha) \to \mathbb{R}^n$ be such a diffeomorphism. We let $\tilde{\varphi}_\alpha := \xi_\alpha \circ \varphi_\alpha : U_\alpha \to \mathbb{R}^n$. A composition of diffeomorphisms is a diffeomorphism, so for all $1 \le \alpha, \beta \le N$, $\tilde{\varphi}_\alpha \circ \tilde{\varphi}_\beta^{-1} : \tilde{\varphi}_\beta(U_\alpha \cap U_\beta) \to \tilde{\varphi}_\alpha(U_\alpha \cap U_\beta)$ is a diffeomorphism. So $\{(U_\alpha, \tilde{\varphi}_\alpha, \rho_\alpha)\}_{1 \le \alpha \le N}$ is clearly a smooth super nice total trivialization atlas. Moreover, if $1 \le \alpha, \beta \le N$ are such that $U_\alpha \cap U_\beta$ is nonempty, then $\tilde{\varphi}_\alpha(U_\alpha \cap U_\beta)$ is $\mathbb{R}^n$ or a bounded open set with Lipschitz continuous boundary. The reason is that $\tilde{\varphi}_\alpha = \xi_\alpha \circ \varphi_\alpha$, and $\varphi_\alpha(U_\alpha \cap U_\beta)$ is $\mathbb{R}^n$ or Lipschitz, ξ_α is a diffeomorphism and being equal to $\mathbb{R}^n$ or Lipschitz is a property that is preserved under diffeomorphisms. Therefore, $\{(U_\alpha, \tilde{\varphi}_\alpha, \rho_\alpha)\}_{1 \le \alpha \le N}$ is a finite super nice total trivialization atlas that is GL compatible with itself. □

A **section** of E is a map $u : M \to E$ such that $\pi \circ u = Id_M$. The collection of all sections of E is denoted by $\Gamma(M, E)$. A section $u \in \Gamma(M, E)$ is said to be smooth if it is smooth as a map from the smooth manifold M to the smooth manifold E. The collection of all smooth sections of $E \to M$ is denoted by $C^\infty(M, E)$. Note that if $\{(U_\alpha, \varphi_\alpha, \rho_\alpha)\}_{\alpha \in I}$ is a total trivialization atlas for the vector bundle $\pi : E \to M$ of rank r, then for $u \in \Gamma(M, E)$ we have $u \in C^\infty(M, E)$ if and only if for all $\alpha \in I$, the local representation of u with respect to the coordinate charts $(U_\alpha, \varphi_\alpha)$ and $(E_{U_\alpha}, (\varphi_\alpha \circ \pi, \rho_\alpha))$ is smooth, that is,

$$\begin{aligned}
u \in C^\infty(M, E) &\iff \forall \alpha \in I \quad x \mapsto (\varphi_\alpha \circ \pi \circ u \circ \varphi_\alpha^{-1}, \rho_\alpha \circ u \circ \varphi_\alpha^{-1}) \text{ is smooth} \\
&\iff \forall \alpha \in I \quad x \mapsto (x, \rho_\alpha \circ u \circ \varphi_\alpha^{-1}) \text{ is smooth} \\
&\iff \forall \alpha \in I \quad x \mapsto \rho_\alpha \circ u \circ \varphi_\alpha^{-1} \text{ is smooth} \\
&\iff \forall \alpha \in I, \forall 1 \le l \le r \quad \rho_\alpha^l \circ u \circ \varphi_\alpha^{-1} \in C^\infty(\varphi_\alpha(U_\alpha))\,.
\end{aligned}$$

A local section of E over an open set $U \subseteq M$ is a map $u : U \to E$ where u has the property that $\pi \circ u = Id_U$ (that is, u is a section of the vector bundle $E_U \to U$). We denote the collection of all local sections on U by $\Gamma(U, E)$ or $\Gamma(U, E_U)$.

Remark 16. *As a consequence of $\rho|_{E_x} : E_x \to \mathbb{R}^r$ being an isomorphism, if u is a section of $E|_U \to U$ and $f : U \to \mathbb{R}$ is a function, then $\rho(fu) = f\rho(u)$. In particular, $\rho(0) = 0$.*

Given a total trivialization triple (U, φ, ρ) we have the following commutative diagram:

$$\begin{array}{ccc} E|_U & \xrightarrow{(\varphi\circ\pi,\rho^j)} & \varphi(U)\times\mathbb{R} \\ \big\downarrow \pi & & \big\downarrow \tilde{\pi} \\ U & \xrightarrow{\varphi} & \varphi(U)\subseteq\mathbb{R}^n \end{array}$$

If s is a section of $E|_U \to U$, then by definition the pushforward of s by ρ^j (the jth component of ρ) is a section of $\varphi(U)\times\mathbb{R}\to\varphi(U)$ which is defined by

$$\rho^j_*(s) = \rho^j \circ s \circ \varphi^{-1} \quad (\text{i.e., } z \in \varphi(U) \mapsto (z, \rho^j \circ s \circ \varphi^{-1}(z)))\,.$$

Let $E \to M$ be a vector bundle of rank r and $U \subseteq M$ be an open set. A (smooth) **local frame** for E over U is an ordered r-tuple $(s_1,\ldots,s_r)$ of (smooth) local sections over U such that for each $x \in U$, $(s_1(x),\ldots,s_r(x))$ is a basis for E_x. Given any vector bundle chart (V,ρ), we can define the associated (smooth) local frame on V as follows:

$$\forall 1 \le l \le r\; \forall x \in V \qquad s_l(x) = \rho|_{E_x}^{-1}(e_l)\,,$$

where $(e_1,\cdots,e_r)$ is the standard basis of $\mathbb{R}^r$. The following theorem states the converse of this observation is also true.

Theorem 35 ([19], p. 258)**.** *Let $E \to M$ be a vector bundle of rank r and let $(s_1,\ldots,s_r)$ be a smooth local frame over an open set $U \subseteq M$. Then (U,ρ) is a vector bundle chart where the map $\rho : E_U \to \mathbb{R}^r$ is defined by*

$$\forall x \in U, \forall u \in E_x \qquad \rho(u) = u^1 e_1 + \ldots + u^r e_r\,,$$

where $u = u^1 s_1(x) + \ldots + u^r s_r(x)$.

Theorem 36 ([19], p. 260)**.** *Let $E \to M$ be a vector bundle of rank r and let $(s_1,\ldots,s_r)$ be a smooth local frame over an open set $U \subseteq M$. If $f \in \Gamma(M,E)$, then f is smooth on U if and only if its component functions with respect to $(s_1,\ldots,s_r)$ are smooth.*

A (smooth) **fiber metric** on a vector bundle E is a (smooth) function which assigns to each $x \in M$ an inner product

$$\langle .,.\rangle_E : E_x \times E_x \to \mathbb{R}\,.$$

Note that the smoothness of the fiber metric means that for all $u, v \in C^\infty(M,E)$ the mapping

$$M \to \mathbb{R}, \qquad x \mapsto \langle u(x), v(x)\rangle_E$$

is smooth. One can show that every (smooth) vector bundle can be equipped with a (smooth) fiber metric ([33], p. 72).

Remark 17. *If (M,g) is a Riemannian manifold, then g can be viewed as a fiber metric on the tangent bundle. The metric g induces fiber metrics on all tensor bundles; it can be shown that ([32]) if (M,g) is a Riemannian manifold, then there exists a unique inner product on each fiber of $T^k_l(M)$ with the property that for all $x \in M$, if $\{e_i\}$ is an orthonormal basis of T_xM with dual basis $\{\eta^i\}$, then the corresponding basis of $T^k_l(T_xM)$ is orthonormal. We denote this inner product by $\langle .,.\rangle_F$ and the corresponding norm by $|.|_F$. If A and B are two tensor fields, then with respect to any local coordinate system*

$$\langle A, B\rangle_F = g^{i_1 r_1}\ldots g^{i_k r_k} g_{j_1 s_1}\cdots g_{j_l s_l} A^{j_1\ldots j_l}_{i_1\ldots i_k} B^{s_1\ldots s_l}_{r_1\ldots r_k}\,.$$

Theorem 37. *Let $\pi : E \to M$ be a vector bundle with rank r equipped with a fiber metric $\langle .,.\rangle_E$. Then given any total trivialization triple (U,φ,ρ), there exists a smooth map $\tilde{\rho} : E_U \to \mathbb{R}^r$ such that with respect to the new total trivialization triple $(U,\varphi,\tilde{\rho})$ the fiber metric trivializes on U, that is,*

$$\forall\, x \in U\ \forall\, u, v \in E_x \qquad \langle u, v\rangle_E = u^1 v^1 + \ldots + u^r v^r\,,$$

where for each $1 \leq l \leq r$, u^l *and* v^l *denote the* l*th components of* u *and* v*, respectively, (with respect to the local frame associated with the bundle chart* $(U, \tilde{\rho})$*).*

Proof. Let $(t_1, \ldots, t_r)$ be the local frame on U associated with the vector bundle chart (U, ρ). That is,

$$\forall\, x \in U,\ \forall 1 \leq l \leq r \qquad t_l(x) = \rho|_{E_x}^{-1}(e_l)\,.$$

Now, we apply the Gram–Schmidt algorithm to the local frame $(t_1, \ldots, t_r)$ to construct an orthonormal frame $(s_1, \ldots, s_r)$ where

$$\forall\, 1 \leq l \leq r \qquad s_l = \frac{t_l - \sum_{j=1}^{l-1} \langle t_l, s_j\rangle_E s_j}{|t_l - \sum_{j=1}^{l-1} \langle t_l, s_j\rangle_E s_j|}\,.$$

$s_l : U \to E$ is smooth because

(1) Smooth local sections over U form a module over the ring $C^\infty(U)$;
(2) The function $x \mapsto \langle t_l(x), s_j(x)\rangle_E$ from U to $\mathbb{R}$ is smooth;
(3) Since $\mathrm{Span}\{s_1, \ldots, s_{l-1}\} = \mathrm{Span}\{t_1, \ldots, t_{l-1}\}$, $t_l - \sum_{j=1}^{l-1} \langle t_l, s_j\rangle_E s_j$ is nonzero on U and $x \mapsto |t_l(x) - \sum_{j=1}^{l-1} \langle t_l(x), s_j(x)\rangle_E s_j(x)|$ as a function from U to $\mathbb{R}$ is nonzero on U and it is a composition of smooth functions.

Thus, for each l, s_l is a linear combination of elements of the $C^\infty(U)$-module of smooth local sections over U, and so it is a smooth local section over U. Now, we let $(U, \tilde{\rho})$ be the associated vector bundle chart described in Theorem 35. For all $x \in U$ and for all $u, v \in E_x$ we have

$$\langle u, v\rangle_E = \langle u^l s_l, v^j s_j\rangle_E = u^l v^j \langle s_l, s_j\rangle_E = u^l v^j \delta_{lj} = u^1 v^1 + \ldots + u^r v^r\,.$$

□

Corollary 2. *As a consequence of Theorem 37, Theorem 34, and Theorem 33 every vector bundle on a compact manifold equipped with a fiber metric admits a nice finite total trivialization atlas (and a super nice finite total trivialization atlas and a finite total trivialization atlas that is GL compatible with itself) such that the fiber metric is trivialized with respect to each total trivialization triple in the atlas.*

5.3. Standard Total Trivialization Triples

Let M^n be a smooth manifold and $\pi : E \to M$ be a vector bundle of rank r. For certain vector bundles there are standard methods to associate with any given smooth coordinate chart $(U, \varphi = (x^i))$ a total trivialization triple (U, φ, ρ). We call such a total trivialization triple the **standard total trivialization** associated with (U, φ). Usually this is done by first associating with (U, φ) a local frame for E_U and then applying Theorem 35 to construct a total trivialization triple.

- $E = T^k_l(M)$: The collection of the following tensor fields on U forms a local frame for E_U associated with $(U, \varphi = (x^i))$.

$$\frac{\partial}{\partial x^{i_1}} \otimes \ldots \otimes \frac{\partial}{\partial x^{i_l}} \otimes dx^{j_1} \otimes \ldots \otimes dx^{j_k}\,.$$

So, given any atlas $\{(U_\alpha, \varphi_\alpha)\}$ of a manifold M^n, there is a corresponding total trivialization atlas for the tensor bundle $T^k_l(M)$, namely $\{(U_\alpha, \varphi_\alpha, \rho_\alpha)\}$ where for each α, ρ_α has n^{k+l} components which we denote by $(\rho_\alpha)^{j_1 \ldots j_l}_{i_1 \ldots i_k}$. For all $F \in \Gamma(M, T^k_l(M))$, we have

$$(\rho_\alpha)^{j_1\ldots j_l}_{i_1\ldots i_k}(F) = (F_\alpha)^{j_1\ldots j_l}_{i_1\ldots i_k}.$$

Here $(F_\alpha)^{j_1\ldots j_l}_{i_1\ldots i_k}$ denotes the components of F with respect to the standard frame for $T^k_l U_\alpha$ described above. When there is no possibility of confusion, we may write $F^{j_1\ldots j_l}_{i_1\ldots i_k}$ instead of $(F_\alpha)^{j_1\ldots j_l}_{i_1\ldots i_k}$.

- $E = \Lambda^k(M)$: This is the bundle whose fiber over each $x \in M$ consists of alternating covariant tensors of order k. The collection of the following forms on U form a local frame for E_U associated with $(U, \varphi = (x^i))$

$$dx^{j_1} \wedge \ldots \wedge dx^{j_k} \quad ((j_1, \ldots, j_k) \text{ is increasing}).$$

- $E = \mathcal{D}(M)$ (the density bundle): The density bundle over M is the vector bundle whose fiber over each $x \in M$ is $\mathcal{D}(T_x M)$. More precisely, if we let

$$\mathcal{D}(M) = \coprod_{x \in M} \mathcal{D}(T_x M),$$

then $\mathcal{D}(M)$ is a smooth vector bundle of rank 1 over M ([19], p. 429). Indeed, for every smooth chart $(U, \varphi = (x^i))$, $|dx^1 \wedge \ldots \wedge dx^n|$ on U is a local frame for $\mathcal{D}(M)|_U$. We denote the corresponding trivialization by $\rho_{\mathcal{D},\varphi}$, that is, given $\mu \in \mathcal{D}(T_y M)$, there exists a number a such that

$$\mu = a(|dx^1 \wedge \ldots \wedge dx^n|_y)$$

and $\rho_{\mathcal{D},\varphi}$ sends μ to a. Sometimes we write $\mathcal{D}$ instead of $\mathcal{D}(M)$ if M is clear from the context. Furthermore, when there is no possibility of confusion we may write $\rho_{\mathcal{D}}$ instead of $\rho_{\mathcal{D},\varphi}$.

Remark 18 (Integration of densities on manifolds). *Elements of $C_c(M, \mathcal{D})$ can be integrated over M. Indeed, for $\mu \in C_c(M, \mathcal{D})$ we may consider two cases*

- ***Case 1:*** *There exists a smooth chart (U, φ) such that $\text{supp}\mu \subseteq U$.*

$$\int_M \mu := \int_{\varphi(U)} \rho_{\mathcal{D},\varphi} \circ \mu \circ \varphi^{-1} \, dV.$$

- ***Case 2:*** *If μ is an arbitrary element of $C_c(M, \mathcal{D})$, then we consider a smooth atlas $\{(U_\alpha, \varphi_\alpha)\}_{\alpha \in I}$ and a partition of unity $\{\psi_\alpha\}_{\alpha \in I}$ subordinate to $\{U_\alpha\}$ and we let*

$$\int_M \mu := \sum_{\alpha \in I} \int_M \psi_\alpha \mu.$$

It can be shown that the above definitions are independent of the choices (charts and partition of unity) involved ([19], pp. 431–432).

5.4. *Constructing New Bundles from Old Ones*

5.4.1. Hom Bundle, Dual Bundle, Functional Dual Bundle

- The construction Hom(.,.) can be applied fiberwise to a pair of vector bundles E and $\tilde{E}$ over a manifold M to give a new vector bundle denoted by $\text{Hom}(E, \tilde{E})$. The fiber of $\text{Hom}(E, \tilde{E})$ at any given point $p \in M$ is the vector space $\text{Hom}(E_p, \tilde{E}_p)$. Clearly, if $\text{rank}\, E = r$ and $\text{rank}\, \tilde{E} = \tilde{r}$, then $\text{rank}\, \text{Hom}(E, \tilde{E}) = r\tilde{r}$.
  If $\{(U_\alpha, \varphi_\alpha, \rho_\alpha)\}$ and $\{(U_\alpha, \varphi_\alpha, \tilde{\rho}_\alpha)\}$ are total trivialization atlases for the vector bundles $\pi : E \to M$ and $\tilde{\pi} : \tilde{E} \to M$, respectively, then $\{U_\alpha, \varphi_\alpha, \hat{\rho}_\alpha\}$ will be a total trivialization atlas for $\pi_{\text{Hom}} : \text{Hom}(E, \tilde{E}) \to M$ where $\hat{\rho}_\alpha : \pi^{-1}_{\text{Hom}}(U_\alpha) \to \text{Hom}(\mathbb{R}^r, \mathbb{R}^{\tilde{r}}) \cong \mathbb{R}^{r\tilde{r}}$ is defined as follows: for $p \in U_\alpha$, $A_p \in \text{Hom}(E_p, \tilde{E}_p)$ is mapped to $[\tilde{\rho}_\alpha|_{\tilde{E}_p}] \circ A \circ [\rho_\alpha|_{E_p}]^{-1}$.

- Let $\pi : E \to M$ be a vector bundle. The **dual bundle** E^* is defined by $E^* = \mathrm{Hom}(E, \tilde{E} = M \times \mathbb{R})$.
- Let $\pi : E \to M$ be a vector bundle and let $\mathcal{D}$ denote the density bundle of M. The **functional dual bundle** $E^\vee$ is defined by $E^\vee = \mathrm{Hom}(E, \mathcal{D})$ (see [24]). Let us describe explicitly what the standard total trivialization triples of this bundle are. Let (U, φ, ρ) be a total trivialization triple for E. We can associate with this triple the total trivialization triple $(U, \varphi, \rho^\vee)$ for $E^\vee$ where $\rho^\vee : E^\vee_U \to \mathbb{R}^r$ is defined as follows: for $p \in U$, $L_p \in \mathrm{Hom}(E_p, \mathcal{D}_p)$ is mapped to $\rho_{\mathcal{D},\varphi} \circ L_p \circ (\rho|_{E_p})^{-1} \in (\mathbb{R}^r)^* \simeq \mathbb{R}^r$. Note that $(\mathbb{R}^r)^* \simeq \mathbb{R}^r$ under the following isomorphism

$$(\mathbb{R}^r)^* \to \mathbb{R}^r, \qquad u \mapsto u(e_1)e_1 + \ldots + u(e_r)e_r .$$

That is, u as an element of $\mathbb{R}^r$ is the vector whose components are $(u(e_1), \ldots, u(e_r))$. In particular, if $z = z_1 e_1 + \ldots + z_r e_r$ is an arbitrary vector in $\mathbb{R}^r$, then

$$u(z) = u(z_1 e_1 + \ldots + z_r e_r) = z_1 u(e_1) + \ldots + z_r u(e_r) = z \cdot u ,$$

where on the LHS u is viewed as an element of $(\mathbb{R}^r)^*$ and on the RHS u is viewed as an element of $\mathbb{R}^r$.
In short, $\rho^\vee : E^\vee_U \to \mathbb{R}^r$ is given by

$$\forall 1 \le l \le r \qquad (\rho^\vee)^l(L_p) = (\rho_{\mathcal{D},\varphi} \circ L_p \circ (\rho|_{E_p})^{-1})(e_l) .$$

5.4.2. Tensor Product of Bundles

Let $\pi : E \to M$ and $\tilde{\pi} : \tilde{E} \to M$ be two vector bundles. Then $E \otimes \tilde{E}$ is a new vector bundle whose fiber at $p \in M$ is $E_p \otimes \tilde{E}_p$. If $\{(U_\alpha, \varphi_\alpha, \rho_\alpha)\}$ and $\{(U_\alpha, \varphi_\alpha, \tilde{\rho}_\alpha)\}$ are total trivialization atlases for the vector bundles $\pi : E \to M$ and $\tilde{\pi} : \tilde{E} \to M$, respectively, then $\{(U_\alpha, \varphi_\alpha, \hat{\rho}_\alpha)\}$ will be a total trivialization atlas for $\pi_{\text{tensor}} : E \otimes \tilde{E} \to M$ where $\hat{\rho}_\alpha : \pi^{-1}_{\text{tensor}}(U_\alpha) \to (\mathbb{R}^r \otimes \mathbb{R}^{\tilde{r}}) \cong \mathbb{R}^{r\tilde{r}}$ is defined as follows: for $p \in U_\alpha$, $a_p \otimes \tilde{a}_p \in E_p \otimes \tilde{E}_p$ is mapped to $\rho_\alpha|_{E_p}(a_p) \otimes \tilde{\rho}_\alpha|_{\tilde{E}_p}(\tilde{a}_p)$.

It can be shown that $\mathrm{Hom}(E, \tilde{E}) \cong E^* \otimes \tilde{E}$ (isomorphism of vector bundles over M).

Remark 19 (Fiber Metric on Tensor Product). *Consider the inner product spaces $(U, \langle ., . \rangle_U)$ and $(V, \langle ., . \rangle_V)$. We can turn the tensor product of U and V, $U \otimes V$ into an inner product space by defining*

$$\langle u_1 \otimes v_1, u_2 \otimes v_2 \rangle_{U \otimes V} = \langle u_1, u_2 \rangle_U \langle v_1, v_2 \rangle_V ,$$

and extending by linearity. As a consequence, if E is a vector bundle (on a Riemannian manifold (M, g)) equipped with a fiber metric $\langle ., . \rangle_E$, then there is a natural fiber metric on the bundle $(T^*M)^{\otimes k}$ and subsequently on the bundle $(T^*M)^{\otimes k} \otimes E$. If $F = F^a_{i_1 \ldots i_k} dx^{i_1} \otimes \ldots \otimes dx^{i_k} \otimes s_a$ and $G = G^b_{j_1 \ldots j_k} dx^{j_1} \otimes \ldots \otimes dx^{j_k} \otimes s_b$ are two local sections of this bundle on a domain U of a total trivialization triple, then at any point in U we have

$$\begin{aligned} \langle F, G \rangle_{(T^*M)^{\otimes k} \otimes E} &= F^a_{i_1 \cdots i_k} G^b_{j_1 \ldots j_k} \langle dx^{i_1}, dx^{j_1} \rangle_{T^*M} \ldots \langle dx^{i_k}, dx^{j_k} \rangle_{T^*M} \langle s_a, s_b \rangle_E \\ &= g^{i_1 j_1} \ldots g^{i_k j_k} h_{ab} F^a_{i_1 \ldots i_k} G^b_{j_1 \ldots j_k} , \end{aligned}$$

where $h_{ab} := \langle s_a, s_b \rangle_E$ (here $\{s_a = \rho^{-1}(e_a)\}_{1 \le a \le r}$ is a local frame for E over U.$\{e_a\}_{1 \le a \le r}$ is the standard basis for $\mathbb{R}^r$ where $r = \text{rank } E$).

5.5. Connection on Vector Bundles, Covariant Derivative

5.5.1. Basic Definitions

Let $\pi : E \to M$ be a vector bundle.

Definition 16. *A **connection** in E is a map*

$$\nabla : C^\infty(M, TM) \times C^\infty(M, E) \to C^\infty(M, E), \quad (X, u) \mapsto \nabla_X u$$

satisfying the following properties:

(1) $\nabla_X u$ *is linear over* $C^\infty(M)$ *in* X

$$\forall f, g \in C^\infty(M) \qquad \nabla_{fX_1 + gX_2} u = f \nabla_{X_1} u + g \nabla_{X_2} u\,.$$

(2) $\nabla_X u$ *is linear over* $\mathbb{R}$ *in* u*:*

$$\forall a, b \in \mathbb{R} \qquad \nabla_X(au_1 + bu_2) = a\nabla_X u_1 + b\nabla_X u_2\,.$$

(3) ∇ *satisfies the following product rule*

$$\forall f \in C^\infty(M) \qquad \nabla_X(fu) = f\nabla_X u + (Xf)u\,.$$

A ***metric connection*** *in a real vector bundle E with a fiber metric is a connection* ∇ *such that*

$$\forall X \in C^\infty(M, TM), \forall u, v \in C^\infty(M, E) \qquad X\langle u, v\rangle_E = \langle \nabla_X u, v\rangle_E + \langle u, \nabla_X v\rangle_E\,.$$

Here is a list of useful facts about connections:

- ([34], p. 183) Using a partition of unity, one can show that any real vector bundle with a smooth fiber metric admits a metric connection;
- ([19], p. 50) If ∇ is a connection in a bundle E, $X \in C^\infty(M, TM)$, $u \in C^\infty(M, E)$, and $p \in M$, then $\nabla_X u|_p$ depends only on the values of u in a neighborhood of p and the value of X at p. More precisely, if $u = \tilde{u}$ on a neighborhood of p and $X_p = \tilde{X}_p$, then $\nabla_X u|_p = \nabla_{\tilde{X}} \tilde{u}|_p$;
- ([19], p. 53) If ∇ is a connection in TM, then there exists a unique connection in each tensor bundle $T^k_l(M)$, also denoted by ∇, such that the following conditions are satisfied:

  (1) On the tangent bundle, ∇ agrees with the given connection.
  (2) On $T^0(M)$, ∇ is given by ordinary differentiation of functions, that is, for all real-valued smooth functions $f : M \to \mathbb{R}$: $\nabla_X f = Xf$.
  (3) $\nabla_X(F \otimes G) = (\nabla_X F) \otimes G + F \otimes (\nabla_X G)$.
  (4) If tr denotes the trace on any pair of indices, then $\nabla_X(\mathrm{tr}F) = \mathrm{tr}(\nabla_X F)$.

  This connection satisfies the following additional property: for any $T \in C^\infty(M, T^k_l(M))$, vector fields Y_i, and differential 1-forms ω^j,

$$\begin{aligned}(\nabla_X T)(\omega^1, \ldots, \omega^l, Y_1, \ldots, Y_k) &= X(T(\omega^1, \ldots, \omega^l, Y_1, \ldots, Y_k)) \\ &\quad - \sum_{j=1}^{l} T(\omega^1, \ldots, \nabla_X \omega^j, \ldots, \omega^l, Y_1, \ldots, Y_k) \\ &\quad - \sum_{i=1}^{k} T(\omega^1, \ldots, \omega^l, Y_1, \ldots, \nabla_X Y_i, \ldots, Y_k)\,.\end{aligned}$$

Definition 17. *Let* ∇ *be a connection in* $\pi : E \to M$*. We define the corresponding* ***covariant derivative*** *on E, also denoted* ∇*, as follows*

$$\nabla : C^\infty(M, E) \to C^\infty(M, Hom(TM, E)) \cong C^\infty(M, T^*M \otimes E), \qquad u \mapsto \nabla u$$

where for all $p \in M$*,* $\nabla u(p) : T_pM \to E_p$ *is defined by*

$$X_p \mapsto \nabla_X u|_p\,,$$

where X on the RHS is any smooth vector field whose value at p is X_p*.*

Remark 20. *Let ∇ be a connection in TM. As it was discussed ∇ induces a connection in any tensor bundle $E = T_l^k(M)$, also denoted by ∇. Some authors (including Lee in [19], p. 53) define the corresponding covariant derivative on $E = T_l^k(M)$ as follows:*

$$\nabla : C^\infty(M, T_l^k(M)) \to C^\infty(M, T_l^{k+1}(M)), \qquad F \mapsto \nabla F$$

where

$$\nabla F(\omega^1, \ldots, \omega^l, Y_1, \ldots, Y_k, X) = (\nabla_X F)(\omega^1, \ldots, \omega^l, Y_1, \ldots, Y_k).$$

This definition agrees with the previous definition of covariant derivative that we had for general vector bundles because

$$T^*M \otimes T_l^k M \cong T^*M \otimes \underbrace{T^*M \otimes \ldots \otimes T^*M}_{k\ factors} \otimes \underbrace{TM \otimes \ldots \otimes TM}_{l\ factors} \cong T_l^{k+1} M.$$

Therefore,

$$C^\infty(M, Hom(TM, T_l^k M)) \cong C^\infty(M, T^*M \otimes T_l^k M) \cong C^\infty(M, T_l^{k+1} M).$$

More concretely, we have the following one-to-one correspondence between $C^\infty(M, Hom(TM, T_l^k M))$ and $C^\infty(M, T_l^{k+1} M)$:

(1) Given $u \in C^\infty(M, T_l^{k+1} M)$, the corresponding element $\tilde{u} \in C^\infty(M, Hom(TM, T_l^k M))$ is given by

$$\forall p \in M \qquad \tilde{u}(p) : T_p M \to T_l^k(T_p M), \quad X \mapsto u(p)(\ldots, \ldots, X).$$

(2) Given $\tilde{u} \in C^\infty(M, Hom(TM, T_l^k M))$, the corresponding element $u \in C^\infty(M, T_l^{k+1} M)$ is given by

$$\forall p \in M \qquad u(p)(\omega^1, \ldots, \omega^l, Y_1, \ldots, Y_k, X) = [\tilde{u}(p)(X)](\omega^1, \ldots, \omega^l, Y_1, \ldots, Y_k).$$

5.5.2. Covariant Derivative on Tensor Product of Bundles

If E an $\tilde{E}$ are vector bundles over M with covariant derivatives $\nabla^E : C^\infty(M, E) \to C^\infty(M, T^*M \otimes E)$ and $\nabla^{\tilde{E}} : C^\infty(M, \tilde{E}) \to C^\infty(M, T^*M \otimes \tilde{E})$, respectively, then there is a uniquely determined covariant derivative ([14], p. 87)

$$\nabla^{E \otimes \tilde{E}} : C^\infty(M, E \otimes \tilde{E}) \to C^\infty(M, T^*M \otimes E \otimes \tilde{E})$$

such that

$$\nabla^{E \otimes \tilde{E}}(u \otimes \tilde{u}) = \nabla^E u \otimes \tilde{u} + \nabla^{\tilde{E}} \tilde{u} \otimes u.$$

The above sum makes sense because of the following isomorphisms:

$$(T^*M \otimes E) \otimes \tilde{E} \cong T^*M \otimes E \otimes \tilde{E} \cong T^*M \otimes \tilde{E} \otimes E \cong (T^*M \otimes \tilde{E}) \otimes E.$$

Remark 21. *Recall that for tensor fields covariant derivative can be considered as a map from $C^\infty(M, T_l^k M) \to C^\infty(M, T_l^{k+1} M)$. Using this, we can give a second description of covariant derivative on $E \otimes \tilde{E}$ when $E = T_l^k M$. In this new description we have*

$$\nabla^{T_l^k M \otimes \tilde{E}} : C^\infty(M, T_l^k M \otimes \tilde{E}) \to C^\infty(M, T_l^{k+1} M \otimes \tilde{E}).$$

Indeed, for $F \in C^\infty(M, T_l^k M)$ and $u \in C^\infty(M, \tilde{E})$

$$\nabla^{T_l^k M \otimes \tilde{E}}(F \otimes u) = \underbrace{(\nabla^{T_l^k M} F)}_{T_l^{k+1} M} \otimes u + \underbrace{\underbrace{F}_{T_l^k M} \otimes \underbrace{\nabla^{\tilde{E}} u}_{T^*M \otimes \tilde{E}}}_{T_l^{k+1} M \otimes \tilde{E}}.$$

In particular, if $f \in C^\infty(M)$ and $u \in C^\infty(M,E)$ we have $\nabla^E(fu) \in C^\infty(M, T^*M \otimes E)$ and it is equal to

$$\nabla^E(fu) = df \otimes u + f\nabla^E u\,.$$

5.5.3. Higher Order Covariant Derivatives

Let $\pi : E \to M$ be a vector bundle. Let ∇^E be a connection in E and ∇ be a connection in TM which induces a connection in T^*M. We have the following chain

$$C^\infty(M,E) \xrightarrow{\nabla^E} C^\infty(M, T^*M \otimes E) \xrightarrow{\nabla^{T^*M\otimes E}} C^\infty(M, (T^*M)^{\otimes 2} \otimes E) \xrightarrow{\nabla^{(T^*M)^{\otimes 2}\otimes E}}$$
$$\cdots \xrightarrow{\nabla^{(T^*M)^{\otimes(k-1)}\otimes E}} C^\infty(M, (T^*M)^{\otimes k} \otimes E) \xrightarrow{\nabla^{(T^*M)^{\otimes k}\otimes E}} \cdots\,.$$

In what follows we denote all the maps in the above chain by ∇^E. That is, for any $k \in \mathbb{N}_0$ we consider ∇^E as a map from $C^\infty(M, (T^*M)^{\otimes k} \otimes E)$ to $C^\infty(M, (T^*M)^{\otimes(k+1)} \otimes E)$. So,

$$(\nabla^E)^k : C^\infty(M,E) \to C^\infty(M, (T^*M)^{\otimes k} \otimes E)\,.$$

As an example, let us consider $(\nabla^E)^k(fu)$ where $f \in C^\infty(M)$ and $u \in C^\infty(M,E)$. We have

$$\begin{aligned}
\nabla^E(fu) &= df \otimes u + f\nabla^E u\,.\\
(\nabla^E)^2(fu) &= \nabla^{T^*M\otimes E}[df \otimes u + f\nabla^E u]\\
&= [\nabla^{T^*M}(df) \otimes u + df \otimes \nabla^E u] + [df \otimes \nabla^E u + f(\nabla^E)^2 u]\\
&= \sum_{j=0}^{2} \binom{2}{j} (\nabla^{T^*M})^j f \otimes (\nabla^E)^{2-j} u\,.
\end{aligned}$$

In general, we can show by induction that

$$(\nabla^E)^k(fu) = \sum_{j=0}^{k} \binom{k}{j} (\nabla^{T^*M})^j f \otimes (\nabla^E)^{k-j} u\,.$$

where $(\nabla^{T^*M})^0 = Id$. Here $(\nabla^{T^*M})^j f$ should be interpreted as applying ∇ (in the sense described in Remark 20) j times; so $(\nabla^{T^*M})^j f$ at each point is an element of $T^j_0 M = (T^*M)^{\otimes j}$.

5.5.4. Three Useful Rules, Two Important Observations

Let $\pi : E \to M$ and $\tilde{\pi} : \tilde{E} \to M$ be two vector bundles over M with ranks r and $\tilde{r}$, respectively. Let ∇ be a connection in TM (which automatically induces a connection in all tensor bundles), ∇^E be a connection in E and $\nabla^{\tilde{E}}$ be a connection in $\tilde{E}$. Let (U, φ, ρ) be a total trivialization triple for E.

(1) $\{\partial_i = \varphi_*^{-1} \frac{\partial}{\partial x^i}\}_{1\le i\le n}$ is a coordinate frame for TM over U.
(2) $\{s_a = \rho^{-1}(e_a)\}_{1\le a\le r}$ is a local frame for E over U ($\{e_a\}_{1\le a\le r}$ is the standard basis for $\mathbb{R}^r$ where $r = \operatorname{rank} E$).
(3) Christoffel Symbols for ∇ on (U, φ, ρ): $\nabla_{\partial_i}\partial_j = \Gamma^k_{ij}\partial_k$.
(4) Christoffel Symbols for ∇^E on (U, φ, ρ): $\nabla_{\partial_i} s_a = (\Gamma_E)^b_{ia} s_b$.

Furthermore, recall that for any 1-form ω,

$$\nabla_X \omega = (X^i \partial_i \omega_k - X^i \omega_j \Gamma^j_{ik}) dx^k\,.$$

Therefore,

$$\nabla_{\partial_i} dx^j = -\Gamma^j_{ik} dx^k\,.$$

- **Rule 1:** For all $u \in C^\infty(M, E)$
$$\nabla^E u = dx^i \otimes \nabla^E_{\partial_i} u \qquad \text{on } U\,.$$
The reason is as follows: Recall that for all $p \in M$, $\nabla^E u(p) \in T^*M \otimes E$. Since $\{dx^i \otimes s_a\}$ is a local frame for $T^*M \otimes E$ on U we have
$$\nabla^E u = R^a_i dx^i \otimes s_a = dx^i \otimes (R^a_i s_a)\,.$$
According to what was discussed in the study of the isomorphism $\text{Hom}(V, W) \cong V^* \otimes W$ in Section 3 we know that at any point $p \in M$, R^a_i is the element in column i and row a of the matrix of $\nabla^E u(p)$ as an element of $\text{Hom}(T_pM, E_p)$. Therefore,
$$\nabla^E_{\partial_i} u = R^a_i s_a\,.$$
Consequently, we have $\nabla^E u = dx^i \otimes (R^a_i s_a) = dx^i \otimes \nabla^E_{\partial_i} u$.
- **Rule 2:** For all $v_1 \in C^\infty(M, E)$ and $v_2 \in C^\infty(M, \tilde{E})$
$$\nabla^{E\otimes\tilde{E}}_{\partial_j}(v_1 \otimes v_2) = (\nabla^E_{\partial_j} v_1) \otimes v_2 + v_1 \otimes (\nabla^{\tilde{E}}_{\partial_j} v_2)\,.$$
- **Rule 3:** For all $u \in C^\infty(M, E)$ and $f \in C^\infty(M)$
$$\nabla^E(fu) = f\nabla^E u + df \otimes u\,.$$
The following two examples are taken from [35].
- **Example 1:** Let $u \in C^\infty(M, E)$. On U we may write $u = u^a s_a$. We have
$$\begin{aligned}
\nabla^E u &= \nabla^E(u^a s_a) \overset{\text{Rule 3}}{=} u^a \nabla^E s_a + du^a \otimes s_a = u^a \nabla^E s_a + (\partial_i u^a dx^i) \otimes s_a \\
&\overset{\text{Rule 1}}{=} u^a dx^i \otimes \nabla^E_{\partial_i} s_a + (\partial_i u^a dx^i) \otimes s_a \\
&= u^a dx^i \otimes \left((\Gamma_E)^b_{ia} s_b\right) + (\partial_i u^a dx^i) \otimes s_a = dx^i \otimes \left(u^a (\Gamma_E)^b_{ia} s_b\right) + dx^i \otimes (\partial_i u^a s_a) \\
&= dx^i \otimes \left(u^b (\Gamma_E)^a_{ib} s_a\right) + dx^i \otimes (\partial_i u^a s_a) \\
&= [\partial_i u^a + (\Gamma_E)^a_{ib} u^b] dx^i \otimes s_a\,.
\end{aligned}$$
That is, $\nabla^E u = (\nabla^E u)^a_i dx^i \otimes s_a$ where
$$(\nabla^E u)^a_i = \partial_i u^a + (\Gamma_E)^a_{ib} u^b\,.$$
- **Example 2:** Let $u \in C^\infty(M, E)$. On U we may write $u = u^a s_a$. We have

$$
\begin{aligned}
(\nabla^E)^2 u &= \nabla^{T^*M\otimes E}([\partial_i u^a + (\Gamma_E)^a_{ib} u^b] dx^i \otimes s_a) \\
&\overset{\text{Rule 3}}{=} [\partial_i u^a + (\Gamma_E)^a_{ib} u^b] \nabla^{T^*M\otimes E}(dx^i \otimes s_a) + d[\partial_i u^a + (\Gamma_E)^a_{ib} u^b] \otimes (dx^i \otimes s_a) \\
&\overset{\text{Rule 1}}{=} [\partial_i u^a + (\Gamma_E)^a_{ib} u^b] dx^j \otimes \nabla^{T^*M\otimes E}_{\partial_j}(dx^i \otimes s_a) + d[\partial_i u^a + (\Gamma_E)^a_{ib} u^b] \otimes (dx^i \otimes s_a) \\
&\overset{\text{Def. of } d}{=} [\partial_i u^a + (\Gamma_E)^a_{ib} u^b] dx^j \otimes \nabla^{T^*M\otimes E}_{\partial_j}(dx^i \otimes s_a) + \partial_j[\partial_i u^a + (\Gamma_E)^a_{ib} u^b] dx^j \otimes dx^i \otimes s_a \\
&\overset{\text{Rule 2}}{=} [\partial_i u^a + (\Gamma_E)^a_{ib} u^b] dx^j \otimes \left[\nabla^{T^*M}_{\partial_j} dx^i \otimes s_a + dx^i \otimes \nabla^E_{\partial_j} s_a\right] + \partial_j[\partial_i u^a + (\Gamma_E)^a_{ib} u^b] dx^j \otimes dx^i \otimes s_a \\
&= [\partial_i u^a + (\Gamma_E)^a_{ib} u^b] dx^j \otimes \left[-\Gamma^i_{jk} dx^k \otimes s_a + dx^i \otimes (\Gamma_E)^c_{ja} s_c\right] + \partial_j[\partial_i u^a + (\Gamma_E)^a_{ib} u^b] dx^j \otimes dx^i \otimes s_a \\
&\overset{i \leftrightarrow k \text{ in the first summand}}{=} [\partial_k u^a + (\Gamma_E)^a_{kb} u^b] dx^j \otimes \left[-\Gamma^k_{ji} dx^i \otimes s_a + dx^k \otimes (\Gamma_E)^c_{ja} s_c\right] + \partial_j[\partial_i u^a + (\Gamma_E)^a_{ib} u^b] dx^j \otimes dx^i \otimes s_a \\
&= \{\partial_j[\partial_i u^a + (\Gamma_E)^a_{ib} u^b] - \Gamma^k_{ji}[\partial_k u^a + (\Gamma_E)^a_{kb} u^b]\} dx^j \otimes dx^i \otimes s_a + [\partial_k u^a + (\Gamma_E)^a_{kb} u^b](\Gamma_E)^c_{ja} dx^j \otimes dx^k \otimes s_c \\
&\overset{i \leftrightarrow k \text{ in the last summand}}{=} \{\partial_j[\partial_i u^a + (\Gamma_E)^a_{ib} u^b] - \Gamma^k_{ji}[\partial_k u^a + (\Gamma_E)^a_{kb} u^b]\} dx^j \otimes dx^i \otimes s_a \\
&\qquad + [\partial_i u^a + (\Gamma_E)^a_{ib} u^b](\Gamma_E)^c_{ja} dx^j \otimes dx^i \otimes s_c \\
&\overset{c \leftrightarrow a \text{ in the last summand}}{=} \{\partial_j[\partial_i u^a + (\Gamma_E)^a_{ib} u^b] - \Gamma^k_{ji}[\partial_k u^a + (\Gamma_E)^a_{kb} u^b]\} dx^j \otimes dx^i \otimes s_a \\
&\qquad + [\partial_i u^c + (\Gamma_E)^c_{ib} u^b](\Gamma_E)^a_{jc} dx^j \otimes dx^i \otimes s_a \,.
\end{aligned}
$$

Considering the above examples we make the following two useful observations that can be proved by induction.

- **Observation 1:** In general $(\nabla^E)^k u = ((\nabla^E)^k u)^a_{i_1\ldots i_k} dx^{i_1} \otimes \ldots \otimes dx^{i_k} \otimes s_a$ $(1 \le a \le r, 1 \le i_1, \ldots, i_k \le n)$ where $((\nabla^E)^k u)^a_{i_1\ldots i_k} \circ \varphi^{-1}$ is a linear combination of $u^1 \circ \varphi^{-1}, \ldots, u^r \circ \varphi^{-1}$ and their partial derivatives up to order k and the coefficients are polynomials in terms of Christoffel symbols (of the linear connection on M and connection in E) and their derivatives (on a compact manifold these coefficients are uniformly bounded provided that the metric and the fiber metric are smooth). That is,

$$((\nabla^E)^k u)^a_{i_1\ldots i_k} \circ \varphi^{-1} = \sum_{|\eta|\le k} \sum_{l=1}^{r} C_{\eta l} \partial^\eta (u^l \circ \varphi^{-1}),$$

where for each η and l, $C_{\eta l}$ is a polynomial in terms of Christoffel symbols (of the linear connection on M and connection in E) and their derivatives.

- **Observation 2:** The highest order term in $((\nabla^E)^k u)^a_{i_1\ldots i_k} \circ \varphi^{-1}$ is $\frac{\partial}{x^{i_1}} \cdots \frac{\partial}{x^{i_k}} (u^a \circ \varphi^{-1})$; that is,

$$((\nabla^E)^k u)^a_{i_1\ldots i_k} \circ \varphi^{-1} = \frac{\partial}{\partial x^{i_1}} \cdots \frac{\partial}{\partial x^{i_k}} (u^a \circ \varphi^{-1}) + \ldots$$

where extra terms contain derivatives of order at most $k-1$ of $u^l \circ \varphi^{-1}$ $(1 \le l \le r)$:

$$((\nabla^E)^k u)^a_{i_1\ldots i_k} \circ \varphi^{-1} = \frac{\partial^k}{\partial x^{i_1} \ldots \partial x^{i_k}} (u^a \circ \varphi^{-1}) + \sum_{|\eta|<k} \sum_{l=1}^{r} C_{\eta l} \partial^\eta (u^l \circ \varphi^{-1}).$$

6. Some Results from the Theory of Generalized Functions

In this section, we collect some results from the theory of distributions that will be needed for our definition of function spaces on manifolds. Our main reference for this part is the exquisite exposition by Marcel De Reus [24].

6.1. Distributions on Domains in Euclidean Space

Let Ω be a nonempty open set in $\mathbb{R}^n$.

(1) Recall that

- $\mathcal{K}(\Omega)$ is the collection of all compact subsets of Ω.
- $C^\infty(\Omega)$ = the collection of all infinitely differentiable (real-valued) functions on Ω.
- For all $K \in \mathcal{K}(\Omega)$, $C_K^\infty(\Omega) = \{\varphi \in C^\infty(\Omega) : \operatorname{supp}\varphi \subseteq K\}$.
- $C_c^\infty(\Omega) = \bigcup_{K\in\mathcal{K}(\Omega)} C_K^\infty(\Omega) = \{\varphi \in C^\infty(\Omega) : \operatorname{supp}\varphi \text{ is compact in } \Omega\}$.

(2) For all $\varphi \in C^\infty(\Omega)$, $j \in \mathbb{N}$ and $K \in \mathcal{K}(\Omega)$ we define

$$\|\varphi\|_{j,K} := \sup\{|\partial^\alpha \varphi(x)| : |\alpha| \le j, x \in K\}.$$

(3) For all $j \in \mathbb{N}$ and $K \in \mathcal{K}(\Omega)$, $\|.\|_{j,K}$ is a seminorm on $C^\infty(\Omega)$. We define $\mathcal{E}(\Omega)$ to be $C^\infty(\Omega)$ equipped with the natural topology induced by the separating family of seminorms $\{\|.\|_{j,K}\}_{j\in\mathbb{N},K\in\mathcal{K}(\Omega)}$. It can be shown that $\mathcal{E}(\Omega)$ is a Frechet space.

(4) For all $K \in \mathcal{K}(\Omega)$ we define $\mathcal{E}_K(\Omega)$ to be $C_K^\infty(\Omega)$ equipped with the subspace topology. This subspace topology on $C_K^\infty(\Omega)$ is the natural topology induced by the separating family of seminorms $\{\|.\|_{j,K}\}_{j\in\mathbb{N}}$. Since $C_K^\infty(\Omega)$ is a closed subset of the Frechet space $\mathcal{E}(\Omega)$, $\mathcal{E}_K(\Omega)$ is also a Frechet space.

(5) We define $D(\Omega) = \bigcup_{K\in\mathcal{K}(\Omega)} \mathcal{E}_K(\Omega)$ equipped with the inductive limit topology with respect to the family of vector subspaces $\{\mathcal{E}_K(\Omega)\}_{K\in\mathcal{K}(\Omega)}$. It can be shown that if $\{K_j\}_{j\in\mathbb{N}_0}$ is an exhaustion by compacts sets of Ω, then the inductive limit topology on $D(\Omega)$ with respect to the family $\{\mathcal{E}_{K_j}\}_{j\in\mathbb{N}_0}$ is exactly the same as the inductive limit topology with respect to $\{\mathcal{E}_K(\Omega)\}_{K\in\mathcal{K}(\Omega)}$.

Remark 22. *Let us mention a trivial but extremely useful consequence of the above description of the inductive limit topology on $D(\Omega)$. Suppose Y is a topological space and the mapping $T : Y \to D(\Omega)$ is such that $T(Y) \subseteq \mathcal{E}_K(\Omega)$ for some $K \in \mathcal{K}(\Omega)$. Since $\mathcal{E}_K(\Omega) \hookrightarrow D(\Omega)$, if $T : Y \to \mathcal{E}_K(\Omega)$ is continuous, then $T : Y \to D(\Omega)$ will be continuous.*

Theorem 38 (Convergence and Continuity for $\mathcal{E}(\Omega)$). *Let Ω be a nonempty open set in $\mathbb{R}^n$. Let Y be a topological vector space whose topology is induced by a separating family of seminorms $\mathcal{Q}$.*

(1) A sequence $\{\varphi_m\}$ converges to φ in $\mathcal{E}(\Omega)$ if and only if $\|\varphi_m - \varphi\|_{j,K} \to 0$ for all $j \in \mathbb{N}$ and $K \in \mathcal{K}(\Omega)$.

(2) Suppose $T : \mathcal{E}(\Omega) \to Y$ is a linear map. Then the following is equivalent

- *T is continuous.*
- *For every $q \in \mathcal{Q}$, there exist $j \in \mathbb{N}$ and $K \in \mathcal{K}(\Omega)$, and $C > 0$ such that*

$$\forall \varphi \in \mathcal{E}(\Omega) \qquad q(T(\varphi)) \le C\|\varphi\|_{j,K}.$$

- *If $\varphi_m \to 0$ in $\mathcal{E}(\Omega)$, then $T(\varphi_m) \to 0$ in Y.*

(3) In particular, a linear map $T : \mathcal{E}(\Omega) \to \mathbb{R}$ is continuous if and only if there exist $j \in \mathbb{N}$ and $K \in \mathcal{K}(\Omega)$, and $C > 0$ such that

$$\forall \varphi \in \mathcal{E}(\Omega) \qquad |T(\varphi)| \le C\|\varphi\|_{j,K}.$$

(4) A linear map $T : Y \to \mathcal{E}(\Omega)$ is continuous if and only if

$$\forall j \in \mathbb{N}, \forall K \in \mathcal{K}(\Omega) \qquad \exists C > 0, k \in \mathbb{N}, q_1, \ldots, q_k \in \mathcal{Q} \quad \text{such that } \forall y \quad \|T(y)\|_{j,K} \le C \max_{1\le i\le k} q_i(y).$$

Theorem 39 (Convergence and Continuity for $\mathcal{E}_K(\Omega)$). *Let Ω be a nonempty open set in $\mathbb{R}^n$ and $K \in \mathcal{K}(\Omega)$. Let Y be a topological vector space whose topology is induced by a separating family of seminorms $\mathcal{Q}$.*

(1) A sequence $\{\varphi_m\}$ converges to φ in $\mathcal{E}_K(\Omega)$ if and only if $\|\varphi_m - \varphi\|_{j,K} \to 0$ for all $j \in \mathbb{N}$.

(2) Suppose $T : \mathcal{E}_K(\Omega) \to Y$ is a linear map. Then the following is equivalent:

- *T is continuous.*

- For every $q \in \mathcal{Q}$, there exists $j \in \mathbb{N}$ and $C > 0$ such that

$$\forall \varphi \in \mathcal{E}_K(\Omega) \qquad q(T(\varphi)) \leq C\|\varphi\|_{j,K}.$$

- If $\varphi_m \to 0$ in $\mathcal{E}_K(\Omega)$, then $T(\varphi_m) \to 0$ in Y.

Theorem 40 (Convergence and Continuity for $D(\Omega)$). *Let Ω be a nonempty open set in $\mathbb{R}^n$. Let Y be a topological vector space whose topology is induced by a separating family of seminorms $\mathcal{Q}$.*

(1) *A sequence $\{\varphi_m\}$ converges to φ in $D(\Omega)$ if and only if there is a $K \in \mathcal{K}(\Omega)$ such that $\text{supp}\varphi_m \subseteq K$ and $\varphi_m \to \varphi$ in $\mathcal{E}_K(\Omega)$.*

(2) *Suppose $T : D(\Omega) \to Y$ is a linear map. Then the following is equivalent*
   - *T is continuous.*
   - *For all $K \in \mathcal{K}(\Omega)$, $T : \mathcal{E}_K(\Omega) \to Y$ is continuous.*
   - *For every $q \in \mathcal{Q}$ and $K \in \mathcal{K}(\Omega)$, there exists $j \in \mathbb{N}$ and $C > 0$ such that*

$$\forall \varphi \in \mathcal{E}_K(\Omega) \qquad q(T(\varphi)) \leq C\|\varphi\|_{j,K}.$$

   - *If $\varphi_m \to 0$ in $D(\Omega)$, then $T(\varphi_m) \to 0$ in Y.*

(3) *In particular, a linear map $T : D(\Omega) \to \mathbb{R}$ is continuous if and only if for every $K \in \mathcal{K}(\Omega)$, there exists $j \in \mathbb{N}$ and $C > 0$ such that*

$$\forall \varphi \in \mathcal{E}_K(\Omega) \qquad |T(\varphi)| \leq C\|\varphi\|_{j,K}.$$

Remark 23. *Let Ω be a nonempty open set in $\mathbb{R}^n$. Here are two immediate consequences of the previous theorems and remark:*

(1) *The identity map*

$$i_{D,\mathcal{E}} : D(\Omega) \to \mathcal{E}(\Omega)$$

is continuous (that is, $D(\Omega) \hookrightarrow \mathcal{E}(\Omega)$).

(2) *If $T : \mathcal{E}(\Omega) \to \mathcal{E}(\Omega)$ is a continuous linear map such that $\text{supp}(T\varphi) \subseteq \text{supp}\varphi$ for all $\varphi \in \mathcal{E}(\Omega)$ (i.e., T is a* **local** *continuous linear map), then T restricts to a continuous linear map from $D(\Omega)$ to $D(\Omega)$. Indeed, the assumption $\text{supp}(T\varphi) \subseteq \text{supp}\varphi$ implies that $T(D(\Omega)) \subseteq D(\Omega)$. Moreover, $T : D(\Omega) \to D(\Omega)$ is continuous if and only if for $K \in \mathcal{K}(\Omega)$ $T : \mathcal{E}_K(\Omega) \to D(\Omega)$ is continuous. Since $T(\mathcal{E}_K(\Omega)) \subseteq \mathcal{E}_K(\Omega)$, this map is continuous if and only if $T : \mathcal{E}_K(\Omega) \to \mathcal{E}_K(\Omega)$ is continuous (see Remark 22). However, since the topology of $\mathcal{E}_K(\Omega)$ is the induced topology from $\mathcal{E}(\Omega)$, the continuity of the preceding map follows from the continuity of $T : \mathcal{E}(\Omega) \to \mathcal{E}(\Omega)$.*

Theorem 41. *Let Ω be a nonempty open set in $\mathbb{R}^n$. Let Y be a topological vector space whose topology is induced by a separating family of seminorms $\mathcal{Q}$. Suppose $T : [D(\Omega)]^{\times r} \to Y$ is a linear map. The following are equivalent: (product spaces are equipped with the product topology)*

(1) *$T : [D(\Omega)]^{\times r} \to Y$ is continuous.*

(2) *For all $K \in \mathcal{K}(\Omega)$, $T : [\mathcal{E}_K(\Omega)]^{\times r} \to Y$ is continuous.*

(3) *For all $q \in \mathcal{Q}$ and $K \in \mathcal{K}(\Omega)$, there exists $j_1, \ldots, j_l \in \mathbb{N}$ such that*

$$\forall (\varphi_1, \ldots, \varphi_r) \in [\mathcal{E}_K(\Omega)]^{\times r} \qquad |q \circ T(\varphi_1, \ldots, \varphi_r)| \leq C(\|\varphi_1\|_{j_1,K} + \ldots + \|\varphi_r\|_{j_r,K}).$$

Theorem 42. *Let Ω be a nonempty open set in $\mathbb{R}^n$.*

(1) *A set $B \subseteq D(\Omega)$ is bounded if and only if there exists $K \in \mathcal{K}(\Omega)$ such that B is a bounded subset of $\mathcal{E}_K(\Omega)$ which is in turn equivalent to the following statement:*

$$\forall j \in \mathbb{N}\, \exists r_j \geq 0 \quad \text{such that} \quad \forall \varphi \in B \quad \|\varphi\|_{j,K} \leq r_j.$$

(2) *If $\{\varphi_m\}$ is a Cauchy sequence in $D(\Omega)$, then it converges to a function $\varphi \in D(\Omega)$. We say $D(\Omega)$ is sequentially complete.*

Remark 24. *Topological spaces whose topology is determined by knowing the convergent sequences and their limits exhibit nice properties and are of particular interest. Let us recall a number of useful definitions related to this topic:*

- *Let X be a topological space and let $E \subseteq X$. The* ***sequential closure*** *of E, denoted $scl(E)$ is defined as follows:*

$$scl(E) = \{x \in X : \text{there is a sequence } \{x_n\} \text{ in } E \text{ such that } x_n \to x\}.$$

Clearly, $scl(E)$ is contained in the closure if E.

- *A topological space X is called a* ***Frechet-Urysohn*** *space if for every $E \subseteq X$ the sequential closure of E is equal to the closure of E.*
- *A subset E of a topological space X is said to be* ***sequentially closed*** *if $E = scl(E)$.*
- *A topological space X is said to be* ***sequential*** *if for every $E \subseteq X$, E is closed if and only if E is sequentially closed. If X is a sequential topological space and Y is any topological space, then a map $f : X \to Y$ is continuous if and only if*

$$\lim_{n\to\infty} f(x_n) = f(\lim_{n\to\infty} x_n)$$

for each convergent sequence $\{x_n\}$ in X.

The following implications hold for a topological space X:

$$X \text{ is metrizable} \to X \text{ is first-countable} \to X \text{ is Frechet-Urysohn} \to X \text{ is sequential}$$

As it was stated, $\mathcal{E}$ and $\mathcal{E}_K$ (For all $K \in \mathcal{K}(\Omega)$) are Frechet and subsequently they are metrizable. However, it can be shown that $D(\Omega)$ is not first-countable and subsequently it is not metrizable. In fact, although according to Theorem 40, the elements of the dual of $D(\Omega)$ can be determined by knowing the convergent sequences in $D(\Omega)$, it can be proved that $D(\Omega)$ is not sequential.

Definition 18. *Let Ω be a nonempty open set in $\mathbb{R}^n$. The topological dual of $D(\Omega)$, denoted $D'(\Omega)$ ($D'(\Omega) = [D(\Omega)]^*$), is called the* ***space of distributions*** *on Ω. Each element of $D'(\Omega)$ is called a* ***distribution*** *on Ω.*

Remark 25. *Every function $f \in L^1_{loc}(\Omega)$ defines a distribution $u_f \in D'(\Omega)$ as follows:*

$$\forall \varphi \in D(\Omega) \qquad u_f(\varphi) := \int_\Omega f\varphi dx. \tag{1}$$

In particular, every function $\varphi \in \mathcal{E}(\Omega)$ defines a distribution u_φ. It can be shown that the map $j : \mathcal{E}(\Omega) \to D'(\Omega)$ which sends φ to u_φ is an injective linear continuous map ([24], p. 11). Therefore, we can identify $\mathcal{E}(\Omega)$ with a subspace of $D'(\Omega)$.

Remark 26. *Let $\Omega \subseteq \mathbb{R}^n$ be a nonempty open set. Recall that $f : \Omega \to \mathbb{R}$ is locally integrable ($f \in L^1_{loc}(\Omega)$) if it satisfies any of the following equivalent conditions:*

(1) *$f \in L^1(K)$ for all $K \in \mathcal{K}(\Omega)$.*
(2) *For all $\varphi \in C^\infty_c(\Omega)$, $f\varphi \in L^1(\Omega)$.*
(3) *For every nonempty open set $V \subseteq \Omega$ such that $\bar{V}$ is compact and contained in Ω, $f \in L^1(V)$.*

(It can be shown that every locally integrable function is measurable ([36], p. 70)).
As a consequence, if we define $Func_{reg}(\Omega)$ to be the set

$$\{f : \Omega \to \mathbb{R} : u_f : D(\Omega) \to \mathbb{R} \text{ defined by Equation (1) is well-defined and continuous}\},$$

then $Func_{reg}(\Omega) = L^1_{loc}(\Omega)$.

Definition 19 (Calculus Rules for Distributions). *Let Ω be a nonempty open set in $\mathbb{R}^n$. Let $u \in D'(\Omega)$.*

- *For all $\varphi \in C^\infty(\Omega)$, φu is defined by*

$$\forall \psi \in C_c^\infty(\Omega) \qquad [\varphi u](\psi) := u(\varphi\psi)\,.$$

It can be shown that $\varphi u \in D'(\Omega)$.

- *For all multiindices α, $\partial^\alpha u$ is defined by*

$$\forall \psi \in C_c^\infty(\Omega) \qquad [\partial^\alpha u](\psi) = (-1)^{|\alpha|} u(\partial^\alpha \psi)\,.$$

It can be shown that $\partial^\alpha u \in D'(\Omega)$.

Furthermore, it is possible to make sense of "change of coordinates" for distributions. Let Ω and Ω' be two open sets in $\mathbb{R}^n$. Suppose $T : \Omega \to \Omega'$ is a C^∞ diffeomorphism. T can be used to move any function on Ω to a function on Ω' and vice versa.

$$T^* : \text{Func}(\Omega', \mathbb{R}) \to \text{Func}(\Omega, \mathbb{R}), \qquad T^*(f) = f \circ T\,,$$
$$T_* : \text{Func}(\Omega, \mathbb{R}) \to \text{Func}(\Omega', \mathbb{R}), \qquad T_*(f) = f \circ T^{-1}\,.$$

T^*f is called the **pullback** of the function f under the mapping T and T_*f is called the **pushforward** of the function f under the mapping T. Clearly, T^* and T_* are inverses of each other and $T_* = (T^{-1})^*$. One can show that T_* sends functions in $L^1_{loc}(\Omega)$ to $L^1_{loc}(\Omega')$ and furthermore T_* restricts to linear topological isomorphisms $T_* : \mathcal{E}(\Omega) \to \mathcal{E}(\Omega')$ and $T_* : D(\Omega) \to D(\Omega')$. Note that for all $f \in L^1_{loc}(\Omega)$ and $\varphi \in C_c^\infty(\Omega')$

$$\begin{aligned} < u_{T_*f}, \varphi >_{D'(\Omega') \times D(\Omega')} &= \int_{\Omega'} (T_*f)(y)\varphi(y)dy = \int_{\Omega'} (f \circ T^{-1})(y)\varphi(y)dy \\ &\overset{x=T^{-1}(y)}{=} \int_\Omega f(x)\varphi(T(x))|\det T'(x)|dx \\ &=< u_f, |\det T'(x)|\varphi(T(x)) >_{D'(\Omega) \times D(\Omega)}\,. \end{aligned}$$

The above observation motivates us to define the pushforward of any distribution $u \in D'(\Omega)$ as follows:

$$\forall \varphi \in D(\Omega') \qquad \langle T_*u, \varphi \rangle_{D'(\Omega') \times D(\Omega')} := \langle u, |\det T'(x)|\varphi(T(x)) \rangle_{D'(\Omega) \times D(\Omega)}\,.$$

It can be shown that $T_*u : D(\Omega') \to \mathbb{R}$ is continuous and so it is in fact an element of $D'(\Omega')$. Similarly, the pullback $T^* : D'(\Omega') \to D'(\Omega)$ is defined by

$$\forall \varphi \in D(\Omega) \qquad \langle T^*u, \varphi \rangle_{D'(\Omega) \times D(\Omega)} := \langle u, |\det(T^{-1})'(y)|\varphi(T^{-1}(y)) \rangle_{D'(\Omega') \times D(\Omega')}\,.$$

It can be shown that $T^*u : D(\Omega) \to \mathbb{R}$ is continuous and so it is in fact an element of $D'(\Omega)$.

Definition 20 (Extension by Zero of a Function). *Let Ω be an open subset of $\mathbb{R}^n$ and V be an open susbset of Ω. We define the linear map $ext^0_{V,\Omega} : Func(V, \mathbb{R}) \to Func(\Omega, \mathbb{R})$ as follows:*

$$ext^0_{V,\Omega}(f)(x) = \begin{cases} f(x) & \text{if } x \in V \\ 0 & \text{if } x \in \Omega \setminus V \end{cases}\,.$$

$ext^0_{V,\Omega}$ restricts to a continuous linear map $D(V) \to D(\Omega)$.

Definition 21 (Restriction of a Distribution). *Let Ω be an open subset of $\mathbb{R}^n$ and V be an open susbset of Ω. We define the restriction map $res_{\Omega,V} : D'(\Omega) \to D'(V)$ as follows:*

$$\langle res_{\Omega,V} u, \varphi\rangle_{D'(V)\times D(V)} := \langle u, ext^0_{V,\Omega}\varphi\rangle_{D'(\Omega)\times D(\Omega)}\,.$$

This is well-defined; indeed, $res_{\Omega,V} : D'(\Omega) \to D'(V)$ is a continuous linear map as it is the adjoint of the continuous map $ext^0_{V,\Omega} : D(V) \to D(\Omega)$. Given $u \in D'(\Omega)$, we sometimes write $u|_V$ instead of $res_{\Omega,V}u$.

Remark 27. *It is easy to see that the restriction of the map $res_{\Omega,V} : D'(\Omega) \to D'(V)$ to $\mathcal{E}(\Omega)$ agrees with the usual restriction of smooth functions.*

Definition 22 (Support of a Distribution). *Let Ω be a nonempty open set in $\mathbb{R}^n$. Let $u \in D'(\Omega)$.*

- *We say u is equal to zero on some open subset V of Ω if $u|_V = 0$.*
- *Let $\{V_i\}_{i\in I}$ be the collection of all open subsets of Ω such that u is equal to zero on V_i. Let $V = \bigcup_{i\in I} V_i$. The support of u is defined as follows:*

$$supp\, u := \Omega \setminus V\,.$$

Note that $suppu$ is closed in Ω but it is not necessarily closed in $\mathbb{R}^n$.

Theorem 43 (Properties of the Support [20,23,24]). *Let Ω and Ω' be nonempty open sets in $\mathbb{R}^n$.*

- *If $f \in L^1_{loc}(\Omega)$, then $suppf = supp\, u_f$.*
- *For all $u \in D'(\Omega)$, $u = 0$ on $\Omega \setminus supp\, u$.*
- *Let $u \in D'(\Omega)$. If $\varphi \in D(\Omega)$ vanishes on an open neighborhood of $supp\, u$, then $u(\varphi) = 0$.*
- *For every closed subset A of Ω and every $u \in D'(\Omega)$, we have $supp\, u \subseteq A$ if and only if $u(\varphi) = 0$ for every $\varphi \in D(\Omega)$ with $supp\, \varphi \subseteq \Omega \setminus A$.*
- *For every $u \in D'(\Omega)$ and $\psi \in \mathcal{E}(\Omega)$, $supp(\psi u) \subseteq supp(\psi) \cap supp(u)$.*
- *Let $u, v \in D'(\Omega)$. If there exists a nonempty open subset U of Ω such that $supp\, u \subseteq U$ and $supp\, v \subseteq U$ and*

$$\langle u|_U, \varphi\rangle_{D'(U)\times D(U)} = \langle v|_U, \varphi\rangle_{D'(U)\times D(U)} \qquad \forall\, \varphi \in C^\infty_c(U)\,,$$

  then $u = v$ as elements of $D'(\Omega)$.
- *Let $u, v \in D'(\Omega)$. Then $supp(u + v) \subseteq supp\, u \cup supp\, v$.*
- *Let $\{u_i\}$ be a sequence in $D'(\Omega)$, $u \in D(\Omega)$, and $K \in \mathcal{K}(\Omega)$ such that $u_i \to u$ in $D'(\Omega)$ and $supp\, u_i \subseteq K$ for all i. Then also $supp\, u \subseteq K$.*
- *For every $u \in D'(\Omega)$ and $\alpha \in \mathbb{N}^n_0$, $supp(\partial^\alpha u) \subseteq supp(u)$.*
- *If $T : \Omega \to \Omega'$ is a diffeomorphism, then $supp(T_* u) = T(supp\, u)$. In particular, if u has compact support, then so has $T_* u$.*

Considering the eighth item in the above theorem, an interesting question that one may ask is the following: Let $\{u_i\}$ be a sequence in $D(\Omega)$ such that $u_i \to u$ in $D'(\Omega)$, and suppose there exists $K \in \mathcal{K}(\Omega)$ such that $\operatorname{supp} u \subseteq K$. Does the fact that the limiting distribution has compact support imply that there exists a compact set $\tilde{K}$ such that $\operatorname{supp} u_i \subseteq \tilde{K}$ for all i? The answer is negative. For example, for each $i \in \mathbb{N}$ let $u_i \in D(\mathbb{R})$ be a nonnegative function such that $u_i = 0$ outside the interval $(i, i+1)$ and $\int_i^{i+1} u_i\, dx = \frac{1}{i}$. Clearly, $u_i \to 0$ in $L^1(\mathbb{R})$ and so $u_i \to 0$ in $D'(\mathbb{R})$. However, there is no compact set $\tilde{K}$ such that $\operatorname{supp} u_i \subseteq \tilde{K}$ for all i.

Theorem 44 ([24], pp. 10 and 20). *Let Ω be a nonempty open set in $\mathbb{R}^n$. Let $\mathcal{E}'(\Omega)$ denote the topological dual of $\mathcal{E}(\Omega)$ equipped with the strong topology. Then*

- *The map that sends $u \in \mathcal{E}'(\Omega)$ to $u|_{D(\Omega)}$ is an injective continuous linear map from $\mathcal{E}'(\Omega)$ into $D'(\Omega)$.*
- *The image of the above map consists precisely of those $u \in D'(\Omega)$ for which $supp\, u$ is compact.*

Due to the above theorem we may identify $\mathcal{E}'(\Omega)$ with distributions on Ω with compact support.

Definition 23 (Extension by Zero of Distributions With Compact Support). *Let Ω be a nonempty open set in $\mathbb{R}^n$ and V be a nonempty open subset of Ω. We define the linear map $ext^0_{V,\Omega} : \mathcal{E}'(V) \to \mathcal{E}'(\Omega)$ as the adjoint of the continuous linear map $res_{\Omega,V} : \mathcal{E}(\Omega) \to \mathcal{E}(V)$; that is,*

$$\langle ext^0_{V,\Omega} u, \varphi\rangle_{\mathcal{E}'(\Omega)\times\mathcal{E}(\Omega)} := \langle u, \varphi|_V\rangle_{\mathcal{E}'(V)\times\mathcal{E}(V)}\,.$$

Suppose Ω' and Ω are two nonempty open sets in $\mathbb{R}^n$ such that $\Omega' \subseteq \Omega$ and $K \in \mathcal{K}(\Omega')$. One can easily show that:

- For all $u \in \mathcal{E}_K(\Omega')$, $\text{res}_{\mathbb{R}^n,\Omega} \circ \text{ext}^0_{\Omega',\mathbb{R}^n} u = \text{ext}^0_{\Omega',\Omega} u$.
- For all $u \in \mathcal{E}_K(\Omega')$, $\text{ext}^0_{\Omega,\mathbb{R}^n} \circ \text{ext}^0_{\Omega',\Omega} u = \text{ext}^0_{\Omega',\mathbb{R}^n} u$.
- For all $u \in \mathcal{E}_K(\Omega)$, $\text{ext}^0_{\Omega',\Omega} \circ \text{res}_{\Omega,\Omega'} u = u$.

We summarize the important topological properties of the spaces of test functions and distributions in Table 1 below.

Table 1. Topological properties of the spaces of test functions.

	$D(\Omega)$	$\mathcal{E}(\Omega)$	$D'(\Omega)$ **Strong**	$\mathcal{E}'(\Omega)$ **Strong**	$D'(\Omega)$ **Weak**	$\mathcal{E}'(\Omega)$ **Weak**
Sequential	No	Yes	No	No	No	No
First-Countable	No	Yes	No	No	No	No
Metrizable	No	Yes	No	No	No	No
Second-Countable	No	Yes	No	No	No	No
Sequentially Complete	Yes	Yes	Yes	Yes	Yes	Yes
Complete	Yes	Yes	Yes	Yes	No	No

6.2. Distributions on Vector Bundles

6.2.1. Basic Definitions, Notation

Let M^n be a smooth manifold (M is not necessarily compact). Let $\pi : E \to M$ be a vector bundle of rank r.

(1) $\mathcal{E}(M,E)$ is defined as $C^\infty(M,E)$ equipped with the locally convex topology induced by the following family of seminorms: let $\{(U_\alpha, \varphi_\alpha, \rho_\alpha)\}_{\alpha\in I}$ be a total trivialization atlas. Then for every $\alpha \in I$, $1 \le l \le r$, and $f \in C^\infty(M,E)$, $\tilde{f}^l_\alpha := \rho^l_\alpha \circ f \circ \varphi_\alpha^{-1}$ is an element of $C^\infty(\varphi_\alpha(U_\alpha))$. For every 4-tuple (l,α,j,K) with $1 \le l \le r$, $\alpha \in I$, $j \in \mathbb{N}$, K a compact subset of U_α (i.e., $K \in \mathcal{K}(U_\alpha)$) we define

$$\|.\|_{l,\alpha,j,K} : C^\infty(M,E) \to \mathbb{R}, \quad f \mapsto \|\rho^l_\alpha \circ f \circ \varphi_\alpha^{-1}\|_{j,\varphi_\alpha(K)}\,.$$

It is easy to check that $\|.\|_{l,\alpha,j,K}$ is a seminorm on $C^\infty(M,E)$ and the locally convex topology induced by the above family of seminorms does not depend on the choice of the total trivialization atlas. Sometimes we may write $\|.\|_{l,\varphi_\alpha,j,K}$ instead of $\|.\|_{l,\alpha,j,K}$.

(2) For any compact subset $K \subseteq M$ we define

$$\mathcal{E}_K(M,E) := \{f \in \mathcal{E}(M,E) : \operatorname{supp} f \subseteq K\}$$

equipped with the subspace topology.

(3) $D(M,E) := C^\infty_c(M,E) = \cup_{K\in\mathcal{K}(M)}\mathcal{E}_K(M,E)$ (union over all compact subsets of M) equipped with the inductive limit topology with respect to the family $\{\mathcal{E}_K(M,E)\}_{K\in\mathcal{K}(M)}$. Clearly, if M is compact, then $D(M,E) = \mathcal{E}(M,E)$ (as topological vector spaces).

Remark 28.

- *If for each* $\alpha \in I$, $\{K_m^\alpha\}_{m\in\mathbb{N}}$ *is an exhaustion by compact sets of* U_α, *then the topology induced by the family of seminorms*

$$\{\|.\|_{l,\alpha,j,K_m^\alpha} : 1 \le l \le r, \alpha \in I, j \in \mathbb{N}, m \in \mathbb{N}\}$$

on $C^\infty(M, E)$ *is the same as the topology of* $\mathcal{E}(M, E)$. *This together with the fact that every manifold has a countable total trivialization atlas shows that the topology of* $\mathcal{E}(M, E)$ *is induced by a countable family of seminorms. So* $\mathcal{E}(M, E)$ *is metrizable.*
- *If* $\{K_j\}_{j\in\mathbb{N}}$ *is an exhaustion by compact sets of* M, *then the inductive limit topology on* $C_c^\infty(M, E)$ *with respect to the family* $\{\mathcal{E}_{K_j}(M, E)\}$ *is the same as the topology on* $D(M, E)$.

Definition 24. *The space of distributions on the vector bundle* E, *denoted* $D'(M, E)$, *is defined as the topological dual of* $D(M, E^\vee)$. *That is,*

$$D'(M, E) = [D(M, E^\vee)]^* .$$

As usual we equip the dual space with the strong topology. Recall that $E^\vee$ *denotes the bundle* $Hom(E, \mathcal{D}(M))$ *where* $\mathcal{D}(M)$ *is the density bundle of* M.

Remark 29. *The reason that space of distributions on the vector bundle* E *is defined as the dual of* $D(M, E^\vee)$ *rather than the dual of the seemingly natural choice* $D(M, E)$ *is well explained in [24,37]. Of course, there are other nonequivalent ways to make sense of distributions on vector bundles (see [37] for a detailed discussion). Furthermore, see Lemma 13 where it is proved that Riemannian density can be used to identify* $D'(M, E)$ *with* $[D(M, E)]^*$.

Remark 30. *Let* U *and* V *be nonempty open sets in* M *with* $V \subseteq U$.

- *As in the Euclidean case, the linear map* $ext^0_{V,U} : \Gamma(V, E^\vee_V) \to \Gamma(U, E^\vee_U)$ *defined by*

$$ext^0_{V,U} f(x) = \begin{cases} f(x) & x \in V \\ 0 & x \in U \setminus V \end{cases}$$

restricts to a continuous linear map from $D(V, E^\vee_V)$ *to* $D(U, E^\vee_U)$.
- *As in the Euclidean case, the restriction map* $res_{U,V} : D'(U, E_U) \to D'(V, E_V)$ *is defined as the adjoint of* $ext^0_{V,U}$:

$$\langle res_{U,V} u, \varphi \rangle_{D'(V,E_V)\times D(V,E^\vee_V)} = \langle u, ext^0_{V,U}\varphi \rangle_{D'(U,E_U)\times D(U,E^\vee_U)} .$$

- *Support of a distribution* $u \in D'(M, E)$ *is defined in the exact same way as for distributions in the Euclidean space. It can be shown that*
  (1) *([24], p. 105) If* $u \in D'(M, E)$ *and* $\varphi \in D(M, E^\vee)$ *vanishes on an open neighborhood of* $supp u$, *then* $u(\varphi) = 0$.
  (2) *([24], p. 104) For every closed subset* A *of* M *and every* $u \in D'(M, E)$, *we have* $supp u \subseteq A$ *if and only if* $u(\varphi) = 0$ *for every* $\varphi \in D(M, E^\vee)$ *with* $supp \varphi \subseteq M \setminus A$.

The strength of the theory of distributions in the Euclidean case is largely due to the fact that it is possible to identify a huge class of ordinary functions with distributions. A question that arises is that whether there is a natural way to identify regular sections of E (i.e., elements of $\Gamma(M, E)$) with distributions. The following theorem provides a partial answer to this question. Recall that compactly supported continuous sections of the density bundle can be integrated over M.

Theorem 45. *Every $f \in \mathcal{E}(M,E)$ defines the following continuous map:*

$$u_f : D(M, E^\vee) \to \mathbb{R}, \quad \psi \mapsto \int_M [\psi, f], \tag{2}$$

where the pairing $[\psi, f]$ defines a compactly supported continuous section of the density bundle:

$$\forall x \in M \quad [\psi, f](x) := [\psi(x)][f(x)] \quad (\psi(x) \in Hom(E_x, \mathcal{D}_x) \text{ evaluated at } f(x) \in E_x).$$

In general, we define $\Gamma_{reg}(M, E)$ as the set

$$\{f \in \Gamma(M, E) : u_f \text{ defined by Equation (2) is well-defined and continuous}\}.$$

Compare this with the definition of $\text{Func}_{reg}(\Omega)$ in Remark 26. Theorem 45 tells us that $\mathcal{E}(M,E)$ is contained in $\Gamma_{reg}(M,E)$. If $u \in D'(M,E)$ is such that $u = u_f$ for some $f \in \Gamma_{reg}(M,E)$, then we say that u is a **regular distribution**.

Now, let (U, φ, ρ) be a total trivialization triple for E and let $(U, \varphi, \rho_{\mathcal{D}})$ and $(U, \varphi, \rho^\vee)$ be the corresponding standard total trivialization triples for $\mathcal{D}(M)$ and $E^\vee$, respectively. The local representation of the pairing $[\psi, f]$ has a very simple expression in terms of the local representations of f and ψ:

$$f \in \Gamma_{reg}(M,E) \Longrightarrow (\vec{f}^1, \ldots, \vec{f}^r) := (f^1 \circ \varphi^{-1}, \ldots, f^r \circ \varphi^{-1}) := \rho \circ f \circ \varphi^{-1} \in [\text{Func}(\varphi(U), \mathbb{R})]^{\times r}$$
$(\vec{f}^1, \ldots, \vec{f}^r)$ is the local representation of f.
$$\psi \in D(M, E^\vee) \Longrightarrow (\tilde{\psi}^1, \ldots, \tilde{\psi}^r) := (\psi^1 \circ \varphi^{-1}, \ldots, \psi^r \circ \varphi^{-1}) := \rho^\vee \circ \psi \circ \varphi^{-1} \in [\text{Func}(\varphi(U), \mathbb{R})]^{\times r}$$
$(\tilde{\psi}^1, \ldots, \tilde{\psi}^r)$ is the local representation of ψ.

Our claim is that the local representation of $[\psi, f]$ (that is, $\rho_{\mathcal{D}} \circ [\psi, f] \circ \varphi^{-1}$) is equal to the Euclidean dot product of the local representations of f and ψ:

$$\rho_{\mathcal{D}} \circ [\psi, f] \circ \varphi^{-1} = \sum_i \vec{f}^i \tilde{\psi}^i.$$

The reason is as follows: Let $y \in \varphi(U)$ and $x = \varphi^{-1}(y)$

$$\begin{aligned}
[\rho_{\mathcal{D}} \circ [\psi, f] \circ \varphi^{-1}](y) &= \rho_{\mathcal{D}}([\psi(x)][f(x)]) = \rho_{\mathcal{D}}([\psi(x)][(\rho|_{E_x})^{-1}(\vec{f}^1(y), \ldots, \vec{f}^r(y))]) \\
&= [\rho_{\mathcal{D}} \circ \psi(x) \circ (\rho|_{E_x})^{-1}](\vec{f}^1(y), \ldots, \vec{f}^r(y)) \\
&= [\rho^\vee(\psi(x))][(\vec{f}^1(y), \ldots, \vec{f}^r(y))] \quad \text{the left bracket is applied to the right bracket} \\
&= \rho^\vee(\psi(x)) \cdot (\vec{f}^1(y), \ldots, \vec{f}^r(y)) \quad \text{dot product! } \rho^\vee(\psi(x)) \text{ viewed as an element of } \mathbb{R}^r \\
&= (\tilde{\psi}^1(y), \ldots, \tilde{\psi}^r(y)) \cdot (\vec{f}^1(y), \ldots, \vec{f}^r(y)).
\end{aligned}$$

6.2.2. Local Representation of Distributions

Let (U, φ, ρ) be a total trivialization triple for $\pi : E \to M$. We know that each $f \in \Gamma(M,E)$ can locally be represented by r components $\vec{f}^1, \ldots, \vec{f}^r$ defined by

$$\forall 1 \le l \le r \quad \vec{f}^l : \varphi(U) \to \mathbb{R}, \quad \vec{f}^l = \rho^l \circ f \circ \varphi^{-1}.$$

These components play a crucial role in our study of Sobolev spaces. Now the question is that whether we can similarly use the total trivialization triple (U, φ, ρ) to locally associate with each distribution $u \in D'(M,E)$, r components $\tilde{u}^1, \ldots, \tilde{u}^r$ belonging to $D'(\varphi(U))$. That is, we want to see whether we can define a nice map

$$D'(U, E_U) = [D(U, E_U^\vee)]^* \to \underbrace{D'(\varphi(U)) \times \ldots \times D'(\varphi(U))}_{r \text{ times}}.$$

(Note that according to Remark 30, if $u \in D'(M,E)$, then $u|_U \in D'(U,E_U)$.) Such a map, in particular, will be important when we want to make sense of Sobolev spaces with negative exponents of sections of vector bundles. Furthermore, it would be desirable to ensure that if u is a regular distribution then the components of u as a distribution agree with the components obtained when u is viewed as an element of $\Gamma(M,E)$.

We begin with the following map at the level of compactly supported smooth functions:

$$\tilde{T}_{E^\vee,U,\varphi} : D(U,E_U^\vee) \to [D(\varphi(U))]^{\times r}, \quad \xi \to \rho^\vee \circ \xi \circ \varphi^{-1} = ((\rho^\vee)^1 \circ \xi \circ \varphi^{-1}, \dots, (\rho^\vee)^r \circ \xi \circ \varphi^{-1}).$$

Note that $\tilde{T}_{E^\vee,U,\varphi}$ has the property that for all $\psi \in C^\infty(U)$ and $\xi \in D(U,E_U^\vee)$

$$\tilde{T}_{E^\vee,U,\varphi}(\psi\xi) = (\psi \circ \varphi^{-1})\tilde{T}_{E^\vee,U,\varphi}(\xi).$$

Theorem 46. *The map $\tilde{T}_{E^\vee,U,\varphi} : D(U,E_U^\vee) \to [D(\varphi(U))]^{\times r}$ is a linear topological isomorphism ($[D(\varphi(U))]^{\times r}$ is equipped with the product topology).*

Proof. Clearly, $\tilde{T}_{E^\vee,U,\varphi}$ is linear. Furthermore, the map $\tilde{T}_{E^\vee,U,\varphi}$ is bijective. Indeed, the inverse of $\tilde{T}_{E^\vee,U,\varphi}$ (which we denote by $T_{E^\vee,U,\varphi}$) is given by

$$T_{E^\vee,U,\varphi} : [D(\varphi(U))]^{\times r} \to D(U,E_U^\vee)$$
$$\forall x \in U \quad T_{E^\vee,U,\varphi}(\xi_1,\dots,\xi_r)(x) = (\rho^\vee|_{E_x^\vee})^{-1} \circ (\xi_1,\dots,\xi_r) \circ \varphi(x).$$

Now, we show that $\tilde{T}_{E^\vee,U,\varphi} : D(U,E_U^\vee) \to [D(\varphi(U))]^{\times r}$ is continuous. To this end, it is enough to prove that for each $1 \le l \le r$ the map

$$\pi^l \circ \tilde{T}_{E^\vee,U,\varphi} : D(U,E_U^\vee) \to D(\varphi(U)), \qquad \xi \mapsto (\rho^\vee)^l \circ \xi \circ \varphi^{-1}$$

is continuous. The topology on $D(U,E_U^\vee)$ is the inductive limit topology with respect to $\{\mathcal{E}_K(U,E_U^\vee)\}_{K\in\mathcal{K}(U)}$, so it is enough to show that for each $K \in \mathcal{K}(U)$, $\pi^l \circ \tilde{T}_{E^\vee,U,\varphi} : \mathcal{E}_K(U,E_U^\vee) \to D(\varphi(U))$ is continuous. Note that $\pi^l \circ \tilde{T}_{E^\vee,U,\varphi}[\mathcal{E}_K(U,E_U^\vee)] \subseteq \mathcal{E}_{\varphi(K)}(\varphi(U))$. Considering that $\mathcal{E}_{\varphi(K)}(\varphi(U)) \hookrightarrow D(\varphi(U))$, it is enough to show that

$$\pi^l \circ \tilde{T}_{E^\vee,U,\varphi} : \mathcal{E}_K(U,E_U^\vee) \to \mathcal{E}_{\varphi(K)}(\varphi(U))$$

is continuous. For all $\xi \in \mathcal{E}_K(U,E_U^\vee)$ and $j \in \mathbb{N}$ we have

$$\|\pi^l \circ \tilde{T}_{E^\vee,U,\varphi}(\xi)\|_{j,\varphi(K)} = \|(\rho^\vee)^l \circ \xi \circ \varphi^{-1}\|_{j,\varphi(K)} = \|\xi\|_{l,\varphi,j,K},$$

which implies the continuity (note that even an inequality in place of the last equality would have been enough to prove the continuity). It remains to prove the continuity of $T_{E^\vee,U,\varphi} : [D(\varphi(U))]^{\times r} \to D(U,E_U^\vee)$. By Theorem 41 it is enough to show that for all $K \in \mathcal{K}(\varphi(U))$, $T_{E^\vee,U,\varphi} : [\mathcal{E}_K(\varphi(U))]^{\times r} \to D(U,E_U^\vee)$ is continuous. It is clear that $T_{E^\vee,U,\varphi}([\mathcal{E}_K(\varphi(U))]^{\times r}) \subseteq \mathcal{E}_{\varphi^{-1}(K)}(U,E_U^\vee)$. Since $\mathcal{E}_{\varphi^{-1}(K)}(U,E_U^\vee) \hookrightarrow D(U,E_U^\vee)$, it is sufficient to show that $T_{E^\vee,U,\varphi} : [\mathcal{E}_K(\varphi(U))]^{\times r} \to \mathcal{E}_{\varphi^{-1}(K)}(U,E_U^\vee)$ is continuous. To this end, by Theorem 41, we just need to show that for all $j \in \mathbb{N}$ and $1 \le l \le r$ there exists $j_1,\dots,j_r$ such that

$$\|T_{E^\vee,U,\varphi}(\xi_1,\dots,\xi_r)\|_{l,\varphi,j,\varphi^{-1}(K)} \le C(\|\xi_1\|_{j_1,K} + \dots \|\xi_r\|_{j_r,K}).$$

However, this obviously holds because

$$\|T_{E^\vee,U,\varphi}(\xi_1,\dots,\xi_r)\|_{l,\varphi,j,\varphi^{-1}(K)} = \|\xi_l\|_{j,K}.$$

□

The adjoint of $T_{E^\vee,U,\varphi}$ is

$$T^*_{E^\vee,U,\varphi} : [D(U, E^\vee_U)]^* \to \left([D(\varphi(U))]^{\times r}\right)^*$$
$$\langle T^*_{E^\vee,U,\varphi} u, (\xi_1, \ldots, \xi_r)\rangle = \langle u, T_{E^\vee,U,\varphi}(\xi_1, \ldots, \xi_r)\rangle .$$

Note that, since $T_{E^\vee,U,\varphi}$ is a linear topological isomorphism, $T^*_{E^\vee,U,\varphi}$ is also a linear topological isomorphism (and in particular it is bijective). For every $u \in [D(U, E^\vee_U)]^*$, $T^*_{E^\vee,U,\varphi} u$ is in $\left([D(\varphi(U))]^{\times r}\right)^*$; we can combine this with the bijective map

$$L : \left([D(\varphi(U))]^{\times r}\right)^* \to [D'(\varphi(U))]^{\times r}, \quad L(v) = (v \circ i_1, \ldots, v \circ i_r)$$

(see Theorem 24) to send $u \in [D(U, E^\vee_U)]^*$ into an element of $[D'(\varphi(U))]^{\times r}$:

$$L(T^*_{E^\vee,U,\varphi} u) = ((T^*_{E^\vee,U,\varphi} u) \circ i_1, \ldots, (T^*_{E^\vee,U,\varphi} u) \circ i_r),$$

where for all $1 \le l \le r$, $(T^*_{E^\vee,U,\varphi} u) \circ i_l \in D'(\varphi(U))$ is given by

$$\begin{aligned}((T^*_{E^\vee,U,\varphi} u) \circ i_l)(\xi) &= (T^*_{E^\vee,U,\varphi} u)(i_l(\xi)) = (T^*_{E^\vee,U,\varphi} u)(0, \ldots, 0, \underbrace{\xi}_{l\text{th position}}, 0, \cdots, 0) \\ &= \langle u, T_{E^\vee,U,\varphi}(0, \ldots, 0, \underbrace{\xi}_{l\text{th position}}, 0, \ldots, 0)\rangle .\end{aligned}$$

If we define $g_{l,\xi,U,\varphi} \in D(U, E^\vee_U)$ by

$$\begin{aligned}g_{l,\xi,U,\varphi}(x) &= T_{E^\vee,U,\varphi}(0, \ldots, 0, \underbrace{\xi}_{l\text{th position}}, 0, \ldots, 0)(x) \\ &= \left(\rho^\vee|_{E^\vee_x}\right)^{-1} \circ (0, \ldots, 0, \underbrace{\xi}_{l\text{th position}}, 0, \cdots, 0) \circ \varphi(x),\end{aligned}$$

then we may write

$$\langle (T^*_{E^\vee,U,\varphi} u) \circ i_l, \xi\rangle_{D'(\varphi(U)) \times D(\varphi(U))} = \langle u, g_{l,\xi,U,\varphi}\rangle_{[D(U,E^\vee_U)]^* \times D(U,E^\vee_U)} .$$

Summary: We can associate with $u \in D'(U, E_U) = (D(U, E^\vee_U))^*$ the following r distributions in $D'(\varphi(U))$:

$$\forall 1 \le l \le r \quad \tilde{u}^l = T^*_{E^\vee,U,\varphi} u \circ i_l ,$$

that is,

$$\forall \xi \in D(\varphi(U)) \quad \langle \tilde{u}^l, \xi\rangle = \langle u, g_{l,\xi,U,\varphi}\rangle ,$$

where $g_{l,\xi,U,\varphi} \in D(U, E^\vee_U)$ is defined by

$$\left(\rho^\vee|_{E^\vee_x}\right)^{-1} \circ (0, \ldots, 0, \underbrace{\xi}_{l\text{th position}}, 0, \ldots, 0) \circ \varphi(x) .$$

In particular,

$$\rho^\vee \circ g_{l,\xi,U,\varphi} \circ \varphi^{-1} = (0, \ldots, 0, \underbrace{\xi}_{l\text{th position}}, 0, \ldots, 0),$$

and so $(\rho^\vee \circ g_{l,\xi,U,\varphi} \circ \varphi^{-1})^l = \xi$.

Let us give a name to the composition of L with $T^*_{E^\vee,U,\varphi}$ that we used above. We set $H_{E^\vee,U,\varphi} := L \circ T^*_{E^\vee,U,\varphi}$:

$$H_{E^\vee,U,\varphi} : [D(U, E^\vee_U)]^* \to (D'(\varphi(U)))^{\times r}, \qquad u \mapsto L(T^*_{E^\vee,U,\varphi} u) = (\tilde{u}^1, \dots, \tilde{u}^r).$$

Remark 31. *Here we make three observations about the mapping $H_{E^\vee,U,\varphi}$.*

(1) *For every $u \in [D(U, E^\vee_U)]^*$*

$$supp[H_{E^\vee,U,\varphi}\, u]^l = supp\tilde{u}^l \subseteq \varphi(supp\, u).$$

Indeed, let $A = \varphi(suppu)$. By Theorem 43, it is enough to show that if $\eta \in D(\varphi(U))$ is such that $supp\eta \subseteq \varphi(U) \setminus A$, then $\tilde{u}^l(\eta) = 0$. Note that

$$\langle \tilde{u}^l, \eta \rangle = \langle u, g_{l,\eta,U,\varphi} \rangle.$$

So, by Remark 30 we just need to show that $g_{l,\eta,U,\varphi} = 0$ on an open neighborhood of suppu. Let $K = supp\eta$. Clearly, $U \setminus \varphi^{-1}(K)$ is an open neighborhood of suppu. We will show that $g_{l,\eta,U,\varphi}$ vanishes on this open neighborhood. Note that

$$g_{l,\eta,U,\varphi}(x) = (\rho^\vee|_{E^\vee_x})^{-1}(0, \dots, 0, \underbrace{\eta \circ \varphi(x)}_{l\text{th position}}, 0, \dots, 0).$$

Since $\rho^\vee|_{E^\vee_x}$ is an isomorphism and $\eta = 0$ on $\varphi(U) \setminus K$, we conclude that $g_{l,\eta,U,\varphi} = 0$ on $\varphi^{-1}(\varphi(U) \setminus K) = U \setminus \varphi^{-1}(K)$.

(2) *Clearly, $H_{E^\vee,U,\varphi} : D'(U, E_U) \to [D'(\varphi(U))]^{\times r}$ preserves addition. Moreover, if $f \in C^\infty(U)$ and $u \in D'(U, E_U)$, then $H_{E^\vee,U,\varphi}(fu) = (f \circ \varphi^{-1}) H_{E^\vee,U,\varphi}(u)$. Recall that $H = L \circ T^*_{E^\vee,U,\varphi}$.*

$$\begin{aligned} \langle T^*_{E^\vee,U,\varphi}(fu), (\xi_1, \dots, \xi_r) \rangle &= \langle fu, T_{E^\vee,U,\varphi}(\xi_1, \dots, \xi_r) \rangle \\ &= \langle u, f T_{E^\vee,U,\varphi}(\xi_1, \dots, \xi_r) \rangle \\ &= \langle u, T_{E^\vee,U,\varphi}[(f \circ \varphi^{-1})(\xi_1, \dots, \xi_r)] \rangle \\ &= \langle T^*_{E^\vee,U,\varphi} u, (f \circ \varphi^{-1})(\xi_1, \dots, \xi_r) \rangle \\ &= \langle (f \circ \varphi^{-1}) T^*_{E^\vee,U,\varphi} u, (\xi_1, \dots, \xi_r) \rangle \end{aligned}$$

(The third equality follows directly from the definition of $T_{E^\vee,U,\varphi}$.) Therefore,

$$T^*_{E^\vee,U,\varphi}(fu) = (f \circ \varphi^{-1}) T^*_{E^\vee,U,\varphi} u.$$

The fact that $L((f \circ \varphi^{-1}) T^*_{E^\vee,U,\varphi} u) = (f \circ \varphi^{-1}) L(T^*_{E^\vee,U,\varphi} u)$ is an immediate consequence of the definition of L.

(3) *Since $T_{E^\vee,U,\varphi}$ and L are both linear topological isomorphisms, $H^{-1}_{E^\vee,U,\varphi} = (L \circ T^*_{E^\vee,U,\varphi})^{-1} : (D'(\varphi(U)))^{\times r} \to D^*(U, E^\vee_U)$ is also a linear topological isomorphism. It is useful for our later considerations to find an explicit formula for this map. Note that*

$$\begin{aligned} H^{-1}_{E^\vee,U,\varphi} &= (L \circ T^*_{E^\vee,U,\varphi})^{-1} = (T^*_{E^\vee,U,\varphi})^{-1} \circ L^{-1} = (T^{-1}_{E^\vee,U,\varphi})^* \circ L^{-1} \\ &= (\tilde{T}_{E^\vee,U,\varphi})^* \circ L^{-1} = (\tilde{T}_{E^\vee,U,\varphi})^* \circ \tilde{L}. \end{aligned}$$

Recall that

$$\tilde{L} : [D^*(\varphi(U))]^{\times r} \to [(D(\varphi(U)))^{\times r}]^*, \quad (v^1, \dots, v^r) \mapsto v^1 \circ \pi_1 + \dots + v^r \circ \pi_r,$$
$$\tilde{T}^*_{E^\vee,U,\varphi} : [(D(\varphi(U)))^{\times r}]^* \to D^*(U, E^\vee_U).$$

Therefore, for all $\xi \in D(U, E_U^\vee)$

$$\begin{aligned} H^{-1}_{E^\vee,U,\varphi}(v^1,\ldots,v^r)(\xi) &= \langle \tilde{T}^*_{E^\vee,U,\varphi}(v^1\circ\pi_1+\ldots+v^r\circ\pi_r),\xi\rangle \\ &= \langle (v^1\circ\pi_1+\ldots+v^r\circ\pi_r),\tilde{T}\xi\rangle \\ &= \langle (v^1\circ\pi_1+\ldots+v^r\circ\pi_r),((\rho^\vee)^1\circ\xi\circ\varphi^{-1},\ldots,(\rho^\vee)^r\circ\xi\circ\varphi^{-1})\rangle \\ &= \sum_i v^i[(\rho^\vee)^i\circ\xi\circ\varphi^{-1}]. \end{aligned}$$

Remark 32. *Suppose* $u \in D'(M,E)$ *is a regular distribution, that is,* $u = u_f$ *where* $f \in \Gamma_{reg}(M,E)$*. We want to see whether the local components of such a distribution agree with its components as an element of* $\Gamma(M,E)$*. With respect to the total trivialization triple* (U,φ,ρ) *we have*

(1) $f \mapsto (\tilde{f}^1,\ldots,\tilde{f}^r),\quad \tilde{f}^l = \rho^l\circ f\circ\varphi^{-1},$

(2) $u_f \mapsto (\tilde{u_f}^1,\ldots,\tilde{u_f}^l).$

The question is whether $u_{\tilde{f}^l} = \tilde{u_f}^l$*? Here we will show that the answer is positive. Indeed, for all* $\xi \in D(\varphi(U))$ *we have*

$$\begin{aligned} \langle \tilde{u_f}^l,\xi\rangle &= \langle u_f, g_{l,\xi,U,\varphi}\rangle = \int_M [g_{l,\xi,U,\varphi},f] = \int_{\varphi(U)} \sum_i (\tilde{g}_{l,\xi,U,\varphi})^i \tilde{f}^i dV = \int_{\varphi(U)} (\tilde{g}_{l,\xi,U,\varphi})^l \tilde{f}^l dV \\ &= \int_{\varphi(U)} \tilde{f}^l \xi dV = \langle u_{\tilde{f}^l},\xi\rangle. \end{aligned}$$

Note that the above calculation in fact shows that the restriction of $H_{E^\vee,U,\varphi}$ *to* $D(U,E_U)$ *is* $\tilde{T}_{E,U,\varphi}$.

7. Spaces of Sobolev and Locally Sobolev Functions in $\mathbb{R}^n$

In this section, we present a brief overview of the basic definitions and properties related to Sobolev spaces on Euclidean spaces.

7.1. Basic Definitions

Definition 25. *Let* $s \geq 0$ *and* $p \in [1,\infty]$*. The Sobolev–Slobodeckij space* $W^{s,p}(\mathbb{R}^n)$ *is defined as follows:*

- *If* $s = k \in \mathbb{N}_0$, $p \in [1,\infty]$,

$$W^{k,p}(\mathbb{R}^n) = \{u \in L^p(\mathbb{R}^n) : \|u\|_{W^{k,p}(\mathbb{R}^n)} := \sum_{|\nu|\leq k} \|\partial^\nu u\|_p < \infty\}.$$

- *If* $s = \theta \in (0,1)$, $p \in [1,\infty)$,

$$W^{\theta,p}(\mathbb{R}^n) = \{u \in L^p(\mathbb{R}^n) : |u|_{W^{\theta,p}(\mathbb{R}^n)} := \Big(\int\int_{\mathbb{R}^n\times\mathbb{R}^n} \frac{|u(x)-u(y)|^p}{|x-y|^{n+\theta p}} dxdy\Big)^{\frac{1}{p}} < \infty\}.$$

- *If* $s = \theta \in (0,1)$, $p = \infty$,

$$W^{\theta,\infty}(\mathbb{R}^n) = \{u \in L^\infty(\mathbb{R}^n) : |u|_{W^{\theta,\infty}(\mathbb{R}^n)} := \operatorname*{ess\,sup}_{x,y\in\mathbb{R}^n, x\neq y} \frac{|u(x)-u(y)|}{|x-y|^\theta} < \infty\}.$$

- *If* $s = k+\theta$, $k \in \mathbb{N}_0$, $\theta \in (0,1)$, $p \in [1,\infty]$,

$$W^{s,p}(\mathbb{R}^n) = \{u \in W^{k,p}(\mathbb{R}^n) : \|u\|_{W^{s,p}(\mathbb{R}^n)} := \|u\|_{W^{k,p}(\mathbb{R}^n)} + \sum_{|\nu|=k} |\partial^\nu u|_{W^{\theta,p}(\mathbb{R}^n)} < \infty\}.$$

Remark 33. *Clearly, for all* $s \geq 0$, $W^{s,p}(\mathbb{R}^n) \subseteq L^p(\mathbb{R}^n)$. *Recall that* $L^p(\mathbb{R}^n) \subseteq L^1_{loc}(\mathbb{R}^n) \subseteq D'(\mathbb{R}^n)$. *So, we may consider elements of* $W^{s,p}(\mathbb{R}^n)$ *as distributions in* $D'(\mathbb{R}^n)$. *Indeed, for* $s \geq 0$, $p \in (1,\infty)$, *and* $u \in D'(\mathbb{R}^n)$ *we define*

$$\begin{cases} \|u\|_{W^{s,p}(\mathbb{R}^n)} := \|f\|_{W^{s,p}(\mathbb{R}^n)} & \text{if } u = u_f \text{ for some } f \in L^p(\mathbb{R}^n) \\ \|u\|_{W^{s,p}(\mathbb{R}^n)} := \infty & \text{otherwise} \end{cases}.$$

As a consequence, we may write

$$W^{s,p}(\mathbb{R}^n) = \{u \in D'(\mathbb{R}^n) : \|u\|_{W^{s,p}(\mathbb{R}^n)} < \infty\}.$$

Remark 34. *Let us make some observations that will be helpful in the proof of a number of important theorems. Let A be a nonempty measurable set in* $\mathbb{R}^n$.

(1) *We may write:*

$$\int\int_{\mathbb{R}^n\times\mathbb{R}^n} \frac{|\partial^\nu u(x) - \partial^\nu u(y)|^p}{|x-y|^{n+\theta p}} dxdy$$
$$= \int\int_{A\times A} \dots dxdy + \int_A \int_{\mathbb{R}^n\setminus A} \dots dxdy + \int_{\mathbb{R}^n\setminus A}\int_A \dots dxdy + \int_{\mathbb{R}^n\setminus A}\int_{\mathbb{R}^n\setminus A} \dots dxdy.$$

In particular, if $\operatorname{supp} u \subseteq A$, *then the last integral vanishes and the sum of the two middle integrals will be equal to* $2\int_A \int_{\mathbb{R}^n\setminus A} \frac{|\partial^\nu u(x)|^p}{|x-y|^{n+\theta p}} dydx$. *Therefore, in this case*

$$\int\int_{\mathbb{R}^n\times\mathbb{R}^n} \frac{|\partial^\nu u(x) - \partial^\nu u(y)|^p}{|x-y|^{n+\theta p}} dxdy =$$
$$\int\int_{A\times A} \frac{|\partial^\nu u(x) - \partial^\nu u(y)|^p}{|x-y|^{n+\theta p}} dxdy + 2\int_A\int_{\mathbb{R}^n\setminus A} \frac{|\partial^\nu u(x)|^p}{|x-y|^{n+\theta p}} dydx.$$

(2) *If A is open,* $K \subseteq A$ *is compact and* $\alpha > n$, *then there exists a number C such that for all* $x \in K$ *we have*

$$\int_{\mathbb{R}^n\setminus A} \frac{1}{|x-y|^\alpha} dy \leq C.$$

(C may depend on A, K, n, and α *but is independent of x.) The reason is as follows: Let* $R = \frac{1}{2}\operatorname{dist}(K, A^c) > 0$. *Clearly, for all* $x \in K$, *the ball* $B_R(x)$ *is inside A. Therefore, for all* $x \in K$, $\mathbb{R}^n \setminus A \subseteq \mathbb{R}^n \setminus B_R(x)$ *which implies that for all* $x \in K$

$$\int_{\mathbb{R}^n\setminus A} \frac{1}{|x-y|^\alpha} dy \leq \int_{\mathbb{R}^n\setminus B_R(x)} \frac{1}{|x-y|^\alpha} dy \overset{z=y-x}{=} \int_{\mathbb{R}^n\setminus B_R(0)} \frac{1}{|z|^\alpha} dz = \sigma(S^{n-1}) \int_R^\infty \frac{1}{r^\alpha} r^{n-1} dr,$$

which converges because $\alpha > n$. *We can let* $C = \sigma(S^{n-1}) \int_R^\infty \frac{1}{r^\alpha} r^{n-1} dr$.

(3) *If A is bounded and* $\alpha < n$, *then there exists a number C such that for all* $x \in A$

$$\int_A \frac{1}{|x-y|^\alpha} dy \leq C.$$

(C depends on A, n, and α *but is independent of x.) The reason is as follows: Since A is bounded there exists* $R > 0$ *such that for all* $x, y \in A$ *we have* $|x-y| < R$. *So, for all* $x \in A$

$$\int_A \frac{1}{|x-y|^\alpha} dy \leq \sigma(S^{n-1}) \int_0^R \frac{1}{r^\alpha} r^{n-1} dr,$$

which converges because $\alpha < n$.

Theorem 47. *Let* $s \geq 0$ *and* $p \in (1,\infty)$. $C_c^\infty(\mathbb{R}^n)$ *is dense in* $W^{s,p}(\mathbb{R}^n)$. *In fact, the identity map* $i_{D,W} : D(\mathbb{R}^n) \to W^{s,p}(\mathbb{R}^n)$ *is a linear continuous map with dense image.*

Proof. The fact that $C_c^\infty(\mathbb{R}^n)$ is dense in $W^{s,p}(\mathbb{R}^n)$ follows from Theorem 7.38 and Lemma 7.44 in [38] combined with Remark 39. Linearity of $i_{D,W}$ is obvious. It remains to prove that this map is continuous. By Theorem 40 it is enough to show that

$$\forall K \in \mathcal{K}(\mathbb{R}^n), \forall \varphi \in \mathcal{E}_K(\mathbb{R}^n) \quad \exists j \in \mathbb{N} \quad \text{s.t.} \quad \|\varphi\|_{W^{s,p}(\mathbb{R}^n)} \preceq \|\varphi\|_{j,K}\,.$$

Let $s = m + \theta$ where $m \in \mathbb{N}_0$ and $\theta \in [0,1)$. If $\theta \neq 0$, by definition $\|\varphi\|_{W^{s,p}(\mathbb{R}^n)} = \|\varphi\|_{W^{m,p}(\mathbb{R}^n)} + \sum_{|\nu|=m} |\partial^\nu \varphi|_{W^{\theta,p}(\mathbb{R}^n)}$. It is enough to show that each summand can be bounded by a constant multiple of $\|\varphi\|_{j,K}$ for some j.

- **Step 1:** If $\theta = 0$,

$$\|\varphi\|_{W^{m,p}(\mathbb{R}^n)} = \sum_{|\nu|\leq m} \|\partial^\nu \varphi\|_{L^p(\mathbb{R}^n)} = \sum_{|\nu|\leq m} \|\partial^\nu \varphi\|_{L^p(K)}$$
$$= \sum_{|\nu|\leq m} (\|\varphi\|_{m,K} |K|^{\frac{1}{p}}) \preceq \|\varphi\|_{m,K}\,,$$

where the implicit constant depends on m, p, and K but is independent of φ.

- **Step 2:** Let A be an open ball that contains K (in particular, A is bounded). As it was pointed out in Remark 34 we may write

$$\int\int_{\mathbb{R}^n\times\mathbb{R}^n} \frac{|\partial^\nu \varphi(x) - \partial^\nu \varphi(y)|^p}{|x-y|^{n+\theta p}} dxdy =$$
$$\int\int_{A\times A} \frac{|\partial^\nu \varphi(x) - \partial^\nu \varphi(y)|^p}{|x-y|^{n+\theta p}} dxdy + 2\int_A \int_{\mathbb{R}^n\setminus A} \frac{|\partial^\nu \varphi(x)|^p}{|x-y|^{n+\theta p}} dydx\,.$$

First note that $\mathbb{R}^n$ is a convex open set; so by Theorem 6 every function $f \in \mathcal{E}_K(\mathbb{R}^n)$ is Lipschitz; indeed, for all $x, y \in \mathbb{R}^n$ we have $|f(x) - f(y)| \preceq \|f\|_{1,K}\|x-y\|$. Hence

$$\int\int_{A\times A} \frac{|\partial^\nu \varphi(x) - \partial^\nu \varphi(y)|^p}{|x-y|^{n+\theta p}} dxdy \leq \int_A \|\partial^\nu \varphi\|_{1,K}^p \int_A \frac{|x-y|^p}{|x-y|^{n+\theta p}} dydx$$
$$= \int_A \|\partial^\nu \varphi\|_{1,K}^p \int_A \frac{1}{|x-y|^{n+(\theta-1)p}} dydx\,.$$

By part 3 of Remark 34 $\int_A \frac{1}{|x-y|^{n+(\theta-1)p}} dy$ is bounded by a constant independent of x; also, clearly, $\|\partial^\nu \varphi\|_{1,K} \leq \|\varphi\|_{m+1,K}$. Considering that $|A|$ is finite we get

$$\int\int_{A\times A} \frac{|\partial^\nu \varphi(x) - \partial^\nu \varphi(y)|^p}{|x-y|^{n+\theta p}} dxdy \preceq \|\varphi\|_{m+1,K}^p\,.$$

Finally, for the remaining integral we have

$$\int_A \int_{\mathbb{R}^n\setminus A} \frac{|\partial^\nu \varphi(x)|^p}{|x-y|^{n+\theta p}} dydx = \int_K \int_{\mathbb{R}^n\setminus A} \frac{|\partial^\nu \varphi(x)|^p}{|x-y|^{n+\theta p}} dydx\,,$$

because the inner integral is zero for $x \notin K$. Now, we can write

$$\int_K \int_{\mathbb{R}^n\setminus A} \frac{|\partial^\nu \varphi(x)|^p}{|x-y|^{n+\theta p}} dydx \preceq \int_K \|\varphi\|_{m,K}^p \int_{\mathbb{R}^n\setminus A} \frac{1}{|x-y|^{n+\theta p}} dydx\,.$$

By part 2 of Remark 34 for all $x \in K$, the inner integral is bounded by a constant. Since $|K|$ is finite we conclude that

$$\int_A \int_{\mathbb{R}^n\setminus A} \frac{|\partial^\nu \varphi(x)|^p}{|x-y|^{n+\theta p}} dydx \preceq \|\varphi\|_{m,K}^p\,.$$

Hence

$$\|u\|_{W^{s,p}(\mathbb{R}^n)} \preceq \|\varphi\|_{m+1,K}\,.$$

□

Definition 26. *Let $s > 0$ and $p \in (1,\infty)$. We define*

$$W^{-s,p'}(\mathbb{R}^n) = (W^{s,p}(\mathbb{R}^n))^* \quad (\frac{1}{p} + \frac{1}{p'} = 1).$$

Remark 35. *Note that since the identity map from $D(\mathbb{R}^n)$ to $W^{s,p}(\mathbb{R}^n)$ is continuous with dense image, the dual space $W^{-s,p'}(\mathbb{R}^n)$ can be viewed as a subspace of $D'(\mathbb{R}^n)$. Indeed, by Theorem 25 the adjoint of the identity map, $i^*_{D,W} : W^{-s,p'}(\mathbb{R}^n) \to D'(\mathbb{R}^n)$ is an injective linear continuous map and we can use this map to identify $W^{-s,p'}(\mathbb{R}^n)$ with a subspace of $D'(\mathbb{R}^n)$. It is a direct consequence of the definition of adjoint that for all $u \in W^{-s,p'}(\mathbb{R}^n)$, $i^*_{D,W}u = u|_{D(\mathbb{R}^n)}$. So, by identifying $u : W^{s,p}(\mathbb{R}^n) \to \mathbb{R}$ with $u|_{D(\mathbb{R}^n)} : D(\mathbb{R}^n) \to \mathbb{R}$, we can view $W^{-s,p'}(\mathbb{R}^n)$ as a subspace of $D'(\mathbb{R}^n)$.*

Remark 36.

- *It is a direct consequence of the contents of pp. 88 and 178 of [8] that for $m \in \mathbb{Z}$ and $1 < p < \infty$*
$$W^{m,p}(\mathbb{R}^n) = H^m_p(\mathbb{R}^n) = F^m_{p,2}(\mathbb{R}^n)\,.$$
- *It is a direct consequence of the contents of pp. 38, 51, 90 and 178 of [8] that for $s \notin \mathbb{Z}$ and $1 < p < \infty$*
$$W^{s,p}(\mathbb{R}^n) = B^s_{p,p}(\mathbb{R}^n)\,.$$

Theorem 48. *For all $s \in \mathbb{R}$ and $1 < p < \infty$, $W^{s,p}(\mathbb{R}^n)$ is reflexive.*

Proof. See the proof of Theorem 64. Additionally, see [39], Section 2.6, p. 198. □

Note that by definition for all $s > 0$ we have $[W^{s,p}(\mathbb{R}^n)]^* = W^{-s,p'}(\mathbb{R}^n)$. Now, since $W^{s,p}(\mathbb{R}^n)$ is reflexive, $[W^{-s,p'}(\mathbb{R}^n)]^*$ is isometrically isomorphic to $W^{s,p}(\mathbb{R}^n)$ and so they can be identified with one another. Thus, for all $s \in \mathbb{R}$ and $1 < p < \infty$ we may write

$$[W^{s,p}(\mathbb{R}^n)]^* = W^{-s,p'}(\mathbb{R}^n)\,.$$

Let $s \geq 0$ and $p \in (1,\infty)$. Every function $\varphi \in C^\infty_c(\mathbb{R}^n)$ defines a linear functional $L_\varphi : W^{s,p}(\mathbb{R}^n) \to \mathbb{R}$ defined by

$$L_\varphi(u) = \int_{\mathbb{R}^n} u\varphi dx\,.$$

L_φ is continuous because by Holder's inequality

$$|L_\varphi(u)| = |\int_{\mathbb{R}^n} u\varphi dx| \leq \|u\|_{L^p(\mathbb{R}^n)}\|\varphi\|_{L^{p'}(\mathbb{R}^n)} \leq \|\varphi\|_{L^{p'}(\mathbb{R}^n)}\|u\|_{W^{s,p}(\mathbb{R}^n)}\,.$$

Furthermore, the map $L : C^\infty_c(\mathbb{R}^n) \to W^{-s,p'}(\mathbb{R}^n)$ which maps φ to L_φ is injective because

$$L_\varphi = L_\psi \to \forall\, u \in W^{s,p}(\mathbb{R}^n) \quad \int_{\mathbb{R}^n} u(\varphi - \psi)dx = 0 \to \int_{\mathbb{R}^n} |\varphi - \psi|^2 dx = 0 \to \varphi = \psi\,.$$

Thus, we may identify φ with L_φ and consider $C^\infty_c(\mathbb{R}^n)$ as a subspace of $W^{-s,p'}(\mathbb{R}^n)$.

Theorem 49. *For all $s > 0$ and $p \in (1,\infty)$, $C^\infty_c(\mathbb{R}^n)$ is dense in $W^{-s,p'}(\mathbb{R}^n)$.*

Proof. The proof given in p. 65 of [1] for the density of $L^{p'}$ in the integer order Sobolev space $W^{-m,p'}$, which is based on reflexivity of Sobolev spaces, works equally well for establishing the density of $C_c^\infty(\mathbb{R}^n)$ in $W^{-s,p'}(\mathbb{R}^n)$. □

Remark 37. *As a consequence of the above theorems, for all $s \in \mathbb{R}$ and $p \in (1,\infty)$, $W^{s,p}(\mathbb{R}^n)$ can be considered as a subspace of $D'(\mathbb{R}^n)$. See Theorem 25 and the discussion thereafter for further insights. Additionally, see Remark 45.*

Next we list several definitions pertinent to Sobolev spaces on open subsets of $\mathbb{R}^n$.

Definition 27. *Let Ω be a nonempty open set in $\mathbb{R}^n$. Let $s \in \mathbb{R}$ and $p \in (1,\infty)$.*

(1) • *If $s = k \in \mathbb{N}_0$,*

$$W^{k,p}(\Omega) = \{u \in L^p(\Omega) : \|u\|_{W^{k,p}(\Omega)} := \sum_{|\nu|\le k} \|\partial^\nu u\|_{L^p(\Omega)} < \infty\}.$$

• *If $s = \theta \in (0,1)$,*

$$W^{\theta,p}(\Omega) = \{u \in L^p(\Omega) : |u|_{W^{\theta,p}(\Omega)} := \Big(\int\int_{\Omega\times\Omega} \frac{|u(x)-u(y)|^p}{|x-y|^{n+\theta p}} dxdy\Big)^{\frac{1}{p}} < \infty\}.$$

• *If $s = k+\theta$, $k \in \mathbb{N}_0$, $\theta \in (0,1)$,*

$$W^{s,p}(\Omega) = \{u \in W^{k,p}(\Omega) : \|u\|_{W^{s,p}(\Omega)} := \|u\|_{W^{k,p}(\Omega)} + \sum_{|\nu|=k} |\partial^\nu u|_{W^{\theta,p}(\Omega)} < \infty\}.$$

• *If $s < 0$,*

$$W^{s,p}(\Omega) = (W_0^{-s,p'}(\Omega))^* \quad (\frac{1}{p} + \frac{1}{p'} = 1),$$

where for all $e \ge 0$ and $1 < q < \infty$, $W_0^{e,q}(\Omega)$ is defined as the closure of $C_c^\infty(\Omega)$ in $W^{e,q}(\Omega)$.

(2) *$W^{s,p}(\bar\Omega)$ is defined as the restriction of $W^{s,p}(\mathbb{R}^n)$ to Ω. That is, $W^{s,p}(\bar\Omega)$ is the collection of all $u \in D'(\Omega)$ such that there is a $v \in W^{s,p}(\mathbb{R}^n)$ with $v|_\Omega = u$. Here $v|_\Omega$ should be interpreted as the restriction of a distribution in $D'(\mathbb{R}^n)$ to a distribution in $D'(\Omega)$. $W^{s,p}(\bar\Omega)$ is equipped with the following norm:*

$$\|u\|_{W^{s,p}(\bar\Omega)} = \inf_{v\in W^{s,p}(\mathbb{R}^n), v|_\Omega = u} \|v\|_{W^{s,p}(\mathbb{R}^n)}.$$

(3)

$$\tilde W^{s,p}(\bar\Omega) = \{u \in W^{s,p}(\mathbb{R}^n) : supp\, u \subseteq \bar\Omega\}.$$

$\tilde W^{s,p}(\bar\Omega)$ is equipped with the norm $\|u\|_{\tilde W^{s,p}(\bar\Omega)} = \|u\|_{W^{s,p}(\mathbb{R}^n)}$.

(4)

$$\tilde W^{s,p}(\Omega) = \{u = v|_\Omega, v \in \tilde W^{s,p}(\bar\Omega)\}. \tag{3}$$

Again $v|_\Omega$ should be interpreted as the restriction of an element in $D'(\mathbb{R}^n)$ to $D'(\Omega)$. So $\tilde W^{s,p}(\Omega)$ is a subspace of $D'(\Omega)$. This space is equipped with the norm $\|u\|_{\tilde W^{s,p}} = \inf \|v\|_{W^{s,p}(\mathbb{R}^n)}$ where the infimum is taken over all v that satisfy the equality in Equation (3). Note that two elements v_1 and v_2 of $\tilde W^{s,p}(\bar\Omega)$ restrict to the same element in $D'(\Omega)$ if and only if $supp(v_1 - v_2) \subseteq \partial\Omega$. Therefore,

$$\tilde W^{s,p}(\Omega) = \frac{\tilde W^{s,p}(\bar\Omega)}{\{v \in W^{s,p}(\mathbb{R}^n) : supp\, v \subseteq \partial\Omega\}}.$$

(5) *For $s \geq 0$ we define*

$$W^{s,p}_{00}(\Omega) = \{u \in W^{s,p}(\Omega) : ext^0_{\Omega,\mathbb{R}^n} u \in W^{s,p}(\mathbb{R}^n)\}.$$

We equip this space with the norm

$$\|u\|_{W^{s,p}_{00}(\Omega)} := \|ext^0_{\Omega,\mathbb{R}^n} u\|_{W^{s,p}(\mathbb{R}^n)}.$$

Note that previously we defined the operator $ext^0_{\Omega,\mathbb{R}^n}$ only for distributions with compact support and functions; this is why the values of s are restricted to be nonnegative in this definition.

(6) *For all $K \in \mathcal{K}(\Omega)$ we define*

$$W^{s,p}_K(\Omega) = \{u \in W^{s,p}(\Omega) : supp\, u \subseteq K\},$$

with $\|u\|_{W^{s,p}_K(\Omega)} := \|u\|_{W^{s,p}(\Omega)}$.

(7)
$$W^{s,p}_{comp}(\Omega) = \bigcup_{K \in \mathcal{K}(\Omega)} W^{s,p}_K(\Omega).$$

This space is normally equipped with the inductive limit topology with respect to the family $\{W^{s,p}_K(\Omega)\}_{K \in \mathcal{K}(\Omega)}$. ***However, in these notes we always consider $W^{s,p}_{comp}(\Omega)$ as a normed space equipped with the norm induced from $W^{s,p}(\Omega)$.***

Remark 38. *Each of these definitions has its advantages and disadvantages. For example, the way we defined the spaces $W^{s,p}(\Omega)$ is well suited for using duality arguments while proving the usual embedding theorems for these spaces on an arbitrary open set Ω is not trivial; on the other hand, duality arguments do not work as well for spaces $W^{s,p}(\bar{\Omega})$ but the embedding results for these spaces on an arbitrary open set Ω automatically follow from the corresponding results on $\mathbb{R}^n$. Various authors adopt different definitions for Sobolev spaces on domains based on the applications in which they are interested. Unfortunately, the notation used in the literature for the various spaces introduced above are not uniform. First note that it is a direct consequence of Remark 36 and the definitions of $B^s_{p,q}(\Omega)$, $H^s_p(\Omega)$ and $F^s_{p,q}(\Omega)$ in [39] p. 310 and [40] that*

$$W^{s,p}(\bar{\Omega}) = \begin{cases} F^s_{p,2}(\Omega) = H^s_p(\Omega) & \text{if } s \in \mathbb{Z} \\ B^s_{p,p}(\Omega) & \text{if } s \notin \mathbb{Z} \end{cases}.$$

With this in mind, we have Table 2 which displays the connection between the notation used in this work with the notation in a number of well-known references.

Table 2. Connection to notation employed in previous literature

This Manuscript	Triebel [39]	Triebel [40]	Grisvard [5]	Bhattacharyya [4]
$W^{s,p}(\Omega)$			$W^s_p(\Omega)$	$W^{s,p}(\Omega)$
$W^{s,p}(\bar{\Omega})$	$W^s_p(\Omega)$	$W^s_p(\Omega)$	$W^s_p(\bar{\Omega})$	$W^{s,p}(\bar{\Omega})$
$\tilde{W}^{s,p}(\bar{\Omega})$	$\tilde{W}^s_p(\Omega)$	$\tilde{W}^s_p(\bar{\Omega})$		
$\tilde{W}^{s,p}(\Omega)$		$\tilde{W}^s_p(\Omega)$		
$W^{s,p}_{00}(\Omega)$			$\tilde{W}^s_p(\Omega)$	$W^{s,p}_{00}(\Omega)$

Remark 39.

- *Note that*

$$\begin{aligned}\|u\|_{W^{k,p}(\Omega)} + \sum_{|\nu|=k} |\partial^\nu u|_{W^{\theta,p}(\Omega)} &\leq \|u\|_{W^{k,p}(\Omega)} + \sum_{|\nu|=k} \|\partial^\nu u\|_{W^{\theta,p}(\Omega)} \\ &= \|u\|_{W^{k,p}(\Omega)} + \sum_{|\nu|=k} \Big(\|\partial^\nu u\|_{L^p(\Omega)} + |\partial^\nu u|_{W^{\theta,p}(\Omega)} \Big) \\ &\preceq \|u\|_{W^{k,p}(\Omega)} + \sum_{|\nu|=k} |\partial^\nu u|_{W^{\theta,p}(\Omega)} \qquad (\text{since } \sum_{|\nu|=k} \|\partial^\nu u\|_{L^p(\Omega)} \leq \|u\|_{W^{k,p}(\Omega)}).\end{aligned}$$

  Therefore, the following is an equivalent norm on $W^{s,p}(\Omega)$

$$\|u\|_{W^{s,p}(\Omega)} := \|u\|_{W^{k,p}(\Omega)} + \sum_{|\alpha|=k} \|\partial^\alpha u\|_{W^{\theta,p}(\Omega)}.$$

- *For* $p \in (1,\infty)$ *and* $a, b > 0$ *we have* $(a^p + b^p)^{\frac{1}{p}} \simeq a + b$; *indeed,*

$$a^p + b^p \leq (a+b)^p \leq (2\max\{a,b\})^p \leq 2^p(a^p + b^p).$$

  More generally, if $a_1, \ldots, a_m$ *are nonnegative numbers, then* $(a_1^p + \ldots + a_m^p)^{\frac{1}{p}} \simeq a_1 + \ldots + a_m$. *Therefore, for any nonempty open set* Ω *in* $\mathbb{R}^n$, $s > 0$, *the following expressions are both equivalent to the original norm on* $W^{s,p}(\Omega)$

$$\|u\|_{W^{s,p}(\Omega)} := \big[\|u\|^p_{W^{k,p}(\Omega)} + \sum_{|\nu|=k} |\partial^\nu u|^p_{W^{\theta,p}(\Omega)}\big]^{\frac{1}{p}},$$

$$\|u\|_{W^{s,p}(\Omega)} := \big[\|u\|^p_{W^{k,p}(\Omega)} + \sum_{|\nu|=k} \|\partial^\nu u\|^p_{W^{\theta,p}(\Omega)}\big]^{\frac{1}{p}},$$

  where $s = k + \theta$, $k \in \mathbb{N}_0$, $\theta \in (0,1)$.

7.2. Properties of Sobolev Spaces on the Whole Space $\mathbb{R}^n$

Theorem 50 (Embedding Theorem I, [39], Section 2.8.1). *Suppose* $1 < p \leq q < \infty$ *and* $-\infty < t \leq s < \infty$ *satisfy* $s - \frac{n}{p} \geq t - \frac{n}{q}$. *Then* $W^{s,p}(\mathbb{R}^n) \hookrightarrow W^{t,q}(\mathbb{R}^n)$. *In particular,* $W^{s,p}(\mathbb{R}^n) \hookrightarrow W^{t,p}(\mathbb{R}^n)$.

Theorem 51 (Multiplication by smooth functions, [12], p. 203). *Let* $s \in \mathbb{R}$, $1 < p < \infty$, *and* $\varphi \in BC^\infty(\mathbb{R}^n)$. *Then the linear map*

$$m_\varphi : W^{s,p}(\mathbb{R}^n) \to W^{s,p}(\mathbb{R}^n), \qquad u \mapsto \varphi u$$

is well-defined and bounded.

A detailed study of the following multiplication theorems can be found in [18].

Theorem 52. *Let* s_i, s *and* $1 \leq p, p_i < \infty$ $(i = 1,2)$ *be real numbers satisfying*

(i) $s_i \geq s \geq 0$,
(ii) $s \in \mathbb{N}_0$,
(iii) $s_i - s \geq n(\dfrac{1}{p_i} - \dfrac{1}{p})$,
(iv) $s_1 + s_2 - s > n(\dfrac{1}{p_1} + \dfrac{1}{p_2} - \dfrac{1}{p}) \geq 0$,

where the strictness of the inequalities in items (iii) and (iv) can be interchanged.

If $u \in W^{s_1,p_1}(\mathbb{R}^n)$ *and* $v \in W^{s_2,p_2}(\mathbb{R}^n)$, *then* $uv \in W^{s,p}(\mathbb{R}^n)$ *and moreover the pointwise multiplication of functions is a continuous bilinear map*

$$W^{s_1,p_1}(\mathbb{R}^n) \times W^{s_2,p_2}(\mathbb{R}^n) \to W^{s,p}(\mathbb{R}^n).$$

Theorem 53 (Multiplication theorem for Sobolev spaces on the whole space, nonnegative exponents). *Assume* s_i, s *and* $1 \le p_i \le p < \infty$ $(i = 1, 2)$ *are real numbers satisfying*

(i) $s_i \ge s$,
(ii) $s \ge 0$,
(iii) $s_i - s \ge n(\frac{1}{p_i} - \frac{1}{p})$,
(iv) $s_1 + s_2 - s > n(\frac{1}{p_1} + \frac{1}{p_2} - \frac{1}{p})$.

If $u \in W^{s_1,p_1}(\mathbb{R}^n)$ *and* $v \in W^{s_2,p_2}(\mathbb{R}^n)$, *then* $uv \in W^{s,p}(\mathbb{R}^n)$ *and moreover the pointwise multiplication of functions is a continuous bilinear map*

$$W^{s_1,p_1}(\mathbb{R}^n) \times W^{s_2,p_2}(\mathbb{R}^n) \to W^{s,p}(\mathbb{R}^n).$$

Theorem 54 (Multiplication theorem for Sobolev spaces on the whole space, negative exponents I). *Assume* s_i, s *and* $1 < p_i \le p < \infty$ $(i = 1, 2)$ *are real numbers satisfying*

(i) $s_i \ge s$,
(ii) $\min\{s_1, s_2\} < 0$,
(iii) $s_i - s \ge n(\frac{1}{p_i} - \frac{1}{p})$,
(iv) $s_1 + s_2 - s > n(\frac{1}{p_1} + \frac{1}{p_2} - \frac{1}{p})$,
(v) $s_1 + s_2 \ge n(\frac{1}{p_1} + \frac{1}{p_2} - 1) \ge 0$.

Then the pointwise multiplication of smooth functions extends uniquely to a continuous bilinear map

$$W^{s_1,p_1}(\mathbb{R}^n) \times W^{s_2,p_2}(\mathbb{R}^n) \to W^{s,p}(\mathbb{R}^n).$$

Theorem 55 (Multiplication theorem for Sobolev spaces on the whole space, negative exponents II). *Assume* s_i, s *and* $1 < p, p_i < \infty$ $(i = 1, 2)$ *are real numbers satisfying*

(i) $s_i \ge s$,
(ii) $\min\{s_1, s_2\} \ge 0$ *and* $s < 0$,
(iii) $s_i - s \ge n(\frac{1}{p_i} - \frac{1}{p})$,
(iv) $s_1 + s_2 - s > n(\frac{1}{p_1} + \frac{1}{p_2} - \frac{1}{p}) \ge 0$,
(v) $s_1 + s_2 > n(\frac{1}{p_1} + \frac{1}{p_2} - 1)$ *(the inequality is strict).*

Then the pointwise multiplication of smooth functions extends uniquely to a continuous bilinear map

$$W^{s_1,p_1}(\mathbb{R}^n) \times W^{s_2,p_2}(\mathbb{R}^n) \to W^{s,p}(\mathbb{R}^n).$$

Remark 40. *Let us discuss further how we should interpret multiplication in the case where negative exponents are involved. Suppose for instance* $s_1 < 0$ *(s_2 may be positive or negative). A moment's thought shows that the relation*

$$W^{s_1,p_1}(\mathbb{R}^n) \times W^{s_2,p_2}(\mathbb{R}^n) \hookrightarrow W^{s,p}(\mathbb{R}^n)$$

in the above theorems can be interpreted as follows: for all $u \in W^{s_1,p_1}(\mathbb{R}^n)$ *and* $v \in W^{s_2,p_2}(\mathbb{R}^n)$, *if* $\{\varphi_i\}$ *in* $C^\infty(\mathbb{R}^n) \cap W^{s_1,p_1}(\mathbb{R}^n)$ *is any sequence such that* $\varphi_i \to u$ *in* $W^{s_1,p_1}(\mathbb{R}^n)$, *then*

(1) *For all* i, $\varphi_i v \in W^{s,p}(\mathbb{R}^n)$ *(multiplication of a smooth function and a distribution);*
(2) $\varphi_i v$ *converges to some element* g *in* $W^{s,p}(\mathbb{R}^n)$ *as* $i \to \infty$*;*
(3) $\|g\|_{W^{s,p}(\mathbb{R}^n)} \preceq \|u\|_{W^{s_1,p_1}(\mathbb{R}^n)} \|v\|_{W^{s_2,p_2}(\mathbb{R}^n)}$ *where the implicit constant does not depend on* u *and* v*;*
(4) $g \in W^{s,p}(\mathbb{R}^n)$ *is independent of the sequence* $\{\varphi_i\}$ *and can be regarded as the product of* u *and* v*.*

In particular, $\varphi_i v \to uv$ *in* $D'(\mathbb{R}^n)$ *and for all* $\psi \in C_c^\infty(\mathbb{R}^n)$

$$\langle uv, \psi \rangle_{D'(\mathbb{R}^n)\times D(\mathbb{R}^n)} = \lim_{i\to\infty} \langle \varphi_i v, \psi \rangle_{D'(\mathbb{R}^n)\times D(\mathbb{R}^n)} = \langle v, \varphi_i \psi \rangle_{D'(\mathbb{R}^n)\times D(\mathbb{R}^n)} .$$

7.3. Properties of Sobolev Spaces on Smooth Bounded Domains

In this section, we assume that Ω is an open bounded set in $\mathbb{R}^n$ with smooth boundary unless a weaker assumption is stated. First we list some facts that can be useful in understanding the relationship between various definitions of Sobolev spaces on domains.

- ([4], p. 584) [Theorem 8.10.13 and its proof] Suppose $s > 0$ and $1 < p < \infty$. Then $W^{s,p}(\Omega) = W^{s,p}(\bar{\Omega})$ in the sense of equivalent normed spaces.
- ([40], pp. 481 and 494) For $s > \frac{1}{p} - 1$, $\tilde{W}^{s,p}(\bar{\Omega}) = \tilde{W}^{s,p}(\Omega)$. That is, for $s > \frac{1}{p} - 1$

$$\{v \in W^{s,p}(\mathbb{R}^n) : \operatorname{supp} v \subseteq \partial\Omega\} = \{0\} .$$

- Let $s > 0$ and $1 < p < \infty$. Then for $s \neq \frac{1}{p}, 1 + \frac{1}{p}, 2 + \frac{1}{p}, \ldots$ (that is, when the fractional part of s is not equal to $\frac{1}{p}$) we have
  (1) ([4], p. 592) [Theorem 8.10.20] $W_{00}^{s,p}(\Omega) = W_0^{s,p}(\Omega)$ in the sense of equivalent normed spaces.
  (2)
  $$\operatorname{ext}^0_{\Omega,\mathbb{R}^n} : (C_c^\infty(\Omega), \|.\|_{s,p}) \to W^{s,p}(\mathbb{R}^n)$$
  is a well-defined bounded linear operator.
  (3)
  $$\operatorname{res}_{\mathbb{R}^n,\Omega} : W^{-s,p'}(\mathbb{R}^n) \to W^{-s,p'}(\Omega) \qquad u \mapsto u|_\Omega$$
  is a well-defined bounded linear operator.

Note that the connection between items (2) and (3) above can be seen as follows: Let $u \in W^{-s,p'}(\mathbb{R}^n)$. $\operatorname{res}_{\mathbb{R}^n,\Omega} u \in W^{-s,p'}(\Omega)$ if and only if $u|_\Omega : (D(\Omega), \|.\|_{s,p}) \to \mathbb{R}$ is continuous, that is, if

$$\sup_{0\neq\varphi\in D(\Omega)} \frac{|\langle u|_\Omega, \varphi \rangle_{D'(\Omega)\times D(\Omega)}|}{\|\varphi\|_{W^{s,p}(\Omega)}} < \infty .$$

We have

$$\begin{aligned} |\langle u|_\Omega, \varphi \rangle_{D'(\Omega)\times D(\Omega)}| &= |\langle u, \operatorname{ext}^0_{\Omega,\mathbb{R}^n}\varphi \rangle_{D'(\mathbb{R}^n)\times D(\mathbb{R}^n)}| = |\langle u, \operatorname{ext}^0_{\Omega,\mathbb{R}^n}\varphi \rangle_{W^{-s,p'}(\mathbb{R}^n)\times W_0^{s,p}(\mathbb{R}^n)}| \\ &\preceq \|u\|_{W^{-s,p'}(\mathbb{R}^n)} \|\operatorname{ext}^0_{\Omega,\mathbb{R}^n}\varphi\|_{W_0^{s,p}(\mathbb{R}^n)} . \end{aligned}$$

So, the desired inequality holds if one can show that for all $\varphi \in D(\Omega)$, $\|\operatorname{ext}^0_{\Omega,\mathbb{R}^n}\varphi\|_{W_0^{s,p}(\mathbb{R}^n)} \preceq \|\varphi\|_{W^{s,p}(\Omega)}$.

Next we recall some facts about extension operators and embedding properties of Sobolev spaces. The existence of extension operator can be helpful in transferring known results for Sobolev spaces defined on $\mathbb{R}^n$ to Sobolev spaces defined on bounded domains.

Theorem 56 (Extension Property I [4], p. 584). *Let* $\Omega \subset \mathbb{R}^n$ *be a bounded open set with Lipschitz continuous boundary. Then for all* $s > 0$ *and for* $1 \leq p < \infty$*, there exists a continuous linear extension operator* $P : W^{s,p}(\Omega) \hookrightarrow W^{s,p}(\mathbb{R}^n)$ *such that* $(Pu)|_\Omega = u$ *and* $\|Pu\|_{W^{s,p}(\mathbb{R}^n)} \leq C\|u\|_{W^{s,p}(\Omega)}$ *for some constant* C *that may depend on* s*,* p*, and* Ω *but is independent of* u*.*

The next theorem states that the claim of Theorem 56 holds for all values of s (positive and negative) if we replace $W^{s,p}(\Omega)$ with $W^{s,p}(\bar{\Omega})$.

Theorem 57 (Extension Property II [40], p. 487, [8], p. 201). *Let $\Omega \subset \mathbb{R}^n$ be a bounded open set with Lipschitz continuous boundary, $p \in (1,\infty)$ and $s \in \mathbb{R}$. Let $R : W^{s,p}(\mathbb{R}^n) \to W^{s,p}(\bar{\Omega})$ be the restriction operator ($R(u) = u|_\Omega$). Then there exists a continuous linear operator $S : W^{s,p}(\bar{\Omega}) \to W^{s,p}(\mathbb{R}^n)$ such that $R \circ S = Id$.*

Corollary 3. *One can easily show that the results of Sobolev multiplication theorems in the previous section (Theorems 52–55) hold also for Sobolev spaces on any Lipschitz domain as long as all the Sobolev spaces involved satisfy $W^{e,q}(\Omega) = W^{e,q}(\bar{\Omega})$ (and so, in particular, existence of an extension operator is guaranteed). Indeed, if $P_1 : W^{s_1,p_1}(\Omega) \to W^{s_1,p_1}(\mathbb{R}^n)$ and $P_2 : W^{s_2,p_2}(\Omega) \to W^{s_2,p_2}(\mathbb{R}^n)$ are extension operators, then $(P_1u)(P_2v)|_\Omega = uv$ and therefore,*

$$\begin{aligned} \|uv\|_{W^{s,p}(\Omega)} = \|uv\|_{W^{s,p}(\bar{\Omega})} \leq \|(P_1u)(P_2v)\|_{W^{s,p}(\mathbb{R}^n)} &\preceq \|P_1u\|_{W^{s_1,p_1}(\mathbb{R}^n)}\|P_2v\|_{W^{s_2,p_2}(\mathbb{R}^n)} \\ &\preceq \|u\|_{W^{s_1,p_1}(\Omega)}\|v\|_{W^{s_2,p_2}(\Omega)} . \end{aligned}$$

Remark 41. *In the above Corollary, we presumed that $(P_1u)(P_2v)|_\Omega = uv$. Clearly, if s_1 and s_2 are both nonnegative, the equality holds. However, what if at least one of the exponents, say s_1, is negative? In order to prove this equality, we may proceed as follows: let $\{\varphi_i\}$ be a sequence in $C^\infty(\mathbb{R}^n) \cap W^{s_1,p_1}(\mathbb{R}^n)$ such that $\varphi_i \to P_1u$ in $W^{s_1,p_1}(\mathbb{R}^n)$. By assumption $W^{s_1,p_1}(\Omega) = W^{s_1,p_1}(\bar{\Omega})$, therefore the restriction operator is continuous and $\{\varphi_i|_\Omega\}$ is a sequence in $C^\infty(\Omega) \cap W^{s_1,p_1}(\Omega)$ that converges to u in $W^{s_1,p_1}(\Omega)$. For all $\psi \in C_c^\infty(\Omega)$ we have*

$$\begin{aligned} \langle [(P_1u)(P_2v)]|_\Omega, \psi\rangle_{D'(\Omega)\times D(\Omega)} &= \langle (P_1u)(P_2v), ext^0_{\Omega,\mathbb{R}^n}\psi\rangle_{D'(\mathbb{R}^n)\times D(\mathbb{R}^n)} \\ &\overset{\text{Remark 40}}{=} \lim_{i\to\infty}\langle \varphi_i(P_2v), ext^0_{\Omega,\mathbb{R}^n}\psi\rangle_{D'(\mathbb{R}^n)\times D(\mathbb{R}^n)} \\ &= \lim_{i\to\infty}\langle (P_2v), \varphi_i ext^0_{\Omega,\mathbb{R}^n}\psi\rangle_{D'(\mathbb{R}^n)\times D(\mathbb{R}^n)} \\ &= \lim_{i\to\infty}\langle (P_2v), ext^0_{\Omega,\mathbb{R}^n}(\varphi_i|_\Omega\psi)\rangle_{D'(\mathbb{R}^n)\times D(\mathbb{R}^n)} \\ &= \lim_{i\to\infty}\langle (P_2v)|_\Omega, \varphi_i|_\Omega\psi\rangle_{D'(\Omega)\times D(\Omega)} \\ &= \lim_{i\to\infty}\langle \varphi_i|_\Omega v, \psi\rangle_{D'(\Omega)\times D(\Omega)} \\ &= \langle uv, \psi\rangle_{D'(\Omega)\times D(\Omega)} . \end{aligned}$$

Theorem 58 (Embedding Theorem II [5]). *Let Ω be a nonempty bounded open subset of $\mathbb{R}^n$ with Lipschitz continuous boundary or $\Omega = \mathbb{R}^n$. If $sp > n$, then $W^{s,p}(\Omega) \hookrightarrow L^\infty(\Omega) \cap C^0(\Omega)$ and $W^{s,p}(\Omega)$ is a Banach algebra.*

Theorem 59 (Embedding Theorem III [18]). *Let Ω be a nonempty bounded open subset of $\mathbb{R}^n$ with Lipschitz continuous boundary. Suppose $1 \leq p, q < \infty$ (p does NOT need to be less than or equal to q) and $0 \leq t \leq s$ satisfy $s - \frac{n}{p} \geq t - \frac{n}{q}$. If $s \notin \mathbb{N}_0$, additionally assume that $s \neq t$. Then $W^{s,p}(\Omega) \hookrightarrow W^{t,q}(\Omega)$. In particular, $W^{s,p}(\Omega) \hookrightarrow W^{t,p}(\Omega)$.*

Theorem 60. *Let Ω be a nonempty bounded open subset of $\mathbb{R}^n$ with Lipschitz continuous boundary. Then $u : \Omega \to \mathbb{R}$ is Lipschitz continuous if and only if $u \in W^{1,\infty}(\Omega)$. In particular, every function in $BC^1(\Omega)$ is Lipschitz continuous.*

Proof. The above theorem is proved in Chapter 5 of [2] for open sets with C^1 boundary. The exact same proof works for open sets with Lipschitz continuous boundary. □

The following theorem (and its corollary) will play an important role in our study of Sobolev spaces on manifolds.

Theorem 61 (Multiplication by smooth functions). *Let Ω be a nonempty bounded open set in $\mathbb{R}^n$ with Lipschitz continuous boundary.*

(1) *Let $k \in \mathbb{N}_0$ and $1 < p < \infty$. If $\varphi \in BC^k(\Omega)$, then the linear map $W^{k,p}(\Omega) \to W^{k,p}(\Omega)$ defined by $u \mapsto \varphi u$ is well-defined and bounded.*
(2) *Let $s \in (0,\infty)$ and $1 < p < \infty$. If $\varphi \in BC^{\lfloor s \rfloor,1}(\Omega)$ (all partial derivatives of φ up to and including order $\lfloor s \rfloor$ exist and are bounded and Lipschitz continuous), then the linear map $W^{s,p}(\Omega) \to W^{s,p}(\Omega)$ defined by $u \mapsto \varphi u$ is well-defined and bounded.*
(3) *Let $s \in (-\infty,0)$ and $1 < p < \infty$. If $\varphi \in BC^{\infty,1}(\Omega)$, then the linear map $W^{s,p}(\Omega) \to W^{s,p}(\Omega)$ defined by $u \mapsto \varphi u$ is well-defined and bounded.*

Note: *According to Theorem 60, when Ω is an open bounded set with Lipschitz continuous boundary, every function in $BC^1(\Omega)$ is Lipschitz continuous. As a consequence, $BC^{\infty,1}(\Omega) = BC^{\infty}(\Omega)$. Of course, as it was discussed after Theorem 6, for a general bounded open set Ω whose boundary is not Lipschitz, functions in $BC^{\infty}(\Omega)$ are not necessarily Lipschitz.*

Proof.

- **Step 1:** $s = k \in \mathbb{N}_0$. The claim is proved in ([29], p. 995).
- **Step 2:** $0 < s < 1$. The proof in p. 194 of [41], with obvious modifications, shows the validity of the claim for the case where $s \in (0,1)$.
- **Step 3:** $1 < s \notin \mathbb{N}$. In this case we can proceed as follows: Let $k = \lfloor s \rfloor$, $\theta = s - k$.

$$\begin{aligned}
\|\varphi u\|_{s,p} &\overset{\text{Remark 39}}{=} \|\varphi u\|_{k,p} + \sum_{|\nu|=k} \|\partial^{\nu}(\varphi u)\|_{\theta,p} \\
&\preceq \|\varphi u\|_{k,p} + \sum_{|\nu|=k}\sum_{\beta\le\nu} \|\partial^{\nu-\beta}\varphi\partial^{\beta}u\|_{\theta,p} \\
&\preceq \|u\|_{k,p} + \sum_{|\nu|=k}\sum_{\beta\le\nu} \|\partial^{\beta}u\|_{\theta,p} \quad \text{(by steps 1 and 2; the implicit constant may depend on } \varphi) \\
&= \|u\|_{s,p} + \sum_{|\nu|=k}\sum_{\beta<\nu} \|\partial^{\beta}u\|_{\theta,p} \\
&\preceq \|u\|_{s,p} + \sum_{|\nu|=k}\sum_{\beta<\nu} \|u\|_{\theta+|\beta|,p} \quad (\partial^{\beta} : W^{\theta+|\beta|,p}(\Omega) \to W^{\theta,p}(\Omega) \text{is continuous}) \\
&\preceq \|u\|_{s,p} + \sum_{|\nu|=k}\sum_{\beta<\nu} \|u\|_{s,p} \quad (\theta+|\beta| < s \Rightarrow W^{s,p}(\Omega) \hookrightarrow W^{\theta+|\beta|,p}(\Omega)) \\
&\preceq \|u\|_{s,p}.
\end{aligned}$$

  Note that the embedding $W^{s,p}(\Omega) \hookrightarrow W^{\theta+|\beta|,p}(\Omega)$ is valid due to the extra assumption that Ω is bounded with Lipschitz continuous boundary (see Theorem 68 and Remark 42).

- **Step 4:** $s < 0$. For this case we use a duality argument. Note that since $\varphi \in C^{\infty}(\Omega)$, φu is defined as an element of $D'(\Omega)$. Furthermore, recall that $W^{s,p}(\Omega)$ is isometrically isomorphic to $[C_c^{\infty}(\Omega), \|.\|_{-s,p'}]^*$ (see the discussion after Remark 10). So, in order to prove the claim, it is enough to show that multiplication by φ is a well-defined continuous operator from $W^{s,p}(\Omega)$ to $A = [C_c^{\infty}(\Omega), \|.\|_{-s,p'}]^*$. We have

$$\begin{aligned}
\|\varphi u\|_A = \sup_{v\in C_c^{\infty}\setminus\{0\}} \frac{|\langle \varphi u, v\rangle_{D'(\Omega)\times D(\Omega)}|}{\|v\|_{-s,p'}} &= \sup_{v\in C_c^{\infty}\setminus\{0\}} \frac{|\langle u, \varphi v\rangle_{D'(\Omega)\times D(\Omega)}|}{\|v\|_{-s,p'}} \\
&\overset{\text{Remark 45}}{=} \sup_{v\in C_c^{\infty}\setminus\{0\}} \frac{|\langle u, \varphi v\rangle_{W^{s,p}(\Omega)\times W_0^{-s,p'}(\Omega)}|}{\|v\|_{-s,p'}} \\
&\le \sup_{v\in C_c^{\infty}\setminus\{0\}} \frac{\|u\|_{s,p}\|\varphi v\|_{-s,p'}}{\|v\|_{-s,p'}} \preceq \sup_{v\in C_c^{\infty}\setminus\{0\}} \frac{\|u\|_{s,p}\|v\|_{-s,p'}}{\|v\|_{-s,p'}} = \|u\|_{s,p}.
\end{aligned}$$

□

Corollary 4. *Let Ω be a nonempty bounded open set in $\mathbb{R}^n$ with Lipschitz continuous boundary. Let $K \in \mathcal{K}(\Omega)$. Suppose $s \in \mathbb{R}$ and $p \in (1,\infty)$. If $\varphi \in C^\infty(\Omega)$, then the linear map $W^{s,p}_K(\Omega) \to W^{s,p}_K(\Omega)$ defined by $u \mapsto \varphi u$ is well-defined and bounded.*

Proof. Let U be an open set such that $K \subset U \subseteq \bar{U} \subseteq \Omega$. Let $\psi \in C_c^\infty(\Omega)$ be such that $\psi = 1$ on K and $\psi = 0$ outside U. Clearly $\psi\varphi \in C_c^\infty(\Omega)$ and thus $\psi\varphi \in BC^{\infty,1}(\Omega)$ (see the paragraph above Theorem 7). So, it follows from Theorem 61 that $\|\psi\varphi u\|_{s,p} \preceq \|u\|_{s,p}$ where the implicit constant in particular may depend on φ and ψ. Now the claim follows from the obvious observation that for all $u \in W^{s,p}_K(\Omega)$, we have $\psi\varphi u = \varphi u$. □

Theorem 62. *Let $\Omega = \mathbb{R}^n$ or Ω be a nonempty bounded open set in $\mathbb{R}^n$ with Lipschitz continuous boundary. Let $K \subseteq \Omega$ be compact, $s \in \mathbb{R}$ and $p \in (1,\infty)$. Then*

(1) $W^{s,p}_K(\Omega) \subseteq W^{s,p}_0(\Omega)$. That is, every element of $W^{s,p}_K(\Omega)$ is a limit of a sequence in $C_c^\infty(\Omega)$;
(2) if $K \subseteq V \subseteq K' \subseteq \Omega$ where and K' is compact and V is open, then for every $u \in W^{s,p}_K(\Omega)$, there exists a sequence in $C^\infty_{K'}(\Omega)$ that converges to u in $W^{s,p}(\Omega)$.

Proof.

(1) Let $u \in W^{s,p}_K(\Omega)$. By Theorems 65 and 66, there exists a sequence $\{\varphi_i\}$ in $C^\infty(\Omega)$ such that $\varphi_i \to u$ in $W^{s,p}(\Omega)$. Let $\psi \in C_c^\infty(\Omega)$ be such that $\psi = 1$ on K. Since $C_c^\infty(\Omega) \subseteq BC^{\infty,1}(\Omega)$, it follows from Theorems 51 and 61 that $\psi\varphi_i \to \psi u$ in $W^{s,p}(\Omega)$. This proves the claim because $\psi\varphi_i \in C_c^\infty(\Omega)$ and $\psi u = u$.
(2) In the above argument, choose $\psi \in C_c^\infty(\Omega)$ such that $\psi = 1$ on K and $\psi = 0$ outside V.

□

Theorem 63 (([40], p. 496), ([39], pp. 317, 330, and 332)). *Let Ω be a bounded Lipschitz domain in $\mathbb{R}^n$. Suppose $1 < p < \infty$, $0 \le s < \frac{1}{p}$. Then $C_c^\infty(\Omega)$ is dense in $W^{s,p}(\Omega)$ (thus $W^{s,p}(\Omega) = W^{s,p}_0(\Omega)$).*

7.4. Properties of Sobolev Spaces on General Domains

In this section, Ω and Ω' are arbitrary nonempty open sets in $\mathbb{R}^n$. We begin with some facts about the relationship between various Sobolev spaces defined on bounded domains.

- Suppose $s \geq 0$ and $\Omega' \subseteq \Omega$. Then for all $u \in W^{s,p}(\Omega)$, we have $\mathrm{res}_{\Omega,\Omega'} u \in W^{s,p}(\Omega')$. Moreover, $\|\mathrm{res}_{\Omega,\Omega'} u\|_{W^{s,p}(\Omega')} \le \|u\|_{W^{s,p}(\Omega)}$. Indeed, if we let $s = k + \theta$

$$\|u\|_{W^{s,p}(\Omega')} = \|u\|_{W^{k,p}(\Omega')} + \sum_{|\nu|=k} \Big(\int\int_{\Omega'\times\Omega'} \frac{|\partial^\nu u(x) - \partial^\nu u(y)|^p}{|x-y|^{n+\theta p}} dxdy\Big)^{\frac{1}{p}}$$
$$= \sum_{|\alpha|\le k} \|\partial^\alpha u\|_{L^p(\Omega')} + \sum_{|\nu|=k} \Big(\int\int_{\Omega'\times\Omega'} \frac{|\partial^\nu u(x) - \partial^\nu u(y)|^p}{|x-y|^{n+\theta p}} dxdy\Big)^{\frac{1}{p}}$$
$$\le \sum_{|\alpha|\le k} \|\partial^\alpha u\|_{L^p(\Omega)} + \sum_{|\nu|=k} \Big(\int\int_{\Omega\times\Omega} \frac{|\partial^\nu u(x) - \partial^\nu u(y)|^p}{|x-y|^{n+\theta p}} dxdy\Big)^{\frac{1}{p}} = \|u\|_{W^{s,p}(\Omega)}.$$

  So, $\mathrm{res}_{\Omega,\Omega'} : W^{s,p}(\Omega) \to W^{s,p}(\Omega')$ is a continuous linear map. Furthermore, as a consequence, for every real number $s \geq 0$

$$W^{s,p}(\bar{\Omega}) \hookrightarrow W^{s,p}(\Omega).$$

  Indeed, if $u \in W^{s,p}(\bar{\Omega})$, then there exists $v \in W^{s,p}(\mathbb{R}^n)$ such that $\mathrm{res}_{\mathbb{R}^n,\Omega} v = u$ and thus $u \in W^{s,p}(\Omega)$. Moreover, for every such v, $\|u\|_{W^{s,p}(\Omega)} = \|\mathrm{res}_{\mathbb{R}^n,\Omega} v\|_{W^{s,p}(\Omega)} \le \|v\|_{W^{s,p}(\mathbb{R}^n)}$. This implies that

$$\|u\|_{W^{s,p}(\Omega)} \le \inf_{v\in W^{s,p}(\mathbb{R}^n), v|_\Omega = u} \|v\|_{W^{s,p}(\mathbb{R}^n)} = \|u\|_{W^{s,p}(\bar{\Omega})}.$$

- Clearly, for all $s \geq 0$
$$W^{s,p}_{00}(\Omega) \hookrightarrow W^{s,p}(\bar{\Omega}).$$
- For every integer $m > 0$ ([5], p. 18)
$$W^{m,p}_{0}(\Omega) \subseteq W^{m,p}_{00}(\Omega) \subseteq W^{m,p}(\bar{\Omega}) \subseteq W^{m,p}(\Omega).$$
- Suppose $s \geq 0$. Clearly, the restriction map $\text{res}_{\mathbb{R}^n,\Omega} : W^{s,p}(\mathbb{R}^n) \to W^{s,p}(\bar{\Omega})$ is a continuous linear map. This combined with the fact that $C_c^\infty(\mathbb{R}^n)$ is dense in $W^{s,p}(\mathbb{R}^n)$ implies that $C_c^\infty(\bar{\Omega}) := \text{res}_{\mathbb{R}^n,\Omega}(C_c^\infty(\mathbb{R}^n))$ is dense in $W^{s,p}(\bar{\Omega})$ for all $s \geq 0$.
- $\tilde{W}^{s,p}(\bar{\Omega})$ is a closed subspace of $W^{s,p}(\mathbb{R}^n)$. Closed subspaces of reflexive spaces are reflexive, hence $\tilde{W}^{s,p}(\bar{\Omega})$ is a reflexive space.

Theorem 64. *Let Ω be a nonempty open set in $\mathbb{R}^n$ and $1 < p < \infty$.*

(1) For all $s \geq 0$, $W^{s,p}(\Omega)$ is reflexive.
(2) For all $s \geq 0$, $W^{s,p}_0(\Omega)$ is reflexive.
(3) For all $s < 0$, $W^{s,p}(\Omega)$ is reflexive.

Proof.

(1) The proof for $s \in \mathbb{N}_0$ can be found in [1]. Let $s = k + \theta$ where $k \in \mathbb{N}_0$ and $0 < \theta < 1$. Let
$$r = \text{card}\{\nu \in \mathbb{N}_0^n : |\nu| = k\}.$$
Define $P : W^{s,p}(\Omega) \to W^{k,p}(\Omega) \times [L^p(\Omega \times \Omega)]^{\times r}$ by
$$P(u) = \left(u, \left(\frac{|\partial^\nu u(x) - \partial^\nu u(y)|}{|x-y|^{\frac{n}{p}+\theta}}\right)_{|\nu|=k}\right).$$
The space $W^{k,p}(\Omega) \times [L^p(\Omega \times \Omega)]^{\times r}$ equipped with the norm
$$\|(f, v_1, \dots, v_r)\| := \|f\|_{W^{k,p}(\Omega)} + \|v_1\|_{L^p(\Omega\times\Omega)} + \dots + \|v_r\|_{L^p(\Omega\times\Omega)}$$
is a product of reflexive spaces and so it is reflexive (see Theorem 9). Clearly, the operator P is an isometry from $W^{s,p}(\Omega)$ to $W^{k,p}(\Omega) \times [L^p(\Omega \times \Omega)]^{\times r}$. Since $W^{s,p}(\Omega)$ is a Banach space, $P(W^{s,p}(\Omega))$ is a closed subspace of the reflexive space $W^{k,p}(\Omega) \times [L^p(\Omega \times \Omega)]^{\times r}$ and thus it is reflexive. Hence $W^{s,p}(\Omega)$ itself is reflexive.

(2) $W^{s,p}_0(\Omega)$ is the closure of $C_c^\infty(\Omega)$ in $W^{s,p}(\Omega)$. Closed subspaces of reflexive spaces are reflexive. Therefore, $W^{s,p}_0(\Omega)$ is reflexive.

(3) A normed space X is reflexive if and only if X^* is reflexive (see Theorem 9). Since for $s < 0$ we have $W^{s,p}(\Omega) = [W^{-s,p'}_0(\Omega)]^*$, the reflexivity of $W^{s,p}(\Omega)$ follows from the reflexivity of $W^{-s,p'}_0(\Omega)$.

□

Theorem 65. *For all $s < 0$ and $1 < p < \infty$, $C_c^\infty(\Omega)$ is dense in $W^{s,p}(\Omega)$.*

Proof. The proof of the density of L^p in $W^{m,p}$ in p. 65 of [1] for integer order Sobolev spaces, which is based on the reflexivity of $W^{-m,p'}_0(\Omega)$, works in the exact same way for establishing the density of $C_c^\infty(\Omega)$ in $W^{s,p}(\Omega)$. □

Theorem 66 (Meyers-Serrin). *For all $s \geq 0$ and $p \in (1, \infty)$, $C^\infty(\Omega) \cap W^{s,p}(\Omega)$ is dense in $W^{s,p}(\Omega)$.*

Next we consider *extension by zero* and its properties.

Lemma 6 ([4], p. 201). *Let Ω be a nonempty open set in $\mathbb{R}^n$ and $u \in W_0^{m,p}(\Omega)$ where $m \in \mathbb{N}_0$ and $1 < p < \infty$. Then*

(1) $\forall\, |\alpha| \leq m$, $\partial^\alpha \tilde{u} = \widetilde{(\partial^\alpha u)}$ *as elements of* $D'(\mathbb{R}^n)$,
(2) $\tilde{u} \in W^{m,p}(\mathbb{R}^n)$ *with* $\|\tilde{u}\|_{W^{m,p}(\mathbb{R}^n)} = \|u\|_{W^{m,p}(\Omega)}$.

Here, $\tilde{u} := ext^0_{\Omega,\mathbb{R}^n} u$ *and* $\widetilde{(\partial^\alpha u)} := ext^0_{\Omega,\mathbb{R}^n}(\partial^\alpha u)$.

Lemma 7 ([6], p. 546). *Let Ω be a nonempty open set in $\mathbb{R}^n$, $K \in \mathcal{K}(\Omega)$, $u \in W_K^{s,p}(\Omega)$ where $s \in (0,1)$ and $1 < p < \infty$. Then $ext^0_{\Omega,\mathbb{R}^n} u \in W^{s,p}(\mathbb{R}^n)$ and*

$$\|ext^0_{\Omega,\mathbb{R}^n}\|_{W^{s,p}(\mathbb{R}^n)} \preceq \|u\|_{W^{s,p}(\Omega)},$$

where the implicit constant depends on n, p, s, K and Ω.

Theorem 67 (Extension by Zero). *Let $s \geq 0$ and $p \in (1,\infty)$. Let Ω be a nonempty open set in $\mathbb{R}^n$ and let $K \in \mathcal{K}(\Omega)$. Suppose $u \in W_K^{s,p}(\Omega)$. Then*

(1) $ext^0_{\Omega,\mathbb{R}^n} u \in W^{s,p}(\mathbb{R}^n)$. *Indeed,* $\|ext^0_{\Omega,\mathbb{R}^n} u\|_{W^{s,p}(\mathbb{R}^n)} \preceq \|u\|_{W^{s,p}(\Omega)}$ *where the implicit constant may depend on s, p, n, K, Ω but it is independent of $u \in W_K^{s,p}(\Omega)$.*
(2) *Moreover,*
$$\|ext^0_{\Omega,\mathbb{R}^n} u\|_{W^{s,p}(\mathbb{R}^n)} \geq \|u\|_{W^{s,p}(\Omega)}.$$

In short, $\|ext^0_{\Omega,\mathbb{R}^n} u\|_{W^{s,p}(\mathbb{R}^n)} \simeq \|u\|_{W^{s,p}(\Omega)}$.

Proof. Let $\tilde{u} = \text{ext}^0_{\Omega,\mathbb{R}^n} u$. If $s \in \mathbb{N}_0$, then both items follow from Lemma 6. So, let $s = m + \theta$ where $m \in \mathbb{N}_0$ and $\theta \in (0,1)$. We have

$$\begin{aligned}
\|\tilde{u}\|_{W^{s,p}(\mathbb{R}^n)} &= \|\tilde{u}\|_{W^{m,p}(\mathbb{R}^n)} + \sum_{|\nu|=m} |\partial^\nu \tilde{u}|_{W^{\theta,p}(\mathbb{R}^n)} \\
&\overset{\text{Lemma 6}}{=} \|u\|_{W^{m,p}(\Omega)} + \sum_{|\nu|=m} |\widetilde{\partial^\nu u}|_{W^{\theta,p}(\mathbb{R}^n)} \\
&\overset{\text{Lemma 7}}{\preceq} \|u\|_{W^{m,p}(\Omega)} + \sum_{|\nu|=m} \|\partial^\nu u\|_{W^{\theta,p}(\Omega)} \\
&\preceq \|u\|_{W^{s,p}(\Omega)}.
\end{aligned}$$

The fact that $\|\tilde{u}\|_{W^{s,p}(\mathbb{R}^n)} \geq \|u\|_{W^{s,p}(\Omega)}$ is a direct consequence of the decomposition stated in item 1 of Remark 34. □

Corollary 5. *Let $s \geq 0$ and $p \in (1,\infty)$. Let Ω and Ω' be nonempty open sets in $\mathbb{R}^n$ with $\Omega' \subseteq \Omega$ and let $K \in \mathcal{K}(\Omega')$. Suppose $u \in W_K^{s,p}(\Omega')$. Then*

(1) $ext^0_{\Omega',\Omega} u \in W^{s,p}(\Omega)$,
(2) $\|ext^0_{\Omega',\Omega} u\|_{W^{s,p}(\Omega)} \simeq \|u\|_{W^{s,p}(\Omega')}$.

Proof.

$$u \in W_K^{s,p}(\Omega') \Longrightarrow \text{ext}^0_{\Omega',\mathbb{R}^n} u \in W^{s,p}(\mathbb{R}^n) \Longrightarrow \text{ext}^0_{\Omega',\mathbb{R}^n} u|_\Omega \in W^{s,p}(\bar{\Omega}).$$

As we know, $W^{s,p}(\bar{\Omega}) \hookrightarrow W^{s,p}(\Omega)$. Furthermore, it is easy to see that $\text{ext}^0_{\Omega',\mathbb{R}^n} u|_\Omega = \text{ext}^0_{\Omega',\Omega} u$. Therefore, $\text{ext}^0_{\Omega',\Omega} u \in W^{s,p}(\Omega)$. Moreover,

$$\|\text{ext}^0_{\Omega',\Omega} u\|_{W^{s,p}(\Omega)} \simeq \|\text{ext}^0_{\Omega,\mathbb{R}^n} \circ \text{ext}^0_{\Omega',\Omega} u\|_{W^{s,p}(\mathbb{R}^n)} = \|\text{ext}^0_{\Omega',\mathbb{R}^n} u\|_{W^{s,p}(\mathbb{R}^n)} \simeq \|u\|_{W^{s,p}(\Omega')}.$$

□

Extension by zero for Sobolev spaces with negative exponents will be discussed in Theorem 71.

Theorem 68 (Embedding Theorem IV). *Let $\Omega \subseteq \mathbb{R}^n$ be an arbitrary nonempty open set.*

(1) *Suppose $1 \le p \le q < \infty$ and $0 \le t \le s$ satisfy $s - \frac{n}{p} \ge t - \frac{n}{q}$. Then $W^{s,p}(\bar{\Omega}) \hookrightarrow W^{t,q}(\bar{\Omega})$.*
(2) *Suppose $1 \le p \le q < \infty$ and $0 \le t \le s$ satisfy $s - \frac{n}{p} \ge t - \frac{n}{q}$. Then $W^{s,p}_K(\Omega) \hookrightarrow W^{t,q}_K(\Omega)$ for all $K \in \mathcal{K}(\Omega)$.*
(3) *For all $k_1, k_2 \in \mathbb{N}_0$ with $k_1 \le k_2$ and $1 < p < \infty$, $W^{k_2,p}(\Omega) \hookrightarrow W^{k_1,p}(\Omega)$.*
(4) *If $0 \le t \le s < 1$ and $1 < p < \infty$, then $W^{s,p}(\Omega) \hookrightarrow W^{t,p}(\Omega)$.*
(5) *If $0 \le t \le s < \infty$ are such that $\lfloor s \rfloor = \lfloor t \rfloor$ and $1 < p < \infty$, then $W^{s,p}(\Omega) \hookrightarrow W^{t,p}(\Omega)$.*
(6) *If $0 \le t \le s < \infty$, $t \in \mathbb{N}_0$, and $1 < p < \infty$, then $W^{s,p}(\Omega) \hookrightarrow W^{t,p}(\Omega)$.*

Proof.

(1) This item can be found in ([39], Section 4.6.1).
(2) For all $u \in W^{s,p}_K(\Omega)$ we have

$$\|u\|_{W^{t,q}(\Omega)} \simeq \|\mathrm{ext}^0_{\Omega,\mathbb{R}^n} u\|_{W^{t,q}(\mathbb{R}^n)} \preceq \|\mathrm{ext}^0_{\Omega,\mathbb{R}^n} u\|_{W^{s,p}(\mathbb{R}^n)} \simeq \|u\|_{W^{s,p}(\Omega)} \,.$$

(3) This item is a direct consequence of the definition of integer order Sobolev spaces.
(4) Proof can be found in [6], p. 524.
(5) This is a direct consequence of the previous two items.
(6) This is true because $W^{s,p}(\Omega) \hookrightarrow W^{\lfloor s \rfloor,p}(\Omega) \hookrightarrow W^{t,p}(\Omega)$.

□

Remark 42. *For an arbitrary open set Ω in $\mathbb{R}^n$ and $0 < t < 1$, the embedding $W^{1,p}(\Omega) \hookrightarrow W^{t,p}(\Omega)$ does NOT necessarily hold (see, e.g., [6], Section 9). Of course, as it was discussed, under the extra assumption that Ω is Lipschitz, the latter embedding holds true. So, if $\lfloor s \rfloor \neq \lfloor t \rfloor$ and $t \notin \mathbb{N}_0$, then in order to ensure that $W^{s,p}(\Omega) \hookrightarrow W^{t,p}(\Omega)$ we need to assume some sort of regularity for the domain Ω (for instance it is enough to assume Ω is Lipschitz).*

Theorem 69 (Multiplication by smooth functions). *Let Ω be any nonempty open set in $\mathbb{R}^n$. Let $p \in (1, \infty)$.*

(1) *If $0 \le s < 1$ and $\varphi \in BC^{0,1}(\Omega)$ (that is, φ is bounded and φ is Lipschitz), then*

$$m_\varphi : W^{s,p}(\Omega) \to W^{s,p}(\Omega), \qquad u \mapsto \varphi u$$

is a well-defined bounded linear map.

(2) *If $k \in \mathbb{N}_0$ and $\varphi \in BC^k(\Omega)$, then*

$$m_\varphi : W^{k,p}(\Omega) \to W^{k,p}(\Omega), \qquad u \mapsto \varphi u$$

is a well-defined bounded linear map.

(3) *If $-1 < s < 0$ and $\varphi \in BC^{\infty,1}(\Omega)$ or $s \in \mathbb{Z}^-$ and $\varphi \in BC^\infty(\Omega)$, then*

$$m_\varphi : W^{s,p}(\Omega) \to W^{s,p}(\Omega), \qquad u \mapsto \varphi u$$

is a well-defined bounded linear map (φu is interpreted as the product of a smooth function and a distribution).

Proof.

(1) Proof can be found in [6], p. 547.
(2) Proof can be found in [29], p. 995.
(3) The duality argument in Step 4 of the proof of Theorem 61 works for this item too.

□

Remark 43. *Suppose $\varphi \in BC^{\infty,1}(\Omega)$. Note that the above theorem says nothing about the boundedness of the mapping $m_\varphi : W^{s,p}(\Omega) \to W^{s,p}(\Omega)$ in the case where s is noninteger such that $|s| > 1$. Of course, if we assume Ω is Lipschitz, then the continuity of m_φ follows from Theorem 61. It is important to note that the proof of that theorem for the case $s > 1$ (noninteger) uses the embedding $W^{k+\theta,p}(\Omega) \hookrightarrow W^{k'+\theta,p}(\Omega)$ with $k' < k$ which as we discussed does not hold for an arbitrary open set Ω. The proof for the case $s < -1$ (noninteger) uses duality to transfer the problem to $s > 1$ and thus again we need the extra assumption of regularity of the boundary of Ω.*

Theorem 70. *Let Ω be a nonempty open set in $\mathbb{R}^n$, $K \in \mathcal{K}(\Omega)$, $p \in (1,\infty)$, and $-1 < s < 0$ or $s \in \mathbb{Z}^-$ or $s \in [0,\infty)$. If $\varphi \in C^\infty(\Omega)$, then the linear map*

$$W_K^{s,p}(\Omega) \to W_K^{s,p}(\Omega), \qquad u \mapsto \varphi u$$

is well-defined and bounded.

Proof. There exists $\psi \in C_c^\infty(\Omega)$ such that $\psi = 1$ on K. Clearly $\psi\varphi \in C_c^\infty(\Omega)$ and if $u \in W_K^{s,p}(\Omega)$, $\psi\varphi u = \varphi u$ on Ω. Thus without loss of generality we may assume that $\varphi \in C_c^\infty(\Omega)$. Since $C_c^\infty(\Omega) \subseteq BC^\infty(\Omega)$ and $C_c^\infty(\Omega) \subseteq BC^{\infty,1}(\Omega)$, the cases where $-1 < s < 0$ or $s \in \mathbb{Z}^-$ follow from Theorem 69. For $s \geq 0$, the proof of Theorem 61 works for this theorem as well. The only place in that proof that the regularity of the boundary of Ω was used was for the validity of the embedding $W^{s,p}(\Omega) \hookrightarrow W^{\theta+|\beta|,p}(\Omega)$. However, as we know (see Theorem 68), this embedding holds for Sobolev spaces with support in a fixed compact set inside Ω for a general open set Ω, that is, for $W_K^{s,p}(\Omega) \hookrightarrow W_K^{\theta+|\beta|,p}(\Omega)$ to be true we do not need to assume Ω is Lipschitz. □

Remark 44. *Note that our proofs for $s < 0$ are based on duality. As a result, it seems that for the case where s is a noninteger less than -1 we cannot have a multiplication by smooth functions result for $W_K^{s,p}(\Omega)$ similar to the one stated in the above theorem (note that there is no fixed compact set K such that every $v \in C_c^\infty(\Omega)$ has compact support in K. Thus, the technique used in Step 4 of the proof of Theorem 61 does not work in this case).*

Theorem 71. *Let $s < 0$ and $p \in (1,\infty)$. Let Ω and Ω' be nonempty open sets in $\mathbb{R}^n$ with $\Omega' \subseteq \Omega$ and let $K \in \mathcal{K}(\Omega')$. Suppose $u \in W_K^{s,p}(\Omega')$. Then*

(1) *If $ext^0_{\Omega',\Omega}u \in W^{s,p}(\Omega)$, then $\|u\|_{W^{s,p}(\Omega')} \preceq \|ext^0_{\Omega',\Omega}u\|_{W^{s,p}(\Omega)}$ (the implicit constant may depend on K).*

(2) *If $s \in (-\infty,-1] \cap \mathbb{Z}$ or $-1 < s < 0$, then $ext^0_{\Omega',\Omega}u \in W^{s,p}(\Omega)$ and $\|ext^0_{\Omega',\Omega}u\|_{W^{s,p}(\Omega)} \simeq \|u\|_{W^{s,p}(\Omega')}$. This result holds for all $s < 0$ if we further assume that Ω is Lipschitz or $\Omega = \mathbb{R}^n$.*

Proof. To be completely rigorous, let $i_{D,W} : D(\Omega') \to W_0^{-s,p'}(\Omega')$ be the identity map and let $i^*_{D,W} : W^{s,p}(\Omega') \to D'(\Omega')$ be its dual with which we identify $W^{s,p}(\Omega')$ with a subspace of $D'(\Omega')$. Previously we defined $ext^0_{\Omega',\Omega}$ for distributions with compact support in Ω'. For any $u \in W_K^{s,p}(\Omega')$ we let

$$\text{ext}^0_{\Omega',\Omega}u := \text{ext}^0_{\Omega',\Omega} \circ i^*_{D,W}u\,,$$

which by definition will be an element of $D'(\Omega)$. Note that (see Remark 45 and the discussion right after Remark 10)

$$\|\text{ext}^0_{\Omega',\Omega}u\|_{W^{s,p}(\Omega)} = \sup_{0\neq\psi\in D(\Omega)} \frac{|\langle \text{ext}^0_{\Omega',\Omega}u,\psi\rangle_{D'(\Omega)\times D(\Omega)}|}{\|\psi\|_{W^{-s,p'}(\Omega)}}$$

$$\|u\|_{W^{s,p}(\Omega')} = \sup_{0\neq\varphi\in D(\Omega')} \frac{|\langle u,\varphi\rangle_{D'(\Omega')\times D(\Omega')}|}{\|\varphi\|_{W^{-s,p'}(\Omega')}}.$$

So, in order to prove the first item we just need to show that

$$\forall 0 \neq \varphi \in D(\Omega') \quad \exists \psi \in D(\Omega) \text{ s.t. } \frac{|\langle u, \varphi \rangle_{D'(\Omega') \times D(\Omega')}|}{\|\varphi\|_{W^{-s,p'}(\Omega')}} \preceq \frac{|\langle \mathrm{ext}^0_{\Omega',\Omega} u, \psi \rangle_{D'(\Omega) \times D(\Omega)}|}{\|\psi\|_{W^{-s,p'}(\Omega)}}.$$

Let $\varphi \in D(\Omega')$. Define $\psi = \mathrm{ext}^0_{\Omega',\Omega}\varphi$. Clearly, $\psi \in D(\Omega)$ and $\psi = \varphi$ on Ω'. Therefore,

$$\langle \mathrm{ext}^0_{\Omega',\Omega} u, \psi \rangle_{D'(\Omega) \times D(\Omega)} = \langle u, \psi|_{\Omega'} \rangle_{D'(\Omega') \times D(\Omega')} = \langle u, \varphi \rangle_{D'(\Omega') \times D(\Omega')}.$$

Moreover, since $-s > 0$

$$\|\psi\|_{W^{-s,p'}(\Omega)} = \|\mathrm{ext}^0_{\Omega',\Omega}\varphi\|_{W^{-s,p'}(\Omega)} \preceq \|\varphi\|_{W^{-s,p'}(\Omega')}.$$

This completes the proof of the first item. For the second item we just need to prove that under the given hypotheses

$$\forall 0 \neq \psi \in D(\Omega) \quad \exists \varphi \in D(\Omega') \text{ s.t. } \frac{|\langle \mathrm{ext}^0_{\Omega',\Omega} u, \psi \rangle_{D'(\Omega) \times D(\Omega)}|}{\|\psi\|_{W^{-s,p'}(\Omega)}} \preceq \frac{|\langle u, \varphi \rangle_{D'(\Omega') \times D(\Omega')}|}{\|\varphi\|_{W^{-s,p'}(\Omega')}}.$$

To this end suppose $\psi \in D(\Omega)$. Choose a compact set $\tilde{K}$ such that $K \subset \mathring{\tilde{K}} \subset \tilde{K} \subset \Omega'$. Fix $\chi \in D(\Omega)$ such that $\chi = 1$ on $\tilde{K}$ and $\mathrm{supp}\,\chi \subset \Omega'$. Clearly, $\psi = \chi\psi$ on a neighborhood of K and if we set $\varphi = \chi\psi|_{\Omega'}$, then $\varphi \in D(\Omega')$. Therefore,

$$\langle \mathrm{ext}^0_{\Omega',\Omega} u, \psi \rangle_{D'(\Omega) \times D(\Omega)} = \langle \mathrm{ext}^0_{\Omega',\Omega} u, \chi\psi \rangle_{D'(\Omega) \times D(\Omega)} = \langle u, \chi\psi|_{\Omega'} \rangle_{D'(\Omega') \times D(\Omega')} = \langle u, \varphi \rangle_{D'(\Omega') \times D(\Omega')}.$$

Furthermore, since $-s > 0$, we have

$$\|\varphi\|_{W^{-s,p'}(\Omega')} \leq \|\mathrm{ext}^0_{\Omega',\Omega}\varphi\|_{W^{-s,p'}(\Omega)} = \|\chi\psi\|_{W^{-s,p'}(\Omega)} \preceq \|\psi\|_{W^{-s,p'}(\Omega)}.$$

The latter inequality is the place where we used the assumption that $s \in (-\infty, -1] \cap \mathbb{Z}$ or $-1 < s < 0$ or Ω is Lipschitz or $\Omega = \mathbb{R}^n$. This completes the proof of the second item. □

Corollary 6. *Let $p \in (1, \infty)$. Let Ω and Ω' be nonempty open sets in $\mathbb{R}^n$ with $\Omega' \subseteq \Omega$ and let $K \in \mathcal{K}(\Omega')$. Suppose $u \in W^{s,p}_K(\Omega)$. It follows from Corollary 5 and Theorem 71 that*

- *If $s \in \mathbb{R}$ is not a noninteger less than -1, then*

$$\|u\|_{W^{s,p}(\Omega)} \simeq \|u\|_{W^{s,p}(\Omega')},$$

- *If Ω is Lipschitz or $\Omega = \mathbb{R}^n$, then for all $s \in \mathbb{R}$*

$$\|u\|_{W^{s,p}(\Omega)} \simeq \|u\|_{W^{s,p}(\Omega')}.$$

Note that on the right hand sides of the above expressions, u stands for $\mathrm{res}_{\Omega,\Omega'} u$. Clearly, $\mathrm{ext}^0_{\Omega',\Omega} \circ \mathrm{res}_{\Omega,\Omega'} u = u$.

Theorem 72. *Let Ω be any nonempty open set in $\mathbb{R}^n$, $K \subseteq \Omega$ be compact, $s > 0$, and $p \in (1, \infty)$. Then the following norms on $W^{s,p}_K(\Omega)$ are equivalent:*

$$\|u\|_{W^{s,p}(\Omega)} := \|u\|_{W^{k,p}(\Omega)} + \sum_{|\nu|=k} |\partial^\nu u|_{W^{\theta,p}(\Omega)},$$

$$[u]_{W^{s,p}(\Omega)} := \|u\|_{W^{k,p}(\Omega)} + \sum_{1 \leq |\nu| \leq k} |\partial^\nu u|_{W^{\theta,p}(\Omega)},$$

where $s = k + \theta$, $k \in \mathbb{N}_0$, $\theta \in (0,1)$. Moreover, if we further assume Ω is Lipschitz, then the above norms are equivalent on $W^{s,p}(\Omega)$.

Proof. Clearly, for all $u \in W^{s,p}(\Omega)$, $\|u\|_{W^{s,p}(\Omega)} \leq [u]_{W^{s,p}(\Omega)}$. So, it is enough to show that there is a constant $C > 0$ such that for all $u \in W_K^{s,p}(\Omega)$ (or $u \in W^{s,p}(\Omega)$ if Ω is Lipschitz)

$$[u]_{W^{s,p}(\Omega)} \leq C\|u\|_{W^{s,p}(\Omega)}\,.$$

For each $1 \leq i \leq k$ we have

$$\sum_{|\nu|=i} |\partial^\nu u|_{W^{\theta,p}(\Omega)} = \|u\|_{W^{i+\theta,p}(\Omega)} - \|u\|_{W^{i,p}(\Omega)}\,.$$

Thus

$$\begin{aligned}[u]_{W^{s,p}(\Omega)} &= \|u\|_{W^{s,p}(\Omega)} + \sum_{1\leq i<k}\sum_{|\nu|=i} |\partial^\nu u|_{W^{\theta,p}(\Omega)} \\ &= \|u\|_{W^{s,p}(\Omega)} + \sum_{1\leq i<k}\left(\|u\|_{W^{i+\theta,p}(\Omega)} - \|u\|_{W^{i,p}(\Omega)}\right).\end{aligned}$$

Therefore, it is enough to show that there exists a constant $C \geq 1$ such that

$$\sum_{1\leq i<k} \|u\|_{W^{i+\theta,p}(\Omega)} \leq (C-1)\|u\|_{W^{s,p}(\Omega)} + \sum_{1\leq i<k} \|u\|_{W^{i,p}(\Omega)}\,.$$

By Theorem 68, for each $1 \leq i < k$, $W_K^{s,p}(\Omega) \hookrightarrow W_K^{i+\theta,p}(\Omega)$ (also, we have $W^{s,p}(\Omega) \hookrightarrow W^{i+\theta,p}(\Omega)$ with the extra assumption that Ω is Lipschitz); so there is a constant C_i such that $\|u\|_{W^{i+\theta,p}(\Omega)} \leq C_i\|u\|_{W^{s,p}(\Omega)}$. Clearly with $C = 1 + \sum_{i=1}^{k-1} C_i$ the desired inequality holds. $\square$

Remark 45. *Let $s \geq 0$ and $1 < p < \infty$. Here we summarize the connection between Sobolev spaces and space of distributions.*

(1) ***Question 1:*** *What does it mean to say $u \in D'(\Omega)$ belongs to $W^{-s,p'}(\Omega)$?*
Answer:

$$\begin{aligned}&u \in D'(\Omega) \text{ is in } W^{-s,p'}(\Omega) \iff u : (D(\Omega), \|.\|_{s,p}) \to \mathbb{R} \text{ is continuous} \\ &\iff u : D(\Omega) \to \mathbb{R} \text{ has a unique continuous extension to } \hat{u} : W_0^{s,p}(\Omega) \to \mathbb{R}\end{aligned}$$

(2) ***Question 2:*** *How should we interpret $W^{-s,p'}(\Omega) \subseteq D'(\Omega)$?*
Answer: *$i : D(\Omega) \to W_0^{s,p}(\Omega)$ is continuous with dense image. Therefore, $i^* : W^{-s,p'}(\Omega) \to D'(\Omega)$ is an injective continuous linear map. If $u \in W^{-s,p'}(\Omega)$, then $i^*u \in D'(\Omega)$ and*

$$\langle i^*u, \varphi\rangle_{D'(\Omega)\times D(\Omega)} = \langle u, i\varphi\rangle_{W^{-s,p'}(\Omega)\times W_0^{s,p}(\Omega)} = \langle u, \varphi\rangle_{W^{-s,p'}(\Omega)\times W_0^{s,p}(\Omega)}\,.$$

So, $i^*u = u|_{D(\Omega)}$ and if we identify with i^*u with u we can write

$$\langle u, \varphi\rangle_{D'(\Omega)\times D(\Omega)} = \langle u, \varphi\rangle_{W^{-s,p'}(\Omega)\times W_0^{s,p}(\Omega)}, \qquad \|u\|_{W^{-s,p'}(\Omega)} = \sup_{0\neq\varphi\in C_c^\infty(\Omega)} \frac{|\langle u, \varphi\rangle_{D'(\Omega)\times D(\Omega)}|}{\|\varphi\|_{W^{s,p}(\Omega)}}\,.$$

(3) ***Question 3:*** *How should we interpret $W^{s,p}(\Omega) \subseteq D'(\Omega)$?*
Answer: *It is a direct consequence of the definition of $W^{s,p}(\Omega)$ that $W^{s,p}(\Omega) \hookrightarrow L^p(\Omega)$ for any open set Ω. So, any $f \in W^{s,p}(\Omega)$ can be identified with the regular distribution $u_f \in D'(\Omega)$ where*

$$\langle u_f, \varphi\rangle = \int f\varphi \qquad \forall\, \varphi \in D(\Omega)\,.$$

(4) ***Question 4:*** *What does it mean to say $u \in D'(\Omega)$ belongs to $W^{s,p}(\Omega)$?*
Answer: *It means there exists $f \in W^{s,p}(\Omega)$ such that $u = u_f$.*

Remark 46. *Let Ω be a nonempty open set in $\mathbb{R}^n$ and $f, g \in C_c^\infty(\Omega)$. Suppose $s \in \mathbb{R}$ and $p \in (1, \infty)$.*

- *If $s \geq 0$, then*

$$\|f\|_{W^{-s,p'}(\Omega)} = \sup_{0 \neq \varphi \in C_c^\infty(\Omega)} \frac{|\langle f, \varphi \rangle_{D'(\Omega) \times D(\Omega)}|}{\|\varphi\|_{W^{s,p}(\Omega)}} = \sup_{0 \neq \varphi \in C_c^\infty(\Omega)} \frac{|\int_\Omega f \varphi \, dx|}{\|\varphi\|_{W^{s,p}(\Omega)}}.$$

So, for all $\varphi \in C_c^\infty(\Omega)$

$$|\int_\Omega f \varphi \, dx| \leq \|f\|_{W^{-s,p'}(\Omega)} \|\varphi\|_{W^{s,p}(\Omega)}.$$

In particular, for g, we have

$$|\int_\Omega f g \, dx| \leq \|f\|_{W^{-s,p'}(\Omega)} \|g\|_{W^{s,p}(\Omega)}.$$

- *If $s < 0$, we may replace the roles of f and g, and also (s, p) and $(-s, p')$ in the above argument to get the exact same inequality: $|\int_\Omega f g \, dx| \leq \|f\|_{W^{-s,p'}(\Omega)} \|g\|_{W^{s,p}(\Omega)}$.*

7.5. Invariance Under Change of Coordinates, Composition

Theorem 73 ([12], Section 4.3). *Let $s \in \mathbb{R}$ and $1 < p < \infty$. Suppose that $T : \mathbb{R}^n \to \mathbb{R}^n$ is a C^∞-diffeomorphism (i.e., T is bijective and T and T^{-1} are C^∞) with the property that the partial derivatives (of any order) of the components of T are bounded on $\mathbb{R}^n$ (the bound may depend on the order of the partial derivative) and $\inf_{\mathbb{R}^n} |\det T'| > 0$. Then the linear map*

$$W^{s,p}(\mathbb{R}^n) \to W^{s,p}(\mathbb{R}^n), \qquad u \mapsto u \circ T$$

is well-defined and is bounded.

Now, let U and V be two nonempty open sets in $\mathbb{R}^n$. Suppose $T : U \to V$ is a bijective map. Similar to [1] we say T is **k-smooth** if all the components of T belong to $BC^k(U)$ and all the components of T^{-1} belong to $BC^k(V)$.

Remark 47. *It is useful to note that if T is 1-smooth, then*

$$\inf_U |\det T'| > 0 \quad \text{and} \quad \inf_V |\det (T^{-1})'| > 0.$$

Indeed, since the first order partial derivatives of the components of T and T^{-1} are bounded, there exist postive numbers M and $\tilde{M}$ such that for all $x \in U$ and $y \in V$

$$|\det T'(x)| < M, \qquad |\det (T^{-1})'(y)| < \tilde{M}.$$

Since $|\det T'(x)| \times |\det (T^{-1})'(T(x))| = 1$, we can conclude that for all $x \in U$ and $y \in V$

$$|\det T'(x)| > \frac{1}{\tilde{M}}, \qquad |\det (T^{-1})'(y)| > \frac{1}{M},$$

which proves the claim.

Remark 48. *Furthermore, it is interesting to note that, as a consequence of the inverse function theorem, if $T : U \to V$ is a bijective map that is C^k ($k \in \mathbb{N}$) with the property that $\det T'(x) \neq 0$ for all $x \in U$, then the inverse of T will be C^k as well, that is, T will automatically be a C^k-diffeomorphism (see, e.g., Appendix C in [19] for more details).*

Remark 49. *Note that since we do not assume that U and V are necessarily convex or Lipschitz, the continuity and boundedness of the partial derivatives of the components of T do not imply that the components of T are Lipschitz. (see the "Warning" immediately after Theorem 6).*

Theorem 74 (([29], p. 1003), ([1], pp. 77–78)). *Let $p \in (1, \infty)$ and $k \in \mathbb{N}$. Suppose that U and V are nonempty open subsets of $\mathbb{R}^n$.*

(1) If $T : U \to V$ is a 1-smooth map, then the map

$$L^p(V) \to L^p(U), \qquad u \mapsto u \circ T$$

is well-defined and is bounded.

(2) If $T : U \to V$ is a k-smooth map, then the map

$$W^{k,p}(V) \to W^{k,p}(U), \qquad u \mapsto u \circ T$$

is well-defined and is bounded.

Theorem 75. *Let $p \in (1, \infty)$ and $k \in \mathbb{Z}^-$ (k is a negative integer). Suppose that U and V are nonempty open subsets of $\mathbb{R}^n$, and $T : U \to V$ is ∞-smooth. Then the map*

$$W^{k,p}(V) \to W^{k,p}(U), \qquad u \mapsto u \circ T$$

is well-defined and is bounded.

Proof. By definition we have (T^*u denotes the pullback of u by T)

$$\begin{aligned} \|T^*u\|_{W^{k,p}(U)} &= \sup_{\varphi \in C_c^\infty(U)} \frac{|\langle T^*u, \varphi\rangle_{D'(U)\times D(U)}|}{\|\varphi\|_{W^{-k,p'}(U)}} \\ &= \sup_{\varphi \in C_c^\infty(U)} \frac{|\langle u, |\det(T^{-1})'|\varphi \circ T^{-1}\rangle_{D'(V)\times D(V)}|}{\|\varphi\|_{W^{-k,p'}(U)}} \\ &\preceq \sup_{\varphi \in C_c^\infty(U)} \frac{\|u\|_{W^{k,p}(V)} \| |\det(T^{-1})'|\varphi \circ T^{-1}\|_{W^{-k,p'}(V)}}{\|\varphi\|_{W^{-k,p'}(U)}} \\ &\overset{|\det(T^{-1})'| \in BC^\infty}{\preceq} \sup_{\varphi \in C_c^\infty(U)} \frac{\|u\|_{W^{k,p}(V)} \|\varphi \circ T^{-1}\|_{W^{-k,p'}(V)}}{\|\varphi\|_{W^{-k,p'}(U)}}. \end{aligned}$$

Since $-k$ is a positive integer, by Theorem 74 we have $\|\varphi \circ T^{-1}\|_{W^{-k,p'}(V)} \preceq \|\varphi\|_{W^{-k,p'}(U)}$. Consequently,

$$\|T^*u\|_{W^{k,p}(U)} \preceq \|u\|_{W^{k,p}(V)}.$$

□

Theorem 76. *Let $p \in (1, \infty)$ and $0 < s < 1$. Suppose that U and V are nonempty open subsets of $\mathbb{R}^n$, $T : U \to V$ is 1-smooth, and T is Lipschitz continuous on U. Then the map*

$$W^{s,p}(V) \to W^{s,p}(U), \qquad u \mapsto u \circ T$$

is well-defined and is bounded.

Proof. Note that

$$\|u \circ T\|_{W^{s,p}(U)} = \|u \circ T\|_{L^p(U)} + |u \circ T|_{W^{s,p}(U)} \overset{\text{Theorem 74}}{\preceq} \|u\|_{L^p(V)} + |u \circ T|_{W^{s,p}(U)}.$$

So, it is enough to show that $|u\circ T|_{W^{s,p}(U)} \preceq |u|_{W^{s,p}(V)}$.

$$|u\circ T|_{W^{s,p}(U)} = \Big(\int\int_{U\times U} \frac{|(u\circ T)(x)-(u\circ T)(y)|^p}{|x-y|^{n+sp}}dxdy\Big)^{\frac{1}{p}}$$

$$\overset{\substack{z=T(x)\\ w=T(y)}}{\preceq} \Big(\int\int_{V\times V} \frac{|u(z)-u(w)|^p}{|T^{-1}(z)-T^{-1}(w)|^{n+sp}} \frac{1}{|\mathrm{det}T'(x)|}\frac{1}{|\mathrm{det}T'(y)|}dzdw\Big)^{\frac{1}{p}}$$

$$\preceq \Big(\int\int_{V\times V} \frac{|u(z)-u(w)|^p}{|T^{-1}(z)-T^{-1}(w)|^{n+sp}}dzdw\Big)^{\frac{1}{p}}.$$

T is Lipschitz continuous on U; so, there exists a constant $C>0$ such that

$$|T(x)-T(y)| \le C|x-y| \Longrightarrow |z-w| \le C|T^{-1}(z)-T^{-1}(w)|.$$

Therefore,

$$|u\circ T|_{W^{s,p}(U)} \preceq \Big(\int\int_{V\times V} \frac{|u(z)-u(w)|^p}{|z-w|^{n+sp}}dzdw\Big)^{\frac{1}{p}} = |u|_{W^{s,p}(V)}.$$

□

Theorem 77. *Let $p\in(1,\infty)$ and $-1<s<0$. Suppose that U and V are nonempty open subsets of $\mathbb{R}^n$, $T:U\to V$ is ∞-smooth, T^{-1} is Lipschitz continuous on V, and $|det(T^{-1})'|$ is in $BC^{0,1}(V)$. Then the map*

$$W^{s,p}(V)\to W^{s,p}(U), \qquad u\mapsto u\circ T$$

is well-defined and is bounded.

Proof. The proof of Theorem 75, with obvious modifications, shows the validity of the above claim. □

Remark 50. *In the previous theorem, by assumption, the first order partial derivatives of the components of T^{-1} are continuous and bounded. Furthermore, it is true that absolute value of a Lipschitz continuous function and the sum and product of bounded Lipschitz continuous functions will be Lipschitz continuous. Consequently, in order to ensure that $|det(T^{-1})'|$ is in $BC^{0,1}(V)$, it is enough to make sure that the first order partial derivatives of the components of T^{-1} are bounded and Lipschitz continuous.*

Theorem 78. *Let $s=k+\theta$ where $k\in\mathbb{N}$, $\theta\in(0,1)$, and let $p\in(1,\infty)$. Suppose that U and V are two nonempty open sets in $\mathbb{R}^n$. Let $T:U\to V$ be a Lipschitz continuous k-smooth map on U such that the partial derivatives up to and including order k of all the components of T are Lipschitz continuous on U as well. Then*

(1) *For each $K\in\mathcal{K}(V)$ the linear map*

$$T^*: W^{s,p}_K(V)\to W^{s,p}_{T^{-1}(K)}(U), \qquad u\mapsto u\circ T$$

is well-defined and is bounded.

(2) *If we further assume that V is Lipschitz (and so U is Lipschitz), the linear map*

$$T^*: W^{s,p}(V)\to W^{s,p}(U), \qquad u\mapsto u\circ T$$

is well-defined and is bounded.

Note: *When U is a Lipschitz domain, the fact that T is k-smooth automatically implies that all the partial derivatives of the components of T up to and including order $k-1$ are Lipschitz continuous (see Theorem 60). So in this case, the only extra assumption, in addition to T*

being k-smooth, is that the partial derivatives of the components of T of order k are Lipschitz continuous on U.

Proof. Recall that $C^\infty(V) \cap W^{s,p}(V)$ is dense in $W^{s,p}(V)$. Our proof consists of two steps: in the first step we addditionally assume that $u \in C^\infty(V)$. Then in the second step we prove the validity of the claim for $u \in W^{s,p}_K(V)$ (or $u \in W^{s,p}(V)$ with the assumption that V is Lipschitz).

- **Step 1:** We have

$$\|u \circ T\|_{W^{s,p}(U)} = \|u \circ T\|_{W^{k,p}(U)} + \sum_{|\nu|=k} |\partial^\nu (u \circ T)|_{W^{\theta,p}(U)}$$
$$\overset{\text{Theorem 74}}{\preceq} \|u\|_{W^{k,p}(V)} + \sum_{|\nu|=k} |\partial^\nu (u \circ T)|_{W^{\theta,p}(U)}\,.$$

Since u and T are both C^k, it can be proved by induction that (see, e.g., [1])

$$\partial^\nu (u \circ T)(x) = \sum_{\beta \le \nu, 1 \le |\beta|} M_{\nu\beta}(x)[(\partial^\beta u) \circ T](x)\,,$$

where $M_{\nu\beta}(x)$ are polynomials of degree at most $|\beta|$ in derivatives of order at most $|\nu|$ of the components of T. In particular, $M_{\nu\beta} \in BC^{0,1}(U)$. Therefore,

$$|\partial^\nu (u \circ T)|_{W^{\theta,p}(U)} \le \|\partial^\nu (u \circ T)\|_{W^{\theta,p}(U)} = \|\sum_{\beta \le \nu, 1 \le |\beta|} M_{\nu\beta}(x)[(\partial^\beta u) \circ T](x)\|_{W^{\theta,p}(U)}$$
$$\overset{\text{Theorem 69}}{\preceq} \sum_{\beta \le \nu, 1 \le |\beta|} \|(\partial^\beta u) \circ T\|_{W^{\theta,p}(U)} = \sum_{\beta \le \nu, 1 \le |\beta|} \|(\partial^\beta u) \circ T\|_{L^p(U)} + |(\partial^\beta u) \circ T|_{W^{\theta,p}(U)}$$
$$\overset{\text{Theorems 74 and 76}}{\preceq} \sum_{\beta \le \nu, 1 \le |\beta|} \|\partial^\beta u\|_{L^p(V)} + |\partial^\beta u|_{W^{\theta,p}(V)} \le \|u\|_{W^{k,p}(V)} + \sum_{\beta \le \nu, 1 \le |\beta|} |\partial^\beta u|_{W^{\theta,p}(V)}\,.$$

(The fact that $\partial^\beta u$ belongs to $W^{\theta,p}(V) \hookrightarrow L^p(V)$ is a consequence of the definition of the Slobodeckij norm combined with our embedding theorems for Sobolev spaces of functions with fixed compact support in an arbitrary domain or embedding theorems for Sobolev spaces of functions on a Lipschitz domain). Hence

$$\|u \circ T\|_{W^{s,p}(U)} \preceq \|u\|_{W^{k,p}(V)} + \sum_{1 \le |\nu| \le k} \sum_{\beta \le \nu, 1 \le |\beta|} |\partial^\beta u|_{W^{\theta,p}(V)}$$
$$\preceq \|u\|_{W^{k,p}(V)} + \sum_{1 \le |\alpha| \le k} |\partial^\alpha u|_{W^{\theta,p}(V)} \overset{\text{Theorem 72}}{\simeq} \|u\|_{W^{s,p}(V)}\,.$$

Note that the last equivalence is due to the assumption that $u \in W^{s,p}_K(V)$ (or $u \in W^{s,p}(V)$ with V being Lipschitz).

- **Step 2:** Now suppose u is an arbitrary element of $W^{s,p}_K(V)$ (or $W^{s,p}(V)$ with V being Lipschitz). There exists a sequence $\{u_m\}_{m \ge 1}$ in $C^\infty(V)$ such that $u_m \to u$ in $W^{s,p}(V)$. In particular, $\{u_m\}$ is Cauchy. By the previous steps we have

$$\|T^* u_m - T^* u_l\|_{W^{s,p}(U)} \preceq \|u_m - u_l\|_{W^{s,p}(V)} \to 0 \qquad (\text{as } m, l \to \infty)\,.$$

Therefore, $\{T^* u_m\}$ is a Cauchy sequence in the Banach space $W^{s,p}(U)$ and subsequently there exists $v \in W^{s,p}(U)$ such that $T^* u_m \to v$ as $m \to \infty$. It remains to show that $v = T^* u$ as elements of $W^{s,p}(U)$. As a direct consequence of the definition of $W^{s,p}$-norm $(s \ge 0)$ we have

$$\|T^*u_m - v\|_{L^p(U)} \leq \|T^*u_m - v\|_{W^{s,p}(U)} \to 0\,,$$
$$\|u_m - u\|_{L^p(V)} \leq \|u_m - u\|_{W^{s,p}(V)} \to 0\,.$$

Note that by Theorem 74, $u_m \to u$ in $L^p(V)$ implies that $T^*u_m \to T^*u$ in $L^p(U)$. Thus $T^*u = v$ as elements of $L^p(U)$ and hence as elements of $W^{s,p}(U)$.
□

Theorem 79. *Let $p \in (1,\infty)$ and $s < -1$ be a **noninteger** number. Suppose that U and V are two nonempty **Lipschitz** open sets in $\mathbb{R}^n$ and $T : U \to V$ is a ∞-smooth map. Then the linear map*

$$T^* : W^{s,p}(V) \to W^{s,p}(U), \qquad u \mapsto u \circ T$$

is well-defined and is bounded.
***Note:** Since V is a Lipschitz domain, the fact that T is ∞-smooth automatically implies that T^{-1} and all the partial derivatives of the components of T^{-1} are Lipschitz continuous (see Theorem 60).*

Proof. The proof is completely analogous to the proof of Theorem 75. We have

$$\begin{aligned}
\|T^*u\|_{W^{s,p}(U)} &= \sup_{\varphi\in C_c^\infty(U)} \frac{|\langle T^*u,\varphi\rangle_{D'(U)\times D(U)}|}{\|\varphi\|_{W^{-s,p'}(U)}} \\
&= \sup_{\varphi\in C_c^\infty(U)} \frac{|\langle u, |\det(T^{-1})'|\varphi\circ T^{-1}\rangle_{D'(V)\times D(V)}|}{\|\varphi\|_{W^{-s,p'}(U)}} \\
&\preceq \frac{\|u\|_{W^{s,p}(V)}\||\det(T^{-1})'|\varphi\circ T^{-1}\|_{W^{-s,p'}(V)}}{\|\varphi\|_{W^{-s,p'}(U)}} \\
&\overset{|\det(T^{-1})'|\in BC^\infty(V)}{\preceq} \frac{\|u\|_{W^{s,p}(V)}\|\varphi\circ T^{-1}\|_{W^{-s,p'}(V)}}{\|\varphi\|_{W^{-s,p'}(U)}}\,.
\end{aligned}$$

Since $-s > 0$, it follows from the hypotheses of this theorem and the result of Theorem 78 that $\|\varphi\circ T^{-1}\|_{W^{-s,p'}(V)} \preceq \|\varphi\|_{W^{-s,p'}(U)}$. Consequently,

$$\|T^*u\|_{W^{s,p}(U)} \preceq \|u\|_{W^{s,p}(V)}\,.$$

□

Lemma 8. *Let U and V be two nonempty open sets in $\mathbb{R}^n$. Suppose $T : U \to V$ ($T = (T^1,\dots,T^n)$) is a C^{k+1}-diffeomorphism for some $k \in \mathbb{N}_0$ and let $B \subseteq U$ be a nonempty bounded open set such that $B \subseteq \bar{B} \subseteq U$. Then*

(1) $T : B \to T(B)$ is a $(k+1)$-smooth map.
(2) $T : B \to T(B)$ and $T^{-1} : T(B) \to B$ are Lipschitz (the Lipschitz constant may depend on B).
(3) For all $1 \leq i \leq n$ and $|\alpha| \leq k$, $\partial^\alpha T^i \in BC^{k,1}(B)$ and $\partial^\alpha (T^{-1})^i \in BC^{k,1}(T(B))$.

Proof. Item 1 is true because $\bar{B}$ is compact and so $T(\bar{B})$ is compact and continuous functions are bounded on compact sets. Items 2 and 3 are direct consequences of Theorem 7. □

Theorem 80. *Let $s \in \mathbb{R}$ and $p \in (1,\infty)$. Suppose that U and V are two nonempty open sets in $\mathbb{R}^n$ and $T : U \to V$ is a C^∞-diffeomorphism (if $s \geq 0$ it is enough to assume T is a $C^{\lfloor s\rfloor+1}$-diffeomorphism). Let $B \subseteq U$ be a nonempty bounded open set such that $B \subseteq \bar{B} \subseteq U$. Let $u \in W^{s,p}(V)$ be such that $\operatorname{supp} u \subseteq T(B)$ (note that if $\operatorname{supp} u$ is compact in V, then such a B exists).*

(1) *If s is NOT a noninteger less than* -1, *then*

$$\|u \circ T\|_{W^{s,p}(U)} \preceq \|u\|_{W^{s,p}(V)}.$$

(The implicit constant may depend on B but otherwise is independent of u.)

(2) *If U and V are Lipschitz or* $\mathbb{R}^n$, *then the above result holds for all* $s \in \mathbb{R}$.

Proof. If s is an integer or $-1 < s < 1$, or if U and V are Lipschitz or $\mathbb{R}^n$ and $s \in \mathbb{R}$ then as a consequence of the above lemma and the preceding theorems we may write

$$\|u \circ T\|_{W^{s,p}(U)} \overset{\text{Corollary 6}}{\simeq} \|u \circ T\|_{W^{s,p}(B)} \preceq \|u\|_{W^{s,p}(T(B))} \overset{\text{Corollary 6}}{\simeq} \|u\|_{W^{s,p}(V)}.$$

For general U and V, if $s = k + \theta$, we let $\hat{B}$ be an open set such that $\bar{\hat{B}}$ is a compact subset of U and $\bar{B} \subseteq \hat{B}$. We can apply the previous lemma to $\hat{B}$ and write

$$\|u \circ T\|_{W^{s,p}(U)} \overset{\text{Corollary 6}}{\simeq} \|u \circ T\|_{W^{s,p}_{\bar{B}}(\hat{B})} \overset{\text{Theorem 78}}{\preceq} \|u\|_{W^{s,p}_{T(\bar{B})}(T(\hat{B}))} \overset{\text{Corollary 6}}{\simeq} \|u\|_{W^{s,p}(V)}.$$

□

Theorem 81 ([42]). *Let* $s \in [1, \infty)$, $1 < p < \infty$, *and let*

$$m = \begin{cases} s, & \text{if } s \text{ is an integer} \\ \lfloor s \rfloor + 1, & \text{otherwise} \end{cases}.$$

If $F \in C^m(\mathbb{R})$ *is such that* $F(0) = 0$ *and* $F, F', \ldots, F^{(m)} \in L^\infty(\mathbb{R})$ *(in particular, note that every* $F \in C_c^\infty(\mathbb{R})$ *with* $F(0) = 0$ *satisfies these conditions), then the map* $u \mapsto F(u)$ *is well-defined and continuous from* $W^{s,p}(\mathbb{R}^n) \cap W^{1,sp}(\mathbb{R}^n)$ *into* $W^{s,p}(\mathbb{R}^n)$.

Corollary 7. *Let s, p, and F be as in the previous theorem. Moreover, suppose* $sp > n$. *Then the map* $u \mapsto F(u)$ *is well-defined and continuous from* $W^{s,p}(\mathbb{R}^n)$ *into* $W^{s,p}(\mathbb{R}^n)$. *The reason is that when* $sp > n$, *we have* $W^{s,p}(\mathbb{R}^n) \hookrightarrow W^{1,sp}(\mathbb{R}^n)$.

7.6. Differentiation

Theorem 82 (([4], pp. 598–605), ([5], Section 1.4)). *Let* $s \in \mathbb{R}$, $1 < p < \infty$, *and* $\alpha \in \mathbb{N}_0^n$. *Suppose* Ω *is a nonempty open set in* $\mathbb{R}^n$. *Then*

(1) *The linear operator* $\partial^\alpha : W^{s,p}(\mathbb{R}^n) \to W^{s-|\alpha|,p}(\mathbb{R}^n)$ *is well-defined and bounded.*

(2) *For* $s < 0$, *the linear operator* $\partial^\alpha : W^{s,p}(\Omega) \to W^{s-|\alpha|,p}(\Omega)$ *is well-defined and bounded.*

(3) *For* $s \geq 0$ *and* $|\alpha| \leq s$, *the linear operator* $\partial^\alpha : W^{s,p}(\Omega) \to W^{s-|\alpha|,p}(\Omega)$ *is well-defined and bounded.*

(4) *If* Ω *is bounded with Lipschitz continuous boundary, and if* $s \geq 0$, $s - \frac{1}{p} \neq$ *integer (i.e., the fractional part of s is not equal to* $\frac{1}{p}$*), then the linear operator* $\partial^\alpha : W^{s,p}(\Omega) \to W^{s-|\alpha|,p}(\Omega)$ *for* $|\alpha| > s$ *is well-defined and bounded.*

Remark 51. *Comparing the first and last items of the previous theorem, we see that not all the properties of Sobolev–Slobodeckij spaces on* $\mathbb{R}^n$ *are fully inherited by Sobolev–Slobodeckij spaces on bounded domains even when the domain has Lipschitz continuous boundary (note that the above difference is related to the more fundamental fact that for* $s > 0$, *even when* Ω *is Lipschitz,* $C_c^\infty(\Omega)$ *is not necessarily dense in* $W^{s,p}(\Omega)$ *and subsequently* $W^{-s,p'}(\Omega)$ *is defined as the dual of* $W_0^{s,p}(\Omega)$ *rather than the dual of* $W^{s,p}(\Omega)$ *itself). For this reason, when working with Sobolev spaces on manifolds, we prefer super nice atlases (i.e., we prefer to work with coordinate charts whose image under the coordinate map is the entire* $\mathbb{R}^n$*). The next best choice would be GGL or GL atlases.*

7.7. Spaces of Locally Sobolev Functions

Material of this section are taken from our manuscript on the properties of locally Sobolev-Slobodeckij functions [17].

Definition 28. *Let $s \in \mathbb{R}$, $1 < p < \infty$. Let Ω be a nonempty open set in $\mathbb{R}^n$. We define*

$$W^{s,p}_{loc}(\Omega) := \{u \in D'(\Omega) : \forall \varphi \in C^{\infty}_c(\Omega) \quad \varphi u \in W^{s,p}(\Omega)\}.$$

$W^{s,p}_{loc}(\Omega)$ is equipped with the natural topology induced by the separating family of seminorms $\{|\cdot|_{\varphi}\}_{\varphi \in C^{\infty}_c(\Omega)}\}$ where

$$\forall u \in W^{s,p}_{loc}(\Omega) \quad \varphi \in C^{\infty}_c(\Omega) \qquad |u|_{\varphi} := \|\varphi u\|_{W^{s,p}(\Omega)}.$$

Theorem 83. *Let $s \in \mathbb{R}$, $1 < p < \infty$, and $\alpha \in \mathbb{N}^n_0$. Suppose Ω is a nonempty bounded open set in $\mathbb{R}^n$ with Lipschitz continuous boundary. Then*

(1) *The linear operator $\partial^{\alpha} : W^{s,p}_{loc}(\mathbb{R}^n) \to W^{s-|\alpha|,p}_{loc}(\mathbb{R}^n)$ is well-defined and continuous.*
(2) *For $s < 0$, the linear operator $\partial^{\alpha} : W^{s,p}_{loc}(\Omega) \to W^{s-|\alpha|,p}_{loc}(\Omega)$ is well-defined and continuous.*
(3) *For $s \geq 0$ and $|\alpha| \leq s$, the linear operator $\partial^{\alpha} : W^{s,p}_{loc}(\Omega) \to W^{s-|\alpha|,p}_{loc}(\Omega)$ is well-defined and continuous.*
(4) *If $s \geq 0$, $s - \frac{1}{p} \neq$ integer (i.e., the fractional part of s is not equal to $\frac{1}{p}$), then the linear operator $\partial^{\alpha} : W^{s,p}_{loc}(\Omega) \to W^{s-|\alpha|,p}_{loc}(\Omega)$ for $|\alpha| > s$ is well-defined and continuous.*

The following statements play a key role in our study of Sobolev spaces on Riemannian manifolds with rough metrics.

Theorem 84. *Let Ω be a nonempty bounded open set in $\mathbb{R}^n$ with Lipschitz continuous boundary or $\Omega = \mathbb{R}^n$. Suppose $u \in W^{s,p}_{loc}(\Omega)$ where $sp > n$. Then u has a continuous version.*

Lemma 9. *Let $\Omega = \mathbb{R}^n$ or Ω be a bounded open set in $\mathbb{R}^n$ with Lipschitz continuous boundary. Suppose $s_1, s_2, s \in \mathbb{R}$ and $1 < p_1, p_2, p < \infty$ are such that*

$$W^{s_1,p_1}(\Omega) \times W^{s_2,p_2}(\Omega) \hookrightarrow W^{s,p}(\Omega).$$

Then

(1) $W^{s_1,p_1}_{loc}(\Omega) \times W^{s_2,p_2}_{loc}(\Omega) \hookrightarrow W^{s,p}_{loc}(\Omega)$,
(2) *For all $K \in \mathcal{K}(\Omega)$, $W^{s_1,p_1}_{loc}(\Omega) \times W^{s_2,p_2}_{K}(\Omega) \hookrightarrow W^{s,p}(\Omega)$. In particular, if $f \in W^{s_1,p_1}_{loc}(\Omega)$, then the mapping $u \mapsto fu$ is a well-defined continuous linear map from $W^{s_2,p_2}_{K}(\Omega)$ to $W^{s,p}(\Omega)$.*

Remark 52. *It can be shown that the locally Sobolev spaces on Ω are metrizable, so the continuity of the mapping*

$$W^{s_1,p_1}_{loc}(\Omega) \times W^{s_2,p_2}_{loc}(\Omega) \to W^{s,p}_{loc}(\Omega), \quad (u,v) \mapsto uv$$

in the above lemma can be interpreted as follows: if $u_i \to u$ in $W^{s_1,p_1}_{loc}(\Omega)$ and $v_i \to v$ in $W^{s_2,p_2}_{loc}(\Omega)$, then $u_i v_i \to uv$ in $W^{s,p}_{loc}(\Omega)$. Furthermore, since $W^{s_2,p_2}_{K}(\Omega)$ is considered as a normed subspace of $W^{s_2,p_2}(\Omega)$, we have a similar interpretation of the continuity of the mapping in item 2.

Lemma 10. *Let $\Omega = \mathbb{R}^n$ or let Ω be a nonempty bounded open set in $\mathbb{R}^n$ with Lipschitz continuous boundary. Let $s \in \mathbb{R}$ and $p \in (1, \infty)$ be such that $sp > n$. Let $B : \Omega \to GL(k, \mathbb{R})$. Suppose for all $x \in \Omega$ and $1 \leq i,j \leq k$, $B_{ij}(x) \in W^{s,p}_{loc}(\Omega)$. Then*

(1) $\det B \in W^{s,p}_{loc}(\Omega)$.
(2) *Moreover, if for each $m \in \mathbb{N}$ $B_m : \Omega \to GL(k, \mathbb{R})$ and for all $1 \leq i,j \leq k$ $(B_m)_{ij} \to B_{ij}$ in $W^{s,p}_{loc}(\Omega)$, then $\det B_m \to \det B$ in $W^{s,p}_{loc}(\Omega)$.*

Theorem 85. *Let $\Omega = \mathbb{R}^n$ or let Ω be a nonempty bounded open set in $\mathbb{R}^n$ with Lipschitz continuous boundary. Let $s \geq 1$ and $p \in (1, \infty)$ be such that $sp > n$.*

(1) *Suppose that $u \in W^{s,p}_{loc}(\Omega)$ and that $u(x) \in I$ for all $x \in \Omega$ where I is some interval in $\mathbb{R}$. If $F : I \to \mathbb{R}$ is a smooth function, then $F(u) \in W^{s,p}_{loc}(\Omega)$.*
(2) *Suppose that $u_m \to u$ in $W^{s,p}_{loc}(\Omega)$ and that for all $m \geq 1$ and $x \in \Omega$, $u_m(x), u(x) \in I$ where I is some open interval in $\mathbb{R}$. If $F : I \to \mathbb{R}$ is a smooth function, then $F(u_m) \to F(u)$ in $W^{s,p}_{loc}(\Omega)$.*
(3) *If $F : \mathbb{R} \to \mathbb{R}$ is a smooth function, then the map taking u to $F(u)$ is continuous from $W^{s,p}_{loc}(\Omega)$ to $W^{s,p}_{loc}(\Omega)$.*

8. Lebesgue Spaces on Compact Manifolds

Let M^n be a compact smooth manifold and $E \to M$ be a smooth vector bundle of rank r.

Definition 29. *A collection $\{(U_\alpha, \varphi_\alpha, \rho_\alpha, \psi_\alpha)\}_{1 \leq \alpha \leq N}$ of 4-tuples is called an* ***augmented total trivialization atlas*** *for $E \to M$ provided that $\{(U_\alpha, \varphi_\alpha, \rho_\alpha)\}_{1 \leq \alpha \leq N}$ is a total trivialization atlas for $E \to M$ and $\{\psi_\alpha\}$ is a partition of unity subordinate to the open cover $\{U_\alpha\}$.*

Let $\{(U_\alpha, \varphi_\alpha, \rho_\alpha, \psi_\alpha)\}_{1 \leq \alpha \leq N}$ be an augmented total trivialization atlas for $E \to M$. Let g be a continuous Riemannian metric on M and $\langle ., . \rangle_E$ be a fiber metric on E (we denote the corresponding norm by $|.|_E$). Suppose $1 \leq q < \infty$.

(1) **Definition A:** The space $L^q(M, E)$ is the completion of $C^\infty(M, E)$ with respect to the following norm:

$$\|u\|_{L^q(M,E)} := \sum_{\alpha=1}^{N} \sum_{l=1}^{r} \|\rho^l_\alpha \circ (\psi_\alpha u) \circ \varphi_\alpha^{-1}\|_{L^q(\varphi_\alpha(U_\alpha))}.$$

Note that for this definition to make sense it is not necessary to have metric on M or fiber metric on E.

(2) **Definition B:** The space $L^q(M, E)$ is the completion of $C^\infty(M, E)$ with respect to the following norm:

$$|u|_{L^q(M,E)} := \left(\int_M |u|^q_E dV_g \right)^{\frac{1}{q}}.$$

(3) **Definition C:** The metric g defines a measure on M. Define the following equivalence relation on $\Gamma(M, E)$:

$$u \sim v \Longleftrightarrow u = v \ a.e.$$

We define

$$L^q(M, E) := \frac{\{u \in \Gamma(M, E) : \|u\|^q_{L^q(M,E)} := \int_M |u|^q_E dV_g < \infty\}}{\sim}.$$

For $q = \infty$ we define

$$L^\infty(M, E) := \frac{\{u \in \Gamma(M, E) : \|u\|_{L^\infty(M,E)} := \text{esssup}|u|_E < \infty\}}{\sim}.$$

Note: We may define negligible sets (sets of measure zero) on a compact manifold using charts (see Chapter 6 in [43]); it can be shown that this definition is independent of the charts and equivalent to the one that is obtained using the metric g. So, it is meaningful to write $u = v$ *a.e* even without using a metric.

Theorem 86. *Definition A is equivalent to Definition B (i.e., the norms are equivalent).*

Proof. Our proof consists of four steps:

- **Step 1:** In the next section it will be proved that different total trivialization atlases and partitions of unity result in equivalent norms (note that $L^q = W^{0,q}$). Therefore, without loss of generality we may assume that $\{(U_\alpha, \varphi_\alpha, \rho_\alpha)\}_{1\leq\alpha\leq N}$ is a total trivialization atlas that trivializes the fiber metric $\langle .,. \rangle_E$ (see Theorem 37 and Corollary 2). So, on any bundle chart (U, φ, ρ) and for any section u we have

$$|u|_E^2 \circ \varphi^{-1} = \langle u, u\rangle_E \circ \varphi^{-1} = \sum_{l=1}^{r} (\rho^l \circ u \circ \varphi^{-1})^2 .$$

- **Step 2:** In this step we show that if there is $1 \leq \beta \leq N$ such that $\operatorname{supp} u \subseteq U_\beta$, then

$$|u|^q_{L^q(M,E)} = \int_M |u|_E^q dV_g \simeq \sum_{l=1}^{r} \|\rho_\beta^l \circ u \circ \varphi_\beta^{-1}\|^q_{L^q(\varphi_\beta(U_\beta))} .$$

We have

$$\begin{aligned}
\int_M |u|_E^q dV_g &= \int_{\varphi_\beta(U_\beta)} (|u|_E \circ \varphi_\beta^{-1})^q \sqrt{\det(g_{ij}\circ\varphi_\beta^{-1})(x)}\, dx^1 \dots dx^n \\
&\simeq \int_{\varphi_\beta(U_\beta)} (|u|_E \circ \varphi_\beta^{-1})^q\, dx^1 \dots dx^n \qquad (\sqrt{\det(g_{ij}\circ\varphi_\beta^{-1})(x)} \text{ is bounded by positive constants}) \\
&= \int_{\varphi_\beta(U_\beta)} \Big(\sqrt{\sum_{l=1}^{r} (\rho_\beta^l \circ u \circ \varphi_\beta^{-1})^2}\Big)^q dx^1 \dots dx^n \\
&\simeq \int_{\varphi_\beta(U_\beta)} [\sum_{l=1}^{r} |\rho_\beta^l \circ u \circ \varphi_\beta^{-1}|]^q\, dx^1 \dots dx^n \qquad (\sqrt{\textstyle\sum a_l^2} \simeq \textstyle\sum |a_l|) \\
&\simeq \int_{\varphi_\beta(U_\beta)} \sum_{l=1}^{r} |\rho_\beta^l \circ u \circ \varphi_\beta^{-1}|^q\, dx^1 \dots dx^n \qquad ((\textstyle\sum a_l)^q \simeq \textstyle\sum a_l^q) \\
&= \sum_{l=1}^{r} \int_{\varphi_\beta(U_\beta)} |\rho_\beta^l \circ u \circ \varphi_\beta^{-1}|^q\, dx^1 \dots dx^n = \sum_{l=1}^{r} \|\rho_\beta^l \circ u \circ \varphi_\beta^{-1}\|^q_{L^q(\varphi_\beta(U_\beta))} .
\end{aligned}$$

- **Step 3:** In this step we will prove that for all $u \in C^\infty(M, E)$

$$|u|^q_{L^q(M,E)} \simeq \sum_\alpha |\psi_\alpha u|^q_{L^q(M,E)} .$$

We have

$$\begin{aligned}
|u|^q_{L^q(M,E)} &= \int_M |u|_E^q dV_g = \sum_\alpha \int_M \frac{\psi_\alpha^q}{\sum_\beta \psi_\beta^q} |u|_E^q dV_g \qquad (\{\tfrac{\psi_\alpha^q}{\sum_\beta \psi_\beta^q}\} \text{ is a partition of unity subordinate to } \{U_\alpha\}) \\
&\simeq \sum_\alpha \int_{U_\alpha} \psi_\alpha^q |u|_E^q dV_g \qquad (\frac{1}{\sum_\beta \psi_\beta^q} \text{ is bounded by positive constants}) \\
&= \sum_\alpha \int_{U_\alpha} |\psi_\alpha u|_E^q dV_g = \sum_\alpha \int_M |\psi_\alpha u|_E^q dV_g \\
&= \sum_\alpha |\psi_\alpha u|^q_{L^q(M,E)} .
\end{aligned}$$

- **Step 4:** Let u be an arbitrary element of $C^\infty(M, E)$. We have

$$|u|^q_{L^q(M,E)} \overset{\text{Step 3}}{\simeq} \sum_\alpha |\psi_\alpha u|^q_{L^q(M,E)} \overset{\text{Step 2}}{\simeq} \sum_\alpha \sum_l \|\rho_\alpha^l \circ (\psi_\alpha u) \circ \varphi_\alpha^{-1}\|^q_{L^q(\varphi_\alpha(U_\alpha))} \simeq \|u\|^q_{L^q(M,E)} .$$

□

9. Sobolev Spaces on Compact Manifolds and Alternative Characterizations

9.1. The Definition

Let M^n be a compact smooth manifold. Let $\pi : E \to M$ be a smooth vector bundle of rank r. Let $\Lambda = \{(U_\alpha, \varphi_\alpha, \rho_\alpha, \psi_\alpha)\}_{1\leq\alpha\leq N}$ be an augmented total trivialization atlas for $E \to M$. For each $1 \leq \alpha \leq N$, let H_α denote the map $H_{E^\vee, U_\alpha, \varphi_\alpha}$ which was introduced in Section 6.

Definition 30.

$$W^{e,q}(M, E; \Lambda) = \{u \in D'(M, E) : \|u\|_{W^{e,q}(M,E;\Lambda)} = \sum_{\alpha=1}^{N}\sum_{l=1}^{r} \|[H_\alpha(\psi_\alpha u)]^l\|_{W^{e,q}(\varphi_\alpha(U_\alpha))} < \infty\}.$$

Remark 53.

(1) *If $u \in W^{e,q}(M, E; \Lambda)$ is a regular distribution, it follows from Remark 32 that*

$$\|u\|_{W^{e,q}(M,E;\Lambda)} = \sum_{\alpha=1}^{N}\sum_{l=1}^{r} \|(\rho_\alpha)^l \circ (\psi_\alpha u) \circ \varphi_\alpha^{-1}\|_{W^{e,q}(\varphi_\alpha(U_\alpha))}.$$

(2) *It is clear that the collection of functions from M to $\mathbb{R}$ can be identified with sections of the vector bundle $E = M \times \mathbb{R}$. For this reason $W^{e,q}(M; \Lambda)$ is defined as $W^{e,q}(M, M \times \mathbb{R}; \Lambda)$. Note that in this case, for each α, ρ_α is the identity map. So, we may consider an augmented total trivialization atlas Λ as a collection of 3-tuples $\{(U_\alpha, \varphi_\alpha, \psi_\alpha)\}_{1\leq\alpha\leq N}$. In particular, if $u \in W^{e,q}(M; \Lambda)$ is a regular distribution, then*

$$\|u\|_{W^{e,q}(M;\Lambda)} = \sum_{\alpha=1}^{N} \|(\psi_\alpha u) \circ \varphi_\alpha^{-1}\|_{W^{e,q}(\varphi_\alpha(U_\alpha))}.$$

(3) *Sometimes, when the underlying manifold M and the augmented total trivialization atlas are clear from the context (or when they are irrelevant), we may write $W^{e,q}(E)$ instead of $W^{e,q}(M, E; \Lambda)$. In particular, for tensor bundles, we may write $W^{e,q}(T^k_l M)$ instead of $W^{e,q}(M, T^k_l M; \Lambda)$.*

Remark 54. *Here is a list of some alternative, not necessarily equivalent, characterizations of Sobolev spaces.*

(1) *Suppose $e \geq 0$.*

$$W^{e,q}(M, E; \Lambda) = \{u \in L^q(M, E) : \|u\|_{W^{e,q}(M,E;\Lambda)} = \sum_{\alpha=1}^{N}\sum_{l=1}^{r} \|(\rho_\alpha)^l \circ (\psi_\alpha u) \circ \varphi_\alpha^{-1}\|_{W^{e,q}(\varphi_\alpha(U_\alpha))} < \infty\}.$$

(2)

$$W^{e,q}(M, E; \Lambda) = \{u \in D'(M, E) : \|u\|_{W^{e,q}(M,E;\Lambda)} = \sum_{\alpha=1}^{N}\sum_{l=1}^{r} \|ext^0_{\varphi_\alpha(U_\alpha),\mathbb{R}^n}[H_\alpha(\psi_\alpha u)]^l\|_{W^{e,q}(\mathbb{R}^n)} < \infty\}.$$

(3)

$$W^{e,q}(M, E; \Lambda) = \{u \in D'(M, E) : [H_\alpha(u|_{U_\alpha})]^l \in W^{e,q}_{loc}(\varphi_\alpha(U_\alpha)),\ \forall 1 \leq \alpha \leq N, \forall 1 \leq l \leq r\}.$$

(4) *$W^{e,q}(M, E; \Lambda)$ is the completion of $C^\infty(M, E)$ with respect to the norm*

$$\|u\|_{W^{e,q}(M,E;\Lambda)} = \sum_{\alpha=1}^{N}\sum_{l=1}^{r} \|(\rho_\alpha)^l \circ (\psi_\alpha u) \circ \varphi_\alpha^{-1}\|_{W^{e,q}(\varphi_\alpha(U_\alpha))}.$$

(5) • *Let g be a smooth Riemannian metric (i.e., a fiber metric on TM). So, g^{-1} is a fiber metric on T^*M.*
• *Let $\langle .,. \rangle_E$ be a smooth fiber metric on E.*
• *Let ∇^E be a metric connection in the vector bundle $\pi : E \to M$.*

For $k \in \mathbb{N}_0$, $W^{k,q}(M, E; g, \nabla^E)$ is the completion of $C^\infty(M, E)$ with respect to the following norm:

$$\|u\|_{W^{k,q}(M,E;g,\nabla^E)} = \Big(\sum_{i=0}^{k} |(\nabla^E)^i u|_{L^q}^q\Big)^{\frac{1}{q}} = \Big(\sum_{i=0}^{k} \int_M |\underbrace{\nabla^E \dots \nabla^E}_{i \text{ times}} u|^q_{(T^*M)^{\otimes i}\otimes E} dV_g\Big)^{\frac{1}{q}}.$$

In particular, if we denote the Levi Civita connection corresponding to the smooth Riemannian metric g by ∇, then $W^{k,q}(M; g)$ is the completion of $C^\infty(M)$ with respect to the following norm

$$\|u\|_{W^{k,q}(M;g)} = \Big(\sum_{i=0}^{k} |\nabla^i u|_{L^q}^q\Big)^{\frac{1}{q}} = \Big(\sum_{i=0}^{k} \int_M |\underbrace{\nabla \dots \nabla}_{i \text{ times}} u|^q_{T^iM} dV_g\Big)^{\frac{1}{q}}.$$

In the subsequent discussions we will study the relation between each of these alternative descriptions of Sobolev spaces and Definition 30.

Remark 55. *As it is discussed for example in [18], Sobolev-Slobodeckij spaces on $\mathbb{R}^n$ with non-integer smoothness degree can be defined using real interpolation. Indeed, for $s \in \mathbb{R} \setminus \mathbb{Z}$ and $\theta = s - \lfloor s \rfloor$,*

$$W^{s,p}(\mathbb{R}^n) = \big(W^{\lfloor s \rfloor,p}(\mathbb{R}^n), W^{\lfloor s \rfloor+1,p}(\mathbb{R}^n)\big)_{\theta,p}.$$

One may use any of the previously mentioned descriptions to define $W^{k,q}(M, E)$ for $k \in \mathbb{Z}$, and then use real interpolation to define $W^{e,q}(M, E)$ for $e \notin \mathbb{Z}$. We postpone the study of this approach to an independent manuscript with focus on the role of interpolation theory in investigation of Bessel potential spaces and Sobolev–Slobodeckij spaces on compact manifolds.

An important question is whether our definition of Sobolev spaces (as topological spaces) depends on the augmented total trivialization atlas Λ. We will answer this question at 3 levels. Although each level can be considered as a generalization of the preceding level, the proofs will be independent of each other. The following theorems show that at least when e is not a noninteger less than -1, the space $W^{e,q}(M, E; \Lambda)$ and its topology are independent of the choice of augmented total trivialization atlas.

Remark 56. *In the following theorems, by the equivalence of two norms $\|.\|_1$ and $\|.\|_2$ we mean there exist constants C_1 and C_2 such that*

$$C_1\|.\|_1 \le \|.\|_2 \le C_2\|.\|_1,$$

where C_1 and C_2 may depend on

$$n, e, q, \varphi_\alpha, U_\alpha, \tilde{\varphi}_\beta, \tilde{U}_\beta, \psi_\alpha, \tilde{\psi}_\beta.$$

Theorem 87 (Equivalence of norms for functions). *Let $e \in \mathbb{R}$ and $q \in (1, \infty)$. Let $\Lambda = \{(U_\alpha, \varphi_\alpha, \psi_\alpha)\}_{1 \le \alpha \le N}$ and $Y = \{(\tilde{U}_\beta, \tilde{\varphi}_\beta, \tilde{\psi}_\beta)\}_{1 \le \beta \le \tilde{N}}$ be two augmented total trivialization atlases for the trivial bundle $M \times \mathbb{R} \to M$. Furthermore, let $\mathcal{W}$ be any vector subspace of $W^{e,q}(M; Y)$ whose elements are regular distributions (e.g., $C^\infty(M)$).*

(1) *If e is not a noninteger less than -1, then W is a subspace of $W^{e,q}(M; \Lambda)$ as well, and the norms produced by Λ and Y are equivalent on $\mathcal{W}$.*

(2) *If e is a noninteger less than -1, further assume that the total trivialization atlases corresponding to Λ and Υ are GLC. Then W is a subspace of $W^{e,\mathcal{q}}(M;\Lambda)$ as well, and the norms produced by Λ and Υ are equivalent on $\mathcal{W}$.*

Proof. Let $u \in \Gamma_{reg}(M)$. Our goal is to show that the following expressions are comparable:

$$\sum_{\alpha=1}^{N} \|(\psi_\alpha u) \circ \varphi_\alpha^{-1}\|_{W^{e,\mathcal{q}}(\varphi_\alpha(U_\alpha))},$$

$$\sum_{\beta=1}^{\tilde{N}} \|(\tilde{\psi}_\beta u) \circ \tilde{\varphi}_\beta^{-1}\|_{W^{e,\mathcal{q}}(\tilde{\varphi}_\beta(\tilde{U}_\beta))}.$$

To this end it suffices to show that for each $1 \leq \alpha \leq N$

$$\|(\psi_\alpha u) \circ \varphi_\alpha^{-1}\|_{W^{e,\mathcal{q}}(\varphi_\alpha(U_\alpha))} \preceq \sum_{\beta=1}^{\tilde{N}} \|(\tilde{\psi}_\beta u) \circ \tilde{\varphi}_\beta^{-1}\|_{W^{e,\mathcal{q}}(\tilde{\varphi}_\beta(\tilde{U}_\beta))}.$$

We have

$$\begin{aligned}
\|(\psi_\alpha u) \circ \varphi_\alpha^{-1}\|_{W^{e,\mathcal{q}}(\varphi_\alpha(U_\alpha))} &= \|\sum_{\beta=1}^{\tilde{N}} \tilde{\psi}_\beta(\psi_\alpha u) \circ \varphi_\alpha^{-1}\|_{W^{e,\mathcal{q}}(\varphi_\alpha(U_\alpha))} \\
&\leq \sum_{\beta=1}^{\tilde{N}} \|\tilde{\psi}_\beta(\psi_\alpha u) \circ \varphi_\alpha^{-1}\|_{W^{e,\mathcal{q}}(\varphi_\alpha(U_\alpha))} \\
&\simeq \sum_{\beta=1}^{\tilde{N}} \|(\tilde{\psi}_\beta \psi_\alpha u) \circ \varphi_\alpha^{-1}\|_{W^{e,\mathcal{q}}(\varphi_\alpha(U_\alpha \cap \tilde{U}_\beta))}.
\end{aligned}$$

The last equality follows from Corollary 6 because $(\tilde{\psi}_\beta \psi_\alpha u) \circ \varphi_\alpha^{-1}$ has support in the compact set $\varphi_\alpha(\operatorname{supp} \psi_\alpha \cap \operatorname{supp} \tilde{\psi}_\beta) \subseteq \varphi_\alpha(U_\alpha \cap \tilde{U}_\beta)$. Note that here we used the assumption that if e is a noninteger less than -1, then $\varphi_\alpha(U_\alpha)$ is Lipschitz or the entire $\mathbb{R}^n$. Clearly,

$$\sum_{\beta=1}^{\tilde{N}} \|(\tilde{\psi}_\beta \psi_\alpha u) \circ \varphi_\alpha^{-1}\|_{W^{e,\mathcal{q}}(\varphi_\alpha(U_\alpha \cap \tilde{U}_\beta))} = \sum_{\beta=1}^{\tilde{N}} \|(\tilde{\psi}_\beta \psi_\alpha u) \circ \tilde{\varphi}_\beta^{-1} \circ \tilde{\varphi}_\beta \circ \varphi_\alpha^{-1}\|_{W^{e,\mathcal{q}}(\varphi_\alpha(U_\alpha \cap \tilde{U}_\beta))}.$$

Since $\tilde{\varphi}_\beta \circ \varphi_\alpha^{-1} : \varphi_\alpha(U_\alpha \cap \tilde{U}_\beta) \to \tilde{\varphi}_\beta(U_\alpha \cap \tilde{U}_\beta)$ is a C^∞-diffeomorphism and $(\tilde{\psi}_\beta \psi_\alpha u) \circ \tilde{\varphi}_\beta^{-1}$ has compact support in the compact set $\tilde{\varphi}_\beta(\operatorname{supp} \psi_\alpha \cap \operatorname{supp} \tilde{\psi}_\beta) \subseteq \tilde{\varphi}_\beta(U_\alpha \cap \tilde{U}_\beta)$, it follows from Theorem 80 that

$$\sum_{\beta=1}^{\tilde{N}} \|(\tilde{\psi}_\beta \psi_\alpha u) \circ \tilde{\varphi}_\beta^{-1} \circ \tilde{\varphi}_\beta \circ \varphi_\alpha^{-1}\|_{W^{e,\mathcal{q}}(\varphi_\alpha(U_\alpha \cap \tilde{U}_\beta))} \preceq \sum_{\beta=1}^{\tilde{N}} \|(\tilde{\psi}_\beta \psi_\alpha u) \circ \tilde{\varphi}_\beta^{-1}\|_{W^{e,\mathcal{q}}(\tilde{\varphi}_\beta(U_\alpha \cap \tilde{U}_\beta))}.$$

Note that here we used the assumption that if e is a noninteger less than -1, then the two total trivialization atlases are GL compatible. As a direct consequence of Corollary 5 and Theorem 71 we have

$$\begin{aligned}
\|(\tilde{\psi}_\beta \psi_\alpha u) \circ \tilde{\varphi}_\beta^{-1}\|_{W^{e,\mathcal{q}}(\tilde{\varphi}_\beta(U_\alpha \cap \tilde{U}_\beta))} &\simeq \|(\tilde{\psi}_\beta \psi_\alpha u) \circ \tilde{\varphi}_\beta^{-1}\|_{W^{e,\mathcal{q}}(\tilde{\varphi}_\beta(\tilde{U}_\beta))} \\
&= \|(\psi_\alpha \circ \tilde{\varphi}_\beta^{-1})[(\tilde{\psi}_\beta u) \circ \tilde{\varphi}_\beta^{-1}]\|_{W^{e,\mathcal{q}}(\tilde{\varphi}_\beta(\tilde{U}_\beta))}.
\end{aligned}$$

Now, note that $\psi_\alpha \circ \tilde{\varphi}_\beta^{-1} \in C^\infty(\tilde{\varphi}_\beta(\tilde{U}_\beta))$ and $(\tilde{\psi}_\beta u) \circ \tilde{\varphi}_\beta^{-1}$ has support in the compact set $\tilde{\varphi}_\beta(\operatorname{supp} \tilde{\psi}_\beta)$. Therefore, by Theorem 70 (for the case where e is not a noninteger less than -1) and Corollary 4 (for the case where e is a noninteger less than -1) we have

$$\|(\psi_\alpha \circ \tilde{\varphi}_\beta^{-1})[(\tilde{\psi}_\beta u) \circ \tilde{\varphi}_\beta^{-1}]\|_{W^{e,q}(\tilde{\varphi}_\beta(\tilde{U}_\beta))} \preceq \|(\tilde{\psi}_\beta u) \circ \tilde{\varphi}_\beta^{-1}\|_{W^{e,q}(\tilde{\varphi}_\beta(\tilde{U}_\beta))}.$$

Hence

$$\|(\psi_\alpha u) \circ \varphi_\alpha^{-1}\|_{W^{e,q}(\varphi_\alpha(U_\alpha))} \preceq \sum_{\beta=1}^{\tilde{N}} \|(\tilde{\psi}_\beta u) \circ \tilde{\varphi}_\beta^{-1}\|_{W^{e,q}(\tilde{\varphi}_\beta(\tilde{U}_\beta))}.$$

□

Theorem 88 (Equivalence of norms for regular sections). *Let $e \in \mathbb{R}$ and $q \in (1, \infty)$. Let $\Lambda = \{(U_\alpha, \varphi_\alpha, \rho_\alpha, \psi_\alpha)\}_{1 \le \alpha \le N}$ and $\Upsilon = \{(\tilde{U}_\beta, \tilde{\varphi}_\beta, \tilde{\rho}_\beta, \tilde{\psi}_\beta)\}_{1 \le \beta \le \tilde{N}}$ be two augmented total trivialization atlases for the vector bundle $E \to M$. Furthermore, let $\mathcal{W}$ be any vector subspace of $W^{e,q}(M, E; \Upsilon)$ whose elements are regular distributions (e.g., $C^\infty(M, E)$).*

(1) *If e is not a noninteger less than -1, then W is a subspace of $W^{e,q}(M, E; \Lambda)$ as well, and the norms produced by Λ and Υ are equivalent on $\mathcal{W}$.*
(2) *If e is a noninteger less than -1, further assume that the total trivialization atlases corresponding to Λ and Υ are GLC. Then W is a subspace of $W^{e,q}(M, E; \Lambda)$ as well, and the norms produced by Λ and Υ are equivalent on $\mathcal{W}$.*

Proof. Let $u \in \Gamma_{reg}(M, E)$. Our goal is to show that the following expressions are comparable:

$$\sum_{\alpha=1}^{N} \sum_{l=1}^{r} \|\rho_\alpha^l \circ (\psi_\alpha u) \circ \varphi_\alpha^{-1}\|_{W^{e,q}(\varphi_\alpha(U_\alpha))},$$

$$\sum_{\beta=1}^{\tilde{N}} \sum_{l=1}^{r} \|\tilde{\rho}_\beta^l \circ (\tilde{\psi}_\beta u) \circ \tilde{\varphi}_\beta^{-1}\|_{W^{e,q}(\tilde{\varphi}_\beta(\tilde{U}_\beta))}.$$

To this end, it is enough to show that for each $1 \le \alpha \le N$ and $1 \le l \le r$

$$\|\rho_\alpha^l \circ (\psi_\alpha u) \circ \varphi_\alpha^{-1}\|_{W^{e,q}(\varphi_\alpha(U_\alpha))} \preceq \sum_{\beta=1}^{\tilde{N}} \sum_{t=1}^{r} \|\tilde{\rho}_\beta^t \circ (\tilde{\psi}_\beta u) \circ \tilde{\varphi}_\beta^{-1}\|_{W^{e,q}(\tilde{\varphi}_\beta(\tilde{U}_\beta))}.$$

We have

$$\begin{aligned}
\|\rho_\alpha^l \circ (\psi_\alpha u) \circ \varphi_\alpha^{-1}\|_{W^{e,q}(\varphi_\alpha(U_\alpha))} &= \|\rho_\alpha^l \circ (\sum_{\beta=1}^{\tilde{N}} \tilde{\psi}_\beta \psi_\alpha u) \circ \varphi_\alpha^{-1}\|_{W^{e,q}(\varphi_\alpha(U_\alpha))} \\
&\le \sum_{\beta=1}^{\tilde{N}} \|\rho_\alpha^l \circ (\tilde{\psi}_\beta \psi_\alpha u) \circ \varphi_\alpha^{-1}\|_{W^{e,q}(\varphi_\alpha(U_\alpha))} \\
&\simeq \sum_{\beta=1}^{\tilde{N}} \|\rho_\alpha^l \circ (\tilde{\psi}_\beta \psi_\alpha u) \circ \varphi_\alpha^{-1}\|_{W^{e,q}(\varphi_\alpha(U_\alpha \cap \tilde{U}_\beta))}.
\end{aligned}$$

The last equality follows from Corollary 6 because $\rho_\alpha^l \circ (\tilde{\psi}_\beta \psi_\alpha u) \circ \varphi_\alpha^{-1}$ has support in the compact set $\varphi_\alpha(\operatorname{supp} \psi_\alpha \cap \operatorname{supp} \tilde{\psi}_\beta) \subseteq \varphi_\alpha(U_\alpha \cap \tilde{U}_\beta)$. Note that here we used the assumption that if e is a noninteger less than -1, then $\varphi_\alpha(U_\alpha)$ is either Lipschitz or equal to the entire $\mathbb{R}^n$. Note that

$$\sum_{\beta=1}^{\tilde{N}} \|\rho_\alpha^l \circ (\tilde{\psi}_\beta \psi_\alpha u) \circ \varphi_\alpha^{-1}\|_{W^{e,q}(\varphi_\alpha(U_\alpha \cap \tilde{U}_\beta))}$$

$$= \sum_{\beta=1}^{\tilde{N}} \|\rho_\alpha^l \circ (\tilde{\psi}_\beta \psi_\alpha u) \circ \tilde{\varphi}_\beta^{-1} \circ \tilde{\varphi}_\beta \circ \varphi_\alpha^{-1}\|_{W^{e,q}(\varphi_\alpha(U_\alpha \cap \tilde{U}_\beta))}$$

$$\overset{\text{Theorem 80}}{\preceq} \sum_{\beta=1}^{\tilde{N}} \|\rho_\alpha^l \circ (\tilde{\psi}_\beta \psi_\alpha u) \circ \tilde{\varphi}_\beta^{-1}\|_{W^{e,q}(\tilde{\varphi}_\beta(U_\alpha \cap \tilde{U}_\beta))}$$

$$= \sum_{\beta=1}^{\tilde{N}} \|(\psi_\alpha \circ \tilde{\varphi}_\beta^{-1})[\rho_\alpha^l \circ (\tilde{\psi}_\beta u) \circ \tilde{\varphi}_\beta^{-1}]\|_{W^{e,q}(\tilde{\varphi}_\beta(U_\alpha \cap \tilde{U}_\beta))}$$

$$= \sum_{\beta=1}^{\tilde{N}} \|(\psi_\alpha \circ \tilde{\varphi}_\beta^{-1})\big[\pi_l \circ \underbrace{\pi' \circ \Phi_\alpha}_{\rho_\alpha} \circ (\tilde{\psi}_\beta u) \circ \tilde{\varphi}_\beta^{-1}\big]\|_{W^{e,q}(\tilde{\varphi}_\beta(U_\alpha \cap \tilde{U}_\beta))}$$

$$= \sum_{\beta=1}^{\tilde{N}} \|(\psi_\alpha \circ \tilde{\varphi}_\beta^{-1})\big[\pi_l \circ \pi' \circ \Phi_\alpha \circ \Phi_\beta^{-1} \circ \Phi_\beta \circ (\tilde{\psi}_\beta u) \circ \tilde{\varphi}_\beta^{-1}\big]\|_{W^{e,q}(\tilde{\varphi}_\beta(U_\alpha \cap \tilde{U}_\beta))}.$$

Let $v_\beta : \tilde{\varphi}_\beta(\tilde{U}_\beta) \to E$ be defined by $v_\beta(x) = (\tilde{\psi}_\beta u) \circ \tilde{\varphi}_\beta^{-1}$. Clearly $\pi(v_\beta(x)) = \tilde{\varphi}_\beta^{-1}(x)$. Therefore,

$$\Phi_\beta(v_\beta(x)) = \big(\pi(v_\beta(x)), \tilde{\rho}_\beta(v_\beta(x))\big) = \big(\tilde{\varphi}_\beta^{-1}(x), \tilde{\rho}_\beta(v_\beta(x))\big).$$

For all $x \in \tilde{\varphi}_\beta(U_\alpha \cap \tilde{U}_\beta)$ we have

$$\begin{aligned}
&\pi' \circ \Phi_\alpha \circ \Phi_\beta^{-1}(\Phi_\beta(v_\beta(x))) \\
&\quad = \pi' \circ \Phi_\alpha \circ \Phi_\beta^{-1}(\tilde{\varphi}_\beta^{-1}(x), \tilde{\rho}_\beta(v_\beta(x))) \\
&\quad \overset{\text{Lemma 4}}{=} \pi' \circ (\tilde{\varphi}_\beta^{-1}(x), \tau_{\alpha\beta}(\tilde{\varphi}_\beta^{-1}(x))\tilde{\rho}_\beta(v_\beta(x))) \\
&\quad = \underbrace{\tau_{\alpha\beta}(\tilde{\varphi}_\beta^{-1}(x))}_{\text{an } r \times r \text{ matrix}} \tilde{\rho}_\beta(v_\beta(x)).
\end{aligned}$$

Let $A_{\alpha\beta} = \tau_{\alpha\beta} \circ \tilde{\varphi}_\beta^{-1}$ on $\tilde{\varphi}_\beta(U_\alpha \cap \tilde{U}_\beta)$. So, we can write

$$\begin{aligned}
&\|\rho_\alpha^l \circ (\psi_\alpha u) \circ \varphi_\alpha^{-1}\|_{W^{e,q}(\tilde{\varphi}_\beta(U_\alpha \cap \tilde{U}_\beta))} \\
&\quad \preceq \sum_{\beta=1}^{\tilde{N}} \|(\psi_\alpha \circ \tilde{\varphi}_\beta^{-1})(x)\big[\pi_l \circ A_{\alpha\beta}(x)\tilde{\rho}_\beta(v_\beta(x))\big]\|_{W^{e,q}(\tilde{\varphi}_\beta(U_\alpha \cap \tilde{U}_\beta))} \\
&\quad = \sum_{\beta=1}^{\tilde{N}} \|(\psi_\alpha \circ \tilde{\varphi}_\beta^{-1})(x)\Big[\sum_{t=1}^{r}(A_{\alpha\beta}(x))_{lt}\tilde{\rho}_\beta^t(v_\beta(x))\Big]\|_{W^{e,q}(\tilde{\varphi}_\beta(U_\alpha \cap \tilde{U}_\beta))} \\
&\quad \leq \sum_{\beta=1}^{\tilde{N}} \sum_{t=1}^{r} \|(\psi_\alpha \circ \tilde{\varphi}_\beta^{-1})(x)(A_{\alpha\beta}(x))_{lt}\tilde{\rho}_\beta^t(v_\beta(x))\|_{W^{e,q}(\tilde{\varphi}_\beta(U_\alpha \cap \tilde{U}_\beta))}.
\end{aligned}$$

Now, note that $(A_{\alpha\beta}(x))_{lt}$ are in $C^\infty(\tilde{\varphi}_\beta(U_\alpha \cap \tilde{U}_\beta))$ and $(\psi_\alpha \circ \tilde{\varphi}_\beta^{-1})(x)\tilde{\rho}_\beta^t(v_\beta(x))$ has support inside the compact set $\tilde{\varphi}_\beta(\operatorname{supp} \tilde{\psi}_\beta \cap \operatorname{supp} \psi_\alpha)$. Therefore, by Theorem 70 (for the case where e is not a noninteger less than -1) and Corollary 4 (for the case where e is a noninteger less than -1), we have

$$\sum_{t=1}^{r} \|(\psi_\alpha \circ \tilde{\varphi}_\beta^{-1})(x)(A_{\alpha\beta}(x))_{lt}\tilde{\rho}_\beta^t(v_\beta(x))\|_{W^{e,q}(\tilde{\varphi}_\beta(U_\alpha \cap \tilde{U}_\beta))} \preceq \sum_{t=1}^{r} \|(\psi_\alpha \circ \tilde{\varphi}_\beta^{-1})(x)\tilde{\rho}_\beta^t(v_\beta(x))\|_{W^{e,q}(\tilde{\varphi}_\beta(U_\alpha \cap \tilde{U}_\beta))}.$$

Therefore,

$$\|\rho_\alpha^l \circ (\psi_\alpha u) \circ \varphi_\alpha^{-1}\|_{W^{e,q}(\varphi_\alpha(U_\alpha))}$$

$$\preceq \sum_{\beta=1}^{\tilde{N}} \sum_{t=1}^{r} \|(\psi_\alpha \circ \tilde{\varphi}_\beta^{-1})(x)\tilde{\rho}_\beta^t(v_\beta(x))\|_{W^{e,q}(\tilde{\varphi}_\beta(U_\alpha \cap \tilde{U}_\beta))}$$

$$\simeq \sum_{\beta=1}^{\tilde{N}} \sum_{t=1}^{r} \|(\psi_\alpha \circ \tilde{\varphi}_\beta^{-1})(x)\tilde{\rho}_\beta^t(v_\beta(x))\|_{W^{e,q}(\tilde{\varphi}_\beta(\tilde{U}_\beta))}$$

(Here we used Corollary 5 and Theorem 71)

$$\preceq \sum_{\beta=1}^{\tilde{N}} \sum_{t=1}^{r} \|\tilde{\rho}_\beta^t(v_\beta(x))\|_{W^{e,q}(\tilde{\varphi}_\beta(\tilde{U}_\beta))}$$

(Here we used Theorem 70 and Corollary 4)

$$= \sum_{\beta=1}^{\tilde{N}} \sum_{t=1}^{r} \|\tilde{\rho}_\beta^t \circ (\tilde{\psi}_\beta u) \circ \tilde{\varphi}_\beta^{-1}\|_{W^{e,q}(\tilde{\varphi}_\beta(\tilde{U}_\beta))}.$$

□

Theorem 89 (Equivalence of norms for distributional sections). *Let $e \in \mathbb{R}$ and $q \in (1,\infty)$. Let $\Lambda = \{(U_\alpha, \varphi_\alpha, \rho_\alpha, \psi_\alpha)\}_{1\leq\alpha\leq N}$ and $\Upsilon = \{(\tilde{U}_\beta, \tilde{\varphi}_\beta, \tilde{\rho}_\beta, \tilde{\psi}_\beta)\}_{1\leq\beta\leq\tilde{N}}$ be two augmented total trivialization atlases for the vector bundle $E \to M$.*

(1) *If e is not a noninteger less than -1, then $W^{e,q}(M,E;\Lambda)$ and $W^{e,q}(M,E;\Upsilon)$ are equivalent normed spaces.*

(2) *If e is a noninteger less than -1, further assume that the total trivialization atlases corresponding to Λ and Υ are GLC. Then $W^{e,q}(M,E;\Lambda)$ and $W^{e,q}(M,E;\Upsilon)$ are equivalent normed spaces.*

Proof. Let $u \in D'(M,E)$. We want to show the following expressions are comparable:

$$\sum_{\alpha=1}^{N} \sum_{l=1}^{r} \|[H_\alpha(\psi_\alpha u)]^l\|_{W^{e,q}(\varphi_\alpha(U_\alpha))},$$

$$\sum_{\beta=1}^{\tilde{N}} \sum_{i=1}^{r} \|[\tilde{H}_\beta(\tilde{\psi}_\beta u)]^i\|_{W^{e,q}(\tilde{\varphi}_\beta(\tilde{U}_\beta))}.$$

To this end it is enough to show that for each $1 \leq \alpha \leq N$ and $1 \leq l \leq r$

$$\|[H_\alpha(\psi_\alpha u)]^l\|_{W^{e,q}(\varphi_\alpha(U_\alpha))} \preceq \sum_{\beta=1}^{\tilde{N}} \sum_{i=1}^{r} \|[\tilde{H}_\beta(\tilde{\psi}_\beta u)]^i\|_{W^{e,q}(\tilde{\varphi}_\beta(\tilde{U}_\beta))}.$$

We have

$$[H_\alpha(\psi_\alpha u)]^l = [H_\alpha(\sum_{\beta=1}^{\tilde{N}} \tilde{\psi}_\beta \psi_\alpha u)]^l \overset{\text{Remark 31}}{=} \sum_{\beta=1}^{\tilde{N}} [H_\alpha(\tilde{\psi}_\beta \psi_\alpha u)]^l.$$

In what follows we will prove that

$$[H_\alpha(\tilde{\psi}_\beta \psi_\alpha u)]^l = \sum_{i=1}^{r} \left((A_{\alpha\beta})_{il}[\tilde{H}_\beta(\tilde{\psi}_\beta \psi_\alpha u)]^i\right) \circ \tilde{\varphi}_\beta \circ \varphi_\alpha^{-1}, \tag{4}$$

for some functions $(A_{\alpha\beta})_{il}$, $(1 \leq i \leq r)$ in $C^\infty(\tilde{\varphi}_\beta(U_\alpha \cap \tilde{U}_\beta))$. For now let us assume the validity of Equation (4) to prove the claim.

$$\|[H_\alpha(\psi_\alpha u)]^l\|_{W^{e,\mathcal{q}}(\varphi_\alpha(U_\alpha))} = \|\sum_{\beta=1}^{\tilde{N}}[H_\alpha(\tilde{\psi}_\beta\psi_\alpha u)]^l\|_{W^{e,\mathcal{q}}(\varphi_\alpha(U_\alpha))}$$

$$\leq \sum_{\beta=1}^{\tilde{N}}\|[H_\alpha(\tilde{\psi}_\beta\psi_\alpha u)]^l\|_{W^{e,\mathcal{q}}(\varphi_\alpha(U_\alpha))}$$

$$\overset{\text{Corollary 6}}{\simeq} \sum_{\beta=1}^{\tilde{N}}\|[H_\alpha(\tilde{\psi}_\beta\psi_\alpha u)]^l\|_{W^{e,\mathcal{q}}(\varphi_\alpha(U_\alpha\cap\tilde{U}_\beta))}$$

(note that by Remark 31 $[H_\alpha(\tilde{\psi}_\beta\psi_\alpha u)]^l$ has support in the compact set $\varphi_\alpha(\operatorname{supp}\psi_\alpha\cap\operatorname{supp}\tilde{\psi}_\beta)$)

$$= \sum_{\beta=1}^{\tilde{N}}\|\sum_{i=1}^{r}\left((A_{\alpha\beta})_{il}[\tilde{H}_\beta(\tilde{\psi}_\beta\psi_\alpha u)]^i\right)\circ\tilde{\varphi}_\beta\circ\varphi_\alpha^{-1}\|_{W^{e,\mathcal{q}}(\varphi_\alpha(U_\alpha\cap\tilde{U}_\beta))}$$

$$\leq \sum_{\beta=1}^{\tilde{N}}\sum_{i=1}^{r}\|\left((A_{\alpha\beta})_{il}[\tilde{H}_\beta(\tilde{\psi}_\beta\psi_\alpha u)]^i\right)\circ\tilde{\varphi}_\beta\circ\varphi_\alpha^{-1}\|_{W^{e,\mathcal{q}}(\varphi_\alpha(U_\alpha\cap\tilde{U}_\beta))}$$

$$\overset{\text{Theorem 80}}{\preceq} \sum_{\beta=1}^{\tilde{N}}\sum_{i=1}^{r}\|(A_{\alpha\beta})_{il}[\tilde{H}_\beta(\tilde{\psi}_\beta\psi_\alpha u)]^i\|_{W^{e,\mathcal{q}}(\tilde{\varphi}_\beta(U_\alpha\cap\tilde{U}_\beta))}$$

$$= \sum_{\beta=1}^{\tilde{N}}\sum_{i=1}^{r}\|(A_{\alpha\beta})_{il}(\psi_\alpha\circ\tilde{\varphi}_\beta^{-1})[\tilde{H}_\beta(\tilde{\psi}_\beta u)]^i\|_{W^{e,\mathcal{q}}(\tilde{\varphi}_\beta(U_\alpha\cap\tilde{U}_\beta))}$$

$$\preceq \sum_{\beta=1}^{\tilde{N}}\sum_{i=1}^{r}\|(\psi_\alpha\circ\tilde{\varphi}_\beta^{-1})[\tilde{H}_\beta(\tilde{\psi}_\beta u)]^i\|_{W^{e,\mathcal{q}}(\tilde{\varphi}_\beta(U_\alpha\cap\tilde{U}_\beta))}$$

$$\simeq \sum_{\beta=1}^{\tilde{N}}\sum_{i=1}^{r}\|(\psi_\alpha\circ\tilde{\varphi}_\beta^{-1})[\tilde{H}_\beta(\tilde{\psi}_\beta u)]^i\|_{W^{e,\mathcal{q}}(\tilde{\varphi}_\beta(\tilde{U}_\beta))}$$

(Here we used Corollary 5 and Theorem 71)

$$\preceq \sum_{\beta=1}^{\tilde{N}}\sum_{i=1}^{r}\|[\tilde{H}_\beta(\tilde{\psi}_\beta u)]^i\|_{W^{e,\mathcal{q}}(\tilde{\varphi}_\beta(\tilde{U}_\beta))}$$

(Here we used Theorem 70 and Corollary 4).

So, it remains to prove Equation (4). Since $\operatorname{supp}[H_\alpha(\tilde{\psi}_\beta\psi_\alpha u)]^l$ is inside the compact set $\varphi_\alpha(\operatorname{supp}\psi_\alpha\cap\operatorname{supp}\tilde{\psi}_\beta)\subseteq\varphi_\alpha(U_\alpha\cap\tilde{U}_\beta)$, it is enough to consider the action of $[H_\alpha(\tilde{\psi}_\beta\psi_\alpha u)]^l$ on elements of $C_c^\infty(\varphi_\alpha(U_\alpha\cap\tilde{U}_\beta))$. $\tilde{\varphi}_\beta\circ\varphi_\alpha^{-1}:\varphi_\alpha(U_\alpha\cap\tilde{U}_\beta)\to\tilde{\varphi}_\beta(U_\alpha\cap\tilde{U}_\beta)$ is a C^∞-diffeomorphism. Therefore, the map

$$C_c^\infty[\tilde{\varphi}_\beta(U_\alpha\cap\tilde{U}_\beta)]\to C_c^\infty[\varphi_\alpha(U_\alpha\cap\tilde{U}_\beta)],\qquad \eta\mapsto\eta\circ\tilde{\varphi}_\beta\circ\varphi_\alpha^{-1}$$

is bijective. In particular, an arbitrary element of $C_c^\infty[\varphi_\alpha(U_\alpha\cap\tilde{U}_\beta)]$ has the form $\eta\circ\tilde{\varphi}_\beta\circ\varphi_\alpha^{-1}$ where η is an element of $C_c^\infty[\tilde{\varphi}_\beta(U_\alpha\cap\tilde{U}_\beta)]$.
For all $\eta\in C_c^\infty[\tilde{\varphi}_\beta(U_\alpha\cap\tilde{U}_\beta)]$ we have (see Section 6.2.2)

$$\langle[H_\alpha(\tilde{\psi}_\beta\psi_\alpha u)]^l,\eta\circ\tilde{\varphi}_\beta\circ\varphi_\alpha^{-1}\rangle=\langle\tilde{\psi}_\beta\psi_\alpha u,g^\alpha_{l,\eta\circ\tilde{\varphi}_\beta\circ\varphi_\alpha^{-1}}\rangle,\qquad(5)$$

where $g^\alpha_{l,\eta\circ\tilde{\varphi}_\beta\circ\varphi_\alpha^{-1}}$ stands for $g_{l,\eta\circ\tilde{\varphi}_\beta\circ\varphi_\alpha^{-1},U_\alpha,\varphi_\alpha}$.

For all $y \in \varphi_\alpha(U_\alpha \cap \tilde{U}_\beta)$ we have $(x = \varphi_\alpha^{-1}(y))$

$$\rho_\alpha^\vee|_{E_x^\vee} \circ g^\alpha_{l,\eta\circ\tilde{\varphi}_\beta\circ\varphi_\alpha^{-1}} \circ \underbrace{\varphi_\alpha^{-1}(y)}_{x} = (0,\ldots,0,\underbrace{\eta\circ\tilde{\varphi}_\beta\circ\varphi_\alpha^{-1}(y)}_{l\text{th position}},0,\ldots,0),$$

$$\tilde{\rho}_\beta^\vee \circ \tilde{g}^\beta_{l,\eta} \circ \underbrace{\tilde{\varphi}_\beta^{-1}(\tilde{\varphi}_\beta\circ\varphi_\alpha^{-1}(y))}_{x} = (0,\ldots,0,\underbrace{\eta\circ\tilde{\varphi}_\beta\circ\varphi_\alpha^{-1}(y)}_{l\text{th position}},0,\ldots,0).$$

Therefore, for all $y \in \varphi_\alpha(U_\alpha \cap \tilde{U}_\beta)$

$$\rho_\alpha^\vee|_{E_x^\vee} \circ g^\alpha_{l,\eta\circ\tilde{\varphi}_\beta\circ\varphi_\alpha^{-1}} \circ \varphi_\alpha^{-1}(y) = \tilde{\rho}_\beta^\vee \circ \tilde{g}^\beta_{l,\eta} \circ \varphi_\alpha^{-1}(y),$$

which implies that on $U_\alpha \cap \tilde{U}_\beta$

$$g^\alpha_{l,\eta\circ\tilde{\varphi}_\beta\circ\varphi_\alpha^{-1}} = [\rho_\alpha^\vee|_{E_x^\vee}]^{-1} \circ [\tilde{\rho}_\beta^\vee|_{E_x^\vee}] \circ \tilde{g}^\beta_{l,\eta}. \tag{6}$$

It follows from Lemma 4 that for all $a \in E_x^\vee$

$$[\tilde{\rho}_\beta^\vee|_{E_x^\vee}] \circ [\rho_\alpha^\vee|_{E_x^\vee}]^{-1} \circ [\tilde{\rho}_\beta^\vee|_{E_x^\vee}](a) = \underbrace{\tau^{\tilde{\beta}\alpha}(x)}_{r\times r}(\tilde{\rho}_\beta^\vee|_{E_x^\vee}(a)).$$

That is,

$$[\rho_\alpha^\vee|_{E_x^\vee}]^{-1} \circ [\tilde{\rho}_\beta^\vee|_{E_x^\vee}](a) = [\tilde{\rho}_\beta^\vee|_{E_x^\vee}]^{-1}[\tau^{\tilde{\beta}\alpha}(x)(\tilde{\rho}_\beta^\vee|_{E_x^\vee}(a))].$$

For $a = \tilde{g}^\beta_{l,\eta}(x)$ we have

$$\tilde{\rho}_\beta^\vee|_{E_x^\vee}(a) = \tilde{\rho}_\beta^\vee|_{E_x^\vee}(\tilde{g}^\beta_{l,\eta}(x)) = (0,\ldots,0,\underbrace{\eta\circ\tilde{\varphi}_\beta(x)}_{l\text{th position}},0,\ldots,0).$$

So,

$$\begin{aligned}
[\rho_\alpha^\vee|_{E_x^\vee}]^{-1} \circ [\tilde{\rho}_\beta^\vee|_{E_x^\vee}] \circ \tilde{g}^\beta_{l,\eta} &= [\tilde{\rho}_\beta^\vee|_{E_x^\vee}]^{-1}[\tau^{\tilde{\beta}\alpha}(x)(\tilde{\rho}_\beta^\vee|_{E_x^\vee}(\tilde{g}^\beta_{l,\eta}(x)))] = [\tilde{\rho}_\beta^\vee|_{E_x^\vee}]^{-1}((\eta\circ\tilde{\varphi}_\beta)\begin{bmatrix}\tau_{1l}^{\tilde{\beta}\alpha}\\ \vdots \\ \tau_{rl}^{\tilde{\beta}\alpha}\end{bmatrix}) \\
&= [\tilde{\rho}_\beta^\vee|_{E_x^\vee}]^{-1}(\begin{bmatrix}(\eta\circ\tilde{\varphi}_\beta)\tau_{1l}^{\tilde{\beta}\alpha}\\ 0 \\ \vdots \\ 0\end{bmatrix} + \cdots + \begin{bmatrix}0\\ \vdots \\ 0 \\ (\eta\circ\tilde{\varphi}_\beta)\tau_{rl}^{\tilde{\beta}\alpha}\end{bmatrix}) \\
&= \tilde{g}^\beta_{1,(\tau_{1l}^{\tilde{\beta}\alpha}\circ\tilde{\varphi}_\beta^{-1})\eta} + \cdots + \tilde{g}^\beta_{r,(\tau_{rl}^{\tilde{\beta}\alpha}\circ\tilde{\varphi}_\beta^{-1})\eta}.
\end{aligned} \tag{7}$$

It follows from (5)–(7) that for all $\eta \in C_c^\infty[\tilde{\varphi}_\beta(U_\alpha \cap \tilde{U}_\beta)]$

$$\begin{aligned}
&\langle [H_\alpha(\tilde{\psi}_\beta \psi_\alpha u)]^l, \eta \circ \tilde{\varphi}_\beta \circ \varphi_\alpha^{-1}\rangle = \langle \tilde{\psi}_\beta \psi_\alpha u, [\rho_\alpha^\vee|_{E_x^\vee}]^{-1} \circ [\tilde{\rho}_\beta^\vee|_{E_x^\vee}] \circ \tilde{g}^\beta_{l,\eta}\rangle \\
&= \langle \tilde{\psi}_\beta \psi_\alpha u, \sum_{i=1}^r \tilde{g}^\beta_{i,(\tau_{il}^{\beta\alpha} \circ \tilde{\varphi}_\beta^{-1})\eta}\rangle \\
&= \sum_{i=1}^r \langle [\tilde{H}_\beta(\tilde{\psi}_\beta \psi_\alpha u)]^i, (\tau_{il}^{\tilde{\beta}\alpha} \circ \tilde{\varphi}_\beta^{-1})\eta\rangle \\
&= \sum_{i=1}^r \langle (\tau_{il}^{\tilde{\beta}\alpha} \circ \tilde{\varphi}_\beta^{-1})[\tilde{H}_\beta(\tilde{\psi}_\beta \psi_\alpha u)]^i, \eta\rangle \\
&= \sum_{i=1}^r \langle (\tau_{il}^{\tilde{\beta}\alpha} \circ \tilde{\varphi}_\beta^{-1})[\tilde{H}_\beta(\tilde{\psi}_\beta \psi_\alpha u)]^i, \eta \circ \tilde{\varphi}_\beta \circ \varphi_\alpha^{-1} \circ (\varphi_\alpha \circ \tilde{\varphi}_\beta^{-1})\rangle \\
&= \sum_{i=1}^r \langle \frac{1}{\det(\varphi_\alpha \circ \tilde{\varphi}_\beta^{-1})}(\tau_{il}^{\tilde{\beta}\alpha} \circ \tilde{\varphi}_\beta^{-1})[\tilde{H}_\beta(\tilde{\psi}_\beta \psi_\alpha u)]^i \circ \tilde{\varphi}_\beta \circ \varphi_\alpha^{-1}, \eta \circ \tilde{\varphi}_\beta \circ \varphi_\alpha^{-1}\rangle .
\end{aligned}$$

For the last equality we used the following identity

$$\langle \frac{1}{\det T^{-1}}(u \circ T), \varphi\rangle = \langle u, \varphi \circ T^{-1}\rangle .$$

Hence

$$[H_\alpha(\tilde{\psi}_\beta \psi_\alpha u)]^l = \sum_{i=1}^r \frac{1}{\det(\varphi_\alpha \circ \tilde{\varphi}_\beta^{-1})}(\tau_{il}^{\tilde{\beta}\alpha} \circ \tilde{\varphi}_\beta^{-1})[\tilde{H}_\beta(\tilde{\psi}_\beta \psi_\alpha u)]^i \circ \tilde{\varphi}_\beta \circ \varphi_\alpha^{-1},$$

and consequently letting

$$(A_{\alpha\beta})_{il} = \frac{1}{\det(\varphi_\alpha \circ \tilde{\varphi}_\beta^{-1})}(\tau_{il}^{\tilde{\beta}\alpha} \circ \tilde{\varphi}_\beta^{-1})$$

leads to (4). □

Remark 57. *Note that the above theorems establish the full independence of* $W^{e,q}(M,E;\Lambda)$ *from* Λ *at least when e is not a noninteger less than* -1*. So, it is justified to write* $W^{e,q}(M,E)$ *instead of* $W^{e,q}(M,E;\Lambda)$ *at least when e is not a noninteger less than* -1*. Additionally, see Remark 61.*

9.2. The Properties

9.2.1. Multiplication Properties

Theorem 90. *Let* M^n *be a compact smooth manifold and* $E \to M$ *be a vector bundle with rank r. Let* $\Lambda = \{(U_\alpha, \varphi_\alpha, \rho_\alpha, \psi_\alpha)\}_{1\leq\alpha\leq N}$ *be an augmented total trivialization atlas for E. Suppose* $e \in \mathbb{R}$, $q \in (1,\infty)$, $\eta \in C^\infty(M)$*. If e is a noninteger less than* -1*, further assume that the total trivialization atlas of* Λ *is GGL. Then the linear map*

$$m_\eta : W^{e,q}(M,E;\Lambda) \to W^{e,q}(M,E;\Lambda), \quad u \mapsto \eta u$$

is well-defined and bounded.

Proof.

$$\|\eta u\|_{W^{e,q}(M,E;\Lambda)} := \sum_{\alpha=1}^{N}\sum_{l=1}^{r} \|(H_\alpha(\psi_\alpha \eta u))^l\|_{W^{e,q}(\varphi_\alpha(U_\alpha))}$$
$$\overset{\text{Remark 31}}{=} \sum_{\alpha=1}^{N}\sum_{l=1}^{r} \|(\eta\circ\varphi_\alpha^{-1})(H_\alpha(\psi_\alpha u))^l\|_{W^{e,q}(\varphi_\alpha(U_\alpha))}$$
$$\preceq \sum_{\alpha=1}^{N}\sum_{l=1}^{r} \|(H_\alpha(\psi_\alpha u))^l\|_{W^{e,q}(\varphi_\alpha(U_\alpha))} = \|u\|_{W^{e,q}(M,E;\Lambda)}\,.$$

For the case where e is not a noninteger less than -1, the last inequality follows from Theorem 70. If e is a noninteger less than -1, then by assumption $\varphi_\alpha(U_\alpha)$ is either entire $\mathbb{R}^n$ or is Lipschitz, and the last inequality is due to Theorem 51 and Corollary 4. □

Theorem 91. *Let M^n be a compact smooth manifold and $E \to M$ be a vector bundle with rank r. Let Λ be an augmented total trivialization atlas for E. Let $s_1, s_2, s \in \mathbb{R}$ and $p_1, p_2, p \in (1,\infty)$. If any of s_1, s_2, or s is a noninteger less than -1, further assume that the total trivialization atlas of Λ is GL compatible with itself.*

(1) *If s_1, s_2, and s are not nonintegers less than -1, and if $W^{s_1,p_1}(\mathbb{R}^n)\times W^{s_2,p_2}(\mathbb{R}^n) \hookrightarrow W^{s,p}(\mathbb{R}^n)$, then*
$$W^{s_1,p_1}(M;\Lambda)\times W^{s_2,p_2}(M,E;\Lambda)\hookrightarrow W^{s,p}(M,E;\Lambda)\,.$$

(2) *If s_1, s_2, and s are not nonintegers less than -1, and if $W^{s_1,p_1}(\Omega)\times W^{s_2,p_2}(\Omega)\hookrightarrow W^{s,p}(\Omega)$, for any open ball Ω, then*
$$W^{s_1,p_1}(M;\Lambda)\times W^{s_2,p_2}(M,E;\Lambda)\hookrightarrow W^{s,p}(M,E;\Lambda)\,.$$

(3) *If any of s_1, s_2, or s is a noninteger less than -1, and if $W^{s_1,p_1}(\Omega)\times W^{s_2,p_2}(\Omega)\hookrightarrow W^{s,p}(\Omega)$ for $\Omega=\mathbb{R}^n$* ***and*** *for any bounded open set Ω with Lipschitz continuous boundary, then*
$$W^{s_1,p_1}(M;\Lambda)\times W^{s_2,p_2}(M,E;\Lambda)\hookrightarrow W^{s,p}(M,E;\Lambda)\,.$$

Proof.

(1) Let $\Lambda_1 = \{(U_\alpha,\varphi_\alpha,\rho_\alpha,\psi_\alpha)\}_{1\le\alpha\le N}$ be any augmented total trivialization atlas which is super nice. Let $\Lambda_2 = \{(U_\alpha,\varphi_\alpha,\rho_\alpha,\tilde{\psi}_\alpha)\}_{1\le\alpha\le N}$ where for each $1\le\alpha\le N$, $\tilde{\psi}_\alpha = \frac{\psi_\alpha^2}{\sum_{\beta=1}^N \psi_\beta^2}$. Note that $\frac{1}{\sum_{\beta=1}^N \psi_\beta^2}\circ\varphi_\alpha^{-1}\in BC^\infty(\varphi_\alpha(U_\alpha))$. For $f\in W^{s_1,p_1}(M;\Lambda)$ and $u\in W^{s_2,p_2}(M,E;\Lambda)$ we have

$$\|fu\|_{W^{s,p}(M,E;\Lambda)} \simeq \|fu\|_{W^{s,p}(M,E;\Lambda_2)} = \sum_{\alpha=1}^{N}\sum_{j=1}^{r}\|[H_\alpha(\tilde{\psi}_\alpha(fu))]^j\|_{W^{s,p}(\varphi_\alpha(U_\alpha))}$$
$$\preceq \sum_{\alpha=1}^{N}\sum_{j=1}^{r}\|((\psi_\alpha f)\circ\varphi_\alpha^{-1})[H_\alpha(\psi_\alpha u)]^j\|_{W^{s,p}(\varphi_\alpha(U_\alpha))}$$
$$\preceq \Big(\sum_{\alpha=1}^{N}\|(\psi_\alpha f)\circ\varphi_\alpha^{-1}\|_{W^{s_1,p_1}(\varphi_\alpha(U_\alpha))}\Big)\Big(\sum_{\alpha=1}^{N}\sum_{j=1}^{r}\|[H_\alpha(\psi_\alpha u)]^j\|_{W^{s_2,p_2}(\varphi_\alpha(U_\alpha))}\Big)$$
$$= \|f\|_{W^{s_1,p_1}(M;\Lambda_1)}\|u\|_{W^{s_2,p_2}(M,E;\Lambda_1)} \simeq \|f\|_{W^{s_1,p_1}(M;\Lambda)}\|u\|_{W^{s_2,p_2}(M,E;\Lambda)}\,.$$

(2) We can use the exact same argument as item 1. Just choose Λ_1 to be "nice" instead of "super nice".

(3) The exact same argument as item 1 works. Just choose $\Lambda_1 = \Lambda$. (The equality $\|fu\|_{W^{s,p}(M,E;\Lambda)} \simeq \|fu\|_{W^{s,p}(M,E;\Lambda_2)}$ holds due to the assumption that $\Lambda = \Lambda_1$ is GL compatible with itself.)

□

Remark 58. *Suppose e is a noninteger less than -1 and $q \in (1,\infty)$. We will prove that if Λ and $\tilde{\Lambda}$ are two augmented total trivialization atlases and each of Λ and $\tilde{\Lambda}$ is GL compatible with itself, then $W^{e,q}(M,E;\Lambda) = W^{e,q}(M,E;\tilde{\Lambda})$ (see Remark 61). Considering this and the fact that we can choose Λ_1 to be super nice (or nice) and GL compatible with itself (see Theorem 34 and Corollary 1), we can remove the assumption "s_1, s_2, and s are not nonintegers less than -1" from part 1 and part 2 of the preceding theorem.*

9.2.2. Embedding Properties

Theorem 92. *Let M^n be a compact smooth manifold. Let $\pi : E \to M$ be a smooth vector bundle of rank r over M. Let Λ be an augmented total trivialization atlas for E. Let $e_1, e_2 \in \mathbb{R}$ and $q_1, q_2 \in (1,\infty)$. If any of e_1 or e_2 is a noninteger less than -1, further assume that the total trivialization atlas in Λ is GGL.*

(1) If e_1 and e_2 are not nonintegers less than -1 and if $W^{e_1,q_1}(\mathbb{R}^n) \hookrightarrow W^{e_2,q_2}(\mathbb{R}^n)$, then $W^{e_1,q_1}(M,E;\Lambda) \hookrightarrow W^{e_2,q_2}(M,E;\Lambda)$.
(2) If e_1 and e_2 are not nonintegers less than -1 and if $W^{e_1,q_1}(\Omega) \hookrightarrow W^{e_2,q_2}(\Omega)$ for all open balls $\Omega \subseteq \mathbb{R}^n$, then $W^{e_1,q_1}(M,E;\Lambda) \hookrightarrow W^{e_2,q_2}(M,E;\Lambda)$.
(3) If any of e_1 or e_2 is a noninteger less than -1 and if $W^{e_1,q_1}(\Omega) \hookrightarrow W^{e_2,q_2}(\Omega)$ for $\Omega = \mathbb{R}^n$ and for any bounded domain $\Omega \subseteq \mathbb{R}^n$ with Lipschitz continuous boundary, then $W^{e_1,q_1}(M,E;\Lambda) \hookrightarrow W^{e_2,q_2}(M,E;\Lambda)$.

Proof.

(1) Let $\Lambda_1 = \{(U_\alpha, \varphi_\alpha, \rho_\alpha, \psi_\alpha)\}_{1\le\alpha\le N}$ be any augmented total trivialization atlas for E which is super nice. We have

$$\begin{aligned}\|u\|_{W^{e_2,q_2}(M,E;\Lambda)} &\simeq \|u\|_{W^{e_2,q_2}(M,E;\Lambda_1)} = \sum_{\alpha=1}^{N}\sum_{l=1}^{r} \|[H_\alpha(\psi_\alpha u)]^l\|_{W^{e_2,q_2}(\varphi_\alpha(U_\alpha))} \\ &\preceq \sum_{\alpha=1}^{N}\sum_{l=1}^{r} \|[H_\alpha(\psi_\alpha u)]^l\|_{W^{e_1,q_1}(\varphi_\alpha(U_\alpha))} \\ &= \|u\|_{W^{e_1,q_1}(M,E;\Lambda_1)} \simeq \|u\|_{W^{e_1,q_1}(M,E;\Lambda)}\,.\end{aligned}$$

(2) We can use the exact same argument as item 1. Just choose Λ_1 to be "nice" instead of "super nice".
(3) The exact same argument as item 1 works. Just choose $\Lambda_1 = \Lambda$.

□

Remark 59. *If we further assume that Λ is GL compatible with itself, then we can remove the assumption "e_1 and e_2 are not nonintegers less than -1" from part 1 and part 2 of the preceding theorem. (see the explanation in Remark 58).*

Theorem 93. *Let M^n be a compact smooth manifold. Let $\pi : E \to M$ be a smooth vector bundle of rank r over M equipped with fiber metric $\langle .,. \rangle_E$ (so it is meaningful to talk about $L^\infty(M,E)$). Suppose $s \in \mathbb{R}$ and $p \in (1,\infty)$ are such that $sp > n$. Then $W^{s,p}(M,E) \hookrightarrow L^\infty(M,E)$. Moreover, every element u in $W^{s,p}(M,E)$ has a continuous version (note that since s is not a noninteger less than -1, the choice of the augmented total trivialization atlas is immaterial).*

Proof. Let $\{(U_\alpha, \varphi_\alpha, \rho_\alpha)\}_{1\leq\alpha\leq N}$ be a nice total trivialization atlas for $E \to M$ that trivializes the fiber metric. Let $\{\psi_\alpha\}_{1\leq\alpha\leq N}$ be a partition of unity subordinate to $\{U_\alpha\}$. We need to show that for every $u \in W^{s,p}(M,E)$

$$|u|_{L^\infty(M,E)} \preceq \|u\|_{W^{s,p}(M,E)} .$$

Note that since $s > 0$, $W^{s,p}(M,E) \hookrightarrow L^p(M,E)$ and we can treat u as an ordinary section of E. We prove the above inequality in two steps:

- **Step 1:** Suppose there exists $1 \leq \beta \leq N$ such that $\mathrm{supp} u \subseteq U_\beta$. We have

$$\begin{aligned}
|u|_{L^\infty(M,E)} &= \operatorname*{ess\,sup}_{x\in M} |u|_E = \operatorname*{ess\,sup}_{x\in U_\beta} |u|_E \\
&= \operatorname*{ess\,sup}_{y\in\varphi_\beta(U_\beta)} \sqrt{\sum_{l=1}^{r} |\rho_\beta^l \circ u \circ \varphi_\beta^{-1}|^2} \qquad \text{(by assumption the triples trivialize the metric)} \\
&\leq \operatorname*{ess\,sup}_{y\in\varphi_\beta(U_\beta)} \sum_{l=1}^{r} |\rho_\beta^l \circ u \circ \varphi_\beta^{-1}| \leq \sum_{l=1}^{r} \operatorname*{ess\,sup}_{y\in\varphi_\beta(U_\beta)} |\rho_\beta^l \circ u \circ \varphi_\beta^{-1}| \\
&= \sum_{l=1}^{r} \|\rho_\beta^l \circ u \circ \varphi_\beta^{-1}\|_{L^\infty(\varphi_\beta(U_\beta))} \\
&\preceq \sum_{l=1}^{r} \|\rho_\beta^l \circ u \circ \varphi_\beta^{-1}\|_{W^{s,p}(\varphi_\beta(U_\beta))} \qquad (sp > n \text{ so } W^{s,p}(\varphi_\beta(U_\beta)) \hookrightarrow L^\infty(\varphi_\beta(U_\beta))) .
\end{aligned}$$

- **Step 2:** Now, suppose u is an arbitrary element of $W^{s,p}(M,E)$. We have

$$\begin{aligned}
|u|_{L^\infty(M,E)} &= |\sum_{\alpha=1}^{N} \psi_\alpha u|_{L^\infty(M,E)} \leq \sum_{\alpha=1}^{N} |\psi_\alpha u|_{L^\infty(M,E)} \\
&\overset{\text{Step 1}}{\preceq} \sum_{\alpha=1}^{N}\sum_{l=1}^{r} \|\rho_\alpha^l \circ \psi_\alpha u \circ \varphi_\alpha^{-1}\|_{W^{s,p}(\varphi_\alpha(U_\alpha))} \simeq \|u\|_{W^{s,p}(M,E)} .
\end{aligned}$$

Next we prove that every element u of $W^{s,p}(M,E)$ has a continuous version. Note that for all $x \in U_\alpha$

$$\psi_\alpha u(x) = \Phi_\alpha^{-1}(x, \rho_\alpha^1 \circ \psi_\alpha u, \ldots, \rho_\alpha^r \circ \psi_\alpha u) .$$

Furthermore, for all $1 \leq l \leq r$ and $1 \leq \alpha \leq N$ we have

$$\rho_\alpha^l \circ \psi_\alpha u \circ \varphi_\alpha^{-1} \in W^{s,p}(\varphi_\alpha(U_\alpha)) .$$

Therefore, $\rho_\alpha^l \circ \psi_\alpha u \circ \varphi_\alpha^{-1}$ has a continuous version which we denote by v_α^l. Suppose A_α^l is the set of measure zero on which $v_\alpha^l \neq \rho_\alpha^l \circ \psi_\alpha u \circ \varphi_\alpha^{-1}$. Let $A_\alpha = \cup_{1\leq l\leq r} A_\alpha^l$. Clearly, A_α is a set of measure zero. Since $\varphi_\alpha : U_\alpha \to \varphi_\alpha(U_\alpha)$ is a diffeomorphism, $B_\alpha := \varphi_\alpha^{-1}(A_\alpha)$ is a set of measure zero in U_α (In general, if M and N are smooth n-manifolds, $F : M \to N$ is a smooth map, and $A \subseteq M$ is a subset of measure zero, then $F(A)$ has measure zero in N. See p. 128 in [19]).
Clearly,

$$(x, v_\alpha^1 \circ \varphi_\alpha, \ldots, v_\alpha^r \circ \varphi_\alpha) = (x, \rho_\alpha^1 \circ \psi_\alpha u, \ldots, \rho_\alpha^r \circ \psi_\alpha u) .$$

on $U_\alpha \setminus B_\alpha$. So,

$$w_\alpha := \Phi_\alpha^{-1}(x, v_\alpha^1 \circ \varphi_\alpha, \ldots, v_\alpha^r \circ \varphi_\alpha) = \Phi_\alpha^{-1}(x, \rho_\alpha^1 \circ \psi_\alpha u, \ldots, \rho_\alpha^r \circ \psi_\alpha u) = \psi_\alpha u$$

on $U_\alpha \setminus B_\alpha$. Note that $w_\alpha : U_\alpha \to E$ is a composition of continuous functions and so it is continuous on U_α. Let $\xi_\alpha \in C_c^\infty(U_\alpha)$ be such that $\xi_\alpha = 1$ on $\mathrm{supp}\psi_\alpha$. So $\xi_\alpha w_\alpha = \psi_\alpha u$ on $M \setminus B_\alpha$. Consequently, if we let $w = \sum_{\alpha=1}^{N} \xi_\alpha w_\alpha$, then w is a continuous function that agrees with $u = \sum_{\alpha=1}^{N} \psi_\alpha u$ on $M \setminus B$ where $B = \cup_{1\leq\alpha\leq N} B_\alpha$. □

9.2.3. Observations Concerning the Local Representation of Sobolev Functions

Let M^n be a compact smooth manifold. Let $E \to M$ be a smooth vector bundle of rank r over M. As it was discussed in Section 6, given a total trivialization triple $(U_\alpha, \varphi_\alpha, \rho_\alpha)$, we can associate with every $u \in D'(M,E)$ and every $f \in \Gamma(M,E)$, a local representation with respect to $(U_\alpha, \varphi_\alpha, \rho_\alpha)$:

$$u \mapsto (\tilde{u}^1, \dots, \tilde{u}^r) \in [D'(\varphi_\alpha(U_\alpha))]^{\times r}, \qquad \tilde{u}^l = [H_\alpha(u|_{U_\alpha})]^l,$$
$$f \mapsto (\tilde{f}^1, \dots, \tilde{f}^r) \in [\mathrm{Func}(\varphi_\alpha(U_\alpha), \mathbb{R})]^{\times r}, \qquad \tilde{f}^l = \rho^l_\alpha \circ (f|_{U_\alpha}) \circ \varphi_\alpha^{-1},$$

and of course, as it was pointed out in Remark 32, the two representations agree when u is a regular distribution. The goal of this section is to list some useful facts about the local representations of elements of Sobolev spaces. In what follows, when there is no possibility of confusion, we may write $H_\alpha(u)$ instead of $H_\alpha(u|_{U_\alpha})$, or $\rho^l_\alpha \circ f \circ \varphi_\alpha^{-1}$ instead of $\rho^l_\alpha \circ (f|_{U_\alpha}) \circ \varphi_\alpha^{-1}$.

Theorem 94. *Let M^n be a compact smooth manifold and $E \to M$ be a vector bundle of rank r. Suppose $\Lambda = \{(U_\alpha, \varphi_\alpha, \rho_\alpha, \psi_\alpha)\}_{\alpha=1}^N$ is an augmented total trivialization atlas for $E \to M$. Let $u \in D'(M,E)$, $e \in \mathbb{R}$, and $q \in (1,\infty)$. If for all $1 \le \alpha \le N$ and $1 \le j \le r$, $[H_\alpha(u)]^j \in W^{e,q}_{loc}(\varphi_\alpha(U_\alpha))$, then $u \in W^{e,q}(M,E;\Lambda)$.*

Proof.

$$\begin{aligned}\|u\|_{W^{e,q}(M,E;\Lambda)} &= \sum_{\alpha=1}^N \sum_{j=1}^r \|[H_\alpha(\psi_\alpha u)]^j\|_{W^{e,q}(\varphi_\alpha(U_\alpha))} \\ &= \sum_{\alpha=1}^N \sum_{j=1}^r \|(\psi_\alpha \circ \varphi_\alpha^{-1}) \cdot ([H_\alpha(u)]^j)\|_{W^{e,q}(\varphi_\alpha(U_\alpha))} .\end{aligned}$$

Now, note that $\psi_\alpha \circ \varphi_\alpha^{-1} : \varphi_\alpha(U_\alpha) \to \mathbb{R}$ is smooth with compact support (its support is in the compact set $\varphi_\alpha(\mathrm{supp}\,\psi_\alpha)$). Therefore, it follows from the assumption that each term on the right hand side of the above equality is finite. □

Remark 60. *Note that, as opposed to what is claimed in some references, it is NOT true in general that if $u \in W^{e,q}(M,E;\Lambda)$, then the components of the local representations of u will be in the corresponding Euclidean Sobolev space; that is, $u \in W^{e,q}(M,E;\Lambda)$ does not imply that for all $1 \le \alpha \le N$ and $1 \le j \le r$, $[H_\alpha(u)]^j \in W^{e,q}(\varphi_\alpha(U_\alpha))$. Consider the following example: $M = S^1$, $e = 0$, $q = 1$, and $f : M \to \mathbb{R}$ defined by $f \equiv 1$. Clearly $f \in W^{0,1}(M) = L^1(S^1)$. Now, consider the atlas $\mathcal{A} = \{(U_1, \varphi_1), (U_2, \varphi_2)\}$ where*

$$U_1 = S^1 \setminus \{(0,1)\}, \qquad \varphi_1(x,y) = \frac{x}{1-y},$$
$$U_2 = S^1 \setminus \{(0,-1)\}, \qquad \varphi_2(x,y) = \frac{x}{1+y} \quad \textit{(stereographic projection)}.$$

Clearly, $f \circ \varphi_1^{-1} = f \circ \varphi_2^{-1} = 1$ and $\varphi_1(U_1) = \varphi_2(U_2) = \mathbb{R}$. So, $f \circ \varphi_1^{-1}$ and $f \circ \varphi_2^{-1}$ do not belong to $L^1(\varphi_1(U_1))$ or $L^1(\varphi_2(U_2))$.

However, the following theorem holds true.

Theorem 95. *Let M^n be a compact smooth manifold and $E \to M$ be a vector bundle of rank r. Let $e \in \mathbb{R}$ and $q \in (1,\infty)$. Suppose $\Lambda = \{(U_\alpha, \varphi_\alpha, \rho_\alpha, \psi_\alpha)\}_{\alpha=1}^N$ is an augmented total trivialization atlas for $E \to M$. If e is a noninteger less than -1 further assume that Λ is GL compatible with* *itself. Let* $u \in W^{e,q}(M,E;\Lambda)$ *be such that* $\operatorname{supp} u \subseteq V \subseteq \bar{V} \subseteq U_\beta$ *for some open set V and some* $1 \le \beta \le N$. *Then for all* $1 \le i \le r$, $[H_\beta(u)]^i \in W^{e,q}(\varphi_\beta(U_\beta))$. *Indeed,*

$$\|[H_\beta(u)]^i\|_{W^{e,q}(\varphi_\beta(U_\beta))} \preceq \|u\|_{W^{e,q}(M,E;\Lambda)}\,.$$

Proof. Let $\Lambda_1 = \{(U_\alpha, \varphi_\alpha, \rho_\alpha, \tilde{\psi}_\alpha)\}_{\alpha=1}^N$ where $\{\tilde{\psi}_\alpha\}_{1\le\alpha\le N}$ is a partition of unity subordinate to the cover $\{U_\alpha\}_{1\le\alpha\le N}$ such that $\tilde{\psi}_\beta = 1$ on a neighborhood of $\bar{V}$ (see Lemma 3). We have

$$\begin{aligned}
\|[H_\beta(u)]^i\|_{W^{e,q}(\varphi_\beta(U_\beta))} &= \|[H_\beta(\tilde{\psi}_\beta u)]^i\|_{W^{e,q}(\varphi_\beta(U_\beta))} \\
&\le \sum_{\alpha=1}^{N}\sum_{j=1}^{r} \|[H_\alpha(\tilde{\psi}_\alpha u)]^j\|_{W^{e,q}(\varphi_\alpha(U_\alpha))} \\
&= \|u\|_{W^{e,q}(M,E;\Lambda_1)} \simeq \|u\|_{W^{e,q}(M,E;\Lambda)}\,.
\end{aligned}$$

□

Corollary 8. *Let* M^n *be a compact smooth manifold and* $E \to M$ *be a vector bundle of rank r. Let* $e \in \mathbb{R}$ *and* $q \in (1,\infty)$. *Suppose* $\Lambda = \{(U_\alpha, \varphi_\alpha, \rho_\alpha, \psi_\alpha)\}_{\alpha=1}^N$ *is an augmented total trivialization atlas for* $E \to M$. *If e is a noninteger less than* -1 *further assume that* Λ *is GL compatible with itself. If* $u \in W^{e,q}(M,E;\Lambda)$, *then for all* $1 \le \alpha \le N$ *and* $1 \le i \le r$, $[H_\alpha(u)]^i$ *(i.e., each component of the local representation of u with respect to* $(U_\alpha, \varphi_\alpha, \rho_\alpha)$*) belongs to* $W^{e,q}_{loc}(\varphi_\alpha(U_\alpha))$. *Moreover, if* $\xi \in C_c^\infty(\varphi_\alpha(U_\alpha))$, *then*

$$\|\xi[H_\alpha(u)]^i\|_{W^{e,q}(\varphi_\alpha(U_\alpha))} \preceq \|u\|_{W^{e,q}(M,E;\Lambda)}\,,$$

where the implicit constant may depend on ξ.

Proof. Define $G : M \to \mathbb{R}$ by

$$G(p) = \begin{cases} \xi \circ \varphi_\alpha & \text{if } p \in U_\alpha \\ 0 & \text{if } p \notin U_\alpha \end{cases}.$$

Clearly, $G \in C^\infty(M)$. So, by Theorem 90, $Gu \in W^{e,q}(M,E;\Lambda)$. Furthermore, since $\xi \in C_c^\infty(\varphi_\alpha(U_\alpha))$, there exists a compact set K such that

$$\operatorname{supp}\xi \subseteq \mathring{K} \subseteq K \subseteq \varphi_\alpha(U_\alpha)\,.$$

Consequently, there exists an open set V_α (e.g., $V_\alpha = \varphi_\alpha^{-1}(\mathring{K})$) such that

$$\operatorname{supp}(Gu) \subseteq \operatorname{supp}(\xi \circ \varphi_\alpha) \subseteq V_\alpha \subseteq \bar{V}_\alpha \subseteq U_\alpha\,.$$

So, by Theorem 95, $[H_\alpha(Gu)]^i \in W^{e,q}(\varphi_\alpha(U_\alpha))$ and

$$\|[H_\alpha(Gu)]^i\|_{W^{e,q}(\varphi_\alpha(U_\alpha))} \preceq \|Gu\|_{W^{e,q}(M,E;\Lambda)} \preceq \|u\|_{W^{e,q}(M,E;\Lambda)}\,.$$

Now, we just need to notice that on $\varphi_\alpha(U_\alpha)$,

$$[H_\alpha(Gu)]^i = (G \circ \varphi_\alpha^{-1})[H_\alpha(u)]^i = \xi[H_\alpha(u)]^i\,.$$

□

9.2.4. Observations Concerning the Riemannian Metric

The Sobolev spaces that appear in this section all have nonnegative smoothness exponents; therefore, the choice of the augmented total trivialization atlas is immaterial and will not appear in the notation.

Corollary 9. *Let (M^n, g) be a compact Riemannian manifold with $g \in W^{s,p}(T^2M)$, $sp > n$. Let $\{(U_\alpha, \varphi_\alpha, \rho_\alpha)\}_{1\leq\alpha\leq N}$ be a standard total trivialization atlas for $T^2M \to M$. Fix some α and denote the components of the metric with respect to $(U_\alpha, \varphi_\alpha, \rho_\alpha)$ by $g_{ij} : U_\alpha \to \mathbb{R}$ ($g_{ij} = (\rho_\alpha)_{ij} \circ g$). As an immediate consequence of Corollary 8 we have*

$$g_{ij} \circ \varphi_\alpha^{-1} \in W^{s,p}_{loc}(\varphi_\alpha(U_\alpha)) .$$

Theorem 96. *Let (M^n, g) be a compact Riemannian manifold with $g \in W^{s,p}(T^2M)$, $sp > n$, $s \geq 1$. Let $\{(U_\alpha, \varphi_\alpha, \rho_\alpha)\}_{1\leq\alpha\leq N}$ be a GGL standard total trivialization atlas for $T^2M \to M$. Fix some α and denote the components of the metric with respect to $(U_\alpha, \varphi_\alpha, \rho_\alpha)$ by $g_{ij} : U_\alpha \to \mathbb{R}$ ($g_{ij} = (\rho_\alpha)_{ij} \circ g$). Then*

(1) *$\det g_\alpha \in W^{s,p}_{loc}(\varphi_\alpha(U_\alpha))$ where $g_\alpha(x)$ is the matrix whose (i,j)-entry is $g_{ij} \circ \varphi_\alpha^{-1}$,*
(2) *$\sqrt{\det g} \circ \varphi_\alpha^{-1} = \sqrt{\det g_\alpha} \in W^{s,p}_{loc}(\varphi_\alpha(U_\alpha))$,*
(3) *$\frac{1}{\sqrt{\det g \circ \varphi_\alpha^{-1}}} \in W^{s,p}_{loc}(\varphi_\alpha(U_\alpha))$.*

Proof.
(1) By Corollary 8, $g_{ij} \circ \varphi_\alpha^{-1}$ is in $W^{s,p}_{loc}(\varphi_\alpha(U_\alpha))$. So, it follows from Lemma 10 that $\det g_\alpha \in W^{s,p}_{loc}(\varphi_\alpha(U_\alpha))$.
(2) This is a direct consequence of item 1 and Theorem 85.
(3) This is a direct consequence of item 1 and Theorem 85.
□

Theorem 97. *Let (M^n, g) be a compact Riemannian manifold with $g \in W^{s,p}(T^2M)$, $sp > n$, $s \geq 1$. Then the inverse metric tensor g^{-1} (which is a $\binom{0}{2}$ tensor field) is in $W^{s,p}(T_2M)$.*

Proof. Let $\{(U_\alpha, \varphi_\alpha, \rho_\alpha)\}_{1\leq\alpha\leq N}$ be a GGL standard total trivialization atlas for $T^2M \to M$. Let $\{\psi_\alpha\}_{1\leq\alpha\leq N}$ be a partition of unity subordinate to $\{U_\alpha\}_{1\leq\alpha\leq N}$. We have

$$\|g^{-1}\|_{W^{s,p}(T_2M)} = \sum_{\alpha=1}^{N} \sum_{i,j} \|\psi_\alpha g^{ij} \circ \varphi_\alpha^{-1}\|_{W^{s,p}(\varphi_\alpha(U_\alpha))} .$$

So, it is enough to show that for all i,j and α, $g^{ij} \circ \varphi_\alpha^{-1}$ is in $W^{s,p}_{loc}(\varphi_\alpha(U_\alpha))$. Let $B = (B_{ij})$ where $B_{ij} = g_{ij} \circ \varphi_\alpha^{-1}$. By assumption, $g \in W^{s,p}(T^2M)$; it follows from Corollary 8 that $B_{ij} \in W^{s,p}_{loc}(\varphi_\alpha(U_\alpha))$. Our goal is to show that the entries of the inverse of B are in $W^{s,p}_{loc}(\varphi_\alpha(U_\alpha))$. Recall that

$$(B^{-1})_{ij} = \frac{(-1)^{i+j}}{\det B} M_{ij} ,$$

where M_{ij} is the determinant of the $(n-1) \times (n-1)$ matrix formed by removing the jth row and ith column of B. Since the entries of B are in $W^{s,p}_{loc}(\varphi_\alpha(U_\alpha))$, it follows from Lemma 10 and Theorem 85 that $\frac{1}{\det B}$ and M_{ij} are in $W^{s,p}_{loc}(\varphi_\alpha(U_\alpha))$. Furthermore, $sp > n$, so $W^{s,p}_{loc}(\varphi_\alpha(U_\alpha))$ is closed under multiplication. Consequently, $(B^{-1})_{ij}$ is in $W^{s,p}_{loc}(\varphi_\alpha(U_\alpha))$. □

Corollary 10. *Let (M^n, g) be a compact Riemannian manifold with $g \in W^{s,p}(T^2M)$, $sp > n$, $s \geq 1$. $\{(U_\alpha, \varphi_\alpha)\}_{1\leq\alpha\leq N}$ be a GGL smooth atlas for M. Denote the standard components of the inverse metric with respect to this chart by $g^{ij} : U_\alpha \to \mathbb{R}$. As an immediate consequence of Theorem 97 and Corollary 8 we have*

$$g^{ij} \circ \varphi_\alpha^{-1} \in W^{s,p}_{loc}(\varphi_\alpha(U_\alpha)) .$$

Furthermore, since

$$\Gamma^k_{ij} \circ \varphi_\alpha^{-1} = \frac{1}{2} g^{kl} (\partial_i g_{jl} + \partial_j g_{il} - \partial_l g_{ij}) \circ \varphi_\alpha^{-1} ,$$

it follows from Corollary 9, Lemma 9, Theorem 83, and the fact that $W^{s,p}(\varphi_\alpha(U_\alpha)) \times W^{s-1,p}(\varphi_\alpha(U_\alpha)) \hookrightarrow W^{s-1,p}(\varphi_\alpha(U_\alpha))$ *that*

$$\Gamma^k_{ij} \circ \varphi_\alpha^{-1} \in W^{s-1,p}_{loc}(\varphi_\alpha(U_\alpha)).$$

9.2.5. A Useful Isomorphism

Let M^n be a compact smooth manifold and $E \to M$ be a vector bundle of rank r. Let $e \in \mathbb{R}$ and $q \in (1,\infty)$. Suppose $\Lambda = \{(U_\alpha, \varphi_\alpha, \rho_\alpha, \psi_\alpha)\}_{\alpha=1}^N$ is an augmented total trivialization atlas for $E \to M$. Given a closed subset $A \subseteq M$, $W^{e,q}_A(M,E;\Lambda)$ is defined to be the subspace of $W^{e,q}(M,E;\Lambda)$ consisting of $u \in W^{e,q}(M,E;\Lambda)$ with $\operatorname{supp} u \subseteq A$. Fix $1 \le \beta \le N$ and suppose $K \subseteq U_\beta$ is compact. Then each element of $W^{e,q}_K(M,E;\Lambda)$ can be identified with an element of $D'(U_\beta, E_{U_\beta})$ under the injective map $u \in W^{e,q}_K(M,E;\Lambda) \subseteq D'(M,E) \mapsto u|_U \in D'(U_\beta, E_{U_\beta})$. So, we can restrict the domain of $H_\beta : [D(U_\beta, E^\vee_{U_\beta})]^* \to (D'(\varphi_\beta(U_\beta)))^{\times r}$ to $W^{e,q}_K(M,E;\Lambda)$ which associates with each element $u \in W^{e,q}_K(M,E;\Lambda)$, the r components of $H_\beta(u) = (\tilde{u}^1_\beta, \cdots, \tilde{u}^r_\beta)$ (here H_β stands for $H_{E^\vee, U_\beta, \varphi_\beta}$).

Lemma 11. *Consider the above setting and further assume that if e is a noninteger less than −1, then the total trivialization atlas in Λ is GL compatible with itself. Then the linear topological isomorphism* $H_\beta : [D(U_\beta, E^\vee_{U_\beta})]^* = D'(U_\beta, E_{U_\beta}) \to (D'(\varphi_\beta(U_\beta)))^{\times r}$ *restricts to a linear topological isomorphism*

$$\hat{H}_\beta : W^{e,q}_K(M,E;\Lambda) \to [W^{e,q}_{\varphi_\beta(K)}(\varphi_\beta(U_\beta))]^{\times r}.$$

Proof. In order to simplify the notation we will use (U,φ,ρ), H, $\hat{H}$, and $\tilde{u}^l$ instead of $(U_\beta, \varphi_\beta, \rho_\beta)$, H_β, $\hat{H}_\beta$, and $\tilde{u}^l_\beta$. In order to prove this claim, we proceed as follows:

(1) First we show that $\operatorname{supp}\tilde{u}^l \subseteq \varphi(K)$.
(2) Next we show that if $u \in W^{e,q}_K(M,E;\Lambda)$, then $\|u\|_{W^{e,q}(M,E;\Lambda)} \simeq \sum_{l=1}^r \|\tilde{u}^l\|_{W^{e,q}(\varphi(U))}$ which proves that:
    (i) $\tilde{u}^l$ is indeed an element of $W^{e,q}(\varphi(U))$;
    (ii) $\hat{H}$ is continuous.

    Note that (i) together with the fact that $\operatorname{supp}\tilde{u}^l \subseteq \varphi(K)$ shows that $\tilde{u}^l$ is indeed an element of $W^{e,q}_{\varphi(K)}(\varphi(U))$ so $\hat{H}$ is well-defined.
(3) We prove that $\hat{H}$ is injective.
(4) In order to prove that $\hat{H}$ is surjective we use our explicit formula for H^{-1} (see Remark 31).

Note that the fact that $\hat{H}$ is bijective combined with the equality $\|u\|_{W^{e,q}(M,E;\Lambda)} \simeq \sum_{l=1}^r \|\tilde{u}^l\|_{W^{e,q}(\varphi(U))}$ implies that $\hat{H}^{-1}$ is continuous as well.
Here are the proofs:

(1) This item is a direct consequence of item 1 in Remark 31.
(2) Define the augmented total trivialization atlas Λ_1 by $\Lambda_1 = \{(U_\alpha, \varphi_\alpha, \rho_\alpha, \tilde{\psi}_\alpha)\}_{\alpha=1}^N$ where $\{\tilde{\psi}_\alpha\}_{1\le\alpha\le N}$ is a partition of unity subordinate to $\{U_\alpha\}_{1\le\alpha\le N}$ such that $\tilde{\psi}_\beta = 1$ on a neighborhood of K. Note that for each α, $\tilde{\psi}_\alpha \ge 0$ and $\sum_{\alpha=1}^N \tilde{\psi}_\alpha = 1$. Thus, the assumption $\tilde{\psi}_\beta = 1$ on K implies that $\tilde{\psi}_\alpha = 0$ on K for all $\alpha \ne \beta$. We have

$$\begin{aligned}\|u\|_{W^{e,q}(M,E;\Lambda)} &\simeq \|u\|_{W^{e,q}(M,E;\Lambda_1)} \simeq \sum_{\alpha=1}^N \sum_{l=1}^r \|(H_\alpha(\tilde{\psi}_\alpha u))^l\|_{W^{e,q}(\varphi_\alpha(U_\alpha))} \\ &= \sum_{l=1}^r \|(H(\tilde{\psi}_\beta u))^l\|_{W^{e,q}(\varphi_\alpha(U_\alpha))} = \sum_{l=1}^r \|[H(u)]^l\|_{W^{e,q}(\varphi_\alpha(U_\alpha))}.\end{aligned}$$

Note that supp$u \subseteq K$ and $\tilde{\psi}_\beta = 1$ on K, so $\tilde{\psi}_\beta u = u|_U$ as elements of $D'(U, E_U)$. Therefore, $H(\tilde{\psi}_\beta u) = H(u) = (\tilde{u}^1, \ldots, \tilde{u}^r)$.

(3) $\hat{H}$ is injective because it is a restriction of the injective map H.

(4) Let $(v^1, \ldots, v^r) \in [W^{e,q}_{\varphi(K)}(\varphi(U))]^{\times r}$. Our goal is to show that $H^{-1}(v^1, \ldots, v^r) \in W^{e,q}_K(M, E; \Lambda) \simeq W^{e,q}_K(M, E; \Lambda_1)$ (this implies that $\hat{H}$ is surjective). By Remark 31, for all $\xi \in D(U, E^\vee_U)$

$$H^{-1}(v^1, \ldots, v^r)(\xi) = \sum_i v^i[(\rho^\vee)^i \circ \xi \circ \varphi^{-1}].$$

First note it follows from Remark 30 that supp$H^{-1}(v^1, \ldots, v^r) \subseteq K$; indeed, if supp$\xi \subseteq U \setminus K$, then $\xi \circ \varphi^{-1} = 0$ on $\varphi(K)$. So, $(\rho^\vee)^i \circ \xi \circ \varphi^{-1} = 0$ on $\varphi(K)$. That is, supp$[(\rho^\vee)^i \circ \xi \circ \varphi^{-1}] \subseteq \varphi(U) \setminus \varphi(K)$. Thus, for all i, $v^i[(\rho^\vee)^i \circ \xi \circ \varphi^{-1}] = 0$ (because, by assumption, supp$v^i \subseteq \varphi(K)$). This shows that if supp$\xi \subseteq U \setminus K$, then $H^{-1}(v^1, \ldots, v^r)(\xi) = 0$. Consequently, supp$H^{-1}(v^1, \ldots, v^r) \subseteq K$.
Furthermore, we have

$$\|H^{-1}(v^1, \ldots, v^r)\|_{W^{e,q}(M,E;\Lambda_1)} \simeq \sum_{l=1}^{r} \|v^l\|_{W^{e,q}(\varphi(U))} < \infty.$$

So, $H^{-1}(v^1, \cdots, v^r) \in W^{e,q}(M, E; \Lambda)$.

□

It is clear that $u \in W^{e,q}(M, E; \Lambda)$ if and only if for all α, $\psi_\alpha u \in W^{e,q}_{K_\alpha}(M, E; \Lambda)$ where K_α can be taken as any compact set such that supp$\psi_\alpha \subseteq K_\alpha \subseteq U_\alpha$. In fact as a direct consequence of the definition of Sobolev spaces and the above mentioned isomorphism we have

$$\begin{aligned} u \in W^{e,q}(M, E; \Lambda) &\Longleftrightarrow \forall 1 \leq \alpha \leq N \quad H_\alpha(\psi_\alpha u) \in [W^{e,q}_{\varphi_\alpha(\text{supp}\psi_\alpha)}(\varphi_\alpha(U_\alpha))]^{\times r} \\ &\Longleftrightarrow \forall 1 \leq \alpha \leq N \quad \psi_\alpha u \in W^{e,q}_{\text{supp}\psi_\alpha}(M, E; \Lambda) \end{aligned}$$

9.2.6. Completeness; Density of Smooth Functions

Our proofs for completeness of Sobolev spaces and density of smooth functions are based on the ideas presented in [24].

Lemma 12. *Let M^n be a compact smooth manifold and $E \to M$ be a vector bundle of rank r. Let $e \in \mathbb{R}$ and $q \in (1, \infty)$. Suppose $\Lambda = \{(U_\alpha, \varphi_\alpha, \rho_\alpha, \psi_\alpha)\}_{\alpha=1}^N$ is an augmented total trivialization atlas for $E \to M$. If e is a noninteger less than -1 further assume that Λ is GL compatible with itself. Let K_α be a compact subset of U_α that contains the support of ψ_α. Let $S : W^{e,q}(M, E; \Lambda) \to \prod_{\alpha=1}^N W^{e,q}_{K_\alpha}(M, E; \Lambda)$ be the linear map defined by $S(u) = (\psi_1 u, \ldots, \psi_N u)$. Then $S : W^{e,q}(M, E; \Lambda) \to S(W^{e,q}(M, E; \Lambda)) \subseteq \prod_{\alpha=1}^N W^{e,q}_{K_\alpha}(M, E; \Lambda)$ is a linear topological isomorphism. Moreover, $S(W^{e,q}(M, E; \Lambda))$ is closed in $\prod_{\alpha=1}^N W^{e,q}_{K_\alpha}(M, E; \Lambda)$.*

Proof. Each component of S is continuous (see Theorem 90), therefore S is continuous. Define $P : \prod_{\alpha=1}^N W^{e,q}_{K_\alpha}(M, E) \to W^{e,q}(M, E)$ by

$$P(v_1, \ldots, v_N) = \sum_i v_i.$$

Clearly, P is continuous. Furthermore, $P \circ S = id$. Now the claim follows from Theorem 23. □

Theorem 98. *Let M^n be a compact smooth manifold and $E \to M$ be a vector bundle of rank r. Let $e \in \mathbb{R}$ and $q \in (1, \infty)$. Suppose $\Lambda = \{(U_\alpha, \varphi_\alpha, \rho_\alpha, \psi_\alpha)\}_{\alpha=1}^N$ is an augmented total trivialization* *atlas for $E \to M$. If e is a noninteger less than -1 further assume that Λ is GL compatible with itself. Then $W^{e,q}(M,E;\Lambda)$ is a Banach space.*

Proof. According to Lemma 11, for each $1 \le \alpha \le N$, $W^{e,q}_{K_\alpha}(M,E;\Lambda)$ is isomorphic to the Banach space $[W^{e,q}_{\varphi_\alpha(K_\alpha)}(\varphi_\alpha(U_\alpha))]^{\times r}$. So $\prod_{\alpha=1}^{N} W^{e,q}_{K_\alpha}(M,E;\Lambda)$ is a Banach space. A closed subspace of a Banach space is Banach. Therefore, $S(W^{e,q}(M,E;\Lambda))$ is a Banach space. Since S is a linear topological isomorphism onto its image, $W^{e,q}(M,E;\Lambda)$ is also a Banach space. □

Theorem 99. *Let M^n be a compact smooth manifold and $E \to M$ be a vector bundle of rank r. Let $e \in \mathbb{R}$ and $q \in (1,\infty)$. Suppose $\Lambda = \{(U_\alpha, \varphi_\alpha, \rho_\alpha, \psi_\alpha)\}_{\alpha=1}^{N}$ is an augmented total trivialization atlas for $E \to M$. If e is a noninteger less than -1 further assume that Λ is GL compatible with itself. Then $D(M,E)$ is dense in $W^{e,q}(M,E;\Lambda)$.*

Proof. Let $K_\alpha = \text{supp}\psi_\alpha$. For each $1 \le \alpha \le N$, let V_α be an open set such that

$$K_\alpha \subseteq V_\alpha \subseteq \bar{V}_\alpha \subseteq U_\alpha\,.$$

Suppose $u \in W^{e,q}(M,E;\Lambda)$ and let $u_\alpha = \psi_\alpha u$. Clearly, $\text{supp}u_\alpha \subseteq K_\alpha$. Furthermore, according to Lemma 11, for each α there exists a linear topological isomorphism

$$\hat{H}_\alpha : W^{e,q}_{\bar{V}_\alpha}(M,E) \to [W^{e,q}_{\varphi_\alpha(\bar{V}_\alpha)}(\varphi_\alpha(U_\alpha))]^{\times r}\,.$$

Note that $\hat{H}_\alpha(u_\alpha) \in [W^{e,q}_{\varphi_\alpha(K_\alpha)}(\varphi_\alpha(U_\alpha))]^{\times r}$. Therefore, by Lemma 62 there exists a sequence $\{(\eta_\alpha)_i\}$ in $[C^\infty_{\varphi_\alpha(\bar{V}_\alpha)}(\varphi_\alpha(U_\alpha))]^{\times r}$ (of course we view each component of $(\eta_\alpha)_i$ as a distribution) that converges to $\hat{H}_\alpha(u_\alpha)$ in $W^{e,q}$ norm as $i \to \infty$. Since $\hat{H}_\alpha$ is a linear topological isomorphism, we can conclude that

$$\hat{H}_\alpha^{-1}((\eta_\alpha)_i) \to u_\alpha, \quad (\text{in } W^{e,q}_{\bar{V}_\alpha}(M,E;\Lambda) \text{ as } i \to \infty)\,.$$

(Note that if a sequence converges in $W^{e,q}_A(M,E;\Lambda)$ where A is a closed subset of M, it also obviously converges in $W^{e,q}(M,E;\Lambda)$.) Let $\xi_i = \sum_{\alpha=1}^{N} \hat{H}_\alpha^{-1}((\eta_\alpha)_i)$. This sum makes sense because, as we will shortly prove, each summand is in $C^\infty_c(U_\alpha, E_\alpha)$ and so by extension by zero can be viewed as an element of $C^\infty(M,E)$. Clearly $\xi_i \to \sum_\alpha u_\alpha = u$ in $W^{e,q}(M,E;\Lambda)$. It remains to show that for each i, ξ_i is in $C^\infty(M,E)$. To this end, it suffices to show that if $\chi = (\chi^1,\dots,\chi^r) \in [C^\infty_c(\varphi_\alpha(U_\alpha))]^{\times r}$, then $\hat{H}_\alpha^{-1}(\chi)$ is in $C^\infty_c(U_\alpha, E_\alpha)$ and so can be considered as an element of $C^\infty(M,E)$ (by extension by zero). Note that $\hat{H}_\alpha^{-1}(\chi)$ is compactly supported in U_α because by definition of $\hat{H}_\alpha$ any distribution in the codomain of $\hat{H}_\alpha^{-1}$ has compact support in $\bar{V}_\alpha$. So, we just need to prove the smoothness of $\hat{H}_\alpha^{-1}(\chi)$. That is, we need to show that there is a smooth section $f \in C^\infty(U_\alpha, E_{U_\alpha})$ such that $u_f = \hat{H}_\alpha^{-1}(\chi)$. It seems that the natural candidate for $f(x)$ should be $(\rho_\alpha|_{E_x})^{-1} \circ \chi \circ \varphi_\alpha(x)$. In fact, if we define f by this formula, then $\hat{H}_\alpha(u_f) = H_\alpha(u_f)$ and by Remark 32 $H_\alpha(u_f)$ is a distribution that corresponds to the regular function $(\tilde{f}^1,\dots,\tilde{f}^r) = \rho_\alpha \circ f \circ \varphi_\alpha^{-1}$. Obviously,

$$\rho_\alpha \circ f \circ \varphi_\alpha^{-1}|_{\varphi_\alpha(x)} = \rho_\alpha \circ (\rho_\alpha|_{E_x})^{-1} \circ \chi \circ \varphi_\alpha \circ \varphi_\alpha^{-1}|_{\varphi_\alpha(x)} = \chi|_{\varphi_\alpha(x)}\,.$$

So, the regular section $f(x) = \rho_\alpha|_{E_x}^{-1} \circ \chi \circ \varphi_\alpha(x)$ corresponds to $\hat{H}_\alpha^{-1}(\chi)$ and we just need to show that f is smooth; this is true because f is a composition of smooth functions. Indeed,

$$f(x) = \rho_\alpha|_{E_x}^{-1} \circ \chi \circ \varphi_\alpha(x) = \Phi_\alpha^{-1}(x, \chi \circ \varphi_\alpha(x)) \Longrightarrow f = \Phi_\alpha^{-1} \circ (Id, \chi \circ \varphi_\alpha)\,,$$

and all the maps involved in the above expression are smooth. □

9.2.7. Dual of Sobolev Spaces

Lemma 13. *Let M^n be a compact smooth manifold and let $\pi : E \to M$ be a vector bundle of rank r equipped with a fiber metric $\langle .,. \rangle_E$. Let $e \in \mathbb{R}$ and $q \in (1, \infty)$. Suppose $\Lambda = \{(U_\alpha, \varphi_\alpha, \rho_\alpha, \psi_\alpha)\}_{\alpha=1}^N$ is an augmented total trivialization atlas for $E \to M$ which trivializes the fiber metric. If e is a noninteger less than -1 further assume that the total trivialization atlas in Λ is GGL.*
Fix a positive smooth density μ on M (for instance we can equip M with a smooth Riemannian metric and consider the corresponding Riemannian density). Let $T : D(M, E) \to D(M, E^\vee)$ be the map that sends ξ to T_ξ where T_ξ is defined by

$$\forall x \in M \quad T_\xi(x) : E_x \to \mathcal{D}_x, \quad a \mapsto \langle a, \xi(x) \rangle_E \, \mu(x) .$$

Then T is a linear bijective continuous map. Moreover, $T : (C^\infty(M, E), \|.\|_{W^{e,q}(M,E;\Lambda)}) \to (C^\infty(M, E^\vee), \|.\|_{W^{e,q}(M,E^\vee;\Lambda^\vee)})$ is a topological isomorphism.

Note: Since M is compact, $D(M, E)$ and $D(M, E^\vee)$ are Frechet spaces. So, by Theorem 17, the continuity of the bijective linear map $T : D(M, E) \to D(M, E^\vee)$ implies the continuity of its inverse. That is, $T : D(M, E) \to D(M, E^\vee)$ is a linear topological isomorphism. As a consequence, the adjoint of T is a well-defined bijective continuous map that can be used to identify $D'(M, E) = [D(M, E^\vee)]^*$ with $[D(M, E)]^*$.

Proof. The fact that T is linear is obvious.

- ***T* is one-to-one:** Suppose $\xi \in D(M, E)$ is such that $T_\xi = 0$. Then

$$\begin{aligned} \forall x \in M \quad T_\xi(x) = 0 &\Longrightarrow \forall x \in M, \forall a \in E_x \quad [T_\xi(x)](a) = 0 \\ &\Longrightarrow \forall x \in M, \forall a \in E_x \quad \langle a, \xi(x) \rangle_E = 0 \\ &\Longrightarrow \forall x \in M \quad \langle \xi(x), \xi(x) \rangle_E = 0 \Longrightarrow \forall x \in M \quad \xi(x) = 0 . \end{aligned}$$

- ***T* is onto:** Let $u \in D(M, E^\vee)$. Our goal is to show that there exists $\xi \in D(M, E)$ such that $u = T_\xi$. Note that

$$\forall x \in M \quad u(x) = T_\xi(x) \Longleftrightarrow \forall x \in M \, \forall a \in E_x \quad \langle a, \xi(x) \rangle_E \, \mu(x) = [u(x)](a) .$$

  Since $\mathcal{D}_x$ is 1-dimensional and both $\mu(x)$ (which is a positive smooth density) and $[u(x)][a]$ belong to $\mathcal{D}_x$,, there exists a number $b(x, a)$ such that

$$[u(x)](a) = b(x, a)\mu(x) .$$

  So, we need to show that there exists $\xi \in D(M, E)$ such that

$$\forall x \in M \, \forall a \in E_x \qquad \langle a, \xi(x) \rangle_E = b(x, a) .$$

  The above equality uniquely defines a functional on E_x which gives us a unique element $\xi(x) \in E_x$ by the Riesz representation theorem. It remains to prove that ξ is smooth. To this end, we will show that for each α, $\xi|_{U_\alpha}$ is smooth. Let $(s_1, \ldots, s_r)$ be a smooth orthonormal frame for E_{U_α}.

$$\forall x \in U_\alpha \quad \xi(x) = \xi^1(x)s_1(x) + \ldots + \xi^r(x)s_r(x) .$$

  It suffices to show that $\xi^1, \ldots, \xi^r$ are smooth functions (see Theorem 36). We have

$$\xi^i(x) = \langle \xi(x), s_i(x) \rangle_E .$$

  It follows from the definition of $\xi(x)$ that

$$[u(x)][s_i(x)] = \langle s_i(x), \xi(x) \rangle_E \, \mu(x) .$$

Therefore, $\xi^i(x)$ satisfies the following equality

$$[u(x)][s_i(x)] = \xi^i(x)\mu(x)\,.$$

That is, if we define a section of $\mathcal{D} \to U_\alpha$ by

$$[u, s_i] : U_\alpha \to \mathcal{D}, \quad x \mapsto [u(x)][s_i(x)]\,,$$

then ξ^i is the component of this section with respect to the smooth frame $\{\mu(x)\}$ on U_α. The smoothness of ξ^i follows from the fact that if N is any manifold, $E \to N$ is a vector bundle and u and v are in $\mathcal{E}(N, E^\vee)$ and $\mathcal{E}(N, E)$, respectively, then $[u, v]$ is in $\mathcal{E}(N, \mathcal{D})$; indeed, the local representation of $[u, v]$ is $\sum_l \tilde{u}^l \tilde{v}^l$ which is a smooth function because $\tilde{u}^l$ and $\tilde{v}^l$ are smooth functions.

- $T : D(M, E) \to D(M, E^\vee)$ **is continuous:**
We make use of Theorem 20. Recall that
(1) The topology on $D(M, E)$ is induced by the seminorms:

$$\forall 1 \leq l \leq r, \forall 1 \leq \alpha \leq N, \forall k \in \mathbb{N}, \forall K \subseteq U_\alpha(\text{compact}) \qquad p_{l,\alpha,k,K}(\xi) = \|\rho^l_\alpha \circ \xi \circ \varphi_\alpha^{-1}\|_{\varphi_\alpha(K),k}\,.$$

(2) The topology on $D(M, E^\vee)$ is induced by the seminorms:

$$\forall 1 \leq l \leq r, \forall 1 \leq \alpha \leq N, \forall k \in \mathbb{N}, \forall K \subseteq U_\alpha(\text{compact}) \qquad q_{l,\alpha,k,K}(\eta) = \|(\rho^\vee_\alpha)^l \circ \eta \circ \varphi_\alpha^{-1}\|_{\varphi_\alpha(K),k}\,.$$

For all $\xi \in D(M, E)$ we have

$$q_{l,\alpha,k,K}(T_\xi) = \|(\rho^\vee_\alpha)^l \circ T_\xi \circ \varphi_\alpha^{-1}\|_{\varphi_\alpha(K),k} = \|(\rho_{\mathcal{D},\varphi_\alpha}) \circ (T_\xi \circ \varphi_\alpha^{-1}) \circ \underbrace{(\varphi_\alpha|_{E_x})^{-1}(e_l)}_{s_l(x)}\|_{\varphi_\alpha(K),k}\,,$$

where $(e_1, \ldots, e_r)$ is the standard basis for $\mathbb{R}^r$. Let $y = \varphi_\alpha(x)$. Note that

$$[T_\xi(\varphi_\alpha^{-1}(y))][s_l(x)] = \langle s_l(x), \xi(x)\rangle_E\, \mu(x)\,.$$

Therefore, if we define the smooth function f_α on U_α by $\mu(x) = f_\alpha(x)|dx^1 \wedge \ldots \wedge dx^n|$, then

$$(\rho_{\mathcal{D},\varphi_\alpha}) \circ (T_\xi \circ \varphi_\alpha^{-1}) \circ s_l(x) = \langle s_l(x), \xi(x)\rangle_E f_\alpha(x) = \xi^l(x) f_\alpha(x) = (\rho^l_\alpha \circ \xi \circ \varphi_\alpha^{-1}(y))(f_\alpha \circ \varphi_\alpha^{-1}(y))\,. \tag{8}$$

So, if we let

$$C = \max_{y \in \varphi_\alpha(K), |\beta| \leq k} |\partial^\beta(f_\alpha \circ \varphi_\alpha^{-1}(y))|\,,$$

then

$$q_{l,\alpha,k,K}(T_\xi) = \|(\rho^l_\alpha \circ \xi \circ \varphi_\alpha^{-1}(y))(f_\alpha \circ \varphi_\alpha^{-1}(y))\|_{\varphi_\alpha(K),k} \leq C\|\rho^l_\alpha \circ \xi \circ \varphi_\alpha^{-1}(y))\|_{\varphi_\alpha(K),k} = C\, p_{l,\alpha,k,K}(\xi)\,.$$

- $T : (C^\infty(M, E), \|.\|_{e,q}) \to (C^\infty(M, E^\vee), \|.\|_{e,q})$ **is a topological isomorphism:**

$$\|\xi\|_{W^{e,q}(M,E;\Lambda)} = \sum_{\alpha=1}^{N}\sum_{l=1}^{r} \|\rho^l_\alpha \circ \psi_\alpha \xi \circ \varphi_\alpha^{-1}\|_{W^{e,q}(\varphi_\alpha(U_\alpha))}\,,$$

$$\|T_\xi\|_{W^{e,q}(M,E^\vee;\Lambda^\vee)} = \sum_{\alpha=1}^{N}\sum_{l=1}^{r} \|(\rho^\vee_\alpha)^l \circ \psi_\alpha T_\xi \circ \varphi_\alpha^{-1}\|_{W^{e,q}(\varphi_\alpha(U_\alpha))}\,.$$

By Equation (8), we have

$$(\rho^\vee_\alpha)^l \circ \psi_\alpha T_\xi \circ \varphi_\alpha^{-1} = \rho_{\mathcal{D},\varphi_\alpha} \circ (\psi_\alpha T_\xi \circ \varphi_\alpha^{-1}) \circ s_l(x) = (\rho^l_\alpha \circ \psi_\alpha \xi \circ \varphi_\alpha^{-1})(f_\alpha \circ \varphi_\alpha^{-1})\,.$$

Therefore,

$$\|T_\xi\|_{W^{e,q}(M,E^\vee;\Lambda^\vee)} = \sum_{\alpha=1}^{N}\sum_{l=1}^{r} \|(\rho_\alpha^l \circ \psi_\alpha\xi \circ \varphi_\alpha^{-1})(f_\alpha \circ \varphi_\alpha^{-1})\|_{W^{e,q}(\varphi_\alpha(U_\alpha))}\,.$$

Now, we just need to notice that $f_\alpha \circ \varphi_\alpha^{-1}$ is a positive function and belongs to $C^\infty(\varphi_\alpha(U_\alpha))$ (so $\frac{1}{f_\alpha\circ\varphi_\alpha^{-1}}$ is also smooth) and $\rho_\alpha^l \circ \psi_\alpha\xi \circ \varphi_\alpha^{-1}$ has support in the compact set $\varphi_\alpha(\mathrm{supp}(\psi_\alpha))$ to conclude that

$$\|\xi\|_{W^{e,q}(M,E;\Lambda)} \simeq \|T_\xi\|_{W^{e,q}(M,E^\vee;\Lambda^\vee)}\,.$$

□

Lemma 14. *Let M^n be a compact smooth manifold and let $\pi : E \to M$ be a vector bundle of rank r equipped with a fiber metric $\langle .,. \rangle_E$. Let $e \in \mathbb{R}$ and $q \in (1,\infty)$. Suppose $\Lambda = \{(U_\alpha, \varphi_\alpha, \rho_\alpha, \psi_\alpha)\}_{\alpha=1}^N$ is an augmented total trivialization atlas for $E \to M$. If e is a noninteger less than -1 further assume that the total trivialization atlas in Λ is GGL. Then $D(M,E) \hookrightarrow W^{e,q}(M,E) \hookrightarrow D'(M,E)$.*

Proof. We refer to [24] for discussion about the case where $e \in \mathbb{Z}$. For $e \in \mathbb{R}\setminus\mathbb{Z}$ we have

$$W^{e,q}(M,E;\Lambda) \hookrightarrow W^{\lfloor e\rfloor,q}(M,E;\Lambda) \hookrightarrow D'(M,E)\,,$$
$$D(M,E) \hookrightarrow W^{\lfloor e\rfloor+1,q}(M,E;\Lambda) \hookrightarrow W^{e,q}(M,E;\Lambda)\,.$$

□

Theorem 100. *Let M^n be a compact smooth manifold and let $\pi : E \to M$ be a vector bundle of rank r equipped with a fiber metric $\langle .,. \rangle_E$. Let $e \in \mathbb{R}$ and $q \in (1,\infty)$. Suppose $\Lambda = \{(U_\alpha, \varphi_\alpha, \rho_\alpha, \psi_\alpha)\}_{\alpha=1}^N$ is an augmented total trivialization atlas for $E \to M$ which trivializes the fiber metric. If e is a noninteger whose magnitude is greater than 1 further assume that the total trivialization atlas in Λ is GL compatible with itself. Fix a positive smooth density μ on M. Consider the L^2 inner product on $D(M,E)$ defined by*

$$\langle u,v\rangle_2 = \int_M \langle u,v\rangle_E \mu\,.$$

Then

(i) $\langle .,. \rangle_2$ extends uniquely to a continuous bilinear pairing $\langle .,. \rangle_2 : W^{-e,q'}(M,E;\Lambda) \times W^{e,q}(M,E;\Lambda) \to \mathbb{R}$ (We are using the same notation (i.e., $\langle .,. \rangle_2$) for the extended bilinear map!)

(ii) The map $S : W^{-e,q'}(M,E;\Lambda) \to [W^{e,q}(M,E;\Lambda)]^*$ defined by $S(u) = l_u$ where

$$l_u : W^{e,q}(M,E;\Lambda) \to \mathbb{R}, \quad l_u(v) = \langle u,v\rangle_2$$

is a well-defined topological isomorphism.

In particular, $[W^{e,q}(M,E;\Lambda)]^*$ can be identified with $W^{-e,q'}(M,E;\Lambda)$.

Proof.

(1) By Theorem 8, in order to prove (i) it is enough to show that

$$\langle .,. \rangle_2 : (C^\infty(M,E), \|.\|_{-e,q'}) \times (C^\infty(M,E), \|.\|_{e,q}) \to \mathbb{R}$$

is a **continuous** bilinear map. Denote the corresponding standard trivialization map for the density bundle $\mathcal{D} \to M$ by $\rho_{\mathcal{D},\varphi_\alpha}$. Let $\Lambda_1 = \{(U_\alpha, \varphi_\alpha, \rho_\alpha, \tilde{\psi}_\alpha)\}_{\alpha=1}^N$ be an augmented total trivialization atlas for E where $\tilde{\psi}_\alpha = \frac{\psi_\alpha^3}{\sum_{\beta=1}^N \psi_\beta^3}$. Note that $\frac{1}{\sum_{\beta=1}^N \psi_\beta^3} \circ \varphi_\alpha^{-1} \in BC^\infty(\varphi_\alpha(U_\alpha))$. Let $K_\alpha = \mathrm{supp}\psi_\alpha$. Recall that on U_α we may write $\mu =$

$h_\alpha |dx^1 \wedge \cdots \wedge dx^n|$ where $h_\alpha = \rho_{\mathcal{D},\varphi_\alpha} \circ \mu$ is smooth. Moreover, for any continuous function $f : M \to \mathbb{R}$,

$$\begin{aligned}
\int_M f\mu &= \sum_{\alpha=1}^{N} \int_M \tilde{\psi}_\alpha f \mu \\
&= \sum_{\alpha=1}^{N} \int_{\varphi_\alpha(U_\alpha)} (\varphi_\alpha^{-1})^* (\tilde{\psi}_\alpha f \mu) \\
&= \sum_{\alpha=1}^{N} \int_{\varphi_\alpha(U_\alpha)} (\tilde{\psi}_\alpha f \circ \varphi_\alpha^{-1})(\varphi_\alpha^{-1})^* \mu \\
&= \sum_{\alpha=1}^{N} \int_{\varphi_\alpha(U_\alpha)} (\tilde{\psi}_\alpha f \circ \varphi_\alpha^{-1})(h_\alpha \circ \varphi_\alpha^{-1})\, dV \\
&\preceq \sum_{\alpha=1}^{N} \int_{\varphi_\alpha(U_\alpha)} (\psi_\alpha^2 f \circ \varphi_\alpha^{-1})(\psi_\alpha h_\alpha \circ \varphi_\alpha^{-1})\, dV \quad (\frac{1}{\sum_{\beta=1}^{N} \psi_\beta^3} \circ \varphi_\alpha^{-1} \in BC^\infty(\varphi_\alpha(U_\alpha)))\,.
\end{aligned}$$

Therefore, we have

$$\begin{aligned}
|\int_M \langle u, v \rangle_E \mu| &= |\sum_{\alpha=1}^{N} \int_M \tilde{\psi}_\alpha \langle u, v \rangle_E \mu| \\
&\preceq |\sum_{\alpha=1}^{N} \int_{\varphi_\alpha(U_\alpha)} (\psi_\alpha^2 \langle u, v \rangle_E \circ \varphi_\alpha^{-1})(\psi_\alpha h_\alpha \circ \varphi_\alpha^{-1}) dV|\,.
\end{aligned}$$

Since by assumption the total trivialization atlas in Λ trivializes the metric, we get

$$\begin{aligned}
|\int_M \langle u, v \rangle_E \mu| &\preceq \sum_{\alpha=1}^{N} \sum_{i=1}^{r} |\int_{\varphi_\alpha(U_\alpha)} (\psi_\alpha \circ \varphi_\alpha^{-1} \tilde{u}_i)(\psi_\alpha \circ \varphi_\alpha^{-1} \tilde{v}_i)(\psi_\alpha h_\alpha \circ \varphi_\alpha^{-1}) dV| \\
&\overset{\text{Remark 46}}{\preceq} \sum_{\alpha=1}^{N} \sum_{i=1}^{r} \|(\psi_\alpha \circ \varphi_\alpha^{-1} \tilde{u}_i)\|_{W^{-e,q'}(\varphi_\alpha(U_\alpha))} \|(\psi_\alpha \circ \varphi_\alpha^{-1} \tilde{v}_i)(\psi_\alpha h_\alpha \circ \varphi_\alpha^{-1})\|_{W^{e,q}(\varphi_\alpha(U_\alpha))} \\
&\preceq \sum_{\alpha=1}^{N} \sum_{i=1}^{r} \|(\psi_\alpha \circ \varphi_\alpha^{-1} \tilde{u}_i)\|_{W^{-e,q'}(\varphi_\alpha(U_\alpha))} \|(\psi_\alpha \circ \varphi_\alpha^{-1} \tilde{v}_i)\|_{W^{e,q}(\varphi_\alpha(U_\alpha))} \\
&\preceq [\sum_{\alpha=1}^{N} \sum_{i=1}^{r} \|(\psi_\alpha \circ \varphi_\alpha^{-1} \tilde{u}_i)\|_{W^{-e,q'}(\varphi_\alpha(U_\alpha))}] [\sum_{\alpha=1}^{N} \sum_{i=1}^{r} \|(\psi_\alpha \circ \varphi_\alpha^{-1} \tilde{v}_i)\|_{W^{e,q}(\varphi_\alpha(U_\alpha))}] \\
&= \|u\|_{W^{-e,q'}(M,E;\Lambda)} \|v\|_{W^{e,q}(M,E;\Lambda)}\,.
\end{aligned}$$

(2) For each $u \in W^{-e,q'}(M, E; \Lambda)$, l_u is continuous because $\langle .,. \rangle_2$ is continuous. So, S is well-defined.

(3) S is a continuous linear map because

$$\begin{aligned}
\forall u \in W^{-e,q'}(M, E; \Lambda) \quad \|S(u)\|_{(W^{e,q}(M,E;\Lambda))^*} &= \sup_{0 \neq v \in W^{e,q}(M,E;\Lambda)} \frac{|S(u)v|}{\|v\|_{W^{e,q}(M,E;\Lambda)}} \\
&= \sup_{0 \neq v \in W^{e,q}(M,E;\Lambda)} \frac{|\langle u, v \rangle_2|}{\|v\|_{W^{e,q}(M,E;\Lambda)}} \leq C \|u\|_{W^{-e,q'}(M,E;\Lambda)}\,,
\end{aligned}$$

where C is the norm of the continuous bilinear form $\langle .,. \rangle_2$.

(4) S is injective: suppose $u \in W^{-e,q'}(M, E; \Lambda)$ is such that $S(u) = 0$, then

$$\forall v \in W^{e,q}(M, E; \Lambda) \quad l_u(v) = \langle u, v \rangle_2 = 0\,.$$

We need to show that $u = 0$.

- **Step 1:** For ξ and η in $D(M,E)$ we have
$$\langle \xi,\eta\rangle_2 = \langle u_\xi, T\eta\rangle_{[D(M,E^\vee)]^*\times D(M,E^\vee)}\,,$$
where T is the map introduced in Lemma 13 (note that if we identify $D(M,E)$ with a subset of $[D(M,E^\vee)]^*$, then we may write ξ instead of u_ξ on the right hand side of the above equality). The reason is as follows:
$$\langle u_\xi, T\eta\rangle_{[D(M,E^\vee)]^*\times D(M,E^\vee)} = \int_M [T_\eta(x)][\xi(x)] \qquad \text{(by definition of } u_\xi\text{)}.$$
Recall that by definition of T_η we have
$$\forall x\in M \quad \forall a\in E_x \qquad [T_\eta(x)][a] = \langle a,\eta(x)\rangle_E\,\mu\,.$$
In particular,
$$[T_\eta(x)][\xi(x)] = \langle \xi(x),\eta(x)\rangle_E\,\mu\,.$$
Therefore,
$$\langle u_\xi, T\eta\rangle_{[D(M,E^\vee)]^*\times D(M,E^\vee)} = \int_M \langle \xi(x),\eta(x)\rangle_E\mu = \langle \xi,\eta\rangle_2\,.$$
- **Step 2:** For $w\in W^{-e,q'}(M,E;\Lambda)$ and $\eta\in D(M,E)\subseteq W^{e,q}(M,E;\Lambda)$ we have
$$\langle w,\eta\rangle_2 = \langle w, T\eta\rangle_{[D(M,E^\vee)]^*\times D(M,E^\vee)}\,.$$
Indeed, let $\{\xi_m\}$ be a sequence in $D(M,E)$ that converges to w in $W^{-e,q'}(M,E;\Lambda)$. Note that $W^{-e,q'}(M,E;\Lambda)\hookrightarrow [D(M,E^\vee)]^*$, so the sequence converges to w in $[D(M,E^\vee)]^*$ as well. By what was proved in the first step, for all m
$$\langle \xi_m,\eta\rangle_2 = \langle \xi_m, T\eta\rangle_{[D(M,E^\vee)]^*\times D(M,E^\vee)}\,.$$
Taking the limit as $m\to\infty$ proves the claim.
- **Step 3:** Finally note that for all $v\in D(M,E)\subseteq W^{e,q}(M,E;\Lambda)$
$$\langle T^*u, v\rangle_{[D(M,E)]^*\times D(M,E)} = \langle u, Tv\rangle_{[D(M,E^\vee)]^*\times D(M,E^\vee)} = \langle u,v\rangle_2 = 0\,.$$
Therefore, $T^*u=0$ as an element of $[D(M,E)]^*$. T is a continuous bijective map, so T^* is injective. It follows that $u=0$ as an element of $[D(M,E^\vee)]^*$ and so $u=0$ as an element of $W^{-e,q'}(M,E;\Lambda)$.

(5) S is surjective. Let $F\in [W^{e,q}(M,E;\Lambda)]^*$. We need to show that there is an element $u\in W^{-e,q'}(M,E;\Lambda)$ such that $S(u)=F$. Since $D(M,E)$ is dense in $W^{e,q}(M,E;\Lambda)$, it is enough to show that there exists an element $u\in W^{-e,q'}(M,E;\Lambda)$ with the property that
$$\forall \xi\in D(M,E) \quad F(\xi) = \langle u,\xi\rangle_2\,.$$
Note that, according to what was proved in Step 2,
$$\langle u,\xi\rangle_2 = \langle u, T\xi\rangle_{[D(M,E^\vee)]^*\times D(M,E^\vee)} = \langle T^*u,\xi\rangle_{[D(M,E)]^*\times D(M,E)}\,.$$
So, we need to show that there exists an element $u\in W^{-e,q'}(M,E;\Lambda)$ such that
$$\forall \xi\in D(M,E) \quad F(\xi) = \langle T^*u,\xi\rangle_{[D(M,E)]^*\times D(M,E)}\,.$$
Since $D(M,E)\hookrightarrow W^{e,q}(M,E;\Lambda)$, $F|_{D(M,E)}$ is an element of $[D(M,E)]^*$. We let
$$u := [T^{-1}]^*(F|_{D(M,E)}) \in [D(M,E^\vee)]^*\,.$$

Clearly, u satisfies the desired equality (note that $[T^{-1}]^* = [T^*]^{-1}$). So, we just need to show that u is indeed an element of $W^{-e,q'}(M,E;\Lambda)$. Note that

$$u \in W^{-e,q'}(M,E;\Lambda) \Longleftrightarrow \forall 1 \le \alpha \le N \quad H_\alpha(\psi_\alpha u) \in [W^{-e,q'}_{\varphi_\alpha(\text{supp}\psi_\alpha)}(\varphi_\alpha(U_\alpha))]^{\times r}.$$

Since $\text{supp}(\psi_\alpha u) \subseteq \text{supp}\psi_\alpha$, it follows from Remark 31 that

$$\forall 1 \le l \le r \quad \text{supp}([H_\alpha(\psi_\alpha u)]^l) \subset \varphi_\alpha(\text{supp}\psi_\alpha).$$

It remains to prove that $[H_\alpha(\psi_\alpha u)]^l \in W^{-e,q'}(\varphi_\alpha(U_\alpha))$. Note that

$$\text{for } e \ge 0 \qquad [W^{e,q}_0(\varphi_\alpha(U_\alpha))]^* = W^{-e,q'}(\varphi_\alpha(U_\alpha)),$$
$$\text{for } e < 0 \qquad [W^{e,q}_0(\varphi_\alpha(U_\alpha))]^* = [W^{e,q}(\varphi_\alpha(U_\alpha))]^* = W^{-e,q'}_0(\varphi_\alpha(U_\alpha)) \subseteq W^{-e,q'}(\varphi_\alpha(U_\alpha)).$$

Consequently, for all e

$$[W^{e,q}_0(\varphi_\alpha(U_\alpha))]^* \subseteq W^{-e,q'}(\varphi_\alpha(U_\alpha)).$$

Therefore, it is enough to show that

$$[H_\alpha(\psi_\alpha u)]^l \in [W^{e,q}_0(\varphi_\alpha(U_\alpha))]^*.$$

To this end, we need to prove that

$$[H_\alpha(\psi_\alpha u)]^l : (C^\infty_c(\varphi_\alpha(U_\alpha)), \|.\|_{e,q}) \to \mathbb{R}$$

is continuous. For all $\xi \in C^\infty_c(\varphi_\alpha(U_\alpha))$ we have

$$\begin{aligned}
[H_\alpha(\psi_\alpha u)]^l(\xi) &= \langle \psi_\alpha u, g_{l,\xi,U_\alpha,\varphi_\alpha}\rangle_{[D(U_\alpha,E^\vee_{U_\alpha})]^* \times D(U_\alpha,E^\vee_{U_\alpha})} = \langle u, \psi_\alpha g_{l,\xi,U_\alpha,\varphi_\alpha}\rangle_{[D(M,E^\vee)]^*\times D(M,E^\vee)} \\
&= \langle [T^{-1}]^* F|_{D(M,E)}, \psi_\alpha g_{l,\xi,U_\alpha,\varphi_\alpha}\rangle_{[D(M,E^\vee)]^*\times D(M,E^\vee)} \\
&= \langle F|_{D(M,E)}, T^{-1}(\psi_\alpha g_{l,\xi,U_\alpha,\varphi_\alpha})\rangle_{D^*(M,E)\times D(M,E)} = F(T^{-1}(\psi_\alpha g_{l,\xi,U_\alpha,\varphi_\alpha})).
\end{aligned}$$

Thus, $[H_\alpha(\psi_\alpha u)]^l$ is the composition of the following maps:

$$\begin{aligned}
(C^\infty_c(\varphi_\alpha(U_\alpha)), \|.\|_{e,q}) &\to [W^{e,q}_{\varphi_\alpha(\text{supp}\psi_\alpha)}(\varphi_\alpha(U_\alpha))]^{\times r} \cap [C^\infty_c(\varphi_\alpha(U_\alpha))]^{\times r} \to W^{e,q}_{\text{supp}\psi_\alpha}(M,E^\vee;\Lambda^\vee) \cap C^\infty(M,E^\vee) \\
&\to (C^\infty(M,E), \|\|_{e,q}) \to \mathbb{R}
\end{aligned}$$

$$\begin{aligned}
\xi \mapsto (0,\cdots,0,\underbrace{(\psi_\alpha \circ \varphi_\alpha^{-1})\xi}_{l\text{th position}},0,\ldots,0) &\mapsto H^{-1}_{E^\vee,U_\alpha,\varphi_\alpha}(0,\ldots,0,(\psi_\alpha\circ\varphi_\alpha^{-1})\xi,0,\cdots,0) = \psi_\alpha g_{l,\xi,U_\alpha,\varphi_\alpha} \\
&\mapsto T^{-1}(\psi_\alpha g_{l,\xi,U_\alpha,\varphi_\alpha}) \mapsto F(T^{-1}(\psi_\alpha g_{l,\xi,U_\alpha,\varphi_\alpha})),
\end{aligned}$$

which is a composition of continuous maps.

(6) $S : W^{-e,q'}(M,E;\Lambda) \to [W^{e,q}(M,E;\Lambda)]^*$ is a continuous bijective map, so by the Banach isomorphism theorem, it is a topological isomorphism.

□

Remark 61.

(1) *The result of Theorem 100 remains valid even if $\Lambda = \{(U_\alpha, \varphi_\alpha, \rho_\alpha, \psi_\alpha)\}$ does not trivialize the fiber metric. Indeed, if e is not a noninteger whose magnitude is greater than 1, then the Sobolev spaces $W^{e,q}$ and $W^{-e,q'}$ are independent of the choice of augmented total trivialization atlas. If e is a noninteger whose magnitude is greater than 1, then by Theorem 37 there exists an augmented total trivialization atlas $\tilde{\Lambda} = \{(U_\alpha, \varphi_\alpha, \tilde{\rho}_\alpha, \psi_\alpha)\}$ that trivializes the metric and*

has the same base atlas as Λ (so it is GL compatible with Λ because by assumption Λ is GL compatible with itself). So, we can replace Λ by $\tilde{\Lambda}$.

(2) *Let Λ be an augmented total trivialization atlas that is GL compatible with itself. Let e be a noninteger less than -1 and $q \in (1,\infty)$. By Theorem 100 and the above observation, $W^{e,q}(M,E;\Lambda)$ is topologically isomorphic to $[W^{-e,q'}(M,E;\Lambda)]^*$. However, the space $W^{-e,q'}(M,E;\Lambda)$ is independent of Λ. So, we may conclude that even when e is a noninteger less than -1, the space $W^{e,q}(M,E;\Lambda)$ is independent of the choice of the augmented total trivialization atlas as long as the corresponding total trivialization atlas is **GL compatible** with itself.*

9.3. On the Relationship between Various Characterizations

Here we discuss the relationship between the characterizations of Sobolev spaces given in Remark 54 and our original definition (Definition 30).

(1) Suppose $e \geq 0$.

$$W^{e,q}(M,E;\Lambda) = \{u \in L^q(M,E) : \|u\|_{W^{e,q}(M,E;\Lambda)} = \sum_{\alpha=1}^{N}\sum_{l=1}^{r} \|(\rho_\alpha)^l \circ (\psi_\alpha u) \circ \varphi_\alpha^{-1}\|_{W^{e,q}(\varphi_\alpha(U_\alpha))} < \infty\}.$$

As a direct consequence of Theorem 92, for $e \geq 0$, $W^{e,q}(M,E;\Lambda) \hookrightarrow L^q(M,E)$ with the original definition of $W^{e,q}(M,E;\Lambda)$. Therefore, the above characterization is completely consistent with the original definition.

(2)

$$W^{e,q}(M,E;\Lambda) = \{u \in D'(M,E) : \|u\|_{W^{e,q}(M,E;\Lambda)} = \sum_{\alpha=1}^{N}\sum_{l=1}^{r} \|\text{ext}^0_{\varphi_\alpha(U_\alpha),\mathbb{R}^n}[H_\alpha(\psi_\alpha u)]^l\|_{W^{e,q}(\mathbb{R}^n)} < \infty\}.$$

It follows from Corollary 6 that

- If e is not a noninteger less than -1, then

$$\|[H_\alpha(\psi_\alpha u)]^l\|_{W^{e,q}(\varphi_\alpha(U_\alpha))} \simeq \|\text{ext}^0_{\varphi_\alpha(U_\alpha),\mathbb{R}^n}[H_\alpha(\psi_\alpha u)]^l\|_{W^{e,q}(\mathbb{R}^n)},$$

- If e is a noninteger less than -1 and $\varphi_\alpha(U_\alpha)$ is $\mathbb{R}^n$ or a bounded open set with Lipschitz continuous boundary, then again the above equality holds.

Therefore, when e is not a noninteger less than -1, the above characterization completely agrees with the original definition. If e is a noninteger less than -1 and the total trivialization atlas corresponding to Λ is GGL, then again the two definitions agree.

(3)

$$W^{e,q}(M,E;\Lambda) = \{u \in D'(M,E) : [H_\alpha(u|_{U_\alpha})]^l \in W^{e,q}_{loc}(\varphi_\alpha(U_\alpha)),\ \forall 1 \leq \alpha \leq N, \forall 1 \leq l \leq r\}.$$

It follows immediately from Theorem 94 and Corollary 8 that the above characterization of the set of Sobolev functions is equivalent to the set given in the original definition provided we assume that if e is a noninteger less than -1, then Λ is GL compatible with itself.

(4) $W^{e,q}(M,E;\Lambda)$ is the completion of $C^\infty(M,E)$ with respect to the norm

$$\|u\|_{W^{e,q}(M,E;\Lambda)} = \sum_{\alpha=1}^{N}\sum_{l=1}^{r} \|(\rho_\alpha)^l \circ (\psi_\alpha u) \circ \varphi_\alpha^{-1}\|_{W^{e,q}(\varphi_\alpha(U_\alpha))}.$$

It follows from Theorem 99 that if e is not a noninteger less than -1 the above characterization of Sobolev spaces is equivalent to the original definition. Furthermore, if e is a noninteger less than -1 and Λ is GL compatible with itself, the two characterizations are equivalent.

Now, we will focus on proving the equivalence of the original definition and the fifth characterization of Sobolev spaces. In what follows instead of $\|.\|_{W^{k,q}(M,E;g,\nabla^E)}$ we just write $|.|_{W^{k,q}(M,E)}$. Furthermore, note that since k is a nonnegative integer, the choice of the augmented total trivialization atlas in Definition 30 is immaterial. Our proof follows the argument presented in [44] and is based on the following five facts:

- **Fact 1:** Let $u \in C^\infty(M,E)$ be such that $\operatorname{supp} u \subseteq U_\beta$ for some $1 \le \beta \le N$. Then

$$|u|^q_{L^q(M,E)} = \int_M |u|^q_E dV_g \simeq \sum_l \| \underbrace{\rho^l_\beta \circ u \circ \varphi_\beta^{-1}}_{u^l} \|^q_{L^q(\varphi_\beta(U_\beta))}.$$

- **Fact 2:** Let $u \in C^\infty(M,E)$ be such that $\operatorname{supp} u \subseteq U_\beta$ for some $1 \le \beta \le N$. Then

$$|u|^q_{W^{k,q}(M,E)} \simeq \sum_{s=0}^{k} \sum_{a=1}^{r} \sum_{1\le j_1,\ldots,j_s \le n} \|((\nabla^E)^s u)^a_{j_1\ldots j_s} \circ \varphi_\beta^{-1}\|^q_{L^q(\varphi_\beta(U_\beta))}.$$

Proof.

$$|u|^q_{W^{k,q}(M,E)} \simeq \sum_{s=0}^{k} |(\nabla^E)^s u|^q_{L^q(M,(T^*M)^{\otimes i}\otimes E)}$$
$$\overset{\text{Fact 1}}{\simeq} \sum_{s=0}^{k} \sum_{a=1}^{r} \sum_{1\le j_1,\ldots,j_s\le n} \| \underbrace{((\nabla^E)^s u)^a_{j_1\ldots j_s}}_{\text{components w.r.t } (U_\beta,\varphi_\beta,\rho_\beta)} \circ \varphi_\beta^{-1}\|^q_{L^q(\varphi_\beta(U_\beta))}.$$

□

- **Fact 3:** Let $u \in C^\infty(M,E)$ be such that $\operatorname{supp} u \subseteq U_\beta$ for some $1 \le \beta \le N$. Then

$$\|u\|_{W^{e,q}(M,E)} \simeq \sum_{l=1}^{r} \|\rho^l_\beta \circ u \circ \varphi_\beta^{-1}\|_{W^{e,q}(\varphi_\beta(U_\beta))}.$$

Proof. Let $\{\psi_\alpha\}$ be a partition of unity such that $\psi_\beta = 1$ on $\operatorname{supp} u$ (note that since elements of a partition of unity are nonnegative and their sum is equal to 1, we can conclude that if $\alpha \neq \beta$ then $\psi_\alpha = 0$ on $\operatorname{supp} u$). We have

$$\|u\|_{W^{e,q}(M,E)} \simeq \sum_{\alpha=1}^{N} \sum_{l=1}^{r} \|\rho^l_\alpha \circ (\psi_\alpha u) \circ \varphi_\alpha^{-1}\|_{W^{e,q}(\varphi_\alpha(U_\alpha))}$$
$$= \sum_{l=1}^{r} \|\rho^l_\beta \circ (\psi_\beta u) \circ \varphi_\beta^{-1}\|_{W^{e,q}(\varphi_\beta(U_\beta))} = \sum_{l=1}^{r} \|\rho^l_\beta \circ u \circ \varphi_\beta^{-1}\|_{W^{e,q}(\varphi_\beta(U_\beta))}.$$

□

- **Fact 4:** Let $u \in C^\infty(M,E)$. Then for any multi-index γ and all $1 \le l \le r$ we have (on any total trivialization triple (U,φ,ρ)):

$$|\partial^\gamma[\rho^l \circ u \circ \varphi^{-1}]| \preceq \sum_{s\le|\gamma|} \underbrace{\sum_{a=1}^{r} \sum_{1\le j_1,\cdots,j_s\le n}}_{\text{sum over all components of } (\nabla^E)^s u} |((\nabla^E)^s u)^a_{j_1\cdots j_s} \circ \varphi^{-1}|.$$

Proof. For any multi-index $\gamma = (\gamma_1,\ldots,\gamma_n)$ we define $\operatorname{seq}\gamma$ to be the following list of numbers:

$$\operatorname{seq}\gamma = \underbrace{1\ldots1}_{\gamma_1 \text{ times}} \underbrace{2\cdots2}_{\gamma_2 \text{ times}} \ldots \underbrace{n\ldots n}_{\gamma_n \text{ times}}.$$

Note that there are exactly $|\gamma| = \gamma_1 + \ldots + \gamma_n$ numbers in seq γ. By Observation 2 in Section 5.5.4 we have

$$((\nabla^E)^{|\gamma|}u)^l_{\text{seq}\,\gamma} \circ \varphi^{-1} = \partial^\gamma[\rho^l \circ u \circ \varphi^{-1}] + \sum_{a=1}^{r} \sum_{\alpha:|\alpha|<|\gamma|} C_{\alpha a}\partial^\alpha[\rho^a \circ u \circ \varphi^{-1}].$$

Thus

$$\partial^\gamma[\rho^l \circ u \circ \varphi^{-1}] = ((\nabla^E)^{|\gamma|}u)^l_{\text{seq}\,\gamma} \circ \varphi^{-1} - \sum_{a=1}^{r} \sum_{\alpha:|\alpha|<|\gamma|} C_{\alpha a}\partial^\alpha[\rho^a \circ u \circ \varphi^{-1}],$$

$$\partial^\alpha[\rho^a \circ u \circ \varphi^{-1}] = ((\nabla^E)^{|\alpha|}u)^a_{\text{seq}\,\alpha} \circ \varphi^{-1} - \sum_{b=1}^{r} \sum_{\beta:|\beta|<|\alpha|} C_{\beta b}\partial^\beta[\rho^b \circ u \circ \varphi^{-1}],$$

$$\vdots$$

where the coefficients $C_{\alpha a}$, $C_{\beta b}$, etc. are polynomials in terms of christoffel symbols and the metric and so they are all bounded on the compact manifold M. Consequently,

$$|\partial^\gamma[\rho^l \circ u \circ \varphi^{-1}]| \preceq \sum_{s\leq|\gamma|} \underbrace{\sum_{a=1}^{r} \sum_{1\leq j_1,\cdots,j_s\leq n}}_{\text{sum over all components of } (\nabla^E)^s u} |((\nabla^E)^s u)^a_{j_1\ldots j_s} \circ \varphi_\beta^{-1}|.$$

□

- **Fact 5:** Let $f \in C^\infty(M, E)$ and $u \in W^{k,q}(M, \tilde{E})$ where $\tilde{E}$ is another vector bundle over M. Then

$$\|f \otimes u\|_{W^{k,q}(M,E\otimes\tilde{E})} \preceq \|u\|_{W^{k,q}(M,\tilde{E})},$$

where the implicit constant may depend on f but it does not depend on u.

Proof. Let $\{(U_\alpha, \varphi_\alpha, \rho_\alpha)\}_{1\leq\alpha\leq N}$ and $\{(U_\alpha, \varphi_\alpha, \tilde{\rho}_\alpha)\}_{1\leq\alpha\leq N}$ be total trivialization atlases for E and $\tilde{E}$, respectively. Let $\{s_{\alpha,a} = \rho_\alpha^{-1}(e_a)\}_{a=1}^{r}$ be the corresponding local frame for E on U_α and $\{t_{\alpha,b} = \tilde{\rho}_\alpha^{-1}(e_b)\}_{b=1}^{\tilde{r}}$ be the corresponding local frame for $\tilde{E}$ on U_α. Let $G : \{1,\ldots,r\} \times \{1,\ldots,\tilde{r}\} \to \{1,\ldots,r\tilde{r}\}$ be an arbitrary but fixed bijective function. Then $\{(U_\alpha, \varphi_\alpha, \hat{\rho}_\alpha)\}$ is a total trivialization atlas for $E \otimes \tilde{E}$ where

$$\hat{\rho}_\alpha(s_{\alpha,a} \otimes t_{\alpha,b}) = e_{G(a,b)} \text{ (as an element of } \mathbb{R}^{r\tilde{r}}),$$

and it is extended by linearity to the $E \otimes \tilde{E}|_{U_\alpha}$. Now we have

$$\begin{aligned}\|f \otimes u\|_{W^{k,q}(M,E\otimes\tilde{E})} &= \sum_{\alpha=1}^{N}\sum_{a=1}^{r}\sum_{b=1}^{\tilde{r}} \|\hat{\rho}_\alpha^{a,b} \circ (\psi_\alpha f \otimes u) \circ \varphi_\alpha^{-1}\|_{W^{k,q}(\varphi_\alpha(U_\alpha))}\\ &= \sum_{\alpha=1}^{N}\sum_{a=1}^{r}\sum_{b=1}^{\tilde{r}} \|(\psi_\alpha \circ \varphi_\alpha^{-1})(f_\alpha^a \circ \varphi_\alpha^{-1})(u_\alpha^b \circ \varphi_\alpha^{-1})\|_{W^{k,q}(\varphi_\alpha(U_\alpha))},\end{aligned}$$

where $f = f_\alpha^a s_{\alpha,a}$ and $u = u_\alpha^b t_{\alpha,b}$ on U_α. Clearly $f_\alpha^a \circ \varphi_\alpha^{-1} \in C^\infty(\varphi_\alpha(U_\alpha))$. Therefore,

$$\|f \otimes u\|_{W^{k,q}(M,E\otimes\tilde{E})} \preceq \sum_{\alpha=1}^{N}\sum_{b=1}^{\tilde{r}} \|(\psi_\alpha \circ \varphi_\alpha^{-1})(u_\alpha^b \circ \varphi_\alpha^{-1})\|_{W^{k,q}(\varphi_\alpha(U_\alpha))} \simeq \|u\|_{W^{k,q}(M,\tilde{E})}.$$

□

- **Part I:** First we prove that $\|u\|_{W^{k,q}(M,E)} \preceq |u|_{W^{k,q}(M,E)}$.

  (1) **Case 1:** Suppose there exists $1 \leq \beta \leq N$ such that $\text{supp}\, u \subseteq U_\beta$. We have

$$\begin{aligned}
\|u\|^q_{W^{k,q}(M,E)} &\overset{\text{Fact 3}}{\simeq} \sum_{l=1}^{r} \|\rho^l_\beta \circ u \circ \varphi^{-1}_\beta\|^q_{W^{k,q}(\varphi_\beta(U_\beta))} \simeq \sum_{l=1}^{r} \sum_{|\gamma|\leq k} \|\partial^\gamma(\rho^l_\beta \circ u \circ \varphi^{-1}_\beta)\|^q_{L^q(\varphi_\beta(U_\beta))} \\
&\overset{\text{Fact 4}}{\preceq} \sum_{l=1}^{r} \sum_{|\gamma|\leq k} \sum_{s\leq|\gamma|} \sum_{a=1}^{r} \sum_{1\leq j_1,\dots,j_s\leq n} \|((\nabla^E)^s u)^a_{j_1\dots j_s} \circ \varphi^{-1}_\beta\|^q_{L^q(\varphi_\beta(U_\beta))} \\
&\preceq \sum_{s=0}^{k} \sum_{a=1}^{r} \sum_{1\leq j_1,\dots,j_s\leq n} \|((\nabla^E)^s u)^a_{j_1\dots j_s} \circ \varphi^{-1}_\beta\|^q_{L^q(\varphi_\beta(U_\beta))} \\
&\overset{\text{Fact 2}}{\simeq} |u|^q_{W^{k,q}(M,E)}\,.
\end{aligned}$$

(2) **Case 2:** Now let u be an arbitrary element of $C^\infty(M,E)$. We have

$$\begin{aligned}
\|u\|_{W^{k,q}(M,E)} &= \|\sum_{\alpha=1}^{N} \psi_\alpha u\|_{W^{k,q}(M,E)} \leq \sum_{\alpha=1}^{N} \|\psi_\alpha u\|_{W^{k,q}(M,E)} \\
&\preceq \sum_{\alpha=1}^{N} |\psi_\alpha u|_{W^{k,q}(M,E)} \qquad \text{(by what was proved in Case 1)} \\
&\preceq \sum_{\alpha=1}^{N} |u|_{W^{k,q}(M,E)} \simeq |u|_{W^{k,q}(M,E)}\,.
\end{aligned}$$

We note that the last inequality holds because

$$\begin{aligned}
|\psi_\alpha u|^q_{W^{k,q}(M,E)} &= \sum_{i=0}^{k} \|(\nabla^E)^i(\psi_\alpha u)\|^q_{L^q(M,(T^*M)^{\otimes i}\otimes E)} \\
&= \sum_{i=0}^{k} \|\sum_{j=0}^{i} \binom{i}{j} \nabla^j \psi_\alpha \otimes (\nabla^E)^{i-j} u\|^q_{L^q(M,(T^*M)^{\otimes i}\otimes E)} \\
&\overset{\text{Fact 5}}{\preceq} \sum_{i=0}^{k} \sum_{j=0}^{i} \|(\nabla^E)^{i-j} u\|^q_{L^q(M,(T^*M)^{\otimes(i-j)}\otimes E)} \\
&\preceq \sum_{s=0}^{k} \|(\nabla^E)^s u\|^q_{L^q(M,(T^*M)^{\otimes s}\otimes E)} \simeq |u|^q_{W^{k,q}(M,E)}\,.
\end{aligned}$$

- **Part II:** Now we show that $|u|_{W^{k,q}(M,E)} \preceq \|u\|_{W^{k,q}(M,E)}$.

  (1) **Case 1:** Suppose there exists $1 \leq \beta \leq N$ such that $\operatorname{supp} u \subseteq U_\beta$.

$$\begin{aligned}
|u|^q_{W^{k,q}(M,E)} &\overset{\text{Fact 2}}{\simeq} \sum_{s=0}^{k} \sum_{a=1}^{r} \sum_{1\leq j_1,\dots,j_s\leq n} \|((\nabla^E)^s u)^a_{j_1\dots j_s} \circ \varphi^{-1}_\beta\|^q_{L^q(\varphi_\beta(U_\beta))} \\
&\overset{\text{Observation 1 in 5.5.4}}{=} \sum_{s=0}^{k} \sum_{a=1}^{r} \sum_{1\leq j_1,\dots,j_s\leq n} \|\sum_{|\eta|\leq s} \sum_{l=1}^{r} (C_{\eta l})^a_{j_1\dots j_s} \partial^\eta(\underbrace{u^l}_{\rho^l_\beta \circ u} \circ \varphi^{-1}_\beta)\|^q_{L^q(\varphi_\beta(U_\beta))} \\
&\preceq \sum_{l=1}^{r} \sum_{|\eta|\leq k} \|\partial^\eta(u^l \circ \varphi^{-1}_\beta)\|^q_{L^q(\varphi_\beta(U_\beta))} = \sum_{l=1}^{r} \|u^l \circ \varphi^{-1}_\beta\|^q_{W^{k,q}(\varphi_\beta(U_\beta))} \\
&\simeq \|u\|^q_{W^{k,q}(M,E)}\,.
\end{aligned}$$

(2) **Case 2:** Now let u be an arbitrary element of $C^\infty(M,E)$.

$$
\begin{aligned}
|u|_{W^{k,q}(M,E)} &= |\sum_{\alpha=1}^{N} \psi_\alpha u|_{W^{k,q}(M,E)} \leq \sum_{\alpha=1}^{N} |\psi_\alpha u|_{W^{k,q}(M,E)} \\
&\overset{\text{Case 1}}{\preceq} \sum_{\alpha=1}^{N} \|\psi_\alpha u\|_{W^{k,q}(M,E)} \overset{\text{Fact 3}}{\simeq} \sum_{\alpha=1}^{N}\sum_{l=1}^{r} \|\rho_\alpha^l \circ (\psi_\alpha u) \circ \varphi_\alpha^{-1}\|_{W^{k,q}(\varphi_\alpha(U_\alpha))} \\
&\simeq \|u\|_{W^{k,q}(M,E)}\,.
\end{aligned}
$$

10. Some Results on Differential Operators

Let M^n be a compact smooth manifold. Let E and $\tilde{E}$ be two vector bundles over M of ranks r and $\tilde{r}$, respectively. A linear operator $P : C^\infty(M,E) \to \Gamma(M,\tilde{E})$ is called **local** if

$$\forall u \in C^\infty(M,E) \qquad \operatorname{supp} Pu \subseteq \operatorname{supp} u\,.$$

If P is a local operator, then it is possible to have a well-defined notion of restriction of P to open sets $U \subseteq M$, that is, if $P : C^\infty(M,E) \to \Gamma(M,\tilde{E})$ is local and $U \subseteq M$ is open, then we can define a map

$$P|_U : C^\infty(U, E_U) \to \Gamma(U, \tilde{E}_U)$$

with the property that

$$\forall u \in C^\infty(M,E) \qquad (Pu)|_U = P|_U(u|_U)\,.$$

Indeed, suppose $u, \tilde{u} \in C^\infty(M,E)$ agree on U, then as a result of P being local we have

$$\operatorname{supp}(Pu - P\tilde{u}) \subseteq \operatorname{supp}(u - \tilde{u}) \subseteq M \setminus U\,.$$

Therefore, if $u|_U = \tilde{u}|_U$, then $(Pu)|_U = (P\tilde{u})|_U$. Thus, if $v \in C^\infty(U, E_U)$ and $x \in U$, we can define $(P|_U)(v)(x)$ as follows: choose any $u \in C^\infty(M,E)$ such that $u = v$ on a neighborhood of x and then let $(P|_U)(v)(x) = (Pu)(x)$.

Recall that for any nonempty set V, $\mathrm{Func}(V, \mathbb{R}^t)$ denotes the vector space of all functions from V to $\mathbb{R}^t$. By the **local representation of** P with respect to the total trivialization triples (U, φ, ρ) of E and $(U, \varphi, \tilde{\rho})$ of $\tilde{E}$ we mean the linear transformation $Q : C^\infty(\varphi(U), \mathbb{R}^r) \to \mathrm{Func}(\varphi(U), \mathbb{R}^{\tilde{r}})$ defined by

$$Q(f) = \tilde{\rho} \circ P(\rho^{-1} \circ f \circ \varphi) \circ \varphi^{-1}\,.$$

Note that $\rho^{-1} \circ f \circ \varphi$ is a section of $E_U \to U$. Furthermore, note that for all $u \in C^\infty(M,E)$

$$\tilde{\rho} \circ (P(u|_U)) \circ \varphi^{-1} = Q(\rho \circ (u|_U) \circ \varphi^{-1})\,. \tag{9}$$

Let us denote the components of $f \in C^\infty(\varphi(U), \mathbb{R}^r)$ by $(f^1, \ldots, f^r)$. Then we can write $Q(f^1, \cdots, f^r) = (h^1, \ldots, h^{\tilde{r}})$ where for all $1 \leq k \leq \tilde{r}$

$$h^k = \pi_k \circ Q(f^1, \ldots, f^r) \overset{\text{Q is linear}}{=} \pi_k \circ Q(f^1, 0, \ldots, 0) + \ldots + \pi_k \circ Q(0, \ldots, 0, f^r)\,.$$

So, if for each $1 \leq k \leq \tilde{r}$ and $1 \leq i \leq r$ we define $Q_{ki} : C^\infty(\varphi(U), \mathbb{R}) \to \mathrm{Func}(\varphi(U), \mathbb{R})$ by

$$Q_{ki}(g) = \pi_k \circ Q(0, \ldots, 0, \underbrace{g}_{i\text{th position}}, 0, \ldots, 0)\,,$$

then we have

$$Q(f^1, \ldots, f^r) = (\sum_{i=1}^{r} Q_{1i}(f^i), \ldots, \sum_{i=1}^{r} Q_{\tilde{r}i}(f^i))\,.$$

In particular, note that the sth component of $\tilde{\rho} \circ Pu \circ \varphi^{-1}$, that is $\tilde{\rho}^s \circ Pu \circ \varphi^{-1}$, is equal to the sth component of $Q(\rho^1 \circ u \circ \varphi^{-1}, \cdots, \rho^r \circ u \circ \varphi^{-1})$ (see Equation (9)) which is equal to

$$\sum_{i=1}^{r} Q_{si}(\rho^i \circ u \circ \varphi^{-1}).$$

Theorem 101. *Let M^n be a compact smooth manifold. Let $P : C^\infty(M,E) \to \Gamma(M,\tilde{E})$ be a local operator. Let $\Lambda = \{(U_\alpha, \varphi_\alpha, \rho_\alpha, \psi_\alpha)\}_{1\leq\alpha\leq N}$ and $\tilde{\Lambda} = \{(U_\alpha, \varphi_\alpha, \tilde{\rho}_\alpha, \psi_\alpha)\}_{1\leq\alpha\leq N}$ be two augmented total trivialization atlases for E and $\tilde{E}$, respectively. Suppose the atlas $\{(U_\alpha, \varphi_\alpha)\}_{1\leq\alpha\leq N}$ is GL compatible with itself. For each $1 \leq \alpha \leq N$, let Q^α denote the local representation of P with respect to the total trivialization triples $(U_\alpha, \varphi_\alpha, \rho_\alpha)$ and $(U_\alpha, \varphi_\alpha, \tilde{\rho}_\alpha)$ of E and $\tilde{E}$, respectively. Suppose $e, \tilde{e} \in \mathbb{R}$, $1 < q, \tilde{q} < \infty$, and for each $1 \leq \alpha \leq N$, $1 \leq i \leq \tilde{r}$, and $1 \leq j \leq r$,*

$$Q^\alpha_{ij} : (C^\infty_c(\varphi_\alpha(U_\alpha)), \|.\|_{e,q}) \to W^{\tilde{e},\tilde{q}}_{loc}(\varphi_\alpha(U_\alpha))$$

is well-defined and continuous and does not increase support. Then

- *$P(C^\infty(M,E)) \subseteq W^{\tilde{e},\tilde{q}}(M,\tilde{E};\tilde{\Lambda})$,*
- *$P : (C^\infty(M,E), \|.\|_{e,q}) \to W^{\tilde{e},\tilde{q}}(M,\tilde{E};\tilde{\Lambda})$ is continuous and so it can be extended to a continuous linear map $P : W^{e,q}(M,E;\Lambda) \to W^{\tilde{e},\tilde{q}}(M,\tilde{E};\tilde{\Lambda})$.*

Proof. First note that

$$\|Pu\|_{W^{\tilde{e},\tilde{q}}(M,\tilde{E};\tilde{\Lambda})} = \sum_{\alpha=1}^{N}\sum_{i=1}^{\tilde{r}} \|\tilde{\rho}^i_\alpha \circ (\psi_\alpha(Pu)) \circ \varphi_\alpha^{-1}\|_{W^{\tilde{e},\tilde{q}}(\varphi_\alpha(U_\alpha))},$$

$$\|u\|_{W^{e,q}(M,E;\Lambda)} = \sum_{\alpha=1}^{N}\sum_{j=1}^{r} \|\rho^j_\alpha \circ (\psi_\alpha u) \circ \varphi_\alpha^{-1}\|_{W^{e,q}(\varphi_\alpha(U_\alpha))}.$$

It is enough to show that for all $1 \leq \alpha \leq N$, $1 \leq i \leq \tilde{r}$

$$\|\tilde{\rho}^i_\alpha \circ (\psi_\alpha(Pu)) \circ \varphi_\alpha^{-1}\|_{W^{\tilde{e},\tilde{q}}(\varphi_\alpha(U_\alpha))} \preceq \sum_{\beta=1}^{N}\sum_{j=1}^{r} \|\rho^j_\beta \circ (\psi_\beta u) \circ \varphi_\beta^{-1}\|_{W^{e,q}(\varphi_\beta(U_\beta))}.$$

We have

$$\begin{aligned}
\|\tilde{\rho}^i_\alpha \circ (\psi_\alpha(Pu)) \circ \varphi_\alpha^{-1}\|_{W^{\tilde{e},\tilde{q}}(\varphi_\alpha(U_\alpha))} &= \|(\psi_\alpha \circ \varphi_\alpha^{-1}) \cdot (\tilde{\rho}^i_\alpha \circ (Pu) \circ \varphi_\alpha^{-1})\|_{W^{\tilde{e},\tilde{q}}(\varphi_\alpha(U_\alpha))} \\
&\leq \sum_{j=1}^{r} \|(\psi_\alpha \circ \varphi_\alpha^{-1}) \cdot Q^\alpha_{ij}(\rho^j_\alpha \circ (\sum_{\beta=1}^{N} \psi_\beta u) \circ \varphi_\alpha^{-1})\|_{W^{\tilde{e},\tilde{q}}(\varphi_\alpha(U_\alpha))} \\
&\quad \text{(see the paragraph above Theorem 101)} \\
&\leq \sum_{\beta=1}^{N}\sum_{j=1}^{r} \|(\psi_\alpha \circ \varphi_\alpha^{-1}) \cdot Q^\alpha_{ij}(\rho^j_\alpha \circ (\psi_\beta u) \circ \varphi_\alpha^{-1})\|_{W^{\tilde{e},\tilde{q}}(\varphi_\alpha(U_\alpha))} \\
&= \sum_{\beta=1}^{N}\sum_{j=1}^{r} \|(\psi_\alpha \circ \varphi_\alpha^{-1}) \cdot Q^\alpha_{ij}(\rho^j_\alpha \circ (\xi\psi_\beta u) \circ \varphi_\alpha^{-1})\|_{W^{\tilde{e},\tilde{q}}(\varphi_\alpha(U_\alpha))},
\end{aligned}$$

where $\xi \in C^\infty_c(U_\alpha)$ is a fixed function such that $\xi = 1$ on $\text{supp}\,\psi_\alpha$. Using the assumption that $Q^\alpha_{ij} : (C^\infty_c(\varphi_\alpha(U_\alpha)), \|.\|_{e,q}) \to W^{\tilde{e},\tilde{q}}_{loc}(\varphi_\alpha(U_\alpha))$ is continuous we get

$$\|\tilde{\rho}^i_\alpha \circ (\psi_\alpha(Pu)) \circ \varphi_\alpha^{-1}\|_{W^{\tilde{e},\tilde{q}}(\varphi_\alpha(U_\alpha))} \preceq \sum_{\beta=1}^{N}\sum_{j=1}^{r} \|\rho^j_\alpha \circ (\xi\psi_\beta u) \circ \varphi_\alpha^{-1}\|_{W^{e,q}(\varphi_\alpha(U_\alpha))}.$$

Note that $\rho^j_\alpha \circ (\xi\psi_\beta u) \circ \varphi_\alpha^{-1} = (\xi\psi_\beta \circ \varphi_\alpha^{-1})(\rho^j_\alpha \circ u \circ \varphi_\alpha^{-1})$ has compact support in $\varphi_\alpha(U_\alpha \cap U_\beta)$. So, it follows from Corollary 6 that

$$\|\rho^j_\alpha \circ (\xi\psi_\beta u) \circ \varphi_\alpha^{-1}\|_{W^{e,q}(\varphi_\alpha(U_\alpha))} \simeq \|\rho^j_\alpha \circ (\xi\psi_\beta u) \circ \varphi_\alpha^{-1}\|_{W^{e,q}(\varphi_\alpha(U_\alpha \cap U_\beta))}.$$

Therefore,

$$\begin{aligned}
&\|\tilde{\rho}^i_\alpha \circ (\psi_\alpha(Pu)) \circ \varphi_\alpha^{-1}\|_{W^{\tilde{e},\tilde{q}}(\varphi_\alpha(U_\alpha))} \\
&\preceq \sum_{\beta=1}^{N}\sum_{j=1}^{r} \|\rho^j_\alpha \circ (\xi\psi_\beta u) \circ \varphi_\alpha^{-1}\|_{W^{e,q}(\varphi_\alpha(U_\alpha \cap U_\beta))} \\
&= \sum_{\beta=1}^{N}\sum_{j=1}^{r} \|\rho^j_\alpha \circ (\xi\psi_\beta u) \circ \varphi_\beta^{-1} \circ \varphi_\beta \circ \varphi_\alpha^{-1}\|_{W^{e,q}(\varphi_\alpha(U_\alpha \cap U_\beta))} \\
&\overset{\text{Theorem 80}}{\preceq} \sum_{\beta=1}^{N}\sum_{j=1}^{r} \|\rho^j_\alpha \circ (\xi\psi_\beta u) \circ \varphi_\beta^{-1}\|_{W^{e,q}(\varphi_\beta(U_\alpha \cap U_\beta))}.
\end{aligned}$$

So, it is enough to prove that $\|\rho^j_\alpha \circ (\xi\psi_\beta u) \circ \varphi_\beta^{-1}\|_{W^{e,q}(\varphi_\beta(U_\alpha \cap U_\beta))}$ can be bounded by $\sum_{\beta=1}^{N}\sum_{j=1}^{r} \|\rho^j_\beta \circ (\psi_\beta u) \circ \varphi_\beta^{-1}\|_{W^{e,q}(\varphi_\beta(U_\beta))}$. Since this can be done in the exact same way as the proof of Theorem 88, we do not repeat the argument here. □

Here we will discuss one simple application of the above theorem. Let (M^n, g) be a compact Riemannian manifold with $g \in W^{s,p}(M, T^2M)$, $sp > n$, and $s \geq 1$. Consider $d : C^\infty(M) \to C^\infty(T^*M)$. The local representations are all assumed to be with respect to charts in a super nice total trivialization atlas that is GL compatible with itself. The local representation of d is $Q : C^\infty(\varphi(U)) \to C^\infty(\varphi(U), \mathbb{R}^n)$ which is defined by

$$\begin{aligned}
Q(f)(a) &= \tilde{\rho} \circ d(\rho^{-1} \circ f \circ \varphi) \circ \varphi^{-1}(a) \\
&= \tilde{\rho} \circ \left(\frac{\partial f}{\partial x^i}|_{\varphi(\varphi^{-1}(a))} dx^i|_{\varphi^{-1}(a)}\right) \\
&= \left(\frac{\partial f}{\partial x^1}|_a, \ldots, \frac{\partial f}{\partial x^n}|_a\right).
\end{aligned}$$

Here we used $\rho = Id$ and the fact that if $g : M \to \mathbb{R}$ is smooth, then

$$(dg)(p) = \frac{\partial(g \circ \varphi^{-1})}{\partial x^i}|_{\varphi(p)} dx^i|_p.$$

Clearly, each component of Q is a continuous operator from $(C^\infty_c(\varphi(U)), \|.\|_{e,q})$ to $W^{e-1,q}(\varphi(U)) \hookrightarrow W^{e-1,q}_{loc}(\varphi(U))$ (see Theorem 82; note that $\varphi(U) = \mathbb{R}^n$). Hence d can be viewed as a continuous operator from $W^{e,q}(M)$ to $W^{e-1,q}(T^*M)$.

Several other interesting applications of Theorem 101 can be found in [16].

11. Conclusions

Sobolev-Slobodeckij spaces play a key role in the study of elliptic differential operators in nonsmooth setting. In this manuscript, we focused on establishing certain fundamental properties of Sobolev-Slobodeckij spaces that are particularly useful in better understanding the behavior of elliptic differential operators on compact manifolds. In particular, we built a general framework for developing multiplication theorems, embedding results, etc. for Sobolev–Slobodeckij spaces on compact manifolds. We paid special attention to spaces with noninteger smoothness order and to general sections of vector bundles. We established in particular that, as long as $1 < q < \infty$ and $e \geq 0$ or $e \in \mathbb{Z}$,

- Various common standard characterizations of $W^{e,q}$ (as discussed in Section 9) are equivalent;
- The local charts definition of $W^{e,q}$ is independent of the chosen atlas;

- Nice properties of $W^{e,q}$ for smooth domains in $\mathbb{R}^n$ (such as embedding properties and multiplication properties) will carry over to $W^{e,q}$ of sections of vector bundles.

Furthermore, we noticed that the local representations of elements of $W^{e,q}$ (for functions on M or, more generally, sections of vector bundles) will not necessarily be in the corresponding Euclidean Sobolev-Slobodeckij space; they should be viewed as elements of *locally* Sobolev-Slobodeckij spaces on the Euclidean space (we have devoted a separate manuscript [17] to the study of the properties of locally Sobolev-Slobodeckij spaces on the Euclidean space). In the same spirit, in Section 10 we observed that *locally* Sobolev-Slobodeckij spaces can be considered as the appropriate target spaces in the study of the local representations of differential operators between Sobolev–Slobodeckij spaces of sections of vector bundles. For the case where $e < -1$ is noninteger, we were not able to prove the validity of these properties in a general setting; however, by introducing notions such as "geometrically Lipschitz atlases", we found sufficient conditions that guarantee the validity of similar results as those we have for the case where $e \in \mathbb{Z}$.

Author Contributions: Conceptualization, A.B. and M.H.; methodology, A.B. and M.H.; formal analysis, A.B. and M.H.; investigation, A.B. and M.H. Both authors contributed equally to this work. All authors have read and agreed to the published version of the manuscript.

Funding: MH was supported in part by NSF Awards [1262982, 1620366, 2012857]. AB was supported by NSF Award [1262982].

Institutional Review Board Statement: Not applicable.

Informed Consent Statement: Not applicable.

Data Availability Statement: Not applicable.

Conflicts of Interest: The authors declare no conflict of interest.

References

1. Adams, R.A.; Fournier, J.J.F. *Sobolev Spaces*, 2nd ed.; Academic Press: New York, NY, USA, 2003.
2. Evans, L. *Partial Differential Equations*, 2nd ed.; American Mathematical Society: Providence, RI, USA, 2010.
3. Renardy, M.; Rogers, R. *An Introduction to Partial Differential Equations*, 2nd ed.; Springer: Berlin/Heildelberg, Germany; New York, NY, USA, 2004.
4. Bhattacharyya, P.K. *Distributions: Generalized Functions with Applications in Sobolev Spaces*; de Gruyter: Berlin, Germany, 2012.
5. Grisvard, P. *Elliptic Problems in Nonsmooth Domains*; Pitman Publishing: Marshfield, MA, USA, 1985.
6. Nezza, E.D.; Palatucci, G.; Valdinoci, E. Hitchhiker's guide to the fractional Sobolev spaces. *Bull. Sci. Math.* **2012**, *136*, 521–573. [CrossRef]
7. Runst, T.; Sickel, W. *Sobolev Spaces of Fractional Order, Nemytskij Operators, and Nonlinear Partial Differential Equations*; Walter de Gruyter: Berlin, Germany, 1996.
8. Triebel, H. *Theory of Function Spaces*; Volume 78 of Monographs in Mathematics; Birkhäuser: Basel, Switzerland, 1983.
9. Aubin, T. *Some Nonlinear Problems in Riemannian Geometry*; Springer: Berlin/Heidelberg, Germany, 1998.
10. Hebey, E. *Sobolev Spaces on Riemannian Manifolds*; Volume 1635 of Lecture Notes in Mathematics; Springer: Berlin, Germany; New York, NY, USA, 1996.
11. Hebey, E. *Nonlinear Analysis on Manifolds: Sobolev Spaces and Inequalities*; American Mathematical Society: Providence, RI, USA, 2000.
12. Triebel, H. *Theory of Function Spaces. II*; Volume 84 of Monographs in Mathematics; Birkhäuser: Basel, Switzerland, 1992.
13. Eichhorn, J. *Global Analysis on Open Manifolds*; Nova Science Publishers: Hauppauge, NY, USA, 2007.
14. Palais, R. *Seminar on the Atiyah-Singer Index Theorem*; Princeton University Press: Princeton, NJ, USA, 1965.
15. Behzadan, A. Some remarks on $W^{s,p}$ interior elliptic regularity estimates. *J. Elliptic Parabol. Equ.* **2021**, *7*, 137–169. [CrossRef]
16. Behzadan, A.; Holst, M. On certain geometric operators between Sobolev spaces of sections of tensor bundles on compact manifolds equipped with rough metrics. *arXiv* **2017**, arXiv:1704.07930v2.
17. Behzadan, A.; Holst, M. Some remarks on the space of locally Sobolev-Slobodeckij functions. *arXiv* **2018**, arXiv:1806.02188.
18. Behzadan, A.; Holst, M. Multiplication in Sobolev spaces, revisited. *Ark. Mat.* **2021**, *59*, 275–306. [CrossRef]
19. Lee, J.M. *Introduction to Smooth Manifolds*, 2nd ed.; Springer: Berlin/Heidelberg, Germany, 2012.
20. Grubb, G. *Distributions and Operators*; Springer: New York, NY, USA, 2009.
21. Folland, G. *Real Analysis: Modern Techniques and Their Applications*, 2nd ed.; Wiley: Hoboken, NJ, USA, 2007.
22. Apostol, T. *Mathematical Analysis*, 2nd ed.; Addison-Wesley: London, UK, 1974.
23. Rudin, W. *Functional Analysis*; McGraw-Hill: New York, NY, USA, 1973.

24. Reus, M.D. An Introduction to Functional Spaces. Master's Thesis, Utrecht University, Utrecht, The Netherlands, 2011.
25. Treves, F. *Topological Vector Spaces, Distributions and Kernels;* Academic Press: New York, NY, USA; London, UK, 1967.
26. Choquet-Bruhat, Y.; DeWitt-Morette, C. *Analysis, Manifolds and Physics, Part 1: Basics*, Revised ed.; North Holland: Amsterdam, The Netherlands, 1982.
27. Bastos, M.A.; Lebre, A.; Samko, S.; Spitkovsky, I.M. *Operator Theory, Operator Algebras and Applications;* Birkhäuser: Basel, Switzerland, 2014.
28. Tu, L. *An Introduction to Manifolds*, 2nd ed.; Springer: Berlin/Heidelberg, Germany, 2011.
29. Driver, B. Analysis Tools with Applications. 2003. Available online: https://mathweb.ucsd.edu/~bdriver/231-02-03/Lecture_Notes/PDE-Anal-Book/analpde2-2p.pdf (accessed on 29 January 2022).
30. Walschap, G. *Metric Structures in Differential Geometry;* Springer: Berlin/Heidelberg, Germany, 2004.
31. Inci, H.; Kappeler, T.; Topalov, P. *On the Regularity of the Composition of Diffeomorphisms;* American Mathematical Society: Providence, RI, USA, 2013; Volume 226.
32. Lee, J.M. *Riemannian Manifolds: An Introduction to Curvature;* Springer: Berlin/Heidelberg, Germany, 1997.
33. Taubes, C. *Differential Geometry: Bundles, Connections, Metrics and Curvature;* Oxford University Press: Oxford, UK, 2011.
34. Moore, J.D. Lectures on Differential Geometry. 2009. Available online: https://web.math.ucsb.edu/~moore/riemanniangeometry.pdf (accessed on 29 January 2022)
35. Gavrilov, A.V. Higher covariant derivatives. *Sib. Math. J.* **2008**, *49*, 997–1007. [CrossRef]
36. Debnath, L.; Mikusinski, P. *Introduction to Hilbert Spaces with Applications*, 3rd ed.; Academic Press: New York, NY, USA, 2005.
37. Grosser, M.; Kunzinger, M.; Oberguggenberger, M.; Steinbauer, R. *Geometric Theory of Generalized Functions with Applications to General Relativity;* Springer: Berlin/Heidelberg, Germany, 2001.
38. Adams, R.A. *Sobolev Spaces;* Academic Press: New York, NY, USA, 1975.
39. Triebel, H. *Interpolation Theory, Function Spaces, Differential Operators;* North-Holland Publishing Company: Amsterdam, The Netherlands, 1977.
40. Triebel, H. Function spaces in Lipschitz domains and on Lipschitz manifolds. Characteristic functions as pointwise multipliers. *Rev. Mat. Complut.* **2002**, *15*, 475–524. [CrossRef]
41. Demengel, F.; Demengel, G. *Function Spaces for the Theory of Elliptic Partial Differential Equations;* Springer: Berlin/Heidelberg, Germany, 2012.
42. Brezis, H.; Mironescu, P. Gagliardo-Nirenberg, composition and products in fractional Sobolev spaces. *J. Evol. Equ.* **2001**, *1*, 387–404. [CrossRef]
43. Lee, J.M. *Introduction to Smooth Manifolds;* Springer: Berlin/Heidelberg, Germany, 2002.
44. Grosse, N.; Schneider, C. Sobolev spaces on Riemannian manifolds with bounded geometry: General coordinates and traces. *Math. Nachr.* **2013**, *286*, 1586–1613. [CrossRef]

mathematics

MDPI

Article

Hardy Inequalities and Interrelations of Fractional Triebel–Lizorkin Spaces in a Bounded Uniform Domain

Jun Cao, Yongyang Jin *, Yuanyuan Li and Qishun Zhang

Department of Applied Mathematics, Zhejiang University of Technology, Hangzhou 310023, China; caojun1860@zjut.edu.cn (J.C.); 2112009013@zjut.edu.cn (Y.L.); zhang15868226102@163.com (Q.Z.)
* Correspondence: yongyang@zjut.edu.cn

Abstract: The interrelations of Triebel–Lizorkin spaces on smooth domains of Euclidean space $\mathbb{R}^n$ are well-established, whereas only partial results are known for the non-smooth domains. In this paper, Ω is a non-smooth domain of $\mathbb{R}^n$ that is bounded and uniform. Suppose $p, q \in [1, \infty)$ and $s \in (n(\frac{1}{p} - \frac{1}{q})_+, 1)$ with $n(\frac{1}{p} - \frac{1}{q})_+ := \max\{n(\frac{1}{p} - \frac{1}{q}), 0\}$. The authors show that three typical types of fractional Triebel–Lizorkin spaces, on Ω: $F^s_{p,q}(\Omega)$, $\mathring{F}^s_{p,q}(\Omega)$ and $\widetilde{F}^s_{p,q}(\Omega)$, defined via the restriction, completion and supporting conditions, respectively, are identical if Ω is E-thick and supports some Hardy inequalities. Moreover, the authors show the condition that Ω is E-thick can be removed when considering only the density property $F^s_{p,q}(\Omega) = \mathring{F}^s_{p,q}(\Omega)$, and the condition that Ω supports Hardy inequalities can be characterized by some Triebel–Lizorkin capacities in the special case of $1 \le p \le q < \infty$.

Keywords: Triebel–Lizorkin space; Hardy inequality; uniform domain; fractional Laplacian

Citation: Cao, J.; Jin, Y.; Li, Y.; Zhang, Q. Hardy Inequalities and Interrelations of Fractional Triebel–Lizorkin Spaces in a Bounded Uniform Domain. *Mathematics* **2022**, *10*, 637. https://doi.org/10.3390/math10040637

Academic Editor: Ioannis K. Argyros

Received: 12 January 2022
Accepted: 15 February 2022
Published: 18 February 2022

1. Introduction

The Triebel–Lizorkin spaces $F^s_{p,q}(\mathbb{R}^n)$ on the Euclidean space $\mathbb{R}^n$, with parameters $s \in \mathbb{R}$ and $p, q \in (0, \infty]$, were introduced in 1970s (see [1–3]). They provide a unified treatment of various kinds of classical concrete function spaces, such as Sobolev spaces, Hölder-Zygmund spaces, Bessel-potential spaces, Hardy spaces and BMO spaces. Nowadays, the theory of $F^s_{p,q}(\mathbb{R}^n)$ is well-established in the literature as has numerous applications (see [4–10] and their references).

When trying to extend the theory of Triebel–Lizorkin space from $\mathbb{R}^n$ to a domain Ω of $\mathbb{R}^n$, one usually meets the fundamental problem of identifying the interrelations among a number of related spaces that are defined from distinct perspectives. In particular, there are three typical ways of defining Triebel–Lizorkin spaces on Ω (see, e.g., [10]). To be precise, let $\mathcal{D}(\Omega) = C^\infty_0(\Omega)$ be the collection of all infinitely differentiable functions in $\mathbb{R}^n$ with compact supports in Ω and $\mathcal{D}'(\Omega)$ the dual space of $\mathcal{D}(\Omega)$. For any $s \in \mathbb{R}$ and $p, q \in (0, \infty]$, recall that

(I) $F^s_{p,q}(\Omega) := \{f \in \mathcal{D}'(\Omega) : \text{there is a } g \in F^s_{p,q}(\mathbb{R}^n) \text{ with } g|_\Omega = f\}$ being the *restriction Triebel–Lizorkin space* endowed with the quasi-norm

$$\|f\|_{F^s_{p,q}(\Omega)} := \inf \|g\|_{F^s_{p,q}(\mathbb{R}^n)}, \tag{1}$$

where the infimum is taken over all $g \in F^s_{p,q}(\mathbb{R}^n)$ satisfying $g|_\Omega = f$. Here, for any $g \in \mathcal{S}'(\mathbb{R}^n)$, $g|_\Omega$ is the *restriction* of g to Ω, defined as a distribution in Ω such that for any $\varphi \in \mathcal{D}(\Omega)$,

$$(g|_\Omega)(\varphi) := g(\varphi);$$

(II) $\mathring{F}^s_{p,q}(\Omega) := \overline{\mathcal{D}(\Omega)}^{\|\cdot\|_{F^s_{p,q}(\Omega)}}$ is the *completion Triebel–Lizorkin space* that is defined as the completion of $\mathcal{D}(\Omega)$ in $F^s_{p,q}(\Omega)$ with respect to the quasi-norm $\|\cdot\|_{F^s_{p,q}(\Omega)}$, as in (1);

(III) $\widetilde{F}^s_{p,q}(\Omega) := \{f \in \mathcal{D}'(\Omega) : \text{there is a } g \in F^s_{p,q}(\mathbb{R}^n) \text{ with } g|_\Omega = f \text{ and } \operatorname{supp} g \subset \overline{\Omega}\}$ being the *supporting Triebel–Lizorkin space* endowed with the quasi-norm

$$\|f\|_{\widetilde{F}^s_{p,q}(\Omega)} = \inf \|g\|_{F^s_{p,q}(\mathbb{R}^n)},$$

where the infimum is taken over all $g \in F^s_{p,q}(\mathbb{R}^n)$ satisfying $g|_\Omega = f$ and $\operatorname{supp} g \subset \overline{\Omega}$.

Note that if $\Omega = \mathbb{R}^n$ is the Euclidean space, it follows easily from their definitions and the density property of $F^s_{p,q}(\mathbb{R}^n)$ that the aforementioned three kinds of Triebel–Lizorkin spaces are identical (see, e.g., [4]). However, if $\Omega \neq \mathbb{R}^n$, the situation becomes much more complex, since in this case the above density property and many other important properties, including the availability of restriction, trace and extension operators may fail (see, e.g., [6,8]). Indeed, it turns out that the interrelations of the aforementioned three kinds of Triebel–Lizorkin spaces depend heavily on the geometry of domain Ω and parameters s, p and q. Let us review some of the known results on this subject.

If Ω is a bounded C^∞-domain, it is known that the following results are almost sharp (see ([8], Chapter 5)).

(A) $F^s_{p,q}(\Omega) = \mathring{F}^s_{p,q}(\Omega)$, if and only if, one of the following two conditions is satisfied:

(a1) $0 < p < \infty$, $-\infty < s < \frac{1}{p}$ and $0 < q < \infty$;

(a2) $1 < p < \infty$, $s = \frac{1}{p}$ and $0 < q < \infty$.

(B) $\mathring{F}^s_{p,q}(\Omega) = \widetilde{F}^s_{p,q}(\Omega)$, if $0 < p < \infty$, $0 < q < \infty$, $s > \sigma_p := n(\frac{1}{p} - 1)_+$ and $s - \frac{1}{p} \notin \mathbb{Z}_+$.

(C) $F^s_{p,q}(\Omega) = \widetilde{F}^s_{p,q}(\Omega)$, if $0 < p \le \infty$, $0 < q \le \infty$ and $\max\{\frac{1}{p} - 1, n(\frac{1}{p} - 1)\} < s < \frac{1}{p}$.

A combination of **(A)**, **(B)** and **(C)** immediately implies the following identities.

$$F^s_{p,q}(\Omega) = \mathring{F}^s_{p,q}(\Omega) = \widetilde{F}^s_{p,q}(\Omega), \tag{2}$$

if $0 < p < \infty$, $0 < q < \infty$ and $\max\{\frac{1}{p} - 1, n(\frac{1}{p} - 1)\} < s < \frac{1}{p}$.

Note the restriction that $s < \frac{1}{p}$ in the above identities can be relaxed if Ω supports some Hardy inequalities. In particular, it is known that

$$\widetilde{F}^s_{p,q}(\Omega) = F^s_{p,q}(\Omega) \cap L^p(\Omega, d(\cdot, \partial\Omega)^{-s}), \tag{3}$$

if $0 < p < \infty$, $0 < q < \infty$ and

$$s > \sigma_{p,q} := n\left(\frac{1}{\min\{p,q\}} - 1\right)_+,$$

where for any $x \in \Omega$, $d(x, \partial\Omega)$ denotes the distance from x to the boundary $\partial\Omega$ of Ω and

$$L^p(\Omega, d(\cdot, \partial\Omega)^{-s}) := \left\{ f : \|f\|_{L^p(\Omega, d(\cdot, \partial\Omega)^{-s})} = \left(\int_\Omega \frac{|f(x)|^p}{d(x, \partial\Omega)^{sp}}\, dx \right)^{1/p} < \infty \right\}$$

denotes the weighted Lebesgue space on Ω. The identity (3) together with **(A)** and **(B)** shows that if Ω supports the Hardy condition $F^s_{p,q}(\Omega) \subset L^p(\Omega, d(\cdot, \partial\Omega)^{-s})$, then identities (2) hold for all

$$1 < p < \infty, 1 \le q < \infty \quad \text{and} \quad 0 < s < \infty. \tag{4}$$

Recall that on the smooth domain, the Hardy inequalities

$$\|f\|_{L^p(\Omega, d(\cdot, \partial\Omega)^{-s})} \le C\|f\|_{\widetilde{F}^s_{p,q}(\Omega)}$$

hold for any $f \in \widetilde{F}^s_{p,q}(\Omega)$ with $0 < p \le \infty$, $0 < q \le \infty$ and $s > \sigma_p$ with σ_p as in **(B)**.

If Ω is a non-smooth domain, there is no comprehensive treatment compared with what is available for smooth domains. Moreover, in the former case we meet much more complicated situations influenced by the geometry of Ω. Let us mention some of the related results.

(i) If $\Omega \subset \mathbb{R}^n$ is a bounded domain such that its boundary $\partial\Omega$ is porous and has upper Minkowski dimension $D \in (0, n]$, Caetano ([11], Proposition 2.5) proved the following identity.

(A′) $F^s_{p,q}(\Omega) = \mathring{F}^s_{p,q}(\Omega)$, if $0 < p < \infty$, $0 < q < \infty$ and $-\infty < s < (n - D)/p$.

Note that for an arbitrary bounded domain Ω, it holds that $D \in [n-1, n]$, and if $D = n - 1$, then the range of s in **(A′)** equal to that in (a1).

(ii) If $\Omega \subset \mathbb{R}^n$ is a domain whose closure $\overline{\Omega}$ is a n-set, and $\partial\Omega$ is a d-set with $n - 1 < d < n$, Ihnatsyeva et al. ([12], Theorem 4.3) obtained the following inclusion.

(B′) $\mathring{F}^s_{p,q}(\Omega) \subset \widetilde{F}^s_{p,q}(\Omega)$, if $1 < p < \infty$, $1 \leq q < \infty$ and $(n - d)/p < s < \infty$.

Note that if $\partial\Omega$ is a d-set with $d < n$, then $\partial\Omega$ is porous (see ([10], Chapter 3)) and has upper Minkowski dimension d (see ([7], Chapter 1)).

(iii) If $\Omega \subset \mathbb{R}^n$ is an arbitrary domain, Triebel ([10], Chapter 2) proved the following identity.

(C′) $F^0_{p,2}(\Omega) = \widetilde{F}^0_{p,2}(\Omega)$, if $1 < p < \infty$.

Moreover, if Ω is a bounded Lipschitz domain, then it is proved in ([9], Proposition 3.1) that identity (2) holds true for all

$$0 < p < \infty, \ \min\{p, 1\} < q < \infty \ \text{and} \ \max\left\{\frac{1}{p} - 1, n\left(\frac{1}{p} - 1\right)\right\} < s < \frac{1}{p}. \tag{5}$$

Motivated by the aforementioned results, it is natural to ask the following.

Main question: Let Ω be a bounded non-smooth domain. Is it possible to extend identity (2) for parameters from (5) to the general fractional case $s \in (0, 1)$?

In this paper, we give an affirmative answer to the above question in the setting that Ω is a bounded uniform domain, which contains a bounded Lipschitz domain as a special case. Recall that a domain $\Omega \subset \mathbb{R}^n$ is called a *uniform domain* (see [13,14]), if there exist constants c_1 and $c_2 > 0$ such that each pair of points $x, y \in \Omega$ can be connected by a rectifiable curve $\Gamma \subset \Omega$ for which

$$\begin{cases} L(\Gamma) \leq c_1|x - y|, \\ \min\{|x - z|, |y - z|\} \leq c_2 d(z, \partial\Omega), \quad \text{for any } z \in \Gamma, \end{cases}$$

where $L(\Gamma)$ denotes the length of Γ.

A closely related notion of uniform domain is the so-called E-thick domain. Recall in [10] that a domain $\Omega \subset \mathbb{R}^n$ is said to be *E-thick*, if there exists $j_0 \in \mathbb{N}$ such that for any interior cube $Q^i \subset Q$ satisfying

$$l(Q^i) \sim 2^{-j} \quad \text{and} \quad d(Q^i, \partial\Omega) \sim 2^{-j} \qquad \text{for some } j \geq j_0 \in \mathbb{N},$$

one finds a complementary exterior cube $Q^e \subset \Omega^c = \mathbb{R}^n \setminus \Omega$ satisfying

$$l(Q^e) \sim 2^{-j} \quad \text{and} \quad d(Q^e, \partial\Omega) \sim d(Q^i, Q^e) \sim 2^{-j},$$

where the implicit constants are independent of Q^i, Q^e and j. It is known that any bounded Lipschitz domain is E-thick and uniform; and if a domain Ω is uniform, then $\overline{\Omega}^c$ is E-thick. Moreover, there exists domain in $\mathbb{R}^n$ that is E-thick but not uniform (see ([10], Remark 3.7)). Note that if Ω is E-thick, then $\partial\Omega$ is a d-set with $d \in [n-1, n)$ (see ([10], Proposition 3.18)).

We also need the following Hardy condition.

$(H)_{s,p,q}$*-condition.* Let $1 \le p,\ q < \infty$, $s \in (0,1)$ and $\Omega \subset \mathbb{R}^n$ be a domain satisfying $\Omega \neq \mathbb{R}^n$. Ω is said to satisfy the $(H)_{s,p,q}$*-condition* if

$$\int_{\mathbb{R}^n} \left| \frac{f(x)}{d(x,\partial\Omega)^s} \right|^p dx < \infty$$

holds for all $f \in F^s_{p,q}(\Omega)$ as in **(I)**.

The main result of the paper is as follows.

Theorem 1. *Let $p,\ q \in [1,\infty)$ and $s \in (n(\frac{1}{p} - \frac{1}{q})_+, 1)$. Assume that Ω is a bounded E-thick uniform domain satisfying the $(H)_{s,p,q}$-condition. Then it holds that*

$$F^s_{p,q}(\Omega) = \mathring{F}^s_{p,q}(\Omega) = \tilde{F}^s_{p,q}(\Omega) \tag{6}$$

with equivalent norms.

We make some remarks on Theorem 1.

Remark 1. *(i) Theorem 1 gives an affirmative answer to the main question. It extends by necessity the identities (2) for parameter s from the range $s \in (\max\{\frac{1}{p} - 1, n(\frac{1}{p} - 1)\}, \frac{1}{p})$ as in (5) to $s \in (n(\frac{1}{p} - \frac{1}{q})_+, 1)$ and for domain Ω from bounded Lipschitz to bounded uniform, E-thick and supporting the $(H)_{s,p,q}$-condition. Moreover, in the proof of Theorem 1, we establish the following two identities:*

(A″) $F^s_{p,q}(\Omega) = \mathring{F}^s_{p,q}(\Omega)$*, if* $1 \le p,\ q < \infty$, $n(\frac{1}{p} - \frac{1}{q})_+ < s < 1$ *and Ω is bounded uniform;*
(C″) $F^s_{p,q}(\Omega) = \widetilde{F}^s_{p,q}(\Omega)$*, if* $1 \le p,\ q < \infty$, $n(\frac{1}{p} - \frac{1}{q})_+ < s < 1$ *and Ω is bounded E-thick,*

which extends by necessity the corresponding identities **(A′)** *and* **(C′)**.

(ii) As in the Sobolev case (see, e.g., [15,16]), the proof of Theorem 1 relies on an intrinsic norm characterization of the restriction space $F^s_{p,q}(\Omega)$ as in **(I)**. *This characterization is established in [17] under the condition $s \in (n(\frac{1}{p} - \frac{1}{q})_+, 1)$, which is shown to be sharp therein. It seems a new method is needed if one considers the case $s \le n(\frac{1}{p} - \frac{1}{q})_+$; see Proposition 1, where a density property is established for a variant of Triebel–Lizorkin space in the full range $s \in (0,1)$. Note that if $1 \le q \le p < \infty$, then $n(\frac{1}{p} - \frac{1}{q})_+ = 0$. In this case, Theorem 1 gives identities (2) for the full range $s \in (0,1)$. We also point out that it is possible to consider the case $s \ge 1$ by using higher order difference. We do not pursue this in the present paper.*

We point out that the most technical part of the proof of Theorem 1 is to prove the first identity

$$F^s_{p,q}(\Omega) = \mathring{F}^s_{p,q}(\Omega), \tag{7}$$

which is also called the density property of $F^s_{p,q}(\Omega)$ and has close relations with other properties, such as zero trace characterization and regularity of the Dirichlet energy integral minimizer (see [18]). As far as we know, if Ω is a non-smooth domain, this density property is only known for some Sobolev spaces, or the case when s is small (see [9,11,15,16,19]). In this paper, we show that the density property (7) holds for bounded uniform domains without the assumption of E-thickness. More precisely, the following result is true.

Theorem 2. *Let $p,\ q \in [1,\infty)$ and $s \in \left(n(\frac{1}{p} - \frac{1}{q})_+, 1\right)$. Assume Ω is a bounded uniform domain satisfying the $(H)_{s,p,q}$-condition. Then the density property (7) holds.*

A few remarks on Theorem 2 are in order.

Remark 2. *(i) Theorem 2 extends by necessity the corresponding density property of $F^s_{p,q}(\Omega)$ by relaxing the restriction $s < (n-D)/p$ as in* **(A')**. *In particular, if $1 \le p = q < \infty$ and $s \in (0,1)$, since in this case $F^s_{p,p} = W^{s,p}$ becomes the fractional Sobolev space, Theorem 2 implies the the following zero trace characterization of fractional Sobolev space: for any $p \in [1,\infty)$ and $s \in (0,1)$, if Ω is a bounded uniform domain supporting the $(H)_{s,p,p}$-condition, then*

$$W^{s,p}(\Omega) = \mathring{W}^{s,p}(\Omega).$$

Recall that the corresponding characterization at the endpoint case $s = 1$ is a well-known result (see, e.g., [15,16]; see also [19] for a very recent result on the fractional case reached using a different method).

(ii) The proofs of Theorems 1 and 2 are based on a localization technique of Whitney decomposition (see Section 2 below). Since this technique has been extended to the more general setting of volume doubling metric measure space (see, e.g., [20]), it is straightforward to establish our results to this setting, once the corresponding intrinsic norm characterization of the restriction space $F^s_{p,q}(\Omega)$ is established.

Finally, we present further discussion on the Hardy $(H)_{s,p,q}$-condition appearing in Theorems 1 and 2. As announced earlier, we prove Theorems 1 and 2 by using a localization technique of Whitney decomposition, together with a smooth partition of unity. This allows us to decompose each $f \in F^s_{p,q}(\Omega)$ into two parts: the interior part v_ε and boundary part w_ε. It is the estimates of the latter part that need the Hardy $(H)_{s,p,q}$-condition. Note that the $(H)_{s,p,q}$-condition is satisfied once we prove the following Hardy's inequality:

$$\left\| \frac{f}{d(\cdot,\partial\Omega)^s} \right\|_{L^p(\Omega)} \lesssim \|f\|_{F^s_{p,q}(\Omega)}, \tag{8}$$

for any $f \in C_c(\Omega)$. Unfortunately, it is known that (8) may not hold in the uniform domains (see [21]). Thus, a characterization of (8) in this setting is necessary. In this paper, we establish a characterization of (8) in terms of capacities, under the additional condition $1 \le q \le p < \infty$. To be precise, for any $1 \le q \le p < \infty$ and $s \in (0,1)$, let Ω be a uniform domain on $\mathbb{R}^n$ and $K \subset \Omega$ be its compact subset. Define the *capacity* $\mathrm{cap}_{s,p,q}(K,\Omega)$ *of* K by setting

$$\mathrm{cap}_{s,p,q}(K,\Omega) := \inf |f|^p_{\mathcal{F}^s_{p,q}(\Omega)}, \tag{9}$$

where the infimum is taken over all real-valued functions $f \in C_c(\Omega)$ such that $f \ge 1$ on K and

$$|f|_{\mathcal{F}^s_{p,q}(\Omega)} := \left[\int_\Omega \left(\int_\Omega \frac{|f(x)-f(y)|^q}{|x-y|^{n+sq}}\, dy \right)^{\frac{p}{q}} dx \right]^{\frac{1}{p}}. \tag{10}$$

The following result gives the capacity characterization of (8) in the setting of a uniform domain.

Theorem 3. *Let $1 \le q \le p < \infty$ and $s \in (0,1)$. Assume that Ω is a uniform domain. The following are equiavalent.*

(i) There is a constant $C_1 > 0$ such that

$$\left\| \frac{f}{d(\cdot,\partial\Omega)^s} \right\|_{L^p(\Omega)} \le C_1 |f|_{\mathcal{F}^s_{p,q}(\Omega)},$$

for any $f \in C_c(\Omega)$.

(ii) *There is a constant $C_2 > 0$ such that*

$$\int_K \frac{1}{d(x,\partial\Omega)^{sp}}\,dx \le C_2 \mathrm{cap}_{s,p,q}(K,\Omega), \qquad (\mathrm{Cap})_{s,p,q}$$

for every compact $K \subset \Omega$.

Based on Theorems 1–3, we immediately obtain the following corollary.

Corollary 1. *Let $1 \le p \le q < \infty$ and $s \in (n(\frac{1}{p} - \frac{1}{q})_+, 1)$. Assume that Ω is a bounded uniform domain satisfying the capacity condition $(\mathrm{Cap})_{s,p,q}$. Then the following two assertions hold.*

(i) $F^s_{p,q}(\Omega) = \mathring{F}^s_{p,q}(\Omega)$ *with equivalent norms.*
(ii) *If, in addition, Ω is E-thick, then $F^s_{p,q}(\Omega) = \mathring{F}^s_{p,q}(\Omega) = \tilde{F}^s_{p,q}(\Omega)$ with equivalent norms.*

We now make some remarks on Theorem 3 and Corollary 1.

Remark 3.

(i) *Theorem 3 is the extension of the corresponding result in [22], where the authors considered the capacity characterization of Hardy's inequalities in the fractional order Sobolev space. Recall that if Ω is domain with $\partial\Omega$ being a d-set satisfying $n - 1 < d < n$, then it is proved in [12] that Hardy's inequalities (8) hold for any $f \in C_c(\Omega)$ with $p \in [1,\infty)$, $q \in [1,\infty]$ and $s > (\frac{n-d}{p}, 1)$. Note that the proof of [12] uses the technique of restriction-extension, whereas the proof of Theorem 3 depends only on the intrinsic norm characterization of $F^s_{p,q}(\Omega)$ defined as in (10).*
(ii) *The restriction $p \le q$ seems technical, which is needed in the proof of Theorem 3 in order to give a dual representation of the capacity in (9). Moreover, since the capacity condition $(\mathrm{Cap})_{s,p,q}$ is difficult to verify, it would be interesting to characterize it in terms of some geometric conditions, which is left for a further study.*

This paper is organized as follows. In Section 2, we collect some necessary technical properties of the Whitney decomposition of the domain Ω that are used out throughout this paper. Section 3.1 is devoted to the proof of Theorem 2. We prove Theorems 1 and 3 in Sections 3.2 and 3.3, respectively.

Notation. Let $\mathbb{N} := \{1,2,\ldots\}$ and $\mathbb{Z}_+ := \mathbb{N} \cup \{0\}$. For any $s \in \mathbb{R}$, let $s_+ := \max\{s,0\}$. For any subset $E \subset \mathbb{R}^n$, $\mathbf{1}_E$ denotes its *characteristic function*. We use C to denote a *positive constant* that is independent of the main parameters involved, whose value may differ from line to line. Constants with subscripts, such as C_1, do not change in different occurrences. For any qualities f, g and h, if $f \le Cg$, we write $f \lesssim g$, and if $f \lesssim g \lesssim f$, we then write $f \sim g$. We also use the following convention: if $f \le Cg$ and $g = h$ or $g \le h$, we write $f \lesssim g \sim h$ or $f \lesssim g \lesssim h$, *rather than* $f \lesssim g = h$ or $f \lesssim g \le h$. Throughout this article, we denote $Q = Q(x,l)$ be the cube with center x and sidelength l whose side parallel to coordinate axes.

2. Preliminaries on Whitney Decomposition

In this section, we collect some basic properties of the Whitney decomposition of domain Ω, with emphasis on those Whitney cubes that are close to the boundary. These properties play an important role in the proofs of our main results. To begin with, we recall the classical form of Whitney decomposition from [23].

Lemma 1 ([23]). *Let $\Omega \subsetneq \mathbb{R}^n$ be a domain. There exists a family of cubes $\{Q_j\}_{j=1}^\infty$ with sides parallel to the coordinate axes and satisfying*

(i) $Q_j^o \cap Q_k^o = \varnothing$, *if $j \ne k$, where Q_j^o denotes the interior of Q_j;*

(ii) *For any* $j \in \mathbb{N}$, $\operatorname{diam} Q_j \le d(Q_j, \partial\Omega) \le 4 \operatorname{diam} Q_j$, *where* $\operatorname{diam} Q_j$ *denotes the diameter of* Q_j;

(iii) $\Omega = \bigcup_{j=1}^{\infty} Q_j^*$, *where* $Q_j^* = (1+\mu)Q_j$ *is the concentric cube of* Q_j *with sidelength* $(1+\mu)l_j$ *and* $\mu \in [0, \frac{1}{4})$;

(iv) *Each* $x \in \Omega$ *is contained in at most* 12^n *cubes* Q_j^*;

(v) *If* Q_i *and* Q_j *touch, namely,* $\overline{Q}_i \cap \overline{Q}_j \neq \emptyset$ *and* $Q_i^o \cap Q_j^o = \emptyset$, *then*

$$\frac{1}{4} \operatorname{diam} Q_i \le \operatorname{diam} Q_j \le 4 \operatorname{diam} Q_i.$$

Throughout this section, for any $\delta > 0$, let Ω_δ be the boundary layer in Ω with length δ defined by setting

$$\Omega_\delta := \{x \in \Omega : d(x, \partial\Omega) < \delta\}. \tag{11}$$

Let $\varepsilon > 0$ and $\{Q_j\}_{j=1}^{\infty}$ be the Whitney decomposition of Ω as in Lemma 1. The following classes of index sets represent three subgroups of $\{Q_j\}_{j=1}^{\infty}$ that are closely related to the boundary layer in Ω.

$$\begin{aligned} &\Lambda_1 := \{j \in \mathbb{N} : d(Q_j, \partial\Omega) \ge \varepsilon\},\ \Lambda_2 := \{j \in \mathbb{N} : d(Q_j, \partial\Omega) < \varepsilon\} \quad \text{and} \\ &\Lambda_3 := \{j \in \mathbb{N} : Q_j \cap (\Omega \setminus \Omega_{14\varepsilon}) \neq \emptyset\} \end{aligned} \tag{12}$$

with $\Omega_{14\varepsilon}$ as in (11).

The following lemma says that a small dilation of the first subgroup $\{Q_j\}_{j\in\Lambda_1}$ of Whitney cubes is contained in the interior of Ω with a positive distance to the boundary $\partial\Omega$.

Lemma 2. *Let* $\varepsilon > 0$ *and* Λ_1 *be the index set as in* (12). *For any* $j \in \Lambda_1$, *let* $Q_j^* := (1+\tilde{\mu})Q_j$ *be the concentric cube of* Q_j *with sidelength* $(1+\tilde{\mu})l_j$ *and* $\tilde{\mu} \in (0, \frac{1}{16})$. *Then it holds that*

$$\bigcup_{j\in\Lambda_1} Q_j^* \subseteq \left\{x \in \Omega : d(x, \partial\Omega) > \frac{3\tilde{\mu}\varepsilon}{8\sqrt{n}}\right\}. \tag{13}$$

Proof. For any $x \in \bigcup_{j\in\Lambda_1} Q_j^*$, there exists $j \in \Lambda_1$ such that $x \in Q_j^* \subseteq \Omega$. By Lemma 1(iii) and the assumption $0 < \tilde{\mu} < \frac{1}{16}$, we obtain $Q_j^* \subseteq (1+4\tilde{\mu})Q_j \subseteq \Omega$. This, together with Lemma 1(ii) and the definition of Λ_1, implies

$$d(x, \partial\Omega) \ge d\left(Q_j^*, \partial\Omega\right) \ge d\left(Q_j^*, (1+4\tilde{\mu})Q_j\right) = \frac{3}{2}\tilde{\mu} l_j \ge \frac{3\tilde{\mu}\varepsilon}{8\sqrt{n}},$$

which proves (13). □

Our next lemma shows that a small dilation of the second subgroup $\{Q_j\}_{j\in\Lambda_2}$ of Whitney cubes is contained in a boundary layer of Ω.

Lemma 3. *Let* $\varepsilon > 0$ *and* Λ_2 *be the index set as in* (12). *For any* $j \in \Lambda_2$, *let* $Q_j^{**} := (1+2\tilde{\mu})Q_j$ *be the concentric cube of* Q_j *with sidelength* $(1+2\tilde{\mu})l_j$ *and* $\tilde{\mu} \in (0, \frac{1}{16})$. *Then it holds that*

$$\bigcup_{j\in\Lambda_2} Q_j^{**} \subseteq \Omega_{3\varepsilon} \tag{14}$$

with $\Omega_{3\varepsilon}$ *as in* (11).

Proof. Let $x \in \bigcup_{j\in\Lambda_2} Q_j^{**}$. By (12), Lemma 1(ii), the assumption $0 < \tilde{\mu} < \frac{1}{16}$ and the definition of Λ_2, we have

$$d(x,\partial\Omega) \le d(Q_j,\partial\Omega) + (1+2\tilde{\mu})l_j\sqrt{n} \le (2+2\tilde{\mu})d(Q_j,\partial\Omega) < 3\varepsilon,$$

which implies (14). □

The following lemma gives a few interesting properties of the third subgroup $\{Q_j\}_{j\in\Lambda_3}$ of Whitney cubes.

Lemma 4. *Let $\varepsilon > 0$ and Λ_3 be the index set as in (12). Then the following assertions hold.*

(i) $\Omega \setminus \Omega_{14\varepsilon} \subseteq \bigcup_{j\in\Lambda_3} Q_j^*$;

(ii) *For any $j \in \Lambda_3$, it holds $Q_j \cap \Omega_{7\varepsilon} = \varnothing$ and $d(Q_j,\partial\Omega) \ge 7\varepsilon$;*

(iii) *For any $k \in \Lambda_2$, let $Q_k^{**} := (1+2\tilde{\mu})Q_k$ be the concentric cube of Q_k with sidelength $(1+2\tilde{\mu})l_k$ and $\tilde{\mu} \in (0,\frac{1}{16})$. Then for any $j \in \Lambda_3$ and any $x \in Q_k^{**}$, and $y \in Q_j$, it holds that*

$$|x-y| \sim D(Q_j,Q_k), \tag{15}$$

where $D(Q_j,Q_k) := d(Q_j,Q_k) + l_j + l_k$ and the implicit constants are independent of ε, j, k, x and y.

Proof. The assertion (i) follows immediately from the definition of the index set Λ_3. To prove (ii), we first show $Q_j \cap \Omega_{7\varepsilon} = \varnothing$ for any $j \in \Lambda_3$. If not, namely, $Q_j \cap \Omega_{7\varepsilon} \ne \varnothing$, then by Lemma 1(ii), we have

$$\operatorname{diam} Q_j \le d(Q_j,\partial\Omega) < 7\varepsilon.$$

This implies $Q_j \cap (\Omega \setminus \Omega_{14\varepsilon}) = \varnothing$, which contradicts the definition of Λ_3. Thus, for any $j \in \Lambda_3$, $Q_j \cap \Omega_{7\varepsilon} = \varnothing$, namely, $d(Q_j,\partial\Omega) \ge 7\varepsilon$, which implies (ii).

We now prove (iii). For any $k \in \Lambda_2$, by Lemma 3, we have $Q_k^{**} \subseteq \Omega_{3\varepsilon}$. Let

$$\Gamma_{3\varepsilon} := \{x \in \Omega : \ d(x,\partial\Omega) = 3\varepsilon\}.$$

From (ii), it follows that for each $j \in \Lambda_3$, it holds that $Q_j \cap \Gamma_{3\varepsilon} = \varnothing$ and

$$d(Q_j,Q_k^{**}) \ge 4\varepsilon.$$

Now let $x_j \in \overline{Q_j}$ and $x_k \in \overline{Q_k^{**}}$ such that

$$d(Q_j,Q_k^{**}) = d(x_j,x_k).$$

Let $x_{\tilde{k}}$ be the intersection point of the segment $\overline{x_jx_k}$ and $\Gamma_{3\varepsilon}$. Denote by $Q_{\tilde{k}}$ the Whitney cube that contains $x_{\tilde{k}}$. It is easy to see that

$$d(Q_j,Q_k^{**}) > d(Q_j,Q_{\tilde{k}}). \tag{16}$$

By the definitions of Λ_1, Λ_2 and Lemma 1(iii), it is clear that $\tilde{k} \in \Lambda_1$. This, together with Lemma 1(iii) implies that

$$\frac{\varepsilon}{4} \le \operatorname{diam} Q_{\tilde{k}} \le 3\varepsilon. \tag{17}$$

Moreover, since $Q_{\tilde{k}} \cap \Omega_{3\varepsilon} \neq \varnothing$, by Lemma 1(ii) again, it follows that $Q_{\tilde{k}} \subseteq \Omega_{6\varepsilon}$; from (ii), it follows that $Q_j \cap \Omega_{7\varepsilon} = \varnothing$. This means that Q_j and $Q_{\tilde{k}}$ are not touched, and by Lemma 1(v), it holds that

$$d(Q_j, Q_{\tilde{k}}) \geq \frac{1}{4} l_j. \tag{18}$$

Thus, for any $x \in Q_k^{**}$ and $y \in Q_j$, we have

$$\begin{aligned} |x-y| &\leq \operatorname{diam}(Q_k^{**}) + d(Q_j, Q_k) + \operatorname{diam} Q_j \\ &\lesssim l_k + d(Q_j, Q_k) + l_j \sim D(Q_j, Q_k). \end{aligned} \tag{19}$$

On the other hand, by $l_k \leq \varepsilon \lesssim l_j$, (16) and (18), it follows that

$$|x-y| \geq d(Q_j, Q_k^{**}) \geq d(Q_j, Q_{\tilde{k}}) \geq \frac{1}{4} l_j \gtrsim l_j + l_k$$

and by (17), we know that

$$d(Q_j, Q_k) \leq d(Q_j, Q_k^{**}) + \operatorname{diam}(Q_k^{**}) \lesssim |x-y|. \tag{20}$$

By combing (19) and (20), we obtain (iii), which completes the proof of Lemma 4. □

The following lemma on the summation of D as in (15) needs the assumption that Ω is bounded and uniform.

Lemma 5 ([17]). *Let Ω be a bounded uniform domain and $\{Q_j\}_{j=1}^{\infty}$ be the Whitney decomposition of Ω as in Lemma 1. Then there exists a positive constant C such that for any $\eta > 0$ and $j_0 \in \mathbb{N}$, it holds that*

$$\sum_{j=1}^{\infty} \frac{l(Q_j)^n}{D(Q_j, Q_{j_0})^{n+\eta}} \leq \frac{C}{l(Q_{j_0})^{\eta}}$$

We end this section by giving properties of two subgroups of Whitney cubes from Λ_2 as in (12). To this end, for any $i \in \Lambda_2$, we make a subdivision of Λ_2 by setting

$$\Lambda_{21}(i) := \{k \in \Lambda_2 : Q_k^{**} \cap Q_i^* \neq \varnothing\} \quad \text{and} \quad \Lambda_{22}(i) := \{k \in \Lambda_2 : Q_k^{**} \cap Q_i^* = \varnothing\}, \tag{21}$$

where $Q_i^* = (1+\widetilde{\mu})Q_i$ and $Q_k^{**} = (1+2\widetilde{\mu})Q_k$ with $\widetilde{\mu} \in (0, 1/16)$. For any $i \in \Lambda_2$ and $k \in \Lambda_{21}(i)$, let

$$\Lambda_{23}(i,k) := \left\{ j \in \Lambda_2 : Q_j^* \cap Q_i^* \neq \varnothing \text{ or } Q_j^* \cap Q_k^{**} \neq \varnothing \right\}. \tag{22}$$

Lemma 6. *Let Ω be a bounded domain, $\varepsilon > 0$ and Λ_2 be as in (12). Then the following two assertions hold.*

(i) *For any $i \in \Lambda_2$, let $\Lambda_{21}(i)$ be the index set as in (21). Then it holds that for any $x \in Q_i^*$,*

$$\bigcup_{k \in \Lambda_{21}(i)} Q_k^{**} \subseteq B(x, 7\varepsilon),$$

where $Q_k^{**} = (1+2\widetilde{\mu})Q_k$ with $\widetilde{\mu} \in (0, 1/(16\sqrt{n}))$;

(ii) *For any $i \in \Lambda_2$ and $k \in \Lambda_{21}(i)$, let $\Lambda_{23}(i,k)$ be the index set as in (22). It holds that there exists a number $N \in \mathbb{N}$, independs of i and k, such that*

$$\operatorname{Card}(\Lambda_{23}(i,k)) \leq N. \tag{23}$$

Moreover, for any $j \in \Lambda_{23}(i,k)$, the sidelengths l_j and l_i of Q_j and Q_i are comparable, namely,

$$l_i \sim l_j \tag{24}$$

with implicit constants are independent on i and j.

Proof. We first prove (i). For any $x \in Q_i^*$ and $y \in \bigcup_{k\in\Lambda_{21}(i)} Q_k^{**}$, there exists $k \in \Lambda_{21}(i)$ such that $y \in Q_k^{**}$ and

$$|x-y| \le (1+\tilde{\mu})\sqrt{n}l_i + (1+2\tilde{\mu})\sqrt{n}l_k.$$

By Lemma 3, it holds that $d(x,\partial\Omega) < 3\varepsilon$ and $d(y,\partial\Omega) < 3\varepsilon$, which combined with Lemma 1(ii) show that $l_i, l_k < \frac{3\varepsilon}{\sqrt{n}}$. Thus, using the assumption $0 < \tilde{\mu} < \frac{1}{16\sqrt{n}}$, we know

$$|x-y| \le 6(1+2\tilde{\mu})\varepsilon \le 7\varepsilon.$$

This implies $\bigcup_{k\in\Lambda_{21}(i)} Q_k^{**} \subseteq B(x,7\varepsilon)$ and hence verifies (i).

We now prove (ii). To this end, we first claim that for any two Whitney cubes Q_j and Q_k, $Q_j^{**} \cap Q_k^{**} \neq \emptyset$ if and only if Q_j and Q_k touch. Indeed, it suffices to show that Q_j and Q_k touch when $Q_j^{**} \cap Q_k^{**} \neq \emptyset$. Otherwise, if $Q_j^{**} \cap Q_k^{**} \neq \emptyset$ and Q_j and Q_k do not touch, then by Lemma 1(v), we have

$$d(Q_j, Q_k) \ge \frac{1}{4}\max\{l_j, l_k\}.$$

This, together with the assumption $\tilde{\mu} \in (0, 1/(16\sqrt{n}))$, implies that

$$d(Q_j^{**}, Q_k^{**}) \ge d(Q_j, Q_k) - \tilde{\mu}\sqrt{n}(l_j + l_k) \ge \frac{1}{8}\max\{l_j, l_k\} > 0,$$

which contradicts the assumption $Q_j^{**} \cap Q_k^{**} \neq \emptyset$ and hence verifies the claim. By this and Lemma 1(iv), we know (23) holds with $N = 2(12)^n$. Moreover, the above claim implies that for each $i \in \Lambda_2$, $k \in \Lambda_{21}(i)$ and $j \in \Lambda_{23}(i,k)$, it holds that either Q_j and Q_i touch; or Q_j and Q_k, and Q_i and Q_k, touch. By Lemma 1(v), we conclude that (24) holds true, which completes the proof of (ii) and hence Lemma 6. □

3. Proofs of Main Results

This section is devoted to the proofs of main results of this paper. We first prove Theorem 2 in Section 3.1; then we prove Theorem 1 in Section 3.2. Finally, Section 3.3 is devoted to the proof of Theorem 3.

3.1. Proof of Theorem 2

In this subsection, we prove the density property of Triebel–Lizorkin space $F^s_{p,q}(\Omega)$ (see Theorem 2) via the intrinsic characterization of $F^s_{p,q}(\Omega)$. To this end, we recall the following definitions of intrinsic Triebel–Lizorkin space $\mathcal{F}^s_{p,q}(\Omega)$ from [17].

Definition 1. *Let Ω be a bounded domain in $\mathbb{R}^n$. For any p, $q \in [1,\infty)$ and $s \in (0,1)$. The intrinsic Triebel–Lizorkin space is defined by*

$$\mathcal{F}^s_{p,q}(\Omega) := \{f \in L^p(\Omega) : \|f\|_{\mathcal{F}^s_{p,q}(\Omega)} < \infty\},$$

where

$$\|f\|_{\mathcal{F}^s_{p,q}(\Omega)} := \|f\|_{L^p(\Omega)} + \left[\int_\Omega \left(\int_\Omega \frac{|f(x)-f(y)|^q}{|x-y|^{n+sq}}\,dy\right)^{\frac{p}{q}} dx\right]^{\frac{1}{p}} \tag{25}$$
$$=: \|f\|_{L^p(\Omega)} + |f|_{\mathcal{F}^s_{p,q}(\Omega)} < \infty.$$

Let $\mathring{\mathcal{F}}^s_{p,q}(\Omega)$ *be the completion of* $\mathcal{D}(\Omega)$ *in* $\mathcal{F}^s_{p,q}(\Omega)$ *with respect to the norm* $\|\cdot\|_{\mathcal{F}^s_{p,q}(\Omega)}$ *as in* (25).

Remark 4. *For any* $p,\ q \in [1,\infty)$ *and* $s \in (n(\frac{1}{p}-\frac{1}{q})_+, 1)$, *let* $F^s_{p,q}(\Omega)$ *be the Triebel–Lizorkin space defined as in* **(I)** *of Introduction. If, in addition,* Ω *is a bounded uniform domain, then it is proved in ([17], Corollary 3.11) that*

$$\mathcal{F}^s_{p,q}(\Omega) = F^s_{p,q}(\Omega) \tag{26}$$

with equivalent norms.

On the other hand, let $\mathring{F}^s_{p,q}(\Omega)$ *be the Triebel–Lizorkin space defined as in* **(II)** *of Introduction. By* (26), *we know that for any* $p,\ q \in [1,\infty)$ *and* $s \in (n(\frac{1}{p}-\frac{1}{q})_+, 1)$, *it holds that*

$$\mathring{\mathcal{F}}^s_{p,q}(\Omega) = \mathring{F}^s_{p,q}(\Omega)$$

with equivalent norms.

Note that Theorem 2 is an immediate consequence of Remark 4 and the following density property fo intrinsic Triebel–Lizorkin spaces $\mathcal{F}^s_{p,q}(\Omega)$.

Proposition 1. *Let* $p,\ q \in [1,\infty)$ *and* $s \in (0,1)$. *Assume* Ω *is a bounded uniform domain satisfying the* $(H)_{s,p,q}$*-condition for all* $f \in \mathcal{F}^s_{p,q}(\Omega)$. *Then it holds that*

$$\mathcal{F}^s_{p,q}(\Omega) = \mathring{\mathcal{F}}^s_{p,q}(\Omega)$$

with equivalent norms, where $\mathcal{F}^s_{p,q}(\Omega)$ *and* $\mathring{\mathcal{F}}^s_{p,q}(\Omega)$ *are defined as in Definition 1.*

Proof. Since Ω is bounded, by an elementary calculation, we know $D(\Omega) \subseteq \mathcal{F}^s_{p,q}(\Omega)$. This immediately implies $\mathring{\mathcal{F}}^s_{p,q}(\Omega) \subset \mathcal{F}^s_{p,q}(\Omega)$. Thus, we only need to prove the converse inclusion $\mathcal{F}^s_{p,q}(\Omega) \subset \mathring{\mathcal{F}}^s_{p,q}(\Omega)$. Since the proof is quite long, we divide it into several steps.

Step 1. Let $\{Q_j\}_{j=1}^\infty$ be the Whitney decomposition of Ω as in Lemma 1 and $\{\psi_j\}_{j=1}^\infty \subset C_0^\infty(\mathbb{R}^n)$ the corresponding partition of unity satisfying the following properties:

(i) $\psi_j \equiv 1$ on Q_j and $\operatorname{supp}\psi_j \subset Q_j^*$, where $Q_j^* := (1+2\widetilde{\mu})Q$ is the concentric cube of Q_j with sidelength $(1+2\widetilde{\mu})l_j$ and $\widetilde{\mu} \in (0, 1/(16n))$;

(ii) For any $x \in \Omega$, it holds that

$$\sum_{j=1}^\infty \psi_j(x) = 1 \tag{27}$$

(iii) There exists a positive constant C such that for all $x \in \mathbb{R}^n$ and $j \in \mathbb{N}$,

$$|\nabla\psi_j(x)| \le \frac{C}{\operatorname{diam} Q_j}. \tag{28}$$

Now let $f \in \mathcal{F}^s_{p,q}(\Omega)$. For any $\varepsilon > 0$ and $x \in \Omega$, by (27) and the definitions of the index sets Λ_1, Λ_2 as in (12), we write

$$f(x) = \sum_{d(Q_j,\partial\Omega)\geq\varepsilon} \psi_j(x)f(x) + \sum_{d(Q_j,\partial\Omega)<\varepsilon} \psi_j(x)f(x) = \sum_{j\in\Lambda_1} \psi_j(x)f(x) + \sum_{j\in\Lambda_2} \psi_j(x)f(x) \tag{29}$$
$$=: v_\varepsilon(x) + w_\varepsilon(x)$$

with v_ε and w_ε being the interior and boundary parts, respectively.

Step 2. We first consider the interior part v_ε by claiming

$$v_\varepsilon \in \mathcal{F}^s_{p,q}(\Omega). \tag{30}$$

Indeed, let $\psi := \sum_{j\in\Lambda_1} \psi_j(x)$. By the property (i) and (13), it holds that $\psi \in C_0^\infty(\Omega)$, which together with the fact that $v_\varepsilon = \psi f$ implies

$$\|v_\epsilon\|_{L^p(\Omega)} = \|\psi f\|_{L^p(\Omega)} \leq \|\psi\|_{L^\infty(\Omega)}\|f\|_{L^p(\Omega)} < \infty, \tag{31}$$

which implies $v_\varepsilon \in L^p(\Omega)$. On the other hand, by (25), we have

$$|\psi f|_{\mathcal{F}^s_{p,q}(\Omega)} = \left[\int_\Omega \left(\int_\Omega \frac{|\psi(x)f(x) - \psi(y)f(y)|^q}{|x-y|^{n+sq}}\,dy\right)^{\frac{p}{q}} dx\right]^{\frac{1}{p}} =: A.$$

Write

$$A \leq \left[\int_\Omega \left(\int_\Omega \frac{|f(x)|^q|\psi(x) - \psi(y)|^q}{|x-y|^{n+sq}}\,dy\right)^{\frac{p}{q}} dx\right]^{\frac{1}{p}} + \left[\int_\Omega \left(\int_\Omega \frac{|f(x) - f(y)|^q|\psi(y)|^q}{|x-y|^{n+sq}}\,dy\right)^{\frac{p}{q}} dx\right]^{\frac{1}{p}}$$
$$=: A_1 + A_2.$$

We first estimate A_1. Since $\psi \in C_0^\infty(\Omega)$, it follows that

$$A_1 \leq \left[\int_\Omega |f(x)|^p \left(\int_\Omega \|\nabla\psi(x)\|^q_{L^\infty(\Omega)}|x-y|^{q(1-s)-n}\,dy\right)^{\frac{p}{q}} dx\right]^{\frac{1}{p}}.$$

Moreover, by the assumption that Ω is bounded, we have

$$A_1 \lesssim \left[\int_\Omega |f(x)|^p \left(\int_0^{\operatorname{diam}\Omega} \rho^{q(1-s)-1}\,d\rho\right)^{\frac{p}{q}} dx\right]^{\frac{1}{p}}$$
$$\lesssim \|f\|_{L^p(\Omega)} < \infty.$$

To bound A_2, it is easy to see that

$$A_2 \leq \left[\int_\Omega \left(\int_\Omega \frac{|f(x) - f(y)|^q}{|x-y|^{n+sq}}\,dy\right)^{\frac{p}{q}} dx\right]^{\frac{1}{p}} \|\psi\|_{L^\infty(\Omega)} \lesssim |f|_{\mathcal{F}^s_{p,q}(\Omega)} < \infty.$$

Combining the estimates of A_1 and A_2, we conclude that $A < \infty$. This, together with $v\varepsilon \in L^p(\Omega)$, implies $v\varepsilon \in \mathcal{F}^s_{p,q}(\Omega)$, and hence verifies the claim (30).

Step 3. Next we prove $v_\varepsilon \in \mathring{\mathcal{F}}^s_{p,q}(\Omega)$. Let $\eta \in C_0^\infty(\mathbb{R}^n)$ satisfying $\eta \geq 0$ in $\mathbb{R}^n$, $\operatorname{supp}\eta \subseteq B(0,1)$ and $\int_{B(0,1)} \eta(x)\,dx = 1$. Let $0 < \delta < \frac{3\tilde{\mu}\varepsilon}{16\sqrt{n}}$ and $\eta^{(\delta)}$ be the mollifier defined by

$$\eta^{(\delta)}(x) := \delta^{-n}\eta(x/\delta)$$

for any $x \in \mathbb{R}^n$. It is easy to see $\eta^{(\delta)} * v_\varepsilon \in \mathcal{D}(\Omega)$, and by the property of the approximations of identity, we have

$$\left\| \left(\eta^{(\delta)} * v_\varepsilon\right) - v_\varepsilon \right\|_{L^p(\Omega)} \to 0$$

as $\delta \to 0$. Then to prove $v_\varepsilon \in \mathring{\mathcal{F}}^s_{p,q}(\Omega)$, it suffices to show $\left| \left(\eta^{(\delta)} * v_\varepsilon\right) - v_\varepsilon \right|_{\mathcal{F}^s_{p,q}(\Omega)} \to 0$ as $\delta \to 0$. From (25), we deduce

$$\begin{aligned}
&\left| \left(\eta^{(\delta)} * v_\varepsilon\right) - v_\varepsilon \right|_{\mathcal{F}^s_{p,q}(\Omega)} \qquad (32)\\
&= \left[\int_\Omega \left(\int_\Omega \frac{\left| \left(\eta^{(\delta)} * v_\varepsilon\right)(x) - v_\varepsilon(x) - \left(\eta^{(\delta)} * v_\varepsilon\right)(y) + v_\varepsilon(y) \right|^q}{|x-y|^{n+sq}} \, dy \right)^{\frac{p}{q}} dx \right]^{\frac{1}{p}} \\
&= \left[\int_\Omega \left(\int_\Omega \frac{\left| \delta^{-n} \int_{B(0,\delta)} [v_\varepsilon(x-z) - v_\varepsilon(y-z)] \eta(z/\delta) \, dz - v_\varepsilon(x) + v_\varepsilon(y) \right|^q}{|x-y|^{n+sq}} \, dy \right)^{\frac{p}{q}} dx \right]^{\frac{1}{p}} \\
&= \left[\int_\Omega \left(\int_\Omega \frac{\int_{B(0,1)} |[v_\varepsilon(x-\delta\tilde{z}) - v_\varepsilon(y-\delta\tilde{z}) - v_\varepsilon(x) + v_\varepsilon(y)] \eta(\tilde{z}) \, d\tilde{z}|^q}{|x-y|^{n+sq}} \, dy \right)^{\frac{p}{q}} dx \right]^{\frac{1}{p}} \\
&\le \int_{B(0,1)} \left[\int_\Omega \left(\int_\Omega \frac{|v_\varepsilon(x-\delta\tilde{z}) - v_\varepsilon(y-\delta\tilde{z}) - v_\varepsilon(x) + v_\varepsilon(y)|^q}{|x-y|^{n+sq}} \, dy \right)^{\frac{p}{q}} dx \right]^{\frac{1}{p}} \eta(\tilde{z}) \, d\tilde{z}.
\end{aligned}$$

Now, let

$$G(x,y) := \frac{v_\varepsilon(x) - v_\varepsilon(y)}{|x-y|^{\frac{n}{q}+s}}.$$

It is easy to see

$$G(x-\delta\tilde{z}, y-\delta\tilde{z}) - G(x,y) = \frac{v_\varepsilon(x-\delta\tilde{z}) - v_\varepsilon(y-\delta\tilde{z}) - v_\varepsilon(x) + v_\varepsilon(y)}{|x-y|^{\frac{n}{q}+s}}.$$

Since

$$\|G(x,y)\|_{L^p_x(L^q_y)(\Omega\times\Omega)} := \left[\int_\Omega \left(\int_\Omega |G(x,y)|^q \, dy \right)^{\frac{p}{q}} dx \right]^{\frac{1}{p}}$$

is a mixed Lebegue norm. By the continuity of translation (see ([24], Theorem 2)), we get

$$\lim_{\delta\to 0} \|G(x-\delta\tilde{z}, y-\delta\tilde{z}) - G(x,y)\|_{L^p_x(L^q_y)(\Omega\times\Omega)} = 0 \qquad (33)$$

for any $\tilde{z} \in B(0,1)$. Now let

$$\psi_{\varepsilon,\delta}(\tilde{z}) := \left[\int_\Omega \left(\int_\Omega \frac{|v_\varepsilon(x-\delta\tilde{z}) - v_\varepsilon(y-\delta\tilde{z}) - v_\varepsilon(x) + v_\varepsilon(y)|^q}{|x-y|^{n+sq}} \, dy \right)^{\frac{p}{q}} dx \right]^{\frac{1}{p}} \eta(\tilde{z}),$$

for any $\tilde{z} \in B(0,1)$. By (13), the assumption $0 < \delta < \frac{3\tilde{\mu}\varepsilon}{16\sqrt{n}}$ and the change of variables, we obtain

$$\begin{aligned}
\psi_{\varepsilon,\delta}(\tilde{z}) &\lesssim \left[\int_\Omega \left(\int_\Omega \frac{|v_\varepsilon(x-\delta\tilde{z}) - v_\varepsilon(y-\delta\tilde{z})|^q}{|x-y|^{n+sq}}\,dy + \int_\Omega \frac{|v_\varepsilon(x)-v_\varepsilon(y)|^q}{|x-y|^{n+sq}}\,dy\right)^{\frac{p}{q}} dx\right]^{\frac{1}{p}} \eta(\tilde{z}) \\
&\lesssim \left[\left(\int_\Omega \left(\int_\Omega \frac{|v_\varepsilon(x-\delta\tilde{z}) - v_\varepsilon(y-\delta\tilde{z})|^q}{|x-y|^{n+sq}}\,dy\right)^{\frac{p}{q}} dx\right)^{\frac{1}{p}} + \left(\int_\Omega \left(\int_\Omega \frac{|v_\varepsilon(x)-v_\varepsilon(y)|^q}{|x-y|^{n+sq}}\,dy\right)^{\frac{p}{q}} dx\right)^{\frac{1}{p}}\right] \eta(\tilde{z}) \\
&\lesssim \left(\int_\Omega \left(\int_\Omega \frac{|v_\varepsilon(x)-v_\varepsilon(y)|^q}{|x-y|^{n+sq}}\,dy\right)^{\frac{p}{q}} dx\right)^{\frac{1}{p}} \eta(\tilde{z}).
\end{aligned}$$

This, together with (25) and (30), shows $\psi_{\varepsilon,\delta} \in L^\infty(B(0,1))$. Now, using (32), (33) and the dominated convergence theorem, we get

$$\begin{aligned}
\lim_{\delta\to 0}\left|(\eta^{(\delta)} * v_\varepsilon) - v_\varepsilon\right|_{\mathcal{F}^s_{p,q}(\Omega)} &= \lim_{\delta\to 0}\int_{B(0,1)} \|G(x-\delta\tilde{z}, y-\delta\tilde{z}) - G(x,y)\|_{L^p_x(L^q_y)(\Omega\times\Omega)}\eta(\tilde{z})\,d\tilde{z} \\
&= \int_{B(0,1)} \lim_{\delta\to 0}\psi_{\varepsilon,\delta}(\tilde{z})\,d\tilde{z} = 0,
\end{aligned} \tag{34}$$

which implies $v_\varepsilon \in \mathring{\mathcal{F}}^s_{p,q}(\Omega)$.

Step 4. We still need to verify the boundary part $w_\varepsilon \in \mathring{\mathcal{F}}^s_{p,q}(\Omega)$. To this end, it suffices to prove that

$$\lim_{\varepsilon\to 0}\|w_\varepsilon\|_{L^p(\Omega)} = 0 \tag{35}$$

and

$$\lim_{\varepsilon\to 0}|w_\varepsilon|_{\mathcal{F}^s_{p,q}(\Omega)} = 0. \tag{36}$$

By Lemma 3, we obtain

$$\begin{aligned}
\int_\Omega |w_\varepsilon(x)|^p\,dx &= \int_{\Omega_{3\varepsilon}} |w_\varepsilon(x)|^p\,dx \\
&\le \int_{\Omega_{3\varepsilon}} \left|f(x)\sum_{j=1}^\infty \psi_j(x)\right|^p dx = \int_{\Omega_{3\varepsilon}} |f(x)|^p\,dx,
\end{aligned}$$

which tends to 0 as $\varepsilon \to 0$ and hence implies (35).

Step 5. We now prove (36). By (29) and the fact that $\operatorname{supp}\psi_j \subseteq Q_j^*$, we write

$$
\begin{aligned}
|w_\varepsilon|_{\mathcal{F}^s_{p,q}(\Omega)} &= \left[\int_\Omega\left(\int_\Omega \frac{\left|f(x)\sum_{j\in\Lambda_2}\psi_j(x)-f(y)\sum_{j\in\Lambda_2}\psi_j(y)\right|^q}{|x-y|^{n+sq}}\,dy\right)^{\frac{p}{q}}dx\right]^{\frac{1}{p}} \\
&= \left\{\left[\int_{\bigcup\limits_{i\in\Lambda_2}Q_i^*}+\int_{\Omega_{14\varepsilon}\setminus\bigcup\limits_{i\in\Lambda_2}Q_i^*}+\int_{\Omega\setminus\Omega_{14\varepsilon}}\right]\left(\int_\Omega\cdots dy\right)^{\frac{p}{q}}dx\right\}^{\frac{1}{p}} \\
&= \left\{\int_{\bigcup\limits_{i\in\Lambda_2}Q_i^*}\left[\left(\int_{\bigcup\limits_{k\in\Lambda_2}Q_k^{**}}+\int_{\Omega\setminus\Omega_{14\varepsilon}}\right)\cdots dy\right]^{\frac{p}{q}}dx+\int_{\Omega_{14\varepsilon}\setminus\bigcup\limits_{i\in\Lambda_2}Q_i^*}\left(\int_{\bigcup\limits_{k\in\Lambda_2}Q_k^*}\cdots dy\right)^{\frac{p}{q}}dx\right. \\
&\quad\left.+\int_{\Omega\setminus\Omega_{14\varepsilon}}\left(\int_{\bigcup\limits_{k\in\Lambda_2}Q_k^*}\cdots dy\right)^{\frac{p}{q}}dx\right\}^{\frac{1}{p}} \\
&\lesssim \left[\int_{\bigcup\limits_{i\in\Lambda_2}Q_i^*}\left(\int_{\bigcup\limits_{k\in\Lambda_2}Q_k^{**}}\frac{\left|f(x)\sum_{j\in\Lambda_2}\psi_j(x)-f(y)\sum_{j\in\Lambda_2}\psi_j(y)\right|^q}{|x-y|^{n+sq}}\,dy\right)^{\frac{p}{q}}dx\right]^{\frac{1}{p}} \\
&\quad+\left[\int_{\bigcup\limits_{i\in\Lambda_2}Q_i^*}\left(\int_{\Omega\setminus\bigcup\limits_{k\in\Lambda_2}Q_k^{**}}\frac{\left|f(x)\sum_{j\in\Lambda_2}\psi_j(x)\right|^q}{|x-y|^{n+sq}}\,dy\right)^{\frac{p}{q}}dx\right]^{\frac{1}{p}} \\
&\quad+\left[\int_{\Omega_{14\varepsilon}\setminus\bigcup\limits_{i\in\Lambda_2}Q_i^*}\left(\int_{\bigcup\limits_{k\in\Lambda_2}Q_k^*}\frac{\left|f(x)\sum_{j\in\Lambda_2}\psi_j(x)-f(y)\sum_{j\in\Lambda_2}\psi_j(y)\right|^q}{|x-y|^{n+sq}}\,dy\right)^{\frac{p}{q}}dx\right]^{\frac{1}{p}} \\
&\quad+\left[\int_{\Omega\setminus\Omega_{14\varepsilon}}\left(\int_{\bigcup\limits_{k\in\Lambda_2}Q_k^*}\frac{\left|f(y)\sum_{j\in\Lambda_2}\psi_j(y)\right|^q}{|x-y|^{n+sq}}\,dy\right)^{\frac{p}{q}}dx\right]^{\frac{1}{p}} \\
&=: \mathrm{I}_1+\mathrm{I}_2+\mathrm{II}+\mathrm{III}.
\end{aligned}
\tag{37}
$$

Step 6. We estimate the above terms in the order of I_2, III, I_1 and II. To estimate I_2, we first write

$$
\mathrm{I}_2 \le \left[\sum_{i\in\Lambda_2}\int_{Q_i^*}|f(x)|^p\left(\int_{\Omega\setminus\bigcup\limits_{k\in\Lambda_2}Q_k^{**}}\frac{1}{|x-y|^{n+sq}}\,dy\right)^{\frac{p}{q}}dx\right]^{\frac{1}{p}}.
$$

From the definitions of Q_i^* and Q_i^{**}, it follows that for any $x\in Q_i^*$ and $y\in\Omega\setminus\bigcup\limits_{i\in\Lambda_2}Q_i^{**}$, $|x-y|\ge\frac{\tilde{\mu}l_i}{2}$, where l_i denotes the sidelength of Q_i. Thus, we have

$$
\mathrm{I}_2 \lesssim \left[\sum_{i\in\Lambda_2}\int_{Q_i^*}|f(x)|^p\left(\int_{\frac{\tilde{\mu}l_i}{2}}^\infty\rho^{-sq-1}\,d\rho\right)^{\frac{p}{q}}dx\right]^{\frac{1}{p}} \lesssim \left[\sum_{i\in\Lambda_2}\int_{Q_i^*}|f(x)|^p\left(\frac{\tilde{\mu}l_i}{2}\right)^{-sp}dx\right]^{\frac{1}{p}}.
$$

Using the properties (ii) and (iv) of Lemma 1, (14) and the $(H)_{s,p,q}$-condition, we obtain

$$\mathrm{I}_2 \lesssim \left[\sum_{i\in\Lambda_2}\int_{Q_i^*}\frac{|f(x)|^p}{d(x,\partial\Omega)^{sp}}\,dx\right]^{\frac{1}{p}} \lesssim \left[\int_{\bigcup_{i\in\Lambda_2} Q_i^*}\frac{|f(x)|^p}{d(x,\partial\Omega)^{sp}}\,dx\right]^{\frac{1}{p}} \lesssim \left\|\frac{f}{d(\cdot,\partial\Omega)^s}\right\|_{L^p(\Omega_{3\varepsilon})} \to 0 \tag{38}$$

as $\varepsilon \to 0$, which is desired. That is

$$\lim_{\varepsilon\to 0}\mathrm{I}_2 = 0. \tag{39}$$

Step 7. To bound III, it is easy to see that

$$\begin{aligned}\mathrm{III} &\lesssim \left[\int_{\Omega\setminus\Omega_{14\varepsilon}}\left(\int_{\bigcup_{k\in\Lambda_2} Q_k^{**}}\frac{|f(x)-f(y)|^q}{|x-y|^{n+sq}}\,dy\right)^{\frac{p}{q}}dx\right]^{\frac{1}{p}} \\ &\quad + \left[\int_{\Omega\setminus\Omega_{14\varepsilon}}\left(\int_{\bigcup_{k\in\Lambda_2} Q_k^{**}}\frac{|f(x)|^q}{|x-y|^{n+sq}}\,dy\right)^{\frac{p}{q}}dx\right]^{\frac{1}{p}} \\ &=: \mathrm{III}_1 + \mathrm{III}_2.\end{aligned}$$

Using the fact that $f \in \mathcal{F}^s_{p,q}(\Omega)$, (14) and (25), we have

$$\mathrm{III}_1 \le \left[\int_\Omega\left(\int_{\Omega_{14\varepsilon}}\frac{|f(x)-f(y)|^q}{|x-y|^{n+sq}}\,dy\right)^{\frac{p}{q}}dx\right]^{\frac{1}{p}} \to 0 \tag{40}$$

as $\varepsilon \to 0$.

Now we estimate III_2. For any $x, y \in \Omega$, let

$$F(x,y) := \frac{f(x)}{|x-y|^{\frac{n}{q}+s}}\mathbf{1}_{\bigcup_{k\in\Lambda_2} Q_k^{**}}(y)\mathbf{1}_{\Omega\setminus\Omega_{14\varepsilon}}(x). \tag{41}$$

It is obvious that

$$\begin{aligned}\mathrm{III}_2 &= \left[\int_\Omega\left(\int_\Omega |F(x,y)|^q\,dy\right)^{\frac{p}{q}}dx\right]^{\frac{1}{p}} = \|F(x,y)\|_{L^p_x(L^q_y)(\Omega\times\Omega)} \\ &\le \sup_{\|V\|_{L^{p'}_x(L^{q'}_y)(\Omega\times\Omega)}\le 1}\left[\int_\Omega\left(\int_\Omega F(x,y)V(x,y)\,dy\right)dx\right].\end{aligned}$$

Let

$$B(F,V) := \left[\int_\Omega\left(\int_\Omega F(x,y)V(x,y)\,dy\right)dx\right].$$

By the definition of F in (41), it holds

$$B(F,V) \le \left[\int_{\Omega\setminus\Omega_{14\varepsilon}}|f(x)|\left(\int_{\bigcup_{k\in\Lambda_2} Q_k^{**}}\frac{|V(x,y)|}{|x-y|^{\frac{n}{q}+s}}\,dy\right)dx\right]. \tag{42}$$

Moreover, since $\|V\|_{L_x^{p'}(L_y^{q'})(\Omega\times\Omega)} \le 1$, we deduce

$$\lim_{\varepsilon\to 0}\left[\int_\Omega\left(\int_{\Omega_{3\varepsilon}}|V(x,y)|^{q'}\,dy\right)^{\frac{p'}{q'}}dx\right]^{\frac{1}{p'}} = 0. \tag{43}$$

Using (42), Lemma 3 and Hölder's inequality, we obtain

$$\begin{aligned}B(F,V) &\lesssim \sum_{i\in\Lambda_3}\int_{Q_i}|f(x)|\left(\sum_{k\in\Lambda_2}\int_{Q_k^*}|V(x,y)|^{q'}\,dy\right)^{\frac{1}{q'}}\left(\sum_{k\in\Lambda_2}\int_{Q_k^*}\left(\frac{1}{|x-y|^{n+sq}}\,dy\right)^{\frac{1}{q}}\right)dx\\ &\lesssim \sum_{i\in\Lambda_3}\int_{Q_i}|f(x)|\left(\sum_{k\in\Lambda_2}\int_{Q_k^*}|V(x,y)|^{q'}\,dy\right)^{\frac{1}{q'}}\left(\sum_{k\in\Lambda_2}\frac{l(Q_k)^n}{D(Q_i,Q_k)^{n+sq}}\right)^{\frac{1}{q}}dx,\end{aligned}$$

which, together with Lemmas 4(iv) and 5, implies

$$\begin{aligned}B(F,V) &\lesssim \sum_{i\in\Lambda_3}\int_{Q_i}|f(x)|\left(\sum_{k\in\Lambda_2}\int_{Q_k^*}|V(x,y)|^{q'}\,dy\right)^{\frac{1}{q'}}\frac{1}{l(Q_i)^s}\,dx\\ &\lesssim \sum_{i\in\Lambda_3}\int_{Q_i}\frac{|f(x)|}{d(x,\partial\Omega)^s}\left(\int_{\Omega_{3\varepsilon}}|V(x,y)|^{q'}\,dy\right)^{\frac{1}{q'}}dx\\ &\sim \int_\Omega\frac{|f(x)|}{d(x,\partial\Omega)^s}\left(\int_{\Omega_{3\varepsilon}}|V(x,y)|^{q'}\,dy\right)^{\frac{1}{q'}}dx\\ &\lesssim \left(\int_\Omega\frac{|f(x)|^p}{d(x,\partial\Omega)^{sp}}\right)^{\frac{1}{p}}\left[\int_\Omega\left(\int_{\Omega_{3\varepsilon}}|V(x,y)|^{q'}\,dy\right)^{\frac{p'}{q'}}dx\right]^{\frac{1}{p'}}.\end{aligned}$$

Combining the former with (43), we get

$$B(F,V)\to 0$$

as $\varepsilon\to 0$. By this and (40), we conclude that

$$\lim_{\varepsilon\to 0}\mathrm{III} = 0. \tag{44}$$

Step 8. Next we consider I_1. Next,

$$\begin{aligned}\mathrm{I}_1 &\le \left[\int_{\bigcup_{i\in\Lambda_2}Q_i^*}\left(\int_{\bigcup_{k\in\Lambda_2}Q_k^{**}}\frac{|f(x)-f(y)|^q\left|\sum_{j\in\Lambda_2}\psi_j(y)\right|^q}{|x-y|^{n+sq}}\,dy\right.\right.\\ &\quad\left.\left.+\int_{\bigcup_{k\in\Lambda_2}Q_k^{**}}\frac{|f(x)|^q\left|\sum_{j\in\Lambda_2}\psi_j(x)-\sum_{j\in\Lambda_2}\psi_j(y)\right|^q}{|x-y|^{n+sq}}\,dy\right)^{\frac{p}{q}}dx\right]^{\frac{1}{p}}\\ &\lesssim \left[\int_{\bigcup_{i\in\Lambda_2}Q_i^*}\left(\int_{\bigcup_{k\in\Lambda_2}Q_k^{**}}\frac{|f(x)-f(y)|^q\left|\sum_{j\in\Lambda_2}\psi_j(y)\right|^q}{|x-y|^{n+sq}}\,dy\right)^{\frac{p}{q}}dx\right]^{\frac{1}{p}}\\ &\quad+\left[\int_{\bigcup_{i\in\Lambda_2}Q_i^*}\left(\int_{\bigcup_{k\in\Lambda_2}Q_k^{**}}\frac{|f(x)|^q\left|\sum_{j\in\Lambda_2}\psi_j(x)-\sum_{j\in\Lambda_2}\psi_j(y)\right|^q}{|x-y|^{n+sq}}\,dy\right)^{\frac{p}{q}}dx\right]^{\frac{1}{p}}\\ &=: \mathrm{I}_{11}+\mathrm{I}_{12}.\end{aligned}$$

For I_{11}, by (25) and Lemma 3, we know that

$$\mathrm{I}_{12} \leq \left[\int_{\underset{i\in\Lambda_2}{\cup} Q_i^*} \left(\int_{\underset{k\in\Lambda_2}{\cup} Q_k^{**}} \frac{|f(x)-f(y)|^q}{|x-y|^{n+sq}} \, dy \right)^{\frac{p}{q}} dx \right]^{\frac{1}{p}} \tag{45}$$

$$\leq \left[\int_{\Omega_{3\varepsilon}} \left(\int_{\Omega_{3\varepsilon}} \frac{|f(x)-f(y)|^q}{|x-y|^{n+sq}} \, dy \right)^{\frac{p}{q}} dx \right]^{\frac{1}{p}},$$

which turns to 0 as $\varepsilon \to 0$.

To bound I_{12}, by the definitions of the index sets $\Lambda_{21}(i)$ and $\Lambda_{22}(i)$ as in (21), we have

$$\mathrm{I}_{12} \leq \left[\sum_{i\in\Lambda_2} \int_{Q_i^*} \left(\int_{\underset{k\in\Lambda_2}{\cup} Q_k^{**}} \frac{|f(x)|^q \left| \sum_{j\in\Lambda_2} \psi_j(x) - \sum_{j\in\Lambda_2} \psi_j(y) \right|^q}{|x-y|^{n+sq}} \, dy \right)^{\frac{p}{q}} dx \right]^{\frac{1}{p}} \tag{46}$$

$$\lesssim \left[\sum_{i\in\Lambda_2} \int_{Q_i^*} \left(\int_{\underset{k\in\Lambda_{21}}{\cup} Q_k^{**}} \frac{|f(x)|^q \left| \sum_{j\in\Lambda_2} \psi_j(x) - \sum_{j\in\Lambda_2} \psi_j(y) \right|^q}{|x-y|^{n+sq}} \, dy \right)^{\frac{p}{q}} dx \right]^{\frac{1}{p}}$$

$$+ \left[\sum_{i\in\Lambda_2} \int_{Q_i^*} \left(\int_{\underset{k\in\Lambda_{22}}{\cup} Q_k^{**}} \frac{|f(x)|^q \left| \sum_{j\in\Lambda_2} \psi_j(x) - \sum_{j\in\Lambda_2} \psi_j(y) \right|^q}{|x-y|^{n+sq}} \, dy \right)^{\frac{p}{q}} dx \right]^{\frac{1}{p}}$$

$$=: \mathrm{I}_{12}^1 + \mathrm{I}_{12}^2.$$

By (28); the definition of the index set Λ_{23} as in (22); Lemmas 5 and 6; and an argument similar to that used in the proof of (38), we obtain

$$\mathrm{I}_{12}^1 \leq \left[\sum_{i\in\Lambda_2} \int_{Q_i^*} |f(x)|^p \left(\sum_{k\in\Lambda_{21}} \int_{Q_k^{**}} \frac{\left| \sum_{j\in\Lambda_{23}} \psi_j(x) - \sum_{j\in\Lambda_{23}} \psi_j(y) \right|^q}{|x-y|^{n+sq}} \, dy \right)^{\frac{p}{q}} dx \right]^{\frac{1}{p}} \tag{47}$$

$$\lesssim \left[\sum_{i\in\Lambda_2} \int_{Q_i^*} |f(x)|^p \left(\sum_{k\in\Lambda_{21}} \int_{Q_k^{**}} \left(\sup_{j\in\Lambda_{23}} \|\nabla \psi_j\|_{L^\infty} \right)^q |x-y|^{q(1-s)-n} \, dy \right)^{\frac{p}{q}} dx \right]^{\frac{1}{p}}$$

$$\lesssim \left[\sum_{i\in\Lambda_2} \int_{Q_i^*} |f(x)|^p \left(\int_0^{7\varepsilon} l_i^{-q} \rho^{q(1-s)-1} \, d\rho \right)^{\frac{p}{q}} dx \right]^{\frac{1}{p}}$$

$$\lesssim \left[\sum_{i\in\Lambda_2} \int_{Q_i^*} |f(x)|^p \left(\int_0^{7\varepsilon} d(x,\partial\Omega)^{-q} \rho^{q(1-s)-1} \, d\rho \right)^{\frac{p}{q}} dx \right]^{\frac{1}{p}}$$

$$\lesssim \left[\sum_{i\in\Lambda_2} \int_{Q_i^*} |f(x)|^p d(x,\partial\Omega)^{-p} \varepsilon^{p(1-s)} \, dx \right]^{\frac{1}{p}}$$

$$\lesssim \left[\int_{\underset{i\in\Lambda_2}{\cup} Q_i^*} \frac{|f(x)|^p}{d(x,\partial\Omega)^{sp}} \, dx \right]^{\frac{1}{p}} \leq \left\| \frac{f}{d(\cdot,\partial\Omega)^s} \right\|_{L^p(\Omega_{3\varepsilon})} \to 0$$

as $\varepsilon \to 0$.

On the other hand, by (21), we know that if $k \in \Lambda_{22}(i)$, then $Q_k^{**} \bigcap Q_i^* = \emptyset$. For any $x \in Q_i^*$ and $y \in Q_k^{**}$, there exists a positive constant c such that $|x-y| \geq cl_i$—that is, $\bigcup_{k\in\Lambda_{22}} Q_k^{**} \subseteq [B(x, Cl_i)]^{\complement}$. This yields that

$$
\begin{aligned}
\mathrm{I}_{12}^2 &\lesssim \left[\sum_{i\in\Lambda_2}\int_{Q_i^*}\left(\int_{\bigcup_{k\in\Lambda_{22}} Q_k^{**}} \frac{|f(x)|^q}{|x-y|^{n+sq}}\,dy\right)^{\frac{p}{q}} dx\right]^{\frac{1}{p}} \\
&\lesssim \left[\sum_{i\in\Lambda_2}\int_{Q_i^*} |f(x)|^p \left(\int_{cl_i}^{+\infty} \rho^{-sq-1}\,d\rho\right)^{\frac{p}{q}} dx\right]^{\frac{1}{p}} \\
&\lesssim \left[\sum_{i\in\Lambda_2}\int_{Q_i^*} |f(x)|^p l_i^{-sp}\,dx\right]^{\frac{1}{p}} \\
&\lesssim \left[\int_{\bigcup_{i\in\Lambda_2} Q_i^*} \frac{|f(x)|^p}{d(x,\partial\Omega)^{sp}}\,dx\right]^{\frac{1}{p}} \lesssim \left\|\frac{f}{d(\cdot,\partial\Omega)^s}\right\|_{L^p(\Omega_{3\varepsilon})} \to 0
\end{aligned}
\tag{48}
$$

as $\varepsilon \to 0$.

Combing (45), (47) and (48), we conclude that

$$
\lim_{\varepsilon\to 0} \mathrm{I}_1 = 0. \tag{49}
$$

Step 9. Finally, we estimate II. Write

$$
\begin{aligned}
\mathrm{II} &\leq \left[\int_{\Omega_{14\varepsilon}}\left(\int_{\bigcup_{k\in\Lambda_2} Q_k^*} \frac{\left|f(x)\sum_{j\in\Lambda_2}\psi_j(x) - f(y)\sum_{j\in\Lambda_2}\psi_j(y)\right|^q}{|x-y|^{n+sq}}\,dy\right)^{\frac{p}{q}} dx\right]^{\frac{1}{p}} \\
&\lesssim \left[\int_{\Omega_{14\varepsilon}}\left(\int_{\bigcup_{k\in\Lambda_2} Q_k^*} \frac{|f(x)|^q\left|\sum_{j\in\Lambda_2}\psi_j(x) - \sum_{j\in\Lambda_2}\psi_j(y)\right|^q}{|x-y|^{n+sq}}\,dy\right)^{\frac{p}{q}} dx\right]^{\frac{1}{p}} \\
&\quad + \left[\int_{\Omega_{14\varepsilon}}\left(\int_{\bigcup_{k\in\Lambda_2} Q_k^*} \frac{|f(x)-f(y)|^q\left|\sum_{j\in\Lambda_2}\psi_j(y)\right|^q}{|x-y|^{n+sq}}\,dy\right)^{\frac{p}{q}} dx\right]^{\frac{1}{p}} \\
&=: \mathrm{II}_1 + \mathrm{II}_2.
\end{aligned}
$$

By an argument similar to that of I_1, it is easy to see that

$$
\lim_{\varepsilon\to 0} \mathrm{II} = 0. \tag{50}
$$

Combining (37), (39), (44), (49) and (50), we obtain $\lim_{\varepsilon\to 0} |w_\varepsilon|_{\mathcal{F}^s_{p,q}(\Omega)} = 0$, which proves (36). This, together with (34) and (35) shows $f \in \mathring{\mathcal{F}}^s_{p,q}(\Omega)$ and hence finishes the proof of Proposition 1. □

3.2. *Proof of Theorem 1*

In this subsection, we prove Theorem 1. To this end, we first recall the following definition of refined localisation Triebel–Lizorkin spaces $F^{s,\mathrm{rloc}}_{p,q}(\Omega)$ from ([10], Definition 2.14).

Definition 2 ([10]). *Let Ω be a bounded domain in $\mathbb{R}^n$. Let $\{Q_j\}_{j=1}^{\infty}$ be the Whitney decomposition of Ω as in Lemma 1, and $\{\psi_j\}_{j=1}^{\infty}$ be the corresponding partition of unity as in (27) and (28). For any $p,q \in [1,\infty]$ and $s \in (0,\infty)$, the refined localisation Triebel–Lizorkin space $F_{p,q}^{s,\mathrm{rloc}}(\Omega)$ is defined by setting*

$$F_{p,q}^{s,\mathrm{rloc}}(\Omega) := \left\{ f \in D'(\Omega) : \|f\|_{F_{p,q}^{s,\mathrm{rloc}}(\Omega)} := \left(\sum_{j=0}^{\infty} \|\psi_j f\|_{F_{p,q}^{s}(\mathbb{R}^n)}^{p} \right)^{\frac{1}{p}} < \infty \right\},$$

where $\|\cdot\|_{F_{p,q}^{s}(\mathbb{R}^n)}$ denotes the classical Triebel–Lizorkin norm on $\mathbb{R}^n$.

Remark 5.

(i) *Let Ω be a bounded domain. For any $p,q \in [1,\infty]$ and $s \in (0,\infty)$, it is well-known that the space $F_{p,q}^{s,\mathrm{rloc}}(\Omega)$ is independent of the choice of the partition of unity $\{\psi\}_{j=1}^{\infty}$ (see ([10], Theorem 2.16)).*

(ii) *Let Ω be a bounded domain. For any $p,q \in [1,\infty]$ and $s \in (0,1)$, it is proved in ([10], Theorem 2.18) (see also ([8], Corollary 5.15)) that $F_{p,q}^{s,\mathrm{rloc}}(\Omega)$ can be characterized by the following intrinsic norm:*

$$\left\| \frac{f}{d(\cdot,\partial\Omega)} \right\|_{L^p(\Omega)} + \left\| \left[\int_0^{cd(\cdot,\partial\Omega)} t^{-sq} (d_{t,u} f)^q \, \frac{dt}{t} \right]^{1/q} \right\|_{L^p(\Omega)} \tag{51}$$

for some $c \in (0,1)$, where for any $u \in (0,1)$, $t \in (0,\infty)$ and $x \in \mathbb{R}^n$,

$$d_{t,u} f(x) := \left[\frac{1}{t^n} \int_{|h| \le t} |f(x+h) - f(x)|^u \, dh \right]^{1/u}.$$

(iii) *Suppose that Ω is a bounded E-thick domain. Let $\widetilde{F}_{p,q}^{s}(\Omega)$ be the Triebel–Lizorkin space defined as in* **(III)** *of Introduction. It is known (see ([10], Proposition 3.10)) that for any $p,q \in [1,\infty]$ and $s \in (0,\infty)$,*

$$\widetilde{F}_{p,q}^{s}(\Omega) = F_{p,q}^{s,\mathrm{rloc}}(\Omega)$$

with equivalent norms.

With the help of Remark 5 and Theorem 2, we now turn to the proof of Theorem 1.

Proof of Theorem 1. Let $p,\ q \in [1,\infty)$ and $s \in \left(n(\frac{1}{p} - \frac{1}{q})_+, 1\right)$. Since Ω is bounded and uniform, it follows from Remark 4 that

$$\mathcal{F}_{p,q}^{s}(\Omega) = F_{p,q}^{s}(\Omega) \quad \text{and} \quad \mathring{\mathcal{F}}_{p,q}^{s}(\Omega) = \mathring{F}_{p,q}^{s}(\Omega) \tag{52}$$

with equivalent norms. Moreover, by $(H)_{s,p,q}$-condition and Proposition 1, we know

$$\mathcal{F}_{p,q}^{s}(\Omega) = \mathring{\mathcal{F}}_{p,q}^{s}(\Omega).$$

This together with (52) implies that

$$F_{p,q}^{s}(\Omega) = \mathring{F}_{p,q}^{s}(\Omega) \tag{53}$$

holds for any $p,\ q \in [1,\infty)$ and $s \in \left(n(\frac{1}{p} - \frac{1}{q})_+, 1\right)$.

On the other hand, since Ω is an E-thick domain, we deduce from Remark 5(iii) that for any $p, q \in [1, \infty]$ and $s \in (0,1)$,

$$\widetilde{F}^s_{p,q}(\Omega) = F^{s,\mathrm{rloc}}_{p,q}(\Omega). \tag{54}$$

Moreover, it is proved in ([25], Theorem 3) that the Triebel–Lizorkin space $F^s_{p,q}(\Omega)$, as in **(I)** of Introduction, can also be characterized by the same intrinsic norm of (51). This, combined with Remark 5(ii), implies that for any $p, q \in [1, \infty]$ and $s \in (0,1)$,

$$F^{s,\mathrm{rloc}}_{p,q}(\Omega) = F^s_{p,q}(\Omega). \tag{55}$$

Taking (53)–(55) together, we conclude that for any $p,\ q \in [1, \infty)$ and $s \in \left(n(\frac{1}{p} - \frac{1}{q})_+, 1\right)$, it holds

$$F^s_{p,q}(\Omega) = \mathring{F}^s_{p,q}(\Omega) = \widetilde{F}^s_{p,q}(\Omega),$$

which completes the proof of Theorem 1. □

3.3. Proof of Theorem 3

In this subsection, we prove Theorem 3.

Proof of Theorem 3. We first prove the implication (i) $\Rightarrow$ (ii). Assume (i) holds. Let $f \in C_c(\Omega)$ satisfy $f(x) \geq 1$ for any $x \in K$. By (i), we know

$$\int_K \frac{1}{d(x, \partial\Omega)^{sp}}\, dx \leq \int_\Omega \frac{|f(x)|^p}{d(x, \partial\Omega)^{sp}}\, dx \leq C_1^p \left[\int_\Omega \left(\int_\Omega \frac{|f(x) - f(y)|^q}{|x-y|^{n+sq}}\, dy\right)^{\frac{p}{q}} dx\right].$$

Taking the infinum over all such functions f and using (9), we obtain

$$\int_K \frac{1}{d(x, \partial\Omega)^{sp}}\, dx \leq C_1^p \mathrm{cap}_{s,p,q}(K, \Omega),$$

which implies (ii) with $C_2 = C_1^p$.

Now we prove the converse implication that (ii) $\Rightarrow$ (i). Suppose (ii) holds. Then, for any $k \in \mathbb{Z}$, let

$$E_k := \{x \in \Omega : |f(x)| > 2^k\} \quad \text{and} \quad A_k := E_k \setminus E_{k+1}.$$

Observe

$$\Omega = \{x \in \Omega : 0 \leq |f(x)| < \infty\} = F \cup \bigcup_{k \in \mathbb{Z}} A_k \tag{56}$$

with

$$F := \{x \in \Omega : f(x) = 0\}. \tag{57}$$

Hence, by (ii) we obtain

$$\int_\Omega \frac{|f(x)|^p}{d(x, \partial\Omega)^{sp}}\, dx \leq \sum_{k \in \mathbb{Z}} 2^{(k+2)p} \int_{A_{k+1}} \frac{1}{d(x, \partial\Omega)^{sp}}\, dx \leq C_2 2^{2p} \sum_{k \in \mathbb{Z}} 2^{kp} \mathrm{cap}_{s,p,q}(\bar{A}_{k+1}, \Omega). \tag{58}$$

Define the function $f_k : \Omega \to [0,1]$ by

$$f_k(x) := \begin{cases} 1, & |f(x)| \geq 2^{k+1}, \\ \frac{|f(x)|}{2^k} - 1, & 2^k < |f(x)| < 2^{k+1}, \\ 0, & |f(x)| \leq 2^k. \end{cases} \tag{59}$$

It is easy to see $f_k \in C_c(\Omega)$, and it satisfies $f_k = 1$ on $\bar{E}_{k+1} \supset \bar{A}_{k+1}$. Hence, we can take f_k as a test function for the capacity. By (9), we have

$$\begin{aligned} \mathrm{cap}_{s,p,q}(\bar{A}_{k+1}, \Omega) &\leq \int_\Omega \left(\int_\Omega \frac{|f_k(x) - f_k(y)|^q}{|x-y|^{n+sq}} \, dy \right)^{\frac{p}{q}} dx \\ &\leq \sup_{\|h\|_{L^{(p/q)'}(\Omega)} \leq 1} \left[\int_\Omega \left(\int_\Omega \frac{|f_k(x) - f_k(y)|^q}{|x-y|^{n+sq}} \, dy \right) h(x) \, dx \right]^{\frac{p}{q}}. \end{aligned} \tag{60}$$

Using (56) and (57), we get

$$\begin{aligned} &\int_\Omega \left(\int_\Omega \frac{|f_k(x) - f_k(y)|^q}{|x-y|^{n+sq}} \, dy \right) h(x) \, dx \\ &= \int_{F \cup \bigcup_{i \in \mathbb{Z}} A_i} \int_{F \cup \bigcup_{j \in \mathbb{Z}} A_j} \frac{|f_k(x) - f_k(y)|^q}{|x-y|^{n+sq}} h(x) \, dy \, dx \\ &= \left(\int_{\bigcup_{i \in \mathbb{Z}} A_i} \int_{\bigcup_{j \in \mathbb{Z}} A_j} + \int_F \int_{\bigcup_{j \in \mathbb{Z}} A_j} + \int_{\bigcup_{i \in \mathbb{Z}} A_i} \int_F \right) \frac{|f_k(x) - f_k(y)|^q}{|x-y|^{n+sq}} h(x) \, dy \, dx \\ &= \left(\sum_{i \geq k} \sum_{j \geq k} \int_{A_i} \int_{A_j} + \sum_{i \geq k} \sum_{j < k} \int_{A_i} \int_{A_j} + \sum_{i < k} \sum_{j \geq k} \int_{A_i} \int_{A_j} + \sum_{i < k} \sum_{j < k} \int_{A_i} \int_{A_j} \right. \\ &\quad \left. + \sum_{j < k} \int_F \int_{A_j} + \sum_{i < k} \int_{A_i} \int_F + \sum_{j \geq k} \int_F \int_{A_j} + \sum_{i \geq k} \int_{A_i} \int_F \right) \frac{|f_k(x) - f_k(y)|^q}{|x-y|^{n+sq}} h(x) \, dy \, dx. \end{aligned} \tag{61}$$

Now for any $x \in A_i = E_i \setminus E_{i+1}$, by the fact that $2^i < |f(x)| \leq 2^{i+1}$ and the definition of f_k as in (59), we claim that the following assertions hold true.

(i) If $i < k$, then $|f(x)| \leq 2^{i+1} \leq 2^k$, this implies $f_k(x) = 0$;

(ii) If $i = k$, then $f_k(x) = \frac{|f(x)|}{2^k} - 1$;

(iii) If $i > k$, then $|f(x)| > 2^i \geq 2^{k+1}$, which implies $f_k(x) = 1$;

(iv) If $i \leq k \leq j$, for any $x \in A_i$ and $y \in A_j$, it holds that

$$|f_k(x) - f_k(y)| \leq 2 \cdot 2^{-j} |f(x) - f(y)|. \tag{62}$$

We only need to verify (iv). Indeed, let $i \leq k \leq j$, $x \in A_i$ and $y \in A_j$. We consider four cases based on the sizes of i, j and k.

If $i = j$, then by (ii), it is easy to see that

$$|f_k(x) - f_k(y)| = 0. \tag{63}$$

If $j = i + 1$ and $k = i$, then by (ii), (iii) and the assumptions $x \in A_i$, $y \in A_j$, we have

$$|f_k(x) - f_k(y)| = \left| 1 - \left(\frac{|f(x)|}{2^k} - 1 \right) \right| = \left| 2 - \frac{|f(x)|}{2^k} \right| = 2^{-k} \left| 2^{k+1} - |f(x)| \right|.$$

Moreover, by the assumption $y \in A_{k+1}$, we know $|f(y)| > 2^{k+1}$. This implies that the above term is bound by

$$2^{-k}||f(y)| - |f(x)|| \le 2^{-k}|f(x) - f(y)| = 2 \cdot 2^{-j}|f(x) - f(y)| \tag{64}$$

If $j = i + 1$ and $k = j$, then by a similar argument, we know

$$|f_k(x) - f_k(y)| = \left|\frac{|f(y)|}{2^k} - 1\right| = 2^{-k}\left||f(y)| - 2^k\right|.$$

By the assumption $y \in A_{k-1}$, it holds that $|f(y)| \le 2^k$, so we obtain

$$2^{-k}||f(y)| - |f(x)|| \le 2^{-k}|f(x) - f(y)| = 2^{-j}|f(x) - f(y)|. \tag{65}$$

Finally, if $j \ge i + 2$, it holds that

$$|f(x) - f(y)| \ge |f(x)| - |f(y)| \ge 2^{j-1}.$$

By the definition of f_k, we have

$$|f_k(x) - f_k(y)| \le 1 \le 2 \cdot 2^{-j}|f(x) - f(y)|. \tag{66}$$

Combining the estimates (63)–(66), we conclude that (62) holds true and hence verifies the claim (iv).

Now by and (i) through (iv), we know that some of the sums in (61) vanish. This, together with (60), implies that

$$\begin{aligned}
\mathrm{cap}_{s,p,q}(\bar{A}_{k+1}, \Omega) \le & \sup_{\|h\|_{L^{(p/q)'}} \le 1} \Bigg[\Bigg(\sum_{i \le k}\sum_{j \ge k}\int_{A_i}\int_{A_j} + \sum_{i \ge k}\sum_{j \le k}\int_{A_i}\int_{A_j} \\
& + \sum_{j \ge k}\int_F\int_{A_j} + \sum_{i \ge k}\int_{A_i}\int_F\Bigg)\frac{|f_k(x) - f_k(y)|^q}{|x-y|^{n+sq}}h(x)\,dy\,dx\Bigg]^{\frac{p}{q}} \\
=: & \sup_{\|h\|_{L^{(p/q)'}} \le 1} (\mathrm{I}_1 + \mathrm{I}_2 + \mathrm{I}_3 + \mathrm{I}_4)^{\frac{p}{q}}.
\end{aligned}$$

By this and (58), we know that

$$\begin{aligned}
\int_\Omega \frac{|f(x)|^p}{d(x,\partial\Omega)^{sp}}\,dx & \le CC_2\sum_{k \in \mathbb{Z}} 2^{kp}\mathrm{cap}_{s,p,q}(\bar{A}_{k+1}, \Omega) \\
& \le CC_2\sum_{k \in \mathbb{Z}} 2^{kp} \sup_{\|h\|_{L^{(p/q)'}} \le 1}\left(\mathrm{I}_1^{\frac{p}{q}} + \mathrm{I}_2^{\frac{p}{q}} + \mathrm{I}_3^{\frac{p}{q}} + \mathrm{I}_4^{\frac{p}{q}}\right).
\end{aligned}$$

We first estimate the sum corresponding to I_1. By the properties (i)–(iv) again, we can show that

$$\begin{aligned}
&CC_2 \sum_{k\in\mathbb{Z}} 2^{kp} \sup_{\|h\|_{L^{(p/q)'}}\le 1} \mathrm{I}_1^{\frac{p}{q}} \\
&\le CC_2 \sum_{k\in\mathbb{Z}} 2^{kp} \sup_{\|h\|_{L^{(p/q)'}}\le 1} \left[2^q \sum_{i\le k}\sum_{j\ge k} \int_{A_i}\int_{A_j} 2^{-jq} \frac{|f(x)-f(y)|^q}{|x-y|^{n+sq}} h(x)\,dy\,dx \right]^{\frac{p}{q}} \\
&= CC_2 \sup_{\|h\|_{L^{(p/q)'}}\le 1} \left[\sum_{k\in\mathbb{Z}} 2^{kq} \sum_{i\le k}\sum_{j\ge k} \int_{A_i}\int_{A_j} 2^{-jq} \frac{|f(x)-f(y)|^q}{|x-y|^{n+sq}} h(x)\,dy\,dx \right]^{\frac{p}{q}} \\
&= CC_2 \sup_{\|h\|_{L^{(p/q)'}}\le 1} \left[\sum_{i\in\mathbb{Z}}\sum_{j\ge i}\sum_{k=i}^{j} \int_{A_i}\int_{A_j} 2^{(k-j)q} \frac{|f(x)-f(y)|^q}{|x-y|^{n+sq}} h(x)\,dy\,dx \right]^{\frac{p}{q}}
\end{aligned}$$

Since $\sum_{k=i}^{j} 2^{(k-j)q} < \sum_{k=-\infty}^{j} 2^{(k-j)q} \le \frac{1}{1-2^{-q}}$ and by $q\ge 1$, it is obvious that $\frac{1}{1-2^{-q}} \le 2$. Thus,

$$\begin{aligned}
&CC_2 \sum_{k\in\mathbb{Z}} 2^{kp} \sup_{\|h\|_{L^{(p/q)'}}\le 1} \mathrm{I}_1^{\frac{p}{q}} \\
&\le CC_2 \left(\frac{1}{1-2^{-q}}\right)^{\frac{p}{q}} \sup_{\|h\|_{L^{(p/q)'}}\le 1} \left[\int_\Omega \left(\int_\Omega \frac{|f(x)-f(y)|^q}{|x-y|^{n+sq}}\,dy \right) h(x)\,dx \right]^{\frac{p}{q}} \\
&\le CC_2 \sup_{\|h\|_{L^{(p/q)'}}\le 1} \left[\int_\Omega \left(\int_\Omega \frac{|f(x)-f(y)|^q}{|x-y|^{n+sq}}\,dy \right)^{\frac{p}{q}} dx \right] \left[\int_\Omega h(x)^{(p/q)'}\,dx \right]^{\frac{p/q}{(p/q)'}} \\
&\le CC_2 |f|^p_{\mathcal{F}^s_{p,q}}(\Omega),
\end{aligned}$$

which is desired. The estimates corresponding to I_2, I_3 and I_4 are similar, the details being omitted. Thus, we conclude that

$$\int_\Omega \frac{|f(x)|^p}{d(x,\partial\Omega)^{sp}}\,dx \le CC_2 |f|^p_{\mathcal{F}^s_{p,q}}(\Omega),$$

which implies (i) by letting $C_1 = CC_2$ and hence completes the proof of Theorem 3. □

Author Contributions: Conceptualization, J.C., Y.J., Y.L. and Q.Z.; methodology, J.C., Y.J., Y.L. and Q.Z.; software, J.C. and Y.L.; validation, J.C. and Y.L.; formal analysis, J.C., Y.J., Y.L. and Q.Z.; investigation, J.C., Y.J., Y.L. and Q.Z.; resources, Y.J. and Q.Z.; data curation, J.C., Y.J., Y.L. and Q.Z.; writing—original draft preparation, Y.L.; writing—review and editing, J.C., Y.J., Y.L. and Q.Z.; visualization, J.C., Y.J., Y.L. and Q.Z.; supervision, J.C. and Y.J.; project administration, J.C., Y.J., Y.L. and Q.Z.; funding acquisition, J.C., Y.J., Y.L. and Q.Z. All authors have read and agreed to the published version of the manuscript.

Funding: This research was funded by the National Natural Science Foundation of China (grant numbers 12071431 and 11771395) and the Zhejiang Provincial Natural Science Foundation of China (grant number LR22A010006).

Institutional Review Board Statement: Not applicable.

Informed Consent Statement: Not applicable.

Data Availability Statement: Not applicable.

Conflicts of Interest: The authors declare no conflict of interest.

References

1. Lizorkin, P.I. Operators connected with fractional differentiation, and classes of differentiable functions. *Trudy Mat. Inst. Steklov.* **1972**, *117*, 212–243.
2. Lizorkin, P.I. Properties of functions in the spaces $\Lambda^r_{p,\theta}$. *Trudy Mat. Inst. Steklov.* **1974**, *131*, 158–181.
3. Triebel, H. Spaces of distributions of Besov type on Euclidean n-space, duality, interpolation. *Ark. Mat.* **1973**, *11*, 13–64. [CrossRef]
4. Triebel, H. *Theory of Function Spaces*; Modern Birkhäuser Classics; Birkhäuser/Springer: Basel, Switzerland, 2010.
5. Triebel, H. *Theory of Function Spaces II*; Monographs in Mathematics, 84; Birkhäuser Verlag: Basel, Switzerland, 1992.
6. Triebel, H. *Theory of Function Spaces III*; Monographs in Mathematics, 100; Birkhäuser Verlag: Basel, Switzerland, 2006.
7. Triebel, H. *Fractals and Spectra, Related to Fourier Analysis and Function Spaces*; l Monographs in Mathematics, 91; Birkhäuser Verlag: Basel, Switzerland, 1997.
8. Triebel, H. *The Structure of Functions*; Monographs in Mathematics, 97; Birkhäuser Verlag: Basel, Switzerland, 2001.
9. Triebel, H. Function paces in Lipschitz domains and on Lipschitz manifolds Characteristic functions as pointwise multipliers. *Rev. Mat. Complut.* **2002**, *15*, 475–524. [CrossRef]
10. Triebel, H. *Function Spaces and Wavelets on Domains*; EMS Tracts in Mathematics, 7; European Mathematical Society (EMS): Zürich, Switzerland, 2008.
11. Caetano, A.M. Approximation by functions of compact support in Besov-Triebel-Lizorkin spaces on irregular domains. *Stud. Math.* **2000**, *142*, 47–63. [CrossRef]
12. Ihnatsyeva, L.; Vähäkangas, A. Hardy inequalities in Triebel-Lizorkin spaces I Aikawa dimension. *Annali di Matematica Pura ed Applicata* **2015**, *194*, 479–493. [CrossRef]
13. Gehring, F.W. Uniform domains and the ubiquitous quasidisk. *Math. Sci. Res. Inst.* **1987**, *89*, 88–103.
14. Gehring, F.W.; Osgood, B.G. Uniform domains and the quasi-hyperbolic metric. *J. Anal. Math.* **1979**, *36*, 50–74. [CrossRef]
15. Edmunds, D.E.; Evans, W.D. *Spectral Theory and Differential Operators*; Oxford University Press: New York, NY, USA, 1987.
16. Edmunds, D.E.; Nekvinda, A. Characterisation of zero trace functions in variable exponent Sobolev spaces. *Math. Nachr.* **2017**, *290*, 2247–2258. [CrossRef]
17. Prats, M.; Saksman, E. A T(1) theorem for fractional Sobolev spaces on domains. *J. Geom. Anal.* **2017**, *27*, 2490–2538. [CrossRef]
18. Hästö, P.A. Counter-examples of regularity in variable exponent Sobolev spaces. *Contemp. Math.* **2005**, *370*, 133–144.
19. Dyda, B.; Kijaczko, M. On density of compactly supported smooth functions in fractional Sobolev spaces. *Annali di Matematica* **2021**, 337. [CrossRef]
20. Heinonen, J.; Koskela, P.; Shanmugalingam, N.; Tyson, J. *Sobolev Spaces on Metric Measure Spaces*; An Approach Based on Upper Gradients; New Mathematical Monographs, 27; Cambridge University Press: Cambridge, UK, 2015.
21. Koskela, P.; Lehrbäck, J. Weighted pointwise Hardy inequalities. *J. Lond. Math. Soc.* **2009**, *79*, 757–779. [CrossRef]
22. Dyda, B.; Vähäkangas, A.V. Characterizations for fractional Hardy inequality. *Adv. Calc. Var.* **2015**, *8*, 173–182. [CrossRef]
23. Stein, E.M. *Singular Integrals and Differentiability Properties of Functions*; Princeton Mathematical Series; Princeton University Press: Princeton, NJ, USA, 1970.
24. Benedek, A.; Panzone, R. The space L^p with mixed norm. *Duke Math. J.* **1961**, *28*, 301–324. [CrossRef]
25. Seeger, A. *A Note on Triebel-Lizorkin Spaces*; Banach Center Publications: Warsaw, Poland, 1989; pp. 391–400.

mathematics

MDPI

Article

Schrödinger Harmonic Functions with Morrey Traces on Dirichlet Metric Measure Spaces

Tianjun Shen [1,†] and Bo Li [2,*,†]

1 Center for Applied Mathematics, Tianjin University, Tianjin 300072, China; shentj@tju.edu.cn
2 College of Data Science, Jiaxing University, Jiaxing 314001, China
* Correspondence: bli@zjxu.edu.cn
† These authors contributed equally to this work.

Abstract: Assume that (X, d, μ) is a metric measure space that satisfies a Q-doubling condition with $Q > 1$ and supports an L^2-Poincaré inequality. Let $\mathcal{L}$ be a nonnegative operator generalized by a Dirichlet form $\mathscr{E}$ and V be a Muckenhoupt weight belonging to a reverse Hölder class $RH_q(X)$ for some $q \geq (Q+1)/2$. In this paper, we consider the Dirichlet problem for the Schrödinger equation $-\partial_t^2 u + \mathcal{L}u + Vu = 0$ on the upper half-space $X \times \mathbb{R}_+$, which has f as its the boundary value on X. We show that a solution u of the Schrödinger equation satisfies the Carleson type condition if and only if there exists a square Morrey function f such that u can be expressed by the Poisson integral of f. This extends the results of Song-Tian-Yan [Acta Math. Sin. (Engl. Ser.) 34 (2018), 787-800] from the Euclidean space $\mathbb{R}^Q$ to the metric measure space X and improves the reverse Hölder index from $q \geq Q$ to $q \geq (Q+1)/2$.

Keywords: Schrödinger equation; Morrey space; Dirichlet problem; metric measure space

MSC: 35J10; 42B35

Citation: Shen, T.; Li, B. Schrödinger Harmonic Functions with Morrey Traces on Dirichlet Metric Measure Spaces. *Mathematics* **2022**, *10*, 1112. https://doi.org/10.3390/math10071112

Academic Editors: Dachun Yang and Wen Yuan

Received: 7 March 2022
Accepted: 25 March 2022
Published: 30 March 2022

1. Introduction

The Dirichlet problem was originally posed for the Laplace equation. In such a case, the problem can be stated as follows. Assume that $\Omega \subset \mathbb{R}^n$ is a domain and f is a continuous map on $\partial\Omega$. Let us find a continuous function u satisfying

$$\begin{cases} -\Delta u(x) = 0, & x \in \Omega, \\ u(x) = f(x), & x \in \partial\Omega. \end{cases}$$

We call f as the boundary value of u. Here, $-\Delta u = 0$ means that

$$\int_\Omega \nabla u \cdot \nabla \phi dx = 0$$

holds for every smooth function ϕ on $\mathbb{R}^n$ with compact support in Ω, where ∇u is the distributional gradient of u. For the upper half-space case, the study of the harmonic extension of a function has become one of the elementary tools of harmonic analysis ever since the seminar work of Stein-Weiss [1]. As we know, for any function $f \in L^p(\mathbb{R}^n)$ with $1 \leq p < \infty$, its Poisson extension $u(x,t) = e^{-t\sqrt{-\Delta}} f(x)$, $(x,t) \in \mathbb{R}^{n+1}_+$, which satisfies

$$\begin{cases} -\partial_t^2 u - \Delta u = 0, & (x,t) \in \mathbb{R}^{n+1}_+, \\ u(x) = f(x), & x \in \mathbb{R}^n. \end{cases}$$

In the study of singular integrals, a natural substitution of the end-point space $L^\infty(\mathbb{R}^n)$ is the space of functions of bounded mean oscillation (BMO). Fefferman-Stein [2] proved that a function f belongs to $\mathrm{BMO}(\mathbb{R}^n)$ if and only if its harmonic extension $u(x,t) = e^{-t\sqrt{-\Delta}}f(x)$ satisfies the following Carleson condition

$$\sup_{x_B, r_B} \int_0^{r_B} \fint_{B(x_B,r_B)} |t\nabla u(x,t)|^2 dx \frac{dt}{t} < \infty, \tag{1}$$

where

$$\fint_{B(x_B,r_B)} := \frac{1}{|B(x_B,r_B)|} \int_{B(x_B,r_B)}.$$

Later, Fabes-Johnson-Neri [3] found that the Carleson condition (1) actually characterizes all harmonic functions $u(x,t)$ on $\mathbb{R}^{n+1}_+$ with BMO traces. Since then, the research on this topic has been widely extended to various settings, including heat equations [4], elliptic equations and systems with complex coefficients [5], degenerate elliptic equations and systems [6], as well as Schrödinger equations [7,8]. We refer the reader to [9–13] and the references therein for more information about this topic.

In this paper, we consider a metric measure space X, which satisfies a Q-doubling condition with $Q > 1$, and supports an L^2-Poincaré inequality. Let $\mathscr{L} = \mathcal{L} + V$ be a Schrödinger operator, where $\mathcal{L}$ is a nonnegative operator generalized by a Dirichlet form $\mathscr{E}$, and the nonnegative potential V is a Muckenhoupt weight belonging to the reverse Hölder class. We study the boundary behavior of Schrödinger harmonic function on $X \times \mathbb{R}_+$. Roughly speaking, we derive that a solution u to the Schrödinger equation

$$-\partial_t^2 u(x,t) + \mathscr{L}u(x,t) = -\partial_t^2 u(x,t) + \mathcal{L}u(x,t) + V(x)u(x,t) = 0$$

satisfies the Carleson type condition analogous to (1) if and only if there exists a square Morrey function f such that $u = e^{-t\sqrt{\mathscr{L}}}f$ holds, where the square Morrey spaces $L^{2,\alpha}(X)$ with $-1/2 < \alpha < 0$ are defined by

$$L^{2,\alpha}(X) = \left\{ f \in L^2_{\mathrm{loc}}(X) : \sup_{B \subset X} \frac{1}{[\mu(B)]^\alpha} \left(\fint_B |f|^2 d\mu \right)^{1/2} < \infty \right\}.$$

We refer the reader to Section 2 for more about the Dirichlet metric measure space, the reverse Hölder classes, the Muckenhoupt weight and the main result. We would like to mention that, when $X = \mathbb{R}^n$, if $V \in RH_q(\mathbb{R}^n)$ for some $q \geq n$, Song-Tian-Yan [8] studied the boundary behavior of Schrödinger harmonic functions. Our result covers more general spaces, such as the Riemannian metric measure space, sub-Riemannian manifold; see [14] (Section 7) for more details.

Regarding their proof, the condition $V \in RH_q(\mathbb{R}^n)$ for some $q \geq n$ is to assure that there exists a pointwise upper bound for the gradient of the Schrödinger Poisson kernel. However, even without the potential V, such bounds are not valid in general metric space unless a group structure or strong nonnegative curvature condition is assumed (see [15,16]). Indeed, for uniformly elliptic operators, the pointwise upper bound of the gradient of heat kernel has already failed; see [14,17] for instance.

To overcome this difficulty, we adopt a Caccioppoli inequality for the Schrödinger Poisson semigroup in a tent domain $B(x_B, r_B) \times (0, r_B)$ from [18], and hence the reverse Hölder index can be improved to $q \geq (n+1)/2$ in the case of Euclidean space setting. At this moment, combined with more delicate analysis, we can remove the C^1-regularity of the Schrödinger harmonic function. Moreover, based on some new observations, we establish a new Calderón reproducing formula, which plays a crucial role in our proof; see Lemma 6 for more details.

The paper is organized as follows. In Section 2, we begin with a brief overview of our settings, i.e., the metric measure space with a Dirichlet form. Next, we recall the definition of the reverse Hölder class and the Muckenhoupt weight and finally state the main result of this paper. In Section 3, we establish some properties for the Schrödinger harmonic functions, which satisfy Carleson-type conditions. In the last two sections, we prove the main result.

Throughout the paper, we denote by the letter C (or c) a positive constant that is independent of the essential parameters but may vary from line to line.

2. Main Result

Before stating the main result, we first briefly describe our Dirichlet metric measure space settings; see [19–22] for more details. Suppose that X is a separable, connected, locally compact and metrisable space. Let μ be a Borel measure that is strictly positive on non-empty open sets and finite on compact sets. We consider a regular and strongly local Dirichlet form $\mathscr{E}$ on $L^2(X,\mu)$ with dense domain $\mathscr{D} \subset L^2(X,\mu)$ (see [20] or [21] for an accurate definition). Suppose that $\mathscr{E}$ admits a "*carré du champ*", which means that, for all $f,g \in \mathscr{D}, \Gamma(f,g)$ is absolutely continuous with respect to the measure μ. Hereafter, for simplicity of notation, let $\langle \nabla_x f, \nabla_x g \rangle$ denote the energy density $\frac{d\Gamma(f,g)}{d\mu}$ and $|\nabla_x f|$ denote the square root of $\frac{d\Gamma(f,f)}{d\mu}$. Assume the space $(X,\mu,\mathscr{E})$ is endowed with the intrinsic (pseudo-)distance on X related to $\mathscr{E}$, which is defined by setting

$$d(x,y) := \sup\{f(x) - f(y) : f \in \mathscr{D}_{\text{loc}} \cap C(X),\ |\nabla_x f| \leq 1 \text{ a.e.}\},$$

where $C(X)$ is the space of continuous functions on X. Suppose d is indeed a distance and induces a topology equivalent to the original topology on X. As a summary of the above situation, we will say that $(X,d,\mu,\mathscr{E})$ is a complete Dirichlet metric measure space.

Let the domain $\mathscr{D}$ be equipped with the norm $(\|f\|_2^2 + \mathscr{E}(f,f))^{1/2}$. We can easily see that it is a Hilbert space and denote it by $W^{1,2}(X)$. Given an open set $U \subset X$, we define the Sobolev spaces $W^{1,p}(U)$ and $W_0^{1,p}(U)$ in the usual sense (see [22–24]). With respect to the Dirichlet form, there exists an operator $\mathcal{L}$ with dense domain $\mathscr{D}(\mathcal{L})$ in $L^2(X,\mu), \mathscr{D}(\mathcal{L}) \subset W^{1,2}(X)$, such that

$$\int_X \mathcal{L}f(x)g(x)d\mu(x) = \mathscr{E}(f,g),$$

for all $f \in \mathscr{D}(\mathcal{L})$ and each $g \in W^{1,2}(X)$.

We denote by $B(x,r)$ the open ball with center x and radius r and set $\lambda B(x,r) := B(x,\lambda r)$. We suppose that μ is doubling, i.e., there exists a constant $C_d > 0$ such that, for every ball $B(x,r) \subset X$,

$$\mu(B(x,2r)) \leq C_d \mu(B(x,r)) < \infty. \tag{2}$$

Note that μ is doubling implies there exists $Q > 1$ such that, for any $0 < r < R < \infty$ and $x \in X$,

$$\mu(B(x,R)) \leq C_d \left(\frac{R}{r}\right)^Q \mu(B(x,r)),$$

and the reverse doubling property holds on a connected space (cf. [25] Remark 8.1.15 or [26] Proposition 5.2), i.e., there exist constants $0 < n \leq Q$ and $0 < c < 1$ such that, for any $0 < r < R < \infty$ and $x \in X$,

$$\mu(B(x,r)) \leq C\left(\frac{r}{R}\right)^n \mu(B(x,R)). \tag{3}$$

There also exist constants $C > 0$ and $0 \leq N \leq Q$ such that

$$\mu(B(y,r)) \leq C\left(1 + \frac{d(x,y)}{r}\right)^N \mu(B(x,r)) \tag{4}$$

uniformly for all $x,y \in X$ and $r > 0$. Indeed, property (4) with $N = Q$ is a direct consequence of the doubling property (2) and the triangle inequality of the metric d. It is worth pointing out that N can be chosen to be zero in the cases of Euclidean space, the Lie group of polynomial growth and metric space with a uniformly distributed measure.

Suppose that $(X, d, \mu, \mathscr{E})$ admits an L^2-Poincaré inequality, namely, there exists a constant $C_P > 0$ such that

$$\left(\fint_B |f - f_B|^2 d\mu \right)^{1/2} \leq C_P r_B \left(\fint_B |\nabla_x f|^2 d\mu \right)^{1/2}, \tag{5}$$

for all balls $B = B(x_B, r_B)$ and $W^{1,2}(B)$ functions f, where f_B denotes the mean (or average) of f over B.

We suppose that V is a non-trivial potential satisfying $0 \leq V \in A_\infty(X) \cap RH_q(X)$, where the Muckenhoupt weight class $A_\infty(X)$ and the reverse Hölder class $RH_q(X)$ are defined as follows (cf. [27,28]).

Definition 1.

(i) *We say that a nonnegative function V on X belongs to the Muckenhoupt weight class $A_\infty(X)$, if there exists a constant $C > 0$ such that*

$$\sup_B \fint_B V d\mu \left(\inf_{x \in B} V \right)^{-1} \leq C,$$

where the infimum is understood as the essential infimum or there exists constant $1 < p < \infty$ and $C > 0$ such that

$$\sup_B \fint_B V d\mu \left(\fint_B V^{\frac{1}{1-p}} d\mu \right)^{p-1} \leq C.$$

(ii) *For any $1 < q \leq \infty$, we say that a nonnegative function V on X belongs to the reverse Hölder class $RH_q(X)$, if there exists a constant $C > 0$ such that*

$$\left(\fint_B V^q d\mu \right)^{1/q} \leq C \fint_B V d\mu,$$

for any ball $B \subset X$, with the usual modification when $q = \infty$.

When $X = \mathbb{R}^n$, it is well known that $A_\infty(\mathbb{R}^n) = \bigcup_{1<q\leq\infty} RH_q(\mathbb{R}^n)$. However, in general metric measure space X, this relationship between the reverse Hölder classes and the Muckenhoupt weight may not hold; see [28] (Chapter 1). We point out that, if the measure μ on X is doubling and the potential V belongs to $A_\infty(X)$, then the induced measure $V d\mu$ is also doubling (cf. [28] Chapter 1).

Let us recall the definition of the critical function $\rho(x)$ associated with the potential V (see [29] Definition 1.3). For all $x \in X$, let

$$\rho(x) := \sup\left\{ r > 0 : r^2 \fint_{B(x,r)} V d\mu \leq 1 \right\}.$$

Since the potential V is non-trivial, it holds that $0 < \rho(x) < \infty$ for every $x \in X$. Additionally, by the results of Yang-Zhou [30] (Lemma 2.1 & Proposition 2.1), the critical function satisfies the following property. If $V \in A_\infty(X) \cap RH_q(X)$ with $q > \max\{1, Q/2\}$, then there exist constants $k_0 \geq 1$ and $C > 0$ such that, for all $x, y \in X$,

$$C^{-1} \rho(x) \left(1 + \frac{d(x,y)}{\rho(x)} \right)^{-k_0} \leq \rho(y) \leq C \rho(x) \left(1 + \frac{d(x,y)}{\rho(x)} \right)^{k_0/(k_0+1)}. \tag{6}$$

In this paper, we consider the Schrödinger operator

$$\mathscr{L} = \mathcal{L} + V.$$

Throughout this paper, we denote, by $\mathscr{P}_t = e^{-t\sqrt{\mathscr{L}}}$, the Schrödinger Poisson semigroup associated with $\mathscr{L}$ and, by $p_t^v(x,y)$, the kernel of $\mathscr{P}_t = e^{-t\sqrt{\mathscr{L}}}$. Due to the perturbation of V, the Schrödinger Poisson kernel and its time derivatives admit the Poisson upper bound with an additional polynomial decay (see [18])—namely, for any $k \in \{0\} \cup \mathbb{N}$ and $K > 0$, there exists a constant $C = C(k,K) > 0$ such that

$$|t^k \partial_t^k p_t^v(x,y)| \leq C \frac{t}{t+d(x,y)} \frac{1}{\mu(B(x,t+d(x,y)))} \left(1 + \frac{t+d(x,y)}{\rho(x)}\right)^{-K}.$$

For more results about the Schrödinger operator and their applications, we refer the reader to [31–44].

Let us recall the definition of $\mathscr{L}_+$-harmonic functions on the upper half-space. A function $u \in W^{1,2}(X \times \mathbb{R}_+)$ is said to be an $\mathscr{L}_+$-harmonic function on $X \times \mathbb{R}_+$, if, for every Lipschitz function ϕ with compact support in $X \times \mathbb{R}_+$, it holds that

$$\int_0^\infty \int_X \partial_t u \partial_t \phi d\mu dt + \int_0^\infty \int_X \langle \nabla_x u, \nabla_x \phi \rangle d\mu dt + \int_0^\infty \int_X V u \phi d\mu dt = 0.$$

Suppose $-1/2 < \alpha < 0$. We define $\mathrm{HL}^{2,\alpha}_{\sqrt{\mathscr{L}}}(X \times \mathbb{R}_+)$ as the class of all $\mathscr{L}_+$-harmonic functions u satisfying

$$\|u\|_{\mathrm{HL}^{2,\alpha}_{\sqrt{\mathscr{L}}}} := \sup_{x_B, r_B} \frac{1}{[\mu(B(x_B,r_B))]^\alpha} \left(\int_0^{r_B} \fint_{B(x_B,r_B)} |t\nabla u(x,t)|^2 d\mu(x) \frac{dt}{t} \right)^{1/2} < \infty.$$

The definition of the Morrey spaces refers to [8,42,45]. For every $-1/2 < \alpha < 0$, the square Morrey space $L^{2,\alpha}(X)$ is defined as

$$L^{2,\alpha}(X) := \left\{ f \in L^2_{\mathrm{loc}}(X) : \sup_{B \subset X} \frac{1}{[\mu(B)]^{2\alpha}} \fint_B |f(x)|^2 d\mu(x) < \infty \right\}.$$

This is a Banach space with respect to the norm

$$\|f\|_{L^{2,\alpha}} := \sup_{B \subset X} \frac{1}{[\mu(B)]^\alpha} \left(\fint_B |f(x)|^2 d\mu(x) \right)^{1/2}.$$

The following theorem is the main result of this paper.

Theorem 1. *Assume that $(X, d, \mu, \mathscr{E})$ is a complete Dirichlet metric measure space that satisfies the doubling condition (2) with $Q > 1$, and admits an L^2-Poincaré inequality (5). Let $0 \leq V \in A_\infty(X) \cap RH_q(X)$ with $q \geq (Q+1)/2$, and $-1/2 < \alpha < 0$.*

(i) *If $f \in L^{2,\alpha}(X)$, then $u(x,t) = \mathscr{P}_t f(x) \in \mathrm{HL}^{2,\alpha}_{\sqrt{\mathscr{L}}}(X \times \mathbb{R}_+)$, and there exists a constant $C > 0$, independent of f, such that*

$$\|u\|_{\mathrm{HL}^{2,\alpha}_{\sqrt{\mathscr{L}}}} \leq C \|f\|_{L^{2,\alpha}}.$$

(ii) *Further assume that $\max\{-1/2, -1/2N\} < \alpha < 0$. If $u \in \mathrm{HL}^{2,\alpha}_{\sqrt{\mathscr{L}}}(X \times \mathbb{R}_+)$, then there exists a function $f \in L^{2,\alpha}(X)$ such that $u(x,t) = \mathscr{P}_t f(x)$. Moreover, there exists a constant $C > 0$, independent of u, such that*

$$\|f\|_{L^{2,\alpha}} \leq C \|u\|_{\mathrm{HL}^{2,\alpha}_{\sqrt{\mathscr{L}}}}.$$

Remark 1.

(i) *In Theorem 1, we assume that the reverse Hölder index q is not less than $(Q+1)/2$. However, the observant readers might notice that, in [29], Shen assumed that the nonnegative potential V belongs to $RH_q(\mathbb{R}^Q)$ for some $q \geq Q/2$. However, we consider the boundary value problem of the Schrödinger equation*
$$-\partial_t^2 u + \mathcal{L}u + Vu = 0$$
on the upper half-space $X \times \mathbb{R}_+$. In order to make sure the above Schrödinger harmonic function is Hölder continuous on $X \times \mathbb{R}_+$, the critical reverse Hölder index $(Q+1)/2$ seems to be the least condition via the natural extension $V(\cdot, t) := V(\cdot)$ for all $t > 0$. One might wonder if there is any possibility of relaxing the requirement $q \geq (Q+1)/2$ in Theorem 1 to $q > 1$ together with $q \geq Q/2$. From the initial value to the solution, this is ensured by the Caccioppoli inequality for the Schrödinger Poisson semigroup; see Proposition 3 for more details. To the contrary, from the solution to the initial value, this is an interesting problem to be solved.

(ii) *The range of α in Theorem 1 (ii) is slightly different from that in (i). This assumption $-1/2N < \alpha < 0$ first appears in Lemma 3 below, which is caused by the time regularity of $\mathrm{HL}^{2,\alpha}_{\sqrt{\mathscr{L}}}$-function*
$$|t\partial_t u(x,t)| \leq C[\mu(B(x,t))]^{\alpha} \|u\|_{\mathrm{HL}^{2,\alpha}_{\sqrt{\mathscr{L}}}}.$$
Since the pointwise upper bound of the time regularity of $\mathrm{HL}^{2,\alpha}_{\sqrt{\mathscr{L}}}$-function has to do with the measure of some ball to the α power, the condition $2\alpha N + 1 > 0$ ensures the series in Lemma 3 is convergent. In fact, for metric measure space X, the nonnegative parameter N arises automatically if we want to calculate the ratio of the volumes of two balls with different centers. However, this would not occur in the cases of Euclidean space, the Lie group of polynomial growth and metric space with a uniformly distributed measure. We remark that N can be chosen to be 0 under these settings, and hence the assumption $-1/2N < \alpha < 0$ is superfluous.

3. Schrödinger Harmonic Functions Satisfying Carleson

In this section, we will establish some properties of $\mathrm{HL}^{2,\alpha}_{\sqrt{\mathscr{L}}}$-function.

Lemma 1. *Assume the Dirichlet metric measure space $(X, d, \mu, \mathscr{E})$ satisfies (2) and (5). Let $V \in A_\infty(X) \cap RH_q(X)$ for some $q > \max\{1, Q/2\}$. If $\mathscr{L}u = \mathcal{L}u + Vu = 0$ holds in a bounded domain $\Omega \subset X$, then there exists a constant $C > 0$ such that, for any ball $B = B(x_B, r_B)$ with $2B \subset \Omega$,*
$$\|u\|_{L^\infty(B)} \leq C \fint_{2B} |u| d\mu.$$
Furthermore, u is locally Hölder continuous in Ω, and there exists a constant $\theta \in (0, \min\{1, 2 - Q/q\})$ such that, for any $x, y \in \frac{1}{2}B$,
$$|u(x) - u(y)| \leq C\left(\frac{d(x,y)}{r_B}\right)^{\theta} \|u\|_{L^\infty(B)} \left(1 + r_B^2 \fint_B V d\mu\right).$$

Proof. For the proof, we refer to [18] (Proposition 2.12). □

Let us extend the potential V to the upper half-space by defining $V(x,t) := V(x)$ for all $t \in \mathbb{R}$. We can easily find that $V(x,t) \in A_\infty(X \times \mathbb{R}) \cap RH_q(X \times \mathbb{R})$ with $q > (Q+1)/2$, if $0 \leq V(x) \in A_\infty(X) \cap RH_q(X)$ with $q > (Q+1)/2$. Therefore, it follows from Lemma 1 that $\mathscr{L}_+$-harmonic functions are locally Hölder continuous on $X \times \mathbb{R}_+$.

Lemma 2. *Suppose the complete Dirichlet metric measure space* $(X, d, \mu, \mathscr{E})$ *satisfies* (2) *and* (5). *Let* $0 \le V \in A_\infty(X) \cap RH_q(X)$ *with* $q > (Q+1)/2$. *If* $u \in \mathrm{HL}^{2,\alpha}_{\sqrt{\mathscr{L}}}(X \times \mathbb{R}_+)$ *with* $-1/2 < \alpha < 0$, *then there exists a constant* $C > 0$ *such that, for all* $x \in X$ *and* $t > 0$,

$$|t\partial_t u(x,t)| \le C[\mu(B(x,t))]^{\alpha} \|u\|_{\mathrm{HL}^{2,\alpha}_{\sqrt{\mathscr{L}}}}.$$

Proof. Let $\epsilon > 0$. Given $-\epsilon < h < \epsilon$, for any $x \in X$ and $t > \epsilon$, set

$$u(x,t;h) := \frac{u(x,t+h) - u(x,t)}{h}.$$

It follows that $u(\cdot,\cdot;h)$ is an $\mathscr{L}_+$-harmonic function on $X \times (\epsilon, \infty)$; see the proof of [18] (Lemma 4.1).

Then, by the mean value property in Lemma 1, we conclude that, for any $t > 2\epsilon$,

$$|u(x,t;h)| \le C\left(\fint_{B(x,t/2)} \fint_{t/2}^{3t/2} |u(y,s;h)|^2 ds d\mu(y) \right)^{1/2}, \tag{7}$$

which, combined with the argument in the proof of Jiang-Li [18] (Lemma 4.1), yields, for each $t > 3\epsilon$, that

$$|tu(x,t;h)| \le C\left(\fint_{B(x,2t)} \int_0^{2t} |s\partial_s u(y,s)|^2 \frac{ds}{s} d\mu(y) \right)^{1/2}.$$

This implies that, for each $t > 3\epsilon$,

$$|t\partial_t u(x,t)| \le C[\mu(B(x,t))]^{\alpha} \|u\|_{\mathrm{HL}^{2,\alpha}_{\sqrt{\mathscr{L}}}}.$$

Letting $\epsilon \to 0$ indicates that the above estimate holds for every $t > 0$. □

Lemma 3. *Assume the complete Dirichlet metric measure space* $(X, d, \mu, \mathscr{E})$ *satisfies* (2) *and* (5). *Suppose* $0 \le V \in A_\infty(X) \cap RH_q(X)$ *with* $q > (Q+1)/2$, *and* $\max\{-1/2, -1/2N\} < \alpha < 0$. *If* $u \in \mathrm{HL}^{2,\alpha}_{\sqrt{\mathscr{L}}}(X \times \mathbb{R}_+)$, *then there exists a constant* $C > 0$ *such that, for any* $x \in X$ *and* $t, \epsilon > 0$,

$$\begin{aligned} &\int_X \frac{|u(y,\epsilon)|^2}{(t + d(x,y))\mu(B(x, t + d(x,y)))} d\mu(y) \\ &\le C(1 + t^{-1}) \|u(\cdot,\epsilon)\|^2_{L^\infty(B(x,2))} + C([\mu(B(x,1))]^{2\alpha} + \epsilon^{2N\alpha}[\mu(B(x,\epsilon))]^{2\alpha}) \|u\|^2_{\mathrm{HL}^{2,\alpha}_{\sqrt{\mathscr{L}}}}. \end{aligned}$$

Proof. By Lemma 1, $u(\cdot,\cdot)$ is locally bounded and locally Hölder continuous in $X \times \mathbb{R}_+$. The integral is split into $B(x,1)$ and $X \backslash B(x,1)$. For the local part $B(x,1)$, it holds that

$$\int_{B(x,1)} \frac{|u(y,\epsilon)|^2}{(t + d(x,y))\mu(B(x, t + d(x,y)))} d\mu(y) \le \frac{C}{t} \|u(\cdot,\epsilon)\|^2_{L^\infty(B(x,1))}.$$

For the global part $X \backslash B(x,1)$, by the annulus argument, we have

$$\begin{aligned} &\int_{X \backslash B(x,1)} \frac{|u(y,\epsilon)|^2}{(t + d(x,y))\mu(B(x, t + d(x,y)))} d\mu(y) \\ &\le C \sum_{j=1}^{\infty} 2^{-j} \fint_{2^{j-1}}^{2^j} \fint_{B(x,2^j)} |u(y,\epsilon)|^2 d\mu(y) d\, s \\ &\le C \sum_{j=1}^{\infty} 2^{-j} \fint_{E_j} |u(y,\epsilon) - u(y,s)|^2 d\mu(y) d\, s \end{aligned}$$

$$
\begin{aligned}
&+C\sum_{j=1}^{\infty}2^{-j}\fint_{E_j}|u(y,s)-u_{E_j}|^2d\mu(y)ds+C\sum_{j=1}^{\infty}2^{-j}|u_{E_j}|^2\\
&=: C(\mathrm{I}_1+\mathrm{I}_2+\mathrm{I}_3),
\end{aligned}
$$

where we denote the cylinder $B(x,2^j)\times[2^{j-1},2^j)$ by E_j for simplicity.

For the term I_1, it holds by Lemma 2 and $-1/2N<\alpha$ that

$$
\begin{aligned}
\mathrm{I}_1&=\sum_{j=1}^{\infty}2^{-j}\fint_{E_j}\left|\int_{\epsilon}^{s}\partial_r u(y,r)dr\right|^2d\mu(y)d\,s\\
&\le C\|u\|^2_{\mathrm{HL}^{2,\alpha}_{\sqrt{\mathscr{L}}}}\sum_{j=1}^{\infty}2^{-j}\fint_{E_j}\left(\int_{\epsilon}^{s}[\mu(B(y,r))]^{\alpha}\frac{dr}{r}\right)^2d\mu(y)d\,s\\
&\le C\|u\|^2_{\mathrm{HL}^{2,\alpha}_{\sqrt{\mathscr{L}}}}\sum_{j=1}^{\infty}2^{-j}\left\{[\mu(B(x,1))]^{2\alpha}+\left(\frac{2^j}{\epsilon}\right)^{-2N\alpha}[\mu(B(x,\epsilon))]^{2\alpha}\right\}\\
&\le C\Big([\mu(B(x,1))]^{2\alpha}+\epsilon^{2N\alpha}[\mu(B(x,\epsilon))]^{2\alpha}\Big)\|u\|^2_{\mathrm{HL}^{2,\alpha}_{\sqrt{\mathscr{L}}}}.
\end{aligned}
$$

Above, in the second inequality, we used the fact that

$$
\begin{aligned}
&\int_{\epsilon}^{s}[\mu(B(y,r))]^{\alpha}\frac{dr}{r}\\
&\le\int_{\epsilon}^{s}[\mu(B(y,r))]^{\alpha}\frac{dr}{r}\Big(\chi_{(0,2^{j-1})}(\epsilon)+\chi_{(2^{j-1},\infty)}(\epsilon)\Big)\\
&\le\int_{\epsilon}^{\infty}[\mu(B(y,r))]^{\alpha}\frac{dr}{r}\chi_{(0,2^{j-1})}(\epsilon)+\int_{2^{j-1}}^{\infty}[\mu(B(y,r))]^{\alpha}\frac{dr}{r}\\
&\le C\left\{\int_{\epsilon}^{\infty}\left(\frac{r}{\epsilon}\right)^{n\alpha}[\mu(B(y,\epsilon))]^{\alpha}\frac{dr}{r}\chi_{(0,2^{j-1})}(\epsilon)+\int_{2^{j-1}}^{\infty}\left(\frac{r}{2^{j-1}}\right)^{n\alpha}[\mu(B(y,2^{j-1}))]^{\alpha}\frac{dr}{r}\right\}\\
&\le C\left\{\left(1+\frac{d(x,y)}{\epsilon}\right)^{-N\alpha}[\mu(B(x,\epsilon))]^{\alpha}\chi_{(0,2^{j-1})}(\epsilon)+\left(1+\frac{d(x,y)}{2^{j-1}}\right)^{-N\alpha}[\mu(B(x,2^j))]^{\alpha}\right\}\\
&\le C\left\{[\mu(B(x,1))]^{\alpha}+\left(\frac{2^j}{\epsilon}\right)^{-N\alpha}[\mu(B(x,\epsilon))]^{\alpha}\right\}.
\end{aligned}
$$

Now, we put $u_s(\cdot):=u(\cdot,s)$. For the term I_2, we use the Poincaré inequality to deduce that

$$
\begin{aligned}
\mathrm{I}_2&\le 2\sum_{j=1}^{\infty}2^{-j}\left(\fint_{2^{j-1}}^{2^j}\fint_{B(x,2^j)}\left|u(y,s)-(u_s)_{B(x,2^j)}\right|^2d\mu(y)ds+\fint_{2^{j-1}}^{2^j}\left|(u_s)_{B(x,2^j)}-u_{E_j}\right|^2ds\right)\\
&\le C\sum_{j=1}^{\infty}2^{-j}\left(2^{2j}\fint_{2^{j-1}}^{2^j}\fint_{B(x,2^j)}|\nabla_y u(y,s)|^2d\mu(y)ds+\fint_{2^{j-1}}^{2^j}\left|(u_s)_{B(x,2^j)}-u_{E_j}\right|^2ds\right).
\end{aligned}\tag{8}
$$

By the Hölder inequality and the Poincaré inequality, it holds that

$$
\begin{aligned}
&\fint_{2^{j-1}}^{2^j}\left|(u_s)_{B(x,2^j)}-u_{E_j}\right|^2d\,s\\
&=\fint_{2^{j-1}}^{2^j}\left|\fint_{B(x,2^j)}u(y,s)d\mu(y)-\fint_{2^{j-1}}^{2^j}\fint_{B(x,2^j)}u(y,r)d\mu(y)dr\right|^2d\,s\\
&=\fint_{2^{j-1}}^{2^j}\left|\fint_{2^{j-1}}^{2^j}\fint_{B(x,2^j)}u(y,s)-u(y,r)d\mu(y)dr\right|^2d\,s
\end{aligned}
$$

$$\leq \fint_{B(x,2^j)} \fint_{2^{j-1}}^{2^j} \fint_{2^{j-1}}^{2^j} |u(y,s) - u(y,r)|^2 drdsd\mu(y)$$
$$\leq C2^{2j} \fint_{B(x,2^j)} \fint_{2^{j-1}}^{2^j} |\partial_s u(y,s)|^2 dsd\mu(y).$$

This, together with (8), gives that

$$\begin{aligned}
\mathrm{I}_2 &\leq C\sum_{j=1}^{\infty} 2^{-j}2^{2j} \fint_{2^{j-1}}^{2^j} \fint_{B(x,2^j)} |\nabla u(y,s)|^2 d\mu(y)d\,s \\
&\leq C\sum_{j=1}^{\infty} 2^{-j} \int_0^{2^j} \fint_{B(x,2^j)} |s\nabla u(y,s)|^2 d\mu(y)\frac{ds}{s} \\
&\leq C\sum_{j=1}^{\infty} 2^{-j} [\mu(B(x,2^j))]^{2\alpha} \|u\|^2_{\mathrm{HL}^{2,\alpha}_{\sqrt{\mathscr{L}}}} \\
&\leq C[\mu(B(x,1))]^{2\alpha} \|u\|^2_{\mathrm{HL}^{2,\alpha}_{\sqrt{\mathscr{L}}}}.
\end{aligned}$$

As $E_j = B(x,2^j) \times [2^{j-1},2^j)$, it holds $E_j, E_{j+1} \subset B(x,2^{j+1}) \times [2^{j-1},2^{j+1}) =: F_{j+1}$. For the term I_3, one writes

$$\begin{aligned}
\mathrm{I}_3 &\leq \sum_{j=1}^{\infty} 2^{-j} \left(|u_{E_1}| + \sum_{i=2}^{j} |u_{E_i} - u_{E_{i-1}}| \right)^2 \\
&\leq \sum_{j=1}^{\infty} 2^{-j} \left(|(u - u(\cdot,\epsilon))_{E_1}| + \|u(\cdot,\epsilon)\|_{L^\infty(B(x,2))} + \sum_{i=2}^{j} (|u_{E_i} - u_{F_i}| + |u_{F_i} - u_{E_{i-1}}|) \right)^2.
\end{aligned}$$

It follows from the Poincaré inequality that

$$\begin{aligned}
|u_{E_i} - u_{F_i}| + |u_{F_i} - u_{E_{i-1}}| &\leq C\left(\fint_{2^{i-2}}^{2^i} \fint_{B(x,2^i)} \left|u(y,s) - u_{F_i}\right|^2 d\mu(y)ds \right)^{1/2} \\
&\leq C\left(\fint_{2^{i-2}}^{2^i} \fint_{B(x,2^i)} \left|u(y,s) - (u_s)_{B(x,2^i)}\right|^2 d\mu(y)ds \right)^{1/2} \\
&\quad + C\left(\fint_{2^{i-2}}^{2^i} \left|(u_s)_{B(x,2^i)} - u_{F_i}\right|^2 ds \right)^{1/2} \\
&\leq C2^i \left(\fint_{2^{i-2}}^{2^i} \fint_{B(x,2^i)} |\nabla u(y,s)|^2 d\mu(y)ds \right)^{1/2} \\
&\leq C\left(\int_0^{2^i} \fint_{B(x,2^i)} |s\nabla u(y,s)|^2 d\mu(y)\frac{ds}{s} \right)^{1/2} \\
&\leq C[\mu(B(x,2^i))]^{\alpha} \|u\|_{\mathrm{HL}^{2,\alpha}_{\sqrt{\mathscr{L}}}} \\
&\leq C[\mu(B(x,1))]^{\alpha} \|u\|_{\mathrm{HL}^{2,\alpha}_{\sqrt{\mathscr{L}}}},
\end{aligned} \tag{9}$$

and from Lemma 2 that

$$\begin{aligned}
|(u - u(\cdot,\epsilon))_{E_1}| &\leq \fint_{B(x,2)\times[1,2)} |u(y,s) - u(y,\epsilon)| d\mu(y)ds \\
&\leq \fint_{B(x,2)\times[1,2)} \left| \int_\epsilon^s \partial_r u(y,r)dr \right| d\mu(y)ds
\end{aligned}$$

$$
\begin{aligned}
&\leq C\|u\|_{\mathrm{HL}^{2,\alpha}_{\sqrt{\mathscr{L}}}} \fint_{B(x,2)\times[1,2)} \left| \int_{\epsilon}^{s} [\mu(B(y,r))]^{\alpha} \frac{dr}{r} \right| d\mu(y)ds \\
&\leq C([\mu(B(x,1))]^{\alpha} + \epsilon^{N\alpha}[\mu(B(x,\epsilon))]^{\alpha})\|u\|_{\mathrm{HL}^{2,\alpha}_{\sqrt{\mathscr{L}}}}.
\end{aligned} \tag{10}
$$

Here, we used the fact that

$$
\begin{aligned}
\left| \int_{\epsilon}^{s} [\mu(B(y,r))]^{\alpha} \frac{dr}{r} \right| &\leq C\left(\int_{\epsilon}^{\infty} [\mu(B(y,r))]^{\alpha} \frac{dr}{r} + \int_{1}^{\infty} [\mu(B(y,r))]^{\alpha} \frac{dr}{r} \right) \\
&\leq C\left(\int_{\epsilon}^{\infty} \left(\frac{r}{\epsilon}\right)^{n\alpha} [\mu(B(y,\epsilon))]^{\alpha} \frac{dr}{r} + \int_{1}^{\infty} r^{n\alpha} [\mu(B(y,1))]^{\alpha} \frac{dr}{r} \right) \\
&\leq C(\epsilon^{N\alpha}[\mu(B(x,\epsilon))]^{\alpha} + [\mu(B(x,1))]^{\alpha}).
\end{aligned}
$$

The above two estimates (9) and (10) yield that

$$
\begin{aligned}
\mathrm{I}_3 &\leq C \sum_{j=1}^{\infty} 2^{-j} \|u(\cdot,\epsilon)\|^2_{L^{\infty}(B(x,2))} \\
&\quad + C \sum_{j=1}^{\infty} 2^{-j} \left([\mu(B(x,1))]^{\alpha} \|u\|_{\mathrm{HL}^{2,\alpha}_{\sqrt{\mathscr{L}}}} + \epsilon^{N\alpha}[\mu(B(x,\epsilon))]^{\alpha} \|u\|_{\mathrm{HL}^{2,\alpha}_{\sqrt{\mathscr{L}}}} \right)^2 \\
&\leq C\left(\|u(\cdot,\epsilon)\|^2_{L^{\infty}(B(x,2))} + [\mu(B(x,1))]^{2\alpha} \|u\|^2_{\mathrm{HL}^{2,\alpha}_{\sqrt{\mathscr{L}}}} + \epsilon^{2N\alpha}[\mu(B(x,\epsilon))]^{2\alpha} \|u\|^2_{\mathrm{HL}^{2,\alpha}_{\sqrt{\mathscr{L}}}} \right).
\end{aligned}
$$

In combination with the estimates of $\mathrm{I}_1, \mathrm{I}_2$ and I_3, we obtain the required conclusion. □

Lemma 4. *Suppose the complete Dirichlet metric measure space $(X, d, \mu, \mathscr{E})$ satisfies (2) with $Q > 1$ and admits (5). Let $0 \leq V \in A_{\infty}(X) \cap RH_q(X)$ with $q > (Q+1)/2$. Assume that w is a solution to $(-\partial_t^2 + \mathscr{L})w = 0$ on $X \times \mathbb{R}$. If there exists $m > 0$ such that*

$$
\int_{\mathbb{R}} \int_X \frac{|w(y,t)|^2}{(1+t+d(x,y))^{m+1}\mu(B(x,1+t+d(x,y)))} d\mu(y)dt < \infty,
$$

then $w \equiv 0$.

Proof. For the proof, we refer to [18] (Corollary 4.5). □

Proposition 1. *Suppose the complete Dirichlet metric measure space $(X, d, \mu, \mathscr{E})$ satisfies (2) with $Q > 1$ and admits (5). Let $0 \leq V \in A_{\infty}(X) \cap RH_q(X)$ with $q > (Q+1)/2$. Assume that $u \in \mathrm{HL}^{2,\alpha}_{\sqrt{\mathscr{L}}}(X \times \mathbb{R}_+)$ with $\max\{-1/2, -1/2N\} < \alpha < 0$. For any $x \in X$ and $s, t > 0$, it holds that*

$$
u(x, t+s) = \mathscr{P}_t(u(\cdot, s))(x).
$$

Proof. For each $t > 0$, let

$$
v(x,t) := u(x,t+s) - \mathscr{P}_t(u(\cdot,s))(x).
$$

As $u(\cdot, \cdot + s)$ is Hölder continuous on $X \times (-s, \infty)$ and $u(\cdot, s)$ is Hölder continuous on X, we see that

$$
v(x,0) := \lim_{t \to 0^+} v(x,t) = \lim_{t \to 0^+} \{u(x,t+s) - \mathscr{P}_t(u(\cdot,s))(x)\} = 0.
$$

We extend $v(x,t)$ to $X \times \mathbb{R}$ as

$$w(x,t) := \begin{cases} v(x,t), & t > 0; \\ 0, & t = 0; \\ -v(x,-t), & t < 0. \end{cases}$$

Then, w is a solution to the Schrödinger equation $(-\partial_t^2 + \mathscr{L})w = 0$ on $X \times \mathbb{R}$. We fix a point $y_0 \in X$. By Lemma 4 and the fact that w is odd with respect to t, it is sufficient to show that there exists $m > 0$ such that

$$\int_0^\infty \int_X \frac{|w(x,t)|^2}{(1+t+d(x,y_0))^{m+1}\mu(B(y_0, 1+t+d(x,y_0)))} d\mu(x)dt < \infty.$$

By Lemma 3, we have

$$\begin{aligned}
&\int_0^\infty \int_X \frac{|u(x,s+t)|^2}{(1+t+d(x,y_0))^{m+1}\mu(B(y_0, 1+t+d(x,y_0)))} d\mu(x)d\,t \\
&\leq \int_0^\infty \frac{1}{(1+t)^m} \int_X \frac{|u(x,s+t)|^2}{(1+d(x,y_0))\mu(B(y_0, 1+d(x,y_0)))} d\mu(x)d\,t \\
&\leq C \int_0^\infty \frac{1}{(1+t)^m} \|u(\cdot, s+t)\|^2_{L^\infty(B(y_0,2))} d\,t \\
&+ C \int_0^\infty \frac{1}{(1+t)^m} \left\{ \left([\mu(B(y_0,1))]^{2\alpha} + (s+t)^{2N\alpha} [\mu(B(y_0, s+t))]^{2\alpha} \right) \|u\|^2_{\mathrm{HL}^{2,\alpha}_{\sqrt{\mathscr{L}}}} \right\} d\,t \\
&\leq C \int_0^\infty \frac{1}{(1+t)^m} \|u(\cdot, s+t)\|^2_{L^\infty(B(y_0,2))} dt \\
&+ C \int_0^\infty \frac{1}{(1+t)^m} \left\{ \left([\mu(B(y_0,1))]^{2\alpha} + s^{2N\alpha} [\mu(B(y_0, s))]^{2\alpha} \right) \|u\|^2_{\mathrm{HL}^{2,\alpha}_{\sqrt{\mathscr{L}}}} \right\} dt.
\end{aligned}$$

It follows from Lemma 2 that

$$\begin{aligned}
\|u(\cdot, s+t)\|_{L^\infty(B(y_0,2))} &\leq \|u(\cdot, s+t) - u(\cdot, s)\|_{L^\infty(B(y_0,2))} + \|u(\cdot, s)\|_{L^\infty(B(y_0,2))} \\
&\leq \left\| \int_s^{s+t} |\partial_r u(\cdot, r)| dr \right\|_{L^\infty(B(y_0,2))} + \|u(\cdot, s)\|_{L^\infty(B(y_0,2))} \\
&\leq C \left(1 + \frac{2}{s}\right)^{-N\alpha} [\mu(B(y_0, s))]^\alpha \|u\|_{\mathrm{HL}^{2,\alpha}_{\sqrt{\mathscr{L}}}} + \|u(\cdot, s)\|_{L^\infty(B(y_0,2))} \\
&= C(\alpha, N, y_0, s, \|u\|_{\mathrm{HL}^{2,\alpha}_{\sqrt{\mathscr{L}}}}, \|u(\cdot, s)\|_{L^\infty(B(y_0,2))}).
\end{aligned}$$

Above, we used the fact that

$$\begin{aligned}
\sup_{x \in B(y_0,2)} \int_s^{s+t} |\partial_r u(x,r)| dr &\leq C \sup_{x \in B(y_0,2)} \int_s^{s+t} [\mu(B(x,r))]^\alpha \|u\|_{\mathrm{HL}^{2,\alpha}_{\sqrt{\mathscr{L}}}} \frac{dr}{r} \\
&\leq C \sup_{x \in B(y_0,2)} \int_s^\infty \left(\frac{r}{s}\right)^{n\alpha} [\mu(B(x,s))]^\alpha \|u\|_{\mathrm{HL}^{2,\alpha}_{\sqrt{\mathscr{L}}}} \frac{dr}{r} \\
&\leq C \left(1 + \frac{2}{s}\right)^{-N\alpha} [\mu(B(y_0, s))]^\alpha \|u\|_{\mathrm{HL}^{2,\alpha}_{\sqrt{\mathscr{L}}}}.
\end{aligned}$$

Therefore, one has

$$\begin{aligned}
&\int_0^\infty \int_X \frac{|u(x,s+t)|^2}{(1+t+d(x,y_0))^{m+1}\mu(B(y_0, 1+t+d(x,y_0)))} d\mu(x)dt \\
&\leq C(\alpha, N, y_0, s, \|u\|_{\mathrm{HL}^{2,\alpha}_{\sqrt{\mathscr{L}}}}, \|u(\cdot, s)\|_{L^\infty(B(y_0,2))}) \int_0^\infty \frac{dt}{(1+t)^m}
\end{aligned}$$

$$\leq C(\alpha, N, y_0, s, \|u\|_{\mathrm{HL}^{2,\alpha}_{\sqrt{\mathscr{L}}}}, \|u(\cdot,s)\|_{L^\infty(B(y_0,2))}) < \infty, \tag{11}$$

provided $m > 1$.

For the remaining term, we need to prove that

$$\mathrm{I} := \int_0^\infty \int_X \frac{|\mathscr{P}_t(u(\cdot,s))(x)|^2}{(1+t+d(x,y_0))^{m+1}\mu(B(y_0,1+t+d(x,y_0)))} d\mu(x)dt < \infty.$$

By the Poisson upper bound and the Hölder inequality, it holds that, for all $t > 0$

$$\begin{aligned}|\mathscr{P}_t(u(\cdot,s))(x)|^2 &\leq C|\mathscr{P}_t(1)(x)| \int_X \frac{t|u(y,s)|^2}{(t+d(x,y))\mu(B(x,t+d(x,y)))} d\mu(y)\\ &\leq C\int_X \frac{t|u(y,s)|^2}{(t+d(x,y))\mu(B(x,t+d(x,y)))} d\mu(y).\end{aligned}$$

Hence, we have

$$\begin{aligned}\mathrm{I} &\leq C\int_0^\infty \int_X\int_X \frac{1}{(1+t+d(x,y_0))^{m+1}\mu(B(y_0,1+t+d(x,y_0)))}\\ &\quad\times \frac{t|u(y,s)|^2}{(t+d(x,y))\mu(B(x,t+d(x,y)))} d\mu(x)d\mu(y)dt\\ &\leq C\left\{\int_0^\infty \int_X \int_{B(y_0,d(y,y_0)/2)} + \int_0^\infty \int_X \int_{B(y_0,d(y,y_0)/2)^\complement}\right\} \cdots d\mu(x)d\mu(y)dt\\ &=: \mathrm{I}_1 + \mathrm{I}_2.\end{aligned}$$

For any $x \in B(y_0, d(y,y_0)/2)$, we have $d(x,y) > d(y,y_0) - d(x,y_0) > d(y,y_0)/2$. Hence, by (4) and Lemma 3, we have

$$\begin{aligned}\mathrm{I}_1 &\leq C\int_0^\infty \frac{1}{(1+t)^m} dt \int_X \frac{t|u(y,s)|^2}{(t+d(y,y_0))\mu(B(y_0,t+d(y,y_0)))} d\mu(y)\\ &\quad\times \int_X \frac{d\mu(x)}{(1+d(x,y_0))\mu(B(y_0,1+d(x,y_0)))}\\ &\leq C\int_0^\infty \frac{1}{(1+t)^m} dt \int_X \frac{t|u(y,s)|^2}{(t+d(y,y_0))\mu(B(y_0,t+d(y,y_0)))} d\mu(y)\\ &\leq C\int_0^\infty \frac{(1+t)\|u(\cdot,s)\|^2_{L^\infty(B(y_0,2))}}{(1+t)^m} dt\\ &\quad + C\int_0^\infty \frac{t[\mu(B(y_0,1))]^{2\alpha}\|u\|^2_{\mathrm{HL}^{2,\alpha}_{\sqrt{\mathscr{L}}}} + ts^{2N\alpha}[\mu(B(y_0,s))]^{2\alpha}\|u\|^2_{\mathrm{HL}^{2,\alpha}_{\sqrt{\mathscr{L}}}}}{(1+t)^m} dt\\ &\leq C(\alpha, N, y_0, s, \|u\|_{\mathrm{HL}^{2,\alpha}_{\sqrt{\mathscr{L}}}}, \|u(\cdot,s)\|_{L^\infty(B(y_0,2))}) < \infty,\end{aligned}$$

provided $m > 2$. For any $x \in B(y_0, d(y,y_0)/2)^\complement$, we have $d(x,y_0) > d(y,y_0)/2$. This, together with Lemma 3, yields that

$$\begin{aligned}\mathrm{I}_2 &\leq C\int_0^\infty \frac{dt}{(1+t)^m} \int_X \frac{|u(y,s)|^2 d\mu(y)}{(1+d(y,y_0))\mu(B(y_0,1+d(y,y_0)))}\\ &\quad\times \int_{B(y_0,d(y,y_0)/2)^\complement} \frac{t d\mu(x)}{(t+d(x,y))\mu(B(x,t+d(x,y)))}\\ &\leq C\int_0^\infty \frac{1}{(1+t)^m} dt \int_X \frac{|u(y,s)|^2}{(1+d(y,y_0))\mu(B(y_0,1+d(y,y_0)))} d\mu(y)\end{aligned}$$

$$\leq C\int_0^\infty \frac{\|u(\cdot,s)\|^2_{L^\infty(B(y_0,2))}+\{[\mu(B(y_0,1))]^{2\alpha}+s^{2N\alpha}[\mu(B(y_0,s))]^{2\alpha}\}\|u\|^2_{\mathrm{HL}^{2,\alpha}_{\sqrt{\mathscr{L}}}}}{(1+t)^m}dt$$
$$\leq C(\alpha,N,y_0,s,\|u\|_{\mathrm{HL}^{2,\alpha}_{\sqrt{\mathscr{L}}}},\|u(\cdot,s)\|_{L^\infty(B(y_0,2))})<\infty,$$

provided $m>1$. Therefore, it holds that

$$\int_0^\infty\int_X \frac{|\mathscr{P}_t(u(\cdot,s))(x)|^2}{(1+t+d(x,y_0))^{m+1}\mu(B(y_0,1+t+d(x,y_0)))}d\mu(x)dt<\infty,$$

which, together with (11), yields that

$$\int_0^\infty\int_X \frac{|w(x,t)|^2}{(1+t+d(x,y_0))^{m+1}\mu(B(y_0,1+t+d(x,y_0)))}d\mu(x)dt<\infty,$$

provided $m>2$. The Liouville theorem (Lemma 4) then implies $w(x,t)\equiv 0$, which means $u(x,t+s)\equiv\mathscr{P}_t(u(\cdot,s))(x)$ and thus finishes the proof. □

Next, for every $u\in\mathrm{HL}^{2,\alpha}_{\sqrt{\mathscr{L}}}(X\times\mathbb{R}_+)$, we will show that $u_s(\cdot)=u(\cdot,s)$ is bounded in $L^{2,\alpha}(X)$ uniformly for all $s>0$. To this end, we introduce a notation

$$|||\mu_{\nabla_t,f}|||_\alpha := \sup_{B\subset X}\frac{1}{[\mu(B)]^\alpha}\left(\int_0^{r_B}\fint_B |t\partial_t\mathscr{P}_tf(x)|^2 d\mu(x)\frac{dt}{t}\right)^{1/2},$$

for any

$$f\in\mathcal{M}_2:=\bigcup_{x_0\in X}\bigcup_{0<\beta\leq 1} L^2(X,(1+d(x,x_0))^{-\beta}\mu(B(x_0,1+d(x,x_0)))^{-1}d\mu(x)),$$

and establish Lemmas 5–7 as follows.

Lemma 5. *Assume the complete Dirichlet metric measure space $(X,d,\mu,\mathscr{E})$ satisfies (2) with $Q>1$ and admits (5). Given a ball $B=B(x_B,r_B)$, a function $f\in\mathcal{M}_2$ and an L^2-function g supported on B, set*

$$F(x,t):=t\partial_t\mathscr{P}_tf(x)\quad\text{and}\quad G(x,t):=t\partial_t\mathscr{P}_tg(x),$$

for any $(x,t)\in X\times\mathbb{R}_+$. If $|||\mu_{\nabla_t,f}|||_\alpha<\infty$, then there exists a constant $C>0$ such that

$$\int_0^\infty\int_X |F(x,t)G(x,t)|d\mu(x)\frac{dt}{t}\leq C[\mu(B)]^{1/2+\alpha}|||\mu_{\nabla_t,f}|||_\alpha\|g\|_{L^2(B)}.$$

Proof. Let us consider the square function $\mathcal{G}(h)$ given by

$$\mathcal{G}(h)(x):=\left(\int_0^\infty |t\partial_t\mathscr{P}_th(x)|^2\frac{dt}{t}\right)^{1/2}.$$

By the spectral theory, the function $\mathcal{G}(h)$ is bounded on $L^2(X)$. Let

$$T(B):=\{(x,t)\in X\times\mathbb{R}_+:\ x\in B,\ 0<t<r_B\}=B\times(0,r_B),$$

and write

$$\int_0^\infty\int_X |F(x,t)G(x,t)|d\mu(x)\frac{dt}{t}$$
$$=\int_{T(2B)}|F(x,t)G(x,t)|d\mu(x)\frac{dt}{t}+\sum_{k=2}^\infty\int_{T(2^kB)\setminus T(2^{k-1}B)}|F(x,t)G(x,t)|d\mu(x)\frac{dt}{t}$$

$$=: \mathrm{A}_1 + \sum_{k=2}^{\infty} \mathrm{A}_k.$$

Using the Hölder inequality and the L^2-boundedness of $\mathcal{G}$, we obtain

$$\mathrm{A}_1 \leq \left(\int_0^{2r_B} \int_{2B} |t\partial_t \mathscr{P}_t f(x)|^2 d\mu(x) \frac{dt}{t} \right)^{1/2} \|\mathcal{G}(g)\|_{L^2} \leq C[\mu(B)]^{1/2+\alpha} |||\mu_{\nabla_t, f}|||_\alpha \|g\|_{L^2(B)}.$$

Let us estimate A_k for $k = 2, 3, \ldots$. Note that, for any $(x,t) \in T(2^k B) \backslash T(2^{k-1}B)$ and $y \in B$, we have $t + d(x,y) \geq 2^{k-2} r_B$. It holds

$$\begin{aligned}
|G(x,t)| &= \left| \int_X t\partial_t p_t^{v}(x,y) g(y) d\mu(y) \right| \\
&\leq C \int_X \frac{t}{t + d(x,y)} \frac{|g(y)|}{\mu(B(x, t + d(x,y)))} d\mu(y) \\
&\leq C \int_X \frac{t}{2^k r_B} \frac{|g(y)|}{\mu(B(x, 2^k r_B))} d\mu(y) \\
&\leq C \frac{t}{2^k r_B} \frac{\|g\|_{L^1(B)}}{\mu(2^k B)},
\end{aligned}$$

which, together with the Hölder inequality and (3), implies that

$$\begin{aligned}
&\int_{T(2^k B)\backslash T(2^{k-1}B)} |F(x,t) G(x,t)| d\mu(x) \frac{dt}{t} \\
&\leq C \left(\int_0^{2^k r_B} ⨍_{2^k B} |t\partial_t \mathscr{P}_t f(x)|^2 d\mu(x) \frac{dt}{t} \right)^{1/2} \|g\|_{L^1(B)} \\
&\leq C[\mu(2^k B)]^{\alpha} |||\mu_{\nabla_t, f}|||_\alpha \|g\|_{L^1(B)} \\
&\leq C 2^{kn\alpha} [\mu(B)]^{1/2+\alpha} |||\mu_{\nabla_t, f}|||_\alpha \|g\|_{L^2(B)}.
\end{aligned}$$

Summing over k leads to

$$\int_0^\infty \int_X |F(x,t) G(x,t)| d\mu(x) \frac{dt}{t} = \sum_{k=1}^{\infty} \mathrm{A}_k \leq C[\mu(B)]^{1/2+\alpha} |||\mu_{\nabla_t, f}|||_\alpha \|g\|_{L^2(B)}.$$

This completes the proof of Lemma 5. □

Lemma 6. *Assume the complete Dirichlet metric measure space $(X, d, \mu, \mathscr{E})$ satisfies (2) with $Q > 1$ and admits (5). Suppose B, f, g, F, G are defined as in Lemma 5. If $|||\mu_{\nabla_t, f}|||_\alpha < \infty$, then we have the equality:*

$$\int_X f(x) g(x) d\mu(x) = 4 \int_0^\infty \int_X F(x,t) G(x,t) d\mu(x) \frac{dt}{t}.$$

Proof. From Lemma 5, we find that

$$\int_0^\infty \int_X |F(x,t) G(x,t)| d\mu(x) \frac{dt}{t} < \infty.$$

By the dominated convergence theorem, the following integral converges absolutely and satisfies

$$\int_0^\infty \int_X F(x,t) G(x,t) d\mu(x) \frac{dt}{t} = \lim_{\delta \to 0} \int_\delta^{1/\delta} \int_X F(x,t) G(x,t) d\mu(x) \frac{dt}{t}.$$

Next, by the commutative property of the semigroup $\{\mathscr{P}_t\}_{t>0}$, we have

$$\int_X F(x,t)G(x,t)d\mu(x) = \int_X f(x)t^2\mathscr{L}\mathscr{P}_{2t}g(x)d\mu(x).$$

This, together with Fubini's theorem, gives

$$\begin{aligned}\int_0^\infty \int_X F(x,t)G(x,t)d\mu(x)\frac{dt}{t} &= \lim_{\delta\to 0}\int_\delta^{1/\delta}\int_X f(x)t^2\mathscr{L}\mathscr{P}_{2t}g(x)d\mu(x)\frac{dt}{t} \\ &= \lim_{\delta\to 0}\int_X f(x)\int_\delta^{1/\delta} t^2\mathscr{L}\mathscr{P}_{2t}g(x)\frac{dt}{t}d\mu(x) \\ &= \lim_{\delta\to 0}\int_X f_1(x)\int_\delta^{1/\delta} t^2\mathscr{L}\mathscr{P}_{2t}g(x)\frac{dt}{t}d\mu(x) \\ &+ \lim_{\delta\to 0}\int_X f_2(x)\int_\delta^{1/\delta} t^2\mathscr{L}\mathscr{P}_{2t}g(x)\frac{dt}{t}d\mu(x) \\ &=: \mathrm{I}_1 + \mathrm{I}_2,\end{aligned}$$

where $f_1(x) := f\chi_{4B}(x)$ and $f_2(x) := f\chi_{(4B)^\complement}(x)$.

We first consider the term I_1. It follows from the spectral theory that

$$g(x) = 4\lim_{\delta\to 0}\int_\delta^{1/\delta} t^2\mathscr{L}\mathscr{P}_{2t}g(x)\frac{dt}{t}$$

in $L^2(X)$. Hence, it holds

$$\mathrm{I}_1 = \lim_{\delta\to 0}\int_X f_1(x)\int_\delta^{1/\delta} t^2\mathscr{L}\mathscr{P}_{2t}g(x)\frac{dt}{t}d\mu(x) = \frac{1}{4}\int_X f_1(x)g(x)d\mu(x).$$

In order to estimate the term I_2, we need to show that, for any $x \in (4B)^\complement$, there exists a constant $C = C(x_B, r_B) > 0$ such that

$$\sup_{\delta>0}\left|\int_\delta^{1/\delta} t^2\mathscr{L}\mathscr{P}_{2t}g(x)\frac{dt}{t}\right| \le C\frac{\|g\|_{L^2(B)}}{(1+d(x,x_B))\mu(B(x_B, 1+d(x,x_B)))}. \tag{12}$$

Recall that $\operatorname{supp} g \subset B$. For any $x \in X\backslash 4B$ and $y \in B$, we have

$$3d(x,x_B)/4 \le d(x,y) \le 5d(x,x_B)/4.$$

Hence, it follows from the Poisson upper bound and (6) that, for any $t > 0$,

$$\begin{aligned}&\left|t^2\mathscr{L}\mathscr{P}_{2t}g(x)\right| \\ &\le C\int_B \frac{2t}{(2t+d(x,y))}\frac{1}{\mu(B(x,2t+d(x,y)))}\left(\frac{2t+d(x,y)}{\rho(y)}\right)^{-2}|g(y)|d\mu(y) \\ &\le C\int_B \frac{t}{(t+d(x,x_B))}\frac{1}{\mu(B(x,t+d(x,x_B)))}\left(\frac{\rho(x_B)\left(1+\frac{r_B}{\rho(x_B)}\right)^{k_0/(k_0+1)}}{t+d(x,x_B)}\right)^2|g(y)|d\mu(y) \\ &\le C(x_B,r_B)\frac{t}{(t+d(x,x_B))^3\mu(B(x_B,t+d(x,x_B)))}\|g\|_{L^1(B)} \\ &\le C(x_B,r_B)\frac{\|g\|_{L^2(B)}}{(1+d(x,x_B))\mu(B(x_B,1+d(x,x_B)))}\frac{t}{(t+d(x,x_B))^2}.\end{aligned}$$

The above estimate, together with the fact

$$\int_0^\infty \frac{t}{(t+d(x,x_B))^2}\frac{dt}{t} \le \int_0^\infty \frac{dt}{(t+r_B)^2} \le C(r_B) < \infty$$

yields that

$$\begin{aligned}\left|\int_\delta^{1/\delta} t^2\mathscr{L}\mathscr{P}_{2t}g(x)\frac{dt}{t}\right| &\le \int_0^\infty \left|t^2\mathscr{L}\mathscr{P}_{2t}g(x)\right|\frac{dt}{t}\\ &\le C(x_B,r_B)\frac{\|g\|_{L^2(B)}}{(1+d(x,x_B))\mu(B(x_B,1+d(x,x_B)))}.\end{aligned}$$

Accordingly, (12) follows readily. Now, we estimate the term I_2. Since $f \in \mathcal{M}_2$, the estimate (12) yields that

$$\sup_{\delta>0}\int_X \left|f_2(x)\int_\delta^{1/\delta} t^2\mathscr{L}\mathscr{P}_{2t}g(x)\frac{dt}{t}\right| d\mu(x) \le C(g,x_B,r_B) < \infty.$$

This allows us to pass the limit inside the integral of I_2. Hence, we conclude

$$\mathrm{I}_2 = \lim_{\delta\to 0}\int_X f_2(x)\int_\delta^{1/\delta} t^2\mathscr{L}\mathscr{P}_{2t}g(x)\frac{dt}{t}d\mu(x) = \frac{1}{4}\int_X f_2(x)g(x)d\mu(x).$$

Combining the previous formulas for I_1 and I_2, we complete the proof. □

Recall that we set $u_s(\cdot) = u(\cdot,s)$ for any $s > 0$.

Lemma 7. *Suppose the complete Dirichlet metric measure space $(X,d,\mu,\mathscr{E})$ satisfies (2) with $Q > 1$ and admits (5). Let $0 \le V \in A_\infty(X)\cap RH_q(X)$ with $q > (Q+1)/2$. Assume that $u \in \mathrm{HL}^{2,\alpha}_{\sqrt{\mathscr{L}}}(X\times\mathbb{R}_+)$ with $\max\{-1/2,-1/2N\} < \alpha < 0$.*

Then, there exists a positive constant C such that, for every $s > 0$,

$$|||\mu_{\nabla_t,u_s}|||_\alpha \le C\|u\|_{\mathrm{HL}^{2,\alpha}_{\sqrt{\mathscr{L}}}}.$$

Proof. Let $B = B(x_B,r_B)$. It holds by Proposition 1 that

$$\frac{1}{[\mu(B)]^\alpha}\left(\int_0^{r_B}\fint_B |t\partial_t\mathscr{P}_t u_s|^2 d\mu\frac{dt}{t}\right)^{1/2} = \frac{1}{[\mu(B)]^\alpha}\left(\int_0^{r_B}\fint_B |t\partial_t u(y,t+s)|^2 d\mu(y)\frac{dt}{t}\right)^{1/2}.$$

If $r_B > s$, by the doubling property (2), we have that

$$\begin{aligned}&\frac{1}{[\mu(B)]^\alpha}\left(\int_0^{r_B}\fint_B |t\partial_t\mathscr{P}_t u_s|^2 d\mu\frac{dt}{t}\right)^{1/2}\\ &\le \frac{1}{[\mu(B)]^\alpha}\left(\frac{1}{\mu(B)}\int_0^{r_B+s}\int_{B(x_B,r_B+s)} |t\partial_t u(y,t)|^2 d\mu(y)\frac{dt}{t}\right)^{1/2}\\ &\le \frac{C}{[\mu(2B)]^\alpha}\left(\int_0^{2r_B}\fint_{2B} |t\partial_t u(y,t)|^2 d\mu(y)\frac{dt}{t}\right)^{1/2}\\ &\le C\|u\|_{\mathrm{HL}^{2,\alpha}_{\sqrt{\mathscr{L}}}}.\end{aligned}$$

Otherwise, $r_B \le s$, Lemma 2 together with elementary integration implies that there exists a positive constant C independent of r_B and s such that

$$\begin{aligned}
&\frac{1}{[\mu(B)]^{\alpha}}\left(\int_0^{r_B}\fint_B |t\partial_t \mathscr{P}_t u_s|^2 d\mu \frac{dt}{t}\right)^{1/2} \\
&\le \frac{C}{[\mu(B)]^{\alpha}}\left(\int_0^{r_B}\fint_B \frac{t^2}{(t+s)^2}[\mu(B(y,t+s))]^{2\alpha}\|u\|^2_{\mathrm{HL}^{2,\alpha}_{\sqrt{\mathscr{L}}}} d\mu(y)\frac{dt}{t}\right)^{1/2} \\
&\le \frac{C}{[\mu(B)]^{\alpha}}\left(\int_0^{r_B}\fint_B \left(\frac{t}{r_B}\right)^2[\mu(B(y,r_B))]^{2\alpha}\|u\|^2_{\mathrm{HL}^{2,\alpha}_{\sqrt{\mathscr{L}}}} d\mu(y)\frac{dt}{t}\right)^{1/2} \\
&\le C\|u\|_{\mathrm{HL}^{2,\alpha}_{\sqrt{\mathscr{L}}}},
\end{aligned}$$

which, together with the case $r_B > s$, means that

$$|||\mu_{\nabla_t,u_s}|||_{\alpha} \le C\|u\|_{\mathrm{HL}^{2,\alpha}_{\sqrt{\mathscr{L}}}},$$

which thus finishes the proof. □

Proposition 2. *Suppose the complete Dirichlet metric measure space $(X,d,\mu,\mathscr{E})$ satisfies (2) with $Q > 1$ and admits (5). Let $0 \le V \in A_\infty(X) \cap RH_q(X)$ with $q > (Q+1)/2$. Assume that $u \in \mathrm{HL}^{2,\alpha}_{\sqrt{\mathscr{L}}}(X \times \mathbb{R}_+)$ with $\max\{-1/2, -1/2N\} < \alpha < 0$. Then, for any $s > 0$, we have $u_s \in L^{2,\alpha}(X)$ and there exists a constant $C > 0$, independent of s, such that*

$$\|u_s\|_{L^{2,\alpha}} \le C\|u\|_{\mathrm{HL}^{2,\alpha}_{\sqrt{\mathscr{L}}}}.$$

Proof. Since $u \in \mathrm{HL}^{2,\alpha}_{\sqrt{\mathscr{L}}}(X \times \mathbb{R}_+)$, it follows from Lemma 3 that $u_s \in \mathcal{M}_2$. Given a ball $B \subset X$, for any L^2 function g supported on B, it follows from Lemmas 5, 6 and 7 that

$$\begin{aligned}
\left|\int_X u_s g d\mu\right| &= 4\left|\int_0^\infty \int_X t\partial_t \mathscr{P}_t u_s t\partial_t \mathscr{P}_t g d\mu \frac{dt}{t}\right| \\
&\le C[\mu(B)]^{1/2+\alpha}|||\mu_{\nabla_t,u_s}|||_{\alpha}\|g\|_{L^2(B)} \\
&\le C[\mu(B)]^{1/2+\alpha}\|u\|_{\mathrm{HL}^{2,\alpha}_{\sqrt{\mathscr{L}}}}\|g\|_{L^2(B)}.
\end{aligned}$$

This together with the L^2-duality argument shows that

$$\begin{aligned}
\frac{1}{[\mu(B)]^{\alpha}}\left(\fint_B |u_s|^2 d\mu\right)^{1/2} &= \frac{1}{[\mu(B)]^{1/2+\alpha}}\sup_{\|g\|_{L^2(B)}\le 1}\left|\int_X u_s g d\mu\right| \\
&\le C\sup_{\|g\|_{L^2(B)}\le 1}\|u\|_{\mathrm{HL}^{2,\alpha}_{\sqrt{\mathscr{L}}}}\|g\|_{L^2(B)} \le C\|u\|_{\mathrm{HL}^{2,\alpha}_{\sqrt{\mathscr{L}}}}.
\end{aligned}$$

Then, by taking the supremum over all the ball B, it holds that

$$\|u_s\|_{L^{2,\alpha}} \le C\|u\|_{\mathrm{HL}^{2,\alpha}_{\sqrt{\mathscr{L}}}},$$

which completes the proof. □

4. From Initial Value to Solution

In this section, we will show that every Morrey function f induces a Carleson type measure $t|\nabla \mathscr{P}_t f|^2 d\mu dt$. In order to estimate the space derivation part $t|\nabla_x \mathscr{P}_t f|^2 d\mu dt$, we introduce a result of Jiang-Li [18] (Proposition 5.2), which establishes a Caccioppoli inequality for the Schrödinger Poisson semigroup in a tent domain $B(x_B, r_B) \times (0, r_B)$.

Proposition 3. *Suppose the complete Dirichlet metric measure space* $(X, d, \mu, \mathscr{E})$ *satisfies* (2) *with* $Q > 1$ *and admits* (5). *Let* $0 \le V \in A_\infty(X) \cap RH_q(X)$ *with* $q > \max\{1, Q/2\}$. *Assume that* g *satisfies for some* $y \in X$ *that*

$$\int_X \frac{|g(x)|}{(1+d(x,y))\mu(B(y,1+d(x,y)))} d\mu(x) < \infty.$$

Then, for any ball $B = B(x_B, r_B)$, *it holds that*

$$\int_0^{r_B} \int_B |t\nabla_x \mathscr{P}_t g|^2 d\mu \frac{dt}{t} \le C \int_0^{2r_B} \int_{2B} \left(|t^2 \partial_t^2 \mathscr{P}_t g||\mathscr{P}_t g| + |\mathscr{P}_t g|^2 \right) d\mu \frac{dt}{t}.$$

Theorem 2. *Assume the complete Dirichlet metric measure space* $(X, d, \mu, \mathscr{E})$ *satisfies* (2) *with* $Q > 1$ *and admits* (5). *Let* $V \in A_\infty(X) \cap RH_q(X)$ *with* $q \ge \max\{1, Q/2\}$. *If* $f \in L^{2,\alpha}(X)$ *with* $-1/2 < \alpha < 0$, *then* $u(x,t) = \mathscr{P}_t f(x) \in \mathrm{HL}^{2,\alpha}_{\sqrt{\mathscr{L}}}(X \times \mathbb{R}_+)$. *Moreover, there exists a constant* $C > 0$ *such that*

$$\|u\|_{\mathrm{HL}^{2,\alpha}_{\sqrt{\mathscr{L}}}} \le C \|f\|_{L^{2,\alpha}}.$$

Proof. For any ball $B = B(x_B, r_B)$, it holds that

$$\left(\int_0^{r_B} ⨍_B |t\nabla \mathscr{P}_t f|^2 d\mu \frac{dt}{t} \right)^{1/2} \le \sum_{k=1}^{\infty} \left(\int_0^{r_B} ⨍_B |t\nabla \mathscr{P}_t f_k|^2 d\mu \frac{dt}{t} \right)^{1/2} =: \sum_{k=1}^{\infty} \mathrm{J}_k,$$

where $f_1 := f\chi_{4B}$ and $f_k := f\chi_{2^{k+1}B \setminus 2^k B}$ for $k \in \{2, 3, 4, \dots\}$.

For the term J_1, we apply the L^2-boundedness of the Riesz operator $\nabla_x \mathscr{L}^{-1/2}$ to obtain that

$$\begin{aligned}
\left(\int_0^{r_B} ⨍_B |t\nabla \mathscr{P}_t f_1|^2 d\mu \frac{dt}{t} \right)^{1/2} &\le \left(\frac{1}{\mu(B)} \int_0^{\infty} \int_X |t\nabla \mathscr{P}_t f_1|^2 d\mu \frac{dt}{t} \right)^{1/2} \\
&\le C \left(\frac{1}{\mu(B)} \int_0^{\infty} \int_X |t\sqrt{\mathscr{L}} \mathscr{P}_t f_1|^2 d\mu \frac{dt}{t} \right)^{1/2} \\
&\le C \left(\frac{1}{\mu(B)} \int_X |f_1|^2 d\mu \right)^{1/2} \\
&\le C [\mu(B)]^{\alpha} \|f\|_{L^{2,\alpha}}.
\end{aligned}$$

Since $f_k \in L^{2,\alpha}(X)$, it is easy to see $f_k \in \mathcal{M}_2$. Hence, f_k satisfies the requirement in Proposition 3, which implies that, for any $k \in \{2, 3, 4, \dots\}$,

$$\mathrm{J}_k \le C \left(\int_0^{2r_B} ⨍_{2B} \left(|t\partial_t \mathscr{P}_t f_k|^2 + |t^2 \partial_t^2 \mathscr{P}_t f_k||\mathscr{P}_t f_k| + |\mathscr{P}_t f_k|^2 \right) \frac{d\mu dt}{t} \right)^{1/2}.$$

Then, for any $x \in 2B$, we apply the Poisson upper bound to obtain

$$\begin{aligned}
&|\mathscr{P}_t f_k(x)| + |t\partial_t \mathscr{P}_t f_k(x)| + |t^2 \partial_t^2 \mathscr{P}_t f_k(x)| \\
&\le C \int_{2^{k+1}B \setminus 2^k B} \frac{t}{(t + d(x,y))} \frac{|f(y)|}{\mu(B(x, t + d(x,y)))} d\mu(y) \\
&\le C 2^{-k} \frac{t}{r_B} ⨍_{2^{k+1}B} |f(y)| d\mu(y) \\
&\le C 2^{-k} \frac{t}{r_B} [\mu(2^{k+1}B)]^{\alpha} \|f\|_{L^{2,\alpha}} \\
&\le C 2^{-k} \frac{t}{r_B} [\mu(B)]^{\alpha} \|f\|_{L^{2,\alpha}},
\end{aligned}$$

which yields

$$J_k \le C2^{-k}[\mu(B)]^{\alpha}\|f\|_{L^{2,\alpha}}.$$

Hence, it follows that

$$\|\mathscr{P}_t f\|_{\mathrm{HL}^{2,\alpha}_{\sqrt{\mathscr{L}}}} = \sup_{B\subset X}\frac{1}{[\mu(B)]^{\alpha}}\left(\int_0^{r_B}\fint_B |t\nabla\mathscr{P}_t f|^2 d\mu\frac{dt}{t}\right)^{1/2} \le \sum_{k=1}^{\infty} J_k \le C\|f\|_{L^{2,\alpha}}.$$

This completes the proof. □

5. From Solution to Initial Value

In this section, we will show that, for every function $u \in \mathrm{HL}^{2,\alpha}_{\sqrt{\mathscr{L}}}(X\times\mathbb{R}_+)$, there is a function $f \in L^{2,\alpha}(X)$ such that $u(x,t) = \mathscr{P}_t f(x)$ with the desired norm control.

Theorem 3. *Suppose the complete Dirichlet metric measure space $(X, d, \mu, \mathscr{E})$ satisfies (2) with $Q > 1$ and admits (5). Assume $0 \le V \in A_\infty(X)\cap RH_q(X)$ with $q \ge (Q+1)/2$, and $\max\{-1/2, -1/2N\} < \alpha < 0$. If $u \in \mathrm{HL}^{2,\alpha}_{\sqrt{\mathscr{L}}}(X\times\mathbb{R}_+)$, then there exists a function $f \in L^{2,\alpha}(X)$ such that $u(x,t) = \mathscr{P}_t f(x)$. Moreover, there exists a constant $C > 0$, independent of u, such that*

$$\|f\|_{L^{2,\alpha}} \le C\|u\|_{\mathrm{HL}^{2,\alpha}_{\sqrt{\mathscr{L}}}}.$$

Proof. Without loss of generality, we may assume $q > (Q+1)/2$ because of the self improvement of the $RH_q(X)$ class. Suppose $u \in \mathrm{HL}^{2,\alpha}_{\sqrt{\mathscr{L}}}(X\times\mathbb{R}_+)$. For any $0 < \epsilon < 1$, by Proposition 2, we have

$$\|u_\epsilon\|_{L^{2,\alpha}} \le C\|u\|_{\mathrm{HL}^{2,\alpha}_{\sqrt{\mathscr{L}}}}. \tag{13}$$

Next, we will fix a point x_0 and look for a function $f \in L^{2,\alpha}(X)$ through $L^2(B(x_0, 2^j))$-boundedness of $\{u_\epsilon\}$ for each $j \in \mathbb{N}$. Indeed, for every $j \in \mathbb{N}$, we use (13) to obtain

$$\int_{B(x_0,2^j)} |u_\epsilon(x)|^2 d\mu(x) \le C[\mu(B(x_0,2^j))]^{1+2\alpha}\|u\|^2_{\mathrm{HL}^{2,\alpha}_{\sqrt{\mathscr{L}}}},$$

which implies that the family $\{u_\epsilon(\cdot)\}_{0<\epsilon<1}$ is uniformly bounded in $L^2(B(x_0, 2^j))$. Then, the Eberlein–Šmulian theorem and the diagonal method imply that there exists a sequence $\epsilon_k \to 0$ $(k \to \infty)$ and a function $g_j \in L^2(B(x_0, 2^j))$ such that $u_{\epsilon_k} \to g_j$ weakly in $L^2(B(x_0, 2^j))$, for any $j \in \mathbb{N}$. Now, we define a function $f(x)$ by

$$f(x) = g_j(x),$$

if $x \in B(x_0, 2^j)$, $j = 1, 2, \ldots$. It is easy to see that f is well defined on $X = \bigcup_{j=1}^{\infty} B(x_0, 2^j)$. We can check that, for any ball $B \subset X$,

$$\int_B |f(x)|^2 d\mu(x) \le C[\mu(B)]^{1+2\alpha}\|u\|^2_{\mathrm{HL}^{2,\alpha}_{\sqrt{\mathscr{L}}}},$$

which implies that

$$\|f\|_{L^{2,\alpha}} \le C\|u\|_{\mathrm{HL}^{2,\alpha}_{\sqrt{\mathscr{L}}}}.$$

Finally, we will show that $u(x,t) = \mathscr{P}_t f(x)$. By Lemma 1, we know that $u(x,\cdot)$ is continuous on $\mathbb{R}_+$. This together with Proposition 1 yields that

$$u(x,t) = \lim_{k\to+\infty} u(x, t+\epsilon_k) = \lim_{k\to+\infty} \mathscr{P}_t u_{\epsilon_k}(x).$$

This reduces to verify that

$$\lim_{k\to+\infty} \mathscr{P}_t u_{\epsilon_k}(x) = \mathscr{P}_t f(x). \tag{14}$$

Indeed, we recall that $p_t^v(x,y)$ is the kernel of $\mathscr{P}_t$, and for any $\ell \in \mathbb{N}$, we write

$$\mathscr{P}_t u_{\epsilon_k}(x) = \int_{B(x,2^\ell t)} p_t^v(x,y) u_{\epsilon_k}(y) d\mu(y) + \int_{X\setminus B(x,2^\ell t)} p_t^v(x,y) u_{\epsilon_k}(y) d\mu(y).$$

Using the Poisson upper bound, the Hölder inequality and (13), we obtain

$$\begin{aligned}\left|\int_{X\setminus B(x,2^\ell t)} p_t^v(x,y) u_{\epsilon_k}(y) d\mu(y)\right| &\le C \sum_{i=\ell}^{\infty} 2^{-i} \fint_{B(x,2^{i+1}t)} |u_{\epsilon_k}(y)| d\mu(y) \\ &\le C \sum_{i=\ell}^{\infty} 2^{-i} [\mu(B(x,2^i t))]^{\alpha} \|u_{\epsilon_k}\|_{L^{2,\alpha}} \\ &\le C 2^{-\ell} [\mu(B(x,t))]^{\alpha} \|u\|_{\mathrm{HL}^{2,\alpha}_{\sqrt{\mathscr{L}}}},\end{aligned}$$

where C is a positive constant independent of k. One has

$$\begin{aligned}0 &\le \lim_{\ell\to+\infty} \lim_{k\to+\infty} \left|\int_{X\setminus B(x,2^\ell t)} p_t^v(x,y) u_{\epsilon_k}(y) d\mu(y)\right| \\ &\le \lim_{\ell\to+\infty} C 2^{-\ell} [\mu(B(x,t))]^{\alpha} \|u\|_{\mathrm{HL}^{2,\alpha}_{\sqrt{\mathscr{L}}}} = 0.\end{aligned}$$

Therefore, it holds that

$$\lim_{k\to+\infty} \mathscr{P}_t u_{\epsilon_k}(x) = \lim_{\ell\to+\infty} \lim_{k\to+\infty} \int_{B(x,2^\ell t)} p_t^v(x,y) u_{\epsilon_k}(y) d\mu(y) = \mathscr{P}_t f(x),$$

which yields (14) readily. Then, we show that

$$u(x,t) = \mathscr{P}_t f(x).$$

The proof of Theorem 3 is complete. □

6. Conclusions

In this article, we solved the Dirichelt problem for the Schrödinger equation on the metric measure space. We obtained that a Schrödinger harmonic function satisfies the Carleson type condition if and only if it is the Poisson extension of a Morrey function. This continues the line of research on the Dirichlet problem with boundary value in L^p space and BMO space, extends the result in Song-Tian-Yan [8] from the Euclidean space to the metric measure space and improves the reverse Hölder index from $q \ge n$ to $q \ge (n+1)/2$.

Author Contributions: Created and conceptualized the idea, T.S. and B.L.; writing—original draft preparation, T.S. and B.L.; writing—review and editing, T.S. and B.L. All authors have read and agreed to the published version of the manuscript.

Funding: This research was funded by National Natural Science Foundation of China (Grant No. 11922114 and 11671039), Scientific Research Project of Jiaxing university (Grant No. CD70521016) and SRT of Jiaxing university (Grant No. CD8517211391).

Institutional Review Board Statement: Not applicable.

Informed Consent Statement: Not applicable.

Data Availability Statement: Not applicable.

Acknowledgments: The two authors would like to thank their advisor Renjin Jiang for proposing this joint work and for the useful discussions and advice on the topic of the paper.

Conflicts of Interest: The authors declare no conflict of interest.

1. Stein, E.; Weiss, G. On the theory of harmonic functions of several variables. I. The theory of H^p-spaces. *Acta Math.* **1960**, *103*, 25–62. [CrossRef]
2. Fefferman, C.; Stein, E. H^p spaces of several variables. *Acta Math.* **1972**, *129*, 137–193. [CrossRef]
3. Fabes, E.; Johnson, R.; Neri, U. Spaces of harmonic functions representable by Poisson integrals of functions in BMO and $L_{p,\lambda}$. *Indiana Univ. Math. J.* **1976**, *25*, 159–170. [CrossRef]
4. Fabes, E.; Neri, U. Characterization of temperatures with initial data in BMO. *Duke Math. J.* **1975**, *42*, 725–734. [CrossRef]
5. Martell, J.; Mitrea, D.; Mitrea, I.; Mitrea, M. The BMO-Dirichlet problem for elliptic systems in the upper half-space and quantitative characterizations of VMO. *Anal. PDE* **2019**, *12*, 605–720. [CrossRef]
6. Auscher, P.; Rosén, A.; Rule, D. Boundary value problems for degenerate elliptic equations and systems. *Ann. Sci. L'École Norm. SupÉrieure. QuatriÈme SÉrie* **2015**, *48*, 951–1000. [CrossRef]
7. Duong, X.; Yan, L.; Zhang, C. On characterization of Poisson integrals of Schrödinger operators with BMO traces. *J. Funct. Anal.* **2014**, *266*, 2053–2085. [CrossRef]
8. Song, L.; Tian, X.; Yan, L. On characterization of Poisson integrals of Schrödinger operators with Morrey traces. *Acta Math. Sin. (Engl. Ser.)* **2018**, *34*, 787–800. [CrossRef]
9. Wang, Y.; Liu, Y.; Sun, C.; Li, P. Carleson measure characterizations of the Campanato type space associated with Schrödinger operators on stratified Lie groups. *Forum Math.* **2020**, *32*, 1337–1373. [CrossRef]
10. Jiang, R.; Xiao, J.; Yang, D. Towards spaces of harmonic functions with traces in square Campanato spaces and their scaling invariants. *Anal. Appl.* **2016**, *14*, 679–703. [CrossRef]
11. Huang, Q.; Zhang, C. Characterization of temperatures associated to Schrödinger operators with initial data in Morrey spaces. *Taiwan. J. Math.* **2019**, *23*, 1133–1151. [CrossRef]
12. Liu, H.; Yang, H.; Yang, Q. Carleson measures and trace theorem for β-harmonic functions. *Taiwan. J. Math.* **2018**, *22*, 1107–1138. [CrossRef]
13. Wang, Y.; Xiao, J. Homogeneous Campanato-Sobolev classes. *Appl. Comput. Harmon. Anal.* **2015**, *39*, 214–247. [CrossRef]
14. Coulhon, T.; Jiang, R.; Koskela, P.; Sikora, A. Gradient estimates for heat kernels and harmonic functions. *J. Funct. Anal.* **2020**, *278*, 108398. [CrossRef]
15. Li, H.-Q. Estimations L^p des opérateurs de Schrödinger sur les groupes nilpotents, (French) [L^p estimates of Schrödinger operators on nilpotent groups]. *J. Funct. Anal.* **1999**, *161*, 152–218. [CrossRef]
16. Lin, C.-C.; Liu, H. $BMO_L(\mathbb{H}^n)$ spaces and Carleson measures for Schrödinger operators. *Adv. Math.* **2011**, *228*, 1631–1688. [CrossRef]
17. Jiang, R.; Lin, F. Riesz transform under perturbations via heat kernel regularity. *J. Math. Pures Appliquées. Neuvième Série* **2020**, *133*, 39–65. [CrossRef]
18. Jiang, R.; Li, B. Dirichlet problem for the Schrödinger equation with boundary value in BMO space. *Sci. China. Math.* **2021**, *64*, 10. [CrossRef]
19. Beurling, A.; Deny, J. Dirichlet spaces. *Proc. Natl. Acad. Sci. USA* **1959**, *45*, 208–215. [CrossRef]
20. Fukushima, M.; Oshima, Y.; Takeda, M. *Dirichlet Forms and Symmetric Markov Processes*; Walter de Gruyter & Co.: Berlin, Germany, 1994.
21. Gyrya, P.; Saloff-Coste, L. Neumann and Dirichlet heat kernels in inner uniform domains. *Astérisque* **2011**, *14*, 1–144.
22. Sturm, K. Analysis on local Dirichlet spaces. III. The parabolic Harnack inequality. *J. Mathématiques Pures Appliquées. Neuvième Série* **1996**, *75*, 273–297.
23. Sturm, K. Analysis on local Dirichlet spaces. II. Upper Gaussian estimates for the fundamental solutions of parabolic equations. *Osaka J. Math.* **1995**, *32*, 275–312.
24. Biroli, M.; Mosco, U. A Saint-Venant type principle for Dirichlet forms on discontinuous media. *Ann. Mat. Pura Applicata. Ser. Quarta* **1995**, *169*, 125–181. [CrossRef]
25. Heinonen, J.; Koskela, P.; Shanmugalingam, N.; Tyson, T. *Sobolev Spaces on Metric Measure Spaces. An Approach Based on upper Gradients*; Cambridge University Press: Cambridge, UK, 2015.
26. Grigor'yan, A.; Hu, J. Upper bounds of heat kernels on doubling spaces. *Mosc. Math. J.* **2014**, *14*, 505–563. [CrossRef]
27. Muckenhoupt, B. Weighted norm inequalities for the Hardy maximal function. *Trans. Am. Math. Soc.* **1972**, *165*, 207–226. [CrossRef]
28. Strömberg, J.; Torchinsky, A. *Weighted Hardy Spaces*; Springer: Cham, Switzerland, 1989.
29. Shen, Z. L^p estimates for Schrödinger operators with certain potentials. *Univ. Grenoble. Ann. L'Institut Fourier* **1995**, *45*, 513–546. [CrossRef]
30. Yang, D.; Zhou, Y. Localized Hardy spaces H^1 related to admissible functions on RD-spaces and applications to Schrödinger operators. *Trans. Am. Math. Soc.* **2011**, *363*, 1197–1239. [CrossRef]

31. Cao, J.; Chang, D.-C.; Yang, D.; Yang, S. Boundedness of second order Riesz transforms associated to Schrödinger operators on Musielak-Orlicz-Hardy spaces. *Commun. Pure Appl. Anal.* **2014**, *13*, 1435–1463. [CrossRef]
32. Chen, P.; Duong, X.; Li, J.; Song, L.; Yan, L. Carleson measures, BMO spaces and balayages associated to Schrödinger operators. *Sci. China. Math.* **2017**, *60*, 2077–2092. [CrossRef]
33. Chen, P.; Duong, X.; Li, J.; Yan, L. Sharp endpoint L^p estimates for Schrödinger groups. *Math. Ann.* **2020**, *378*, 667–702. [CrossRef]
34. Guliyev, V.S. Function spaces and integral operators associated with Schrödinger operators: An overview. *Proc. Inst. Math. Mechanics. Natl. Acad. Sci. Azerbaijan* **2014**, *40*, 178–202.
35. Guliyev, V.S.; Guliyev, R.V.; Omarova, M.N.; Ragusa, M.A. Schrödinger type operators on local generalized Morrey spaces related to certain nonnegative potentials. *Discret. Contin. Dyn. Syst. Ser. B. A J. Bridg. Math. Sci.* **2020**, *25*, 671–690. [CrossRef]
36. Guliyev, V.S.; Omarova, M.N.; Ragusa, M.A.; Scapellato, A. Regularity of solutions of elliptic equations in divergence form in modified local generalized Morrey spaces. *Anal. Math. Phys.* **2021**, *11*, 13. [CrossRef]
37. Jiang, R.; Yang, D.; Yang, D. Maximal function characterizations of Hardy spaces associated with magnetic Schrödinger operators. *Forum Math.* **2012**, *24*, 471–494. [CrossRef]
38. Pan, G.; Tang, L.; Zhu, H. Global weighted estimates for higher order Schrödinger operators with discontinuous coefficients. *J. Fourier Anal. Appl.* **2021**, *27*, 85. [CrossRef]
39. Song, L.; Yan, L. Riesz transforms associated to Schrödinger operators on weighted Hardy spaces. *J. Funct. Anal.* **2010**, *259*, 1466–1490. [CrossRef]
40. Wu, L.; Yan, L. Heat kernels, upper bounds and Hardy spaces associated to the generalized Schrödinger operators. *J. Funct. Anal.* **2016**, *270*, 3709–3749. [CrossRef]
41. Yang, D.; Yang, D.; Zhou, Y. Endpoint properties of localized Riesz transforms and fractional integrals associated to Schrödinger operators. *Potential Anal.* **2009**, *30*, 271–300. [CrossRef]
42. Yang, D.; Yang, D.; Zhou, Y. Localized Morrey-Campanato spaces on metric measure spaces and applications to Schrödinger operators. *Nagoya Math. J.* **2010**, *198*, 77–119. [CrossRef]
43. Yang, D.; Yang, S. Second-order Riesz transforms and maximal inequalities associated with magnetic Schrödinger operators. *Can. Math. Bull.* **2015**, *58*, 432–448. [CrossRef]
44. Yang, D.; Yang, S. Regularity for inhomogeneous Dirichlet problems of some Schrödinger equations on domains. *J. Geom. Anal.* **2016**, *26*, 2097–2129. [CrossRef]
45. Morrey, C.B. On the solutions of quasi-linear elliptic partial differential equations. *Trans. Am. Math. Soc.* **1938**, *43*, 126–166. [CrossRef]

MDPI
St. Alban-Anlage 66
4052 Basel
Switzerland
Tel. +41 61 683 77 34
Fax +41 61 302 89 18
www.mdpi.com

Mathematics Editorial Office
E-mail: mathematics@mdpi.com
www.mdpi.com/journal/mathematics

www.ingramcontent.com/pod-product-compliance
Lightning Source LLC
LaVergne TN
LVHW070250100526
838202LV00015B/2202